RIVERSIDE COMMUNITY COLLEGE
1916

ITIAL EQUAT
2

DATE DUE

JA 7 '94			
AP 29 '94			
MY 27 '94			
JE 17 '94			
RENEW			
MR 3 '95			
MY 12 '95			
MR 1 '96			
MR 18 '99			
JE 11 '03			
JE 9 '04			
MR 18 05			
MY 9 07			

DEMCO 38-296

D1118585

HANDBOOK OF
DIFFERENTIAL EQUATIONS

Second Edition

HANDBOOK OF
DIFFERENTIAL EQUATIONS

Second Edition

Daniel Zwillinger

Department of Mathematical Sciences
Rensselaer Polytechnic Institute
Troy, New York

ACADEMIC PRESS, INC.

Harcourt Brace Jovanovich, Publishers

Boston San Diego New York
London Sydney Tokyo Toronto

Riverside Community College
Library
4800 Magnolia Avenue
Riverside, California 92506

JUL '93

This book is printed on acid-free paper. ⊚

Copyright © 1992, 1989 by Academic Press, Inc.

All rights reserved.
No part of this publication may be reproduced or
transmitted in any form or by any means, electronic
or mechanical, including photocopy, recording, or
any information storage and retrieval system, without
permission in writing from the publisher.

This book was typeset by the author using TEX.

Figure 3.3 originally appeared as Figure 1 in H. Abelson, "The Bifurcation Interpreter: A Step Towards the Automatic Analysis of Dynamical Systems," *Comp. & Maths with Appls.*, **20**, No. 8, 1990, pages 13–35. Reprinted courtesy of Pergamon Press, Inc.

Figure 5.4 originally appeared as Figure 4 in S. De Souza-Machado, R. W. Rollins, D. T. Jacobs, and J. L. Hartman, "Studying Chaotic Systems Using Microcomputer Simulations and Lyapunov Exponents," *Am. J. Phys.*, **58**, No. 4, April 1990, pages 321–329. Reprinted courtesy of American Association of Physics Teachers.

Figure 118.3 originally appeared as Figure 9.23 in W. E. Boyce and R. C. DiPrima, *Elementary Differential Equations and Boundary Value Problems*, Fourth Edition, 1986. Reprinted courtesy of John Wiley & Sons.

Figure 155.3 originally appeared as Figure 9.23 in J. R. Rice, "Parallel Methods for Partial Differential Equations," in *The Characteristics of Parallel Algorithms*, L. H. Jamieson, D. B. Gannon, and R. J. Douglas (eds.), 1987, pages 209–231. Reprinted courtesy of MIT Press.

ACADEMIC PRESS, INC.
1250 Sixth Avenue, San Diego, CA 92101

United Kingdom Edition published by
ACADEMIC PRESS LIMITED
24–28 Oval Road, London NW1 7DX

Library of Congress Cataloging-in-Publication Data

Zwillinger, Daniel. 1957–
        Handbook of differential equations/Daniel Zwillinger. — 2nd ed.
            p.   cm.
        Includes bibliographical references and index.
        ISBN 0-12-784391-4
        1. Differential equations.   I. Title.
    QA371.Z88   1992
    515′.35—dc20   91–58716
                                                             CIP

Printed in the United States of America
92 93 94 95   9 8 7 6 5 4 3 2 1

Contents

I.A   Definitions and Concepts

I.B   Transformations

II   Exact Analytical Methods

II.A   Exact Methods for ODEs

---

* These methods are also applicable to partial differential equations.

---

* These methods are also applicable to partial differential equations.

II.B   Exact Methods for PDEs

III   Approximate Analytical Methods

IV.A   Numerical Methods: Concepts

IV.B   Numerical Methods for ODEs

* These methods are also applicable to partial differential equations.

IV.C   Numerical Methods for PDEs

---

* These methods are also applicable to partial differential equations.

Preface

When I was a graduate student in applied mathematics at the California Institute of Technology, we solved many differential equations (both ordinary differential equations and partial differential equations). Given a differential equation to solve, I would think of all the techniques I knew that might solve that equation. Eventually the number of techniques I knew became so large that I began to forget some.

Then I would have to consult books on differential equations to familiarize myself with a technique that I only vaguely remembered. This was a slow process and often unrewarding; I might spend twenty minutes reading about a technique only to realize that it did not apply to the equation I was trying to solve.

Eventually I created a list of the different techniques that I knew. Each technique had a brief description of how the method was used and to what types of equations it applied. As I learned more techniques they were added to the list. This book is a direct outgrowth of that list.

At Caltech we were taught the usefulness of approximate analytic solutions and the necessity of being able to solve differential equations numerically when exact or approximate solution techniques could not be found. Hence, approximate analytical solution techniques and numerical solution techniques were also added to the list.

Given a differential equation to analyze, most people spend only a small amount of time using analytical tools and then use a computer to see what the solution "looks like." Because this procedure is so prevalent, this second edition has expanded the section on numerical methods. New sections on finite difference formulas, grid generation, lattice gases, multigrid methods, parallel computers, and software availability have been added. Other parts of this book have also been improved; additional sections include: chaos, existence, uniqueness, and stability theorems, inverse problems, normal forms, and exact partial differential equations.

In writing this book, I have assumed that the reader is familiar with differential equations and their solutions. The object of this book is not to teach novel techniques, but to provide a handy reference to many popular techniques. All the techniques included are elementary in the usual mathematical sense; since this book is designed to be functional it does not include many abstract methods of limited applicability. This handbook has been designed to serve as both a reference book and as a complement to a text on differential equations. Each technique described is accompanied by several current references; these allow each topic to be studied in more detail.

It is hoped that this book will be used by students taking courses in differential equations (at either the undergraduate or graduate level). It will introduce the student to more techniques than they usually see in a differential equations class and will illustrate the different types of techniques. Furthermore, it should act as a concise reference for the techniques that a student has learned. This book should also be useful for the practicing engineer or scientist who solves differential equations on an occasional basis.

A feature of this book is that it has sections dealing with stochastic differential equations and delay differential equations as well as ordinary differential equations and partial differential equations. Stochastic differential equations and delay differential equations are often only studied in advanced texts and courses; yet, the techniques used to analyze these equations are easy to understand and easy to apply.

Had this book been available when I was a graduate student, it would have saved me much time. It has saved me time in solving problems that arose from my own work in industry (the Jet Propulsion Laboratory, Sandia Laboratories, EXXON Research and Engineering, and the MITRE Corporation).

Parts of the text have been utilized in differential equations classes at the Rensselaer Polytechnic Institute. The students' comments have been used to clarify the text. Unfortunately, there may still be some errors in the text; the author would greatly appreciate receiving notice of any such errors. Please send these comments care of Academic Press.

Boston, Mass. 1991   Daniel Zwillinger

Introduction

This book is a compilation of the most important and widely applicable methods for solving and approximating differential equations. As a reference book, it provides convenient access to these methods and contains examples showing their use.

The book is divided into four parts. The first part is a collection of transformations and general ideas about differential equations. This section of the book describes the techniques needed to determine if a partial differential equation is well-posed, what the "natural" boundary conditions are, and many other things. At the beginning of this section is a list of definitions for many of the terms describing differential equations and their solutions.

The second part of the book is a collection of exact analytical solution techniques for differential equations. The techniques are listed (nearly) alphabetically. First is a collection of techniques for ordinary differential equations, then a collection of techniques for partial differential equations. Those techniques that can be used for both ordinary differential equations and partial differential equations have a star ($*$) next to the method name. For nearly every technique the following are given:

- · the types of equations to which the method is applicable
- · the idea behind the method
- · the procedure for carrying out the method
- · at least one simple example of the method
- · any cautions that should be exercised
- · notes for more advanced users
- · references to the literature for more discussion or more examples

The material for each method has deliberately been kept short to simplify use. Proofs have been intentionally omitted.

It is hoped that, by working through the simple example(s) given, the method will be understood. Enough insight should be gained from working the example(s) to apply the method to other equations. Further references are given for each method so that the principle may be studied in more detail, or more examples seen. Note that not all of the references listed at the end of a method may be referred to in the text.

The author has found that computer languages that perform symbolic manipulations (such as MACSYMA) are very useful for performing the calculations necessary to analyze differential equations. For some of the exact analytical techniques, illustrative MACSYMA programs† are given.

Not all differential equations have exact analytical solutions; sometimes an approximate solution will have to do. Other times, an approximate solution may be *more* useful than an exact solution. For instance, an exact solution in terms of a slowly converging infinite series may be laborious to approximate numerically. The same problem may have a simple approximation that indicates some characteristic behavior or allows a numerical value to be obtained.

The third part of this book deals with approximate analytical solution techniques. For the methods in this part of the book, the format is similar to that used for the exact solution techniques. We classify a method as an approximate method if it gives some information about the solution, but will not give the solution of the original equation(s) at all values of the independent variable(s). The methods in this section describe, for example, how to determine perturbation expansions for the solutions to a differential equation.

When an exact or an approximate solution technique cannot be found, it may be necessary to find the solution numerically. Other times, a numerical solution may convey more information than an exact or approximate analytical solution. The fourth part of this book is concerned with the most important methods for finding numerical solutions of common types of differential equations. Although there are many techniques available for numerically solving differential equations, this book has only tried to illustrate the main techniques for each class of problem. At the beginning of the fourth section is a brief introduction to the terms used in numerical methods.

When possible, a short FORTRAN program† has been given. Once again, those techniques that can be used for both ordinary differential

---

† We make no warranties, express or implied, that these programs are free of error. The author and publisher disclaim all liability for direct or consequential damages resulting from your use of the programs.

equations and partial differential equations have a star next to the method name.

This book is not designed to be read at one sitting. Rather, it should be consulted as needed. Occasionally we have used "ODE" to stand for "ordinary differential equation" and "PDE" to stand for "partial differential equation".

This book contains many references to other books. While some books cover only one or two topics well, some books cover all their topics well. The following books are recommended as a first source for detailed understanding of the differential equation techniques they cover: each is broad in scope and easy to read.

References

[1] C. M. Bender and S. A. Orszag, *Advanced Mathematical Methods for Scientists and Engineers*, McGraw–Hill, New York, 1978.

[2] W. E. Boyce and R. C. DiPrima, *Elementary Differential Equations and Boundary Value Problems*, Fourth Edition, John Wiley & Sons, New York, 1986.

[3] E. Butkov, *Mathematical Physics*, Addison–Wesley Publishing Co., Reading, MA, 1968.

[4] C. R. Chester, *Techniques in Partial Differential Equations*, McGraw–Hill Book Company, New York, 1970.

[5] L. Collatz, *The Numerical Treatment of Differential Equations*, Springer–Verlag, New York, 1966.

[6] C. W. Gear, *Numerical Initial Value Problems in Ordinary Differential Equations*, Prentice–Hall Inc., Englewood Cliffs, NJ, 1971.

[7] E. L. Ince, *Ordinary Differential Equations*, Dover Publications, Inc., New York, 1964.

[8] L. V. Kantorovich and V. I. Krylov, *Approximate Methods of Higher Analysis*, Interscience Publishers, New York, 1958.

How to Use This Book

This book has been designed to be easy to use when solving, or approximating, the solutions to differential equations. This introductory section outlines the procedure for using this book to analyze a given differential equation.

First, determine if the differential equation has been studied in the literature. A list of many such equations may be found in the "Look Up" Section beginning on page 148. If the equation you wish to analyze is contained on one of the lists in that section, then see the indicated reference. This technique is the single most useful technique in this book.

Alternately, if the differential equation that you wish to analyze does not appear on those lists, or if the references do not yield the information you desire, then the analysis that must be performed depends on the type of the differential equation.

Before any other analysis is performed, it must be verified that the equation is well-posed. This means that a solution of the differential equation(s) exists, is unique, and depends continuously on the "data". See pages 14, 50, 80, and 94.

Given an Ordinary Differential Equation

[1] It may be useful to transform the differential equation to a canonical form, or to a form that appears in the "Look Up" Section. For some common transformations, see pages 101–133.

[2] If the equation has some sort of special form, then there may be a specialized solution technique that may work. See the techniques on pages: 230, 233, 338.

[3] If the equation is a

    (A) Bernoulli equation, see page 194.

    (B) Chaplygin equation, see page 438.

    (C) Clairaut equation, see page 196.

    (D) Euler equation, see page 235.

    (E) Lagrange equation, see page 311.

    (F) Riccati equation, see page 332.

[4] If the equation does not depend explicitly on the independent variable, see pages 190 and 350.

[5] If the equation does not depend explicitly on the dependent variable, (undifferentiated) see pages 216 and 349.

[6] If one solution of the equation is known, it may be possible to lower the order of the equation, see page 330.

[7] Are discontinuous terms present? See page 219.

[8] See all of the exact solution techniques, on pages 185–360.

[9] If an approximate solution is desired, see the section "Looking for an Approximate Solution," on page xix.

Given a Partial Differential Equation

Partial differential equations are treated in a different manner from ordinary differential equations; in particular, the *type* of the equation dictates the solution technique. First, determine the type of the partial differential equation; it may be hyperbolic, elliptic, parabolic, or of mixed type (see page 33).

[1] It may be useful to transform the differential equation to a canonical form, or to a form that appears in the "Look Up" Section. For transformations, see the following pages: 118, 139, 144, 390, 400.

[2] The simplest technique for working with partial differential equations, which does not always work, is to "freeze" all but one of the independent variables, and then analyze the resulting partial differential equation or ordinary differential equation. Then the other variables may be added back in, one at a time.

[3] If every term is linear in the dependent variable, then separation of variables may work, see page 419.

[4] If the boundary of the domain must be determined as part of the problem, see the technique on page 262.

[5] See all of the exact solution techniques, on pages 365–432. In addition, many of the techniques that can be used for ordinary differential equations are also applicable to partial differential equations. These techniques are indicated by a star in the method name.

[6] If the equation is hyperbolic:

(A) In principle, the differential equation may be solved using the method of characteristics, see page 368. Often, though, the calculations are impossible to perform analytically.

(B) See the section on the exact solution to the wave equation, page 429.

[7] If an approximate solution is desired, see the section "Looking for an Approximate Solution," on page xix.

Given a System of Differential Equations

[1] First, verify that the system of equations is consistent, see page 39.

[2] Note that many of the methods for a single differential equation may be generalized to handle systems.

[3] By using differential resultants, it may be possible to obtain a single equation, see page 46.

[4] The following methods are for systems of equations:

(A) The method of generating functions, see page 265.

(B) The methods for constant coefficient differential equations, see pages 360 and 384.

(C) Finding integrable combinations, see page 283.

[5] If the system is hyperbolic, then the method of characteristics will work (in principle), see page 368.

[6] See also the method for Pfaffian equations (page 326) and the method for matrix Riccati equations (page 335).

Given a Stochastic Differential Equation

[1] To determine the transition probability density, see the discussion of the Fokker–Planck equation on page 254.

[2] To obtain the moments, without solving the complete problem, see pages 491 and 494.

[3] If the noise appearing in the differential equation is not "white noise," the section on stochastic limit theorems might be useful (see page 545).

[4] To numerically simulate the solutions of a stochastic differential equation, see the technique on page 695.

Given a Delay Equation

See the techniques on page 209.

Looking for an Approximate Solution

[1] If exact bounds on the solution are desired, see the methods on pages: 470, 476, 484.

[2] If the solution has singularities that are to be recovered, see page 503.

[3] If the differential equation(s) can be formulated as a contraction mapping, then approximations may be obtained in a natural way, see page 54.

Looking for a Numerical Solution

[1] It is extremely important that the differential equation(s) be well-posed before a numerical solution is attempted. See the theorem on page 648 for an indication of the problems that can arise.

[2] The numerical solution technique must be stable if the numerical solution is to approximate the true solution of the differential equation. See pages 613, 618 and 621.

[3] It is often easiest to use commercial software packages when looking for a numerical solution, see pages 570 and 586.

[4] If the problem is "stiff," then a method for dealing with "stiff" problems will probably be required, see page 690.

[5] If a low accuracy solution is acceptable, then a Monte Carlo solution technique may be used, see pages 721 and 752.

[6] To determine a grid on which to approximate the solution numerically, see page 606.

[7] To find an approximation scheme that works on a parallel computer, see page 676.

Other Things to Consider

[1] Does the differential equation undergo bifurcations? See page 16.

[2] Is the solution bounded? See pages 476 and 484.

[3] Is the differential equation well-posed? See pages 14 and 94.

[4] Does the equation exhibit symmetries? See pages 314 and 404.

[5] Is the system chaotic? See page 26.

[6] Are some terms in the equation discontinuous? See page 219.

[7] Are there generalized functions in the differential equation? See pages 268 and 279.

[8] Are fractional derivatives involved? See page 258.

[9] Does the equation involve a small parameter? See the perturbation methods (pages 507, 510, 518, 524, 528, 532) or pages 463, 558.

[10] Is the general form of the solution known? See page 354.

[11] Are there multiple time or space scales in the problem? See pages 463 and 524.

I.A

Definitions and Concepts

1. Definition of Terms

Adiabatic Invariants   When the parameters of a physical system vary slowly under the effect of an external perturbation, some quantities are constant to any order of the variable describing the slow rate of change. Such a quantity is called an adiabatic invariant. This does not mean that these quantities are exactly constant, but rather that their variation goes to zero faster then any power of the small parameter.

Analytic   A function is analytic at a point if the function has a power series expansion valid in some neighborhood of that point.

Asymptotic equivalence   Two functions, $f(x)$ and $g(x)$, are said to be *asymptotically equivalent* as $x \to x_0$ if $f(x)/g(x) \sim 1$ as $x \to x_0$, that is: $f(x) = g(x)[1 + o(1)]$ as $x \to x_0$. See Erdélyi [4] for details.

Asymptotic expansions   Given a function $f(x)$ and an asymptotic series $\{g_k(x)\}$ at x_0, the formal series $\sum_{k=0}^{\infty} a_k g_k(x)$, where the $\{a_k\}$ are given constants, is said to be an *asymptotic expansion* of $f(x)$ if $f(x) - \sum_{k=0}^{n} a_k g_k(x) = o(g_n(x))$ as $x \to x_0$ for every n; this is expressed as $f(x) \sim \sum_{k=0}^{\infty} a_k g_k(x)$. Partial sums of this formal series are called *asymptotic*

1

approximations to $f(x)$. Note that the formal series need not converge. See Erdélyi [4] for details.

Asymptotic series   A sequence of functions, $\{g_k(x)\}$, forms an *asymptotic series* at x_0 if $g_{k+1}(x) = o(g_k(x))$ as $x \to x_0$.

Autonomous   An ordinary differential equation is autonomous if the independent variable does not appear explicitly in the equation. For example, $y_{xxx} + (y_x)^2 = y$ is autonomous while $y_x = x$ is not (see page 190).

Bifurcation   The solution of an equation is said to undergo a bifurcation if, at some critical value of a parameter, the number of solutions to the equation changes. For instance, in a quadratic equation with real coefficients, as the constant term changes the number of real solutions can change from 0 to 2 (see page 16).

Boundary data   Given a differential equation, the value of the dependent variable on the boundary may be given in many different ways. For

> **Dirichlet boundary conditions** The dependent variable is prescribed on the boundary. This is also called a boundary condition of the first kind.

> **Homogeneous boundary conditions** The dependent variable vanishes on the boundary.

> **Mixed boundary conditions** A linear combination of the dependent variable and its normal derivative are given on the boundary, or, one type of boundary data is given on one part of the boundary while another type of boundary data is given on a different part of the boundary. This is also called a boundary condition of the third kind.

> **Neumann boundary conditions** The normal derivative of the dependent variable is given on the boundary. This is also called a boundary condition of the second kind. Sometimes the boundary data also includes values of the dependent variable at points interior to the boundary.

Boundary layer   A boundary layer is a small region, near a boundary, in which a function undergoes a large change (see page 510).

Boundary value problem   An ordinary differential equation, where not all of the data is given at one point, is a boundary value problem. For example, the equation $y'' + y = 0$ with the data $y(0) = 1$, $y(1) = 1$ is a boundary value problem.

Characteristics   A hyperbolic partial differential equation can be decomposed into ordinary differential equations along curves known as characteristics. These characteristics are, themselves, determined to be the solutions of ordinary differential equations. See page 368.

Cauchy problem   The Cauchy problem is an initial value problem for a partial differential equation. For this type of problem there are initial conditions, but no boundary conditions.

Commutator   If $L[\cdot]$ and $H[\cdot]$ are two linear differential operators, then the commutator of $L[\cdot]$ and $H[\cdot]$ is defined to be the differential operator given by $[L, H] := L \circ H - H \circ L = -[H, L]$. For example, the commutator of the operators $L[\cdot] = x\dfrac{d}{dx}$ and $H[\cdot] = 1 + \dfrac{d}{dx}$ is

$$[L, H] = \left(x\frac{d}{dx}\right)\left(1 + \frac{d}{dx}\right) - \left(1 + \frac{d}{dx}\right)\left(x\frac{d}{dx}\right) = -\frac{d}{dx}.$$

See Goldstein [6] for details.

Complete   A set of functions is said to be complete on an interval if any other function that satisfies appropriate boundedness and smoothness conditions can be expanded as a linear combination of the original functions. Usually the expansion is assumed to converge in the "mean square," or L_2 sense. For example, the functions $\{u_n(x)\} := \{\sin(n\pi x), \cos(n\pi x)\}$ are complete on the interval $[0, 1]$ since any $C^1[0, 1]$ function, $f(x)$, can be written as

$$f(x) = a_0 + \sum_{n=1}^{\infty}\Big(a_n \cos(n\pi x) + b_n \sin(n\pi x)\Big)$$

for some set of $\{a_n, b_n\}$. See Courant and Hilbert [3] for details.

Complete system   The system of nonlinear partial differential equations: $\{F_k(x_1, \ldots, x_r, y, p_1, \ldots, p_r) = 0 \mid k = 1, \ldots, s\}$, in one dependent variable, $y(\mathbf{x})$, where $p_i = dy/dx_i$, is called a complete system if each $\{F_j, F_k\}$, for $1 \le j, k \le r$, is a linear combination of the $\{F_k\}$. Here $\{\,,\}$ represents the Lagrange bracket. See Iyanaga and Kawada [8], page 1304.

Conservation form   A hyperbolic partial differential equation is said to be in conservation form if each term is a derivative with respect to some variable. That is, it is an equation for $u(\mathbf{x}) = u(x_1, x_2, \ldots, x_n)$ that has the form $\dfrac{\partial f_1(u, \mathbf{x})}{\partial x_1} + \cdots + \dfrac{\partial f_n(u, \mathbf{x})}{\partial x_n} = 0$ (see page 43).

Consistency   There are two types of consistency:

Genuine consistency This occurs when the exact solution to an equation can be shown to satisfy some approximations that have been made in order to simplify the equation's analysis.

Apparent consistency This occurs when the approximate solution to an equation can be shown to satisfy some approximations that have been made in order to simplify the equation's analysis.

When simplifying an equation to find an approximate solution, the derived solution must always show apparent consistency. Even then, the approximate solution may not be close to the exact solution, unless there is genuine consistency. See Lin and Segel [9].

Coupled systems of equations A set of differential equations is said to be coupled if there is more than one dependent variable, and each equation involves more than one dependent variable. For example, the system $\{y' + v = 0, \ v' + y = 0\}$ is a coupled system for $\{y(x), v(x)\}$.

Degree The degree of an ordinary differential equation is the greatest number of times the dependent variable appears in any single term. For example, the degree of $y' + (y'')^2 y + 1 = 0$ is 3, while the degree of $y'' y' y^2 + x^5 y = 1$ is 4. The degree of $y' = \sin y$ is infinite. If all the terms in a differential equation have the same degree, then the equation is called equidimensional-in-y (see page 233).

Delay Equation A delay equation, also called a differential delay equation, is an equation that depends on the "past" as well the "present." For example, $y''(t) = y(t - \tau)$ is a delay equation when $\tau > 0$.

Determined A truncated system of differential equations is said to be determined if the inclusion of any higher order terms cannot affect the topological nature of the local behavior about the singularity.

differential form A first order differential equation is said to be in differential form if it is written $P(x, y)dx + Q(x, y)dy = 0$.

Dirichlet problem The Dirichlet problem is a partial differential equation with Dirichlet data given on the boundaries. That is, the dependent variable is prescribed on the boundary.

Eigenvalues, Eigenfunctions Given a linear operator $L[\cdot]$ with boundary conditions $B[\cdot]$ there will sometimes exist non-trivial solutions to the equation $L[y] = \lambda y$ (the solutions may or may not be required to also satisfy $B[y] = 0$). When such a solution exists, the value of λ is called an eigenvalue. Corresponding to the eigenvalue λ there will exist solutions $\{y(\cdot; \lambda)\}$; these are called eigenfunctions. See Stakgold [12] for details.

Elliptic operator   The differential operator $\sum_{i,j=1}^{n} a_{ij} \dfrac{\partial^2}{\partial x_i \partial x_j}$ is an elliptic differential operator if the quadratic form $\mathbf{x}^{\mathrm{T}} A \mathbf{x}$, where $A = (a_{ij})$, is positive definite whenever $\mathbf{x} \neq \mathbf{0}$. If the $\{a_{ij}\}$ are functions of some variable, say t, and the operator is elliptic for all values of t of interest, then the operator is called *uniformly elliptic*. See page 33.

Euler–Lagrange equation   If $u = u(x)$ and $J[u] = \int f(u', u, x)\, dx$ then the condition for the vanishing of the variational derivative of J with respect to u, $\dfrac{\delta J}{\delta u} = 0$, is given by the Euler–Lagrange equation:

$$\left(\frac{\partial}{\partial u} - \frac{d}{dx} \frac{\partial}{\partial u'} \right) f = 0.$$

If $w = w(x)$ and $J = \int g(w'', w', w, x)\, dx$, then the Euler–Lagrange equation is

$$\left(\frac{\partial}{\partial w} - \frac{d}{dx} \frac{\partial}{\partial w'} + \frac{d^2}{dx^2} \frac{\partial}{\partial w''} \right) g = 0.$$

If $v = v(x, y)$ and $J = \iint h(v_x, v_y, v, x, y)\, dx\, dy$, then the Euler–Lagrange equation is

$$\left(\frac{\partial}{\partial v} - \frac{d}{dx} \frac{\partial}{\partial v_x} - \frac{d}{dy} \frac{\partial}{\partial v_y} \right) h = 0.$$

See page 88 for more details.

First integral: ODE   When a given differential equation is of order n and, by a process of integration, an equation of order $n - 1$ involving an arbitrary constant is obtained, then this new equation is known as a first integral of the given equation. For example, the equation $y'' + y = 0$ has the equation $(y')^2 + y^2 = C$ as a first integral.

First Integral: PDE   A function $u(x, y, z)$ is called a first integral of the vector field $\mathbf{V} = (P, Q, R)$ (or of its associated system: $\dfrac{dx}{P} = \dfrac{dy}{Q} = \dfrac{dz}{R}$) if, at every point in the domain, $\mathbf{V}$ is orthogonal to grad u, i.e.,

$$\mathbf{V} \cdot \nabla u = P\frac{\partial u}{\partial x} + Q\frac{\partial u}{\partial y} + R\frac{\partial u}{\partial z} = 0.$$

Conversely, any solution of this partial differential equation is a first integral of $\mathbf{V}$. Note that if $u(x, y, z)$ is a first integral of $\mathbf{V}$, then so is $f(u)$.

Fréchet derivative, Gâteaux derivative   The Gâteaux derivative of the operator $N[\cdot]$, at the "point" $u(\mathbf{x})$, is the linear operator defined by

$$L[z(\mathbf{x})] = \lim_{\varepsilon \to 0} \frac{N[u + \varepsilon z] - N[u]}{\varepsilon}.$$

For example, if $N[u] = u^3 + u'' + (u')^2$ then $L[z] = 3u^2 z + z'' + 2u'z'$. If, in addition,

$$\lim_{||h|| \to 0} \frac{||N[u + h] - N[u] - L[u]h||}{||h||} = 0$$

(as is true in our example), then $L[u]$ is also called the Fréchet derivative of $N[\cdot]$. See Olver [11] for details.

Fuchsian equation   A Fuchsian equation is an ordinary differential equation whose only singularities are regular singular points.

Fundamental matrix   The vector ordinary differential equation $\mathbf{y}' = A\mathbf{y}$ for $y = y(\mathbf{x})$, where A is a matrix, has the fundamental matrix $\Phi(\mathbf{x})$ if Φ satisfies $\Phi' = A\Phi$ and the determinant of Φ is non-vanishing (see page 97).

General solution   Given an n-th order linear ordinary differential equation, the general solution contains all n linearly independent solutions, with a constant multiplying each one. For example, the differential equation $y'' + y = 1$ has the general solution $y(x) = 1 + A \sin x + B \cos x$, where A and B are arbitrary constants.

Green's function   A Green's function is the solution of a linear differential equation which has a delta function appearing either in the equation or in the boundary conditions (see page 268).

Harmonic functions   A function $\phi(\mathbf{x})$ is harmonic if it satisfies Laplace's equation: $\nabla^2 \phi = 0$.

Hodograph   In a partial differential equation, if the independent variables and dependent variables are switched, then the space of independent variables is called the hodograph space (in two dimensions, the hodograph plane). See page 390 for more details.

Homogeneous equations – 1   An equation is said to be homogeneous if all terms depend linearly on the dependent variable or it's derivatives. For example, the equation $y_{xx} + xy = 0$ is homogeneous while the equation $y_{xx} + y = 1$ is not.

Homogeneous equations – 2   A first order ordinary differential equation is said to be homogeneous if the forcing function is a ratio of homogeneous polynomials (see page 276).

Ill-posed problems   A problem that is not well-posed is said to be ill-posed. Typical ill-posed problems are the Cauchy problem for the Laplace equation, the initial/boundary value problem for the backwards heat equation, and the Dirichlet problem for the wave equation. See page 94 for more details.

Initial value problem   A ordinary differential equation with all of the data given at one point is an initial value problem. For example, the equation $y'' + y = 0$ with the data $y(0) = 1$, $y'(0) = 1$ is an initial value problem.

Involutory transformation   An involutory transformation T is one which, when applied twice, does not change the original system; i.e., T^2 is equal to the identity function.

L_2 functions   A function $f(x)$ is said to belong to L_2 if $\int_0^\infty |f(x)|^2 \, dx$ is finite.

Lagrange bracket   If $\{F_j\}$ and $\{G_j\}$ are sets of functions of the independent variables $\{u, v, \ldots\}$ then the Lagrange bracket of u and v is defined to be

$$\{u, v\} = \sum_j \left(\frac{\partial F_j}{\partial u} \frac{\partial G_j}{\partial v} - \frac{\partial F_j}{\partial v} \frac{\partial G_j}{\partial u} \right) = - \{v, u\} \, .$$

See Goldstein [6] for details.

Lagrangian derivative   The Lagrangian derivative (also called the material derivative) is defined by $\dfrac{DF}{Dt} := \dfrac{\partial F}{\partial t} + \mathbf{v} \cdot \nabla F$, where $\mathbf{v}$ is a given vector. See Iyanaga and Kawada [8], page 669.

Laplacian   The Laplacian is the differential operator usually denoted by ∇^2 (in many books it is represented as Δ). It is defined by $\nabla^2 \phi = \text{div}(\text{grad}\,\phi)$, when ϕ is a scalar. The *vector Laplacian* of a vector is the differential operator denoted by $\vec{\maltese}$ (in most books it is represented as ∇^2). It is defined by $\vec{\maltese}\mathbf{v} = \text{grad}(\text{div}\,\mathbf{v}) - \text{curl}\,\text{curl}\,\mathbf{v}$, when $\mathbf{v}$ is a vector. See Moon and Spencer [10] for details.

Leibniz's rule   Leibniz's rule states that

$$\frac{d}{dt} \left(\int_{f(t)}^{g(t)} h(t, \zeta) \, d\zeta \right) = g'(t)h(t, g(t)) - f'(t)h(t, f(t)) + \int_{f(t)}^{g(t)} \frac{\partial h}{\partial t}(t, \zeta) \, d\zeta.$$

Lie Algebra   A Lie algebra is a vector space equipped with a Lie bracket (often called a commutator) $[x, y]$ that satisfies three axioms:

(A) $[x, y]$ is bilinear (i.e., linear in both x and y separately),

(B) the Lie bracket is anti-commutative (i.e., $[x, y] = -[y, x]$),

(C) the Jacobi identity, $[x, [y, z]] + [y, [z, x]] + [z, [x, y]] = 0$, holds.

See Olver [11] for details.

Limit cycle   A limit cycle is a solution to a differential equation that is a periodic oscillation of finite amplitude. See page 63 for more details.

Linear differential equations   A differential equation is said to be linear if the dependent variable only appears with an exponent of 0 or 1. For example, the equation $x^3 y''' + y' + \cos x = 0$ is a linear equation, while the equation $yy' = 1$ is *nonlinear*.

Linearize   To linearize a nonlinear differential equation means to approximate the equation by a linear differential equation in some region. For example, in regions where $|y|$ is "small," the nonlinear ordinary differential equation $y'' + \sin y = 0$ could be linearized to $y'' + y = 0$.

Lipschitz condition   If $f(x, y)$ is a bounded continuous function in a domain D, then $f(x, y)$ satisfies a Lipschitz condition in y in D if

$$|f(x, y_1) - f(x, y_2)| \leq K_y |y_1 - y_2|$$

for some finite constant K_y, independent of x, y_1, and y_2 in D. If, for some finite constant K_x, $f(x, y)$ satisfies

$$|f(x_1, y) - f(x_2, y)| \leq K_x |x_1 - x_2|$$

independent of x_1, x_2, and y in D, then $f(x, y)$ satisfies a Lipschitz condition in x in D. If both of these conditions are satisfied and $K = \max(K_x, K_y)$, then $f(x, y)$ satisfies a Lipschitz condition in D, with Lipschitz constant K. This also extends to higher dimensions. See Coddington and Levinson [2] for details.

Maximum principle   There are many "maximum principles" in the literature. The most common is "a harmonic function attains its absolute maximum on the boundary." See page 484 for more details.

Mean value theorem   This is a statement about the solution of Laplace's equation. It states, "If $\nabla^2 u = 0$ (in N dimensions), then $u(\mathbf{z}) = \int_S u \, dS / \int_S dS$ where S is the boundary of a N-dimensional sphere centered at z." For example, in $N = 2$, we have, "In 2 dimensions, the value of a solution to Laplace's equation at a point is the average of the values on any circle about that point." See Iyanaga and Kawada [8], page 624.

Natural Hamiltonian A natural Hamiltonian is one having the form $H = T + V$, where $T = \frac{1}{2}\sum_{k=1}^{n} p_k^2$ and V is a function of the position variables only (i.e., $V = V(\mathbf{q}) = V(q_1, \ldots, q_n)$).

Neumann problem The Neumann problem is a partial differential equation with Neumann data given on the boundaries. That is, the normal derivative of the dependent variable is given on the boundary. See Iyanaga and Kawada [8], page 999.

Metaparabolic equations A metaparabolic equation has the form $L[u] + M[u_t] = 0$, where $u = u(\mathbf{x}, t)$, $L[\cdot]$ is a linear differential operator in x of degree n, $M[\cdot]$ is a linear differential operator in x of degree m, and $m < n$. If, conversely, $m > n$, then the equation is called *pseudoparabolic*. See Gilbert and Jensen [5] for details.

Near identity transformation A near-identity transformation is a transformation in a differential equation from the old variables $\{a, b, c, \ldots\}$ to the new variables $\{\alpha, \beta, \gamma, \ldots\}$ via

$$a = \alpha + A(\alpha, \beta, \gamma, \ldots),$$
$$b = \beta + B(\alpha, \beta, \gamma, \ldots),$$
$$c = \gamma + C(\alpha, \beta, \gamma, \ldots),$$
$$\vdots$$

where $\{A, B, C, \ldots\}$ are strictly nonlinear functions (i.e., there are no linear or constant terms). Very frequently $\{A, B, C, \ldots\}$ are taken to be homogeneous polynomials (of, say, degree N) in the variables $\alpha, \beta, \gamma, \ldots$, with unknown coefficients. For example, in two variables we might take

$$A(\alpha, \beta) = \sum_{j=0}^{n} A_{j,n-j}\alpha^j \beta^{n-j},$$

$$B(\alpha, \beta) = \sum_{j=0}^{n} B_{j,n-j}\alpha^j \beta^{n-j},$$

for some given value of n (see page 70).

Normal form An ordinary differential equation is said to be in normal form if it can be solved explicitly for the highest derivative; i.e., $y^{(n)} = G(x, y, y', \ldots, y^{(n-1)})$. A system of partial differential equations (with dependent variables $\{u_1, u_2, \ldots, u_m\}$ and independent variables $\{x, y_1, y_2, \ldots, y_k\}$) is said to be in normal form if it has the form

$$\frac{\partial^r u_j}{\partial x^r} = F_j\left(x, y_1, \ldots, y_k, u_1, \ldots, u_m, \frac{\partial u_1}{\partial x}, \ldots, \frac{\partial^{r-1} u_m}{\partial x^{r-1}}, \ldots, \frac{\partial u_1}{\partial y_1}, \ldots, \frac{\partial^r u_m}{\partial y_k^r}\right),$$

for $j = 1, 2, \ldots, m$. See page 70 or Iyanaga and Kawada [8], page 988.

Nonlinear   A differential equation that is not linear in the dependent variable is nonlinear.

Order of a differential equation   The order of a differential equation is the greatest number of derivatives in any term in the differential equation. For example, the partial differential equation $u_{xxxx} = u_{tt} + u^5$ is of fourth order while the ordinary differential equation $v_x + x^2 v^3 + v = 3$ is of first order.

Orthogonal   Two vectors, $\mathbf{x}$ and $\mathbf{y}$, are said to be orthogonal, with respect to the matrix W if $\mathbf{x}^{\mathrm{T}} W \mathbf{y} = 0$ (often, W is taken to be the identity matrix). Two functions, say $f(x)$ and $g(x)$, are said to be orthogonal with respect to a weighting function $w(x)$ if $(f(x), g(x)) := \int f(x) w(x) \bar{g}(x)\, dx = 0$ over some appropriate range of integration. Here, an overbar indicates the complex conjugate.

Padé approximant   A Padé approximant is a ratio of polynomials. The polynomials are usually chosen so that the Taylor series of the ratio is a prescribed function. See page 503 for more details.

Particular solution   Given a linear differential equation, $L[y] = f(\mathbf{x})$, the general solution can be written as $y = y_p + \sum_i C_i y_i$ where y_p, the particular solution, is any solution that satisfies $L[y] = f(\mathbf{x})$. The y_i are homogeneous solutions that satisfy $L[y] = 0$, and the $\{C_i\}$ are arbitrary constants. If $L[\cdot]$ is an n-th order differential operator then there will be n linearly independent homogeneous solutions.

Poisson bracket   If f and g are functions of $\{p_j, q_j\}$ then the Poisson bracket of f and g is defined to be

$$[f, g] = \sum_j \left(\frac{\partial f}{\partial q_j} \frac{\partial g}{\partial p_j} - \frac{\partial f}{\partial p_j} \frac{\partial g}{\partial q_j} \right) = \sum_j \frac{\partial(f, g)}{\partial(q_j, p_j)} = -[g, f].$$

The Poisson bracket is invariant under a change of independent variables. See Goldstein [6] or Olver [11] for details.

Quasilinear equations   A partial differential equation is said to be quasilinear if it is linear in the first partial derivatives. That is, it has the form $\displaystyle \sum_{k=1}^{n} A_k(u, \mathbf{x})\, \frac{\partial u}{\partial x_k} = B(u, \mathbf{x})$ when the dependent variable is $u(\mathbf{x}) = u(x_1, x_2, \ldots, x_n)$ (see page 368).

Radiation condition   The radiation condition states that a wave equation has no waves incoming from an infinite distance, only outgoing waves. For example, the equation $u_{tt} = \nabla^2 u$ might have the radiation condition

$u(x,t) \simeq A_- \exp(ik(t - x))$ as $x \to -\infty$ and $u(x,t) \simeq A_+ \exp(ik(t + x))$ as $x \to +\infty$. This is also called the Sommerfeld radiation condition. See Butkov [1] for details.

Riemann's P function Riemann's differential equation (see page 156) is the most general second order linear ordinary differential equation with three regular singular points. If these singular points are taken to be a, b, and c, and the exponents of the singularities are taken to be α, α'; β, β'; γ, γ' (where $\alpha + \alpha' + \beta + \beta' + \gamma + \gamma' = 1$), then the solution to Riemann's differential equation may written in the form of Riemann's P function as

$$y(x) = P \left\{ \begin{array}{ccc} a & b & c \\ \alpha & \beta & \gamma & x \\ \alpha' & \beta' & \gamma' \end{array} \right\}.$$

Robbins Problem An elliptic partial differential equation with mixed boundary conditions is called a Robbins problem. See Iyanaga and Kawada [8], page 999.

Schwarzian derivative If $y = y(x)$ then the Schwarzian derivative of y with respect to x is defined by $\{x, y\} := -\dfrac{1}{(y')^2} \left[\dfrac{y'''}{y''} - \dfrac{3}{2} \left(\dfrac{y''}{y'} \right)^2 \right]$. Note that $\{x, y\} = -\left(\dfrac{dx}{dy} \right)^2 \{y, x\}$ and, if $y = y(z)$, then $\{s, z\} = \{s, y\} \left(\dfrac{dy}{dz} \right)^2 + \{y, z\}$. See Ince [7] for details.

Shock A shock is a narrow region in which the dependent variable undergoes a large change. Also called a "layer" or a "propagating discontinuity." (See page 368.

Singular points Given the homogeneous n-th order linear ordinary differential equation

$$y^{(n)} + q_{(n-1)}(x)y^{(n-1)} + q_{(n-2)}(x)y^{(n-2)} + \cdots + q_0(x)y = 0,$$

the point x_0 is classified as being an

Ordinary point: if each of the $\{q_i\}$ are analytic at $x = x_0$.

Singular point: if it is not an ordinary point.

Regular singular point: if it is not an ordinary point and $(x - x_0)^i q_i(x)$ is analytic for $i = 0, 1, \ldots, n$.

Irregular singular point: if it is not an ordinary point and not a regular singular point.

The point at infinity is classified by changing variables to $t = 1/x$ and then analyzing the point $t = 0$. See page 342 for more details.

Singular solutions   A singular solution is a solution of a differential equation that is not derivable from the general solution by any choice of the arbitrary constants appearing in the general solution. Only nonlinear equations have singular solutions. See page 540 for more details.

Stability   The solution to a differential equation is said to be stable if small perturbations in the initial conditions, boundary conditions, or coefficients in the equation itself lead to "small" changes in the solution. There are many different types of stability that are useful.

> **Stable** A solution $\mathbf{y}(x)$ of the system $\mathbf{y}' = f(\mathbf{y}, x)$ that is defined for $x > 0$ is said to be stable if, given any $\varepsilon > 0$, there exists a $\delta > 0$ such that any solution $\mathbf{w}(x)$ of the system satisfying $|\mathbf{w}(0) - \mathbf{y}(0)| < \delta$ also satisfies $|\mathbf{w}(x) - \mathbf{y}(x)| < \varepsilon$.
>
> **Asymptotic stability** The solution $\mathbf{u}(x)$ is said to be asymptotically stable if, in addition to being stable, $|\mathbf{w}(x) - \mathbf{u}(x)| \to 0$ as $x \to \infty$.
>
> **Relative stability** The solution $\mathbf{u}(x)$ is said to be relatively stable if $|\mathbf{w}(0) - \mathbf{u}(0)| < \delta$ implies that $|\mathbf{w}(x) - \mathbf{u}(x)| < \varepsilon\mathbf{u}(x)$.

See page 80 or Coddington and Levinson [2] (Chapter 13) for details.

Stefan problem   A Stefan problem is one in which the boundary of the domain must be solved as part of the problem. For instance, when a jet of water leaves an orifice, not only must the fluid mechanics equations be solved in the stream, but the boundary of the stream must also be determined. Stefan problems are also called free boundary problems (see page 262).

Superposition principle   If $u(\mathbf{x})$ and $v(\mathbf{x})$ are solutions to a linear differential equation (ordinary differential equation or partial differential equation), then the superposition principle states that $\alpha u(\mathbf{x}) + \beta v(\mathbf{x})$ is also a solution, where α and β are any constants (see page 352).

Total differential equations   A total differential equation is an equation of the form: $\sum_k a_k(\mathbf{x}) \, dx_k = 0$. See page 326 for more details.

Trivial solution   The trivial solution is the identically zero solution.

Turning points   Given the equation $y'' + p(x)y = 0$, points at which $p(x) = 0$ are called turning points. The asymptotic behavior of $y(x)$ can change at these points. See Wasow [13] for details.

Weak solution   A weak solution to a differential equation is a function that satisfies only an integral form of the defining equation. For example, a weak solution of the differential equation $a(x)y'' - b(x) = 0$ only needs to satisfy $\int_S [a(x)y'' - b(x)] \, dx = 0$ where S is some appropriate region. For this example, the weak solution may not be twice differentiable everywhere. See Zauderer [14] for details.

Well-posed problems A problem is said to be well-posed if a unique, stable solution exists that depends continuously on the data. See page 94 for more details.

Wronskian Given the functions $\{y_1, y_2, \ldots, y_n\}$, the Wronskian is the determinant

$$\begin{vmatrix} y_1 & y_2 & \cdots & y_n \\ y_1' & y_2' & \cdots & y_n' \\ \vdots & \vdots & \ddots & \vdots \\ y_1^{(n)} & y_2^{(n)} & \cdots & y_n^{(n)} \end{vmatrix}.$$

If the Wronskian does not vanish, then the functions are linearly independent. See page 97.

References

[1] E. Butkov, *Mathematical Physics*, Addison–Wesley Publishing Co., Reading, MA, 1968, page 617.

[2] E. A. Coddington and N. Levinson, *Theory of Ordinary Differential Equations*, McGraw–Hill Book Company, New York, 1955.

[3] R. Courant and D. Hilbert, *Methods of Mathematical Physics*, Interscience Publishers, New York, 1953, pages 51–54.

[4] A. Erdélyi, *Asymptotic Expansions*, Dover Publications, Inc., New York, 1956.

[5] R. P. Gilbert and J. Jensen, "A Computational Approach for Constructing Singular Solutions of One-Dimensional Pseudoparabolic and Metaparabolic Equations," *SIAM J. Sci. Stat. Comput.*, **3**, No. 1, March 1982, pages 111–125.

[6] H. Goldstein, *Classical Mechanics*, Addison–Wesley Publishing Co., Reading, MA, 1950.

[7] E. L. Ince, *Ordinary Differential Equations*, Dover Publications, Inc., New York, 1964, page 394.

[8] S. Iyanaga and Y. Kawada, *Encyclopedic Dictionary of Mathematics*, MIT Press, Cambridge, MA, 1980.

[9] C. C. Lin and L. A. Segel, *Mathematics Applied to Deterministic Problems in the Natural Sciences*, Macmillan, New York, 1974, page 188.

[10] P. Moon and D. E. Spencer, "The Meaning of the Vector Laplacian," *J. Franklin Institute*, **256**, 1953, pages 551–558.

[11] P. J. Olver, *Applications of Lie Groups to Differential Equations*, Graduate Texts in Mathematics # 107, Springer–Verlag, New York, 1986.

[12] I. Stakgold, *Green's Functions and Boundary Value Problems*, John Wiley & Sons, New York, 1979, Chapter 7 (pages 411–466).

[13] W. Wasow, *Linear Turning Point Theory*, Applied Mathematical Sciences, **54**, Springer–Verlag, New York, 1985.

[14] E. Zauderer, *Partial Differential Equations of Applied Mathematics*, John Wiley & Sons, New York, 1983, pages 288–294.

2.   Alternative Theorems

Applicable to   Linear ordinary differential equations.

Idea

It is often possible to determine when a linear ordinary differential equation has a unique solution. Also, when the solution is not unique, it is sometimes possible to describe the degrees of freedom that make it non-unique.

Procedure

Alternative theorems describe, in some way, the type of solutions to expect from linear differential equations. The most common alternative theorems for differential equations were derived by Fredholm.

Suppose we wish to analyze the n-th order linear inhomogeneous ordinary differential equation with boundary conditions

$$L[u] = f(x),$$
$$B_i[u] = 0, \qquad \text{for } i = 1, 2, \ldots, n, \tag{2.1}$$

for $u(x)$ on the interval $x \in [a, b]$. First we must analyze the homogeneous equation and the adjoint homogeneous equation. That is, consider the two problems

$$L[u] = 0,$$
$$B_i[u] = 0, \qquad \text{for } i = 1, 2, \ldots, n, \tag{2.2}$$

and

$$L^*[v] = 0,$$
$$B_i^*[v] = 0, \qquad \text{for } i = 1, 2, \ldots, n, \tag{2.3}$$

where $L^*[\cdot]$ is the adjoint of $L[\cdot]$, and $\{B_i^*[\cdot]\}$ are the adjoint boundary conditions (see page 74). Then Fredholm's alternative theorem states that

[1] If the system in (2.2) has only the trivial solution, that is $u(x) \equiv 0$, then

  (A) the system in (2.3) has only the trivial solution;

  (B) the system in (2.1) has a unique solution.

[2] Conversely, if the system in (2.2) has k linearly independent solutions, say $\{u_1, u_2, \ldots, u_k\}$, then:

(A) the system in (2.3) has k linearly independent solutions, say $\{v_1, v_2, \ldots, v_k\}$;

(B) the system in (2.1) has a solution if and only if the forcing function appearing in (2.1), f, is orthogonal to all solutions to the adjoint system. That is $(f, v_i) := \int_a^b f(x)v_i(x)\,dx = 0$ for $i = 1, 2, \ldots, k$;

(C) the solution to (2.1), if (B) is satisfied, is given by $u(x) = \bar{u}(x) + \sum_{j=i}^{k} c_j u_j(x)$ for any arbitrary constants $\{c_j\}$ where $\bar{u}(x)$ is any solution to (2.1).

Example 1

Given the ordinary differential equation for $u(x)$

$$u' + u = f(x),$$
$$u(0) = 0,$$
$$(2.4)$$

we form the homogeneous system

$$u' + u = 0,$$
$$u(0) = 0.$$
$$(2.5)$$

Since (2.5) has only the trivial solution, we know that the solution to (2.4) is unique. By the method of integrating factors (see page 305) the solution to (2.4) is found to be $u(x) = \int_0^x f(t)e^{t-x}dt$.

Example 2

Given the ordinary differential equation for $u(x)$

$$u' + u = f(x),$$
$$u(0) - eu(1) = 0,$$
$$(2.6)$$

we form the homogeneous system

$$u' + u = 0,$$
$$u(0) - eu(1) = 0.$$
$$(2.7)$$

In this case, (2.7) has the single non-trivial solution $u(x) = e^{-x}$. Hence, the solution to (2.6) is *not* unique. To find out what restrictions must be placed on $f(x)$ for (2.6) to have a solution, consider the corresponding adjoint homogeneous equation

$$v' - v = 0,$$
$$-ev(0) + v(1) = 0.$$
$$(2.8)$$

Since (2.8) has a single non-trivial solution, $v(x) = e^x$, we conclude that (2.6) has a solution if and only if

$$\int_0^1 f(t)e^t \, dt = 0. \tag{2.9}$$

If (2.9) is satisfied, then the solution of (2.6) will be given by

$$u(x) = Ce^{-x} + \int_0^x f(t)e^{t-x}dt$$

where C is an arbitrary constant.

Notes
[1] Epstein [1] discusses the Fredholm theorems in the general setting of a Banach space and a Hilbert space.
[2] A generalized Green's function is a Green's function (see page 268) for a differential equation that does not have a unique solution. See Greenberg [2] for more details.

References
[1] B. Epstein, *Partial Differential Equations An Introduction*, McGraw–Hill Book Company, New York, 1962, pages 83 and 111.
[2] M. D. Greenberg, *Application of Green's Functions in Science and Engineering*, Prentice–Hall Inc., Englewood Cliffs, NJ, 1971.
[3] R. Haberman, *Elementary Applied Partial Differential Equations*, Prentice–Hall Inc., Englewood Cliffs, NJ, 1983, pages 307–314.
[4] I. Stakgold, *Green's Functions and Boundary Value Problems*, John Wiley & Sons, New York, 1979, pages 82–90, 207–214, and 319–323.

3.   Bifurcation Theory

Applicable to   Nonlinear differential equations.

Idea

Given a nonlinear differential equation that depends on a set of parameters, the number of distinct solutions may change as the parameters change. Points where the number of solutions change are called *bifurcation points*.

Procedure

While bifurcations occur in all types of equations, we restrict our discussion to ordinary differential equations. Consider the autonomous system

$$\frac{d\mathbf{x}}{dt} = \mathbf{f}(\mathbf{x}; \boldsymbol{\alpha}), \qquad (3.1)$$

where $\mathbf{x}$ and $\mathbf{f}$ are n-dimensional vectors and $\boldsymbol{\alpha}$ is a set of parameters. Define the Jacobian matrix by

$$J(\mathbf{x}; \boldsymbol{\alpha}) := \frac{d\mathbf{f}}{d\mathbf{x}} = \left(\frac{\partial f_i}{\partial x_j}(\mathbf{x}; \boldsymbol{\alpha}) \mid i, j = 1, \ldots, n \right). \qquad (3.2)$$

Note that $J(\mathbf{x}; \boldsymbol{\alpha})\mathbf{z}$ is the *Fréchet derivative* of $\mathbf{f}$, at the point $\mathbf{x}$ (see page 6).

Using the solution $\mathbf{x}(t, \boldsymbol{\alpha})$ of (3.1), the values of $\boldsymbol{\alpha}$ where one or more of the eigenvalues of J are zero are defined to be bifurcation points. At such points the number of (real) solutions to (3.1) may change, and the stability of the solutions might also change.

If any of the eigenvalues have positive real parts, then that solution is unstable. If we are only concerned with the steady state solutions of (3.1), as is often the case, then the bifurcation points will satisfy the simultaneous equations

$$\mathbf{f}(\mathbf{x}; \boldsymbol{\alpha}) = \mathbf{0}, \qquad \det J = 0. \qquad (3.3)$$

Define the eigenvalues of the Jacobian matrix in (3.2) to be $\{\lambda_i \mid i = 1, \ldots, n\}$. We now presume that (3.1) depends on the single parameter α. Suppose that the change in stability is at the point $\alpha = \widehat{\alpha}$, where the real part of a complex conjugate pair of eigenvalues ($\lambda_1 = \overline{\lambda}_2$) pass through zero:

$$\text{Re}\,\lambda_1(\widehat{\alpha}) = 0, \qquad \text{Im}\,\lambda_1(\widehat{\alpha}) > 0, \qquad \text{Re}\,\lambda_1'(\widehat{\alpha}) \neq 0,$$

and, for all values of α near $\widehat{\alpha}$, $\text{Re}\,\lambda_i(\alpha) < 0$ for $i = 3, \ldots, n$.

Then, under certain smoothness conditions, it can be shown that a small amplitude periodic solution exists for α near $\widehat{\alpha}$. Let ε measures the amplitude of the periodic solution. Then, there are functions $\mu(\varepsilon)$ and $\tau(\varepsilon)$, defined for all sufficiently small, real ε, such that $\mu(0) = \tau(0) = 0$ and that the system with $\alpha = \widehat{\alpha} + \mu(\varepsilon)$ has a unique small amplitude solution of period $T = 2\pi\,(1 + \tau(\varepsilon))\,/\,\text{Im}\,\lambda_1(\widehat{\alpha})$. When expanded, we have $\mu(\varepsilon) = \mu_2\varepsilon^2 + O(\varepsilon^3)$. The sign of μ_2 indicates where the oscillations occur, i.e., for $\alpha < \widehat{\alpha}$ or for $\alpha > \widehat{\alpha}$.

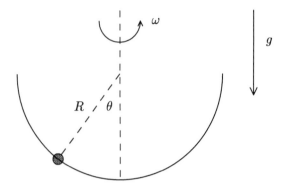

Figure 3.1 A bead on a spinning semi-circular wire.

Example 1

The nonlinear ordinary differential equation

$$\frac{du}{dt} = g(u) = u^2 - \lambda_1 u - \lambda_2 \tag{3.4}$$

has steady states that satisfy $g(u) = u^2 - \lambda_1 u - \lambda_2 = 0$. These steady state solutions have bifurcation points given by

$$\frac{dg}{du} = 2u - \lambda_1 = 0.$$

Solving these last two equations simultaneously, it can be shown that the bifurcation points, of the steady state solutions, are along the curve $4\lambda_2 + \lambda_1^2 = 0$. Further analysis shows that (3.4) will have two real steady state solutions when $4\lambda_2 + \lambda_1^2 > 0$, and it will have no real steady state solutions when $4\lambda_2 + \lambda_1^2 < 0$.

Example 2

Consider a frictionless bead that is free to slide on a semi-circular hoop of wire of radius R that is spinning at an angular rate ω (see Figure 3.1). The equation for $\theta(t)$, the angle of the bead from the vertical, is given by

$$\frac{d^2\theta}{dt^2} + g\sin\theta\left(1 - \frac{\omega^2 R}{g}\cos\theta\right) = 0, \tag{3.5}$$

where g is the magnitude of the gravitational force. We define the parameter ν by $\nu = g/\omega^2 R$. We will only analyze the case $\nu \geq 0$.

The three possible steady solutions of (3.5) are given by:

$$\theta(t) = \theta_1 = 0, \qquad \text{for } \nu \geq 0,$$
$$\theta(t) = \theta_2 = \arccos \nu, \qquad \text{for } \nu \leq 1,$$
$$\theta(t) = \theta_3 = -\arccos \nu, \qquad \text{for } \nu \leq 1.$$

Therefore, for $\nu > 1$ (which corresponds to slow rotation speeds), the only steady solution is $\theta(t) = \theta_1$. For $\nu \leq 1$, however, there are three possible solutions. The solution $\theta(t) = \theta_1$ will be shown to be unstable for $\nu < 1$.

To determine which solution is stable in a region where there are multiple solutions, a stability analysis must be performed. This is accomplished by assuming that the true solution is slightly perturbed from the given solution, and the rate of change of the perturbation is obtained. If the perturbation grows, then the solution is unstable. Conversely, if the perturbation decays (stays bounded), then the solution is stable (neutrally stable).

First we perform a stability analysis for the solution $\theta(t) = \theta_1$. Define

$$\theta(t) = \theta_1 + \varepsilon\phi(t), \tag{3.6}$$

where ε is a small number and $\phi(t)$ is an unknown function. Using (3.6) in (3.5) and expanding all terms for $\varepsilon \ll 1$ results in

$$\frac{d^2\phi}{dt^2} + g\frac{\nu - 1}{\nu}\phi = O(\varepsilon). \tag{3.7}$$

The leading order terms in (3.7) represent the Fréchet derivative of (3.5) at the "point" $\theta(t) = \theta_1$, applied to the function $\phi(t)$. The solution of this differential equation for $\phi(t)$, to leading order in ε, is

$$\phi(t) = A\cos\alpha t + B\sin\alpha t, \tag{3.8}$$

where A and B are arbitrary constants and $\alpha = \sqrt{g\left(\dfrac{\nu - 1}{\nu}\right)}$. If $\nu > 1$, then α is real, and the solutions for $\phi(t)$ remain bounded. Conversely, if $\nu < 1$ then α becomes imaginary, and the solution in (3.8) becomes unbounded as t increases. Hence the solution $\theta(t) = \theta_1$ is unstable for $\nu < 1$.

Now we perform a stability analysis for the solution $\theta(t) = \theta_2$. Writing $\theta(t) = \theta_2 + \varepsilon\psi(t)$, and using this form in (3.5) leads to the equation for $\psi(t)$:

$$\frac{d^2\psi}{dt^2} + g\frac{1 - \nu^2}{\nu}\psi = O(\varepsilon). \tag{3.9}$$

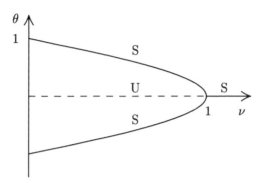

Figure 3.2 Bifurcation diagram for the equation in 3.6. A branch with the label "S" ("U") is a stable (unstable) branch.

The leading order terms in (3.9) represent the Fréchet derivative of (3.5) at the "point" $\theta(t) = \theta_2$, applied to the function $\psi(t)$. The solution of this differential equation for $\psi(t)$ is $\psi(t) = A\cos\beta t + B\sin\beta t$, where A and B are arbitrary constants and $\beta = \sqrt{g\left(\dfrac{1-\nu^2}{\nu}\right)}$. If $\nu < 1$, then β is real, and the solutions for $\psi(t)$ remain bounded. Therefore, the solution $\theta(t) = \theta_2$ is stable for $\nu < 1$. In an exactly analogous manner, $\theta(t) = \theta_3$ is stable for $\nu < 1$.

From what we have found, we can construct the *bifurcation diagram* shown in Figure 3.2. In this diagram, the unstable steady solutions are indicated by a dashed line and the letter U, the stable steady solutions are indicated by the solid line and the letter S. In words this diagram states:

(A) For no rotation ($\omega = 0$ or $\nu = \infty$), the only solution is $\theta(t) = \theta_1 = 0$.

(B) As the frequency or rotation increases (and so ν decreases), the solution $\theta(t) = \theta_1$ becomes unstable at the bifurcation point $\nu = 1$.

(C) For $\nu < 1$, the are two stable solutions, $\theta(t) = \theta_2$ and $\theta(t) = \theta_3$. In this example, there is no way to know in advance which of these two solutions will occur (physically, the bead can slide up either side of the wire).

The formula in (3.3) can applied to (3.5) to determine the location of the bifurcation point, without performing all of the above analysis. If we define $x_1 = \theta$ and $x_2 = \dfrac{d\theta}{dt}$, then (3.5) can be written as the system of ordinary differential equations

$$\frac{d}{dt}\begin{pmatrix} x_1 \\ x_2 \end{pmatrix} = \mathbf{f}(\mathbf{x}) = \begin{pmatrix} x_2 \\ -g\sin x_1\left[1 - \dfrac{\cos x_1}{\nu}\right] \end{pmatrix},$$

which has the Jacobian matrix

$$J = \frac{d\mathbf{f}}{d\mathbf{x}} = \begin{pmatrix} 0 & 1 \\ -g\cos x_1 + \frac{g}{\nu}\left(\cos^2 x_1 - \sin^2 x_1\right) & 0 \end{pmatrix}.$$

If $\nu > 1$, then no choice of (x_1, x_2) will allow both $\mathbf{f}$ and $\det J$ to be zero simultaneously. For $\nu = 1$, however, $x_1 = x_2 = 0$ make both $\mathbf{f}$ and $\det J$ equal to zero. Hence, a bifurcation occurs at $\nu = 1$.

Example 3

Abelson [1] has developed a computer program in LISP that automatically explores the steady-state orbits of one-parameter families of periodically-driven oscillators. The program generates both textual descriptions and schematic diagrams.

For example, consider Duffing's equation in the form $\ddot{x} + 0.1\dot{x} + x^3 = p\cos t$ where the parameter p is in the range $[1, 25]$ and only those solutions with $-5 \leq \dot{x} \leq 5$ and $-10 \leq \ddot{x} \leq 10$ are considered. The program produced the graphical output shown in Figure 3.3, along with the following textual description:

> The system was explored for values of p between 1 and 25, and 10 classes of stable periodic orbits were identified.
>
> Class A is already present at the start of the parameter range $p = 1$ with a family of order-1 orbits A_0. Near $p = 2.287$, there is a supercritical-pitchfork bifurcation, and A_0 splits into symmetric families $A_{1,0}$ and $A_{1,1}$, each of order 1. $A_{1,0}$ vanishes at a fold bifurcation near $p = 3.567$. $A_{1,1}$ vanishes similarly.
>
> Class B appears appears around $p = 3.085$ with a family of order-1 orbits B_0 arising from a fold bifurcation. As the parameter p increases, B_0 undergoes a period doubling cascade, reaching order2 near $p = 4.876$, and order 4 near $p = 5.441$. Although the cascade was not traced past the order 4 orbit, there is apparently another period-doubling near $p = 5.52$, and a chaotic orbit was observed at $p = 5.688$.
>
> $\vdots$
>
> Class J appear around $p = 23.96$ as a family of order-5 orbits J_0 arising from a fold bifurcation. J_0 is present at the end of the parameter range at $p = 25$.

This program is capable of recognizing the following types of bifurcations: fold bifurcations, supercritical and subcritical flip bifurcations, supercritical and subcritical Niemark bifurcations, supercritical and subcritical pitchfork bifurcations, and transcritical bifurcations.

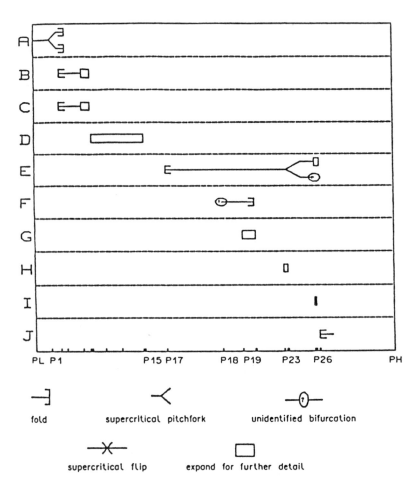

Figure 3.3. Graphical output generated automatically from the Bifurcation Interpreter in Abelson [1]. For Duffing's equation, the evolution of 10 classes of families of periodic orbits and their bifurcations has been traced. The p values along the horizontal axis indicate the parameter value at which the bifurcations occur.

Notes

[1] There are many different types of bifurcations. See Figure 3.4 for diagrams of some of the following bifurcations:

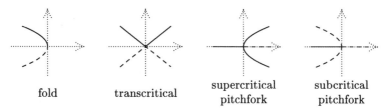

		supercritical	subcritical
fold	transcritical	pitchfork	pitchfork

Figure 3.4 Diagrams of some types of bifurcations. Unstable solutions are indicated by dashed lines, stable solutions are indicated by solid lines.

(A) *Hopf bifurcation*: a stable steady solution bifurcates into a stable oscillatory solution. That is, there are no stable steady solutions in that particular region of parameter space. This occurs by having some of the eigenvalues of the Jacobian in (3.2) become purely imaginary.

(B) *fold bifurcation*: on one side of the bifurcation point a stable and an unstable periodic point (of the same order) coexist. On the other side of the bifurcation point, both periodic points have vanished.

(C) *flip bifurcation (supercritical)*: a stable periodic point of order n transitions to a stable periodic point of order $2n$ and an unstable periodic point of order n.

(D) *flip bifurcation (subcritical)*: an unstable periodic point of order $2n$ and a stable periodic point of order n transition to an unstable periodic point of order n.

(E) *Niemark bifurcation (supercritical)*: a stable periodic transitions to an unstable periodic point and a stable limit cycle.

(F) *Niemark bifurcation (subcritical)*: a stable periodic point and unstable limit cycle transition to an unstable periodic point.

(G) *pitchfork bifurcation (supercritical)*: a stable periodic point transitions to two stable periodic points and an unstable periodic point, all of the same order.

(H) *pitchfork bifurcation (subcritical)*: a stable periodic point and two unstable periodic points transition to an unstable periodic point.

(I) *transcritical bifurcation*: a stable periodic point and an unstable periodic point exchange stabilities; on the other side of the bifurcation point, the extrapolated stable point is now unstable, and vice-versa.

[2] For a differential equation that is not autonomous, bifurcations can also occur from time dependent solutions to other time dependent solutions.

[3] For the general finite dimensional mapping, $G(\mathbf{x})$, from R^m to R^n, the Jacobian $J(\mathbf{x}) := \dfrac{\partial G}{\partial \mathbf{x}}$ need not be square. In this case, the critical points (which include the bifurcation points) are in the set C, with

$$C := \{\mathbf{x} \mid \mathbf{x} \in \mathrm{R}^m,\ \mathrm{rank}\ J(\mathbf{x}) < \min(m, n)\}.$$

The *regular points* are $\mathrm{R}^m - C$. The critical values are the values in the set $G(C) := \{\mathbf{y} \mid \mathbf{y} \in \mathrm{R}^n, \mathbf{y} = G(\mathbf{x})$ for some $\mathbf{x} \in C\}$. The regular values are $\mathrm{R}^n - G(C)$.

[4] Sacks [11] describes the program POINCARE which classifies bifurcation points and constructs representative phase diagrams for each type of behavior. The program is available from the author.

References

[1] H. Abelson, "The Bifurcation Interpreter: A Step Towards the Automatic Analysis of Dynamical Systems," *Comp. & Maths. with Appls.*, **20**, No. 8, 1990, pages 13–35.

[2] B. Eaton and K. Gustafson, "Calculation of Critical Branching Points in Two-Parameter Bifurcation Problems," *J. Comput. Physics*, **50**, 1983, pages 171–177.

[3] J. Guckenheimer, "Patterns of Bifurcations," in P. J. Holmes (ed.), *New Approaches to Nonlinear Problems in Dynamics*, SIAM, Philadelphia, 1980, pages 71–104.

[4] B. D. Hassard, *Theory and Applications of Hopf Bifurcation*, Cambridge University Press, New York, 1981.

[5] M. Holodniok and M. Kubiček, "New Algorithms for the Evaluation of Complex Bifurcation Points in Ordinary Differential Equations. A Comparative Numerical Study," *Appl. Math. and Comp.*, **15**, 1984, pages 261–274.

[6] G. Iooss and D. D. Joseph, *Elementary Stability and Bifurcation Theory*, Second Edition, Springer–Verlag, New York, 1989.

[7] A. D. Jepson and A. Spence "Numerical Methods for Bifurcation Problems," in A. Iserles and M. J. D. Powell (eds.), *The State of the Art in Numerical Analysis*, Clarendon Press, Oxford, 1987, pages 273–298.

[8] M. Kubiček and M. Marek, *Computational Methods in Bifurcation Theory and Dissipative Structures*, Springer–Verlag, New York, 1983.

[9] J. E. Marsden and M. McCracken, "The Hopf Bifurcation and Its Applications," Springer–Verlag, New York, 1976.

[10] R. H. Rand and D. Armbruster, *Perturbation Methods, Bifurcation Theory and Computer Algebra*, Springer–Verlag, New York, 1987.

[11] E. Sacks, "Automatic Analysis of One-Parameter Planar Ordinary Differential Equations by Intelligent Numeric Simulation," *Artificial Intelligence*, **48**, 1991, pages 27–56.

[12] R. Seydel, "From Equilibrium to Chaos: Practical Bifurcation and Stability Analysis," American Elsevier Publishing Company, New York, 1988.

[13] E. F. Wood, J. A. Kempf, and R. K. Mehra, "BISTAB: A Portable Bifurcation and Stability Analysis Package," *Appl. Math. and Comp.*, **15**, 1984, pages 343–355.

4.   A Caveat for Partial Differential Equations

To solve partial differential equations correctly, a good understanding of the nature of the partial differential equation is required. This requires more than a knowledge of the "physics" of the problem: a thorough understanding of the type of partial differential equation is needed. From Collatz [1] (page 260):

> That an investigation of the situation is absolutely essential is revealed even by quite simple examples; they show that formal calculation applied to partial differential equations can lead to false results very easily and that approximate methods can converge in a disarmingly innocuous manner to values bearing no relation to the correct solution.

Example

Suppose we wish to solve the following wave equation

$$u_{xx} = u_{tt},$$
$$u(x,0) = \cos x, \qquad \text{for } |x| < \pi/2,$$
$$\frac{\partial u(x,0)}{\partial t} = \cos x, \qquad \text{for } |x| < \pi/2, \qquad (4.1.a\text{--}d)$$
$$u\left(\pm\frac{\pi}{2}, t\right) = \sin t, \qquad \text{for } t > 0.$$

We will attempt to solve (4.1) by looking for a series solution of the form

$$u(x,t) = \sum_{n,m=0}^{\infty} a_{mn} x^m t^n. \qquad (4.2)$$

Using (4.2) in (4.1.a), we find that

$$a_{m,n+2} = \frac{(m+2)(m+1)}{(n+2)(n+1)} a_{m+2,n}. \qquad (4.3)$$

To satisfy (4.1.b), we require $a_{k,1} = 0$. To satisfy (4.1.c), we also require

$$a_{k,0} = \begin{cases} 0, & k = \text{odd}, \\ (-1)^q/(2q)!, & k = \text{even} = 2q. \end{cases} \qquad (4.4)$$

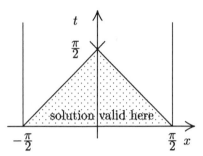

Figure 4. Depiction of the characteristics and the range of validity of the solution found for equation (4.1).

Evaluating (4.2) at $x = 0$ and using (4.3) and (4.4), we find that

$$u(0, t) = \sum_{k=0}^{\infty} a_{0,k} t^k = \sum_{q=0}^{\infty} \frac{(-1)^q}{(2q)!} t^{2q} = \cos t \qquad (4.5)$$

Now the conclusion in (4.5) is correct, but *only* for $0 \le t \le \pi/2$. This is because the characteristics (see page 368), $t = \pi/2 \pm x$, emanating from the points $(\pi/2, 0)$, $(-\pi/2, 0)$ do not allow $u(0, t)$ to be determined directly for $t > \pi/2$.

See Figure 4 for a graphical representation of the characteristics of (4.1) and the region of validity for the solution in (4.5).

Notes
[1] The above example is from Collatz [1].

References
[1] L. Collatz, *The Numerical Treatment of Differential Equations*, Springer–Verlag, New York, 1966.

5.   Chaos in Dynamical Systems

Applicable to   Nonlinear differential equations.

Yields
    Information on whether or not a system is chaotic.

Idea

Chaos is a phenomenon that can appear in solutions to nonlinear differential equations. Chaos is easily defined and can be easily (numerically) found in some equations.

Procedure

For simplicity, we focus on deterministic systems modeled by coupled, autonomous, first order, ordinary differential equations of the form

$$\frac{dx_i}{dt} = g_i(\mathbf{x}; \mathbf{q}) \quad \text{for } i = 1, 2, \ldots, n \tag{5.1}$$

where $\mathbf{x} = (x_1, x_2, \ldots, x_n)$ is the state-space vector and $\mathbf{q} = (q_1, q_2, \ldots, q_m)$ is a set of parameters. This equation determines a set of solutions, each specified by their initial values. We can specify the solution corresponding to the initial condition $\mathbf{p}$ by $\mathbf{x}(t; \mathbf{p})$.

Consider a set of initial conditions contained in a vanishing small volume V. Under the action of (5.1), the volume will change as a function of t. Precisely,

$$\frac{dV}{dt} = \iint_V \cdots \int \left(\sum_{i=1}^{m} \frac{\partial g_i}{\partial x_i} \right) dx_1 \cdots dx_n.$$

The summation term is the generalized divergence of $\mathbf{g}$ and is called the Lie derivative. Dissipative systems are characterized by contracting volumes; this is equivalent to $dV/dt < 0$. Conservative or Hamiltonian systems, in which (5.1) are Hamilton's equations, obey Liouville's theorem: $dV/dt = 0$.

Any trajectory of a dissipative system as $t \to \infty$ will approach a bounded region of phase space called an attractor. An attractor has zero volume in phase space. Attractors include points, limit cycles, and tori. For example, consider an unforced damped pendulum. The attractor for this is a point in phase space, the stable configuration with the pendulum hanging straight down. In this case, starting the pendulum swinging with slightly different initial conditions will lead to close paths in phase space and the same final state.

For nonlinear systems exhibiting chaos, the separation of two nearby trajectories increases exponentially with time. This is referred to as *sensitive dependence on initial conditions*. For dissipative systems, a stretching in one direction has to be accompanied by a more-than-compensating contraction in other directions, so that the volume of an arbitrary droplet of initial conditions will contract with time. The phase-space trajectories for a chaotic system asymptotically approach a *strange attractor*, an attractor with a fractional dimension (i.e., a fractal).

Lyapunov exponents are a measure of the rate of divergence (or convergence) of initially infinitesimally separated trajectories. The ith Lyapunov exponent, λ_i, can be found by considering the evolution of a vanishingly small set of initial conditions which form a hyperellipsoid. We define:

$$\lambda_i := \lim_{\substack{t \to \infty \\ \rho_i(0) \to 0}} \left[\frac{1}{t} \left(\frac{\rho_i(t)}{\rho_i(0)} \right) \right] \tag{5.2}$$

where $\rho_i(t)$ is the length of the ith principal axis of the hyperellipsoid at time t, for $i = 1, 2, \ldots, n$. An attractor is chaotic if it has at least one positive Lyapunov exponent.

The Lyapunov exponents can be determined by analyzing the linearized equations corresponding to (5.1). For illustrative purposes, we specialize to $n = 3$ for the rest of this section. Consider the two close initial points: $\mathbf{p}_0 = (x_0, y_0, z_0)$ and $\mathbf{p}_1 = \mathbf{p}_0 + \delta\mathbf{x} = (x_0 + \delta x, y_0 + \delta y, z_0 + \delta z)$. We want to find the evolution of the difference $\mathbf{a}(t) := \mathbf{x}(t; \mathbf{p}_1) - \mathbf{x}(t; \mathbf{p}_0)$. From Taylor series:

$$\frac{da_1}{dt} = \frac{d[x_1(t; \mathbf{p}_1) - x_1(t; \mathbf{p}_0)]}{dt} = \frac{d[g_1(\mathbf{x}(t; \mathbf{p}_0 + \delta\mathbf{x})) - g_1(\mathbf{x}(t; \mathbf{p}_0))]}{dt}$$

$$\approx \frac{\partial g_1}{\partial x}\delta x + \frac{\partial g_1}{\partial y}\delta y + \frac{\partial g_1}{\partial z}\delta z$$

$$= \frac{\partial g_1}{\partial x}a_1 + \frac{\partial g_1}{\partial y}a_2 + \frac{\partial g_1}{\partial z}a_3,$$

where the partial derivatives are evaluated at $\mathbf{x}(t; \mathbf{p}_0)$. In general:

$$\frac{d\mathbf{a}}{dt} = M(\mathbf{x})\mathbf{a} = \begin{pmatrix} \dfrac{\partial g_1}{\partial x} & \dfrac{\partial g_1}{\partial y} & \dfrac{\partial g_1}{\partial z} \\[2mm] \dfrac{\partial g_2}{\partial x} & \dfrac{\partial g_2}{\partial y} & \dfrac{\partial g_2}{\partial z} \\[2mm] \dfrac{\partial g_3}{\partial x} & \dfrac{\partial g_3}{\partial y} & \dfrac{\partial g_3}{\partial z} \end{pmatrix} \mathbf{a},$$

where M is the Jacobian of the vector $\mathbf{g}$. The Lyapunov exponents are related to the eigenvalues of the matrix M.

In special situations, analytical methods can be used to obtain the Lyapunov spectra, while numerical methods must be used in general. When there is a stationary solution given by $\dfrac{d\mathbf{x}}{dt} = \mathbf{g}(\mathbf{x}) = \mathbf{0}$, the Jacobian matrix is time-independent, and we can analytically obtain the (possibly complex)

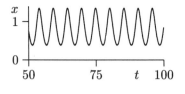
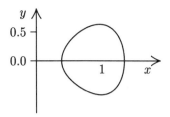

Figure 5.1 Duffing equation with $\Gamma = .20$. (Period 1 solution.)

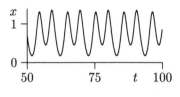
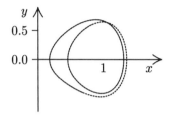

Figure 5.2 Duffing equation with $\Gamma = .28$. (Period 2 solution.)

eigenvalues, from which the Lyapunov exponents may be found. In general, there are no stationary solutions and the equations $\dfrac{d\mathbf{x}}{dt} = \mathbf{g}$ and $\dfrac{d\mathbf{a}}{dt} = M(\mathbf{x})\mathbf{a}$ must be numerically solved simultaneously. See Wolf *et al.* [15] for a numerical technique for computing Lyapunov exponents.

Example

Consider the Duffing equation: $\ddot{x} + k\dot{x} - x + x^3 = \Gamma \cos \omega t$. This can be converted to an autonomous system as follows:

$$\frac{d\mathbf{x}}{dt} = \frac{d}{dt} \begin{pmatrix} x \\ y \\ z \end{pmatrix} = \begin{pmatrix} y \\ -ky + x - x^3 + \Gamma \cos z \\ \omega \end{pmatrix}. \qquad (5.3)$$

In Figure 5.1–Figure 5.3 we show the different behavior of this system ($x(t)$ versus t and $x(t)$ versus $y(t)$) when $k = .3$, $\omega = 1.2$, and Γ takes on the values 0.20, 0.28 and 0.50. For the numerical simulations shown, the initial conditions used were $\mathbf{x}_0 = (1.3, 0, 0)$, and we began plotting the results when $t = 50$ to remove any initial transients. From deeper analysis it can be shown that the system has a period 1 (2, 4, 5, 2, 1) solution when $\Gamma = 0.20$ (0.28, 0.29, 0.37, 0.65, 0.73). The solution is chaotic when $\Gamma = .50$.

A different set of parameters is shown in Figure 5.4. This figure has a plot of the three Lyapunov exponents of (5.3) when $\omega = 1.0$, $k = 0.5$ and Γ is varied from 0.2 to 0.9. At low values of Γ, the system is periodic

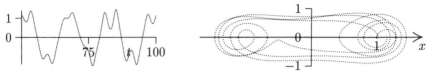

Figure 5.3 Duffing equation with $\Gamma = .50$. (Chaotic solution.)

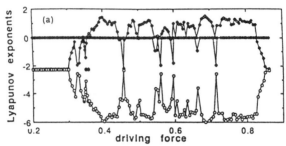

Figure 5.4. The three Lyapunov exponents for Duffing's equation with $\omega = 1.0$ and $k = 0.5$ when Γ is varied from 0.2 to 0.9. (From De Souza-Machado *et al.* [4].)

since the largest Lyapunov exponent is zero. The system follows a period doubling route to chaos at $\Gamma \approx 0.36$, when the largest Lyapunov exponent becomes greater than zero. The system remains chaotic until the driving force gets very large ($\Gamma > 0.84$) except for windows of periodicity, which occur throughout the chaotic regime.

Notes
[1] There are many software packages for numerically computing Lyapunov exponents. See, for example, Rollins [12] and Parker and Chua [11].

[2] There are at least three scenarios in which the regular behavior of a system becomes chaotic. A standard route is via a series of period doubling bifurcations, see page 16. Two other routes to chaos that are fairly well understood are via intermittent behavior and through quasiperiodic solutions.

[3] Many equations have been shown to be chaotic:
(A) Hale and Sternberg [6] have shown that the differential delay equation
$$\frac{dx(t)}{dt} = ax(t) + b\frac{x(t-\tau)}{1 + x^n(t-\tau)}$$ is chaotic for certain parameter regimes.

(B) The equations defining the Lorenz attractor are
$$\dot{x} = 10y - 10x,$$
$$\dot{y} = -y - xz + 28x,$$
$$\dot{z} = xy - \tfrac{8}{3}z.$$

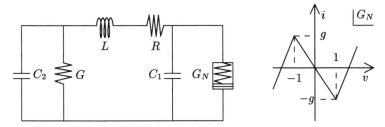

Figure 5.5 The canonical piecewise-linear circuit, and the voltage-current characteristic of the nonlinear resistor G_N.

(C) The Rössler equations are

$$\dot{x} = -(y + z),$$
$$\dot{y} = x + ay,$$
$$\dot{z} = b + xz - cz.$$

When $a = 0.343$, $b = 1.82$, and $c = 9.75$, this generates the "Rössler funnel." When $a = 0.2$, $b = 0.2$, and $c = 5.7$, this generates "the simple Rössler attractor."

[4] For any autonomous electronic circuit to exhibit chaos, it must composed of at least three energy storage devices. (Otherwise, the Poincaré–Bendixson theorem states that the limiting set will be a point or a limit cycle, not a strange attractor.) A simple circuit with three energy storage devices that produces chaos was given in Matsumoto [9].

The circuit given in Chua and Lin [3], see Figure 5.5, is almost as simple as that given by Matsumoto and can simulate (by choosing different values for the nonlinear resistor) all possible chaotic phenomena in a rather large, three-dimensional state space. This circuit contains only six two-terminal elements: five of them are linear resistors, capacitors, and inductors; and only one element (G_N) is a three-segment, piecewise linear resistor.

[5] The papers by Ablowitz and Herbst [1], Yamaguti and Ushiki [16], and Lorenz [7] describe and illustrate how numerical discretizations of a differential equation can lead to discrete equations exhibiting chaos.

[6] Different types of dynamical systems can have greater or lesser degrees of randomness. A simple classification of the amount of randomness in dynamical systems is as follows:

(A) *Ergodic systems*: this is the "weakest" level of randomness, in which phase averages equal time averages.

(B) *Mixing systems*: here, no time averaging is required to reach "equilibrium."

(C) *K-systems*: systems with positive Kolmogorov entropy. This means that a connected neighborhood of trajectories must exhibit a positive average rate of exponential divergence.

(D) *C-systems*: every trajectory has a positive Lyapunov exponent.

(E) *Bernoulli systems*: these systems are as random as a fair coin toss.
See Tabor [14] for details.

[7] In this section we have focused on chaos appearing in coupled, first-order, ordinary differential equations. Chaos can also appear in partial differential equations and stochastic equations.

[8] The journal *Nonlinear Dynamics* has many articles about chaos.

References

[1] M. J. Ablowitz and B. M. Herbst, "On Homoclinic Structure and Numerically Induced Chaos for the Nonlinear Schrödinger Equation," *SIAM J. Appl. Math.*, **50**, No. 2, April 1990, pages 339–351.

[2] B. S. Berger and M. Rokni, "Least Squares Approximation of Lyapunov Exponents," *Quart. Appl. Math.*, **47**, No. 3, 1989, pages 505–508.

[3] L. O. Chua and G.-N. Lin, "Canonical Realization of Chua's Circuit Family," *IEEE Trans. Circ. & Syst.*, **37**, No. 7, July 1990, pages 885–902.

[4] S. De Souza-Machado, R. W. Rollins, D. T. Jacobs, and J. L. Hartman, "Studying Chaotic Systems Using Microcomputer Simulations and Lyapunov Exponents," *Am. J. Phys.*, **58**, No. 4, April 1990, pages 321–329.

[5] R. L. Devaney, *An Introduction to Chaotic Dynamical Systems*, Second Edition, Addison–Wesley Publishing Co., Reading, MA, 1989.

[6] J. K. Hale and N. Sternberg, "Onset of Chaos in Differential Delay Equations," *J. Comput. Physics*, **77**, 1988, pages 221–239.

[7] E. N. Lorenz, "Computational Chaos—A Prelude to Computational Instability," *Physica D*, **35**, 1989, pages 299–317.

[8] M. Marek and I. Schreiber, *Chaotic Behavior of Deterministic Dissipative Systems*, Cambridge University Press, New York, 1991.

[9] T. Matsumoto, "Chaos in Electronic Circuits," *Proc. IEEE*, **75**, No. 8, August 1987.

[10] T. S. Parker and L. O. Chua, "Chaos: A Tutorial for Engineers," *Proc. IEEE*, **75**, No. 8, August 1987, pages 982–1008.

[11] T. S. Parker and L. O. Chua, *Practical Numerical Algorithms for Chaotic Systems*, Springer–Verlag, New York, 1989.

[12] R. W. Rollins, *Chaotic Dynamics Workbench*, Physics Academic Software, AIP, New York, 1990.

[13] R. Seydel, *From Equilibrium to Chaos: Practical Bifurcation and Stability Analysis*, American Elsevier Publishing Company, New York, 1988.

[14] M. Tabor, *Chaos and Integrability in Nonlinear Dynamics*, John Wiley & Sons, New York, 1989.

[15] A. Wolf, J. B. Swift, H. L. Swinney, and J. A. Vastano, "Determining Lyapunov Exponents from a Time Series," *Physica D*, **16**, 1985, pages 285–317.

[16] M. Yamaguti and S. Ushiki, "Chaos in Numerical Analysis of Ordinary Differential Equations," *Physica D*, **3**, 1981, pages 618–626.

6.   Classification of Partial Differential Equations

Applicable to   Partial differential equations.

Yields

Knowledge of the type of equation under consideration.

Procedure

Most partial differential equations are of three basic types: elliptic, hyperbolic, and parabolic.

Elliptic equations are often called potential equations. They result from potential problems, where the potential might be temperature, voltage, or a similar quantity. Elliptic equations are also the steady solutions of diffusion equations, and they require boundary values in order to determine the solution.

Hyperbolic equations are sometimes called wave equations, since they often describe the propagation of waves. They require initial conditions (where the waves start from) as well as boundary conditions (to describe how the wave and boundary interact; for instance, the wave might be scattered or adsorbed). These equations can be solved, in principle, by the method of characteristics (see page 368).

Parabolic equations are often called diffusion equations since they describe the diffusion and convection of some substance (such as heat). The dependent variable usually represents the density of the substance. These equations require initial conditions (what the initial concentration of substance is) as well as boundary conditions (to specify, for instance, whether the substance can cross the boundary or not).

The above classification is most useful for second order partial differential equations. For second order equations only characteristic curves need to be considered. For equations of higher degree, characteristic surfaces must be considered, see Whitham [8] (pages 139–141) or Zauderer [10] for more details. After two special cases, we specialize the rest of this section to second order partial differential equations.

Special Case 1

The most general second order linear partial differential equation with constant coefficients

$$\sum_{i,j=1}^{n} a_{ij} \frac{\partial^2 u}{\partial x_i \partial x_j} + \sum_{i=1}^{n} b_i \frac{\partial u}{\partial x_i} + cu = d,$$

may be placed in the form

$$u_{\xi_1 \xi_1} + \cdots + u_{\xi_n \xi_n} + \lambda u = 0,$$

if the equation is elliptic, or may be placed in the form

$$u_{\xi_1 \xi_1} - u_{\xi_2 \xi_2} - \cdots - u_{\xi_n \xi_n} + \lambda u = 0,$$

if the equation is hyperbolic, for some value of λ. See Garabedian [3] for details.

Special Case 2

The (real valued) second order partial differential equation in n dimensions

$$\sum_{i,j=1}^{n} a_{ij}(\mathbf{x}) \frac{\partial^2 u}{\partial x_i \partial x_j} + f\left(\mathbf{x}, u, \frac{\partial u}{\partial x_1}, \ldots, \frac{\partial u}{\partial x_n}\right) = 0, \qquad (6.1)$$

for $u(\mathbf{x}) = u(x_1, \ldots, x_n)$, where $a_{ij} = a_{ji}$, may be classified at the point $\mathbf{x}_0$ as follows. Let A be the matrix $(a_{ij}(\mathbf{x}_0))$. By means of a linear transformation, the quadratic form $\mathbf{g}^T A \mathbf{g}$ may be reduced to the form

$$\lambda_1 g_1^2 + \lambda_2 g_2^2 + \cdots + \lambda_n g_n^2.$$

The values of $\{\lambda_i\}$, which are the eigenvalues of A, determine the nature of the partial differential equation in (6.1). Since A has been assumed to be symmetric, all of the eigenvalues will be real. The classification at the point $\mathbf{x}_0$ is then given by:

(A) If all of the $\{\lambda_i\}$ are of the same sign, then (6.1) is elliptic at $\mathbf{x}_0$.

(B) If any of the $\{\lambda_i\}$ are zero, then (6.1) is parabolic at $\mathbf{x}_0$.

(C) If none of the $\{\lambda_i\}$ are zero and they are not all of the same sign, then (6.1) is hyperbolic at $\mathbf{x}_0$.

(D) If none of the $\{\lambda_i\}$ are zero and there are at least two that are positive and at least two that are negative, then (6.1) is *ultrahyperbolic* at $\mathbf{x}_0$.

If an equation is parabolic along a smooth curve in a domain D, and the equation is hyperbolic on one side of the curve and elliptic on the other side of the curve, then the equation is of *mixed type*. The smooth curve is called the *curve of parabolic degeneracy*.

Little more can be said in general. We further specialize here and restrict ourselves to second order equations in two independent variables.

Second Order Equations

Consider partial differential equations of second order in two independent variables, of the form

$$A(x,y)\frac{\partial^2 u}{\partial x^2} + B(x,y)\frac{\partial^2 u}{\partial x \partial y} + C(x,y)\frac{\partial^2 u}{\partial y^2} = \Psi\left(u, \frac{\partial u}{\partial x}, \frac{\partial u}{\partial y}, x, y\right), \quad (6.2)$$

where Ψ need not be a linear function.

If $\left\{\begin{array}{l} B^2 - 4AC > 0 \\ B^2 - 4AC = 0 \\ B^2 - 4AC < 0 \end{array}\right\}$ at some point (x,y), then (6.2) is $\left\{\begin{array}{l} \text{hyperbolic} \\ \text{parabolic} \\ \text{elliptic} \end{array}\right\}$ at that point. If an equation is of the same type at all points in the domain, then the equation is simply said to be of that type.

A partial differential equation of second order can be transformed into a canonical form for each of the three types mentioned above. The procedures are as follows.

Hyperbolic Equations

For hyperbolic equations we look for a new set of independent variables $\zeta = \zeta(x,y)$ and $\eta = \eta(x,y)$ in which (6.2) may be written in the standard form

$$u_{\zeta\eta} = \phi\left(u, u_\eta, u_\zeta, \eta, \zeta\right). \quad (6.3)$$

Utilizing this change of variables, we can calculate

$$u_x = u_\eta \eta_x + u_\zeta \zeta_x,$$
$$u_y = u_\eta \eta_y + u_\zeta \zeta_y,$$
$$u_{xx} = u_{\eta\eta}\eta_x\eta_x + 2u_{\eta\zeta}\eta_x\zeta_x + u_{\zeta\zeta}\zeta_x\zeta_x + u_\eta\eta_{xx} + u_\zeta\zeta_{xx},$$
$$u_{xy} = u_{\eta\eta}\eta_x\eta_y + 2u_{\eta\zeta}(\eta_x\zeta_y + \eta_y\zeta_x) + u_{\zeta\zeta}\zeta_x\zeta_y + u_\eta\eta_{xy} + u_\zeta\zeta_{xy},$$
$$u_{yy} = u_{\eta\eta}\eta_y\eta_y + 2u_{\eta\zeta}\eta_y\zeta_y + u_{\zeta\zeta}\zeta_y\zeta_y + u_\eta\eta_{yy} + u_\zeta\zeta_{yy},$$

to find that (6.2) transforms into

$$\overline{A}u_{\zeta\zeta} + \overline{B}u_{\zeta\eta} + \overline{C}u_{\eta\eta} = \Phi\left(u, u_\eta, u_\zeta, \eta, \zeta\right), \quad (6.4)$$

where

$$\overline{A} = A\zeta_x^2 + B\zeta_x\zeta_y + C\zeta_y^2,$$
$$\overline{B} = A\zeta_x\eta_x + B(\zeta_x\eta_y + \zeta_y\eta_x) + 2C\zeta_y\eta_y,$$
$$\overline{C} = A\eta_x^2 + B\eta_x\eta_y + C\eta_y^2.$$

Setting $\overline{A} = \overline{C} = 0$, we can find the following partial differential equations for ζ and η

$$\frac{\zeta_x}{\zeta_y} = \frac{-B + \sqrt{B^2 - 4AC}}{2A},$$

$$\frac{\eta_x}{\eta_y} = \frac{-B - \sqrt{B^2 - 4AC}}{2A}.$$

(6.5.a–b)

These equations may be readily solved (in principle) by the method of characteristics. For example, to solve equation (6.5.a) we only need to solve

$$-\frac{dy}{dx} = \frac{-B + \sqrt{B^2 - 4AC}}{2A}$$

for $Q(x, y) = R$, where R is an arbitrary constant. Then ζ will be given by $\zeta = Q(x, y)$.

After ζ and η are determined, then the original equation must be transformed into the new coordinates (see page 139). The resulting equation will then be in standard form.

Note that another standard form for hyperbolic equations (in two independent variables) is obtained from (6.3) by the change of variables

$$\alpha = \eta - \zeta, \qquad \beta = \eta + \zeta. \tag{6.6}$$

This results in the equation

$$u_{\alpha\alpha} - u_{\beta\beta} = \phi\left(u, u_\alpha - u_\beta, u_\alpha + u_\beta, \tfrac{1}{2}(\beta + \alpha), \tfrac{1}{2}(\beta - \alpha)\right).$$

Example 1

Suppose we have the equation

$$y^2 u_{xx} - x^2 u_{yy} = 0. \tag{6.7}$$

We recognize this equation to be hyperbolic away from the lines $x = 0$ and $y = 0$. To find the new variables ζ and η, we must solve the differential equations in (6.5). For this equation we have $\{A = y^2, B = 0, C = -x^2\}$, therefore (6.5) becomes

$$\frac{\zeta_x}{\zeta_y} = -\frac{x}{y}, \qquad \frac{\eta_x}{\eta_y} = \frac{x}{y},$$

with the solutions $\zeta = y^2 - x^2$, $\eta = y^2 + x^2$. In these new variables, equation (6.7) becomes

$$u_{\zeta\eta} = \frac{\zeta}{2(\zeta^2 - \eta^2)} u_\eta - \frac{\eta}{2(\zeta^2 - \eta^2)} u_\zeta. \tag{6.8}$$

If the change of independent variable in (6.6) is made, then equation (6.8) becomes

$$u_{\alpha\alpha} - u_{\beta\beta} = \frac{1}{2\beta} u_\beta - \frac{1}{2\alpha} u_\alpha.$$

Parabolic Equations

For parabolic equations we look for a new set of variables $\zeta = \zeta(x, y)$ and $\eta = \eta(x, y)$ in which (6.2) can be written in one of the standard forms

$$u_{\zeta\zeta} = \phi\left(u, u_\eta, u_\zeta, \eta, \zeta\right), \tag{6.9.a}$$

or

$$u_{\eta\eta} = \phi\left(u, u_\eta, u_\zeta, \eta, \zeta\right). \tag{6.9.b}$$

Utilizing (6.4), we see that we need to determine ζ and η in such a way that

$$\overline{B} = 0 = \overline{C}, \qquad \text{corresponding to (6.9.a),} \tag{6.10.a}$$

or

$$\overline{B} = 0 = \overline{A}, \qquad \text{corresponding to (6.9.b).} \tag{6.10.b}$$

If $A \neq 0$, then case (6.10.a) corresponds to the single equation

$$\frac{\zeta_x}{\zeta_y} = -\frac{B}{2A}, \tag{6.11.a}$$

while, if $C \neq 0$, then case (6.10.b) corresponds to the equation

$$\frac{\zeta_x}{\zeta_y} = -\frac{B}{2C}. \tag{6.11.b}$$

In either case, we have only to solve a single equation to determine ζ. The variable η can then be chosen to be anything linearly independent of ζ. As before, once ζ and η are determined, then the equation needs to be written in terms of these new variables.

Example 2

Suppose we have the equation

$$y^2 u_{xx} - 2xy u_{xy} + x^2 u_{yy} + u_y = 0. \tag{6.12}$$

Since $\{A = y^2,\ B = -2xy,\ C = x^2\}$, we find that $B^2 - 4AC =)$ and so this equation is parabolic. In this case we choose to make $\overline{B} = \overline{C} = 0$. From equation (6.11) we must solve $\dfrac{\zeta_x}{\zeta_y} = \dfrac{x}{y}$, which has the solution $\zeta = y^2 + x^2$. We choose $\eta = x$. Using these values of η and ζ, we find that (6.12) becomes

$$u_{\eta\eta} = \frac{2(\zeta + \eta)}{\zeta - \eta^2} u_\zeta + \frac{1}{\zeta - \eta^2} u_\eta.$$

Elliptic Equations

For elliptic equations we look for a new set of variables $\alpha = \alpha(x, y)$ and $\beta = \beta(x, y)$ in which (6.2) can be written in the standard form

$$u_{\alpha\alpha} + u_{\beta\beta} = \phi(u, u_\alpha, u_\beta, \alpha, \beta).$$

The easiest way in which to find α and β is to determine variables $\zeta = \zeta(x, y)$ and $\eta = \eta(x, y)$ that satisfy (6.5) and then form $\alpha = (\eta + \zeta)/2$, $\beta = (\eta - \zeta)/2i$ (where, as usual, $i = \sqrt{-1}$). Note that in this case, the differential equations in (6.5) are complex. However, since ζ and η are conjugate complex functions, the quantities α and β will be real.

Example 3

Suppose we have the equation

$$y^2 u_{xx} + x^2 u_{yy} = 0.$$

We recognize this equation to be elliptic away from the lines $x = 0$ and $y = 0$. To find the new variables ζ and η, we must solve the differential equations in (6.5). For this equation we have $\{A = y^2,\ B = 0,\ C = x^2\}$, therefore (6.5) becomes

$$\frac{\zeta_x}{\zeta_y} = -\frac{ix}{y}, \qquad \frac{\eta_x}{\eta_y} = \frac{ix}{y},$$

with the solutions $\zeta = y^2 - ix^2$, $\eta = y^2 + ix^2$. Forming α and β results in

$$\alpha = \frac{\eta + \zeta}{2} = y^2, \qquad \beta = \frac{\eta - \zeta}{2i} = x^2.$$

In these new variables, equation (6.7) becomes

$$u_{\alpha\alpha} + u_{\beta\beta} = -\frac{1}{2\alpha} u_\alpha - \frac{1}{2\beta} u_\beta.$$

Notes

[1] Equations of mixed type are discussed in Smirnoff [6] and also in Haack and Wendland [4].

[2] Given a partial differential equation in the form of (6.1), the characteristic surfaces are defined by the characteristic equation

$$\sum_{i,j=1}^{n} a_{ij}(\mathbf{x}) \left(\frac{\partial u}{\partial x_i} \right) \left(\frac{\partial u}{\partial x_j} \right) = 0.$$

The solutions to this equation are the only surfaces across which $u(\mathbf{x})$ may have discontinuities in its second derivatives.

[3] See the Notes section of the characteristics section (page 368) for how to determine when a system of partial differential equations is hyperbolic.

References

[1] A. V. Bitsadze, *Equations Of The Mixed Type*, The MacMillan Company, New York, 1964.

[2] S. J. Farlow, *Partial Differential Equations for Scientists and Engineers*, John Wiley & Sons, New York, 1982, pages 174–182 and 331–339.

[3] P. R. Garabedian, *Partial Differential Equations*, Wiley, New York, 1964, pages 70–76.

[4] E. R. Haack and W. N. Wendland, *Lectures on Partial and Pfaffian Differential Equations*, translated by E. R. Dawson and W. N. Everitt, Pergamon Press, New York, 1972, Chapter 6 (pages 162–182).

[5] P. Moon and D. E. Spencer, *Partial Differential Equations*, D. C. Heath, Lexington, MA, 1969, pages 137–146.

[6] M. M. Smirnoff, *Equations of Mixed Type*, Amer. Math. Soc., Providence, Rhode Island, 1978.

[7] I. Stakgold, *Green's Functions and Boundary Value Problems*, John Wiley & Sons, New York, 1979, pages 467–482.

[8] G. B. Whitham, *Linear and Nonlinear Waves*, Wiley Interscience, New York, 1974, Chapter 5 (pages 113–142).

[9] E. C. Young, *Partial Differential Equations*, Allyn and Bacon, Inc., Boston, MA, 1972, pages 60–70.

[10] E. Zauderer, *Partial Differential Equations of Applied Mathematics*, John Wiley & Sons, New York, 1983, pages 78–85 and 91–97.

7. Compatible Systems

Applicable to   Systems of differential equations.

Yields

Knowledge of whether the equations are consistent.

Procedure 1

The two equations $f(x, y, z, p, q) = 0$ and $g(x, y, z, p, q) = 0$ for $z = z(x, y)$ (where, as usual, $p = z_x, q = z_y$) are said to be *compatible* if every solution of the first equation is also a solution of the second equation, and conversely. These two equations will be compatible if $\{f, g\} = 0$, where

$$\{f, g\} := \frac{\partial(f, g)}{\partial(x, p)} + p \frac{\partial(f, g)}{\partial(z, p)} + \frac{\partial(f, g)}{\partial(y, q)} + q \frac{\partial(f, g)}{\partial(z, q)},$$

and where $\dfrac{\partial(u, v)}{\partial(a, b)} = \begin{vmatrix} u_a & v_a \\ u_b & v_b \end{vmatrix} = u_a v_b - v_a u_b$ is the usual Jacobian.

Example

Suppose we have the two following nonlinear partial differential equations for $z(x, y)$

$$x z_x = y z_y, \qquad z(x z_x + y z_y) = 2xy. \qquad (7.1)$$

From (7.1) we identify

$$f(x, y, z, p, q) = xp - yq, \qquad g(x, y, z, p, q) = z(xp + yq) - 2xy. \quad (7.2)$$

Using (7.2) we can easily calculate

$$\frac{\partial(f, g)}{\partial(x, p)} = 2xy, \qquad \frac{\partial(f, g)}{\partial(z, p)} = -x^2 p,$$

$$\frac{\partial(f, g)}{\partial(y, q)} = -2xy, \qquad \frac{\partial(f, g)}{\partial(z, q)} = xyp.$$

Therefore, computing $\{f, g\}$, we find it to be zero. Hence, the two equations in (7.1) have identical solution sets.

Since the equations in (7.1) are compatible, we can combine them without changing the solution sets. Solving the equations in (7.1) simultaneously for p and q to obtain $\{z_x = p = y/z, \ z_y = q = x/z\}$. These last two equations can be easily solved to obtain $z^2 = C + 2xy$, where C is an arbitrary constant.

Procedure 2

The conditions for consistency of a system of simultaneous partial differential equations of the first order, if the number of equations is an exact multiple of the number of dependent variables involved, is given in Forsyth [3]. To write the consistency conditions, let the unknown dependent variables be $\{z_i \mid i = 1, \ldots, m\}$, let the independent variables be $\{x_j \mid j = 1, \ldots, n\}$, and define $p_{ij} = \partial z_i / \partial x_j$. We presume the system has rm equations (with $r \leq n$), and that these equations can be solved with respect to the p_{ij}. That is

$$p_{ij} = \frac{\partial z_i}{\partial x_j} = f_{ij}\left(\{x_l\}, \{z_k\}, \{p_{\lambda\mu}\}\right),$$

for† $i = \langle 1, m \rangle$, $j = \langle 1, n \rangle$, $l = \langle 1, n \rangle$, $\lambda = \langle 1, m \rangle$, $\mu = \langle r + 1, n \rangle$. Then, for consistency, the following conditions must be satisfied

$$
\frac{\partial f_{ij}}{\partial x_a} - \frac{\partial f_{ia}}{\partial x_j} + \sum_{\lambda=1}^{m} \left(f_{\lambda a} \frac{\partial f_{ij}}{\partial z_\lambda} - f_{\lambda j} \frac{\partial f_{ia}}{\partial z_\lambda} \right)
$$
$$
+ \sum_{s=1}^{m} \sum_{\mu=r+1}^{n} \left(\frac{\partial f_{ij}}{\partial p_{s\mu}} \frac{\partial f_{sa}}{\partial x_\mu} - \frac{\partial f_{ia}}{\partial p_{s\mu}} \frac{\partial f_{sj}}{\partial x_\mu} \right) \tag{7.3}
$$
$$
+ \sum_{s=1}^{m} \sum_{\mu=r+1}^{n} \sum_{\lambda=1}^{m} \left[\left(\frac{\partial f_{ij}}{\partial p_{s\mu}} \frac{\partial f_{sa}}{\partial z_\lambda} - \frac{\partial f_{ia}}{\partial p_{s\mu}} \frac{\partial f_{sj}}{\partial z_\lambda} \right) p_{\lambda\mu} \right] = 0,
$$

where $i = \langle 1, m \rangle$, $a = \langle j + 1, r \rangle$, $j = \langle 1, r - 1 \rangle$, and

$$
\sum_{s=1}^{m} \left(\frac{\partial f_{ij}}{\partial p_{s\mu}} \frac{\partial f_{sa}}{\partial p_{k\tau}} - \frac{\partial f_{ia}}{\partial p_{s\mu}} \frac{\partial f_{sj}}{\partial p_{k\tau}} + \frac{\partial f_{ij}}{\partial p_{s\tau}} \frac{\partial f_{sa}}{\partial p_{k\mu}} - \frac{\partial f_{ia}}{\partial p_{s\tau}} \frac{\partial f_{sj}}{\partial p_{k\mu}} \right) = 0, \tag{7.4}
$$

where $i, k = \langle 1, m \rangle$, $a = \langle j + 1, r \rangle$, $\mu, \tau = \langle r + 1, n \rangle$, $j = \langle 1, r - 1 \rangle$.

Special Case 1

In the special case of $m = 1$, we have one dependent variable (which we call z) and r equations. Let $p_j = \partial z / \partial x_j = f_j(z, x_1, \ldots, x_n, p_{r+1}, \ldots, p_n)$. In this case, (7.4) is automatically satisfied while (7.3) becomes

$$
\frac{df_j}{dx_a} - \frac{df_a}{dx_j} + \sum_{\mu=r+1}^{n} \left(\frac{\partial f_j}{\partial p_\mu} \frac{df_a}{dx_\mu} - \frac{\partial f_a}{\partial p_\mu} \frac{df_j}{dx_\mu} \right) = 0
$$

for $a = \langle 1, j - 1 \rangle$, $j = \langle 1, r \rangle$, where we have defined $\dfrac{d}{dx_s} = \dfrac{\partial}{\partial x_s} + p_s \dfrac{\partial}{\partial z}$.

Special Case 2

In the special case of $r = n$, the system of mn equations becomes $p_{ij} = F_{ij}(z_1, \ldots, z_m, x_1, \ldots, x_n)$ and the consistency conditions become

$$
\frac{\partial f_{ij}}{\partial x_a} - \frac{\partial f_{ia}}{\partial x_j} + \sum_{\lambda=1}^{m} \left(f_{\lambda a} \frac{\partial f_{ij}}{\partial z_\lambda} - f_{\lambda j} \frac{\partial f_{ia}}{\partial z_\lambda} \right) = 0
$$

for $i = \langle 1, m \rangle$, $a = \langle 1, j - 1 \rangle$ and $j = \langle 1, n \rangle$. These are known as Mayer's system of completely integrable equations.

---

† To simplify notation, define $\langle a, b \rangle$ to be the sequence of numbers $a, a + 1, a + 2, \ldots, b$.

Special Case 3

Consider the special case of $r = 1$, with $\{F_1 = 0, F_2 = 0, \ldots, F_m = 0\}$, where each $F_j = p_j - f_j(z, x_1, \ldots, x_n, p_{r+1}, \ldots, p_n)$ is analytical in each of its arguments. A necessary and sufficient condition for the set of equations to be consistent is that $[F_i, F_j] = 0$, for all combinations of i and j. Here, $[,]$ represents the usual Poisson bracket.

Notes

[1] Jacobi's method (see page 397) takes a given partial differential equation and creates a compatible equation and then uses elimination between these two equations.

[2] If it is known that a linear homogeneous ordinary differential equation of order n has solutions in common with a linear homogeneous ordinary differential equation of order m (with $m < n$), then it is possible to determine a differential equation of lower degree that has, as its solutions, these common solutions.

If the linear homogeneous ordinary differential equations $L_1(u) = 0$ and $L_2(u) = 0$ are defined by

$$L_1 := p_0 D^n + p_1 D^{n-1} + \ldots + p_{n-1} D + p_n,$$
$$L_2 := q_0 D^m + q_1 D^{m-1} + \ldots + q_{m-1} D + q_m,$$

where D represents d/dx and each of the functions $\{p_i, q_i\}$ depends on x, define the ordinary differential equation $R_1(u) = 0$ by

$$R_1 := r_0 D^{n-m} + r_1 D^{n-m-1} + \ldots + r_{n-m-1} D + r_{n-m},$$

where the $\{r_i\}$ are defined by

$$p_0 = r_0 q_0,$$

$$p_1 = r_1 q_0 + r_0 \left[\binom{n-m}{1} q_0' + q_1 \right],$$

$$p_2 = r_2 q_0 + r_1 \left[\binom{n-m-1}{1} q_0' + q_1 \right]$$
$$+ r_0 \left[\binom{n-m}{2} q_0'' + \binom{n-m}{1} q_1' + q_2 \right],$$

$$\vdots$$

$$p_{n-m} = r_{n-m} q_0 + r_{n-m-1} \left[\binom{1}{1} q_0' + q_1 \right]$$
$$+ r_{n-m-2} \left[\binom{2}{2} q_0'' + \binom{2}{1} q_1' + \binom{2}{0} q_2 \right] + \ldots,$$
$$= r_{n-m} q_0 + r_{n-m-1} \left[q_0' + q_1 \right] + r_{n-m-2} \left[q_0'' + 2q_1' + q_2 \right] + \ldots.$$

Then the order of the operator $L_3 := L_1 - R_1 L_2$ will be depressed as much as is possible (the order of L_3 will not exceed $m - 1$). Note that only a finite number of rational operations and differentiations are required to determine the $\{r_i\}$.

From the definition of L_3 we see that all solutions common to both $L_1(u) = 0$ and to $L_2(u) = 0$ will also be solutions to $L_3(u) = 0$. If L_3 is identically zero, then we have found a factorization of L_1 (see page 246). See Ince [4] or Valiron [6] for details.

[3] Differential resultants can also be used to derive consistency conditions. See Berkovich and Tsirulik [2] for details.

[4] Wolf [7] describes an algorithm which determines if an overdetermined system of two equations for one function has any solution. An implementation in FORMAC is mentioned.

References

[1] W. F. Ames, "Ad Hoc Exact Techniques for Nonlinear Partial Differential Equations," in W. F. Ames (ed.), *Nonlinear Partial Differential Equations in Engineering*, Academic Press, New York, 1967, pages 54–65.

[2] L. M. Berkovich and V. G. Tsirulik, "Differential Resultants and Some of Their Applications," *Differentsial'nye Uravneniya*, **22**, No. 5, May 1986, pages 750–757.

[3] A. R. Forsyth, *Theory of Differential Equations*, Part IV, Dover Publications, Inc., New York, 1959, pages 411–419.

[4] E. L. Ince, *Ordinary Differential Equations*, Dover Publications, Inc., New York, 1964, pages 126–128.

[5] I. N. Sneddon, *Elements of Partial Differential Equations*, McGraw–Hill Book Company, New York, 1957, pages 67–68.

[6] G. Valiron, *The Geometric Theory of Ordinary Differential Equations and Algebraic Functions*, Math Sci Press, Brookline, MA, 1950, pages 320–322.

[7] T. Wolf, "An Analytic Algorithm for Decoupling and Integrating Systems of Nonlinear Partial Differential Equations," *J. Comput. Physics*, **60**, 1985, pages 437–446.

8.   Conservation Laws

Applicable to   Partial differential equations.

Yields

Quantities that remain invariant during the evolution of the partial differential equation.

Procedure

Given an evolution equation, which is a partial differential equation of the form

$$u_t = F(u, u_x, u_{xx}, \ldots), \tag{8.1}$$

a conservation law is a partial differential equation of the form

$$\frac{\partial}{\partial t} T\Big(u(x,t)\Big) + \frac{\partial}{\partial x} X\Big(u(x,t)\Big) = 0, \tag{8.2}$$

which is satisfied by all solutions of (8.1). We define $T(\ )$ to be the *conserved density* and $X(\ )$ to be the *flux*. An alternative statement of (8.2) is that

$$\int T\Big(u(x,t)\Big)\, dx \tag{8.3}$$

is independent of t, for solutions of (8.1) such that the integral converges.

More generally, a partial differential equation of order m in the n independent variables $\mathbf{x} = (x_1, x_2, \ldots, x_n)$ and a single dependent variable u is in *conservation form* if it can be written as

$$\sum_{i=1}^{n} \frac{\partial}{\partial x_i} F_i(\mathbf{x}, u, \partial u, \partial^2 u, \ldots, \partial^{m-1} u) = 0 \tag{8.4}$$

Here $\partial^j u$ represents all j-th order partial derivatives of u with respect to $\mathbf{x}$.

Example 1

The Korteweg–de Vries equation

$$u_t = u_{xxx} + u u_x \tag{8.5}$$

has an infinite set of conservation laws. The first few, in order of increasing rank, have the conserved densities

$$T = u,$$
$$T = u^2,$$
$$T = u^3 - 3u_x^2,$$
$$T = 5u^4 - 60uu_x^2 - 36u_x u_{xxx},$$
$$\vdots$$

To demonstrate, for instance, that $T = u^2$ is a conserved density, we compute

$$\frac{\partial T}{\partial t} = \frac{\partial(u^2)}{\partial t} = 2uu_t = 2uu_{xxx} + 2u^2 u_x,$$

where we have used the defining equation (8.5) to replace the u_t term. Now we must determine a flux X such that (8.2) is satisfied. In this case, we find $X = u_x^2 - 2uu_{xx} - \frac{2}{3}u^3$.

Example 2

The Schrödinger equation

$$-\frac{\partial^2 u}{\partial x^2} + V(x)u = i\frac{\partial u}{\partial t}$$

can be expressed in the form of (8.2) with

$$T = i\nu(x)u,$$

$$X = \nu(x)\frac{\partial u}{\partial x} - \nu'(x)u,$$

where $\nu(x)$ is defined by $\nu''(x) = V(x)\nu(x)$.

Notes

[1] Conservation laws allow estimates of the accuracy of a numerical solution scheme (since (8.3) must be invariant in time).

[2] Not all partial differential equations have an infinite number of conservation laws; there may be none or a finite number.

[3] A conservation law for an evolution equation is called trivial if T is, itself, the x derivative of some expression. If (8.1) has an infinite sequence of nontrivial conservation laws, then the equation is formally integrable. Infinite sequences of nontrivial conservation laws are given by Cavalcante and Tenenblat [2] for the following equations: Burgers, KdV, mKdV, sine–Gordon, sinh–Gordon.

[4] If a given partial differential equation is not written in conservation form, there are a number of ways of attempting to put it in a conserved form. See Bluman, Reid, and Kumei [1] for a short list of techniques.

[5] If (8.4) is satisfied, then there exists an $(n-1)$-exterior differential form $\mathbf{F}$ such that (8.4) can be written $d\mathbf{F} = 0$. This implies that there is an $(n-2)$-form ϕ such that $\mathbf{F} = d\phi$. This, in turn, means that there exists an antisymmetric tensor of rank n, ψ, such that

$$F_i(\mathbf{x}, u, \partial u, \partial^2 u, \dots, \partial^{m-1}u) = \sum_{i<j\leq n}(-1)^j\frac{\partial\psi_{ij}}{\partial x_j} + \sum_{1\leq j<i}(-1)^{i-1}\frac{\partial\psi_{ji}}{\partial x_j},$$

for $i = 1, 2, \dots, n$.

[6] A computer program in REDUCE for determining conservation laws is given in Ito and Kako [7]. In Gerdt, Shvachka and Zharkov [4] is the description of a computer program in FORMAC that determines conservation laws, determines Lie–Bäcklund symmetries, and also attempts to determine when an evolution equation is formally integrable.

[7] Torriani [11] shows how the terms appearing in the expression of the densities and the fluxes for the Korteweg-de Vries equation may be found by combinatorial methods.

References

[1] G. W. Bluman, G. J. Reid, and S. Kumei, "New Classes of Symmetries for Partial Differential Equations," *J. Math. Physics*, **29**, No. 4, April 1988, pages 806–811.

[2] J. A. Cavalcante and K. Tenenblat, "Conservation Laws for Nonlinear Evolution Equations," *J. Math. Physics*, **29**, No. 4, April 1988, pages 1044–1049.

[3] C. W. Gear, "Maintaining Solution Invariants in the Numerical Solution of ODEs," *SIAM J. Sci. Stat. Comput.*, **7**, No. 3, July 1986, pages 734–743.

[4] V. P. Gerdt, A. B. Shvachka, and A. Y. Zharkov, "Computer Algebra Applications for Classification of Integrable Non-Linear Evolution Equations," *J. Symbolic Comp.*, **1**, 1985, pages 101–107.

[5] D. Greenspan, "Conservative Numerical Methods for $\ddot{x} = f(x)$," *J. Comput. Physics*, **56**, No. 4, October 1984, pages 28–41.

[6] N. H. Ibragimov, "Group Theoretical Nature of Conservation Theorems," *Letters Math. Physics*, **1**, 1977, pages 423–428.

[7] M. Ito and F. Kako, "A REDUCE program for finding Conserved Densities of Partial Differential Equations with Uniform Rank," *Comput. Physics Comm.*, **38**, 1985, pages 415–419.

[8] E. Littinsky, "Polynomial Integrals of Evolution Equations," *Comm. Math. Physics*, **121**, 1989, pages 669–682.

[9] I. G. Marchuk, "A Special Class of Green's Formulas," *Differentsial'nye Uravneniya*, **22**, No. 2, February 1986, pages 315–326.

[10] P. J. Olver, *Applications of Lie Groups to Differential Equations*, Graduate Texts in Mathematics #107, Springer–Verlag, New York, 1986, Chapter 4 (pages 246–291).

[11] H. H. Torriani, "Conservation Laws for the Korteweg-de Vries Equation and the Theory of Partitions," *Physics Letters*, **113A**, No. 7, 6 January 1986, pages 345–348.

[12] M. Vinokur, "An Analysis of Finite-Difference and Finite-Volume Formulations of Conservations Laws," *J. Comput. Physics*, **81**, 1989, pages 1–52.

9.   Differential Resultants

Applicable to   Two polynomial ordinary differential equations.

Yields

One ordinary differential equation in one independent variable.

Idea

Given two polynomial equations (in, say, x and y) the classical method of resultants is as follows: the equations can always be written as the system of linear equations $A\mathbf{w} = \mathbf{0}$, where $A = A(y)$ and $\mathbf{w} = \mathbf{w}(x) \neq \mathbf{0}$. Since this system must have $\det A = 0$, a polynomial equation only in y may be determined. The technique for polynomial differential equations is very similar.

Procedure

Resultants have classically been used to eliminate one variable between two polynomial equations. For example, given the two equations

$$x^3 - 3y^2x^2 + x + 5y^2 = 0, \tag{9.1}$$

$$x^3 + 5y^2x^2 - x + 3y^2 = 0, \tag{9.2}$$

(9.1) and (9.2) may be multiplied by powers of x to obtain the system of equations:

$$
\begin{aligned}
x^5 - 3y^2x^4 + x^3 + 5y^2x^2 &= 0, \\
x^4 - 3y^2x^3 + x^2 + 5y^2x &= 0, \\
x^3 - 3y^2x^2 + x + 5y^2 &= 0, \\
x^3 + 5y^2x^2 - x + 3y^2 &= 0, \\
x^4 + 5y^2x^3 - x^2 + 3y^2x &= 0, \\
x^5 + 5y^2x^4 - x^3 + 3y^2x^2 &= 0.
\end{aligned}
$$

This system can be written in matrix form as

$$
\begin{pmatrix}
1 & -3y^2 & 1 & 5y^2 & 0 & 0 \\
0 & 1 & -3y^2 & 1 & 5y^2 & 0 \\
0 & 0 & 1 & -3y^2 & 1 & 5y^2 \\
0 & 0 & 1 & 5y^2 & -1 & 3y^2 \\
0 & 1 & 5y^2 & -1 & 3y^2 & 0 \\
1 & 5y^2 & -1 & 3y^2 & 0 & 0
\end{pmatrix}
\begin{pmatrix}
x^5 \\
x^4 \\
x^3 \\
x^2 \\
x \\
1
\end{pmatrix}
=
\begin{pmatrix}
0 \\
0 \\
0 \\
0 \\
0 \\
0
\end{pmatrix}. \tag{9.3}
$$

This last equation is a 6×6 system of the form $A\mathbf{w} = \mathbf{0}$. Since $\mathbf{w} \neq \mathbf{0}$ (since, at least, the last component of $\mathbf{w}$ is non-zero), the determinant of A must vanish. Taking the determinant of the matrix in (9.3), we find that y must satisfy the equation

$$32y^2(289y^8 + 16y^4 + 1) = 0. \tag{9.4}$$

All the different values of y, from the solutions of (9.1) and (9.2), must satisfy (9.4).

Differential resultants are the analogue of resultants applied to differential systems. There are two steps analogous to multiplying the original equations by powers of x. They are

(A) differentiating one of the equations,

(B) multiplying one of the equations by some term which may involve the independent and/or the dependent variables.

While there are algorithms published on how to proceed in any given case, as in Mishina and Proskuryakov [3], they are generally written in the language of abstract algebra.

Example

Suppose we have the following two coupled differential equations for $\{y(x), z(x)\}$

$$A: \quad 3yz + z - y_x = 0,$$
$$B: \quad -z_x + z^2 + y^2 + y = 0.$$

We seek a single differential equation only involving $z(x)$. Note that we could solve equation (A) for $y(x)$ (by integrating factors) and then substitute this result in equation (B), but this creates an algebraic mess. This, in turn, makes it difficult to obtain a single simple equation for $z(x)$.

If we form the equations $\{A, B, yA, yB, y_x B, \partial_x B, y\partial_x A\}$, then we obtain the system

$$
\begin{pmatrix}
0 & 0 & -1 & 0 & 0 & 3z & z \\
1 & 0 & 0 & 0 & 0 & 1 & z^2 - z_x \\
3z & 0 & 0 & -1 & 0 & z & 0 \\
1 & 1 & 0 & 0 & 0 & z^2 - z_x & 0 \\
0 & 0 & z^2 - z_x & 1 & 1 & 0 & 0 \\
0 & 0 & 1 & 2 & 0 & 0 & 2zz_x - z_{xx} \\
0 & 0 & 0 & 1 & 2 & 2zz_x - z_{xx} & 0
\end{pmatrix}
\begin{pmatrix}
y^2 \\
y^3 \\
y_x \\
yy_x \\
y^2 y_x \\
y \\
1
\end{pmatrix}
=
\begin{pmatrix}
0 \\
0 \\
0 \\
0 \\
0 \\
0 \\
0
\end{pmatrix}.
$$
(9.5)

Taking the determinant of the matrix appearing in (9.5) we conclude that $z(x)$ is a solution of the single ordinary differential equation

$$z_{xx}^2 + (-16z_x + 12z^2 - 3)zz_{xx} + 64z^2 z_x^2 + (23 - 96z^2)z^2 z_x$$
$$+ (36z^4 - 17z^2 + 2)z^2 = 0.$$

Notes

[1] This technique applies directly to systems of partial differential equations and to higher order equations.

[2] There are specific technical requirements for when the classical method of resultants (when applied to polynomials) will work. There are similar requirements for when differential resultants will work. See Mishina and Proskuryakov [3] for details.

Rubel [6] proves the following theorem which indicates that elimination is not always possible, at least for algebraic differential equations (ADEs, see page 644):

> There exists a system of two ADEs, in the two dependent variables u and v which possesses a real-valued $C^{n,m}$ solution $\bar{u}, \bar{v}$ on a certain open interval I, but which has no solution u, v on I for which v satisfies an ADE that does not involve u or any derivative of u.

[3] By taking equations pairwise a system of, say, 10 equations in 10 different independent variables could, if fortunate, be reduced to a single equation in a single independent variable.

[4] The two differential equations considered do not both have to be polynomial for this reduction scheme to work. The two equations only have to be polynomials in one of the dependent variables (the one that will be removed).

[5] Any linear second order ordinary differential equation system can be interpreted as the resultant of the elimination of a dependent variable from a pair of conjugate first order Hamilton's equations. See Tolstoy [8] for details.

References

[1] L. M. Berkovich and V. G. Tsirulik, "Differential Resultants and Some of Their Applications," *Differentsial'nye Uravneniya*, **22**, No. 5, May 1986, pages 750–757.

[2] M. Bôcher, *Introduction to Higher Algebra*, The MacMillan Company, New York, 1907.

[3] A. P. Mishina and I. V. Proskuryakov, *Higher Algebra*, Pergamon Press, New York, 1965.

[4] J. F. Ritt, *Differential Algebra*, Dover Publications, Inc., New York, 1966.

[5] L. A. Rubel, "A Counterexample to Elimination in Systems of Algebraic Differential Equations," *Mathematika*, **30**, 1983, pages 74–76.

[6] L. A. Rubel, "An Elimination Theory for Systems of Algebraic Differential Equations," *Houston J. Math.*, **8**, No. 2, 1982, pages 289–295.

[7] A. Seidenberg, *An Elimination Theory for Differential Algebra*, University of California Press, **3**, No. 2, 1956, pages 31–66.

[8] I. Tolstoy, "Remarks on the Linearization of Differential Equations," *J. Inst. Maths. Applics*, **20**, 1977, pages 53–60.

[9] B. L. Van Der Waerden, *Modern Algebra*, Frederick Ungar Publishing, New York, 1940.

10.   Existence and Uniqueness Theorems

Applicable to   Differential equations of all types.

Yields

Knowledge of whether a solution exists and, if so, if the solution is unique.

Idea

There are theorems available for most cases of interest.

Procedure

Corresponding to the difficulty of the subjects involved, there are more theorems applicable to: ordinary differential equations than partial differential equations, linear equations than nonlinear equations, and initial value problems than boundary value problems. In the following we indicate some of the simple theorems that are frequently useful.

The last theorem is applicable to partial differential equations, the rest are applicable to ordinary differential equations. The first and last theorems are for vector systems, the other theorems are for scalar equations (which may sometimes be derived from the vector result).

Theorem: Consider the initial value problem: $\dfrac{d\mathbf{x}}{dt} = \mathbf{F}(t,\mathbf{x})$, with $\mathbf{x}(t_0) = \mathbf{x}_0$, where $\mathbf{x} = \mathbf{x}(t) = (x_1(t),\, x_2(t),\, \ldots,\, x_n(t))$. If each of the functions $\{F_i\}$ and $\left\{\dfrac{\partial F_i}{\partial x_j}\right\}$ are continuous in a region R of $(t,\mathbf{x})$ space containing the point $\mathbf{x}_0$, then there is an interval $|t - t_0| < h$ in which there exists a unique solution to the problem.

Theorem: Consider the initial value problem: $y' = f(x,y)$ with $y(x_0) = y_0$. Let the functions f be continuous in some rectangle $a < x < b,\ c < y < d$ containing the point (x_0, y_0). Assume that $f(x,y)$ satisfies a Lipschitz condition in y. Then, in some interval $x_0 - h < x < x_0 + h$ contained in $a < x < b$, there is a unique solution to the given problem.

Theorem: Consider the initial value problem: $y' = f(x,y)$ with $y(x_0) = y_0$. Let the functions f and $\partial f/\partial y$ be continuous in some rectangle $a < x < b,\ c < y < d$ containing the point (x_0, y_0). Then, in some interval $x_0 - h < x < x_0 + h$ contained in $a < x < b$, there is a unique solution to the given problem.

Theorem: Consider the initial value problem: $y'' = f(x, y, y')$ with $y(x_0) = y_0$, $y'(x_0) = y_0'$. Let the functions f, f_y, and $f_{y'}$ be continuous in an open region R of three-dimensional (x, y, y') space. If the point (x_0, y_0, y_0') is in R, then there exists some interval about x_0 for which there is a unique solution to the given problem.

Theorem: Consider the initial value problem:

$$y^{(n)} + p_1(x)y^{(n-1)} + \ldots + p_{n-1}(x)y' + p_1(x)y = q(x),$$

with

$$y(x_0) = y_0, \quad y'(x_0) = y_0', \quad \ldots \quad y^{(n-1)}(x_0) = y_0^{(n-1)}.$$

If the functions $\{p_i(x)\}$ and $q(x)$ are continuous on the open interval $a < x < b$, then there exists a unique solution to the problem.

Theorem: Consider the initial value problem:

$$x' = f(x, y, t), \qquad y' = g(x, y, t)$$

with $x(t_0) = x_0$, $y(t_0) = y_0$. If f and g satisfy a Lipschitz condition (with respect to x and y) in the region $\{|t - t_0| \le A,\ |x - x_0| \le B,\ |y - y_0| \le C\}$, then the problem has a unique solution in some interval $a < t < b$ about the point t_0.

Theorem: Consider the boundary value problem:

$$x'' = f(t, x, x'), \qquad 0 < t < 1,$$
$$x(0) = A, \quad x(1) = B.$$

If f and f_x are continuous and $f_x \ge 0$, then there exists a unique solution.

Theorem: Consider the initial value problem

$$y'' + f(x, y, y') = 0,$$
$$B_1[y] := y'(a) + Ay(a) - C_1 = 0, \qquad (10.1)$$
$$B_2[y] := y'(b) + By(b) - C_2 = 0,$$

where f satisfies a Lipschitz condition, and f_y and $f_{y'}$ are bounded for x in the interval $[a, b]$ and for values of (y, y') of interest. Consider the two comparison equations

$$u_1'' + h_1(x, u_1, u_1') = 0, \quad B_1[u_1] = 0, \qquad B_2[u_1] = 0,$$
$$u_2'' + h_2(x, u_2, u_2') = 0, \quad B_1[u_2] = 0, \qquad B_2[u_2] = 0,$$

with $h_1(x, y, y') \le f(x, y, y') \le h_2(x, y, y')$. We assume that the u_1 and u_2 problems have unique solutions. Then there exists at least one solution to (10.1) in the given region, and every solution has the property $u_1(x) \le y(x) \le u_2(x)$. (This theorem is one of the major results of the theory of differential inequalities.)

Cauchy–Kowalewski Theorem: If the vector $\mathbf{u} = (u_1, u_2, \ldots, u_n)^{\mathrm{T}}$ satisfies

$$\mathbf{u}_t = A(\mathbf{u})\mathbf{u}_x, \qquad \mathbf{u}(0, x) = \mathbf{h}(x),$$

where $u_k = u_k(x, t)$, $A(\mathbf{u})$ is an analytic matrix, and $\mathbf{h}(x)$ is an analytic function, then a neighborhood of $t = 0$ can be found in which there is a unique solution $\mathbf{u}$, with each u_k being analytic.

Example 1

The first order initial value problem

$$y' = |y|^{1/3}, \qquad y(x_0) = 0 \tag{10.2}$$

has a right-hand side that is not Lipschitz continuous at $y = 0$. This equation, in fact, has an infinite number of solutions. Let x_1 and x_2 be any two numbers such that $x_1 < x_0 < x_2$. Then the following function

$$f(x) = \begin{cases} -\left(\frac{2}{3}\right)^{3/2}(x_1 - x)^{3/2}, & \text{if } x < x_1, \\ 0, & \text{if } x_1 < x < x_2, \\ \left(\frac{2}{3}\right)^{3/2}(x - x_2)^{3/2}, & \text{if } x_2 < x, \end{cases}$$

is a solution to (10.2).

Example 2

The nonlinear second order equation

$$\left(u'^3\right)' + 24(1 - u) = 0, \qquad u(0) = 1, \quad u'(0) = 0,$$

has at least three solutions: $u(t) = 1$, $u(t) = 1 - t^2$, and $u(t) = 1 + t^2$.

Notes

[1] Differential equations with discontinuities (see page 219) and delay equations (see page 209) do not meet the requirements of the above theorems. They must be investigated separately.

[2] It is often possible to determine when a linear ordinary differential equation has a unique solution. When the solution is not unique, it is sometimes possible to describe the degrees of freedom that make it non-unique. See the section on alternative theorems, starting on page 14.

[3] Fixed point theorems are a specific method that can be used to prove the existence of a solution, see page 54. The section on well-posedness of differential equations contains some results on existence and uniqueness; see page 94.

[4] Bobisud and O'Regan [1] consider existence questions for some second order initial value problems of the form $y'' + F(t, y, y') = 0$, where F is allowed to be suitably singular. For example, $F(t, y, y') = t^{-1/2}y^{-1/2}$ is allowed.

[5] The existence of solutions to a differential equation can be critically dependent on the size of the coefficients in the equation. For example, Coddington and Levinson [3] show that the problem

$$\varepsilon y'' = -y' - (y')^3,$$
$$y(0) = A, \quad y(1) = B \qquad (A \neq B)$$

does not have a solution for small enough $\varepsilon > 0$.

[6] The classical problem

$$-\nabla^2 u = u^p \quad \text{in } \Omega, \qquad u = 0 \quad \text{on } \partial\Omega,$$

where Ω is a bounded domain in $\mathbb{R}^N$, with smooth boundary $\partial\Omega$, has the interesting existence result (see Peletier [8]):

(A) if $p < \dfrac{N+2}{N-2}$, then existence of a solution is assured for any domain Ω;

(B) if $p \geq \dfrac{N+2}{N-2}$, then there exists no solution in any star-shaped domain.

Similar results are available for the equation $u_t = \nabla^2 u + u^p$; existence of a global positive solution depends on whether p is greater than $1 + 2/N$ (see Fujita [5]).

[7] A classic result of Lewy [7] is that the equation

$$-u_x - iu_y + 2(ix - y)u_z = F(x, y, z),$$

where $F(x, y, z)$ is of class C^∞, has no H^1-solution, no matter what open (x, y, z) set is taken as the domain of existence.

References

[1] L. E. Bobisud and D. O'Regan, "Existence of Solutions to Some Singular Initial Value Problems," *J. Math. Anal. Appl.*, **133**, 1988, pages 214–230.

[2] W. E. Boyce and R. C. DiPrima, *Elementary Differential Equations and Boundary Value Problems*, Fourth Edition, John Wiley & Sons, New York, 1986.

[3] E. A. Coddington and N. Levinson, "A Boundary Value Problem for a Nonlinear Differential Equation with a Small Parameter," *Proc. Amer. Math. Soc.*, **3**, 1952, pages 73–81.

[4] M. Fečkan, "A New Method for the Existence of Solution of Nonlinear Differential Equations," *J. Differential Equations*, **89**, 1991, pages 203–223.

[5] H. Fujita, "On the Blowing Up of Solutions of the Cauchy Problem for $u_t = \Delta u + u^{1+\alpha}$," *J. Fac. Sci. Univ. Tokyo Sect. A. Math*, **16**, 1966, pages 105–113.

[6] H. A. Levine, "The Role of Critical Exponents in Blowup Theorems," *SIAM Review*, **32**, No. 2, 1990, pages 262–288.

[7] H. Lewy, "An Example of a Smooth Linear Partial Differential Equation without Solution," *Annals of Math.*, **66**, No. 1, July 1957, pages 155–158.

[8] L. A. Peletier, "Elliptic Equations with Nearly Critical Growth," in C. M. Dafermos, G. Ladas, and G. Papanicolaou (eds.), *Equadiff 1987*, Marcel Dekker, New York, 1987, pages 561–574.

[9] M. Plum, "Computer-Assisted Existence Proofs for Two-Point Boundary Value Problems," *Computing*, **46**, 1991, pages 19–34.

11.   Fixed Point Existence Theorems

Applicable to   Differential equations of all types.

Yields

A statement about the existence of the solution.

Idea

If the statement concerning the existence of a solution to a differential equation can be interpreted as a statement concerning fixed points in a Banach space, then a fixed point theorem might be useful.

Procedure

The Schrauder fixed point theorem states:

> Let X be a non-empty convex set in a Banach space and let Y be a compact subset of X. Suppose $Y = f(X)$ maps X continuously into Y. Then there is a fixed point $x^* = f(x^*)$.

By interpreting a given differential equation as a continuous function in a Banach space, the above theorem indicates the existence of a solution.

Example

Suppose we wish to determine whether a solution exists to the nonlinear boundary value problem

$$u'' = -e^{-u(x)},$$
$$u(0) = u(1) = 0,$$

(11.1)

on the interval $x \in [0, 1]$. We first note that the problem

$$v'' = -\phi(x),$$
$$v(0) = v(1) = 0,$$

has the solution

$$v(x) = \int_0^1 G(x,z)\phi(z)\,dz,$$

where $G(x,z)$ is the Green's function (see page 271)

$$G(x,z) = \begin{cases} (1-x)z, & \text{for } 0 \le z \le x, \\ (1-z)x, & \text{for } x \le z \le 1. \end{cases}$$

Hence, we can write equation (11.1) in the form of an equivalent integral equation

$$u(x) = f(u(x)) \equiv \int_0^1 G(x,z)e^{-u(z)}\,dz. \tag{11.2}$$

To apply Schrauder's fixed-point theorem to (11.2), we need to carefully define the Banach space B and the sets X and Y. If we define

$$B = \text{rm space of continuous functions on } (0,1),$$
$$X = \{u(x) \mid 0 \le u(x) \le 1, u(x) \text{ is continuous}\},$$
$$Y = f(X),$$

then we can apply the theorem. Note that in this example, X is not compact but Y is. Note also that the bounds in X were derived after some analysis of (11.1). Finally, then, we conclude that (11.1) has a solution.

Notes

[1] In the example above we used a a fairly standard linearization trick that can be described in more generality. Suppose that an expression $D(f,g)$ (which could involve derivatives of f and/or g) is linear in f. Suppose also that the linear differential equation

$$D(f,g) = 0$$

has a unique solution $f = T[g]$ for each g in some function space. Then to find a solution, in that function space, of the (possibly nonlinear) equation

$$D(f,f) = 0$$

is equivalent to finding a fixed point of the mapping T. Thus a particular nonlinear differential equation can be studied by means of a more general linear differential equation, together with a fixed point problem.

[2] Once a differential equation has been formulated as a fixed point statement, numerical methods that search for fixed points in a function space can be used. See, for example, Allgower [1].

[3] Interval techniques (see page 470) may also be used to bound the solution of a fixed point statement. See Moore [7] for details.

[4] A *contraction mapping* is a functional iteration, say $y_{n+1} = N[y_n]$, that converges to the solution of the fixed point equation $y = F[y]$. The Picard iteration (see page 535) is such a mapping.

[5] Another fixed point theorem that is of use in differential equations is Krasnoselskii's theorem (see Franklin [3] for details):

> Consider the fixed-point equation $\mathbf{x} = \mathbf{f}(\mathbf{x}) + \mathbf{g}(\mathbf{x})$ for $\mathbf{x}$ in a Banach space $\mathcal{B}$. Let X be a non-empty closed convex set in $\mathcal{B}$. Let $\mathbf{f}(\mathbf{x})$ map X continuously into a compact subset $Y \subset X$. Let $\mathbf{g}(\mathbf{x})$ be a contraction mapping on X (note that the range of $\mathbf{g}$ need not be compact). If it is assumed that $\mathbf{y} + \mathbf{g}(\mathbf{x}) \in X$ for $\mathbf{y} \in Y$ and $\mathbf{x} \in X$, then there is a fixed point of $\mathbf{x} = \mathbf{f}(\mathbf{x}) + \mathbf{g}(\mathbf{x})$.

[6] Another fixed point theorem that is of use in differential equations is the Tihonov fixed-point theorem (see Iyanaga and Kawada [6] for details):

> Let R be a locally compact topological linear space, A a compact convex subset of R, and T a continuous mapping sending A into itself. Then T has fixed points.

[7] Some theorems regarding existence of solutions for differential equations may be found on page 50.

References
[1] E. L. Allgower, "Application of a Fixed Point Search Algorithm to Nonlinear Boundary Value Problems Having Several Solutions," in S. Karamardian (ed.), *Fixed Points: Algorithms and Applications*, Academic Press, New York, 1977.

[2] T. A. Burton, *Stability and Periodic Solutions of Ordinary and Functional Differential Equations*, Academic Press, New York, 1985, Chapter 3 (pages 164–196).

[3] J. Franklin, *Methods of Mathematical Economics*, Springer–Verlag, New York, 1980, page 277.

[4] J. K. Hale, *Oscillations in Nonlinear Systems*, McGraw–Hill Book Company, New York, 1963, Appendix (pages 171–172).

[5] P. Hartman, *Ordinary Differential Equations*, John Wiley & Sons, New York, 1964, Chapter 12 (pages 404–449).

[6] S. Iyanaga and Y. Kawada, *Encyclopedic Dictionary of Mathematics*, MIT Press, Cambridge, MA, 1980, pages 542–543.

[7] R. E. Moore, *Interval Analysis*, Prentice–Hall Inc., Englewood Cliffs, NJ, 1966, Chapter 15 (pages 97–102).

[8] D. R. Smart, *Fixed Point Theorems*, Cambridge University Press, New York, 1974, Chapter 6 (pages 41–52).

[9] I. Stakgold, *Green's Functions and Boundary Value Problems*, John Wiley & Sons, New York, 1979, pages 243–259.

12. Hamilton–Jacobi Theory

Applicable to   Conservative dynamical systems.

Yields

A reformulation of a system of ordinary differential equations.

Idea

A change of variables may lead to more tractable equations.

Procedure

A conservative dynamical system has a Lagrangian L defined by $L = T - V$, where $T(V)$ is the kinetic (potential) energy. If the generalized coordinates in this system are $\mathbf{q} = (q_1, q_2, \ldots, q_n)$, then the equations of motion are given by

$$\frac{d}{dt}\left(\frac{\partial L}{\partial \dot{q}_i}\right) - \frac{\partial L}{\partial q_i} = 0, \quad \text{for } i = 1, 2, \ldots, n, \tag{12.1}$$

where a dot denotes differentiation with respect to t. The equations in (12.1) are called Lagrange's equations. If we define the generalized momenta by $p_i = \dfrac{\partial L}{\partial q_i}$ and the Hamiltonian by $H = \mathbf{p}^T\dot{\mathbf{q}} - L$, then Lagrange's equations become

$$\dot{q}_i = \frac{\partial H}{\partial p_i},$$
$$\dot{p}_i = -\frac{\partial H}{\partial q_i}, \tag{12.2}$$
$$\frac{\partial L}{\partial t} = -\frac{\partial H}{\partial t}.$$

These equations are called Hamilton's equations. If we change from the $(H, \mathbf{p}, \mathbf{q})$ variables to the $(J, \mathbf{P}, \mathbf{Q})$ variables via the canonical transformation defined by the generating function $S(\mathbf{P}, \mathbf{q}, t)$ (see page 105), then

$$p_i = \frac{\partial S}{\partial q_i},$$
$$Q_i = \frac{\partial S}{\partial P_i}, \tag{12.3}$$
$$J(\mathbf{P}, \mathbf{Q}, t) = H\Big(\mathbf{p}(\mathbf{P}, \mathbf{Q}, t), \mathbf{q}(\mathbf{P}, \mathbf{Q}, t), t\Big) + \frac{\partial S}{\partial t}.$$

In these new variables, Hamilton's equations may be written

$$\dot{Q}_i = \frac{\partial J}{\partial P_i},$$
$$\dot{P}_i = -\frac{\partial J}{\partial Q_i}.$$

(12.4)

If the canonical transformation is chosen so that $J = 0$, then (12.4) says that $\mathbf{P}$ and $\mathbf{Q}$ are constants. To have J vanish identically we require (from (12.3))

$$H\left(\frac{\partial S}{\partial q_1}, \frac{\partial S}{\partial q_2}, \ldots, \frac{\partial S}{\partial q_n}, q_1, q_2, \ldots, q_n, t\right) + \frac{\partial S}{\partial t} = 0$$

This last equation is known as the Hamilton–Jacobi equation. The procedure is to solve the Hamilton–Jacobi equation for the generating function S, make a canonical change of variables using this generating function, and then solve Hamilton's equation in these new coordinates. This will yield a solution to Lagrange's equations.

Example

Suppose we want to solve the linear constant coefficient ordinary differential equation

$$\ddot{q} + \omega^2 q = 0. \tag{12.5}$$

This differential equation comes from the Hamiltonian $H = \frac{1}{2}\left(p^2 + \omega^2 q^2\right)$, which, in turn, corresponds to the following Hamilton–Jacobi equation

$$\frac{1}{2}\left[\left(\frac{\partial S}{\partial q}\right)^2 + \omega^2 q^2\right] + \frac{\partial S}{\partial t} = 0. \tag{12.6}$$

To solve for $S(q, t)$, we use separation of variables (see page 419), and look for a solution in the form $S(q, t) = a(q) + b(t)$, for some unknown functions $a(q)$ and $b(t)$. Using this form for S in (12.6) and making the usual argument about which terms must depend upon which variables, we determine that $a(q)$ and $b(t)$ must satisfy

$$\dot{b} = -\alpha,$$

$$\left(\frac{da}{dq}\right)^2 + \omega^2 q^2 = 2\alpha,$$

where α is a separation constant. Hence, $S = -\alpha t + \int \sqrt{2\alpha - \omega^2 q^2}\, dq$. If we call $\alpha = P$, then we can compute from (12.3)

$$Q = \frac{\partial S}{\partial P} = -t + \int (2P - \omega^2 q^2)^{-1/2}\, dq = -t + \frac{1}{\omega}\arcsin\left(\frac{\omega q}{\sqrt{2P}}\right),$$

which may be inverted to yield $q = \dfrac{\sqrt{2P}}{\omega}\sin\left[\omega(t+Q)\right]$, which is the solution to (12.5).

Notes

[1] Lagrange's equations can be interpreted as the Euler–Lagrange equations for the functional $J = \int L \, dt$. See page 88 for more details.

[2] The functions f and g are said to be *in involution* or to *Poisson commute* if the Poisson bracket $[f, g]$ is identically equal to zero. Liouville's theorem states that a function F is a first integral of a system with Hamiltonian function H if and only if H and F are in involution. See Abraham, Marsden, and Ratiu [1] for details.

[3] Poisson's theorem states that the Poisson bracket of two first integrals of a Hamiltonian system is again a first integral. See Goldstein [2] for details.

[4] Any function $A(\mathbf{p}, \mathbf{q})$ defined along the trajectories of (12.2) satisfies

$$\frac{dA}{dt} = [A, H] = \sum_j \left(\frac{\partial A}{\partial q_j} \frac{\partial H}{\partial p_j} - \frac{\partial A}{\partial p_j} \frac{\partial H}{\partial q_j} \right)$$

where the square brackets denote the Poisson bracket.

[5] A general form for a non-conservative system is often taken to be

$$
\begin{aligned}
\dot{q}_i &= \frac{\partial C}{\partial p_i} + \frac{\partial D}{\partial q_i} \\
\dot{p}_i &= -\frac{\partial C}{\partial q_i} + \frac{\partial D}{\partial p_i}
\end{aligned}
\tag{12.7}
$$

Where $C(\mathbf{p}, \mathbf{q})$ and $D(\mathbf{p}, \mathbf{q})$ are called the conservative and dissipation functions. For $D = 0$, this reduces to (12.3). For $C = 0$, this becomes a gradient system. Any function $A(\mathbf{p}, \mathbf{q})$ defined along the trajectories of (12.7) satisfies

$$\frac{dA}{dt} = \nabla A \cdot \nabla D + [A, C]$$

Choosing $A = C$ and $A = D$ we obtain the evolution equations for the conservative and dissipative functions

$$
\begin{aligned}
\frac{dC}{dt} &= \nabla C \cdot \nabla D, \\
\frac{dD}{dt} &= \nabla D \cdot \nabla D + [D, C].
\end{aligned}
$$

Note that $\nabla^2 D$ equals the divergence of the vector field of (12.7) and that the system is dissipative when $\nabla^2 D < 0$.

[6] Given the equations of motion: $\ddot{q}_i = f_i(\mathbf{q}, \dot{\mathbf{q}}, t)$, the inverse problem of classical mechanics is to determine whether these equations are equivalent to the Euler–Lagrange equations based on a Lagrangian L. That is, a matrix $w = w(\mathbf{q}, \dot{\mathbf{q}}, t)$ is desired so that

$$w_{ij}(\ddot{q}_j - f_j) = \frac{d}{dt}\left(\frac{\partial L}{\partial \dot{q}_i}\right) - \frac{\partial L}{\partial q_i}.$$

The necessary and sufficient conditions for the existence of w and L are called the Helmholtz conditions, they are:

$$\frac{\partial w_{ij}}{\partial \dot{x}_k} = \frac{\partial w_{ik}}{\partial \dot{x}_j}, \qquad w_{ij} = w_{ji},$$

$$\mathcal{D}w_{ij} = -\frac{1}{2}w_{ik}\frac{\partial f_k}{\partial \dot{x}_j} - \frac{1}{2}w_{jk}\frac{\partial f_k}{\partial \dot{x}_i}$$

$$\frac{1}{2}\mathcal{D}\left(w_{ik}\frac{\partial f_k}{\partial \dot{x}_j} - w_{jk}\frac{\partial f_k}{\partial \dot{x}_i}\right) = w_{ik}\frac{\partial f_k}{\partial x_j} - w_{jk}\frac{\partial f_k}{\partial x_i}$$

where $\mathcal{D} := \dfrac{\partial}{\partial t} + \sum_m \left(\dot{x}_m\dfrac{\partial}{\partial x_m} + f_m\dfrac{\partial}{\partial \dot{x}_m}\right)$. See Hojman and Shepley [4].

References

[1] R. Abraham, J. E. Marsden, and T. Ratiu, *Manifolds, Tensor Analysis, and Applications*, Addison–Wesley Publishing Co., Reading, MA, 1983, page 471.

[2] H. Goldstein, *Classical Mechanics*, Addison–Wesley Publishing Co., Reading, MA, 1950, Chapter 9 (pages 273–317).

[3] D. ter Haar, *Elements of Hamiltonian Mechanics*, Pergamon Press, New York, 1971, Chapter 6 (pages 121–145).

[4] S. A. Hojman and L. C. Shepley, "No Lagrangian? No Quantization!," *J. Math. Physics*, **32**, No. 1, January 1991, pages 142–146.

[5] A. H. Nayfeh, *Perturbation Methods*, John Wiley, New York, 1973, pages 179–189.

[6] J. M. Sanz-Serna, "Runge–Kutta Schemes for Hamiltonian Systems," *BIT*, **28**, 1988, pages 877–883.

13.   Inverse Problems

Applicable to   Inverse problems.

Yields

Information about parameters appearing in a differential equation.

Idea

There are theorems that are used to determine which inverse problems may be solved.

Procedure

The field of inverse problems is filled with specialized theorems that are useful for specific applications.

Example 1

Consider the eigenvalue problem

$$-u'' + q(x)u = \lambda u, \qquad \text{for } 0 \le x \le 1,$$
$$u(0)\cos\alpha + u'(0)\sin\alpha = 0, \qquad\qquad (13.1)$$
$$u(1)\cos\beta + u'(1)\sin\beta = 0,$$

where λ is a complex parameter, $q(x)$ is a real-valued function that is integrable on the interval $[0,1]$, and α and β are values in the interval $[0,\pi)$.

One common inverse problem consists of determining the function $q(x)$ from the eigenvalues of (13.1). There are many different results in this area. For example:

> **Theorem:** Suppose that $(\alpha, \beta, q(x))$ give rise to the eigenvalues $\{\lambda_n\}$ and suppose that $(\overline{\alpha}, \overline{\beta}, \overline{q}(x))$ give rise to the eigenvalues $\{\overline{\lambda}_n\}$. If $\lambda_n = \overline{\lambda}_n$ for $n = 0, 1, \ldots$; $q(x) = \overline{q}(x)$ for $x \in (0, \frac{1}{2})$; and $\alpha = \overline{\alpha}$, then $q(x) = \overline{q}(x)$ almost everywhere on the interval $(0, 1)$.

Another typical theorem is the following:

> **Theorem:** Let $\lambda_0 < \lambda_1 < \lambda_2 < \ldots$ be the eigenvalues of the problem $-y'' + q(x)y = \lambda y$ with $y'(0) = y'(\pi) = 0$, where $q(x)$ is a real-valued continuous function. If $\lambda_n = n^2$ for $n = 0, 1, 2, \ldots$, then $q(x) = 0$.

Example 2

One common technique to show uniqueness for an inverse problem is to investigate a mapping between the solutions of two equations with different values for the parameter(s) of interest. We have, for example (see Rundell [11]):

Theorem: Let $u(x)$ and $v(x)$ satisfy

$$u_t = u_{xx} - a(x)u, \qquad u_x(0, t) = 0,$$
$$v_t = v_{xx} - \overline{a}(x)v, \qquad v_x(0, t) = 0,$$

for $0 \le x \le 1$ and $0 \le t < T$. If $u(0, t) = v(0, t)$, then $v(x, t) = u(x, t) + \int_0^x K(x, s)u(s, t)\, ds$, where $K(x, s)$ satisfies the Goursat problem

$$K_{ss} - K_{tt} = (a(s) - \overline{a}(x))\, K(x, s), \qquad \text{for } 0 \le s \le x \le 1,$$
$$K_s(x, 0) = 0 \qquad \text{for } 0 \le x \le 1,$$
$$K(x, x) = \tfrac{1}{2} \int_0^x (a(r) - \overline{a}(r))\, dr \qquad \text{for } 0 \le x \le 1.$$

In this case it is possible to show that if $\int_0^x K(x, s)f(s)\, ds = 0$ for some positive function $f(x)$, then $a = \overline{a}$.

Notes

[1] The numerical methods used to solve inverse problems tend to result in ill-conditioned systems.

References

[1] G. Anger, *Inverse Problems in Differential Equations*, Plenum Publishing Corp., New York, 1990.

[2] D. C. Barnes, "The Inverse Eigenvalue Problem with Finite Data," *SIAM J. Math. Anal.*, **22**, No. 3, May 1991, pages 732–753.

[3] J. R. Cannon and Y. Lin, "An Inverse Problem of Finding a Parameter in a Semi-Linear Heat Equation," *J. Math. Anal. Appl.*, **145**, No. 2, 1990, pages 470–484.

[4] R. D. R. Castillo, "On Boundary Conditions of an Inverse Sturm–Liouville Problem," *SIAM J. Appl. Math.*, **50**, No. 6, 1990, pages 1745–1751.

[5] D. Colton, R. Ewing, and W. Rundell (eds.), *Inverse Problems in Partial Differential Equations*, SIAM, Philadelphia, 1990.

[6] G. Eskin, "Inverse Spectral Problem for the Schroedinger Equation with Periodic Vector Potential," *Comm. Math. Physics*, **125**, No. 2, 1989, pages 263–300.

[7] A. A. M. Hassan and I. H. Abdel-Halim, "Some Inverse Eigenvalue Problems for the Laplacian Operator, II ," *J. Inst. Math. Comput. Sci. Math. Ser.*, **2**, No. 2, 1989, pages 125–146.

[8] B. M. Levitan and I. S. Sargsjan, *Sturm–Liouville and Dirac Operators*, Kluwer Academic Publishers, Dordrecht, The Netherlands, 1991, Chapter 6 (pages 139–182) and Chapter 12 (pages 324–340).

[9] M. Pilant and W. Rundell, "Determining a Coefficient in a First-Order Hyperbolic Equation," *SIAM J. Appl. Math.*, **51**, No. 2, April 1991, pages 494–506.

[10] D. N. G. Roy, *Methods of Inverse Problems in Physics*, CRC Press, Boca Raton, Florida, 1990.

[11] W. Rundell, "The Use of Integral Operators in Undetermined Coefficient Problems for Partial Differential Equations," *Appl. Analysis*, **18**, 1984, pages 309–324.

14.   Limit Cycles

Applicable to   Systems of nonlinear autonomous differential equations.

Yields

Knowledge of whether or not there exist limit cycles.

Idea

Knowing that limit cycles exist for a differential system allows global characterizations of the differential system.

Procedure

A non-constant solution of the system $\dfrac{d\mathbf{x}}{dt} = \mathbf{f}(\mathbf{x})$ is called a cycle (or a limit cycle) if there is a positive number T (called the period of the cycle) such that $\mathbf{x}(t + T) = \mathbf{x}(t)$ for all t. It is easy to show that inside of every cycle is at least one critical point (i.e., a point where $\mathbf{f}(\mathbf{x}) = \mathbf{0}$, see page 451).

In many systems it is not only true that there are finitely many cycles but also that all solutions tend to one of these cycles. This knowledge permits a concise characterization of the phase plane.

Example 1

The nonlinear autonomous system

$$\frac{dx}{dt} = -y + x(1 - x^2 - y^2),$$

$$\frac{dy}{dt} = x + y(1 - x^2 - y^2),$$

becomes, under the change of variables, $\{x = r\cos\theta,\ y = r\sin\theta\}$ the uncoupled system

$$\frac{dr}{dt} = r(1 - r^2), \qquad \frac{d\theta}{dt} = 1.$$

These new equations have the solution

$$r(t) = \frac{1}{\sqrt{1 + Ae^{-2t}}}, \qquad \theta(t) = t + B,$$

where A and B are arbitrary constants. Hence, the solution of the original system is

$$x(t) = \frac{\cos(t + B)}{\sqrt{1 + Be^{-2t}}}, \qquad y(t) = \frac{\sin(t + B)}{\sqrt{1 + Be^{-2t}}}.$$

This states that all solutions tend to the circle $x^2(t) + y^2(t) = 1$ as $t \to \infty$.

Of course, in most circumstances it is not possible to construct explicitly the limit cycle. Generally theorems (such as those below) are used to prove the existence of a limit cycle.

Example 2

The Van der Pol equation

$$\frac{d^2x}{dt^2} - \mu\left(1 - x^2\right)\frac{dx}{dt} + x = 0$$

with $\mu > 0$ has limit cycles. For this equation, there is negative damping for small values of x and positive damping for large values of x. Hence the value of x increases when x is small and it decreases when x is large.

Notes

[1] Given a limit cycle Γ and a positive number a define *the annulus centered on* Γ to be $\{\mathbf{x} \mid \text{distance from } \mathbf{x} \text{ to } \Gamma \text{ is less than } a\}$ where the distance from $\mathbf{x}$ to Γ is defined to be $\min_{\mathbf{u} \in \Gamma} |\mathbf{x} - \mathbf{u}|$.

A cycle Γ is called *isolated* if there is a positive number a for which the annulus centered on Γ contains no other limit cycles. A cycle is *non-isolated* if every annulus centered of Γ contains at least one other limit cycle. The system

$$\frac{dx}{dt} = x \sin\left(x^2 + y^2\right) - y, \qquad \frac{dy}{dt} = y \sin\left(x^2 + y^2\right) + x$$

has infinitely many isolated cycles while the system $\{x' = y, \; y' = -x\}$ has infinitely many non-isolated cycles.

[2] Part of Hilbert's 16th problem asked for the maximum number of limit cycles of the system $\{x' = A(x,y), y' = B(x,y)\}$ when A and B are polynomials. If A and B are polynomials of degree n, then the maximum number is known as the Hilbert number or the Hilbert function, H_n. It is known that $H_0 = 0$, $H_1 = 0$, $H_2 \geq 4$, $H_3 \geq 6$, $\dfrac{n-1}{2} \leq H_n$ if n is odd, and $H_n < \infty$.

The example that shows that $H_2 \geq 4$ (found by Songling [9] is

$$x' = ax - y - 10x^2 + (5+b)xy + y^2,$$
$$y' = x + x^2 + (8c - 25 - 9b)xy,$$

where $a = -10^{-200}$, $b = -10^{-13}$, and $c = -10^{-52}$.

[3] If $f(x)$ and $g(x)$ are continuous, have continuous derivatives, and satisfy the conditions:

(A) $xg(x) > 0$ for $x \neq 0$,

(B) $f(x)$ is negative in the interval $a < x < b$ (with $a < 0$ and $b > 0$) and positive outside of this interval,

(C) $\int_0^\infty f(x)\,dx = \int_{-\infty}^0 f(x)\,dx = \infty$,

then every nontrivial solution of Liénard's equation

$$\frac{d^2 x}{dt^2} + f(x)\frac{dx}{dt} + g(x) = 0 \tag{14.1}$$

is either a limit cycle or a spiral which tends toward a limit cycle as $t \to \infty$. See Birkhoff and Rota [1] for details.

[4] Liénard's theorem states:

If $f(x)$ and $g(x)$ are continuous, and satisfy the conditions:

(A) $F(x) := \int_0^x f(x)\,dx$ is an odd function,

(B) $F(x)$ is zero only at $x = 0$, $x = a$, $x = -a$, for some $a > 0$,

(C) $F(x) \rightarrow \infty$ monotonically for $x > a$,

(D) $g(x)$ is an odd function, and $g(x) > 0$ for $x > 0$,

then (14.1) has a unique limit cycle.

For details see Jordan and Smith [4]. Note that Van der Pol's equation (see example 2) satisfies Liénard's theorem and, hence, has a unique limit cycle.

[5] Bendixson's theorem states:

If $\dfrac{\partial F}{\partial x} + \dfrac{\partial G}{\partial y}$ is continuous and is always positive or always negative in a certain region of the phase plane, then the autonomous system

$$\frac{dx}{dt} = F(x, y), \qquad \frac{dy}{dt} = G(x, y)$$

has no limit cycles in that region.

For details see Simmons [8]. For example, the equation for the Lewis regulator

$$\frac{d^2 x}{dt^2} + (1 - |x|)\frac{dx}{dt} + x = 0$$

which is equivalent to

$$\frac{dx}{dt} = F(x, y) = y, \qquad \frac{dy}{dt} = G(x, y) = -x - (1 - |x|)y$$

has $\dfrac{\partial F}{\partial x} + \dfrac{\partial G}{\partial y} = 1 - |x|$. Hence, the Lewis regulator has no limit cycles in the strip $-1 < x < 1$.

[6] The Levinson–Smith theorem states:

For the differential equation

$$x'' + f(x, x')x' + g(x) = 0 \tag{14.2}$$

if the following conditions are satisfied:

(A) $xg(x) > 0$ for all $x > 0$,

(B) $\int_0^\infty g(x)\, dx = \infty$,

(C) $f(0,0) < 0$,

(D) there exists an $x_0 > 0$ such that $f(x, x') \geq 0$ for $|x| > x_0$, for every x',

(E) there exists a constant $M > 0$, such that $f(x, x') \geq -M$ for $|x| \leq x_0$,

(F) there exists an $x_1 > x_0$ such that $\int_{x_0}^{x_1} f(x, v(x))\, dx \geq 10Mx_0$, where $v(x)$ is any arbitrary positive and monotonically decreasing function of x,

then (14.2) has at least one limit cycle.

See Hagedorn [3] for details.

References

[1] G. Birkhoff and G.-C. Rota, *Ordinary Differential Equations*, John Wiley & Sons, New York, 1978, pages 135–137.

[2] T. R. Blows and N. G. Lloyd, "The Number of Limit Cycles of Certain Polynomial Differential Equations," *Proc. Roy. Soc. Edinburgh*, **98A**, 1984, pages 215–239.

[3] P. Hagedorn, *Non-Linear Oscillations*, Clarendon Press, Oxford, 1982, page 143.

[4] D. W. Jordan and P. Smith, *Nonlinear Ordinary Differential Equations*, Clarendon Press, Oxford, Second Edition, 1987.

[5] D. E. Koditschek and K. S. Narendra, "Limit Cycles of Planar Quadratic Differential Equations," *J. Differential Equations*, **54**, 1984, pages 181–195.

[6] Y. Kuang, "Finiteness of Limit Cycles in Planar Autonomous Systems," *Appl. Anal.*, **32**, 1989, No. 3-4, pages 253–264.

[7] S. Shahshahani, "Periodic Solutions of Polynomial First Order Differential Equations," *Nonlinear Analysis*, **5**, No. 2, 1981, pages 157–165.

[8] G. F. Simmons, *Differential Equations with Applications and Historical Notes*, McGraw–Hill Book Company, New York, 1972, pages 338–352.

[9] S. Songling, "A Concrete Example of the Existence of Four Limit Cycles for Plane Quadratic Systems," *Sci. Sinica*, **23**, 1980, pages 153–158.

[10] Y. Yan-Qian, *Theory of Limit Cycles*, Translations of Mathematical Monographs, Volume 66, Amer. Math. Soc., Providence, Rhode Island, 1986.

15. Natural Boundary Conditions for a PDE

Applicable to   Partial differential equations.

Yields

A proper set of boundary conditions.

Idea

Given a partial differential equation it is not always clear what the "correct" boundary conditions are. This is especially true for nonlinear partial differential equations. However, most partial differential equations that arise in mathematical physics have been obtained from a variational principle (see page 88).

If we *start* with the variational principle, then "natural" boundary conditions will be generated while deriving the equation we started with.

These boundary condition are, in a sense, the most appropriate boundary conditions for the original equation if there is no physical reason for imposing other conditions.

Procedure

The variational principle that is most often used is $\delta J = 0$, where δ represents a variation, J is a functional given by

$$J[\phi] = \iint\limits_R L(\phi, \phi_t, \phi_\mathbf{x}) \, dt \, d\mathbf{x},$$

$L(\ )$ is a linear or nonlinear functional and $\phi(\mathbf{x}, t)$ is the unknown function to be determined. This variational principle states that the integral $J[\phi]$ should be stationary to small changes in ϕ. If we let $h(\mathbf{x}, t)$ be a continuously differentiable function, that is "small" in magnitude, then we can form

$$J[\phi + h] - J[\phi] = \iint\limits_R \left\{ L_{\phi_t} h_t + L_{\phi_{x_j}} h_{x_j} + L_\phi \right\} dt \, d\mathbf{x} + O(\|h\|^2),$$

where subscripts on L denote partial derivatives. The variational principle requires that $\delta J := J[\phi + h] - J[\phi] = 0$, or that

$$\iint\limits_R \left\{ L_{\phi_t} h_t + L_{\phi_{x_j}} h_{x_j} + L_\phi \right\} dt \, d\mathbf{x} = 0. \tag{15.1}$$

If R is assumed to be a parallelpiped, then let D_t (D_{x_j}) denote the two parts of the boundary of R on which t (x_j) is constant. By integrating by parts, equation (15.1) can be written as

$$\iint\limits_R \left\{ -\frac{\partial}{\partial t} L_{\phi_t} - \frac{\partial}{\partial x_j} L_{\phi_{x_j}} + L_\phi \right\} h \, dt \, d\mathbf{x} = 0, \tag{15.2}$$

where we have assumed that

$$L_{\phi_t} \bigg|_{D_t} = 0, \qquad L_{\phi_{x_j}} \bigg|_{D_{x_j}} = 0. \tag{15.3}$$

Now $h(\mathbf{x}, t)$ was assumed to be arbitrary, so from (15.2) we conclude that

$$\frac{\partial}{\partial t} L_{\phi_t} + \frac{\partial}{\partial x_j} L_{\phi_{x_j}} - L_\phi = 0. \tag{15.4}$$

We conclude: if we can write a given partial differential equation in the form of (15.4), for some operator $L(\ )$, then (15.3) gives the "natural" boundary conditions.

Example

Given the partial differential equation

$$\phi_{tt} - \alpha^2 \nabla^2 \phi + \beta^2 \phi = 0, \tag{15.5}$$

where $\nabla^2 \phi = \sum_{j=1}^{N} \phi_{x_j x_j}$, we find that

$$L(\phi, \phi_t, \phi_\mathbf{x}) = \tfrac{1}{2}\phi_t^2 - \tfrac{1}{2}\alpha^2 \sum_{j=1}^{N} \phi_{x_j}^2 - \tfrac{1}{2}\beta^2 \phi^2 \tag{15.6}$$

makes (15.4) and (15.5) identical. Therefore, the "natural" boundary conditions for (15.5) are, using (15.6) in (15.3),

$$\phi_t\bigg|_{D_t} = 0, \qquad \phi_{x_j}\bigg|_{D_{x_j}} = 0. \tag{15.7}$$

Equation (15.7) states that the partial differential equation in (15.5) requires both initial and boundary conditions. This was to be expected since (15.5) is a hyperbolic equation.

For example, if $N = 1$ and R is the region $[0, T] \times [0, \infty)$, then $D_t = \{t = 0\} \cup \{t = T\}$ and $D_{x_1} = \{x_1 = 0\} \cup \{x_1 = \infty\}$. Hence, the natural boundary conditions for (15.5) require that $\{\phi_t(0, x_1), \phi_t(T, x_1), \phi_{x_1}(t, 0), \phi_{x_1}(t, \infty)\}$ be specified.

Notes

[1] Finding the operator $L(\ )$, or equivalently finding the variational principle δJ, is a non-trivial task in general. Also, very often one wants a vector variational principle that will encompass, simultaneously, several separate equations.

[2] See the section on variational equations (page 88) for more examples.

References

[1] L. V. Kantorovich and V. I. Krylov, *Approximate Methods of Higher Analysis*, Interscience Publishers, New York, 1958, Chapter 4 (pages 241–357).

[2] G. B. Whitham, *Linear and Nonlinear Waves*, Wiley Interscience, New York, 1974.

16.   Normal Forms: Near-Identity Transformations

Applicable to   Systems of ordinary differential equations.

Yields

A reformulation of the differential equations.

Idea

Find a change of variables, in the form of an infinite series, so that the original system of differential equations goes into a "normal" (or "simple" or "canonical") form. The normal form is the simplest member of an equivalence class of differential equations, all exhibiting the same qualitative behavior. Normal forms are often useful for stability analyses.

Procedure

Start with the system $\mathbf{x}' = \mathbf{f}(\mathbf{x})$ such that (without loss of generality) $\mathbf{x} = \mathbf{0}$ is a critical point. Expand this system to obtain

$$\mathbf{x}' = A\mathbf{x} + \mathbf{H}(\mathbf{x}),$$

where $\mathbf{H}(\mathbf{x})$ has *strictly nonlinear* functions (i.e., there are no linear or constant terms).

If $\mathbf{H}(\mathbf{x})$ has nonlinear terms of at least degree n, then make a near-identity transformation using polynomials of degree n with unknown coefficients. By appropriately choosing the unknown coefficients in the near-identity transformation, the original differential equations, when written in the new variables, will have increased the degree of the nonlinear terms by one.

If the critical point is "hyperbolic" (all eigenvalues have non-zero real parts) then the nonlinear terms can always be removed (i.e., one order at a time). Also, the topological nature does not change. See Guckenheimer and Holmes [6], Section 3.3.

Mathematically, we can summarize the procedure as follows:

[1] We are given the system of ordinary differential equations $\mathbf{x}' = \mathbf{f}(\mathbf{x}) = A\mathbf{x} + \mathbf{H}(\mathbf{x})$, which we wish to analyze near the point $\mathbf{x} = \mathbf{0}$.

[2] We make the near-identity transformation from $\mathbf{x}$ to $\mathbf{u}$ via $\mathbf{x} = \mathbf{u} + \mathbf{g}(\mathbf{u})$, where $\mathbf{g}(\ )$ is a strictly nonlinear function.

[3] This change of variables produces the new equation

$$\mathbf{u}' = [I + J]^{-1}\mathbf{f}(\mathbf{u} + \mathbf{g}(\mathbf{u})) = A\mathbf{u} + \mathbf{K}(\mathbf{u}), \qquad (16.1)$$

where I is the identity matrix and $J = \dfrac{\partial \mathbf{g}}{\partial \mathbf{u}}$ is the Jacobian of the transformation.

[4] The function $\mathbf{g}(\ )$ is chosen to eliminate the nonlinear terms in the equation for $\mathbf{u}$ that are of least order.

This procedure can be iterated.

Example 1

Suppose we have the system of equations

$$\frac{dx}{dt} = x + y^2,$$

$$\frac{dy}{dt} = y + xy.$$

Defining $\mathbf{x} = (\,x \quad y\,)^{\mathrm{T}}$, this system has the form

$$\frac{d\mathbf{x}}{dt} = \begin{pmatrix} 1 & 0 \\ 0 & 1 \end{pmatrix}\mathbf{x} + \begin{pmatrix} y^2 \\ xy \end{pmatrix} = \begin{pmatrix} 1 & 0 \\ 0 & 1 \end{pmatrix}\mathbf{x} + \mathbf{H}(\mathbf{x}), \qquad (16.2)$$

where $\mathbf{H}(\mathbf{x})$ has quadratic nonlinearities. We now choose to make the near-identity change of variables (of second order)

$$\begin{aligned} x &= u + a_{02}u^2 + a_{11}uv + a_{20}v^2, \\ y &= v + b_{02}u^2 + b_{11}uv + b_{20}v^2, \end{aligned} \qquad (16.3)$$

where u and v are functions of t. Combining (16.2) and (16.3) we find

$$\frac{du}{dt} = u + (1 - a_{02})v^2 - a_{11}uv - a_{20}u^2 + \text{higher order terms},$$

$$\frac{dv}{dt} = v - b_{02}v^2 + (1 - b_{11})uv - b_{20}u^2 + \text{higher order terms}, \qquad (16.4)$$

where "higher order terms" means terms that are of order $O(u^3, u^2v, uv^2, v^3)$. To eliminate the second order terms in (16.4) we take $\{a_{02} = 1,\ a_{11} = 0,\ a_{20} = 0,\ b_{02} = 0,\ b_{11} = 1,\ b_{20} = 0\}$. With these values, the transformation in (16.3) becomes

$$x = u + u^2,$$

$$y = v + uv$$

so that the original differential equations in (16.2) becomes

$$\frac{du}{dt} = u + \text{higher order terms},$$

$$\frac{dv}{dt} = v - \text{higher order terms}.$$

This new system now has cubic (or higher order) nonlinearities.

Example 2

The system of ordinary differential equations for $x(t)$ and $y(t)$:

$$\begin{aligned} x' &= -y + F(x, y), \\ y' &= x + G(x, y), \end{aligned}$$

(16.5)

where $F(\,)$ and $G(\,)$ are strictly nonlinear, has the normal form

$$\begin{aligned} \theta' &= 1 + D_1 r^2 + D_2 r^4 + D_3 r^6 + \dots, \\ r' &= B_1 r^3 + B_2 r^5 + B_3 r^7 + \dots, \end{aligned}$$

where $u = r\cos\theta$, $v = r\sin\theta$, and $\{u, v\}$ are related, via a near-identity transformation, to $\{x, y\}$. In this example, the linear equations are not sufficient to determine the local behavior. Knowledge of B_1 is needed to determine stability (unless it is zero, in which case B_2 is needed, etc.).

For example, if equation (16.5) has the form

$$\begin{aligned} x' &= y + F_{xx}\frac{x^2}{2} + F_{xy}xy + F_{yy}\frac{y^2}{2} + F_{xxx}\frac{x^3}{6} + F_{xxy}\frac{x^2 y}{2} \\ &\quad + F_{xyy}\frac{xy^2}{2} + F_{yyy}\frac{y^3}{6} + \dots, \\ y' &= G_{xx}\frac{x^2}{2} + G_{xy}xy + F_{yy}\frac{y^2}{2} + G_{xxx}\frac{x^3}{6} + G_{xxy}\frac{x^2 y}{2} \\ &\quad + G_{xyy}\frac{xy^2}{2} + G_{yyy}\frac{y^3}{6} + \dots, \end{aligned}$$

then we find (see Takens [10] for details)

$$\begin{aligned} 16 B_1 &= G_{yyy} + G_{xxy} + F_{xyy} + F_{xxx} + F_{yy}G_{yy} - F_{xx}G_{xx} - G_{xx}G_{xy} \\ &\quad - G_{yy}G_{xy} + F_{xx}F_{xy} + F_{xy}F_{yy}. \end{aligned}$$

Example 3

The system of ordinary differential equations for $x(t)$ and $y(t)$:

$$\begin{aligned} x' &= y + F(x, y), \\ y' &= G(x, y), \end{aligned}$$

(16.6)

where $F(\,)$ and $G(\,)$ are strictly nonlinear, has the normal form

$$u' = v + \sum_{n=2}^{\infty} b_n u^n, \qquad v' = \sum_{n=2}^{\infty} a_n u^n,$$

(16.7)

where $\{u, v\}$ are related, via a near-identity transformation, to $\{x, y\}$. For example, if equation (16.6) has the form

$$x' = -y + F_{xx}\frac{x^2}{2} + F_{xy}xy + F_{yy}\frac{y^2}{2} + F_{xxx}\frac{x^3}{6} + F_{xxy}\frac{x^2 y}{2}$$

$$+ F_{xyy}\frac{xy^2}{2} + F_{yyy}\frac{y^3}{6} + \cdots,$$

$$y' = x + G_{xx}\frac{x^2}{2} + G_{xy}xy + F_{yy}\frac{y^2}{2} + G_{xxx}\frac{x^3}{6} + G_{xxy}\frac{x^2 y}{2}$$

$$+ G_{xyy}\frac{xy^2}{2} + G_{yyy}\frac{y^3}{6} + \cdots,$$

then we find that

$$u' = v + \tfrac{1}{2}\left(G_{xy} + F_{xx}\right)u^2 + \tfrac{1}{12}\left(G_{xy}G_{yy} - F_{xx}G_{yy}\right.$$
$$\left. + 2F_{xy}G_{xy} + 2G_{xxy} - F_{yy}G_{xx} - G_{xx} + 2F_{xxx}\right)u^3 + \cdots, \quad (16.8)$$
$$v' = \tfrac{1}{2}G_{xx}u^2 + \tfrac{1}{6}\left(3F_{xy}G_{xx} + G_{xxx} - F_{xx}G_{xy}\right)u^3 + \cdots,$$

where C is an arbitrary constant. See Takens [10] for details.

Another normal form for (16.6) is given by

$$U' = V,$$

$$U' = \sum_{n=2}^{\infty} a_n U^n + \sum_{n=2}^{\infty} n b_n U^{n-1},$$

where $\{U, V\}$ are related, via a near-identity transformation, to $\{x, y\}$. See Guckenheimer and Holmes [6] for details.

Notes

[1] If $a_2 \neq 0$, then the flow of the system in (16.7) is topologically equivalent to the flow of the system $\{u' = v, \ v' = a_2 u^2\}$, which can be integrated in terms of elliptic integrals. If $a_2 = 0$, then other conclusions are possible; see Rand and Keith [8] for details.

[2] To avoid computing the matrix inverse in equation (16.1), it is sufficient to expand $(I + J)^{-1}$ into $I - J + J^2 - \ldots + (-J)^{n-1}$ if only the nonlinear terms of order n are to be removed.

[3] The concept of normal forms does not require that the transformations used be near-identity ones; but they are the ones most often used in practice.

[4] The computations needed for this technique quickly become unmanageable unless a computer algebra system is used. MACSYMA programs for performing the necessary computations are given in Rand and Keith [9] and in Chow *et al.* [2].

References

[1] M. Ashkenazi and S.-N. Chow, "Normal Forms Near Critical Points for Differential Equations and Maps," *IEEE Trans. Circ. & Syst.*, **35**, No. 7, July 1988, pages 850–862.

[2] S.-N. Chow, B. Byron, and D. Wang, "Computation of Normal Forms," *J. Comput. Appl. Math.*, **29**, No. 2, 1990, pages 129–143.

[3] L. O. Chua and H. Kokubu, "Normal Forms for Nonlinear Vector Fields—Part I: Theory and Algorithm," *IEEE Trans. Circ. & Syst.*, **35**, No. 7, July 1988, pages 863–880.

[4] L. O. Chua and H. Oka, "Normal Forms for Constrained Nonlinear Differential Equations—Part I: Theory," *IEEE Trans. Circ. & Syst.*, **35**, No. 7, July 1988, pages 881–901.

[5] E. Freire, E. Gamero, and E. Ponce, "An Algorithm for Symbolic Computation of Hopf Bifurcation," in E. Kaltofen and S. M. Watt (eds.), *Computers and Mathematics*, Springer–Verlag, New York, 1990, pages 109–118.

[6] J. Guckenheimer and P. Holmes, *Nonlinear Oscillations, Dynamical Systems, and Bifurcations of Vector Fields*, Springer–Verlag, New York, 1983.

[7] R. H. Rand and D. Armbruster, *Perturbation Methods, Bifurcation Theory and Computer Algebra*, Springer–Verlag, New York, Applied Mathematical Sciences #65, 1987, Chapter 3, pages 50–88.

[8] R. H. Rand and W. L. Keith, "Determinacy of Degenerate Equilibria with Linear Part $x' = y$, $y' = 0$ Using MACSYMA," *Appl. Math. and Comp.*, **21**, 1987, pages 1–19.

[9] R. H. Rand and W. L. Keith, "Normal Forms and Center Manifolds Calculations on MACSYMA," in R. Pavelle (ed.), *Applications of Computer Algebra*, Kluwer Academic Publishers, Dordrecht, The Netherlands, 1985.

[10] F. Takens, "Singularities of Vector Fields," *Publ. Math. Inst. Hautes Etudes Sci.*, **43**, 1974, pages 47–100.

17.   Self-Adjoint Eigenfunction Problems

Applicable to   Linear differential operators.

Yields

Information that may be used to show completeness of a set of functions.

Procedure

Many of the differential equations of mathematical physics are related to self-adjoint eigenfunction problems. As a special subcase, Sturm–Liouville equations are often self-adjoint eigenfunction problems (Sturm–Liouville problems are discussed in more detail on page 82).

Let $L[\cdot]$ be the n-th order linear operator defined by

$$L[y] = p_n(x)\frac{d^n y}{dx^n} + p_{n-1}(x)\frac{d^{n-1}y}{dx^{n-1}} + \cdots + p_0(x)y,$$

where the $\{p_i(x)\}$ are complex valued and analytic, and $p_n(x) \neq 0$ on the interval $x \in [a,b]$. Define n boundary conditions by

$$B_j[y] := \sum_{k=1}^{n}\left(M_{jk}\frac{d^{(k-1)}y}{dx^{(k-1)}}(a) + N_{jk}\frac{d^{(k-1)}y}{dx^{(k-1)}}(b)\right) = 0, \qquad j = 1, \ldots, n,$$

where the $\{M_{jk}, N_{jk}\}$ are given complex constants.

The problem we will consider here is

$$L[y] = \lambda y, \qquad B[y] = 0, \tag{17.1}$$

where $\{B[y] = 0\}$ is a shorthand notation for $\{B_j[y] = 0 \mid j = 1,\ldots,n\}$. The system in (17.1) will always have the trivial solution, $y(x) = 0$. But, for certain values of λ, called *eigenvalues*, the system in (17.1) will have non-trivial solutions. Corresponding to a specific eigenvalue, λ_n, will be one or more *eigenfunctions*; that is, non-trivial solutions to (17.1) when $\lambda = \lambda_n$.

We represent the complex conjugate of g by $\bar{g}$. Define the *inner product* of $f(x)$ and $g(x)$ by $(f,g) = \int_a^b f(t)\bar{g}(t)\,dt$, and the *norm* of $f(x)$ by $\|f\| := \sqrt{(f,f)}$. If $(f,g) = 0$, then f and g are said to be *orthogonal*. If $\{f_1, f_2, \ldots, f_n\}$ are a set of functions with $(f_i, f_j) = 0$ when $i \neq j$, then the $\{f_i(x)\}$ are an *orthogonal family*.

The *adjoint* operator to $L[\cdot]$, called $L^*[\cdot]$, is defined by

$$L^*[y] := (-1)^n\frac{d^{(n)}[\bar{p}_n(x)y]}{dx^{(n)}} + (-1)^{n-1}\frac{d^{(n-1)}[\bar{p}_{n-1}(x)y]}{dx^{(n-1)}} + \cdots + \bar{p}_0 y.$$

Let $u(x)$ be a solution to the system $\{L[u] = 0, B[u] = 0\}$, and let $v(x)$ be a solution to the adjoint system $\{L^*[u] = 0, B^*[u] = 0\}$, where $\{B^*[y] = 0\}$ is a shorthand notation for $\{B_j^*[y] = 0 \mid j = 1,\ldots,n\}$ and the $B_i^*[\cdot]$ are, for the moment, unspecified. Using the definitions of $u(x)$ and $v(x)$, we can calculate

$$vL[u] - uL^*[v] = \frac{d}{dx}J(u,v), \tag{17.2}$$

where $J(u, v)$ is called the *bilinear concomitant* and is defined by

$$J(u, v) = \sum_{m=1}^{n} \sum_{j+k=m-1} (-1)^k \left(\frac{d^k}{dx^k}(p_m u) \right) \left(\frac{d^j v}{dx^j} \right). \tag{17.3}$$

Integrating (17.2) results in

$$\int_a^b (vL[u] - uL^*[v]) \, dx = J(u, v) \Big|_a^b = J\Big(u(b), v(b)\Big) - J\Big(u(a), v(a)\Big). \tag{17.4}$$

We now define the $B_i^*[\cdot]$ to be those boundary conditions for which the right-hand side of (17.4) vanishes.

If $L = L^*$, then L is said to be *formally self-adjoint*. If $L = L^*$ and $B = B^*$, then L is said to be *self-adjoint*. Note that, if $L[\cdot]$ is formally self-adjoint, then $n = 2r$ and $L[\cdot]$ must be of the form

$$L[u] = \frac{d^r}{dx^r} \left(b_r(x) \frac{d^r u}{dx^r} \right) + \cdots + \frac{d}{dx} \left(b_1(x) \frac{du}{dx} \right) + b_0(x)u. \tag{17.5}$$

As we now record, self-adjoint operators have some very useful properties. If $L[\cdot]$ is self-adjoint, then

(A) The eigenvalues λ_n of (17.1) are real.

(B) The eigenvalues are enumerable (with no cluster point).

(C) The eigenfunctions $y_n(x)$ corresponding to distinct eigenvalues are orthogonal.

(D) If $f(x)$ is any analytic function that satisfies the boundary conditions in (17.1) (i.e., $B_j[f] = 0$, for $j = 1, \ldots, n$), then, on the interval $[a, b]$,

$$f(x) = \sum_{k=0}^{\infty} \frac{(f, y_k)}{(y_k, y_k)} y_k(x). \text{ That is, the } \{y_k(x)\} \text{ are complete. It is this}$$

last statement that is of particular importance in solving differential equations; the method suggested by this statement, the method of eigenfunction expansions, is described on page 223.

Example 1

Suppose we have the linear differential operator

$$L[y] = \frac{d^2}{dx^2}\left(r_2(x)\frac{d^2y}{dx^2}\right) + \frac{d}{dx}\left(r_1(x)\frac{dy}{dx}\right) + r_0(x). \qquad (17.6)$$

Because of the form of the operator, we know that $L[\cdot]$ will be formally self-adjoint (see (17.5)). For this operator we can evaluate $J(u, v)$ at the upper and lower limits (from (17.3)) to find

$$J(u,v)\Big|_a^b = \left[v(r_2u'')' - v'r_2u'' + r_2v''u' - u(r_2v'')' + r_1(vu' - uv')\right]\Big|_a^b. \quad (17.7)$$

To determine whether $L[\cdot]$ is self-adjoint or not, we need to specify $B[y]$. Since (17.6) is a fourth order operator, four boundary conditions are required. We will consider three separate cases.

Case 1

If $B[y]$ is defined by

$$\begin{aligned} B_1[y] &= y(a), \\ B_2[y] &= y''(a), \\ B_3[y] &= y(b), \\ B_4[y] &= y''(b), \end{aligned} \qquad (17.8)$$

then $J(u, v)$ can be evaluated and (17.7) can be simplified to yield

$$r_2v''u' + r_1vu'\Big|_a^b. \qquad (17.9)$$

If we choose $B = B^*$ (i.e., $B_i^*[y] = B_i[y]$), then the quantity in (17.9) is identically zero. Hence, $L[\cdot]$, as defined by (17.6) and (17.8) is self-adjoint.

Case 2

If $B[y]$ is defined by

$$\begin{aligned} B_1[y] &= y(a), \\ B_2[y] &= y'(a), \\ B_3[y] &= y(b), \\ B_4[y] &= y'(b), \end{aligned} \qquad (17.10)$$

then $J(u, v)$ can be evaluated and (17.7) can be simplified to yield

$$v(r_2u'')' - v'r_2u''\Big|_a^b. \qquad (17.11)$$

Once again, if we choose $B = B^*$ then the quantity in (17.11) is identically zero. Hence, $L[\cdot]$, as defined by (17.6) and (17.10) is self-adjoint.

Case 3

If $B[y]$ is defined by

$$
\begin{aligned}
B_1[y] &= y(a), \\
B_2[y] &= y'(a), \\
B_3[y] &= y''(a), \\
B_4[y] &= y'''(a),
\end{aligned}
\tag{17.12}
$$

then $J(u, v)$ can be evaluated and (17.7) can be simplified to yield

$$
v(r_2u'')' - v'r_2u'' + r_2v''u' - u(r_2v'')' + r_1(vu' - uv') \Big|_{x=b}. \tag{17.13}
$$

If, in this case, we choose $B = B^*$, then the quantity in (17.13) does *not* vanish. If $B = B^*$, then no information has been given at the boundary $x = b$, and the quantity in (17.13) is indeterminate. Hence, $L[\cdot]$, as defined by (17.6) and (17.12), is not self-adjoint. An initial value problem can never be self-adjoint.

Example 2

The operator

$$
L[y] = \frac{d}{dx}\left(a_2(x)\frac{dy}{dx}\right) + a_1(x)\frac{dy}{dx} + a_0(x),
$$

with the boundary conditions

$$
\begin{aligned}
B_1[y] &= y(a), \\
B_2[y] &= y'(b),
\end{aligned}
$$

is self-adjoint. See the section on Sturm–Liouville theory (page 82).

Example 3

A third order linear ordinary differential equation is formally self-adjoint if it has the form

$$
\frac{d^2}{dx^2}\left(P(x)\frac{dy}{dx}\right) + \frac{d}{dx}\left(P(x)\frac{d^2y}{dx^2}\right) + \frac{d}{dx}\left(Q(x)y\right) + Q(x)\frac{dy}{dx} = 0. \tag{17.14}
$$

The general third order linear ordinary differential equation

$$
A(x)\frac{d^3y}{dx^3} + B(x)\frac{d^2y}{dx^2} + C(x)\frac{dy}{dx} + D(x) = 0,
$$

will be formally self-adjoint if and only if $B = \frac{3}{2}A'$ and $D = \frac{1}{2}\left(C - \frac{1}{3}B'\right)'$. The self-adjoint third order equation (17.14) has the first integral

$$
P\left(2yy'' - (y')^2\right) + P'yy' + Qy^2 = \text{constant}.
$$

Example 4

The general fourth order linear ordinary differential equation

$$A(x)y'''' + B(x)y''' + C(x)y'' + D(x)y' + E(x)y = 0,$$

will be formally self-adjoint if and only if $B = 2A'$ and $D = \left(C - \frac{1}{2}B'\right)'$.

Notes

[1] Some of the conditions above can be relaxed, and the main results for self-adjoint operators will still be true. See, for instance, Coddington and Levinson [3].

[2] For partial differential equations there are many results analogous to those mentioned above for ordinary differential equations. We enumerate some of them for the Helmholtz equation in two dimensions. For the equation

$$\nabla^2 \phi + \lambda\phi = 0,$$

in a region R, with the boundary conditions

$$a\phi + b\nabla\phi \cdot \mathbf{n} = 0,$$

given on the entire boundary of R (here $\mathbf{n}$ represents the unit normal):

(A) All the eigenvalues $\{\lambda_i\}$ are real.

(B) There are an infinite number of eigenvalues. There is an eigenvalue of least magnitude, but no largest one.

(C) The eigenfunctions $\{\phi_i(x,y)\}$ form a complete set: any analytic function can be represented in the form $f(x,y) = \sum_i a_i\phi_i(x,y)$, for some set of constants $\{a_i\}$.

(D) Eigenfunctions belonging to different eigenvalues are orthogonal. That is $\iint_R \phi_i\bar{\phi}_j \, dx \, dy = 0$, if $\lambda_i \neq \lambda_j$.

(E) An eigenfunction ϕ is related to it's eigenvalue λ by the Rayleigh quotient

$$\lambda = \frac{-\oint \phi\nabla\phi \cdot \mathbf{n} \, ds + \iint_R |\nabla\phi|^2 \, dx \, dy}{\iint_R \phi^2 \, dx \, dy}.$$

Many other partial differential equations have very similar properties. See Haberman [5] for details.

[3] Partial differential equations can also be self-adjoint. The elliptic equation $au_{xx} + cu_{yy} + du_x + eu_y + fu = g(x,y)$ is said to be essentially self-adjoint when $N_x = M_y$, where

$$N := d - \frac{a_x}{a}, \qquad M := e - \frac{c_l}{c}.$$

In this case, an integrating factor is given by e^ϕ where $\phi_x = N$, $\phi_y = M$. Multiplying the original equation by this factor puts the equation in self-adjoint form. For example, the equation

$$u_{xx} + u_{yy} + x^2 u_x + y^2 u_y + u = 0$$

has $N = x^2$, $M = y^2$ which leads to $\phi = \frac{1}{3}\left(x^3 + y^3\right)$. Multiplying the equation by e^ϕ results in the self-adjoint form of the equation:

$$\left[\exp\left(\tfrac{1}{3}\left(x^3 + y^3\right)\right) u_x\right]_x + \left[\exp\left(\tfrac{1}{3}\left(x^3 + y^3\right)\right) u_y\right]_y + \exp\left(\tfrac{1}{3}\left(x^3 + y^3\right)\right) u = 0.$$

References

[1] G. Birkhoff and G.-C. Rota, *Ordinary Differential Equations*, John Wiley & Sons, New York, 1978, Chapters 10–11.
[2] E. Butkov, *Mathematical Physics*, Addison–Wesley Publishing Co., Reading, MA, 1968, Chapter 9 (pages 332–404).
[3] E. A. Coddington and N. Levinson, *Theory of Ordinary Differential Equations*, McGraw–Hill Book Company, New York, 1955, Chapter 7.
[4] N. Dunford and J. T. Schwartz, *Linear Operators*, Interscience Publishers, New York, 1958.
[5] R. Haberman, *Elementary Applied Partial Differential Equations*, Prentice–Hall Inc., Englewood Cliffs, NJ, 1983, pages 214–219.
[6] E. L. Ince, *Ordinary Differential Equations*, Dover Publications, Inc., New York, 1964, Chapters 9–11 (pages 204–278).
[7] I. Stakgold, *Green's Functions and Boundary Value Problems*, John Wiley & Sons, New York, 1979, Chapter 3.

18.   Stability Theorems

Applicable to   Differential equations of all types.

Yields

Knowledge of whether or not there are stable solutions.

Idea

There are theorems available for most cases of interest.

Procedure

There are many theorems that can be used to determine whether the solutions to a differential equation are stable. In the following we indicate some of the simple theorems that are frequently useful.

> **Theorem:** Consider the equation $\mathbf{y}' = A\mathbf{y} + \mathbf{f}(t, \mathbf{y})$, where A is a real constant matrix whose eigenvalues all have negative real parts. Let $\mathbf{f}$ be real, continuous for small $|\mathbf{y}|$ and $t \geq 0$, and $\mathbf{f}(t, \mathbf{y}) = o(|\mathbf{y}|)$ as $|\mathbf{f}| \to 0$, uniformly for $t \geq 0$. Then the identically zero solution is asymptotically stable.

> **Theorem:** If all solutions of $\mathbf{y}' = A\mathbf{y}$, where A is a constant matrix, are bounded as $t \to \infty$, the same is true of the solutions of $\mathbf{y}' = (A+B(t))\mathbf{y}$, provided that $\int^{\infty} \|B(t)\|\, dt < \infty$.

> **Theorem:** If all solutions of $\mathbf{y}' = A(t)\mathbf{y}$, where A is a periodic matrix, are bounded, the same is true of the solutions of $\mathbf{y}' = (A(t) + B(t))\mathbf{y}$, provided that $\int^{\infty} \|B(t)\|\, dt < \infty$.

> **Theorem:** If all solutions of $\mathbf{y}' = A(t)\mathbf{y}$, where $\lim_{t \to \infty} \int^t \mathrm{tr}(A)\, dt > -\infty$, are bounded, the same is true of the solutions of $\mathbf{y}' = (A(t) + B(t))\mathbf{y}$, provided that $\int^{\infty} \|B(t)\|\, dt < \infty$.

Notes

[1] Stability is required if a differential equation is to be well-posed; see page 94.
[2] Floquet theory and Lyapunov functions are two techniques that can determine if an equation has stable or unstable solutions; see pages 448 and 476.
[3] Note that solutions to the equation $\mathbf{y}' = A(t)\mathbf{f}(\mathbf{y})$ can be increasing even if all the eigenvalues of $A(t)$ have negative real parts for any fixed value of t. For example, consider the matrix $A(t) = \begin{pmatrix} -\dfrac{1}{4(1+t)} & \dfrac{1}{(1+t)^2} \\ -\dfrac{1}{4} & -\dfrac{1}{4(1+t)} \end{pmatrix}$.

This matrix has the eigenvalues $\lambda_{1,2} = \dfrac{-1 \pm 2i}{4(1+t)}$, yet the general solution to $\mathbf{y}' = A(t)\mathbf{f}(\mathbf{y})$ is given by

$$\mathbf{y}(t) = \alpha \begin{pmatrix} (1+t)^{-3/4} \\ -\frac{1}{2}(1+t)^{1/4} \end{pmatrix} + \beta \begin{pmatrix} (1+t)^{-3/4} \log(1+t) \\ (1+t)^{1/4}\left(1 - \frac{1}{2}\log(1+t)\right) \end{pmatrix},$$

where α and β are arbitrary constants.
[4] There are many different technical definitions of stability. For the equation

$$\mathbf{y}' = \mathbf{f}(t, \mathbf{y}) \tag{18.1}$$

defined when $t \geq t_0$, the solution is said to be:

(A) *stable* if, for each $\varepsilon > 0$ there is a corresponding $\delta = \delta(\varepsilon) > 0$ such that any solution $\widehat{\mathbf{y}}(t)$ of (18.1) which satisfies the inequality $|\widehat{\mathbf{y}}(t_0) - \mathbf{y}(t_0)| < \delta$ exists and satisfies the inequality $|\widehat{\mathbf{y}}(t) - \mathbf{y}(t)| < \delta$ for all $t \geq t_0$. A solution that is not stable is said to be *unstable*.

(B) *asymptotically stable* if, in addition to the above stability requirements, $|\widehat{\mathbf{y}}(t) - \mathbf{y}(t)| \to 0$ as $t \to \infty$, whenever $|\widehat{\mathbf{y}}(t_0) - \mathbf{y}(t_0)|$ is sufficiently small.

(C) *uniformly stable* if, for each $\varepsilon > 0$ there is a corresponding $\delta = \delta(\varepsilon) > 0$ such that any solution $hatbfy(t)$ of (18.1) which satisfies the inequality $|\widehat{\mathbf{y}}(t_0) - \mathbf{y}(t_0)| < \delta$ for some $t_1 \geq t_0$ exists and satisfies the inequality $|\widehat{\mathbf{y}}(t) - \mathbf{y}(t)| < \varepsilon$ for all $t \geq t_1$.

(D) *uniformly asymptotically stable* if, in addition to the requirements for asymptotic stability, there is a $\delta_0 > 0$, and for each $\varepsilon > 0$ a corresponding $T = T(\varepsilon) > 0$ such that if $|\widehat{\mathbf{y}}(t_1) - \mathbf{y}(t_1)| < \delta_0$ for some $t_1 \geq t_0$, then $|\widehat{\mathbf{y}}(t) - \mathbf{y}(t)| < \varepsilon$ for all $t \geq t_1 + T$.

(E) *strongly stable* if, for each $\varepsilon > 0$ there is a corresponding $\delta = \delta(\varepsilon) > 0$ such that any solution $hatbfy(t)$ of (18.1) which satisfies the inequality $|\widehat{\mathbf{y}}(t_0) - \mathbf{y}(t_0)| < \delta$ for some $t_1 \geq t_0$ exists and satisfies the inequality $|\widehat{\mathbf{y}}(t) - \mathbf{y}(t)| < \varepsilon$ for all $t \geq t_0$.

References

[1] R. Bellman, *Stability Theory of Differential Equations*, McGraw–Hill Book Company, New York, 1953.

[2] W. E. Boyce and R. C. DiPrima, *Elementary Differential Equations and Boundary Value Problems*, Fourth Edition, John Wiley & Sons, New York, 1986.

[3] E. A. Coddington and N. Levinson, *Theory of Ordinary Differential Equations*, McGraw–Hill Book Company, New York, 1955.

19.   Sturm–Liouville Theory

Applicable to   Second order linear ordinary differential operators.

Yields

Information about whether an operator is self-adjoint.

Procedure

Many of the differential equations of mathematical physics are Sturm–Liouville equations. Sturm–Liouville equations arise naturally, for instance, when separation of variables (see page 419) is applied to the wave equation, the potential equation, or the diffusion equation.

The Sturm–Liouville operator, $\mathcal{L}$, is defined by

$$\mathcal{L} := \frac{1}{s(x)} \left(-\frac{d}{dx} \left[p(x) \frac{d}{dx} \right] + q(x) \right), \tag{19.1}$$

where p, p', q, s are real and continuous, and $s(x) > 0$ and $p(x) > 0$ on the interval (a, b). The Sturm–Liouville equation is defined by

$$\mathcal{L}[y(x)] = -\lambda y(x), \tag{19.2}$$

or, equivalently,

$$-\frac{d}{dx} \left[p(x) \frac{dy}{dx} \right] + q(x)y + \lambda s(x)y = 0, \tag{19.3}$$

for $x \in [a, b]$. The parameter λ is an eigenvalue of the equation. Given a specific set of boundary conditions, there may be specific values of λ for which (19.2) has a non-trivial solution. For different types of boundary conditions, different types of behavior are possible.

Many facts are known about Sturm–Liouville systems:

[1] $\mathcal{L}$, as defined by (19.1), is formally self-adjoint (see page 74), with the inner product, $(f, g)_s := \int s(x) f(x) \bar{g}(x) \, dx$.

[2] $\mathcal{L}$ is self-adjoint (see page 74) when

   (A) The boundary conditions are *unmixed* (or separated). That is, they are of the form

$$\begin{aligned} \alpha_1 y(a) + \beta_1 y'(a) &= 0, \\ \alpha_2 y(b) + \beta_2 y'(b) &= 0. \end{aligned} \tag{19.4}$$

   (B) The boundary conditions are *periodic*. That is, they are of the form

$$\begin{aligned} y(a) &= y(b), \\ y'(a) &= y'(b). \end{aligned}$$

[3] When the boundary conditions are given as in (19.4), and, in addition, $p(x) > 0$, $q(x) > 0$, $\alpha_1/\beta_1 > 0$, $\alpha_2/\beta_2 > 0$, then

(A) $\mathcal{L}$ is a positive definite operator (i.e., $(\mathcal{L}u, u) > 0$, for all $u \neq 0$).

(B) The eigenvalues are simple (i.e., each eigenvalue has a single eigen-function associated with it).

[4] When the operator $\mathcal{L}$ is not self-adjoint then

(A) If λ is a complex eigenvalue of $\mathcal{L}$, then $\bar{\lambda}$ is an eigenvalue of $\mathcal{L}^*$, the adjoint of $\mathcal{L}$.

(B) Eigenfunctions of $\mathcal{L}$ are orthogonal to those of $\mathcal{L}^*$.

If the interval $[a, b]$ is finite and $p(x)$ and $s(x)$ are positive at the endpoints, then the problem is said to be *regular*. Otherwise, it is said to be *singular*. For singular Sturm–Liouville problems, problems are subdivided into two cases, the limit-circle case and limit-point case. Consider (19.2) when one of the endpoints is regular and the other singular. Define the s-norm of a function $u(x)$ by

$$||u||_s = (u, u)_s = \int_a^b s(x)|u(x)|^2 \, dx.$$

If, for any particular complex number λ, the solution to (19.2) satisfies

[1] $||y||_s < \infty$, then $\mathcal{L}$ is said to be of the *limit-circle type* at infinity. In this case, all solutions of (19.2) will satisfy $||y||_s < \infty$, for any value of λ.

[2] $||y||_s = \infty$, then $\mathcal{L}$ is said to be of the *limit-point type* at infinity.

If both endpoints are singular, we introduce an intermediate point l, $a < l < b$, and then classify $\mathcal{L}$ as being of the limit point type or the limit circle type at each endpoint according to the behavior of solutions in $a < x < l$ and in $l < x < b$ (the classification is independent of the choice of l).

For a given real λ, the problem in (19.2) is

- *oscillatory* at $x = a$ if and only if every solution has infinitely many zeros clustering at a.
- *nonoscillatory* at $x = a$ if and only if no solution has infinitely many zeros clustering at a.

The classification is mutually exclusive for a fixed λ, but can vary with λ.

If $\mathcal{L}$ is in the limit-point case at infinity, then there is the following completeness theorem:

Theorem: If $g(\lambda) = \int_0^\infty f(x)\Psi(x, \lambda) \, dx$ then

$f(x) = \int_{-\infty}^\infty g(\lambda) \; \Psi(x, \lambda) \, d\rho(\lambda)$ for a (computable) density function $\rho(\lambda)$.

A completeness theorem is required for a proof that a separation of variables calculation (see page 419) has been done correctly.

The following theorem and corollaries may help decide the type of the operator $\mathcal{L}$:

> **Theorem:**   Let M be a positive differentiable function, and let k_1 and k_2 be two positive constants such that for large x,
>
> $$q(x) \geq -k_1 M(x),$$
>
> $$\int_x^\infty (p(t)M(t))^{-1/2}\, dt = \infty,$$
>
> $$|p^{1/2}(x)M'(x)M^{-3/2}(x)| < k_2,$$
>
> then $\mathcal{L}$ is in the limit-point case at infinity.

> **Corollary:**   If $q(x) \geq -k$, where k is a positive constant, and $\int_n^\infty p^{-1/2}(t)\, dt = \infty$ (where n is any finite number), then $\mathcal{L}$ is in the limit-point case at infinity.

> **Corollary:**   If $p(x) = 1$ for $0 < x < \infty$ and $q(x) \geq -kx^2$ for some positive constant k, then $\mathcal{L}$ is in the limit-point case at infinity.

Classification of Sturm–Liouville Problems

Pruess *et al.* [7] have devised a classification scheme and taxonomy for Sturm–Liouville problems on the interval (a, b). They define:

Category 1: Problem (19.2) is nonoscillatory at $x = a$ and $x = b$.
> *The spectrum is simple, purely discrete, and bounded below.*

Category 2: Problem (19.2) is nonoscillatory at one endpoint. At the other endpoint it is nonoscillatory for $\lambda \in (-\infty, t_0)$ and oscillatory for $\lambda \in (t_0, \infty)$.
> *The spectrum is simple, bounded below. The point spectrum (if any) is in $(-\infty, t_0)$ while (t_0, ∞) is the continuous spectrum.*

Category 3: Problem (19.2) is nonoscillatory at one endpoint. At the other endpoint it is limit circle and oscillatory.
> *The spectrum is simple, unbounded both above and below, and is purely discrete.*

Category 4: Problem (19.2) is nonoscillatory at one endpoint. At the other endpoint it is limit point and oscillatory.
> *The spectrum is simple and purely continuous; the continuous spectrum is the entire real line.*

Category 5: Problem (19.2) is limit circle and oscillatory at $x = a$. It is limit point and oscillatory at $x = b$.
> *The spectrum is simple, unbounded both above and below, and purely discrete.*

Category 6: Problem (19.2) is limit point and oscillatory at $x = a$. It is limit point and oscillatory at $x = b$.

The nature of the spectrum is unknown; a continuous spectrum is likely.

Category 7: Problem (19.2) is limit point and oscillatory at one endpoint. At the other endpoint it is limit circle and oscillatory.

The spectrum is simple and purely continuous; the continuous spectrum is the entire real line.

Category 8: Problem (19.2) is limit circle and oscillatory at one endpoint. At the other endpoint it is nonoscillatory for $\lambda \in (-\infty, t_0)$ and oscillatory for $\lambda \in (t_0, \infty)$.

The spectrum is simple; the point spectrum (if any) is unbounded below but bounded above by t_0. The continuous spectrum is in (t_0, ∞).

Category 9: Problem (19.2) is limit point and oscillatory at one endpoint. At the other endpoint it is nonoscillatory for $\lambda \in (-\infty, t_0)$ and oscillatory for $\lambda \in (t_0, \infty)$.

The spectrum may be nonsimple.

Category 10:   At $x = a$ the problem in (19.2) is nonoscillatory for $\lambda \in (-\infty, t_0)$ and oscillatory for $\lambda \in (t_0, \infty)$. At $x = b$ it is nonoscillatory for $\lambda \in (-\infty, t_1)$ and oscillatory for $\lambda \in (t_1, \infty)$.

The spectrum may be nonsimple. The point spectrum (if any) is in the interval $(-\infty, \min(t_0, t_1))$ and is bounded below. The continuous spectrum is in $(\min(t_0, t_1), \infty)$.

Example 1

The differential equation and boundary conditions

$$-(xy')' = \lambda xy,$$
$$u(1) = 0,$$
$$u(2) = 0,$$

correspond to the Sturm–Liouville operator in (19.1) with $p(x) = x$, $q(x) = 0$, $s(x) = x$. This is a regular Sturm–Liouville problem on the interval $[1, 2]$. The eigenvalues and eigenfunctions are readily computed (see Stakgold [8], page 423). If we define $\lambda_n = r_n^2$, then the r_n are determined from

$$\frac{J_0(r_n)}{J_0(2r_n)} = \frac{N_0(r_n)}{N_0(2r_n)},$$

and the corresponding eigenfunction is given by

$$y_n(x) = \frac{r_n \pi J_0(2r_n)}{\sqrt{2}\sqrt{J_0(r_n)^2 - J_0(2r_n)^2}}[J_0(r_n)N_0(r_n x) - J_0(r_n x)N_0(r_n)].$$

Example 2

The differential equation with boundary conditions

$$-(x^2 y')' - \lambda u = 0,$$
$$u(1) = 0,$$
$$u(e) = 0,$$

for $x \in [1, e]$ is a regular Sturm–Liouville problem with unmixed boundary conditions, so the eigenfunctions are complete. In this case we find

$$\lambda_n = n^2 \pi^2 + \tfrac{1}{2},$$
$$y_n = x^{-1/2} \sin(n\pi \log x).$$

Notes

[1] For transformations of (19.3), see page 128.
[2] The regular Sturm–Liouville equation, written in the form

$$\frac{d^2 z}{dt^2} - r(t)z + \lambda z = 0,$$

with the boundary conditions $z(0) = z(L) = 0$ has the asymptotic eigenvalues and eigenfunctions

$$z_n(t) = \sqrt{\frac{2}{L}} \sin\left(\frac{n\pi}{L}t\right) + O\left(\frac{1}{n}\right),$$
$$\lambda_n = \frac{n^2 \pi^2}{L^2} + O(1).$$

as $n \to \infty$. (See the Prüfer method on page 122.)

References

[1] G. Birkhoff and G.-C. Rota, *Ordinary Differential Equations*, John Wiley & Sons, New York, 1978, Chapters 10–11.
[2] E. Butkov, *Mathematical Physics*, Addison–Wesley Publishing Co., Reading, MA, 1968, pages 337–341.
[3] E. A. Coddington and N. Levinson, *Theory of Ordinary Differential Equations*, McGraw–Hill Book Company, New York, 1955, Chapters 7–12.
[4] N. Dunford and J. Schwartz, *Linear Operators, Part II: Spectral Theory*, John Wiley & Sons, New York, 1958.
[5] E. L. Ince, *Ordinary Differential Equations*, Dover Publications, Inc., New York, 1964, pages 217–218 and 235–241.
[6] B. M. Levitan and I. S. Sargsjan, *Sturm–Liouville and Dirac Operators*, Kluwer Academic Publishers, Dordrecht, The Netherlands, 1991, Chapter 6 (pages 139–182) and Chapter 12 (pages 324–340).

[7] S. Pruess, C. T. Fulton, and Y. Xie, "The Automatic Classification of Sturm–Liouville Problems," (submitted for publication).

[8] I. Stakgold, *Green's Functions and Boundary Value Problems*, John Wiley & Sons, New York, 1979, Chapter 7 (pages 411–466).

[9] E. Zauderer, *Partial Differential Equations of Applied Mathematics*, John Wiley & Sons, New York, 1983, pages 136–159.

20.   Variational Equations

Applicable to   Differential equations that arise from variational principles.

Yields

A variational principle.

Procedure

Most differential equations that arise in mathematical physics have been obtained from a variational principle. The variational principle that is most often used is $\delta J = 0$, where δ represent a variation, and J is a functional given by

$$J[u] = \iint\limits_{R} L(\mathbf{x}, \partial_{x_j}) u(\mathbf{x})\, d\mathbf{x}. \tag{20.1}$$

Here, $L(\ )$ is a linear or nonlinear function of its arguments and $u(\mathbf{x})$ is the unknown function to be determined. This variational principle states that the integral $J[u]$ should be stationary to small changes in $u(\mathbf{x})$. If we let $h(\mathbf{x})$ be a "small," continuously differentiable function, then we can form

$$J[u+h] - J[u] = \iint\limits_{R} \left\{ L(\mathbf{x}, \partial_{x_j})(u(\mathbf{x}) + h(\mathbf{x})) - L(\mathbf{x}, \partial_{x_j}) u(\mathbf{x}) \right\}\, d\mathbf{x}. \tag{20.2}$$

By integration by parts, (20.2) can often be written as

$$J[u+h] - J[u] = \iint\limits_{R} N(\mathbf{x}, \partial_{x_j}) u(\mathbf{x})\, d\mathbf{x} + O(||h||^2),$$

plus some boundary terms (see page 67). The variational principle requires that $\delta J := J[u+h] - J[u]$ vanishes to leading order, or that

$$N(\mathbf{x}, \partial_{x_j}) u(\mathbf{x}) = 0. \tag{20.3}$$

Equation (20.3) is called the first variation of (20.1), or the *Euler–Lagrange equation* corresponding to (20.1). (This is sometimes called the *Euler equation*.) A functional in the form of (20.1) determines an Euler–Lagrange equation. Conversely, given an Euler–Lagrange equation, a corresponding functional can sometimes be obtained.

Many approximate and numerical techniques utilize the functional associated with a given system of Euler–Lagrange equations. See, for example, the Rayleigh–Ritz method (page 554) and the finite element method (page 656).

The following collection of examples assume that the dependent variable in the given differential equation has natural boundary conditions (see page 67). If the dependent variable did not have these specific boundary conditions, then the boundary terms that were discarded in going from (20.2) to (20.3) would have to be satisfied in addition to the Euler–Lagrange equation.

Example 1

The Euler–Lagrange equation for the functional

$$J[y] = \int_R F\left(x, y, y', \dots, y^{(n)}\right) dx,$$

where $y = y(x)$ is

$$\frac{\partial F}{\partial y} - \frac{d}{dx}\left(\frac{\partial F}{\partial y'}\right) + \frac{d^2}{dx^2}\left(\frac{\partial F}{\partial y''}\right) - \cdots + (-1)^n \frac{d^n}{dx^n}\left(\frac{\partial F}{\partial y^{(n)}}\right) = 0.$$

For this equation the natural boundary conditions are given by

$$y(x_0) = y_0, \quad y'(x_0) = y_0', \quad \dots, \quad y^{(n-1)}(x_0) = y_0^{(n-1)},$$
$$y(x_1) = y_1, \quad y'(x_1) = y_1', \quad \dots, \quad y^{(n-1)}(x_1) = y_1^{(n-1)}.$$

Example 2

The Euler–Lagrange equation for the functional

$$J[u] = \iint_R F(x, y, u, u_x, u_y, u_{xx}, u_{xy}, u_{yy}) \, dx \, dy,$$

where $u = u(x, y)$ is

$$\frac{\partial F}{\partial u} - \frac{\partial}{\partial x}\left(\frac{\partial F}{\partial u_x}\right) - \frac{\partial}{\partial y}\left(\frac{\partial F}{\partial u_y}\right) + \frac{\partial^2}{\partial x^2}\left(\frac{\partial F}{\partial u_{xx}}\right) +$$
$$\frac{\partial^2}{\partial x \partial y}\left(\frac{\partial F}{\partial u_{xy}}\right) + \frac{\partial^2}{\partial y^2}\left(\frac{\partial F}{\partial u_{yy}}\right) = 0.$$

(20.4)

Example 3

The Euler–Lagrange equation for the functional

$$J[u] = \iint_R \left[a \left(\frac{\partial u}{\partial x} \right)^2 + b \left(\frac{\partial u}{\partial x} \right)^2 + cu^2 + 2fu \right] dx \, dy,$$

is

$$\frac{\partial}{\partial x} \left(a \frac{\partial u}{\partial x} \right) + \frac{\partial}{\partial y} \left(b \frac{\partial u}{\partial y} \right) - cu = f.$$

Example 4

For the $2m$-th order ordinary differential equation (in formally self-adjoint form)

$$\sum_{k=0}^{m} (-1)^k \frac{d^k}{dx^k} \left(p_k(x) \frac{d^k u}{dx^k} \right) = f(x),$$

$$u(a) = u'(a) = \cdots = u^{(m-1)}(a) = 0,$$

$$u(b) = u'(b) = \cdots = u^{(m-1)}(b) = 0,$$

a corresponding functional is

$$J[u] = \int_a^b \left(\sum_{k=0}^{m} p_k(x) \left(\frac{d^k u}{dx^k} \right)^2 - 2f(x)u(x) \right) dx.$$

Example 5

Consider the system of n second order ordinary differential equations for the unknowns $\{u_k(x) \mid k = 1, \ldots, n\}$

$$-\sum_{k=1}^{n} \left[\frac{d}{dx} \left(p_{jk}(x) \frac{du_k}{dx} \right) + q_{jk}(x) u_k \right] = f_j(x), \tag{20.5}$$

$$u_j(a) = u_j(b) = 0,$$

for $j = 1, 2, \ldots, n$. If $p_{jk} = p_{kj}$, $q_{jk} = q_{kj}$, if the matrix $\{p_{jk}\}$ is bounded and positive definite, and if the matrix $\{q_{jk}\}$ is bounded and non-negative definite, then a functional corresponding to (20.5) is

$$J[u] = \int_a^b \left(\sum_{j,k=1}^{n} \left[p_{jk}(x) \frac{du_j}{dx} \frac{du_k}{dx} + q_{jk}(x) u_j u_k \right] - \sum_{j=1}^{n} f_j(x) u_j(x) \right) dx.$$

Example 6
    If $A_{ij}(x)$ is a symmetric and positive definite matrix, so that the partial differential equation for $u(\mathbf{x}) = u(x_1, \ldots, x_m)$

$$-\sum_{i,j=1}^{m} \frac{\partial}{\partial x_i}\left(A_{ij}\frac{\partial u}{\partial x_j}\right) + C(\mathbf{x})u = f(\mathbf{x}),$$

is elliptic in Ω, $C(\mathbf{x}) > 0$, and there are Dirichlet boundary conditions

$$u\Big|_{\partial\Omega} = 0, \tag{20.6}$$

then a corresponding functional is

$$J[u] = \int_{\Omega}\left(\sum_{i,j=1}^{m} A_{ij}\frac{\partial u}{\partial x_i}\frac{\partial u}{\partial x_j} + Cu^2 - 2fu\right)dx, \tag{20.7}$$

where (20.7) is to be minimized over those functions that satisfy (20.6).

Example 7
    If $A_{ij}(x)$ is a symmetric and positive definite matrix, so that the partial differential equation for $u(\mathbf{x}) = u(x_1, \ldots, x_m)$,

$$-\sum_{i,j=1}^{m} \frac{\partial}{\partial x_j}\left(A_{ij}\frac{\partial u}{\partial x_j}\right) + C(\mathbf{x})u = f(\mathbf{x}),$$

is elliptic in Ω, $C(\mathbf{x}) > 0$, and there are the boundary conditions

$$\left[\sum_{i,j=1}^{m} A_{ij}\frac{\partial u}{\partial x_j}\cos(\nu, x_i) + \sigma u\right]_{\partial\Omega} = 0, \tag{20.8}$$

where ν is normal to $\partial\Omega$ and σ is a positive function on $\partial\Omega$, then a corresponding functional is

$$J[u] = \int_{\Omega}\left(\sum_{i,j=1}^{m} A_{ij}\frac{\partial u}{\partial x_i}\frac{\partial u}{\partial x_j} + Cu^2 - 2fu\right)dx + \int_{\partial\Omega}\sigma u^2\, dS, \tag{20.9}$$

where (20.9) is to be minimized over those functions for which (20.8) is satisfied.

Notes

[1] Note that two different functionals can yield the same set of Euler–Lagrange equations. For example, $\delta \int J \, dx = \delta \int (J + y + xy') \, dx$. The reason that $\delta \int (y + xy') \, dx = 0$ is because the integrand is an exact differential ($\int (y + xy') \, dx = \int d(xy)$). Hence, this integral is path independent; its value is determined by the boundary conditions.

   The Euler–Lagrange equations for the two functionals $\int \int u_{xx} u_{yy} \, dx \, dy$ and $\int \int (u_{xy})^2 \, dx \, dy$ are also the same.

[2] If a differential equation can be derived from a variational principle, then admittance of a Lie group is a necessary condition to find conservation laws by Noether's theorem.

[3] Even if the boundary conditions given with a differential equation are not natural, a variational principle may sometimes be found. Consider

$$J[u] = \int_{x_1}^{x_2} F(x, u, u') \, dx - g_1(x, u) \bigg|_{x=x_1} + g_2(x, u) \bigg|_{x=x_2}$$

where $g_1(x, u)$ and $g_2(x, u)$ are unspecified functions. The necessary conditions for u to minimize $J[u]$ are (see Mitchell and Wait [5])

$$\frac{\partial F}{\partial u} - \frac{d}{dx} \frac{\partial F}{\partial u'} = 0,$$

$$\frac{\partial F}{\partial u'} + \frac{\partial g_1}{\partial u} \bigg|_{x=x_1} = 0, \qquad \frac{\partial F}{\partial u'} + \frac{\partial g_2}{\partial u} \bigg|_{x=x_2} = 0.$$

If g_1 and g_2 are identically zero, then we recover the natural boundary conditions. However, we may choose g_1 and g_2 to suit other boundary conditions. For example, the problem

$$u'' + f(x) = 0,$$

$$u' + \alpha u \bigg|_{x=x_1} = 0, \qquad u' + \beta u \bigg|_{x=x_2} = 0$$

corresponds to the functional

$$J[u] = \int_{x_1}^{x_2} \left[\tfrac{1}{2} \left(u' \right)^2 - f(x)u \right] \, dx + \frac{\beta u^2}{2} \bigg|_{x=x_2} - \frac{\alpha u^2}{2} \bigg|_{x=x_1}.$$

[4] This technique can be used in higher dimensions. For example, consider the functional

$$J[u] = \iint\limits_{R} F(x, y, u, u_x, u_y, u_{xx}, u_{xy}, u_{yy}) \, dx \, dy$$

$$+ \int_{\partial R} G(x, y, u, u_\sigma, u_{\sigma\sigma}, u_n) \, d\sigma,$$

where $\partial/\partial\sigma$ and $\partial/\partial n$ are partial differential operators in the directions of the tangent and normal to the curve ∂R. Necessary conditions for $J[u]$ to have a minimum are the Euler–Lagrange equations (given in (20.4)) together with the boundary conditions:

$$\left[\frac{\partial F}{\partial u_x} - \frac{\partial}{\partial x} \frac{\partial F}{\partial u_{xx}} \right] y_\sigma - \left[\frac{\partial F}{\partial u_y} - \frac{\partial}{\partial y} \frac{\partial F}{\partial u_{yy}} \right] x_\sigma$$

$$- \left[\frac{\partial}{\partial \sigma} \left(\frac{\partial F}{\partial u_{xx}} - \frac{\partial F}{\partial u_{yy}} \right) \right] x_\sigma y_\sigma + \frac{1}{2} \left[\frac{\partial}{\partial \sigma} \frac{\partial F}{\partial u_{xy}} \left(x_\sigma^2 - y_\sigma^2 \right) \right]$$

$$+ \frac{1}{2} \left[\left(\frac{\partial}{\partial x} \frac{\partial F}{\partial u_{xy}} \right) x_\sigma - \left(\frac{\partial}{\partial y} \frac{\partial F}{\partial u_{xy}} \right) y_\sigma \right]$$

$$+ G_u - \frac{\partial}{\partial \sigma} \frac{\partial G}{\partial u_\sigma} + \frac{\partial^2}{\partial \sigma^2} \frac{\partial G}{\partial u_{\sigma\sigma}} = 0,$$

$$\frac{\partial G}{\partial u_n} + \frac{\partial F}{\partial u_{xx}} y_\sigma^2 + \frac{\partial F}{\partial u_{yy}} x_\sigma^2 + \frac{\partial F}{\partial u_{xy}} x_\sigma y_\sigma = 0,$$

where $x_\sigma = \dfrac{dx}{d\sigma}$ and $y_\sigma = \dfrac{dy}{d\sigma}$. See Mitchell and Wait [5] for details.

References

[1] E. Butkov, *Mathematical Physics*, Addison–Wesley Publishing Co., Reading, MA, 1968, pages 573–588.

[2] L. Collatz, *The Numerical Treatment of Differential Equations*, Springer–Verlag, New York, 1966, pages 540–541.

[3] S. J. Farlow, *Partial Differential Equations for Scientists and Engineers*, John Wiley & Sons, New York, 1982, pages 362–369.

[4] L. V. Kantorovich and V. I. Krylov, *Approximate Methods of Higher Analysis*, Interscience Publishers, New York, 1958, Chapter 4 (pages 241–357).

[5] A. R. Mitchell and R. Wait, *The Finite Element Method in Differential Equations*, Wiley, New York, 1977, pages 27–31.

[6] W. Yourgrau and S. Mandelstam, *Variational Principles in Dynamics and Quantum Theory*, Dover Publications, Inc., New York, 1979.

21.   Well-Posedness of Differential Equations

Applicable to   Ordinary differential equations and partial differential equations.

Yields

Knowledge of whether the equation is intrinsically well-posed.

Idea

Before an attempt is made to determine or approximate the solution of a differential equation, it should be checked to determine if the differential equation problem is intrinsically well-posed.

Procedure

A well-posed differential equation is one in which

(A) The solution exists.
(B) The solution is unique.
(C) The solution is stable (i.e., the solution depends continuously on the boundary conditions and initial conditions).

If the differential equation is not well-posed, it is called an ill-posed or improperly posed problem. For such problems: there may not be a solution, there may be more than one solution, or whatever solution is determined (by an approximate scheme) may be unrelated to the actual solution.

For partial differential equations, the third condition (concerning stability) is generally the easiest to check.

Example

Consider the initial value problem for the unknown function $u(x, t)$,

$$u_{tt} = u_{xxxx},$$
$$u(x, 0) = g(x). \tag{21.1}$$

We will show that the solution to this problem is not stable. Suppose that (21.1) has a solution, say $u_0(x, t)$. Assume that ε is a fixed number, much smaller than one in magnitude, and define $u_1(x, t)$ by

$$u_1(x, t) = u_0(x, t) + \varepsilon e^{ikx} e^{\sigma t},$$

where k and σ are also constants. At $t = 0$, $u_1(x, 0)$ differs from $g(x)$ by a quantity that has magnitude ε, an arbitrarily small amount.

However, using $u_1(x,t)$ in (21.1), we determine that $u_1(x,t)$ will satisfy the equation if $\sigma = \pm k^2$. Therefore, at any fixed value of t, say $t = T$, there exists a solution $u_0(x,T)$ and an approximation to the solution $u_1(x,T) = u_0(x,T) + \varepsilon e^{ikx} e^{k^2 T}$. The approximation satisfies the same differential equation that the true solution satisfies. But since k is arbitrary, the approximate solution can be arbitrarily larger than the true solution, by taking k arbitrarily large. Since two different expressions satisfy the same differential equation and initially were arbitrarily close, and are arbitrarily different in magnitude at any future time, we conclude that the problem is ill-posed.

Note that, with the proper boundary conditions and initial conditions, the equation in (21.1) would have a unique solution. But the solution would be unstable since the equation is intrinsically ill-posed as an initial value problem. Hence, there would be, for instance, no easy way to numerically approximate the solution.

Notes

[1] For a discussion of some existence and uniqueness theorems, see page 50. For a discussion of some stability theorems, see page 80.

[2] A standard example of an ill-posed problem is Laplace's equation with initial data. For example, the equation $\nabla^2 u = 0$ with the initial data $\dfrac{\partial u}{\partial y}(x,0) = \dfrac{1}{n}\sin nx$ has the solution $u(x,y) = \dfrac{1}{n^2}\sin nx \sinh ny$. As $n \to \infty$, the initial data are becoming arbitrarily small in magnitude, while the solution (for $y > 0$) is becoming arbitrarily large.

[3] Certain classes of equations have been well studied. We can therefore state:

(A) For Laplace's equation and elliptic equations in general, the Dirichlet problem is well-posed. Also, the Neumann problem does not have a unique solution, but is otherwise well-posed.

(B) For the two-dimensional wave equation and hyperbolic equations in general, both are well-posed as an initial value problem. Both are, generally, ill-posed as boundary value problems.

(C) For the heat equation and diffusion equations in general, both are well-posed when given Dirichlet data and the time variable is increasing; both are ill-posed when the time variable is decreasing. See Beck, Blackwell, and St. Clair [2] for numerical schemes related to a specific ill-posed problem.

[4] A backwards heat equation (a parabolic equation with decreasing time) is ill-posed. It may be made well-posed, however, by requiring the solution to satisfy a suitable constraint. Typically, one asks for non-negative solutions or for solutions that satisfy an a priori bound, which is obtained from physical considerations.

[5] Payne [10] contains the following non-exhaustive list of methods that have been proposed and used in treating various types of improperly posed Cauchy problems:

(A) Function theoretic methods.
(B) Eigenfunction methods.
(C) Logarithmic convexity methods.
(D) Weighted energy methods.
(E) Lagrange identity methods.
(F) Quasireversibility methods.
(G) Restriction of data methods.
(H) Numerical and programming methods.
(I) Concavity methods.
(J) Stochastic and probabilistic methods.
(K) Method of generalized inverse in reproducing kernel spaces.
(L) Comparison methods.

Payne illustrates several of these methods on a backwards heat equation.

[6] As Fichera [4] shows, finding the correct boundary conditions for a *degenerate* problem (one in which the type changes) can be difficult in general. Fichera shows, for example, that the first order equation for $u(x, y)$

$$a(x, y)u_x + b(x, y)u_y + cu = f$$

in the rectangle $R = \{-\alpha \leq x \leq \alpha, -\beta \leq y \leq \beta\}$, when a and b satisfy

$$a(-\alpha, y) \geq 0, \qquad a(\alpha, y) \leq 0,$$
$$b(x, -\beta) \geq 0, \qquad b(x, \beta) \leq 0,$$

has *no* boundary conditions! However, the equation,

$$-a(x, y)u_x - b(x, y)u_y + cu = f,$$

in R, with the same conditions on a and b, requires that u be given on the entire boundary of R!

References

[1] I. K. Argyros, "On the Cardinality of Solutions of Multilinear Differential Equations and Applications," *Int. J. Math. & Math. Sci.*, **9**, No. 4, 1986, pages 757–766.

[2] J. V. Beck, B. Blackwell, and C. R. St. Clair, Jr., *Inverse Heat Problems*, Wiley, New York, 1985.

[3] B. L. Buzbee and A. Carasso, "On the Numerical Computation of Parabolic Problems for Preceding Times," *Math. of Comp.*, **27**, No. 122, April 1973, pages 237–266.

[4] G. Fichera, "On a Unified Theory of Boundary Value Problems for Elliptic–Parabolic Equations of Second Order," in R. E. Langer (ed.), *Boundary Problems in Differential Equations*, Univ. of Wisconsin Press, Madison, Wisconsin, 1960, pages 97–120.

[5] P. R. Garabedian, *Partial Differential Equations*, Wiley, New York, 1964, pages 450–457.

[6] M. M. Lavrent'ev, V. G. Romanov, and S. P. Shishatskii, *Ill-Posed Problems of Mathematical Physics and Analysis*, Amer. Math. Soc., Providence, Rhode Island, 1986.

[7] V. A. Morozov, *Methods for Solving Incorrectly Posed Problems*, Springer–Verlag, New York, 1984.

[8] L. E. Payne, *Improperly Posed Problems in Partial Differential Equations*, SIAM, Philadelphia, 1975.

[9] N. N. Pavlov, "Smoothing of Input Data in the Solution of Ill-Posed Problems," *U.S.S.R. Comput. Maths. Math. Phys.*, **29**, No. 5, 1989, pages 110–114.

[10] E. Zauderer, *Partial Differential Equations of Applied Mathematics*, John Wiley & Sons, New York, 1983, pages 103–113.

22. Wronskians and Fundamental Solutions

Let $L[\cdot]$ be the linear n-th order ordinary differential operator

$$L[y] = \frac{d^n y}{dx^n} + a_1(x)\frac{d^{(n-1)}y}{dx^{(n-1)}} + \cdots + a_n(x)y.$$

The vector equation associated with the linear equation $L[y] = 0$ is given by (see page 118)

$$\mathbf{y}' = A(x)\mathbf{y}, \tag{22.1}$$

where $\mathbf{y} = (y, y', y'', \ldots, y^{(n-1)})^{\mathrm{T}}$ and A is the matrix

$$A = \begin{pmatrix} 0 & 1 & 0 & 0 & \cdots & 0 \\ 0 & 0 & 1 & 0 & \cdots & 0 \\ 0 & 0 & 0 & 1 & & 0 \\ \vdots & \vdots & \vdots & & \ddots & \\ 0 & 0 & 0 & 0 & & 1 \\ -a_n & -a_{n-1} & -a_{n-3} & -a_{n-2} & \cdots & -a_1 \end{pmatrix}. \tag{22.2}$$

If $\{y_1, y_2, \ldots, y_n\}$ is any set of n solutions to the equation $L[y] = 0$, then the matrix

$$\Phi(x) = \begin{pmatrix} y_1 & y_2 & \cdots & y_n \\ y_1' & y_2' & \cdots & y_n' \\ \vdots & \vdots & \ddots & \vdots \\ y_1^{(n-1)} & y_2^{(n-1)} & \cdots & y_n^{(n-1)} \end{pmatrix}$$

is a *solution matrix* for (22.1). It is also called a *fundamental solution*. This matrix satisfies the differential equation $\Phi' = A\Phi$.

The determinant of this matrix, $\det \Phi(x)$, is called the *Wronskian* of $L[y] = 0$ with respect to $\{y_1, y_2, \ldots, y_n\}$ and is denoted by $W(y_1, y_2, \ldots, y_n)$. Note that the Wronskian is a function of x.

If $\Phi(x)$ satisfies $\Phi' = A\Phi$, then $|\Phi(x)|' = |\Phi| \operatorname{tr} A(t)$, and hence

$$\det \Phi(x) = \det \Phi(x_0) \exp\left(\int_{x_0}^{x} \operatorname{tr} A(s) \, ds \right),$$

where $\operatorname{tr} A$ denotes the trace of the matrix A. For the matrix in (22.2) we have $\operatorname{tr} A = -a_1$ so that

$$W(y_1, \ldots, y_n)(x) = W(y_1, \ldots, y_n)(x_0) \exp\left(-\int_{x_0}^{x} a_1(s) \, ds \right). \qquad (22.3)$$

This is sometimes called *Liouville's formula*.

From (22.3) we conclude that either $W(x)$ vanishes for all values for x, or it is never equal to zero. If the Wronskian never vanishes, then the set $\{y_1, y_2, \ldots, y_n\}$ is said to be *linearly independent*. A set of n linearly independent solutions to $L[y] = 0$ is called a *basis* or a *fundamental set*.

Alternately, given a set of n linearly independent continuous functions, $\{y_1, y_2, \ldots, y_n\}$, it is possible to find a unique homogeneous differential equation of order n (with the coefficient of $y^{(n)}$ being one) for which the set forms a fundamental set. This differential equation is given by

$$(-1)^n \frac{W(y, y_1, y_2, \ldots, y_n)}{W(y_1, y_2, \ldots, y_n)} = 0. \qquad (22.4)$$

Example 1

Given the second order linear ordinary differential equation

$$y'' + y = 0, \tag{22.5}$$

the set $\{\sin x, \cos x\}$ forms a fundamental set because each element in this set satisfies (22.5) and also the Wronskian is given by

$$W(\sin x, \cos x) = \begin{vmatrix} \sin x & \cos x \\ \cos x & -\sin x \end{vmatrix} = -1,$$

which does not vanish. Since the Wronskian is constant, we have verified that $a_1(x) = 0$ in (22.5) (the $a_1(x)$ term in this equation corresponds to the first derivative term).

Example 2

If we choose the two functions $y_1 = \sin x$ and $y_2 = x$, we can determine the linear second order equation that has these solutions as its fundamental set by constructing (22.4). Here $n = 2$ so we find

$$(-1)^2 \frac{W(y, x, \sin x)}{W(x, \sin x)} = \frac{\begin{vmatrix} y & x & \sin x \\ y' & 1 & \cos x \\ y'' & 0 & -\sin x \end{vmatrix}}{\begin{vmatrix} x & \sin x \\ 1 & \cos x \end{vmatrix}},$$

$$= \frac{(x\cos x - \sin x)y'' + (x\sin x)y' - (\sin x)y}{(x\cos x - \sin x)},$$

$$= y'' + \frac{x\sin x}{(x\cos x - \sin x)}y' - \frac{\sin x}{(x\cos x - \sin x)}y.$$

Notes

[1] Given the linear partial differential equation

$$L[u] = \sum_{i,j=1}^{n} a_{ij}(\mathbf{x})\frac{\partial^2 u}{\partial x_i \partial x_j} + \sum_{i=1}^{n} b_i \frac{\partial u}{\partial x_i} + cu$$

for $u(\mathbf{x})$, let $\Gamma = \Gamma(\mathbf{x}, \boldsymbol{\xi}) = \Gamma(\boldsymbol{\xi}, \mathbf{x})$ be the geodesic distance between the points $\mathbf{x}$ and $\boldsymbol{\xi}$. (For a rectangular coordinate system, $\Gamma(\mathbf{x}, \boldsymbol{\xi}) = ||\mathbf{x} - \boldsymbol{\xi}|| = \sqrt{(x_1 - \xi_1)^2 + \cdots + (x_n - \xi_n)^2}$.) A fundamental solution, $S(\mathbf{x}, \boldsymbol{\xi})$, satisfies

$L[S] = 0$ and, near $\mathbf{x} = \boldsymbol{\xi}$, has the form $S = \dfrac{U}{\Gamma^m} + V \log \Gamma + W$, where U, V, and W are analytic functions and $m = (n-2)/2$. For example, for Laplace's equation in n dimensions with $n > 2$, $\nabla^2 u = 0$, a fundamental solution is given by

$$S = \frac{1}{r^{n-2}}, \quad \text{with} \quad r = \sqrt{(x_1 - \xi_1)^2 + \cdots + (x_n - \xi_n)^2}.$$

See Garabedian [3] for details.

[2] The canonical form of a self-adjoint third order linear homogeneous differential equation is $y''' + 2Ay' + A'y = 0$, (see pages 78 and 135). A fundamental set of solutions for this equation are $\{u^2, uv, v^2\}$, where $u(x)$ and $v(x)$ are any two linearly independent solutions of the second order differential equation $u'' + \frac{1}{2}Au = 0$.

References

[1] W. E. Boyce and R. C. DiPrima, *Elementary Differential Equations and Boundary Value Problems*, Fourth Edition, John Wiley & Sons, New York, 1986, pages 113–126.

[2] E. A. Coddington and N. Levinson, *Theory of Ordinary Differential Equations*, McGraw–Hill Book Company, New York, 1955, pages 67–84.

[3] P. R. Garabedian, *Partial Differential Equations*, Wiley, New York, 1964, pages 152–153.

[4] E. L. Ince, *Ordinary Differential Equations*, Dover Publications, Inc., New York, 1964, pages 116–121.

[5] E. D. Rainville and P. E. Bedient, *Elementary Differential Equations*, The MacMillan Company, New York, 1964, pages 84–86.

[6] G. F. Simmons, *Differential Equations with Applications and Historical Notes*, McGraw–Hill Book Company, New York, 1972, pages 76–80.

I.B

Transformations

23.   Canonical Forms

Applicable to   The ordinary differential equations:

$$\frac{d^2y}{dx^2} + 2\left(\frac{e}{x} + f\right)\frac{dy}{dx} + \left(\frac{p}{x^2} + \frac{2q}{x} + r\right)y = 0, \tag{23.1}$$

$$\frac{d^2y}{dx^2} + 2(e + fx)\frac{dy}{dx} + (px^2 + 2qx + r)y = 0, \tag{23.2}$$

$$(\alpha + \beta x)\frac{d^2y}{dx^2} + (b + mx)\frac{dy}{dx} + (c + nx)y = 0, \tag{23.3}$$

$$\frac{d^2y}{dx^2} = F\left(\frac{dy}{dx}, y, x\right). \tag{23.4}$$

Idea

Each of these equations has certain canonical forms. When approximations and numerical values for these equations are reported in the literature, it is generally for the canonical forms.

Procedure 1

By changing the dependent and independent variables from $y = y(x)$ to $v = v(z)$, equation (23.1) will take the form of one of the following four canonical forms:

$$\frac{d^2v}{dz^2} + \left(\frac{A}{z^2} + \frac{2}{z} + B\right)v = 0,$$

$$\frac{d^2v}{dz^2} + \left(\frac{A}{z^2} + \frac{2}{z}\right)v = 0,$$

$$\frac{d^2v}{dz^2} + \left(\frac{A}{z^2} + 1\right)v = 0,$$

$$\frac{d^2v}{dz^2} + \frac{A}{z^2}v = 0,$$

where A and B are constants. The transformation is given by

$$y(x) = \nu z^\lambda e^{\mu z} v(z),$$
$$x = \kappa z,$$

for some choice of the constants $\{\nu, \lambda, \mu, \kappa\}$.

Procedure 2

By changing the dependent and independent variables from $y = y(x)$ to $v = v(z)$, equation (23.2) will take the form of one the following four canonical forms:

$$\frac{d^2v}{dz^2} + \left(z^2 + J\right)v = 0,$$

$$\frac{d^2v}{dz^2} - vz = 0,$$

$$\frac{d^2v}{dz^2} + v = 0,$$

$$\frac{d^2v}{dz^2} = 0,$$

where J is a constant. The transformation is given by

$$y(x) = \nu e^{\mu z + \xi z^2} v(z),$$
$$x = \kappa z + \eta,$$

for some choice of the constants $\{\nu, \mu, \xi, \kappa, \eta\}$.

Procedure 3

By changing the dependent and independent variables, equation (23.3) can be reduced to Weiler's canonical form (this is also known as a Kummer equation)

$$z\frac{d^2v}{dz^2} + (b - z)\frac{dv}{dz} - av = 0. \tag{23.5}$$

The transformations that give (23.5) have several different forms depending on the numerical values of the coefficients in (23.3), see Bateman [2] for details.

Procedure 4

A critical point is called a moving critical point if its location depends on the initial conditions for the differential equation (and so the location of the critical point is not fixed solely by the coefficients of the differential equation). For example, the nonlinear differential equation $y'' = (y')^2\dfrac{2y - 1}{y^2 + 1}$ has the general solution $y(x) = \tan\left[\log(Ax + B)\right]$, where A and B are arbitrary constants. The initial conditions determine A and B and thus determine the location of the singularities of $y(x)$.

Given an ordinary differential equation in the form of equation (23.4), if $F(y', y, x)$ is rational in y', algebraic in y, and analytic in x, and if all of the critical points are fixed, then a change of variables of the form

$$y(x) = \frac{az(x) + b}{cz(x) + d},$$

where a, b, c, d, and w are some functions of x, will transform the equation to one of fifty standard forms. Each of these fifty differential equations is for the unknown function $z(x)$.

Of these standard forms, six have solutions in terms of the Painlevé transcendents and all the others have first integrals that are equations of first order or have elementary integrals. The equations that define the six Painlevé transcendents are

$$\frac{d^2y}{dx^2} = 6y^2 + x,$$

$$\frac{d^2y}{dx^2} = 2y^3 + xy + \alpha,$$

$$\frac{d^2y}{dx^2} = \frac{1}{y}\left(\frac{dy}{dx}\right)^2 - \frac{1}{x}\frac{dy}{dx} + \frac{1}{x}(\alpha y^2 + \beta) + \gamma y^3 + \frac{\delta}{y},$$

$$\frac{d^2y}{dx^2} = \frac{1}{2y}\left(\frac{dy}{dx}\right)^2 + \frac{3y^3}{2} + 4xy^2 + 2(x^2 - \alpha)y + \frac{\beta}{y},$$

$$\frac{d^2y}{dx^2} = \left(\frac{1}{2y} + \frac{1}{y-1}\right)\left(\frac{dy}{dx}\right)^2 - \frac{1}{x}\frac{dy}{dx} + \frac{(y-1)^2}{x^2}\left(\alpha y + \frac{\beta}{y}\right) + \frac{\gamma y}{x}$$
$$+ \frac{\delta y(y+1)}{y-1},$$

$$\frac{d^2y}{dx^2} = \frac{1}{2}\left(\frac{1}{y} + \frac{1}{y-1} + \frac{1}{y-x}\right)\left(\frac{dy}{dx}\right)^2 - \left(\frac{1}{x} + \frac{1}{x-1} + \frac{1}{y-x}\right)\frac{dy}{dx}$$
$$+ \frac{y(y-1)(y-x)}{x^2(x-1)^2}\left(\alpha + \frac{\beta x}{y^2} + \frac{\gamma(x-1)}{(y-1)^2} + \frac{\delta x(x-1)}{(y-x)^2}\right).$$

In the above equations, all of the parameters are assumed to be constant.

Notes

[1] The first three transformations may be found in Bateman [2].

[2] The transformations for equation (23.4) may be found in Ince [4].

[3] Even though the Painlevé equations do not have elementary solutions in general, some choices of the parameters will lead to equations solvable in terms of elementary functions. For example, $y = -1/x$ is a solution of the second Painlevé equation when $\alpha = 1$, and $y = -1/x + 3x^2/(x^3 + 4)$ is a solution of the same equation when $\alpha = -2$. See Airault [1] for details.

[4] An ordinary differential equation is said to have the Painlevé property if all of its solutions are free of moving critical points. Rand and Winternitz [7], describe a MACSYMA program for determining whether a nonlinear ordinary differential equation has the Painlevé property (the differential equation must be a polynomial in both the dependent and independent variables and in all derivatives).

References

[1] H. Airault, "Rational Solutions of Painlevé Equations," *Stud. Appl. Math.*, **61**, 1979, pages 31–53.

[2] H. Bateman, *Partial Differential Equations of Mathematical Physics*, Dover Publications, Inc., New York, 1944, pages 75–79.

[3] L. M. Berkovitch, "Canonical Forms of Ordinary Linear Differential Equations," *Arch. Math. (Brno)*, **24**, No. 1, 1988, pages 25–42.

[4] E. L. Ince, *Ordinary Differential Equations*, Dover Publications, Inc., New York, 1964, Chapter 14 (pages 317–355).

[5] D. Irvine and M. A. Savageau, "Efficient Solution of Nonlinear Ordinary Differential Equations Expressed in S-System Canonical Form," *SIAM J. Numer. Anal.*, **27**, No. 3, 1990, pages 704–735.

[6] F. Neuman, "Transformation and Canonical Forms of Functional–Differential Equations," *Proc. Roy. Soc. Edin.* **115A**, 1990, pages 349–357.

[7] D. W. Rand and P. Winternitz, "ODEPAINLEVE — A Macsyma Package for Painleve Analysis of Ordinary Differential Equations," *Comput. Physics Comm.*, **42**, 1986, pages 359–383.

24.   Canonical Transformations

Applicable to   A system of ordinary differential equations that arise from a Hamiltonian.

Yields
A different system of ordinary differential equations that arise from a different Hamiltonian.

Procedure
A Hamiltonian $H(\mathbf{p}, \mathbf{q})$ (with $\mathbf{p} = (p_1, \ldots, p_n)$ and $\mathbf{q} = (q_1, \ldots, q_n)$) defines the system of ordinary differential equations

$$\dot{p}_i = -\frac{\partial H}{\partial q_i} = -H_{q_i},$$

$$\dot{q}_i = \frac{\partial H}{\partial p_i} = H_{p_i},$$

where a dot denotes differentiation with respect to the independent variable t (see page 57). The $\{p_i, q_i\}$ are called the *coordinates* of the Hamiltonian. The transformation to the new system of coordinates $\{P_i, Q_i\}$ via

$$\begin{aligned} p_i &= p_i(\mathbf{P}, \mathbf{Q}), \\ q_i &= q_i(\mathbf{P}, \mathbf{Q}), \end{aligned} \tag{24.1}$$

is (commonly) said to be canonical if Hamilton's equations remain invariant. That is, there exists a new Hamiltonian $K(\mathbf{P}, \mathbf{Q})$ such that the equations

$$\begin{aligned} \dot{P}_i &= -K_{Q_i}, \\ \dot{Q}_i &= K_{P_i}, \end{aligned} \tag{24.2}$$

are valid.
Canonical transformations can be defined implicitly by a *generating function*. For instance, for almost arbitrary $S(\mathbf{p}, \mathbf{Q}, t)$, a canonical transformation is given by

$$\begin{aligned} P_i &= -S_{Q_i}, \\ q_i &= -S_{p_i}, \\ K(\mathbf{P}, \mathbf{Q}) &= H(\mathbf{p}, \mathbf{q}) + S_t, \end{aligned} \tag{24.3.a–c}$$

where equations (24.3.a) and (24.3.b) must be solved to obtain explicit expressions for $\mathbf{q}(\mathbf{P}, \mathbf{Q})$, $\mathbf{p}(\mathbf{P}, \mathbf{Q})$. Also, for the S_t term, the derivative is taken with respect to the explicit dependence of S on t.

Other functional forms for the generating function are also possible. For example, a function of the form $\overline{S}(\mathbf{q}, \mathbf{P}, t)$ gives rise to the canonical transformation

$$p_i = \overline{S}_{q_i},$$
$$Q_i = \overline{S}_{P_i}, \qquad\qquad (24.4.a\text{--}c)$$
$$K(\mathbf{P}, \mathbf{Q}) = H(\mathbf{p}, \mathbf{q}) + \overline{S}_t.$$

Example

Given the Hamiltonian

$$H = \tfrac{1}{2}\left(p^2 + a^2(t)q^2\right), \qquad\qquad (24.5)$$

Hamilton's equations are $\{\dot{p} = -a^2 q, \ \dot{q} = p\}$, which can be combined to yield

$$\ddot{q} + a^2 q = 0. \qquad\qquad (24.6)$$

Hence, the Hamiltonian in (24.5) defines the second order ordinary differential equation in (24.6). Now consider the canonical transformation induced by the generating function $\overline{S}(q, P, t) = q^2 P$. From (24.4) we find

$$p = 2qP,$$
$$Q = q^2,$$
$$K(Q, P) = \frac{1}{2}\left(p^2 + a^2 q^2\right) = \frac{Q}{2}\left(4P^2 + a^2\right).$$

The equations corresponding to the new Hamiltonian are

$$\dot{P} = -\tfrac{1}{2}\left(4P^2 + a^2\right),$$
$$\dot{Q} = 4PQ. \qquad\qquad (24.7.a\text{--}b)$$

Equation (24.7.a) is a nonlinear *first* order ordinary differential equation for $P(t)$. After $P(t)$ is determined, (24.7.b) can be used to determine $Q(t)$ by quadrature. Hence, this change of variable has changed a second order linear ordinary differential equation into two successive first order ordinary differential equations.

Notes

[1] Canonical transformations are sometimes called contact transformations. See page 206 for the correct definition of a contact transformation.

[2] Technically, and in more generality, a transformation of the $2n$ variables $\{x_j, p_j \mid j = 1, \ldots, n\}$ to the $2n$ variables $\{X_j, P_j \mid j = 1, \ldots, n\}$ is a canonical transformation if the differential form $\sum_{j=1}^{n}(P_j dX_j - p_j dx_j)$ is exact, i.e., there exists a function $U = U(\mathbf{x}, \mathbf{p})$ such that

$$\sum_{j=1}^{n}(P_j dX_j - p_j dx_j) = dU. \tag{24.8}$$

[3] The section on Hamilton–Jacobi theory (see page 57) utilizes canonical transformations to derive the Hamilton–Jacobi equation.

[4] Tolstoy [8] shows that any given nonlinear ordinary differential equation may be transformed, in principle, by a variable transformation into a linear differential equation, or a system of such equations. This is the reverse of the process that was seen in the example.

[5] The set of all canonical transformations forms a group.

[6] Fouling transformations are canonical transformations in which the $\mathbf{p}$ coordinates in configuration space are preserved; i.e., $\mathbf{P} = \mathbf{p}$, $\mathbf{Q} = \mathbf{Q}(\mathbf{p}, \mathbf{q})$. See Gelman and Saletan [4] for details.

[7] A transformation, given by (24.1), which allows (24.2) to be written, and may or may not satisfy (24.8) is technically called a canonoid transformation. The lack of distinction between canonical and canonoid has occasionally led to ambiguity in the literature. See Negri, Oliveira, and Teixeira [7] or Currie and Saletan [3] for details.

References

[1] C. Carathéodory, *Calculus of Variations and Partial Differential Equations of the First Order*, Holden–Day, Inc., San Francisco, 1965, Chapter 6, (pages 79–101).

[2] C. R. Chester, *Techniques in Partial Differential Equations*, McGraw–Hill Book Company, New York, 1970, pages 197–206.

[3] D. G. Currie and E. J. Saletan, "Canonical Transformations and Quadratic Hamiltonians," *Nuovo Cimento B*, **9**, No. 1, 1972, pages 143–153.

[4] Y. Gelman and E. J. Saletan, "q-Equivalent Particle Hamiltonians. II: The Two-Dimensional Classical Oscillator," *Nuovo Cimento B*, **18**, No. 1, 1973, pages 53–71.

[5] H. Goldstein, *Classical Mechanics*, Addison–Wesley Publishing Co., Reading, MA, 1950, Chapter 8 (pages 237–272).

[6] D. ter Haar, *Elements of Hamiltonian Mechanics*, Pergamon Press, New York, 1971, pages 98–103.

[7] L. J. Negri, L. C. Oliveira, and J. M. Teixeira, "Canonoid Transformations and Constants of the Motion," *J. Math. Physics*, **28**, No. 10, October 1987, pages 2369–2372.

[8] I. Tolstoy, "Remarks on the Linearization of Differential Equations," *J. Inst. Maths. Applics*, **20**, 1977, pages 53–60.

25.   Darboux Transformation

Applicable to   Linear second order ordinary differential equations, a single equation or a system.

Yields

A reformulation of the problem.

Procedure

Given the equation

$$y'' = (f(x) + \kappa)y \tag{25.1}$$

for $y(x)$, we say that the transformation

$$z(x) = A(x, \lambda)y + B(x, \lambda)y'$$

is a Darboux transformation if $z(x)$ satisfies a differential equation of the form

$$z'' = (g(x) + \lambda)z. \tag{25.2}$$

For example, if $w(x)$ is a solution of (25.1), then a Darboux transformation is given by

$$z = y' - y\frac{w'}{w}. \tag{25.3}$$

In this case, if y satisfies (25.1), then $z(x)$ satisfies (25.2) with

$$f(x) = g(x) - 2[\log w(x)]''.$$

That is to say, this transformation changes the potential function appearing in (25.1) from $f(x)$ by $\delta f = -2[\log w(x)]''$, where $w(x)$ is an arbitrary solution of (25.1). The usefulness of this technique is that equation (25.2) might be easy to solve for $z(x)$; then $y(x)$ may be found from (25.3) by a single integration.

For the system of second order ordinary differential equations

$$\mathbf{y}'' = D(x)\mathbf{y}, \tag{25.4}$$

where $D(x)$ is the matrix

$$D(x) = \begin{pmatrix} d_{11}(x) & d_{12}(x) & \cdots & d_{1n}(x) \\ d_{21}(x) & d_{22}(x) & \cdots & d_{2n}(x) \\ \vdots & \vdots & \ddots & \vdots \\ d_{n1}(x) & d_{n2}(x) & \cdots & d_{nn}(x) \end{pmatrix},$$

we say that

$$\mathbf{z}(x) = A(x)\mathbf{y} + B(x)\mathbf{y}', \tag{25.5}$$

where A and B are matrices, is a Darboux transformation if $\mathbf{z}$ satisfies an equation of the form

$$\mathbf{z}'' = F(x)\mathbf{z}, \tag{25.6}$$

where $F(x)$ is some new matrix

$$F(x) = \begin{pmatrix} f_{11}(x) & f_{12}(x) & \cdots & f_{1n}(x) \\ f_{21}(x) & f_{22}(x) & \cdots & f_{2n}(x) \\ \vdots & \vdots & \ddots & \vdots \\ f_{n1}(x) & f_{n2}(x) & \cdots & f_{nn}(x) \end{pmatrix}.$$

Sometimes Darboux transformations of this type can be used to decouple systems of differential equations. See Humi [1] for details.

Example 1

If the solution of the differential equation

$$y'' = (f(x) + \lambda)y \tag{25.7}$$

is known for all values of λ (call it y_λ), and $w(x) = y_\mu(x)$ is the solution when $\lambda = \mu$, then the general solution of the differential equation

$$z'' = \left(w(x)\frac{d^2}{dx^2}\left(\frac{1}{w(x)}\right) + \lambda - \mu \right) z \tag{25.8}$$

for $z(x)$ is given by (see (25.3))

$$z = y_\lambda' - y_\lambda \frac{w'(x)}{w(x)}, \tag{25.9}$$

for $\lambda \neq \mu$. In particular, if we take $f(x) = 0$ in (25.7), then $y_0(x) = Ax + B$ when $\lambda = 0$ and $y_\lambda(x) = e^{\pm\sqrt{\lambda}x}$ for $\lambda \neq 0$. If we take $\mu = 0$ and $w(x) = x$, then equation (25.8) becomes

$$z'' = \left(\frac{2}{x^2} + \lambda \right) z,$$

with the solution given by equation (25.9); i.e.,

$$z(x) = e^{\pm\sqrt{\lambda}x}\left(\pm\sqrt{\lambda} - \frac{1}{x} \right).$$

Example 2

This example is from Humi [1]. Suppose we wish to decouple a system of symmetric equations in the form of (25.4) with

$$D(x) = \begin{pmatrix} u_1(x) + \lambda & d(x) \\ d(x) & u_2(x) + \lambda \end{pmatrix}.$$

If we apply a Darboux transformation, we can hope to obtain the form of (25.6) with $F(x)$ given by

$$F(x) = \begin{pmatrix} v_1(x) + \lambda & 0 \\ 0 & v_2(x) + \lambda \end{pmatrix}. \tag{25.10}$$

If we choose $B = I$ in (25.5), then to obtain (25.6) we require

$$A'' + D' + AD = FA,$$
$$2A' + D = F.$$

In our case, with $D(x)$ given by (25.7) and $F(x)$ given by (25.10) we require that the elements of the matrix $A(x)$ satisfy

$$2a'_{12} = 2a'_{21} = -d,$$
$$2a'_{11} + u_1(x) = v_1(x), \tag{25.11}$$
$$2a'_{22} + u_2(x) = v_2(x).$$

It is a simple matter to integrate these equations to obtain

$$a_{12}(x) = c(x), \quad a_{21}(x) = c(x),$$
$$a_{11}(x) = \frac{1}{2c}\left(\frac{1}{2}d(x) + \alpha + I\right),$$
$$a_{22}(x) = \frac{1}{2c}\left(\frac{1}{2}d(x) - \alpha + I\right),$$

where α is an arbitrary constant and

$$c(x) = -\frac{1}{2}\int^x d(t)\,dt,$$
$$I(x) = \int^x c(t)[u_2(t) - u_1(t)]\,dt.$$

This solution is valid if the consistency constraint

$$u_1 + u_2 = 2c^2 - \left(\frac{d}{2c}\right)' + \frac{1}{2}\left(\frac{d}{2c}\right)^2 + \frac{1}{2c^2}(\alpha + I)^2 \tag{25.12}$$

is satisfied. This constraint was derived in the solution of (25.11).

Stated another way, we can choose d and $u_1 - u_2$ as arbitrary functions and then use (25.12) to compute the corresponding $u_1 + u_2$ for which the resulting system of equations can be decoupled by the use of a Darboux transformation.

References

[1] M. Humi, "Separation of Coupled Systems of Differential Equations by Darboux Transformation," *J. Phys. A: Math. Gen.*, **18**, 1985, pages 1085–1091.

[2] E. L. Ince, *Ordinary Differential Equations*, Dover Publications, Inc., New York, 1964, page 182.

[3] B. G. Konopelchenko, "On Exact Solutions of Nonlinear Integrable Equations via Integral Linearising Transforms and Generalised Bäcklund–Darboux Transformations," *J. Phys. A: Math. Gen.*, **23**, 190, pages 3761–3768.

[4] G. L. Lamb, *Elements of Soliton Theory*, John Wiley & Sons, New York, 1980, pages 38–41.

[5] D. Levi, "Toward a Unification of the Various Techniques used to Integrate Nonlinear Partial Differential Equations: Bäcklund and Darboux Transformations vs. Dressing Method," *Rep. Math. Phys.*, **23**, No. 1, 1986, pages 41–56.

[6] I. V. Poplavskii, "Generalized Darboux–Crum–Krein Transformations," *Theo. Math. Physics*, **69**, No. 3, 1986, pages 1278–1282.

[7] M. A. Sall, "Darboux Transformations for Non-Abelian and Nonlocal Equations of the Toda Chain Type," *Theo. Math. Physics*, **53**, 1982, pages 1092–1099.

[8] S. Stanek and J. Vosmansky, "Transformations of Linear Second Order Ordinary Differential Equations," *Archivum Mathematicum (BRNO)*, **22**, No. 1, 1986, pages 55–60.

[9] W. M. Zheng, "The Darboux Transformation and Solvable Double-Well Potential Models for Schrödinger Equations," *J. Math. Physics*, **25**, No. 1, January 1984, pages 88–90.

26.   An Involutory Transformation

Applicable to   Nonlinear partial differential equations of a certain form.

Yields

A reformation of the partial differential equation.

Idea

Inverting the dependent and independent variables might lead to a more tractable equation.

Procedure

   Suppose we have a partial differential equation of the form

$$\Phi\left(u; \frac{\partial}{\partial x}; \frac{\partial}{\partial t}\right) := \Phi(u, u_x, u_{xx}, \ldots; u_t, u_{tt}, \ldots) = 0, \qquad (26.1)$$

for $u = u(x, t)$. We introduce the *inverse transformation*

$$T = \begin{cases} u' = x, \\ x' = u, \\ t' = t. \end{cases}$$

Since applying T twice is equivalent to not applying T, the transformation is involutory (i.e., $T^2 = I =$ the identity). Noting that

$$\frac{\partial}{\partial x} = \frac{1}{\partial u'/\partial x'} \frac{\partial}{\partial x'} := D',$$

$$\frac{\partial}{\partial t} = \frac{\partial}{\partial t'} - \frac{\partial u'/\partial t'}{\partial u'/\partial x'} \frac{\partial}{\partial x'} := \partial',$$

then, under T, equation 1 becomes

$$\Phi\left(x; D'; \partial'\right) = 0. \qquad (26.2)$$

This transformation may be used to change classes of nonlinear equations with Dirichlet boundary conditions to linear form. For example, the class

$$\frac{\partial u'}{\partial t'} - \gamma(u')\frac{\partial}{\partial x'}\left(\sum_{i=1}^{N}\alpha_i(u', t')D'^i x'\right) = 0,$$

$$\begin{array}{lll} u' = \Psi_1(t') & \text{on} & x' = \Phi_1(t'), \\ u' = \Psi_2(t') & \text{on} & x' = \Phi_2(t'), \\ u' = \Theta(x') & \text{at} & t' = 0, \end{array}$$

transforms, under T, to

$$\frac{\partial u}{\partial t} + \gamma(u)\frac{\partial}{\partial x}\left(\sum_{i=1}^{N}\alpha_i(x, t)\frac{\partial^i u}{\partial x^i}\right) = 0,$$

$$\begin{array}{lll} u = \Phi_1(t) & \text{on} & x = \Psi_1(t), \\ u = \Phi_2(t) & \text{on} & x = \Psi_2(t), \\ u = \Theta^{-1}(x) & \text{at} & t = 0. \end{array}$$

Example

Given the equation and initial/boundary conditions

$$\frac{\partial u'}{\partial t'} = \frac{\kappa}{\left(\dfrac{\partial u'}{\partial x'}\right)^2} \frac{\partial^2 u'}{\partial x'^2},$$

$$
\begin{array}{llll}
u' = 0 & \text{on} & x' = \Phi_1(t'), & \\
u' = L & \text{on} & x' = \Phi_2(t'), & (26.3) \\
u' = \Theta(x') & \text{at} & t' = 0, &
\end{array}
$$

the transformed equation and initial/boundary conditions become

$$\frac{\partial u}{\partial t} = \kappa \frac{\partial^2 u}{\partial x^2},$$

$$
\begin{array}{llll}
u = \Phi_1(t) & \text{on} & x = 0, & \\
u = \Phi_2(t) & \text{on} & x = L, & (26.4) \\
u = \Theta^{-1}(x) & \text{at} & t = 0. &
\end{array}
$$

The equation in (26.4) can be easily solved (by use of, say, Fourier transforms) to yield

$$
u(x,t) = \frac{2}{L} \sum_{n=1}^{\infty} \exp\left(-\frac{\kappa n^2 \pi^2 t}{L^2}\right) \sin\left(\frac{n\pi x}{L}\right) \left[\int_0^L \Theta^{-1}(\sigma) \sin\left(\frac{n\pi\sigma}{L}\right) d\sigma\right]
$$
$$
+ \frac{n\kappa x}{L} \int_0^1 \exp\left(\frac{\kappa n^2 \pi^2 t}{L^2}\right) [\Psi_1(\tau) - (-1)^n \Psi_2(\tau)] \, d\tau
$$

This last relation, can be implicitly solved for $x = x(u,t)$; which (under T) is the solution to (26.3) (i.e., $u' = u'(x',t')$).

Notes

[1] The Hodograph transformation is a different way in which the dependent and independent variables are interchanged. See page 390 for details.

References

[1] C. Rogers, "Inverse Transformations and the Reduction of Nonlinear Dirich-
    let Problems," *J. Phys. A: Math. Gen.*, **17**, 1984, pages L681–L685.

27. Liouville Transformation − 1

Applicable to   The general Sturm–Liouville equation

$$-[p(x)y']' + r(x)y = \lambda\rho(x)y, \qquad \text{for } a \leq x \leq b,$$
$$y'(a) + \alpha y(a) = 0, \tag{27.1}$$
$$y'(b) + \beta y(b) = 0.$$

Procedure

The Liouville transformation (version 1) is to change the independent
variable from $x \in [a, b]$ to $t \in [0, \pi]$ by

$$t = \frac{1}{J} \int_a^x \left(\frac{\rho(x)}{p(x)}\right)^{1/2} dx, \tag{27.2}$$

where J is defined by

$$J = \frac{1}{\pi} \int_a^b \left(\frac{\rho(x)}{p(x)}\right)^{1/2} dx, \tag{27.3}$$

and to change the dependent variable from $y(x)$ to $u(t)$ by

$$u(t) = f(x)y(x) = [\rho(x)p(x)]^{1/4}y(x), \tag{27.4}$$

where we have defined $f(x) := [\rho(x)p(x)]^{1/4}$. With this change of variable,
(27.1) becomes

$$\frac{d^2u}{dt^2} + [k^2 - q(t)]u = 0, \qquad \text{for } 0 \leq t \leq \pi,$$
$$u'(0) + hu(0) = 0,$$
$$u'(\pi) + Hu(\pi) = 0,$$

which is in *Liouville normal form*. The definitions of $\{k, q(t), h, H\}$ are as follows

$$k^2 = J^2\lambda,$$

$$m(t) = \frac{r(x)}{\rho(x)},$$

$$q(t) = \frac{f_{tt}}{f} + J^2 m(t),$$

$$h = \frac{1}{f^2(a)}[\alpha Jp(a) - f(a)f_t(a)],$$

$$H = \frac{1}{f^2(b)}[\beta Jp(b) - f(b)f_t(b)].$$

Note that $q(t)$ may also be written as

$$q(t) = \frac{r}{p} + (p\rho)^{-1/4}\frac{d^2}{dt^2}[(p\rho)^{1/4}],$$

$$= \frac{r}{p} + \frac{p}{4\rho}\left[\left(\frac{p'}{p}\right)' + \left(\frac{\rho'}{\rho}\right)' + \frac{3}{4}\left(\frac{p'}{p}\right)^2 + \frac{1}{2}\left(\frac{p'}{p}\right)\left(\frac{\rho'}{\rho}\right) - \frac{1}{4}\left(\frac{\rho'}{\rho}\right)^2\right]$$

Example

If we have the equation and boundary conditions

$$-(xy')' + \frac{1}{x}y = \lambda xy, \qquad \text{for } \pi \le x \le 2\pi,$$

$$y'(\pi) = 0,$$

$$y'(2\pi) = 0.$$

Then we identify

$$p(x) = x, \qquad r(x) = \frac{1}{x}, \qquad \rho(x) = x,$$

$$a = \pi, \qquad b = 2\pi, \qquad \alpha = 0, \qquad \beta = 0.$$

A simple calculation results in: $J = 1$, $t = x - \pi$, $f(x) = \sqrt{x} = \sqrt{t+1}$,
$m(t) = \dfrac{1}{x^2} = \dfrac{1}{(t+1)^2}$, $q(t) = \dfrac{3}{4(t+1)^2}$, $k^2 = \lambda$, $h = -\dfrac{1}{2}$, and $H =$
$-\dfrac{1}{2(\pi+1)}$. Hence, we obtain

$$u'' + \left(\lambda - \frac{3}{4(t+1)^2}\right)u = 0, \qquad \text{for } 0 \le t \le \pi,$$

$$u'(0) - \tfrac{1}{2}u(0) = 0, \qquad\qquad (27.5)$$

$$u'(\pi) - \frac{1}{2(\pi+1)}u(\pi) = 0.$$

Equation (27.5) is in Liouville normal form.

Notes

[1] The standard assumptions required on (27.1) are that, on the interval $[a, b]$:
p and q are real-valued, $p > 0$, q does not vanish, and p and q have continuous
second derivatives. Boundedness conditions are also required for the new
functions.

[2] The transformation

$$t = \int_{x_0}^{x} \sqrt{\frac{|q(z)|}{p(z)}}\, dz,$$

$$u(t) = [p(x)|q(x)|]^{1/4}\, y(x),$$

(27.6)

when applied to (27.1) with $\rho = 0$, results in

$$\frac{d^2 u}{dt^2} + [\pm 1 + R(t)]\, u(t) = 0,$$

(27.7)

where

$$R(t) = p^{1/4}|q|^{-3/4} \frac{dp(x)}{dx} \frac{d}{dx} [p(x)|q(x)|]^{-1/4} \Bigg|_{x=x(t)},$$

and the plus (minus) sign is taken in (27.7) if $q(x) > 0$ ($q(x) < 0$). This is
also called the Liouville transformation, See Eastham [5].

[3] The two different transformations, the one in theeqb and (27.4), and the one
in (27.6), are each sometimes called the Liouville–Green transformation.

References

[1] G. Birkhoff and G.-C. Rota, *Ordinary Differential Equations*, John Wiley
    & Sons, New York, 1978, pages 265–267.

[2] W. E. Boyce, "Random Eigenvalue Problems," in A. T. Bharucha-Reid
    (ed.), *Probabilistic Methods in Applied Mathematics*, Academic Press,
    New York, 1968, pages 1–73, pages 20–21.

[3] J. S. Cassell, "An Extension of the Liouville–Green Asymptotic Formula for
    Oscillatory Second-Order Differential Equations," *Proc. Roy. Soc .Edin.*,
    100A, 1985, pages 181–190.

[4] M. S. P. Eastham, "Asymptotic Formulae of Liouville–Green Type for Higher-
    Order Differential Equations," *J. London Math. Soc.*, **2**, No. 28, 1983,
    pages 507–518.

[5] M. S. P. Eastham, "The Liouville–Green Asymptotic Theory for Second-
    Order Differential Equations: A New Approach and Extensions," in W.
    N. Everitt and R. T. Lewis (eds.), *Ordinary Differential Equations and
    Operators*, Springer–Verlag, New York, 1983.

[6] E. Hille, *Lectures on Ordinary Differential Equations*, Addison–Wesley
    Publishing Co., Reading, MA, 1969, page 340.

[7] H. C. Howard and V. Marić, "An Extension of the Liouville–Green Approx-
    imation," *J. Math. Anal. Appl.*, **143**, No. 2, 1989, pages 548–559.

[8] W. D. Lakin and D. A. Sanchez, *Topics in Ordinary Differential Equa-
    tions*, Dover Publications, Inc., New York, 1970, pages 36–41.

[9] G. Valiron, *The Geometric Theory of Ordinary Differential Equations and Algebraic Functions*, Math Sci Press, Brookline, MA, 1950, page 511.

28. Liouville Transformation – 2

Applicable to The second order linear ordinary differential equation

$$\frac{d^2y}{dt^2} + \lambda m^4(t)y = 0 \tag{28.1}$$

on the finite interval $0 \leq t \leq T$, where λ is a constant and $m(t) > 0$.

Procedure

The Liouville transformation (version 2) is to change the dependent and independent variables in (28.1) by

$$x = \frac{1}{J} \int_0^t m^2(z)\, dz,$$

$$J = \frac{1}{\pi} \int_0^T m^2(z)\, dz,$$

$$w(x) = m(t)y(t).$$

This transformation changes (28.1) into

$$\frac{d^2w}{dx^2} + \left[\lambda J^2 + Q(x) \right] w = 0, \tag{28.2}$$

for $0 \leq x \leq \pi$, where $Q(x)$ is defined by

$$Q(x) = \frac{1}{m(t)} \frac{d^2 m(t)}{dx^2} = -\frac{J^2}{m(t)^3} \frac{d^2}{dt^2}\left(\frac{1}{m(t)} \right). \tag{28.3}$$

The inverse transformation, which takes (28.2) into (28.1), is given by

$$t = J \int_0^x \frac{d\zeta}{[m^*(\zeta)]^2},$$

$$J = T \left(\int_0^\pi \frac{d\zeta}{[m^*(\zeta)]^2} \right)^{-1},$$

where $m^*(x) = m(t)$ is any positive solution of the differential equation

$$\frac{d^2 m^*}{dx^2} = Q(x)m^*(x).$$

Example

Suppose we have (essentially) Airy's equation

$$\frac{d^2y}{dt^2} + \lambda t y = 0. \tag{28.4}$$

Comparing (28.4) to (28.1) shows that $m(t) = t^{1/4}$. Using this value for $m(t)$ produces

$$J = \frac{2}{3\pi}T^{3/2},$$

$$x = \pi\left(\frac{t}{T}\right)^{3/2},$$

$$w(x) = t^{1/4}y(t).$$

Under this change of variables, (28.4) becomes

$$\frac{d^2w}{dx^2} + \left(\frac{4\lambda}{9\pi^2}T^3 - \frac{5}{36}\frac{1}{x^2}\right)w = 0. \tag{28.5}$$

For large values of x, an approximation to (28.5) might be obtained by discarding the second term in the parentheses.

Notes

[1] The function $Q(x)$ defined in (28.3) will be a constant if and only if $m(t) = (\alpha t^2 + \beta t + \delta)^{-1/2}$. In this case, $Q(x) = -J^2(\alpha\delta - 4\beta^2)$.

[2] This transformation is useful when followed by some sort of asymptotic analysis. When the magnitude of λ is large compared to $Q(x)$, then the first order approximation to (28.2) will be to discard the $Q(x)$ term.

References

[1] W. Magnus and S. Winkler, *Hill's Equation*, Dover Publications, Inc., New York, 1966, page 51.

29.   Reduction of Linear ODEs
to a First Order System

Applicable to   Linear ordinary differential equations.

Yields

A first order vector system.

Idea

By introducing variables to represent the derivatives in an n-th order linear ordinary differential equation, a first order system of differential equations may be obtained.

Procedure

Given the linear ordinary differential equation

$$\frac{d^n y}{dx^n} = a_{n-1}(x)\frac{d^{(n-1)}y}{dx^{(n-1)}} + \cdots + a_1(x)\frac{dy}{dx} + a_0(x)y + b(x) \qquad (29.1)$$

for $y(x)$, introduce the variables $\{z_1, z_2, \ldots, z_n\}$ defined by

$$z_1 = \frac{dy}{dx}, \quad z_2 = \frac{d^2 y}{dx^2}, \quad \ldots, \quad z_n = \frac{d^n y}{dx^n}.$$

Using these new variables, (29.1) may be written as

$$\frac{d}{dx}\mathbf{y} = A(x)\mathbf{y} + \mathbf{b}(x),$$

where

$$\mathbf{y} = (y, y^{(1)}, \ldots, y^{(n-1)})^{\mathrm{T}} = (y, z_1, z_2, \ldots, z_{n-1})^{\mathrm{T}},$$
$$\mathbf{b} = (0, 0, \ldots, 0, b(x))^{\mathrm{T}},$$

and A is the matrix

$$\begin{pmatrix} 0 & 1 & 0 & 0 & \cdots & 0 \\ 0 & 0 & 1 & 0 & \cdots & 0 \\ 0 & 0 & 0 & 1 & & 0 \\ \vdots & \vdots & \vdots & & \ddots & \\ 0 & 0 & 0 & 0 & & 1 \\ a_0(x) & a_1(x) & a_2(x) & a_3(x) & \cdots & a_{n-1}(x) \end{pmatrix}. \qquad (29.2)$$

If the initial conditions for equation (29.1) were in the form

$$y(x_0) = c_0, \ y'(x_0) = c_1, \ y''(x_0) = c_2, \ \ldots, \ y^{(n-1)}(x_0) = c_{n-1},$$

then the initial condition for equation (29.2) is

$$\mathbf{y}(x_0) = (c_0, c_1, c_2, \ldots, c_{n-1})^{\mathrm{T}}.$$

To solve an equation in the form of (29.2), see the section on vector ordinary differential equations (page 360).

Example

Given the linear ordinary differential equation with initial conditions

$$\frac{d^2y}{dx^2} + x^2\frac{dy}{dx} + (\log x)y = \sin x,$$

$$y(0) = 3, \qquad y'(0) = 4,$$

it may easily be changed into the equivalent first order system

$$\frac{d}{dx}\begin{pmatrix} y \\ y' \end{pmatrix} = \begin{pmatrix} 0 & 1 \\ -\log x & -x^2 \end{pmatrix}\begin{pmatrix} y \\ y' \end{pmatrix} + \begin{pmatrix} 0 \\ \sin x \end{pmatrix},$$

or, equivalently,

$$\frac{d\mathbf{y}}{dt} = A(x)\mathbf{y} + \mathbf{b},$$

where $\mathbf{y} = (y, y')^{\mathrm{T}}$, $\mathbf{b} = (0, \sin x)^{\mathrm{T}}$, and A is the matrix

$$A = \begin{pmatrix} 0 & 1 \\ -\log x & -x^2 \end{pmatrix}.$$

Notes
[1] Many packaged computer programs require the input to be in the form of a first order vector system.
[2] The *method of elimination* is the opposite of the method presented here. In the method of elimination, a system of simultaneous equations is converted into a single equation of higher order. See Finizio and Ladas [1] for details.

References
[1] N. Finizio and G. Ladas, *Ordinary Differential Equations with Modern Applications*, Wadsworth Publishing Company, Belmont, Calif, 1982, pages 162–170.

30.   Prüfer Transformation

Applicable to   Linear, homogeneous, second order differential equations.

Yields

An equivalent system of two first order differential equations.

Idea

This standard transformation changes an equation from Liouville normal form to two successive ordinary differential equations.

Procedure

Suppose we have the Sturm–Liouville equation

$$\frac{d}{dx}\left(P(x)\frac{du}{dx}\right) + Q(x)u = 0, \tag{30.1}$$

defined on $a < x < b$, with $P > 0$, $P \in C^1$, and Q continuous. If we think of this single second order equation as two first order equations for the unknowns $\{u, u'\}$, then we can change the dependent variables from $\{u, u'\}$ to $R(x)$ and $\theta(x)$ by

$$P(x)u'(x) = R(x)\cos\theta(x),$$
$$u(x) = R(x)\sin\theta(x). \tag{30.2}$$

Using (30.2) in (30.1) we obtain two sequential first order ordinary differential equations for the unknowns $R(x)$ and $\theta(x)$

$$\frac{d\theta}{dx} = Q(x)\sin^2\theta + \frac{1}{P(x)}\cos^2\theta,$$
$$\frac{dR}{dx} = \left[\frac{1}{P(x)} - Q(x)\right]R(x)\sin 2\theta. \tag{30.3.a–b}$$

If (30.3.a) can be integrated, then (30.3.b) can be solved for

$$R(x) = R(a)\exp\left(\int_a^x \left[\frac{1}{P(t)} - Q(t)\right]\sin 2\theta(t)\, dt\right). \tag{30.4}$$

Example

If we have the linear second order homogeneous ordinary differential equation

$$xu'' - u' + x^3 u = 0, \tag{30.5}$$

then we can write (30.5) in Liouville normal form as

$$\frac{d}{dx}\left(\frac{1}{x}u'\right) + xu = 0,$$

from which we can identify: $P(x) = 1/x$, $Q(x) = x$. Therefore, from (30.3.a), we have

$$\frac{d\theta}{dx} = x\sin^2\theta + \frac{1}{1/x}\cos^2\theta$$
$$= x.$$

This equation can be solved to yield $\theta(x) = \dfrac{x^2}{2} + C$, where C is an arbitrary constant. From equation (30.4) we then find $R(x) = R(a)$. Therefore, we conclude that

$$u(x) = R(a)\sin\left(\frac{x^2}{2} + C\right)$$

$$= u(a)\frac{\sin(x^2/2 + C)}{\sin(a^2/2 + C)},$$

is the solution to (30.5).

Notes

[1] The Prüfer transformation is often used to obtain information about the zeros of $u(x)$.

References

[1] D. Adamová, J. Holřejší, and I. Úlehla, "The Atkinson–Prüfer Transformation and the Eigenvalue Problem for Coupled Systems of Schrödinger Equations," *J. Phys. A: Math. Gen.*, **17**, 1984, pages 2621–2631.

[2] P. B. Bailey, "Sturm–Liouville Eigenvalues via a Phase Function," *SIAM J. Appl. Math.*, **14**, 1966, pages 242–249.

[3] R. Bellman and G. M. Wing, *An Introduction to Invariant Imbedding*, John Wiley & Sons, New York, 1975, pages 140–142.

[4] D. C. Benson, "A Prüfer Transformation for Liénard's Equation," *SIAM J. Numer. Anal.*,**15**, No. 4, July 1984, pages 656–669.

[5] G. Birkhoff and G.-C. Rota, *Ordinary Differential Equations*, John Wiley & Sons, New York, 1978, pages 257–266.

[6] I. Ulehla and J. Horejsi, "Generalized Prufer Transformation and the Eigenvalue Problem for Radial Dirac Equations," *Phys. Lett. A*, **113**, No. 7, 1986, pages 355–358.

31.   Modified Prüfer Transformation

Applicable to   Linear, homogeneous, second order ordinary differential equations.

Yields

An equivalent system of two first order ordinary differential equations.

Idea

This standard transformation changes an equation from Liouville normal form to two successive ordinary differential equations.

Procedure

Suppose we have an ordinary differential equation in Liouville normal form

$$u'' + Q(x)u = 0, \tag{31.1}$$

defined on $a < x < b$, with $Q > 0$. We define the *modified amplitude* $R(x)$ and the *modified phase* $\phi(x)$ by

$$u(x) = \frac{R(x)}{Q^{1/4}} \sin \phi(x),$$
$$u'(x) = R(x)Q^{1/4} \cos \phi(x). \tag{31.2}$$

Using (31.2) in (31.1) we determine the modified Prüfer system corresponding to (31.1) to be

$$\frac{d\phi}{dx} = -Q^{1/2} - \frac{1}{4}\frac{Q'}{Q}\sin 2\phi,$$
$$\frac{1}{R}\frac{dR}{dx} = \frac{1}{4}\frac{Q'}{Q}\cos 2\phi. \tag{31.3.a-b}$$

The modified Prüfer transformation is usually used to obtain asymptotic information about the solution to (31.1).

Example

If $u(x)$ satisfies

$$u'' + \left(1 - \frac{M}{x^2}\right)u = 0, \tag{31.4}$$

for $0 < x < \infty$, then the exact solution is $u(x) = \sqrt{x}Z_n(x)$, where $Z_n(x)$ is a Bessel function and $n = \pm\sqrt{M + \frac{1}{4}}$. Comparing (31.4) to (31.1) we identify $Q(x) = 1 - \frac{M}{x^2}$, so that (31.3) becomes

$$\frac{d\phi}{dx} = -\sqrt{1 - \frac{M}{x^2}} + \frac{M\sin 2\phi}{2(x^3 - Mx)},$$
$$\frac{1}{R}\frac{dR}{dx} = -\frac{M\cos 2\phi}{2(x^3 - Mx)}.$$

For $M = O(1)$ and $x \gg 1$, the above expressions can be expanded to yield

$$\frac{d\phi}{dx} \simeq 1 - \frac{1}{2}\frac{M}{x^2} + O\left(\frac{1}{x^3}\right),$$

$$\frac{1}{R}\frac{dR}{dx} \simeq O\left(\frac{1}{x^3}\right),$$

which can be integrated (and then simplified) to yield

$$\phi(x) \simeq \phi_\infty + x - \frac{M}{2x} + O\left(\frac{1}{x^2}\right),$$

$$R(x) \simeq R_\infty + O\left(\frac{1}{x^2}\right). \tag{31.5}$$

Using (31.5) and $Q(x)$ in (31.2.a) provides an approximation to $u(x)$ for large values of x. This, in turn, provides an approximation to the n-th Bessel function.

Notes

[1] The modified Prüfer transformation is often used with $Q(x) = \lambda - q(x)$ when λ is large in magnitude compared to $q(x)$.

References

[1] G. Birkhoff and G.-C. Rota, *Ordinary Differential Equations*, John Wiley & Sons, New York, 1978, pages 267–277.

[2] B. A. Hargrave, "Numerical Approximation of Eigenvalues of Sturm–Liouville Systems," *J. Comput. Physics*, **20**, 1976, pages 381–396.

32.   Transformations of Second Order Linear ODEs – 1

Applicable to   The second order linear ordinary differential equation

$$y'' + a(x)y' + b(x)y = 0. \tag{32.1}$$

Transformation 1

If the dependent and independent variables in (32.1) are changed by

$$t = \int_{x_0}^{x} \exp\left(-\int_{x_0}^{r} a(z)\, dz\right) dr,$$
$$w(t) = y(x),$$

then (32.1) becomes

$$\frac{d^2 w}{dt^2} + b(x(t)) \exp\left[-2 \int_{x_0}^{x} a(z)\, dz\right] w = 0. \tag{32.2}$$

Example

For the ordinary differential equation

$$y'' - \frac{3x}{1 - x^2} y' + \frac{7}{1 - x^2} y = 0,$$

the change of variables becomes $t = x/\sqrt{1 - x^2}$ and the equation corresponding to (32.2) is $\dfrac{d^2 w}{dt^2} + \dfrac{7}{(1 + t^2)^2} w = 0.$

Transformation 2

If in equation (32.1) the expression

$$\frac{b' + 2ab}{b^{3/2}} \tag{32.3}$$

is found to be a constant, then the change of independent variable given by

$$z = C \int \sqrt{b(x)}\, dx, \tag{32.4}$$

where C is an arbitrary constant, will reduce equation (32.1) to an equation with constant coefficients. Moreover, if the expression in (32.3) is not constant, then no change of independent variable alone will reduce equation (32.1) to an equation with constant coefficients.

Example

Given the equation

$$xy'' + (8x^2 - 1)y' + 20x^3y = 0, \qquad (32.5)$$

we note that $a(x) = 8x - 1/x$ and $b(x) = 20x^2$. Hence, the expression in (32.3) becomes

$$\frac{b' + 2ab}{b^{3/2}} = \frac{40x + 40x^2(8x - x^{-1})}{20^{3/2}x^3} = \frac{320x^3}{20^{3/2}x^3} = \text{constant}.$$

Therefore, if the independent variable is changed by $z = C\int \sqrt{20x}\,dx$, then equation (32.5), written in terms of z, will be a constant coefficient differential equation. A natural choice for C is $C = 2/\sqrt{20}$ so that the transformation becomes $z = x^2$. Using this new variable in (32.5) results in the equation

$$\frac{d^2y}{dz^2} + 4\frac{dy}{dz} + 5y = 0,$$

which has the solution $y = e^{-2z}(A\cos z + B\sin z)$, where A and B are arbitrary constants. Hence, the general solution to (32.5) is

$$y = \left(A\cos x^2 + B\sin x^2\right)\exp\left(-2x^2\right).$$

Transformation 3

If the dependent variable in (32.1) is changed by

$$y(x) = u(x)\exp\left(-\frac{1}{2}\int^x a(z)\,dz\right),$$

then (32.1) becomes

$$u'' + I(x)u = 0, \qquad (32.6)$$

where

$$I(x) = \left(b - \frac{1}{4}a^2 - \frac{1}{2}\frac{da}{dx}\right). \qquad (32.7)$$

Equation (32.6) is said to be the *normal form* for equation (32.1). The quantity $I(x)$ is the *invariant* of (32.1).

Two ordinary differential equations which have the same name normal form (i.e., $I(x)$ is the same) are said to be *equivalent*. This is because if $y_1(x)$ and $y_2(x)$ satisfy

$$\begin{aligned} y_1'' + p_1y_1' + q_1y_1 &= 0, \\ y_2'' + p_2y_2' + q_2y_2 &= 0, \end{aligned} \qquad (32.8)$$

and if both equations have the same invariant, then

$$y_1(x) = y_2(x) \exp\left(-\frac{1}{2}\int^x \left(p_1(z) - p_2(z)\right) dz\right). \tag{32.9}$$

Conversely, if y_1 and y_2 are solutions to (32.8), and if $y_1(x) = f(x)y_2(x)$ for some $f(x)$, then the invariants of the two equations in (32.8) are the same.

Example

Suppose we wish to solve the equation

$$\frac{d^2y}{dx^2} - \frac{2}{x}\frac{dy}{dx} + \left(a^2 + \frac{2}{x^2}\right)y = 0, \tag{32.10}$$

in which a is a constant. We find that (comparing (32.10) with (32.1), and using (32.7))

$$I(x) = \left(a^2 + \frac{2}{x^2}\right) - \frac{1}{4}\frac{4}{x^2} - \frac{1}{2}\frac{2}{x^2} = a^2.$$

Now, we know the solution of

$$\frac{d^2v}{dx^2} + a^2v = 0 \tag{32.11}$$

to be $v(x) = A\cos ax + B\sin ax$, where A and B are arbitrary constants. Since the equations in (32.10) and (32.11) have the same invariant, one can be transformed into the other. Using (32.9), we find

$$y(x) = v(x)\exp\left(\int \frac{dx}{x}\right) = xv,$$

and, hence, the solution of (32.10) is $y(x) = Ax\cos ax + Bx\sin cx$.

Transformation 4

If, instead of (32.1), both sides of

$$y'' + a(x)y' + b(x)y = c(x) \tag{32.12}$$

are multiplied by

$$p(x) = \exp\left(\int_{x_0}^x a(z)\,dz\right),$$

then (32.12) is put in the *formally self-adjoint* form

$$\frac{d}{dx}\left(p(x)\frac{dy}{dx}\right) + q(x)y = r(x), \tag{32.13}$$

where

$$q(x) = p(x)b(x),$$
$$r(x) = p(x)c(x).$$

See the method on page 128 for transformations of an equation in the form of (32.13).

Notes

[1] Note that the invariant of the adjoint of (32.1) is equal to the invariant of (32.1). That is to say, invariants are preserved under the operation of taking the adjoint.

[2] If (32.6) has the two linearly independent solutions $u(x)$ and $v(x)$ and if we define $s(x) := u(x)/v(x)$, then $\{s, x\} = 2I(x)$, where $\{\ ,\ \}$ denotes the Schwarzian derivative.

[3] Kamran and Olver [5] completely solve the equivalence problem, that is, determining when two second order linear differential operators are the same under a change of variable.

References

[1] L. M. Berkovich, "Canonical Forms of Ordinary Linear Differential Equations," *Arch. Math. (BRNO)*, **24**, No. 1, 1988, pages 25–42.

[2] W. E. Boyce and R. C. DiPrima, *Elementary Differential Equations and Boundary Value Problems*, Fourth Edition, John Wiley & Sons, New York, 1986, pages 141–143.

[3] J. M. Hill, *Solution of Differential Equations by Means of One-Parameter Groups*, Pitman Publishing Co., Marshfield, MA, 1982, pages 42–43.

[4] E. L. Ince, *Ordinary Differential Equations*, Dover Publications, Inc., New York, 1964, page 394.

[5] N. Kamran and P. J. Olver, "Equivalence of Differential Operators," *SIAM J. Math. Anal.*, **20**, No. 5, September 1989, pages 1172–1185.

[6] G. Murphy, *Ordinary Differential Equations*, D. Van Nostrand Company, Inc., New York, 1960, pages 88–89.

[7] H. T. H. Piaggio, *An Elementary Treatise on Differential Equations and Their Applications*, G. Bell & Sons, Ltd, London, 1926, pages 91–92.

[8] E. D. Rainville, *Intermediate Differential Equations*, The MacMillan Company, New York, 1964, pages 7–10 and 15–23.

33.   Transformations of Second Order Linear ODEs – 2

Applicable to   The second order linear ordinary differential equation in formally self-adjoint form

$$L[y] := \frac{d}{dx}\left(p(x)\frac{dy}{dx}\right) + q(x)y = 0. \tag{33.1}$$

Transformation 1

If the dependent variable in (33.1) is changed from x to s by $s = \int \frac{dx}{p(x)}$, and if $p(x) > 0$ for $x > x_0$, and $\int_{x_0}^{\infty} \frac{dx}{p(x)} = \infty$, then (33.1) becomes

$$\frac{d^2 y}{ds^2} + p(x)q(x)y = 0.$$

Note that, as $x \to \infty$, we have $s \to \infty$.

Transformation 2

If the dependent variable in (33.1) is changed from $y(x)$ to $w(x)$ by

$$w(x) = \sqrt{p(x)}y(x),$$

then (33.1) becomes

$$\frac{d^2 w}{dx^2} + \left[\frac{q}{p} - \frac{1}{2}\frac{d}{dx}\left(\frac{p'}{p}\right) - \frac{1}{4}\left(\frac{p'}{p}\right)^2 \right] w = 0.$$

Transformation 3

If the independent and dependent variables are changed in (33.1) by

$$y(x) = \mu(x)w(t),$$
$$t = \int^x \eta(z)\, dz,$$

then (33.1) becomes

$$\frac{\eta}{\mu}\frac{d}{dt}\left(p\mu^2\eta\frac{dw}{dt} \right) + L[\mu]w = 0. \tag{33.2}$$

Note that the operator $L[\cdot]$ is defined by (33.1). If $\eta(z)$ is chosen to be

$$\eta(z) = \frac{1}{p(z)\mu^2(z)},$$

then (33.2) simplifies to $\dfrac{1}{p\mu^3}\dfrac{d^2 w}{dt^2} + L[\mu]w = 0$.

References

[1] R. Courant and D. Hilbert, *Methods of Mathematical Physics*, Interscience
     Publishers, New York, 1953, page 292.

34.   Transformation of an ODE
to an Integral Equation

Applicable to   Second order linear ordinary differential equations.

Yields

An equivalent integral equation.

Idea

An ordinary differential equation may sometimes be formulated as an
integral equation.

Procedure

There is a standard transformation that will allow a linear second
order initial value ordinary differential equation to be written as a Volterra
integral equation. Given the differential equation with initial conditions
for $y(x)$,

$$\frac{d^2 y}{dx^2} + A(x)\frac{dy}{dx} + B(x)y = g(x),$$

$$y(a) = \alpha, \quad y'(a) = \beta,$$

an equivalent Volterra integral equation is

$$y(x) = f(x) + \int_a^x K(x,\zeta)y(\zeta)\,d\zeta,$$

where

$$f(x) = \int_a^x (x - \zeta)g(\zeta)\,d\zeta + (x - a)\Big(A(a)\alpha + \beta\Big) + \alpha,$$

$$K(x,\zeta) = (\zeta - x)\Big(B(\zeta) - A'(\zeta)\Big) - A(\zeta).$$

There is also standard transformation that will allow a linear second
order boundary value ordinary differential equation to be written as a

Fredholm integral equation. Given the differential equation and boundary conditions for $w(x)$,

$$\frac{d^2w}{dx^2} + C(x)\frac{dw}{dx} + D(x)w = j(x),$$

$$w(a) = \gamma, \quad w(b) = \delta,$$

an equivalent Fredholm integral equation is

$$w(x) = h(x) + \int_a^b H(x,\zeta)w(\zeta)\,d\zeta,$$

where

$$h(x) = \gamma + \int_a^x (x-\zeta)j(\zeta)\,d\zeta + \frac{x-a}{b-a}\left[\delta - \gamma - \int_a^b (b-\zeta)j(\zeta)\,d\zeta\right],$$

$$H(x,\zeta) = \begin{cases} \dfrac{x-b}{b-a}\left[C(\zeta) - (a-\zeta)\left(C'(\zeta) - D(\zeta)\right)\right], & \text{for } x > \zeta, \\[2mm] \dfrac{x-a}{b-a}\left[C(\zeta) - (b-\zeta)\left(C'(\zeta) - D(\zeta)\right)\right], & \text{for } x < \zeta. \end{cases}$$

Example

If $y(x)$ satisfies

$$\begin{aligned} y'' + y &= x, \\ y(0) = 0, \quad y'(0) &= 0, \end{aligned} \tag{34.1}$$

then $y(x)$ satisfies the following Volterra integral equation

$$y(x) = \frac{x^3}{6} + \int_0^x (\zeta - x)y(\zeta)\,d\zeta. \tag{34.2}$$

The solution to (34.1), $y = x - \sin x$, satisfies (34.2).

Notes

[1] There are many other ways in which an ordinary differential equation may be transformed into an integral equation. For example, if $y(x)$ satisfies the nth order ordinary differential equation

$$y^{(n)}(x) = f(x) + \sum_{j=1}^{n} C_j(x) y^{(j-1)}(x)$$

and $u(x) := y^{(n)}(x)$, then $u(x)$ satisfies the integral equation

$$u(x) = F(x) + \int_a^x K(x,t) u(t)\, dt,$$

$$K(x,t) = \sum_{j=1}^{n} C_j(x) \frac{(t-x)^{j-1}}{(j-1)!},$$

where $F(x)$ is $f(x)$ plus a polynomial in $(x-a)$ generated by the initial conditions. See Squire [3] for more details on this technique, as well as two other techniques.

[2] Bose's [1] shows that every solution of the n-th order linear homogeneous differential equation

$$y^{(n)} = a_{n-1}(x) y^{(n-1)} + \cdots + a_0(x) y$$

satisfies the integral equation

$$y(x) = y(x_0) + \int_{x_0}^x h(u)\, du + \int_{x_0}^x \left\{ \int_{x_0}^u G(u,v) a_0(v) y(v)\, dv \right\} du,$$

where $h(x)$ is the unique solution to

$$h^{(n-1)} = a_{n-1}(x) h^{(n-2)} + \ldots + a_1(x) h,$$

$$h(x_0) = y'(x_0), \quad h'(x_0) = y''(x_0), \cdots, h^{(n-2)}(x_0) = y^{(n-1)}(x_0),$$

$$(34.3)$$

and $G(x,u)$ is the Green's function associated with (34.3).

References

[1] A. K. Bose, "An Integral Equation Associated with Linear Homogeneous Differential Equations," *Int. J. Math. & Math. Sci.*, **9**, No. 2, 1986, pages 405–408.

[2] A. J. Jerri, *Introduction to Integral Equations with Applications*, Marcel Dekker, New York, 1985, pages 60–67.

[3] W. Squire, *Integration for Engineers and Scientists*, American Elsevier Publishing Company, New York, 1970, pages 223–227.

35. Miscellaneous ODE Transformations

Transformation 1

If $y(x)$ is defined by the ordinary differential equation

$$\frac{d^2y}{dx^2} = f(x)y, \tag{35.1}$$

and the dependent variable is changed by

$$w(\zeta) = \sqrt{\zeta'(x)}\, y(x), \tag{35.2}$$

(for arbitrary $\zeta = \zeta(x)$, or $x = x(\zeta)$), then (35.1) becomes

$$\frac{d^2w}{d\zeta^2} = \left[\dot{x}^2 f(x) + \sqrt{\dot{x}}\frac{d^2}{d\zeta^2}\left(\dot{x}^{-1/2}\right)\right] w,$$
$$= \left[\dot{x}^2 f(x) - \tfrac{1}{2}\{x,\zeta\}\right] w, \tag{35.3}$$

where dots denote differentiation with respect to ζ, and $\{x,\zeta\}$ is the Schwarzian derivative of x with respect to ζ. If we choose $\zeta(x)$ by

$$\zeta(x) = \int^x \sqrt{f(z)}\, dz, \tag{35.4}$$

so that $w(\zeta) = y(x)f^{1/4}(x)$, then (35.3) becomes

$$\frac{d^2w}{d\zeta^2} = [1 + \phi(\zeta)]w, \tag{35.5}$$

with
$$\phi(\zeta) = \frac{4ff'' - 5(f')^2}{16f^3} = -\frac{1}{f^{3/4}}\frac{d^2}{dx^2}\left(\frac{1}{f^{1/4}}\right).$$

This is called the Liouville transformation by Olver [7], and the Liouville–Green transformation by Lakin and Sanchez [6]. By neglecting $\phi(\zeta)$ in equation (35.5), and solving for $w(\zeta)$, we obtain the first term in the WKB approximation (see page 558).

Example
If we apply this transformation to Airy's equation
$$y'' = xy,$$
for $x > 0$, then we find (using $f(x) = x$)
$$\zeta(x) = \int^x \sqrt{z}\,dz = \frac{2}{3}x^{3/2},$$
$$w(\zeta) = \sqrt{\zeta'(x)}y(x) = x^{-1/4}y(x).$$
And so (35.5) becomes
$$\frac{d^2w}{d\zeta^2} - \left(1 + \frac{5}{36\zeta^2}\right)w = 0.$$

This leads to the approximation $w'' - w = 0$ when $\zeta \gg 1$ (which corresponds to $x \gg 1$).

Transformation 2
This transformation removes the $(n-1)$-th derivative term in an n-th order ordinary differential equation. If $y(x)$ satisfies
$$(-1)^n (py^{(n)})^{(n)} + L[y] = \lambda qy, \tag{35.6}$$
for $0 \le x \le 1$, where $L[y]$ is a linear differential operator of degree less than or equal to $2n - 2$, and if the dependent and independent variables are changed from $y(x)$ to $w(t)$ by
$$w(t) = (q^{2n-1}p)^{1/4n}y(x),$$
$$t = \frac{1}{K}\int_0^x \left(\frac{q}{p}\right)^{1/2n}dx,$$
$$K = \int_0^1 \left(\frac{q}{p}\right)^{1/2n}dx,$$
then (35.6) is transformed into
$$\frac{d^{2n}w}{dt^{2n}} + H[w] = K^{2n}\lambda w,$$
where $H[w]$ is another linear differential operator of degree less than or equal to $2n - 2$. See Boyce [1].

Transformation 3

The general third order linear homogeneous ordinary differential equation

$$y''' + p_1(x)y'' + p_2(x)y' + p_3(x)y = 0,$$

can be changed to the canonical form

$$w''' + 2Aw' + (A' + b)w = 0, \qquad (35.7)$$

by the change of variables

$$w(x) = y(x) \exp\left(-\int_{x_0}^x p_1(t)\, dt\right).$$

If we write

$$P_2 = p_2 - p_1^2 - p_1',$$
$$P_3 = p_3 - 3p_1 p_2 + 2p_1^3 - p_1'',$$

then $A(x)$ and $b(x)$ may be written as

$$A(x) = \tfrac{3}{2}P_2,$$
$$b(x) = P_3 - \tfrac{3}{2}P_2'.$$

See Greguš [4] for details.

Transformation 4

The general fourth order linear homogeneous ordinary differential equation

$$A(x)y'''' + B(x)y''' + C(x)y'' + D(x)y' + E(x)y = 0,$$

for $y(x)$ can be changed to the canonical form

$$w'''' + a(t)w'' + b(t)w' + c(t)w = 0,$$

for $w(t)$, by the transformation

$$w(t) = \alpha(x)y(x), \qquad t = \beta(x),$$

where $\{\alpha(x), \beta(x)\}$ are chosen to satisfy

$$\alpha\beta'^3 = \exp\left[-\frac{1}{2}\int_{x_0}^x \frac{B(z)}{A(z)}\, dz\right].$$

Notes

[1] If the transformation given by (35.2) is applied to the equation

$$\frac{d^2 y}{dx^2} = [f(x) + g(x)]y,$$

with ζ defined by (35.4), then we obtain

$$\frac{d^2 w}{d\zeta^2} = \left(1 + \phi + \frac{g}{f}\right) w.$$

[2] The differential equation adjoint to (35.7) has the form: $z''' + 2Az' + (A' - b)z = 0$. Hence, the equation in (35.7) will be self-adjoint if and only if $b(x) = 0$.

[3] Olver [8] proves that any one-dimensional, first order Hamiltonian differential operator can be put into constant coefficient form by a suitable change of variables.

References

[1] W. E. Boyce, "Random Eigenvalue Problems," in A. T. Bharucha-Reid (ed.), *Probabilistic Methods in Applied Mathematics*, Academic Press, New York, 1968, pages 1–73.

[2] A. González-López, "On the Linearization of Second-Order Ordinary Differential Equations," *Lett. Math. Phys.*, **17**, No. 4, 1989, pages 341–349.

[3] C. Grissom, G. Thompson, and G. Wilkens, "Linearization of Second Order Ordinary Differential Equations via Cartan's Equivalence Method," *J. Differential Equations*, **77**, No. 1, 1989, pages 1–15.

[4] M. Greguš, *Third Order Linear Differential Equations*, D. Reidel Publishing Co., Boston, 1987, pages 1–2.

[5] J. M. Hill, *Solution of Differential Equations by Means of One-Parameter Groups*, Pitman Publishing Co., Marshfield, MA, 1982, pages 44–45.

[6] W. D. Lakin and D. A. Sanchez, *Topics in Ordinary Differential Equations*, Dover Publications, Inc., New York, 1970, pages 36–41.

[7] F. W. J. Olver, *Asymptotics and Special Functions*, Academic Press, New York, 1974, pages 190–192.

[8] P. J. Olver, "Darboux' Theorem for Hamiltonian Differential Operators," *J. Differential Equations*, **71**, 1988, pages 10–33.

36.   Reduction of PDEs to a First Order System

Applicable to   Nonlinear partial differential equations.

Yields

A first order system of partial differential equations.

Idea

By introducing variables to represent the derivatives in a partial differential equation, a first order system may be obtained.

Procedure

Sometimes it is advantageous to reduce a partial differential equation of high order for a single unknown function to a system of several first order equations. This might be done, for instance, to utilize a specific numerical package that requires a partial differential equation to be input as a first order system. This can always be done by introducing an appropriate set of derivatives as unknowns.

The general procedure is to introduce new variables as the derivatives of the desired function, and then "discover" relations among these functions. The derivation for the following second order equation may be found in Garabedian [1].

Suppose we have the second order partial differential equation, with boundary conditions

$$\begin{aligned}
u_{xx} &= G\left(x, y, u, u_x, u_y, u_{xy}, u_{yy}\right), \\
u(0, y) &= f(y), \\
u_x(0, y) &= g(y),
\end{aligned} \tag{36.1}$$

for the unknown $u(x, y)$. We introduce new variables, $\{u_1, \ldots, u_8\}$, which are assumed to depend upon the new independent variables ζ and η, by the definitions

$$\begin{array}{lll}
u_1 = x, & u_4 = u_x, & u_7 = u_{xy}, \\
u_2 = y, & u_5 = u_y, & u_8 = u_{yy}, \\
u_3 = u, & u_6 = u_{xx}.
\end{array}$$

If we specify the new independent variables by requiring

$$\frac{\partial u_1}{\partial \zeta} = \frac{\partial u_2}{\partial \eta}, \qquad \frac{\partial u_2}{\partial \zeta} = 0,$$
$$u_1(0, \eta) = 0, \qquad u_2(0, \eta) = \eta,$$

then $u_1 = x = \zeta$ and $u_2 = y = \eta$ The purpose of introducing these new independent variables is to eliminate explicit dependence on x and y.

With these new variables, the equation in (36.1) can be written as the system

$$\frac{\partial u_1}{\partial \zeta} = \frac{\partial u_2}{\partial \eta}, \qquad \frac{\partial u_2}{\partial \zeta} = 0, \qquad \frac{\partial u_3}{\partial \zeta} = u_4 \frac{\partial u_2}{\partial \eta},$$

$$\frac{\partial u_4}{\partial \zeta} = u_6 \frac{\partial u_2}{\partial \eta}, \qquad \frac{\partial u_5}{\partial \zeta} = \frac{\partial u_4}{\partial \eta},$$

$$\frac{\partial u_6}{\partial \zeta} = G_x \frac{\partial u_2}{\partial \eta} + u_4 G_u \frac{\partial u_2}{\partial \eta} + u_6 G_{u_x} \frac{\partial u_2}{\partial \eta} + G_{u_y} \frac{\partial u_4}{\partial \eta} + G_{u_{xy}} \frac{\partial u_6}{\partial \eta} + G_{u_{yy}} \frac{\partial u_7}{\partial \eta},$$

$$\frac{\partial u_7}{\partial \zeta} = \frac{\partial u_6}{\partial \eta}, \qquad \frac{\partial u_8}{\partial \zeta} = \frac{\partial u_7}{\partial \eta}.$$

$$(36.2)$$

Most of the above equations are consistency requirements; i.e., $(u_x)_y = (u_y)_x$ implies that $(u_5)_\zeta = (u_4)_\eta$. The initial conditions for the variables $\{u_1, \ldots, u_8\}$ are given by

$$
\begin{aligned}
u_1(0, \eta) &= 0, \\
u_2(0, \eta) &= \eta, \\
u_3(0, \eta) &= f(\eta), \\
u_4(0, \eta) &= g(\eta), \\
u_5(0, \eta) &= f'(\eta), \\
u_6(0, \eta) &= G(0, \eta, f(\eta), g(\eta), f'(\eta), g'(\eta), f''(\eta)), \\
u_7(0, \eta) &= g'(\eta), \\
u_8(0, \eta) &= f''(\eta).
\end{aligned}
$$

$$(36.3)$$

Note that (36.2) is in the general form of a linear first order system

$$\frac{\partial u_j}{\partial \zeta} = \sum_{k=1}^{8} a_{jk}(u_1, \ldots, u_8) \frac{\partial u_k}{\partial \eta},$$

for $j = 1, 2, \ldots, 8$.

To convert the system in (36.2) back to the system in (36.1) may require the use of the boundary conditions in (36.3).

Notes

[1] Systems of high order partial differential equations can also be made into first order systems, by the introduction of enough terms. For instance, the system of equations for $u(x, y)$ and $v(x, y)$

$$F_1(x, y, u, u_x, u_y, v, v_x, v_y) = 0$$
$$F_2(x, y, u, u_x, u_y, v, v_x, v_y) = 0$$

can be written as a first order system, but the resulting system has 12 dependent variables. See Garabedian [1] for details.

References

[1] P. R. Garabedian, *Partial Differential Equations*, Wiley, New York, 1964, pages 7–11.

37. Transforming Partial Differential Equations

Applicable to   Partial differential equations.

Idea

Changing variables in a partial differential equation is a straightforward process.

Procedure

The general procedure is simple: Construct a new function, which depends upon new variables, and then differentiate with respect to the old variables to see how the derivatives transform.

If a differential equation can be written in terms of coordinate-free expressions (for example, in terms of the gradient operator), then a change of variables can be avoided by simply using the metric of the new coordinate system. At the end of this section are representations of coordinate-free expressions for an orthogonal coordinate system. Moon and Spencer [3] list the metric coefficients for 43 different orthogonal coordinate systems. (These consist of 11 general systems, 21 cylindrical systems, and 11 rotational systems.)

In an orthogonal coordinate system, let $\{\mathbf{a}_i\}$ denote the unit vectors in each of the three coordinate directions, and let $\{u_i\}$ denote distance along each of these axes. The coordinate system may be designated by the *metric coefficients* $\{g_{11}, g_{22}, g_{33}\}$, defined by

$$g_{ii} = \left(\frac{\partial x_1}{\partial u_i}\right)^2 + \left(\frac{\partial x_2}{\partial u_i}\right)^2 + \left(\frac{\partial x_3}{\partial u_i}\right)^2 \tag{37.1}$$

where $\{x_1, x_2, x_3\}$ represent rectangular coordinates. Using the metric coefficients defined in (37.1), we define $g = g_{11}g_{22}g_{33}$.

Operations for orthogonal coordinate systems are sometimes written in terms of $\{h_i\}$ functions, instead of the $\{g_{ii}\}$ terms. Here, $h_i = \sqrt{g_{ii}}$, so that $\sqrt{g} = h_1 h_2 h_3$. For example:

(A) Cylindrical Polar Coordinates

$$x_1 = r \cos \phi, \qquad x_2 = r \sin \phi, \qquad x_3 = z$$

$$h_1 = 1, \qquad h_2 = r, \qquad h_3 = 1 \tag{37.2}$$

(B) Elliptic Cylinder Coordinates

$$x_1 = u_1 u_2, \qquad x_2 = \sqrt{(u_1^2 - c^2)(1 - u_2^2)}, \qquad x_3 = u_3$$

$$h_1 = \sqrt{\frac{u_1^2 - c^2 u_2^2}{u_1^2 - c^2}}, \qquad h_2 = \sqrt{\frac{u_1^2 - c^2 u_2^2}{1 - u_2^2}}, \qquad h_3 = 1$$

Example 1

Suppose we have the equation

$$f_{xx} + f_{yy} + x f_y = 0, \tag{37.3}$$

and we would like to transform the equation from the $\{x, y\}$ variables to the $\{u, v\}$ variables, where

$$u = x, \qquad v = \frac{x}{y}.$$

Note that the inverse transformation is given by $x = u$, $y = u/v$.

We define $g(u, v)$ to be equal to the function $f(x, y)$ when written in the new variables. That is

$$f(x, y) = g(u, v) = g\left(x, \frac{x}{y}\right). \tag{37.4}$$

Now we create the needed derivative terms, carefully applying the chain rule. For example, by differentiating (37.4) with respect to x we obtain

$$f_x(x, y) = g_u \frac{\partial}{\partial x}(u) + g_v \frac{\partial}{\partial x}(v)$$

$$= g_1 \frac{\partial}{\partial x}(x) + g_2 \frac{\partial}{\partial x}\left(\frac{x}{y}\right)$$

$$= g_1 + g_2 \frac{1}{y}$$

$$= g_1 + \frac{v}{u} g_2,$$

where we have used a subscript of "1" ("2") to indicate a derivative with respect to the first (second) argument of the function $g(u,v)$. That is, $g_1(u,v) = g_u(u,v)$. Use of this "slot notation" tends to minimize errors.

In a like manner we find

$$f_y(x,y) = g_u \frac{\partial}{\partial y}(u) + g_v \frac{\partial}{\partial y}(v)$$

$$= g_1 \frac{\partial}{\partial y}(x) + g_2 \frac{\partial}{\partial y}\left(\frac{x}{y}\right)$$

$$= -\frac{x}{y^2} g_2$$

$$= -\frac{v^2}{u} g_2.$$

The second order derivatives can be calculated similarly:

$$f_{xx}(x,y) = \frac{\partial}{\partial x}(f_x(x,y))$$

$$= \frac{\partial}{\partial x}\left(g_1 + \frac{1}{y} g_2\right)$$

$$= g_{11} + \frac{2v}{u} g_{12} + \frac{v^2}{u^2} g_{22},$$

$$f_{xy}(x,y) = \frac{\partial}{\partial x}\left(-\frac{x}{y^2} g_2\right)$$

$$= -\frac{u^2}{v^2} g_2 - \frac{u^3}{v^3} g_{12} - \frac{u^2}{v^2} g_{22},$$

$$f_{yy}(x,y) = \frac{\partial}{\partial y}\left(-\frac{x}{y^2} g_2\right)$$

$$= \frac{2v^3}{u^2} g_2 + \frac{v^4}{u^2} g_{22}.$$

Finally, then, we can determine what equation (37.3) looks like in the new variables:

$$0 = f_{xx} + f_{yy} + x f_y$$

$$= \left(g_{11} + \frac{2v}{u} g_{12} + \frac{v^2}{u^2} g_{22}\right) + \left(\frac{2v^3}{u^2} g_2 + \frac{v^4}{u^2} g_{22}\right) + (u)\left(-\frac{v^2}{u} g_2\right)$$

$$= \frac{v^2(2v - u^2)}{u^2} g_v + g_{uu} + \frac{2v}{u} g_{uv} + \frac{v^2(1 + v^2)}{u^2} g_{vv}.$$

Example 2

As a simple example of using coordinate free representations, consider the diffusion equation in rectilinear coordinates:

$$u_t = \kappa \left(u_{xx} + u_{yy} + u_{zz} \right) = \kappa \nabla^2 u. \tag{37.5}$$

To convert to cylindrical polar coordinates we use (37.2) and (37.1) to transform (37.5) to;

$$u_t = \kappa \nabla^2 u = \kappa \left(\frac{\partial^2 u}{\partial r^2} + \frac{1}{r} \frac{\partial u}{\partial r} + \frac{1}{r^2} \frac{\partial^2 u}{\partial \phi^2} + \frac{\partial^2 u}{\partial z^2} \right)$$

Notes

[1] A MACSYMA program that will perform changes of variables in partial differential equations is described in Sternberg [4].

[2] When ϕ represents a scalar and $\mathbf{E} = E_1 \mathbf{a}_1 + E_2 \mathbf{a}_2 + E_3 \mathbf{a}_3$ represents a vector, we have:

$$\operatorname{grad} \phi = \nabla \phi = \frac{\mathbf{a}_1}{\sqrt{g_{11}}} \frac{\partial \phi}{\partial u_1} + \frac{\mathbf{a}_2}{\sqrt{g_{22}}} \frac{\partial \phi}{\partial u_2} + \frac{\mathbf{a}_3}{\sqrt{g_{33}}} \frac{\partial \phi}{\partial u_3}, \tag{37.6}$$

$$\operatorname{div} \mathbf{E} = \nabla \cdot \mathbf{E}$$
$$= \frac{1}{\sqrt{g}} \left\{ \frac{\partial}{\partial u_1} \left(\frac{g E_1}{g_{11}} \right) + \frac{\partial}{\partial u_2} \left(\frac{g E_2}{g_{22}} \right) + \frac{\partial}{\partial u_3} \left(\frac{g E_3}{g_{33}} \right) \right\}, \tag{37.7}$$

$$\operatorname{curl} \mathbf{E} = \nabla \times \mathbf{E} = \mathbf{a}_1 \frac{\Gamma_1}{\sqrt{g_{11}}} + \mathbf{a}_2 \frac{\Gamma_2}{\sqrt{g_{22}}} + \mathbf{a}_3 \frac{\Gamma_3}{\sqrt{g_{33}}}, \tag{37.8}$$

$$\nabla^2 \phi = \frac{1}{\sqrt{g}} \left\{ \frac{\partial}{\partial u_1} \left[\frac{\sqrt{g}}{g_{11}} \frac{\partial \phi}{\partial u_1} \right] + \frac{\partial}{\partial u_2} \left[\frac{\sqrt{g}}{g_{22}} \frac{\partial \phi}{\partial u_2} \right] + \frac{\partial}{\partial u_3} \left[\frac{\sqrt{g}}{g_{33}} \frac{\partial \phi}{\partial u_3} \right] \right\},$$
$$= \frac{1}{h_1 h_2 h_3} \left\{ \frac{\partial}{\partial u_1} \left[\frac{h_2 h_3}{h_1} \frac{\partial \phi}{\partial u_1} \right] + \frac{\partial}{\partial u_2} \left[\frac{h_3 h_1}{h_2} \frac{\partial \phi}{\partial u_2} \right] + \frac{\partial}{\partial u_3} \left[\frac{h_1 h_2}{h_3} \frac{\partial \phi}{\partial u_3} \right] \right\}, \tag{37.9}$$

$$\operatorname{grad} \operatorname{div} \mathbf{E} = \nabla (\nabla \cdot \mathbf{E}) = \frac{\mathbf{a}_1}{\sqrt{g_{11}}} \frac{\partial \Upsilon}{\partial x_1} + \frac{\mathbf{a}_2}{\sqrt{g_{22}}} \frac{\partial \Upsilon}{\partial x_2} + \frac{\mathbf{a}_3}{\sqrt{g_{33}}} \frac{\partial \Upsilon}{\partial x_3}, \tag{37.10}$$

$$\text{curl}\,\text{curl}\,\mathbf{E} = \nabla \times (\nabla \times \mathbf{E})$$

$$= \mathbf{a}_1 \sqrt{\frac{g_{11}}{g}} \left[\frac{\partial \Gamma_3}{\partial x_2} - \frac{\partial \Gamma_2}{\partial x_3} \right] + \mathbf{a}_2 \sqrt{\frac{g_{22}}{g}} \left[\frac{\partial \Gamma_1}{\partial x_3} - \frac{\partial \Gamma_3}{\partial x_1} \right] \qquad (37.11)$$

$$+ \mathbf{a}_3 \sqrt{\frac{g_{33}}{g}} \left[\frac{\partial \Gamma_2}{\partial x_1} - \frac{\partial \Gamma_1}{\partial x_2} \right],$$

$$\maltese \mathbf{E} = \text{grad}\,\text{div}\,\mathbf{E} - \text{curl}\,\text{curl}\,\mathbf{E}$$

$$= \nabla(\nabla \cdot \mathbf{E}) - \nabla \times (\nabla \times \mathbf{E})$$

$$= \mathbf{a}_1 \left\{ \frac{1}{\sqrt{g_{11}}} \frac{\partial \Upsilon}{\partial x_1} + \sqrt{\frac{g_{11}}{g}} \left[\frac{\partial \Gamma_2}{\partial x_3} - \frac{\partial \Gamma_3}{\partial x_2} \right] \right\}$$

$$+ \mathbf{a}_2 \left\{ \frac{1}{\sqrt{g_{22}}} \frac{\partial \Upsilon}{\partial x_2} + \sqrt{\frac{g_{22}}{g}} \left[\frac{\partial \Gamma_3}{\partial x_1} - \frac{\partial \Gamma_1}{\partial x_3} \right] \right\} \qquad (37.12)$$

$$+ \mathbf{a}_3 \left\{ \frac{1}{\sqrt{g_{33}}} \frac{\partial \Upsilon}{\partial x_3} + \sqrt{\frac{g_{33}}{g}} \left[\frac{\partial \Gamma_1}{\partial x_2} - \frac{\partial \Gamma_2}{\partial x_1} \right] \right\},$$

where Υ and $\mathbf{\Gamma} = (\Gamma_1, \Gamma_2, \Gamma_3)$ are defined by

$$\Upsilon = \frac{1}{\sqrt{g}} \left\{ \frac{\partial}{\partial x_1} \left[E_1 \sqrt{\frac{g}{g_{11}}} \right] + \frac{\partial}{\partial x_2} \left[E_2 \sqrt{\frac{g}{g_{22}}} \right] + \frac{\partial}{\partial x_3} \left[E_3 \sqrt{\frac{g}{g_{33}}} \right] \right\}, \qquad (37.13)$$

$$\Gamma_1 = \frac{g_{11}}{\sqrt{g}} \left\{ \frac{\partial}{\partial x_2} \left(\sqrt{g_{33}} E_3 \right) - \frac{\partial}{\partial x_3} \left(\sqrt{g_{22}} E_2 \right) \right\},$$

$$\Gamma_2 = \frac{g_{22}}{\sqrt{g}} \left\{ \frac{\partial}{\partial x_3} \left(\sqrt{g_{11}} E_1 \right) - \frac{\partial}{\partial x_1} \left(\sqrt{g_{33}} E_3 \right) \right\}, \qquad (37.14)$$

$$\Gamma_3 = \frac{g_{22}}{\sqrt{g}} \left\{ \frac{\partial}{\partial x_1} \left(\sqrt{g_{22}} E_2 \right) - \frac{\partial}{\partial x_2} \left(\sqrt{g_{11}} E_1 \right) \right\}.$$

References

[1] E. Butkov, *Mathematical Physics*, Addison–Wesley Publishing Co., Reading, MA, 1968, pages 34–39.

[2] D. Harper, "Vector 33: A REDUCE Program for Vector Algebra and Calculus in Orthogonal Curvilinear Coordinates," *Comput. Physics Comm.*, **54**, 1989, pages 295–305.

[3] P. Moon and D. E. Spencer, *Field Theory For Engineers*, D. Van Nostrand Company, Inc., New York, 1961, Chapter 3.

[4] P. Moon and D. E. Spencer, *Field Theory Handbook*, Springer–Verlag, New York, 1961.

[5] S. Sternberg, "Change of Variables in Partial Differential Equations," Department of Mathematics, University of New Mexico, preprint (or DTIC document number AD-A214 702).

38.   Transformations of Partial Differential Equations

Euler Transformation

Given the first order partial differential equation in two independent variables, $F(x, y, z, p, q) = 0$ (with, as usual, $p = z_x$, $q = z_y$), and $z_{xx} \neq 0$ the transformation

$$
\begin{cases}
x = Z_X \\
y = Y \\
z = X Z_X - Z \\
p = X \\
q = -Z_Y
\end{cases}
\iff
\begin{cases}
X = z_x \\
Y = y \\
Z = x z_x - z \\
P = x \\
Q = -z_y
\end{cases}
\tag{38.1}
$$

is known as the Euler transformation. Note that $Z_Y + z_y = 0$. Under this transformation, the original equation transforms into $F(Z_X, Y, X Z_X - Z, X, -Z_Y) = 0$.

As an example, the equation $G(xp - z, y, p, q) = 0$ becomes, under the Euler transformation, $G(Z, Y, X, -Z_Y) = 0$. As another example, the Clairaut partial differential equation $F = z - (x z_x + y z_y + f(z_x, z_y)) = 0$ is transformed into $F = Z - Y Z_Y + f(X, -Z_Y) = 0$. Note that this latter equation is really an ordinary differential equation for $Z = Z(Y)$ (the variable X acts as a parameter).

Kirchoff Transformation

Given the elliptic partial differential equation

$$
\text{div}[K(\psi)\, \text{grad}\, \psi] = \nabla \cdot [K(\psi)\nabla \psi] = 0,
\tag{38.2}
$$

for $\psi = \psi(\mathbf{x})$, the Kirchoff transformation introduces the new dependent variable, $\Phi(\mathbf{x})$, defined by $\Phi = \int_{\psi_0}^{\psi} K(t)\, dt$, where ψ_0 is some arbitrary reference value. This transforms (38.2) into Laplace's equation $\nabla^2 \Phi = 0$. See Ames [1] for details.

Transformations of Parabolic Differential Equations I
The parabolic partial differential equation

$$u_t = \alpha^2 u_{xx} - \delta u_x + \varepsilon u,$$

where $\{\alpha, \delta, \varepsilon\}$ are constants, may be transformed into the simple diffusion equation

$$\phi_t = \alpha^2 \phi_{xx},$$

by means of the transformation

$$u(x,t) = \phi(x,t) \exp\left[\frac{\delta}{2\alpha^2}x + \left(\varepsilon - \frac{\delta^2}{4\alpha^2}\right)t\right]. \tag{38.3}$$

Transformations of Parabolic Differential Equations II
The nonlinear parabolic partial differential equation

$$c_t = (D(c)c_x)_x$$

may be transformed into the following equation with a simpler nonlinearity:

$$D(c)v_t = v^2 v_{cc}.$$

The transformation is given by $v(c,t) = D(c)c_x$. See Hill [5].

Removing First Derivative Terms
Linear elliptic equations and hyperbolic equations of second order, all of whose coefficients of the derivative terms are constants, can be transformed so that the first derivative terms no longer appear. For example, we presume that $u(\mathbf{x})$ satisfies

$$\sum_{k=1}^{n} \lambda_k \frac{\partial^2 u}{\partial^2 x_k} + \sum_{k=1}^{n} b_k \frac{\partial u}{\partial x_k} + c(\mathbf{x})u = 0. \tag{38.4}$$

Note that scaling of the $\{x_k\}$ allows (38.4) to be written with each $\{\lambda_k\|$ equal to 0, 1, or -1. If we presume that no λ_k is equal to zero, and we define

$$w(\mathbf{x}) = u(\mathbf{x}) \exp\left[\frac{1}{2}\sum_{k=1}^{n}\left(\frac{b_k}{\lambda_k}\right)x_k\right],$$

then $w(\mathbf{x})$ satisfies

$$\sum_{k=1}^{n} \lambda_k \frac{\partial^2 w}{\partial^2 x_k} + \left(c(\mathbf{x}) - \frac{1}{4}\sum_{i=1}^{n}\frac{b_k^2}{\lambda}\right)w = 0,$$

Von Mises Transformation

For fluid flow with constant viscosity, the Navier–Stokes equations (see page 148) sometimes take the form

$$u\frac{\partial u}{\partial x} + v\frac{\partial v}{\partial y} = \nu\frac{\partial^2 u}{\partial y^2},$$

$$\frac{\partial u}{\partial x} + \frac{\partial v}{\partial y} = 0,$$

$$(38.5.a\text{--}b)$$

these are called the *boundary layer equations*. A standard procedure for analyzing the Navier–Stokes equations (and equations derived from them) is to introduce the stream function Ψ, defined by

$$u = \frac{\partial \Psi}{\partial y}, \qquad v = -\frac{\partial \Psi}{\partial x}.$$

With this definition, (38.5.b) is automatically satisfied. In the Von Mises transformation, Ψ and x are treated as the independent variables, instead of y and x. This transforms (38.5.a) into

$$\frac{\partial u}{\partial x} = \frac{\partial}{\partial \Psi}\left[\nu u \frac{\partial u}{\partial \Psi}\right]$$

For applications, see Schlichting [8].

Notes

[1] If the boundary data of are the Neuman type, then the Kirchoff transformation may introduce nonlinearities in the boundary data for the Φ problem.

[2] The Kirchoff transformation is frequently useful in free boundary problems, where $K(\psi)$ changes value across the (unknown) boundary.

References

[1] W. F. Ames, *Nonlinear Partial Differential Equations*, Academic Press, New York, 1967, pages 6–7 and 21–23.

[2] H. Bateman, *Partial Differential Equations of Mathematical Physics*, Dover Publications, Inc., New York, 1944, pages 75–79.

[3] S. J. Farlow, *Partial Differential Equations for Scientists and Engineers*, John Wiley & Sons, New York, 1982, page 58.

[4] P. R. Garabedian, *Partial Differential Equations*, Wiley, New York, 1964, pages 74–75.

[5] J. M. Hill, *Solution of Differential Equations by Means of One-Parameter Groups*, Pitman Publishing Co., Marshfield, MA, 1982, page 148.

[6] E. Kamke, *Differentialgleichungen Lösungsmethoden und Lösungen*, Volume II, Chelsea Publishing Company, New York, 1947, pages 100–101.

[7] L. Rosenhead, *Laminar Boundary Layers*, Clarendon Press, Oxford, 1963.

[8] H. Schlichting, *Boundary Layer Theory*, McGraw–Hill Book Company, New York, 1955.

II

Exact Analytical Methods

39. Introduction to Exact Analytical Methods

The methods in this section of the book are for the exact solution of differential equations. The methods have been separated into two parts:

[1] Methods which can be used for ordinary differential equations and, sometimes, partial differential equations. When a method in this part can be used for a partial differential equation, there is a star (∗) alongside the method name.

[2] Methods which can only be used for partial differential equations.

Since many of the common methods for partial differential equations are also useful as methods for ordinary differential equations, the first part of this section should not be overlooked when attempting to find the solution of a partial differential equation.

Listed below are, in the author's opinion, those methods that are the most useful when solving ordinary differential equations and partial differential equations. These are the methods that might be tried first.

Most Useful Methods for ODEs
- Look Up Technique
- Look Up ODE Forms
- Computer-Aided Solution
- Constant Coefficient Linear Equations
- Eigenfunction Expansions*
- Green's Functions*
- Integral Transforms: Infinite Intervals*
- Integrating Factors*
- Series Solution*
- Method of Undetermined Coefficients*

Most Useful Methods for PDEs
- Look Up Technique
- Eigenfunction Expansions*
- Green's Functions*
- Integral Transforms: Infinite Intervals*
- Method of Characteristics
- Conformal Mappings
- Lie Groups: PDEs
- Separation of Variables
- Similarity Methods

40.   Look Up Technique

Applicable to   Equations of certain forms.

Yields

A reference to the literature, which may yield an analytical solution, an approximate analytical solution, or an approximate numerical solution.

Idea

Many functions of mathematical physics have been well studied. If a differential equation can be transformed to a known form, then information about the solution may be obtained by looking in the right reference.

Procedure

Compare the differential equation that you are trying to analyze with the lists on the following pages. If the equation you are investigating appears, see the references cited for that equation.

The equations listed in this section include

[1] Ordinary differential equations
  (A) First order equations
  (B) Second order equations
  (C) Higher order equations
[2] Partial differential equations
  (A) Linear equations
  (B) Second order nonlinearity
  (C) Higher order and variable order nonlinearities
[3] Systems of differential equations
  (A) Systems of ordinary differential equations
  (B) Systems of partial differential equations

Notes

[1] Realize that the same equation may look different when written in different variables. Some scaling of any given equation may be required to make it look like one of the forms listed.

[2] Carslaw and Jaeger [27] have a large collection of exact analytical solutions for parabolic partial differential equations.

[3] In Murphy [86] and in Kamke's two books ([64] and [65]) are long listings of ordinary differential equations and partial differential equations and their exact solutions.

[4] The references follow the listings of differential equations.

Ordinary Differential Equations

First Order Equations

Abel equation of the first kind (see Murphy [106], page 23)

$$y' = f_0(x) + f_1(x)y + f_2(x)y^2 + f_3(x)y^3.$$

Abel equation of the second kind (see Murphy [106], page 25)

$$[g_0(x) + g_1(x)y]\, y' = f_0(x) + f_1(x)y + f_2(x)y^2 + f_3(x)y^3.$$

Bernoulli equation (see page 194)

$$y' = a(x)y^n + b(x)y.$$

Binomial equation (see Hille [69], page 675)
$$(y')^m = f(x,y).$$

Briot and Bouquet's equation (see Ince [74], page 295)
$$xy' - \lambda y = a_{10}x + a_{20}x^2 + a_{11}yx + a_{02}y^2 + \ldots.$$

Clairaut's equation (see page 196)
$$f(xy' - y) = g(y').$$

Elliptic functions (see Gradshteyn and Ryzhik [60], page 917)
$$y' = \sqrt{(1 - y^2)(1 - k^2 y^2)}.$$

Euler equation (see Valiron [138], page 201)
$$y' = \pm \sqrt{\frac{ay^4 + by^3 + cy^2 + dy + e}{ax^4 + bx^3 + cx^2 + dx + e}}.$$

Euler equation (see Valiron [138], page 212)
$$y' + y^2 = \alpha x^m.$$

Heisenberg equation of motion (see Iyanaga and Kawada [76], page 1083)
$$\frac{dA(t)}{dt} = \frac{i}{\hbar} [H, A(t)].$$

Jacobi equation (see Ince [74], page 22)
$$(a_1 + b_1 x + c_1 y)(xy' - y) - (a_2 + b_2 x + c_2 y)y' + (a_3 + b_3 x + c_3 y) = 0.$$

Lagrange's equation (see page 311)
$$y = xf(y') + g(y').$$

Löwner's equation (see Iyanaga and Kawada [76], page 1345)
$$y' = -y\frac{1 + \kappa(x)y}{1 - \kappa(x)y}.$$

Riccati equation (see page 332)
$$y' = a(x)y^2 + b(x)y + c(x).$$

Weierstrass function (see Rainville [113], page 312)
$$y' = \sqrt{4y^3 - g_2 y - g_3}.$$

Unnamed equation (see Boyd)
$$y' = -ae^{-b/y}.$$

Unnamed equation (see Goldstein and Braun [58], page 42)
$$g(y)y' = f(x) + h(x)\,G\left(\int f(x)\,dx - \int g(y)\,dy\right).$$

Second Order Equations

Airy equation (see Abramowitz and Stegun [2], Section 10.4.1)

$$y'' = xy.$$

Anger functions (see Gradshteyn and Ryzhik [60], page 989)

$$y'' + \frac{y'}{x} + \left(1 - \frac{\nu^2}{x^2}\right)y = \frac{x - \nu}{\pi x^2}\sin\nu\pi.$$

Baer equation (see Moon and Spencer [103], page 156)

$$(x - a_1)(x - a_2)y'' + \tfrac{1}{2}\left[2x - (a_1 + a_2)\right]y' - \left[p^2 x + q^2\right]y = 0.$$

Baer wave equation (see Moon and Spencer [103], page 157)

$$(x - a_1)(x - a_2)y'' + \tfrac{1}{2}\left[2x - (a_1 + a_2)\right]y' - \left[k^2 x^2 - p^2 x + q^2\right]y = 0.$$

Bessel equation (see Abramowitz and Stegun [2], Section 9.1.1)

$$x^2 y'' + xy' + (x^2 - n^2)y = 0.$$

Bessel equation – modified (see Abramowitz and Stegun [2], Section 9.6.1)

$$x^2 y'' + xy' - (x^2 + n^2)y = 0.$$

Bessel equation – spherical (see Abramowitz and Stegun [2], Section 10.1.1)

$$x^2 y'' + 2xy' + \left[x^2 - n(n + 1)\right]y = 0.$$

Bessel equation – modified spherical (see Abramowitz and Stegun [2], Section 10.2.1)

$$x^2 y'' + 2xy' - \left[x^2 + n(n + 1)\right]y = 0.$$

Bessel equation – wave (see Moon and Spencer [103], page 154)

$$x^2 y'' + xy' + \left[a^2 x^4 + b^2 x^2 - c^2\right]y = 0.$$

Bôcher equation (see Moon and Spencer [103], page 127)

$$y'' + \frac{1}{2}\left[\frac{m_1}{x - a_1} + \ldots + \frac{m_{n-1}}{x - a_{n-1}}\right]y'$$

$$+ \frac{1}{4}\left[\frac{A_0 + A_1 x + \ldots + A_l x^l}{(x - a_1)^{m_1}(x - a_2)^{m_2}\cdots(x - a_{n-1})^{m_{n-1}}}\right]y = 0.$$

Coulomb wave functions (see Abramowitz and Stegun [2], Section 14.1.1)

$$y'' + \left[1 - \frac{2\eta}{x} - \frac{L(L + 1)}{x^2}\right]y = 0.$$

Duffing's equation (see Bender and Orszag [19], page 547)

$$y'' + y + \varepsilon y^3 = 0.$$

Eckart equation (see Barut, Inomata, and Wilson [17])

$$y'' + \left[\frac{\alpha \eta}{1 + \eta} + \frac{\beta \eta}{(1 + \eta)^2} + \gamma \right] y = 0, \qquad \eta = e^{\delta x}.$$

Ellipsoidal wave equation (see Arscott [12])

$$y'' - (a + bk^2 \operatorname{sn}^2 x + qk^4 \operatorname{sn}^4 x) y = 0.$$

Complete elliptic integral (see Gradshteyn and Ryzhik [60], page 907)

$$\frac{d}{dx} \left[x(1 - x^2) \frac{dy}{dx} \right] - xy = 0.$$

Confluent equation – general (see Abramowitz and Stegun [2], Section 13.1.35)

$$y'' + \left[\frac{2a}{x} + 2f' + \frac{bh'}{h} - h' - \frac{h''}{h'} \right] y' + \left[\left(\frac{bh'}{h} - h' - \frac{h''}{h'} \right) \left(\frac{a}{x} + f' \right) \right.$$

$$\left. + \frac{a(a - 1)}{x^2} + \frac{2af'}{x} + f'' + (f')^2 - \frac{a(h')^2}{h} \right] y = 0.$$

Complete elliptic integral (see Gradshteyn and Ryzhik [60], page 907)

$$(1 - x^2) \frac{d}{dx} \left(x \frac{dy}{dx} \right) + xy = 0.$$

Emden equation (see Leach [90])

$$(x^2 y')' + x^2 y^n = 0.$$

Emden equation – modified (see Leach [92])

$$y'' + a(x)y' + y^n = 0.$$

Emden–Fowler equation (see Rosenau [118])

$$(x^p y')' \pm x^\sigma y^n = 0.$$

Integrals of the error function (see Abramowitz and Stegun [2], Section 7.2.2)

$$y'' + 2xy' - 2ny = 0.$$

Gegenbauer functions (see Infeld and Hull [75])

$$(1 - x^2)y'' - (2m + 3)xy' + \lambda y = 0.$$

Halm's equation (see Hille [69], page 357)
$$(1 + x^2)^2 y'' + \lambda y = 0.$$

Heine equation (see Moon and Spencer [103], page 157)
$$y'' + \frac{1}{2}\left[\frac{1}{x - a_1} + \frac{2}{x - a_2} + \frac{2}{x - a_3}\right]y + \frac{1}{4}\left[\frac{A_0 + A_1 x + A_2 x^2 + A_3 x^3}{(x - a_1)(x - a_2)^2(x - a_3)^2}y\right] = 0.$$

Hermite polynomials (see Abramowitz and Stegun [2], Section 22.6.21)
$$y'' - xy' + ny = 0.$$

Heun's equation (see Valent [137])
$$y'' + \left[\frac{\gamma}{x} - \frac{\delta}{1 - x} - \frac{(1 + \alpha + \beta - \gamma - \delta)k^2}{1 - k^2 x}\right]y' + \frac{\alpha\beta k^2 x + s}{x(1 - x)(1 - k^2 x)}y = 0.$$

Hill's equation (see Ince [74], page 384)
$$y'' + (a_0 + 2a_1 \cos 2x + 2a_2 \cos 4x + \dots)y = 0.$$

Hypergeometric equation (see Abramowitz and Stegun [2], Section 15.5.1)
$$x(1 - x)y'' + [c - (a + b + 1)x]y' - aby = 0.$$

Hyperspherical differential equation (see Iyanaga and Kawada [76], page 1185)
$$(1 - x^2)y'' - 2axy' + by = 0.$$

Ince equation (see Athorne [13])
$$y'' + \frac{\alpha + \beta \cos 2t + \gamma \cos 4t}{(1 + a \cos 2t)^2}y = 0.$$

Jacobi's equation (see Iyanaga and Kawada [76], page 1480)
$$x(1 - x)y'' + [\gamma - (\alpha + 1)x]y' + n(\alpha + n)y = 0.$$

Kelvin functions (see Abramowitz and Stegun [2], Section 9.9.3)
$$x^2 y'' + xy' - (ix^2 + \nu^2)y = 0.$$

Kummer's equation (see Abramowitz and Stegun [2], Section 13.1.1)
$$xy'' + (b - x)y' - ay = 0.$$

Lagerstrom equation (see Rosenblat and Shepherd [119])
$$y'' + \frac{k}{x}y' + \varepsilon yy' = 0.$$

Laguerre equation (see Iyanaga and Kawada [76], page 1481)

$$xy'' + (\alpha + 1 - x)y' + \lambda y = 0.$$

Lamé equation (see Moon and Spencer [103], page 157)

$$y'' + \frac{1}{2}\left[\frac{1}{x - a_1} + \frac{1}{x - a_2} + \frac{1}{x - a_3}\right]y' + \frac{1}{4}\left[\frac{A_0 + A_1 x}{(x - a_1)(x - a_2)(x - a_3)}\right]y = 0.$$

Lamé equation (see Ward [143])

$$y'' + (h - n(n + 1)k^2 \operatorname{sn}^2 x)y = 0.$$

Lamé equation – wave (see Moon and Spencer [103], page 157)

$$y'' + \frac{1}{2}\left[\frac{1}{x} + \frac{1}{x - a} + \frac{1}{x - b}\right]y' + \frac{1}{4}\left[\frac{(a^2 + b^2)q - p(p + 1)x + \kappa x^2}{x(x - a)(x - b)}\right]y = 0.$$

Lane–Emden equation (see Seshadri and Na [124], page 193)

$$y'' + \frac{2}{x}y' + y^k = 0.$$

Legendre equation (see Abramowitz and Stegun [2], Section 8.1.1)

$$(1 - x^2)y'' - 2xy' + \left[n(n + 1) - \frac{m^2}{1 - x^2}\right]y = 0.$$

Legendre equation – wave (see Moon and Spencer [103], page 155)

$$(1 - x^2)y'' - 2xy' - \left[k^2 a^2 (x^2 - 1) - p(p + 1) - \frac{q^2}{x^2 - 1}\right]y = 0.$$

Lewis regulator (see Hagedorn [62], page 152)

$$y'' + (1 - |y|)y' + y = 0.$$

Liénard's equation (see Villari [140])

$$y'' + f(x)y' + y = 0.$$

Liouville's equation (see Goldstein and Braun [58], page 98)

$$y'' + g(y)(y')^2 + f(x)y' = 0.$$

Lommel functions (see Gradshteyn and Ryzhik [60], page 986)

$$x^2 y'' + xy' + (x^2 - \nu^2)y = x^{\mu+1}.$$

Magnetic pole equation (see Infeld and Hull [75])

$$y'' - \left[\frac{m(m + 1) + \frac{1}{4} - (m + \frac{1}{2})\cos x}{\sin^2 x} + \left(\lambda + \frac{1}{2}\right)\right]y = 0.$$

Mathieu equation (see Abramowitz and Stegun [2], Section 20.1.1)

$$y'' + (a - 2q \cos 2x)y = 0.$$

Mathieu equation – associated (see Ince [74], page 503)

$$y'' + [(1 - 2r) \cot x] y' + (a + k^2 \cos^2 x)y = 0.$$

Mathieu equation – modified (see Abramowitz and Stegun [2], Section 20.1.2)

$$y'' - (a - 2q \cosh 2x)y = 0.$$

Morse–Rosen equation (see Barut, Inomata, and Wilson [17])

$$y'' + \left[\frac{\alpha}{\cosh^2 ax} + \beta \tanh ax + \gamma \right] y = 0.$$

Neumann's polynomials (see Gradshteyn and Ryzhik [60], page 990)

$$x^2 y'' + 3xy' + (x^2 + 1 - n^2)y = x \cos^2 \frac{n\pi}{2} + n \sin^2 \frac{n\pi}{2}.$$

Painlevé transcendent – first (see Ince [74], page 345)

$$y'' = 6y^2 + x.$$

Painlevé transcendent – second (see Ince [74], page 345)

$$y'' = 2y^3 + xy + \alpha.$$

Painlevé transcendent – third (see Ince [74], page 345)

$$y'' = \frac{1}{y} (y')^2 - \frac{1}{x} y' + \frac{1}{x} (\alpha y^2 + \beta) + \gamma y^3 + \frac{\delta}{y}.$$

Painlevé transcendent – fourth (see Ince [74], page 345)

$$y'' = \frac{1}{2y} (y')^2 + \frac{3y^3}{2} + 4xy^2 + 2(x^2 - \alpha)y + \frac{\beta}{y}.$$

Painlevé transcendent – fifth (see Ince [74], page 345)

$$y'' = \left(\frac{1}{2y} + \frac{1}{y-1} \right) (y')^2 - \frac{1}{x} y' + \frac{(y-1)^2}{x^2} \left(\alpha y + \frac{\beta}{y} \right) + \frac{\gamma y}{x} + \frac{\delta y(y+1)}{y-1}.$$

Painlevé transcendent – sixth (see Ince [74], page 345)

$$y'' = \frac{1}{2} \left(\frac{1}{y} + \frac{1}{y-1} + \frac{1}{y-x} \right) (y')^2 - \left(\frac{1}{x} + \frac{1}{x-1} + \frac{1}{y-x} \right) y'$$
$$+ \frac{y(y-1)(y-x)}{x^2(x-1)^2} \left[\alpha + \frac{\beta x}{y^2} + \frac{\gamma(x-1)}{(y-1)^2} + \frac{\delta x(x-1)}{(y-x)^2} \right].$$

Parabolic cylinder equation (see Abramowitz and Stegun [2], Section 19.1.1)

$$y'' + (ax^2 + bx + c)y = 0.$$

Poisson–Boltzmann equation (see Chambré [36])

$$y'' + \frac{k}{x}y' = -\delta e^y.$$

Pöschl–Teller equation – first (see Barut, Inomata, and Wilson [16])

$$y'' - \left[a^2 \left(\frac{\kappa(\kappa - 1)}{\sin^2 ax} + \frac{\lambda(\lambda - 1)}{\cos^2 ax}\right) - b^2\right] y = 0.$$

Pöschl–Teller equation – second (see Barut, Inomata, Wilson [17])

$$y'' - \left[a^2 \left(\frac{\kappa(\kappa - 1)}{\sinh^2 ax} + \frac{\lambda(\lambda - 1)}{\cosh^2 ax}\right) - b^2\right] y = 0.$$

Polytropic differential equation (see Iyanaga and Kawada [76], page 908)

$$(x^2 y')' = -x^2 y^n.$$

Rayleigh equation (see Birkhoff and Rota [21], page 134)

$$y'' - \mu \left[1 - (y')^2\right] y' + y = 0.$$

Riccati–Bessel equation (see Abramowitz and Stegun [2], Section 10.3.1)

$$x^2 y'' + \left[x^2 - n(n + 1)\right] y = 0.$$

Riemann's differential equation (see Abramowitz and Stegun [2], Section 15.6.1)

$$y'' + \left[\frac{1 - \alpha - \alpha'}{x - a} + \frac{1 - \beta - \beta'}{x - b} + \frac{1 - \gamma - \gamma'}{x - c}\right] y'$$

$$+ \left[\frac{\alpha\alpha'(a - b)(a - c)}{x - a} + \frac{\beta\beta'(b - c)(b - a)}{x - b} + \frac{\gamma\gamma'(c - a)(c - b)}{x - c}\right]$$

$$\times \frac{y}{(x - a)(x - b)(x - c)} = 0.$$

Spheroidal wave functions (oblate) (see Abramowitz and Stegun [2], Section 21.6.4)

$$[(1 - x^2)y']' + \left(\lambda + c^2 x^2 - \frac{m^2}{1 - x^2}\right) y = 0.$$

Spheroidal wave functions radial (see Abramowitz and Stegun [2], Section 21.6.3)

$$[(1 + x^2)y']' - \left(\lambda - c^2 x^2 - \frac{m^2}{x^2 + 1}\right) y = 0.$$

Struve functions (see Abramowitz and Stegun [2], Section 12.1.1)

$$x^2y'' + xy' + (x^2 - \nu^2)y = \frac{4\left(\frac{z}{2}\right)^{\nu+1}}{\sqrt{\pi}\Gamma(\nu + \frac{1}{2})}.$$

Symmetric top equation (see Infeld and Hull [75])

$$y'' - \left[\frac{M^2 - \frac{1}{4} + K^2 - 2MK\cos x}{\sin^2 x} + \left(\sigma + K^2 + \frac{1}{4}\right)\right]y = 0.$$

Tchebycheff equation (see Abramowitz and Stegun [2], Section 22.6.9)

$$(1 - x^2)y'' - xy' + n^2y = 0.$$

Thomas–Fermi equation (see Bender and Orszag [19], page 25)

$$y'' = y^{3/2}x^{-1/2}.$$

Titchmarsh's equation (see Hille [69], page 617)

$$y'' + \left(\lambda - x^{2n}\right)y = 0.$$

Ultraspherical equation (see Abramowitz and Stegun [2], Section 22.6.5)

$$(1 - x^2)y'' - (2\alpha + 1)xy' + n(n + 2\alpha)y = 0.$$

Van der Pol equation (see Birkhoff and Rota [21], page 134)

$$y'' - \mu(1 - y^2)y' + y = 0.$$

Wangerin equation (see Moon and Spencer [103], page 157)

$$y'' + \frac{1}{2}\left[\frac{1}{x - a_1} + \frac{1}{x - a_2} + \frac{2}{x - a_3}\right]y' + \frac{1}{4}\left[\frac{A_0 + A_1x + A_2x^2}{(x - a_1)(x - a_2)(x - a_3)^2}\right]y = 0.$$

Weber functions (see Gradshteyn and Ryzhik [60], page 989)

$$y'' + \frac{y'}{x} + \left(1 - \frac{\nu^2}{x^2}\right)y = -\frac{1}{\pi x^2}\left[x + \nu + (x - \nu)\cos\nu\pi\right].$$

Weber equation (see Moon and Spencer [103], page 153)

$$y'' + \left(a^2 - \frac{b^2}{4}x^2\right)y = 0.$$

Whittaker's equation (see Abramowitz and Stegun [2], equation 13.1.31)

$$y'' + \left(-\frac{1}{4} + \frac{\kappa}{x} + \frac{\frac{1}{4} - \mu^2}{x^2}\right)y = 0.$$

Whittaker–Hill equation (see Urwin and Arscott [135])

$$y'' + (A + B\cos 2x + C\cos 4x)y = 0.$$

Unnamed equation (see Chrisholm and Common [41])

$$y'' + (a_0 + a_1 y)y' + b_0 + b_1 y + b_2 y^2 + b_3 y^3 = 0.$$

Unnamed equation (see Gilding [57])

$$y'' = -\lambda y^p.$$

Unnamed equation (see Latta [89])

$$(1 - x^2)y'' - 2axy' + (b + cx^2)y = 0.$$

Unnamed equation (see Rubel [121])

$$xyy'' + yy' - x(y')^2 = 0.$$

Unnamed equation (see Setoyanagi [125])

$$y'' + (ax^p + bx^q)y = 0.$$

Unnamed equation (see Tsukamoto [134])

$$y'' + e^{at} y^b = 0.$$

Higher Order Equations

Products of Airy functions (see Abramowitz and Stegun [2], equation 10.4.57)

$$y''' - 4xy' - 2y = 0.$$

Blasius equation (see Meyer [98], page 127)

$$y''' + yy'' = 0.$$

Falkner–Skan equation (see Cebeci and Keller [35])

$$y''' + yy'' + \beta \left[1 - (y')^2\right] = 0.$$

Generalized hypergeometric equation (see Miller [101], page 271)

$$\left(x\frac{d}{dx} + a_1\right) \cdots \left(x\frac{d}{dx} + a_p\right) y - \frac{d}{dx}\left(x\frac{d}{dx} + b_1\right) \cdots \left(x\frac{d}{dx} + b_q\right) y = 0.$$

Laplace equations (see Valiron [138], pages 306–315)

$$(a_0 x + b_0)y^{(n)} + (a_1 x + b_1)y^{(n-1)} + \ldots + (a_n x + b_n)y = 0.$$

Sixth order Onsager equation (see Viecelli [139])

$$(e^x (e^x y_x)_{xx})_{xxx} = f(x).$$

Orr–Sommerfeld equation (see Herron [66])

$$\frac{1}{i\alpha R}\left(\frac{d^2}{dx^2}-\alpha^2\right)^2 y - \left[(f(x)-c)\left(\frac{d^2}{dx^2}-\alpha^2\right)-f''(x)\right]y = 0.$$

Unnamed equation (see Hershenov [67])

$$y''' + \alpha xy' + \beta y = 0.$$

Partial Differential Equations

Linear Equations

Biharmonic equation (see Kantorovich and Krylov [81], pages 595–615)

$$\nabla^4 u = 0.$$

Linear Boussinesq equation (see Whitham [146], page 9)

$$u_{tt} - \alpha^2 u_{xx} = \beta^2 u_{xxtt}.$$

Busemann equation (see Chaohao [236])

$$(1-x^2)u_{xx} - 2xyu_{xy} + (1-y^2)u_{yy} + 2a(xu_x + yu_y) - a(a+1)u = 0.$$

Chaplygin's equation (see Landau and Lifshitz [87], page 432)

$$u_{xx} + \frac{y^2}{1-y^2/c^2}u_{yy} + yu_y = 0.$$

Diffusion equation (see Morse and Feshback [105], page 271)

$$\nabla \cdot (\kappa(\mathbf{x},t)\nabla u) = u_t.$$

Euler–Darboux equation (see Miller [100])

$$u_{xy} + \frac{1}{x-y}(\alpha u_x - \beta u_y) = 0.$$

Euler–Poisson–Darboux equation (see Ames [7], Section 3.3)

$$u_{xy} + \frac{N}{x+y}(u_x + u_y) = 0.$$

Helmholtz equation (see Morse and Feshback [105], page 271)

$$\nabla^2 u + k^2 u = 0.$$

Klein–Gordon equation (see Morse and Feshback [105], page 272)

$$\nabla^2 u - \frac{1}{c^2}u_{tt} = \mu^2 u.$$

Kramers equation (see Duck, Marshall, and Watson [48])

$$P_t = P_{xx} - uP_x + \frac{\partial}{\partial u}\left[(u - F(x))P\right].$$

Lambropoulos' equation (see Wilcox [147])

$$u_{xy} + axu_x + byu_y + cxyu + u_t = 0.$$

Laplace's equation (see Morse and Feshback [105], page 271)

$$\nabla^2 u = 0.$$

Lavrent'ev–Bitsadze equation (see Chang [38])

$$u_{xx} + (\operatorname{sgn} y)u_{yy} = f(x, y).$$

Onsager equation (see Wood and Morton [148])

$$\left(e^x \left(e^x u_{xx}\right)_{xx}\right)_{xx} + B^2 u_{yy} = F(x, y).$$

Poisson equation (see Morse and Feshback [105], page 271)

$$\nabla^2 u = -4\pi\rho(\mathbf{x}).$$

Schröedinger equation (see Morse and Feshback [105], page 272)

$$-\frac{\hbar^2}{2m}\nabla^2 u + V(\mathbf{x})u = i\hbar u_t.$$

Spherical harmonics in three dimensions (see Humi [73])

$$\left[\frac{1}{\sin\theta}\frac{\partial}{\partial\theta}\left(\sin\theta\frac{\partial}{\partial\theta}\right) + \frac{1}{\sin^2\theta}\frac{\partial^2}{\partial\phi^2} + l(l+1)\right]Y_{l,m} = 0.$$

Spherical harmonics in four dimensions (see Humi [73])

$$u_{xx} + 2(\cot x)u_x + \frac{1}{\sin^2 x}\left(u_{yy} + (\cot y)u_y + \frac{1}{\sin^2 y}u_{zz}\right) + (n^2 - 1)u = 0.$$

Tricomi equation (see Manwell [95])

$$u_{yy} = yu_{xx}.$$

Wave equation (see Morse and Feshback [105], page 271)

$$u_{tt} = c^2\nabla^2 u.$$

Weinstein equation – generalized (see Akin [5])

$$\nabla^2 u + \frac{p}{x_{n-1}}u_{x_{n-1}} + \frac{q}{x_n}u_{x_n} = 0.$$

Second Order Nonlinearity

Benjamin–Bona–Mahony equation (see Avrin and Goldstein [14])

$$u_t - u_{xxx} + uu_x = 0.$$

Boussinesq equation (see Calogero and Degasperis [31], page 54)

$$u_{tt} - u_{xx} - u_{xxxx} + 3(u^2)_{xx} = 0.$$

Burgers' equation (see Benton and Platzman [20])

$$u_t + uu_x = \nu u_{xx}.$$

Burgers equation – non-planar (see Sachdev and Nair [122])

$$u_t + uu_x + \frac{Ju}{2t} = \frac{\delta}{2}u_{xx}.$$

Ernst equation (see Calogero and Degasperis [31], page 62)

$$(\text{Re}\, u)\left(u_{rr} + \frac{u_r}{r} + u_{zz}\right) = u_r^2 + u_z^2.$$

Fisher's equation (see Kaliappan [78])

$$u_t = Du_{xx} + u - u^2.$$

Kadomtsev–Petviashvili equation (see Latham [88])

$$(u_t + u_{xxx} - 6uu_x)_x \pm u_{yy} = 0.$$

Generalized Kadomtsev–Petviashvili–Burgers equation (see Brugarino [28])

$$\left(u_t + \frac{J}{2t}u + J_1 uu_x + J_2 u_{xx} + J_3 u_{xxx}\right)_x + J_4(t)u_{yy} = 0.$$

Khokhlov–Zabolotskaya equation (see Chowdhury and Nasker [40])

$$u_{xt} - (uu_x)_x = u_{yy}.$$

Korteweg–de Vries equation (KdV) (see Lamb [86], Chapter 4)

$$u_t + u_{xxx} - 6uu_x = 0.$$

KdV equation – cylindrical (see Calogero and Degasperis [31], page 50)

$$u_t + u_{xxx} - 6uu_x + \frac{u}{2t} = 0.$$

KdV equation – generalized (see Boyd [26])

$$u_t + uu_x - u_{xxxxx} = 0.$$

KdV equation – spherical (see Calogero and Degasperis [31], page 51)

$$u_t + u_{xxx} - 6uu_x + \frac{u}{t} = 0.$$

KdV equation – transitional (see Calogero and Degasperis [31], page 50)

$$u_t + u_{xxx} - 6f(t)uu_x = 0.$$

KdV equation – variable coefficient (see Nimala, Vedan, and Baby [110])

$$u_t + \alpha t^n uu_x + \beta t^m u_{xxx} = 0.$$

Korteweg–de Vries–Burgers equation (KdVB) (see Canosa and Gazdag [32])

$$u_t + 2uu_x - \nu u_{xx} + \mu u_{xxx} = 0.$$

Kuramoto–Sivashinksy equation (see Michelson [99])

$$u_t + \nabla^4 u + \nabla^2 u + \tfrac{1}{2}|\nabla^2 u|^2 = 0.$$

Lin–Tsien equation (see Ames and Nucci [9])

$$2u_{tx} + u_x u_{xx} - u_{yy} = 0.$$

Regularized long-wave equation (RLW) (see Calogero and Degasperis [31], page 49)

$$u_t + u_x - 6uu_x - u_{txx} = 0.$$

Thomas equation (see Rosales [116])

$$u_{xy} + \alpha u_x + \beta u_y + \gamma u_x u_y = 0.$$

Unnamed equation (see Rosen [117])

$$u_{tt} + 2uu_t - u_{xx} = 0.$$

Higher Order and Variable Order Nonlinearities

Generalized Benjamin–Bona–Mahony equation (see Goldstein and Wichnoski [59])

$$u_t - \nabla^2 u_t + \nabla \cdot \phi(u)) = 0.$$

Born–Infeld equation (see Whitham [146], page 617)

$$\left(1 - u_t^2\right) u_{xx} + 2u_x u_t u_{xt} - \left(1 + u_x^2\right) u_{tt} = 0.$$

Boussinesq equation – modified (see Clarkson [42])

$$\tfrac{1}{3}u_{tt} - u_t u_{xx} - \tfrac{3}{2}u_x^2 u_{xx} + u_{xxxx} = 0.$$

Boussinesq equation – modified (see Clarkson [43])
$$u_{tt} - u_t u_{xx} - \tfrac{1}{2} u_x^2 u_{xx} + u_{xxxx} = 0.$$

Buckmaster equation (see Hill and Hill [68])
$$u_t = \left(u^4\right)_{xx} + \left(u^3\right)_x.$$

Generalized Burgers–Huxley equation (see Wang, Zhu and Lu [142])
$$u_t - \alpha u^\delta u_x - u_{xx} = \beta u \left(1 - u^\delta\right)\left(u^\delta - \gamma\right).$$

Cahn–Hilliard equation (see Novick-Cohen and Segel [111])
$$u_t = \nabla \cdot \left[M(u)\nabla \left(\frac{\partial f}{\partial u} - K\nabla^2 u\right)\right].$$

Calogero–Degasperis–Fokas equation (see Gerdt, Shvachka, and Zharkov [55])
$$u_{xxx} - \tfrac{1}{8} u_x^3 + u_x \left(A e^u + B e^{-u}\right) = 0.$$

Caudrey–Dodd–Gibbon–Sawada–Kotera equation (see Aiyer, Fuchsteiner, and Oevel [4])
$$u_t + u_{xxxxx} + 30 u u_{xxx} + 30 u_x u_{xx} + 180 u^2 u_x = 0.$$

Clairaut's equation (see Iyanaga and Kawada [76], page 1446)
$$u = x u_x + y u_y + f(u_x, u_y).$$

Eckhaus partial differential equation (see Kundu [85])
$$i u_t + u_{xx} + 2 \left(|u|^2\right)_x u + |u|^4 u = 0.$$

Fisher equation – generalized (see Wang [141])
$$u_t - u_{xx} - \frac{m}{u} u_x^2 = u \left(1 - u^\alpha\right).$$

Gardner equation (see Tabor [130], page 289)
$$u_t = 6(u + \varepsilon^2 u^2)u_x + u_{xxx}.$$

Ginzburg–Landau equation (see Katou [83])
$$u_t = (1 + ia)u_{xx} + (1 + ic)u - (1 + id)|u|^2 u.$$

Hamilton–Jacobi equation (see page 57)
$$V_t + H(t, \mathbf{x}, V_{x_1}, \ldots, V_{x_n}) = 0.$$

Harry Dym equation (see Calogero and Degasperis [31], page 53)
$$u_t = u_{xxx} u^3.$$

Hirota equation (see Calogero and Degasperis [31], page 56)

$$u_t + iau + ib(u_{xx} - 2\eta|u^2|u) + cu_x + d(u_{xxx} - 6\eta|u|^2 u_x) = 0.$$

Kadomtsev–Petviashvili equation – modified (see Clarkson [42])

$$u_{xt} = u_{xxx} + 3u_{yy} - 6u_x^2 u_{xx} - 6u_y u_{xx}.$$

Klein–Gordon equation – quasilinear (see Nayfeh [108], page 76)

$$u_{tt} - \alpha^2 u_{xx} + \gamma^2 u = \beta u^3.$$

Klein–Gordon equation – nonlinear (see Matsuno [97])

$$\nabla^2 u + \lambda u^p = 0.$$

KdV equation – deformed (see Dodd and Fordy [47])

$$u_t + \left(u_{xx} - 2\eta u^3 - \frac{3}{2} \frac{u u_x^2}{\eta + u^2} \right)_x = 0.$$

KdV equation – generalized (see Rammaha [115])

$$u_t + u u_x + p|u|^{p-1} u_x = 0.$$

KdV equation – modified (mKdV) (see Calogero and Degasperis [31], page 51)

$$u_t + u_{xxx} \pm 6u^2 u_x = 0.$$

KdV equation – modified modified (see Dodd and Fordy [47])

$$u_t + u_{xxx} - \tfrac{1}{8}u_x^3 + u_x \left(A e^{au} + B + C e^{-au} \right) = 0.$$

KdV equation – Schwarzian (see Weiss [145])

$$\frac{u_t}{u_x} + \{u; x\} = \lambda.$$

Kupershmidt equation (see Fuchssteiner, Oevel, and Wiwianka [52])

$$u_t = u_{xxxxx} + \tfrac{5}{2}u_{xxx}u + \tfrac{25}{4}u_{xx}u_x + \tfrac{5}{4}u^2 u_x.$$

Liouville equation (see Matsuno [97])

$$\nabla^2 u + e^{\lambda u} = 0.$$

Liouville equation (see Calogero and Degasperis [31], page 60)

$$u_{xt} = e^{\eta u}.$$

Molenbroek's equation (see Cole and Cook [44], page 34)

$$\nabla^2\phi = M_\infty^2 \left\{ \phi_x^2\phi_{xx} + 2\phi_x\phi_y\phi_{xy} + \phi_y^2\phi_{yy} + \frac{\gamma-1}{2}\left(\phi_x^2 + \phi_y^2 - 1\right)\right.$$

$$\left. \times \left(\phi_{xx} + \phi_{yy} + \varepsilon\frac{\phi_y}{y}\right)\right\}.$$

Monge–Ampère equation (see Moon and Spencer [104], page 171)

$$\left(u_{xy}\right)^2 - u_x u_y = f\left(x, y, u, u_x, u_y\right).$$

Monge–Ampère equation (see Gilbarg and Trudinger [56])

$$\begin{vmatrix} u_{x_1 x_1} & u_{x_1 x_2} & \cdots & u_{x_1 x_n} \\ u_{x_2 x_1} & u_{x_2 x_2} & \cdots & u_{x_2 x_n} \\ \vdots & \vdots & \ddots & \vdots \\ u_{x_n x_1} & u_{x_n x_2} & \cdots & u_{x_n x_n} \end{vmatrix} = f(u, \mathbf{x}, \nabla u).$$

Phi–four equation (see Calogero and Degasperis [31], page 60)

$$u_{tt} - u_{xx} - u + u^3 = 0.$$

Plateau's equation (see Bateman [18], page 501)

$$(1 + u_x^2)u_{xx} - 2u_x u_y u_{xy} + (1 + u_y^2)u_{yy} = 0.$$

Porous-medium equation (see Elliot, Herrero, King, and Ockendon [49])

$$u_t = \nabla \cdot (u^m \nabla u).$$

Rayleigh wave equation (see Hall [65])

$$u_{tt} - u_{xx} = \varepsilon(u_t - u_t^3).$$

Sawada–Kotera equation (see Matsuno [96], page 7)

$$u_t + 45u^2 u_x + 15u_x u_{xx} + 15u u_{xxx} + u_{xxxxx} = 0.$$

Schröedinger equation – logarithmic (see Cazenave [34])

$$iu_t + \nabla^2 u + u\log|u|^2 = 0.$$

Schröedinger equation – nonlinear (see Calogero and Degasperis [31], page 56)

$$iu_t + u_{xx} \pm 2|u|^2 u = 0.$$

Schröedinger equation – derivative nonlinear (see Calogero and Degasperis [31], page 56)

$$iu_t + u_{xx} \pm i\left(|u|^2 u\right)_x = 0.$$

Sine–Gordon equation (see Calogero and Degasperis [31], page 59)

$$u_{xx} - u_{yy} \pm \sin u = 0.$$

Sine–Gordon equation – damped (see Levi, Hoppensteadt and Miranker [91])

$$u_{tt} + \sigma u_t - u_{xx} + \sin u = 0.$$

Sine–Gordon equation – double (see Calogero and Degasperis [31], page 60)

$$u_{xt} \pm \left[\sin u + \eta \sin \left(\frac{u}{2} \right) \right] = 0.$$

Sinh–Gordon equation (see Grauel [61])

$$u_{xt} = \sinh u.$$

Sinh–Poisson equation (see Ting, Chen, and Lee [131])

$$\nabla^2 u + \lambda^2 \sinh u = 0.$$

Strongly damped wave equation (see Ang and Dinh [11])

$$u_{tt} - \nabla^2 u - \nabla^2 u_t + f(u) = 0.$$

Wadati–Konno–Ichikawa–Schimizu equation (see Calogero and Degasperis [31], page 53)

$$iu_t + \left[\left(1 + |u|^2 \right)^{-1/2} u \right]_{xx} = 0.$$

Zoomeron equation (see Calogero and Degasperis [31], page 58)

$$\left(\frac{\partial^2}{\partial t^2} - \frac{\partial^2}{\partial x^2} \right) \left(\frac{u_{xt}}{u} \right) + 2 \left(u^2 \right)_{xt} = 0.$$

Unnamed equation (see Aguirre and Escobedo [3])

$$u_t - \nabla^2 u = u^p.$$

Unnamed equation (see Bluman and Kumei [22])

$$u_t - \frac{\partial}{\partial x} \left[\frac{au_x}{(u+b)^2} \right] = 0.$$

Unnamed equation (see Calogero [29])

$$u_{xt} + uu_{xx} + F(u_x) = 0.$$

Unnamed equation (see Calogero [30])

$$u_t = u_{xxx} + 3(u_{xx}u^2 + 3u_x^2 u) + 3u_x u^4.$$

Unnamed equation (see Daniel and Sahadevan [236])

$$u_t = u_{xxx} + u^2 u_{xx} + 3uu_x^2 + \tfrac{1}{3}u^4 u_x.$$

Unnamed equation (see Fujita [53])

$$\nabla^2 u + e^u = 0.$$

Unnamed equation (see Fujita [53])

$$u_t = \nabla^2 u + e^u.$$

Unnamed equation (see Fung and Au [54])

$$u_t + u_{xxx} - 6u^2 u_x + 6\lambda u_x = 0.$$

Unnamed equation (see Kaliappan [78])

$$u_t = Du_{xx} + u - u^k.$$

Unnamed equation (see Lin [45])

$$\nabla^2 u + Ae^{-u} = 0.$$

Unnamed equation (see Lindquist [94])

$$\nabla \cdot (|\nabla u|^p \nabla u) = f.$$

Unnamed equation (see Roy and Chowdhury [120])

$$-iu_t + u_{xx} + \frac{2|u_x|^2 u}{1 - uu^*} = 0.$$

Unnamed equation (see Shivaji [127])

$$-\nabla^2 u = \lambda \exp\left(\frac{\alpha u}{\alpha + u}\right).$$

Unnamed equation (see Trubek [132])

$$\nabla^2 u + Ke^{2u} = 0.$$

Unnamed equation (see Trubek [133])

$$\nabla^2 u + Ku^\sigma = 0.$$

Unnamed equation (see Utepbergenov [136])

$$z^2 u_{zz} + \nabla^2 u + a(z)u = 0.$$

Systems of Differential Equations

Systems of ODEs

Bonhoeffer-van der Pol (BVP) oscillator (see Rajasekar and Lakshmanan [114])

$$x' = x - \frac{x^3}{3} - y + I(t),$$
$$y' = c(x + a - by).$$

Brusselator (see Hairer, Nørsett, and Wanner [64], page 112)

$$u' = A + u^2v - (B+1)u,$$
$$v' = Bu - u^2v.$$

Full Brusselator (see Hairer, Nørsett, Wanner [64], page 114)

$$u' = 1 + u^2v - (w+1)u,$$
$$v' = uw - u^2v,$$
$$w' = -uw + \alpha.$$

Hamilton's differential equations (see Iyanaga and Kawada [76], page 1005)

$$\frac{dx_i}{dt} = H_{p_i}(t, \mathbf{x}, \mathbf{p}),$$
$$\frac{dp_i}{dt} = -H_{x_i}(t, \mathbf{x}, \mathbf{p}).$$

Jacobi elliptic functions (see Hille [69], page 66)

$$u' = vw,$$
$$v' = -uw,$$
$$w' = -k^2uv.$$

Kowalevski's top (see Haine and Horozov [63])

$$\frac{d\mathbf{m}}{dt} = \lambda\mathbf{m} \times \mathbf{m} + \boldsymbol{\gamma} \times \mathbf{l},$$
$$\frac{d\boldsymbol{\gamma}}{dt} = \lambda\boldsymbol{\gamma} \times \mathbf{m}.$$

Lorenz equations (see Sparrow [128])

$$x' = \sigma(y - x),$$
$$y' = rx - y - xz,$$
$$z' = xy - bz.$$

Lorenz equations – complex (see Flessas [51])

$$x' = \sigma(y - x),$$
$$y' = rx - ay - xz,$$
$$z' = -bz + \tfrac{1}{2}(x^*y + xy^*).$$

Lotka–Volterra equations (see Boyce and DiPrima [25], page 494)

$$u' = u(a - bv),$$
$$v' = v(-c + du).$$

Nahm's equations (see Steeb and Louw [129])

$$U_t = [V, W],$$
$$V_t = [W, U],$$
$$W_t = [U, V].$$

Toda molecule equation – cylindrical (see Hirota and Nakamura [72])

$$\left(\partial_{rr} + r^{-1}\partial_r\right) \log V_n - V_{n+1} + 2V_n - V_{n-1} = 0$$

Systems of PDEs

Dispersive long-wave equation (see Boiti, Leon, and Pempinelli [24])

$$u_t = (u^2 - u_x + 2w)_x,$$
$$w_t = (2uw + w_x)_x.$$

Beltrami equation (see Iyanaga and Kawada [76], page 1087)

$$f_{\bar{z}} = \mu(z)f_z.$$

Boomeron equation (see Calogero and Degasperis [31], page 57)

$$u_t = \mathbf{b} \cdot \mathbf{v}_x,$$
$$\mathbf{v}_{xt} = u_{xx}\mathbf{b} + \mathbf{a} \times \mathbf{v}_x - 2\mathbf{v} \times [\mathbf{v} \times \mathbf{b}].$$

Carleman equation (see Kaper and Leaf [82])

$$u_t + u_x = v^2 - u^2,$$
$$v_t - v_x = u^2 - v^2.$$

Cauchy–Riemann equations (see Levinson and Redheffer [92])

$$u_x - v_y = 0,$$
$$u_y + v_x = 0.$$

Chiral field equation (see Calogero and Degasperis [31], page 61)

$$(U^*U_x)_t + (U^*U_t)_x = 0.$$

Davey–Stewartson equations (see Champagne and Winternitz [37])

$$iu_t + u_{xx} + au_{yy} + bu|u|^2 - uw = 0,$$
$$w_{xx} + cw_{yy} + d\left(|u|^2\right)_{yy} = 0.$$

Dirac equation in $1+1$ dimensions (see Alvarez, Pen–Yu, and Vazquez [6])

$$u_t + v_x + imu + 2i\lambda \left(|u|^2 - |v|^2\right) u = 0,$$
$$v_t + u_x + imv + 2i\lambda \left(|v|^2 - |u|^2\right) v = 0.$$

Drinfel'd–Sokolov–Wilson equation (see Hirota, Grammaticos, and Ramani [71])

$$u_t = 3ww_x,$$
$$w_t = 2w_{xxx} + 2uw_x + u_xw.$$

Klein–Gordon–Maxwell equations (see Deumens [46])

$$\nabla^2 s - (|\mathbf{a}|^2 + 1)s = 0,$$
$$\nabla^2 \mathbf{a} - \nabla(\nabla \cdot \mathbf{a}) - s^2 \mathbf{a} = \mathbf{0}.$$

Euler equations (see Landau and Lifshitz [87], page 3)

$$\frac{\partial \mathbf{v}}{\partial t} + (\mathbf{v} \cdot \mathrm{grad})\mathbf{v} = -\frac{1}{\rho} \, \mathrm{grad} \, P.$$

Fitzhugh–Nagumo equations (see Sherman and Peskin [126])

$$u_t = u_{xx} + u(u - a)(1 - u) + w,$$
$$w_t = \varepsilon u.$$

Gross–Neveu model (see Calogero and Degasperis [31], page 62)

$$iu_x^{(n)} = v^{(n)} \sum_{m=1}^{N} \left(v^{(m)*}u^{(m)} + u^{(m)*}v^{(m)}\right),$$
$$iv_t^{(n)} = u^{(n)} \sum_{m=1}^{N} \left(v^{(m)*}u^{(m)} + u^{(m)*}v^{(m)}\right).$$

Heisenberg ferromagnet equation (see Calogero and Degasperis [31], page 56)

$$\mathbf{s}_t = \mathbf{s} \times \mathbf{s}_{xx}.$$

Hirota–Satsuma equation (see Weiss [144])

$$u_t = \tfrac{1}{2}u_{xxx} + 3uu_x - 6ww_x,$$
$$w_t = -w_{xxx} - 3uw_x.$$

KdV equation – super (see Kersten and Gragert [84])

$$u_t = 6uu_x - u_{xxx} + 3ww_{xx},$$
$$w_t = 3u_x w + 6uw_x - 4w_{xxx}.$$

Matrix Liouville equation (see Andreev [10])

$$\left(U_x U^{-1}\right)_t = U$$

Von Kármán equations (see Ames and Ames [8])

$$\nabla^4 u = E\left[w_{xy}^2 - w_{xx}w_{yy}\right],$$
$$\nabla^4 w = a + b\left[u_{yy}w_{xx} + u_{xx}w_{yy} - 2u_{xy}w_{xy}\right].$$

Kaup's equation (see Dodd and Fordy [47])

$$f_x = 2fgc(x - t),$$
$$g_t = 2fgc(x - t).$$

Landau–Lifshitz equation (see Barouch, Fokas, and Papageorgiou [15])

$$U_t = U \cdot U_{xx} + U \cdot JU.$$

Maxwell's equations (see Jackson [77], page 177)

$$\nabla \cdot \mathbf{D} = 4\pi\rho, \quad \nabla \times \mathbf{H} = \frac{4\pi}{c}\mathbf{J},,$$
$$\nabla \cdot \mathbf{B} = 0, \qquad \nabla \times \mathbf{E} + \frac{1}{c}\frac{\partial \mathbf{B}}{\partial t} = 0.$$

Reduced Maxwell–Bloch equations (see Calogero and Degasperis [31], page 59)

$$E_t - v = 0, \qquad q_x + Ev = 0,$$
$$r_x + \omega v = 0, \qquad v_x - \omega r - Eq = 0.$$

Nambu–Jona Lasinio–Vaks–Larkin model (see Calogero and Degasperis [31], page 62)

$$iu_x^{(n)} = v^{(n)}\sum_{m=1}^{N} v^{(m)*}u^{(m)},$$
$$iv_t^{(n)} = u^{(n)}\sum_{m=1}^{N} u^{(m)*}v^{(m)}.$$

Navier's equation (see Eringen and Suhubi [50])

$$(\lambda + 2\mu)\nabla\nabla \cdot \mathbf{u} - \mu\nabla \times \nabla \times \mathbf{u} = \rho\frac{\partial^2 \mathbf{u}}{\partial t^2}.$$

Navier–Stokes equations (see Landau and Lifshitz [87], page 49)

$$\mathbf{u}_t + (\mathbf{u} \cdot \nabla)\mathbf{u} = -\frac{\nabla P}{\rho} + \nu\nabla^2\mathbf{u}.$$

Pohlmeyer–Lund–Regge model (see Calogero and Degasperis [31], page 61)

$$u_{xx} - u_{yy} \pm \sin u \cos u + \left(\frac{\cos u}{\sin^3 u} \right) \left(v_x^2 - v_y^2 \right) = 0,$$

$$\left(v_x \cot^2 u \right)_x = \left(v_y \cot^2 u \right)_y.$$

Prandtl's boundary layer equations (see Iyanaga and Kawada [76], page 672)

$$u_t + u u_x + v u_y = U_t + U U_x + \frac{\mu}{\rho} u_{yy},$$

$$u_x + v_y = 0.$$

Toda equation – 3 + 1-dimensional (see Hirota [70])

$$\nabla^2 \log V_n - V_{n+1} + 2 V_n - V_{n-1} = 0$$

Vector Poisson equation (see Moon and Spencer [102])

$$\maltese \mathbf{A} = - \operatorname{curl} \mathbf{E}.$$

sigma-model (see Calogero and Degasperis [31], page 61)

$$\mathbf{v}_{xt} + (\mathbf{v}_x \mathbf{v}_t) \, \mathbf{v} = 0.$$

Massive Thirring model (see Calogero and Degasperis [31], page 62)

$$i u_x + v + u |v|^2 = 0,$$

$$i v_t + u + v |u|^2 = 0.$$

Veselov–Novikov equation (see Bogdanov [23])

$$\left(\partial_t + \partial_z^3 + \partial_{\bar{z}}^3 \right) v + \partial_z (uv) + \partial_{\bar{z}}(vw) = 0,$$

$$\partial_{\bar{z}} u = 3 \partial_z v,$$

$$\partial_z w = 3 \partial_{\bar{z}} v.$$

Yang–Mills equation (see Calogero and Degasperis [31], page 62)

$$(U^* U_t)_t - (U^* U_x)_{\bar{x}} = 0.$$

Anti-self-dual Yang–Mills equation (see Ablowitz, Costa, and Teneblat [1])

$$\frac{\partial}{\partial \bar{x}_1} \left(\Omega^{-1} \frac{\partial \Omega}{\partial x_1} \right) + \frac{\partial}{\partial \bar{x}_2} \left(\Omega^{-1} \frac{\partial \Omega}{\partial x_2} \right) = 0.$$

Unnamed equation (see Salingaros [123])

$$\nabla \times \mathbf{u} = k \mathbf{u}.$$

References

[1] M. J. Ablowitz, D. G. Costa, and K. Tenenblat, "Solutions of Multidimensional Extensions of the Anti-Self-Dual Yang–Mills Equation," *Stud. Appl. Math.*, **77**, 1987, pages 37–46.

[2] M. Abramowitz and I. A. Stegun, *Handbook of Mathematical Functions*, National Bureau of Standards, Washington, DC, 1964.

[3] J. Aguirre and M. Escobedo, "A Cauchy Problem for $u_t - \Delta u = u^p$ with $0 < p < 1$. Asymptotic behavior of Solutions," *Ann. Fac. Sci. Toulouse Math.*, **8**, No. 2, 1986/87, pages 175–203.

[4] R. N. Aiyer, B. Fuchssteiner, and W. Oevel, "Solitons and Discrete Eigenfunctions of the Recursion Operator of Non-Linear Evolution Equations: I. The Caudrey–Dodd–Gibbon–Sawada–Kotera equation," *J. Phys. A: Math. Gen.*, **19**, 1986, pages 3755–3770.

[5] Ö. Akin, "The Integral Representation of the Positive Solutions of the Generalized Weinstein Equation on a Quarter-Space," *SIAM J. Appl. Math.*, **19**, No. 6, November 1988, pages 1348–1354.

[6] A. Alvarez, K. Pen–Yu, and L. Vazquez, "The Numerical Study of a Nonlinear One-Dimensional Dirac Equation," *Appl. Math. and Comp.*, **13**, 1983, pages 1–15.

[7] W. F. Ames, "Ad Hoc Exact Techniques for Nonlinear Partial Differential Equations," in W. F. Ames (ed.), *Nonlinear Partial Differential Equations in Engineering*, Academic Press, New York, 1967.

[8] K. A. Ames and W. F. Ames, "On Group Analysis of the Von Kármán Equations," *Nonlinear Analysis*, **6**, No. 8, 1982, pages 845–853.

[9] W. F. Ames and M. C. Nucci, "Analysis of Fluid Equations by Group Methods," *Journal of Engineering Mathematics*, **20**, 1985, pages 181–187.

[10] V. A. Andreev, "Matrix Liouville Equation," *Theo. and Math. Physics*, **83**, No. 1, April 1990, pages 366–372.

[11] D. D. Ang and A. P. N. Dinh, "On the Strongly Damped Wave Equation: $u_{tt} - \Delta u - \Delta u_t + f(u) = 0$," *SIAM J. Appl. Math.*, **19**, No. 6, November 1988, pages 1409–1418.

[12] F. M. Arscott, "The Land Beyond Bessel: A Survey of Higher Special Functions," in W. N. Everitt and B. D. Sleeman (ed.), *Ordinary and Partial Differential Equations*, Springer–Verlag, New York, 1981, pages 26–45.

[13] C. Athorne, "On a Subclass of Ince Equations," *J. Phys. A: Math. Gen.*, **23**, 1990, pages L137–L139.

[14] J. Avrin and J. A. Goldstein, "Global Existence for the Benjamin–Bona–Mahony Equation in Arbitrary Dimensions," *Nonlinear Analysis*, **9**, No. 8, 1985, pages 861–865.

[15] E. Barouch, A. S. Fokas, and V. G. Papageorgiou, "The Bi-Hamiltonian Formulation of the Landau–Lifshitz Equation," *J. Math. Physics*, **29**, No. 12, December 1988, pages 2628–2633.

[16] A. O. Barut, A. Inomata, and R. Wilson, "A New Realisation of Dynamical Groups and Factorisation Method," *J. Phys. A: Math. Gen.*, **20**, 1987, pages 4075–4083.

[17] A. O. Barut, A. Inomata, and R. Wilson, "Algebraic Treatment of Second Pöschl–Teller, Morse–Rosen and Eckart Equations," *J. Phys. A: Math. Gen.*, **20**, 1987, pages 4083–4096.

[18] H. Bateman, *Partial Differential Equations of Mathematical Physics*, Cambridge University Press, New York, 1959.

[19] C. M. Bender and S. A. Orszag, *Advanced Mathematical Methods for Scientists and Engineers*, McGraw–Hill, New York, 1978.

[20] E. R. Benton and G. W. Platzman, "A Table of Solutions of the One–Dimensional Burgers Equation," *Quart. Appl. Math.*, 1972, pages 195–212.

[21] G. Birkhoff and G.-C. Rota, *Ordinary Differential Equations*, John Wiley & Sons, New York, 1978.

[22] G. Bluman and S. Kumei, "On the Remarkable Nonlinear Diffusion Equation $(\partial/\partial x)[a(u+b)^{-2}(\partial u/\partial x)] - (\partial u/\partial t) = 0$," *J. Math. Physics*, **21**, No. 5, May 1980, pages 1019–1023.

[23] L. V. Bogdanov, "Veselov–Novikov Equation as a Natural Two-Dimensional Generalization of the Korteweg–de Vries Equation," *Theo. and Math. Physics*, **70**, No. 2, August 1987, pages 219–233.

[24] M. Boiti, J. J.-P. Leon, and F. Pempinelli, "Integrable Two-Dimensional Generalisation of the Sine– and Sinh–Gordon Equations," *Inverse Prob.*, **3**, 1987, pages 37–49.

[25] W. E. Boyce and R. C. DiPrima, *Elementary Differential Equations and Boundary Value Problems*, Fourth Edition, John Wiley & Sons, New York, 1986.

[26] J. P. Boyd, "Solitons from Sine Waves: Analytical and Numerical Methods for Non–Integrable Solitary and Cnoidal Waves," *Physica D*, **21**, 1986, pages 227–246.

[27] J. P. Boyd, "An Analytical Solution for a Nonlinear Differential Equation with Logarithmic Decay," *Advances in Appl. Math.*, **9**, No. 3, 1988, pages 358–363.

[28] T. Brugarino, "Similarity Solutions of the Generalized Kadomtsev–Petviashvili–Burgers equations," *Nuovo Cimento B*, **92**, No. 2, 1986, pages 142–156.

[29] F. Calogero, "A Solvable Nonlinear Wave Equation," *Stud. Appl. Math.*, **70**, No. 3, June 1984, pages 189–199.

[30] F. Calogero, "The Evolution Partial Differential Equation $u_t = u_{xxx} + 3(u_{xx}u^2 + 3u_x^2u) + 3u_xu^4$," *J. Math. Physics*, **28**, No. 3, March 1987, pages 538–555.

[31] F. Calogero and A. Degasperis, *Spectral Transform and Solitons: Tools to Solve and Investigate Nonlinear Evolution Equations*, North–Holland Publishing Co., New York, 1982.

[32] J. Canosa and J. Gazdag, "The Korteweg–de Vries–Burgers Equation," *J. Comput. Physics*, **23**, 1977, pages 393–403.

[33] H. S. Carslaw and J. C. Jaeger, *Conduction of Heat in Solids*, Clarendon Press, Oxford, 1984.

[34] T. Cazenave, "Stable Solutions of the Logarithmic Schrödinger Equation," *Nonlinear Analysis*, **7**, No. 10, 1983, pages 1127–1140.

[35] T. Cebeci and H. B. Keller, "Shooting and Parallel Shooting Methods for Solving Falkner–Skan Boundary Layer Equation," *J. Comput. Physics*, **71**, 1971, pages 289–300.

[36] P. L. Chambré, "On the Solution of the Poisson–Boltzmann Equation with Application to the Theory of Thermal Explosions," *J. Chem. Physics*, **20**, No. 11, November 1952, pages 1795–1797.

[37] B. Champagne and P. Winternitz, "On the Infinite-Dimensional Group of the Davey–Stewartson Equations," *J. Math. Physics*, **29**, No. 1, January 1988, pages 1–8.

[38] C. C. Chang, "On Generalized Lavrent'ev–Bitsadze Problems," *Mixed Type Equations*, Teubner, Leipzig, 1987, pages 55–63.

[39] G. Chaohao, "The Mixed PDE for Amplifying Spiral Waves," *Lett. Math. Physics*, **16**, 1988, pages 69–76.

[40] A. R. Chowdhury and M. Nasker, "Towards the Conservation Laws and Lie Symmetries for the Khokhlov–Zabolotskaya Equation in Three Dimensions," *J. Phys. A: Math. Gen.*, **19**, 1986, pages 1775–1781.

[41] J. S. R. Chrisholm and A. K. Common, "A Class of Second-Order Differential Equations and Related First-Order Systems," *J. Phys. A: Math. Gen.*, **20**, 1987, pages 5459–5472.

[42] P. A. Clarkson, "The Painlevé Property, a Modified Boussinesq Equation and a Modified Kadomtsev–Petviashvili Equation," *Physica D*, **19**, 1986, pages 447–450.

[43] P. A. Clarkson, "New Similarity Solutions for the Modified Boussinesq Equation," *J. Phys. A: Math. Gen.*, **22**, No. 13, 1989, pages 2355–2367.

[44] J. D. Cole and P. Cook, *Transonic Aerodynamics*, North–Holland Publishing Co., New York, 1986.

[45] M. Daniel and R. Sahadevan, "On the Weak Painlevé Property and Linearization of the Evolution Equation $u_t = u_{xxx} + u^2 u_{xx} + 3u u_x^2 + \frac{1}{3} u^4 u_x$," *Phys. Lett. A*, **130**, No. 1, 1988, pages 19–21.

[46] E. Deumens, "The Klein–Gordon–Maxwell Nonlinear System of Equations" *Physica D*, **18**, 1986, pages 371–373.

[47] R. Dodd and A. Fordy, "The Prolongation Structures of Quasi-Polynomial Flows," *Proc. R. Soc. A.*, **385**, 1983, pages 389–429.

[48] P. W. Duck, T. W. Marshall, and E. J. Watson, "First-Passage Times for the Uhlenbeck–Ornstein Process," *J. Phys. A: Math. Gen.*, **19**, 1986, pages 3545–3558.

[49] C. M. Elliot, M. A. Herrero, J. R. King, and J. R. Ockendon, "The Mesa Problem: Diffusion Patterns for $u_t = \nabla \cdot (u^m \nabla u)$ as $m \to \infty$," *IMA J. Appl. Mathematics*, **7**, No. 2, pages 147–154.

[50] A. C. Eringen and E. S. Suhubi, *Elastodynamics*, Academic Press, New York, Volume 2, 1975, Chapter 5.

[51] G. P. Flessas, "New Exact Solutions of the Complex Lorenz Equations," *J. Phys. A: Math. Gen.*, **22**, 1989, pages L137–L141.

[52] B. Fuchssteiner, W. Oevel, and W. Wiwianka, "Computer-Algebra Methods for Investigation of Hereditary Operators of High Order Soliton Equations," *Comput. Physics Comm.*, **44**, 1987, pages 47–55.

[53] H. Fujita, "On the Nonlinear Equations $\Delta u + e^u = 0$ and $\partial v/\partial t = \Delta v + e^v$," *Bull. Amer. Math. Soc.*, **75**, 1969, pages 132–135.

[54] P. C. W. Fung and C. Au, "A series of New Analytical Solutions to the Nonlinear Equation $y_t + y_{xxx} - 6y^2 y_x + 6\lambda y_x = 0$. I," *J. Math. Physics*, **25**, No. 5, May 1984, pages 1370–1371.

[55] V. P. Gerdt, A. B. Shvachka, and A. Y. Zharkov, "Computer Algebra Applications for Classification of Integrable Non-linear Evolution Equations," *J. Symbolic Comp.*, **1**, 1985, pages 101–107.

[56] D. Gilbarg and N. S. Trudinger, *Elliptic Partial Differential Equations of Second Order*, Springer–Verlag, Berlin, 1983, page 441.

[57] B. H. Gilding, "The First Boundary Value Problem for $-u'' = \lambda u^p$," *J. Math. Anal. Appl.*, **128**, No. 2, 1987, pages 419–442.

[58] M. E. Goldstein and W. H. Braun, *Advanced Methods for the Solution of Differential Equations*, NASA SP-316, U.S. Government Printing Office, Washington, D.C., 1973.

[59] J. A. Goldstein and B. J. Wichnoski, "On the Benjamin–Bona–Mahony Equation in Higher Dimensions," *Nonlinear Analysis*, **4**, No. 4, 1980, pages 665–675.

[60] I. S. Gradshteyn and I. M. Ryzhik, *Tables of Integrals, Series, and Products*, Academic Press, New York, 1980.

[61] A. Grauel, "Sinh–Gordon Equation, Painlevé Property and Bäcklund Transformation," *Physica A*, **132**, 1985, pages 557–568.

[62] P. Hagedorn, *Non-Linear Oscillations*, Clarendon Press, Oxford, 1982.

[63] L. Haine and E. Horozov, "A Lax Pair for Kowalevski's Top," *Physica D*, **29**, 1987, pages 173–180.

[64] E. Hairer, S. P. Nørsett, and G. Wanner, *Solving Ordinary Differential Equations I*, Springer–Verlag, New York, 1987.

[65] W. S. Hall, "The Rayleigh Wave Equation — An Analysis," *Nonlinear Analysis*, **2**, No. 2, 1978, pages 129–156.

[66] I. H. Herron, "The Orr–Sommerfeld Equations on Infinite Intervals," *SIAM Review*, **29**, No. 4, 1987, pages 597–620.

[67] J. Hershenov, "Solutions of the Differential Equation $u''' + \lambda^2 z u' + (\alpha - 1)\lambda^2 u = 0$," *Stud. Appl. Math.*, **55**, 1976, pages 301–314.

[68] J. M. Hill and D. L. Hill, "High-Order Nonlinear Evolution Equations," *IMA J. Appl. Mathematics*, **45**, 1990, pages 243–265.

[69] E. Hille, *Lectures on Ordinary Differential Equations*, Addison–Wesley Publishing Co., Reading, MA, 1969.

[70] R. Hirota, "Exact Solutions of the Spherical Toda Molecule Equation," *J. Phys. Soc. Japan*, **57**, No. 1, 1988, pages 66–70.

[71] R. Hirota, B. Grammaticos, and A. Ramani, "Soliton Structure of the Drinfel'd–Sokolov–Wilson Equation," *J. Math. Physics*, **27**, No. 6, June 1986, pages 1499–1505.

[72] R. Hirota and A. Nakamura, "Exact Solutions of the Cylindrical Toda Molecule Equation," *J. Phys. Soc. Japan*, **56**, No. 9, 1987, pages 3055–3061.

[73] M. Humi, "Factorisation of Separable Partial Differential Equations" *J. Phys. A: Math. Gen.*, **20**, 1987, pages 4577–4585.

[74] E. L. Ince, *Ordinary Differential Equations*, Dover Publications, Inc., New York, 1964.

[75] L. Infeld and T. E. Hull, "The Factorization Method," *Rev. Mod. Physics*, **23**, No. 1, January 1951, pages 21–68.

[76] S. Iyanaga and Y. Kawada, *Encyclopedic Dictionary of Mathematics*, MIT Press, Cambridge, MA, 1980.

[77] J. D. Jackson, *Classical Electrodynamics*, John Wiley & Sons, New York, 1962.

[78] P. Kaliappan, "An Exact Solution for Travelling Waves of $u_t = Du_{xx} + u - u^k$," *Physica D*, **11**, 1984, pages 368–374.

[79] E. Kamke, *Differentialgleichungen Lösungsmethoden und Lösungen*, Chelsea Publishing Company, New York, 1948.

[80] E. Kamke, *Differentialgleichungen Lösungsmethoden und Lösungen*, Volume II, Chelsea Publishing Company, New York, 1947.

[81] L. V. Kantorovich and V. I. Krylov, *Approximate Methods of Higher Analysis*, Interscience Publishers, New York, 1958.

[82] H. G. Kaper and G. K. Leaf, "Initial Value Problems for the Carleman Equation," *Nonlinear Analysis*, **4**, No. 2, 1980, pages 343–362.

[83] K. Katou, "Asymptotic Spatial Patterns on the Complex Time–Dependent Ginzburg–Landau Equation," *J. Phys. A: Math. Gen.*, **19**, 1986, pages L1063–L1066.

[84] P. H. M. Kersten and P. K. H. Gragert, "Symmetries for the Super-KdV Equation," *J. Phys. A: Math. Gen.*, **21**, 1988, pages L579–L584.

[85] A. Kundu, "Comments on: 'The Eckhaus PDE $i\psi_t + \psi_{xx} + 2\left(|\psi|^2\right)_x \psi + |\psi|^4\psi = 0$'," *Inverse Problems*, **4**, No. 4, November 1988, pages 1143–1144.

[86] G. L. Lamb, *Elements of Soliton Theory*, John Wiley & Sons, New York, 1980.

[87] L. D. Landau and E. M. Lifshitz, *Fluid Mechanics*, Pergamon Press, New York, 1959.

[88] G. A. Latham, "Solutions of the KP Equation Associated to Rank-Three Commuting Differential Operators over a Singular Elliptic Curve," *Physica D*, **41**, 1990, pages 55–66.

[89] G. E. Latta, "Some Differential Equations of the Mathieu Type, and Related Integral Equations," *J. Math. and Physics*, **42**, 1963, pages 139–146.

[90] P. G. L. Leach, "First Integrals for the Modified Emden Equation $\ddot{q}+\alpha(t)\dot{q}+q^n = 0$," *J. Math. Physics*, **26**, No. 10, October 1985, pages 2510–2514.

[91] M. Levi, F. C. Hoppensteadt and W. L. Miranker, "Dynamics of the Josephson Junction," *Quart. Appl. Math.*, July 1978, pages 167–198.

[92] N. Levinson and R. M. Redheffer, *Complex Variables*, Holden–Day, Inc., San Francisco, 1970.

[93] S.-S. Lin, "Symmetry Breaking for $\Delta u + 2\delta e^{-u} = 0$ on a Disk with General Boundary Conditions," *Proc. Roy. Soc. Edinburgh Sect. A*, **113**, No. 1-2, 1989, pages 105–117.

[94] P. Lindquist, "Stability for the Solutions of div$(|\nabla u|^{p-2}\nabla u) = f$ with Varying p," *J. Math. Anal. Appl.*, **127**, 1987, pages 93–102.

[95] A. R. Manwell, *The Tricomi Equation with Applications to the Theory of Plane Transonic Flow*, Pitman Publishing Co., Marshfield, MA, 1979.

[96] Y. Matsuno, *Bilinear Transformation Method*, Academic Press, New York, 1984.

[97] Y. Matsuno, "Exact Solutions for the Nonlinear Klein–Gordon and Liouville Equations in Four-Dimensional Euclidean space," *J. Math. Physics*, **28**, No. 10, October 1987, pages 2317–2322.

[98] G. H. Meyer, *Initial Value Methods for Boundary Value Problems: Theory and Application of Invariant Imbedding*, Academic Press, New York, 1973.

[99] D. Michelson, "Steady Solutions of the Kuramoto–Sivashinsky Equation," *Physica D*, **19**, 1986, pages 89–111.

[100] W. Miller, Jr., "Symmetries of Differential Equations. The Hypergeometric and Euler–Darboux Equations," *SIAM J. Math. Anal.*, **4**, No. 2, May 1973, pages 314–328.

[101] W. Miller, Jr., "Symmetry and Separation of Variables," Addison–Wesley Publishing Co., Reading, MA, 1977.

[102] P. Moon and D. E. Spencer, "The Meaning of the Vector Laplacian," *J. Franklin Institute*, **256**, 1953, pages 551–558.

[103] P. Moon and D. E. Spencer, *Field Theory For Engineers*, D. Van Nostrand Company, Inc., New York, 1961.

[104] P. Moon and D. E. Spencer, *Partial Differential Equations*, D. C. Heath, Lexington, MA, 1969.

[105] P. M. Morse and H. Feshback, *Methods of Theoretical Physics*, McGraw–Hill Book Company, New York, 1953.

[106] G. Murphy, *Ordinary Differential Equations*, D. Van Nostrand Company, Inc., New York, 1960.

[107] C. Naim, "On Generalized Heat Polynomials" *Int. J. Math. & Math. Sci.*, **11**, No. 2, 1988, pages 393–400.

[108] A. H. Nayfeh, *Perturbation Methods*, John Wiley, New York, 1973.

[109] J. Nikolaus and M. G. Schmidt, "Die Lösungen mit kleinster Wachstumsordnung der Differentialgleichung $c_3 w''' + (b_2 x + c_2)w'' + b_1 w' - w = 0$," *Resultate Math.*, **10**, No. 1-2, 1986, pages 137–142l.

[110] N. Nimala, M. J. Vedan, and B. V. Baby, "A Variable Coefficient Korteweg–de Vries Equation: Similarity Analysis and Exact Solution. II," *J. Math. Physics*, **27**, No. 11, November 1986, pages 2644–2646.

[111] A. Novick-Cohen and L. A. Segel, "Nonlinear Aspects of the Cahn–Hilliard Equation," *Physica D*, **10**, 1984, pages 277–298.

[112] V. I. Oliker and L. D. Prussner, "On the Numerical Solution of the Equation $\dfrac{\partial^2 z}{\partial x^2}\dfrac{\partial^2 z}{\partial y^2} - \left(\dfrac{\partial^2 z}{\partial x \partial y}\right)^2 = f$ and its Discretizations, I," *Numerische Mathematik*, **54**, 1988, pages 271–293.

[113] E. D. Rainville, *Special Functions*, Chelsea Publishing Company, New York, 1960.

[114] S. Rajasekar and M. Lakshmanan, "Period-Doubling Bifurcations, Chaos, Phase-Locking and Devil's Staircase in a Bonhoeffer–Van Der Pol Oscillator," *Physica D*, **32**, 1988, pages 146–152.

[115] M. A. Rammaha, "On the Asymptotic Behavior of Solutions of Generalized Korteweg-de Vries Equations," *J. Math. Anal. Appl.*, **140**, No. 1, 1989, pages 228–240.

[116] R. R. Rosales, "Exact Solutions of Some Nonlinear Evolution Equations," *Stud. Appl. Math.*, **59**, 1978, pages 117–151.

[117] G. Rosen, "Solutions of a Certain Nonlinear Wave Equation," *J. Math. and Physics*, **45**, 1966, pages 235–265.

[118] P. Rosenau, "A Note on Integration of the Emden–Fowler Equation." *Int. J. Non-Lin. Mech.*, **19**, No. 4, 1984, pages 303–308.

[119] S. Rosenblat and J. Shepherd, "On the Asymptotic Solution of the Lagerstrom Model Equation," *SIAM J. Appl. Math.*, **29**, No. 1, July 1975, pages 110–120.

[120] S. Roy and A. R. Chowdhury, "Prolongation Theory. A New Nonlinear Schrödinger Equations," *Int. J. Theo. Physics*, **26**, No. 7, 1987, pages 707–714.

[121] L. A. Rubel, "A Differential Equation Satisfied by all n-Nomials," *Nieuw Arch. Wisk.*, **6**, No. 3, 1988, pages 263–267.

[122] P. L. Sachdev and K. R. C. Nair, "Generalized Burgers Equations and Euler-Painlevé Transcendents. II," *J. Math. Physics*, **28**, No. 5, May 1987, pages 997–1004.

[123] N. A. Salingaros, "On Solutions of the Equation $\nabla \times \mathbf{a} = k\mathbf{a}$: II. The Magnetic Force–Free Model," *J. Phys. A: Math. Gen.*, **19**, 1986, pages L705–L708.

[124] R. Seshadi and T. Y. Na, *Group Invariance in Engineering Boundary Value Problems*, Springer–Verlag, New York, 1985.

[125] M. Setoyanagi, "Liouvillian Solutions of the Differential Equation $y'' + S(x)y = 0$ with $S(x)$ Binomial," *Proc. Am. Math. Soc.*, **100**, No. 4, August 1987, pages 607–612.

[126] A. S. Sherman and C. S. Peskin, "A Monte Carlo Method for Scalar Reaction Diffusion Equations," *SIAM J. Sci. Stat. Comput.*, **7**, No. 4, October 1986, pages 1360–1372.

[127] R. Shivaji, "A Note on the Persistence of an S-Shaped Bifurcation Curve," *Appl. Anal.*, **24**, No. 3, 1987, pages 175–179.

[128] C. Sparrow, *The Lorenz Equations: Bifurcations, Chaos, and Strange Attractors*, Springer–Verlag, New York, 1982.

[129] W.-H. Steeb and J. A. Louw, "Nahm's Equations, Singular Point Analysis, and Integrability," *J. Math. Physics*, **27**, No. 10, 1986, pages 2458–2460.

[130] M. Tabor, *Chaos and Integrability in Nonlinear Dynamics*, John Wiley & Sons, New York, 1989.

[131] A. C. Ting, H. H. Cheb, and Y. C. Lee, "Exact Solutions of a Nonlinear Boundary Value Problem: the Vortices of the Two-Dimensional Sinh–Poisson Equation," *Physica D*, 1987, pages 37–66.

[132] J. Trubek, *Asymptotic Behavior of solutions to $\nabla^2 u + K e^{2u} = 0$ and $\nabla^2 u + K u^\sigma = 0$ on Euclidean Spaces*, PhD thesis, Northeastern Univ., Mass., 1988. See *AMS Notices*, RI, November 1988, page 1321.

[133] J. Trubek, "Asymptotic Behavior of Solutions to $\nabla^2 u + K u^\sigma = 0$ on $\mathbb{R}^n$ for $n \geq 3$," *Proc. Amer. Math. Soc.*, **106**, No. 4, 1989, pages 953–959.

[134] I. Tsukamoto, "On Solutions of $x'' = -e^{\alpha \lambda t} x^{1+\alpha}$," *Tokyo J. Math.*, **12**, No. 1, 1989, pages 181–203.

[135] K. M. Urwin and F. M. Arscott, "Theory of the Whittaker–Hill Equation," *Proc. R. Soc. Edin.*, **69**, 1970, pages 28–44.

[136] M. Utepbergenov, "Integral Representations of Solutions of Equations with Strong Degeneration, *Studies in Multidimensional Elliptic Systems of Partial Differential Equations (Russian)*, Akad. Nauk SSSR Sibirsk Otdel., Inst Mat., Novosibirsk, 1986.

[137] G. Valent, "An Integral Transform Involving Heun Functions and a Related Eigenvalue Problem," *SIAM J. Math. Anal.*, **17**, No. 3, May 1986, pages 688–703.

[138] G. Valiron, *The Geometric Theory of Ordinary Differential Equations and Algebraic Functions*, Math Sci Press, Brookline, MA, 1950.

[139] J. A. Viecelli, "Exponential Difference Operator Approximation for the Sixth Order Onsager Equation," *J. Comput. Physics*, **50**, 1983, pages 162–170.

[140] G. Villari, "Periodic Solutions of Liénard's Equation," *J. Math. Anal. Appl.*, **86**, 1982, pages 379–386.

[141] X. Y. Wang, "Exact and Explicit Solitary Wave Solutions for the Generalised Fisher Equation," *Phys. Lett. A*, **131**, No. 4–5, 1988, pages 277–279.

[142] X. Y. Wang, Z. S. Zhu and Y. K. Lu, "Solitary Wave Solutions of the Generalized Burgers–Huxley Equation," *J. Phys. A: Math. Gen.*, **23**, 1990, pages 271–274.

[143] R. S. Ward, "The Nahn Equations, Finite-Gap Potentials and Lamé Functions," *J. Phys. A: Math. Gen.*, **20**, 1987, pages 2679–2683.

[144] J. Weiss, "Modified Equations, Rational Solutions, and the Painlevé Property for the Kadomtsev–Petviashvili and Hirota–Satsuma Equations," *J. Math. Physics*, **26**, No. 9, September 1985, pages 2174–2180.

[145] J. Weiss, "Periodic Fixed Points of Bäcklund Transformation and the Korteweg-de Vries Equation," *J. Math. Physics*, **27**, No. 11, November 1986, pages 2647–2656.

[146] G. B. Whitham, *Linear and Nonlinear Waves*, Wiley Interscience, New York, 1974.

[147] R. Wilcox, "Closed-Form Solution of the Differential Equation $\left(\dfrac{\partial^2}{\partial x \partial y}+\right.$ $ax\dfrac{\partial}{\partial x}+ by\dfrac{\partial}{\partial y}+ cxy+ \dfrac{\partial}{\partial t}\Big)P = 0$ by Normal-Ordering Exponential Operators," *J. Math. Physics*, **11**, No. 4, April 1970, pages 1235–1237.

[148] H. G. Wood and J. B. Morton, "Onsager's Pancake Approximation for the Fluid Dynamics of a Gas Centrifuge," *J. Fluid Mech.*, **101**, Part 1, 1980, pages 1–31.

41.   Look Up ODE Forms

Applicable to   Ordinary differential equations.

Yields

An idea of whether or not a differential equation has a closed-form solution.

Idea

An experienced differential equations practitioner can look at many second order ordinary differential equations and readily guess whether or not there is a closed form solution. This is because there are many familiar forms that often appear.

Procedure

Having a listing of familiar differential equation forms will make it possible to recognize these forms. We have tabulated below many of the familiar forms that appear for second order ordinary differential equations. All of the equations below are from Abramowitz and Stegun [1]; the section in which each appears is referenced.

In the table, ( ) represents a term that contains constants. Such a term may or may not be correlated with other terms of the form ( ). For example, equation 22.6.5 in [1] is

$$\left(1 - x^2\right) y'' - (2\alpha + 1)xy' + n(n + 2\alpha)y = 0$$

where α is a real constant and n is an integer. Isolating the x dependence, we list this equation as

$$\left(1 - x^2\right) y'' + (\,)xy' + (\,)y = 0,$$

and disregard the fact that the hidden values have constraints on them and, in fact, are related.

Notes

[1] Realize that the same equation may look different when written in different variables. Some scaling of any given equation may be required to make it look like one of the forms listed.

References

[1] M. Abramowitz and I. A. Stegun, *Handbook of Mathematical Functions*, National Bureau of Standards, Washington, DC, 1964.

Equations of the form: $y'' + c(x)y = 0$

$$c(x) = (\,)$$
See 22.6.10

$$c(x) = -x$$
See 10.4.1

$$c(x) = (\,) - x^2$$
See 22.6.20

$$c(x) = (\,) + (\,)x + (\,)x^2$$
See 19.1.1

$$c(x) = (\,)x^{(\,)}$$
See 9.1.51

$$c(x) = (\,) + \frac{(\,)}{x^2}$$
See 9.1.49

$$c(x) = \frac{(\,)}{x} + \frac{(\,)}{x^2}$$
See 9.1.50

$$c(x) = (\,) - \frac{(\,)}{x} - \frac{(\,)}{x^2}$$
See 14.1.1

$$c(x) = (\,) - x^2 + \frac{(\,)}{x^2}$$
See 13.1.1 and 22.6.8

$$c(x) = (\,)e^{2x} - (\,)$$
See 9.1.54

$$c(x) = \frac{(\,)}{1 - x^2} + \frac{(\,) + x^2}{4(1 - x^2)^2}$$
See 22.6.7

$$c(x) = \frac{(\,)}{1 - x^2} + \frac{1}{(1 - x^2)^2}$$
See 22.6.14

$$c(x) = \frac{(\,)}{(1 - x)^2} + \frac{(\,)}{(1 + x)^2} + \frac{(\,)}{1 - x^2}$$
See 22.6.3

$$c(x) = \frac{(\,)}{x} + \frac{(\,)}{x^2} + (\,)$$
See 22.6.17

$$c(x) = (\,) + \frac{(\,)}{\sin^2 x}$$
See 22.6.8

$$c(x) = (\,) + \frac{(\,)}{\sin^2 \frac{x}{2}} + \frac{(\,)}{\cos^2 \frac{x}{2}}$$
See 22.6.4

Equations of the form: $y'' + b(x)y' + c(x)y = 0$

$b(x) = -x, \;\; c(x) = (\;)$	See 22.6.21
$b(x) = -2x, \;\; c(x) = (\;)$	See 22.6.19
$b(x) = 2x, \;\; c(x) = -(\;)x$	See 7.2.2
$b(x) = 2x, \;\; c(x) = x^2 - (\;)$	See 10.1.1
$b(x) = 2x, \;\; c(x) = (\;) - x^2$	See 10.2.1
$b(x) = (\;) - x, \;\; c(x) = (\;)$	See 22.6.15
$b(x) = (\;)x, \;\; c(x) = (\;) + x^{(\;)}$	See 9.1.53
$b(x) = \dfrac{(\;)}{x}, \;\; c(x) = (\;)$	See 9.1.52
$b(x) = (\;), \;\; c(x) = (\;) - (\;)\cos x$	See 20.1.1

Equations of the form: $xy'' + b(x)y' + c(x)y = 0$

$b(x) = (\;) - x, \;\; c(x) = (\;)$	See 13.1.1
$b(x) = (\;) + x, \;\; c(x) = (\;) + \dfrac{(\;)}{x}$	See 22.6.16

Equations of the form: $\left(1 - x^2\right) y'' + b(x)y' + c(x)y = 0$

$b(x) = (\;), \;\; c(x) = (\;) - (\;)x^2$	See 20.1.8
$b(x) = -x, \;\; c(x) = (\;)$	See 22.6.9
$b(x) = -x, \;\; c(x) = (\;) - (\;)x^2$	See 20.1.7
$b(x) = -2x, \;\; c(x) = (\;)$	See 22.6.13
$b(x) = -2x, \;\; c(x) = (\;) + \dfrac{(\;)}{1 - x^2}$	See 8.1.1
$b(x) = -3x, \;\; c(x) = (\;)$	See 22.6.11 and 22.6.12
$b(x) = (\;)x, \;\; c(x) = (\;)$	See 22.6.5 and 22.6.6
$b(x) = (\;) + (\;)x, \;\; c(x) = (\;)$	See 22.6.1 and 22.6.2

Equations of the form: $x^2 y'' + b(x)y' + c(x)y = 0$

$b(x) = x, \quad c(x) = x^2 - (\ )$	See 9.1.1
$b(x) = x, \quad c(x) = (\ ) - x^2$	See 9.6.1
$b(x) = 2x, \quad c(x) = (\ ) + x^2$	See 10.1.1
$b(x) = 2x, \quad c(x) = (\ ) - x^2$	See 10.2.1

Equations of the form: $x(1 - x)y'' + b(x)y' + c(x)y = 0$

$b(x) = (\ ) - (\ )x, \quad c(x) = (\ )$	See 15.5.1

II.A

Exact Methods for ODEs[*]

42. An N-th Order Equation

Applicable to The equation $\dfrac{d^n y}{dx^n} = f(x)$.

Yields

Two exact forms of the solution are available.

Idea

The explicit solution can be written analytically.

---

[*] Some of the methods in this section can be used for partial differential equations as well. These methods are indicated by a star ($*$).

Procedure

The geberal solution of the ordinary differential equation for $y(x)$

$$\frac{d^n y}{dx^n} = f(x)$$

can be found integrating with respect to x a total of n times. This produces

$$
\begin{aligned}
y(x) &= \int_{x_0}^{x} dx \int_{x_0}^{x} dx \cdots \int_{x_0}^{x} f(x)\, dx + C_1 \frac{(x - x_0)^{n-1}}{(n-1)!} \\
&\quad + C_2 \frac{(x - x_0)^{n-2}}{(n-2)!} + \cdots + C_{n-1}(x - x_0) + C_n,
\end{aligned}
\tag{42.1}
$$

for any x_0, where the $\{C_j\}$ represent arbitrary constants. This solution can also be written as

$$
\begin{aligned}
y(x) &= \frac{1}{(n-1)!} \int_{x_0}^{x} (x - t)^{n-1} f(t)\, dt + C_1 \frac{(x - x_0)^{n-1}}{(n-1)!} \\
&\quad + C_2 \frac{(x - x_0)^{n-2}}{(n-2)!} + \cdots + C_{n-1}(x - x_0) + C_n
\end{aligned}
\tag{42.2}
$$

in which there are no repeated integrals. Sometimes the form in (42.2) is more useful than the form in (42.1).

Example

The ordinary differential equation

$$
\begin{aligned}
y^{(4)} &= \sin x, \\
y(0) &= 0, \quad y'(0) = 0, \\
y''(0) &= 0, \quad y'''(0) = 0,
\end{aligned}
$$

has the solution

$$y(x) = \int_0^x dx \int_0^x dx \int_0^x dx \int_0^x \sin x\, dx.
\tag{42.3}$$

This solution may also be written as

$$y(x) = \frac{1}{6} \int_0^x (x - t)^3 \sin t\, dt.
\tag{42.4}$$

For some it is easier to evaluate the expression in (42.4) (by expanding out $(x - t)^3$ and integrating the four terms) to determine that

$$y(x) = \sin x - x + \frac{x^3}{6}$$

than it is to evaluate the expression in (42.3).

Notes

[1] When the answer is to be computed numerically, the solution represented by
(42.2) is more useful than the form in (42.1). It is much faster to numerically
approximate a one-dimensional integral than a multi-dimensional integral.

References

[1] E. L. Ince, *Ordinary Differential Equations*, Dover Publications, Inc., New
York, 1964, page 42.

43.   Use of the Adjoint Equation*

Applicable to   Linear differential equations.

Yields

A linear differential equation of lower order.

Idea

For every solution we can find of the adjoint equation, we can reduce
the order of the original equation by one.

Procedure

If we have the n-th order linear differential operator $L[\cdot]$ (shown operating on the function $u(x)$)

$$L[u(x)] = a_0(x)\frac{d^n u}{dx^n} + a_1(x)\frac{d^{n-1}u}{dx^{n-1}} + \cdots + a_{n-1}(x)\frac{du}{dx} + a_n(x)u, \quad (43.1)$$

then the adjoint of $L[\cdot]$ is defined to be $L^*[\cdot]$, where $L^*[\cdot]$ is given by (shown
operating on the function $w(x)$)

$$L^*[w(x)] = (-1)^n \frac{d^n}{dx^n}[a_0(x)w] + (-1)^{n-1}\frac{d^{n-1}}{dx^{n-1}}[a_1(x)w] + \cdots$$
$$+ (-1)^1 \frac{d}{dx}[a_{n-1}(x)w] + (-1)^0[a_n(x)]w,$$

see page 75 for details. The *bilinear concomitant* of $L[\cdot]$ is defined to be

$$B(u, w) = \sum_{k=1}^{n-1}\sum_{m=k}^{n-1}(-1)^{m-k}u^{(n-m-1)}(a_k w)^{(m-k)},$$

and satisfies the equation

$$wL[u] - uL^*[w] = \frac{d}{dx}B(u, w), \tag{43.2}$$

for all $u(x)$ and $w(x)$.

Suppose we wish to solve the equation $L[u] = f(x)$. If we can find a solution to $L^*[w] = 0$, call it $w^*(x)$, then we have (substituting into (43.2))

$$w^* L[u] - uL^*[w^*] = \frac{d}{dx}B(u, w^*),$$

or

$$w^*(x)f(x) = \frac{d}{dx}B(u, w^*),$$

or

$$B(u, w^*) = \int^x w^*(x)f(x)\, dx. \tag{43.3}$$

Therefore, to find $u(x)$ we can solve (43.3) instead of $L[u] = f(x)$. In other words, $w^*(x)$ is an integrating factor for the equation $L[u] = f(x)$. The original differential equation, $L[u] = f(x)$, is of degree n while equation (43.3) is of degree $n - 1$.

Special Case

For $n = 2$ the adjoint equation is important enough to write separately. If the linear operator $L[\cdot]$ is defined by $L[u(x)] = R(x)u'' + S(x)u' + T(x)u$, then the adjoint is $L^*[w(x)] = Rw'' + (2R' - S)w' + (R'' - S' + T)w$, and the bilinear concomitant is $B(u, v) = uSw + u'Rw - u(Rw)'$.

Example

Suppose we wish to solve the equation $L[u] = 1$, where

$$L[u] = (x^2 - x)u'' + (2x^2 + 4x - 3)u' + 8xu.$$

The adjoint, in this case, is the operator

$$L^*[w] = (x^2 - x)w'' + (-2x^2 + 1)w' + (4x - 2)w,$$

and the bilinear concomitant is given by

$$B(u, w) = u(x^2 - x)w + u'(2x^2 + 2x - 2)w - u(x^2 - x)w'. \tag{43.4}$$

A solution to $L^*[w] = 0$, obtained by the method of undetermined coefficients, is $w^*(x) = x^2$ Using this solution in (43.3) we obtain (with $f(x) = 1$)

$$B(u, w^*) = \int^x w^*(x)f(x)dx = \int^x x^2 dx = \frac{x^3}{3} + C,$$

where C is an arbitrary constant. Using $w = w^* = x^2$ in (43.4) produces

$$B(u, w^*) = (x^4 - x^3)u' + 2x^4 u.$$

Equating these last two equations yields a first order equation for u

$$(x^4 - x^3)u' + 2x^4 u = \frac{x^3}{3} + C. \tag{43.5}$$

Note that equation (43.5) is a first order equation (the original differential equation was of second order). Since (43.5) is a first order linear equation, it can be solved by the use of integrating factors. Multiplying by $\dfrac{x-1}{x^3}e^{2x}$ and integrating results in

$$(x - 1)^2 e^{2x} u(x) = \int^x \left[\frac{x-1}{3} e^{2x} + Ce^{2x} \frac{x-1}{x^3} \right] dx$$

$$= \frac{2x - 3}{12} e^{2x} + \frac{C}{2x^2} e^{2x} + D,$$

where D is another arbitrary constant. Hence, the final solution is

$$u(x) = \frac{1}{(x-1)^2} \left[\frac{2x - 3}{12} + \frac{C}{2x^2} + De^{-2x} \right].$$

Notes

[1] If an operator and its adjoint are identical, then the operator is said to be formally self-adjoint (see page 74). In this case, the adjoint method does not help to find a solution of the original differential equation. .

[2] Similar results hold for linear partial differential equations. For the partial differential operator

$$L[u] = \sum_{i,j=1}^{n} a_{ij}(\mathbf{x}) \frac{\partial^2 u}{\partial x_i \partial x_j} + \sum_{i=1}^{n} b_i(\mathbf{x}) \frac{\partial u}{\partial x_i} + c(\mathbf{x})u,$$

the adjoint operator is defined by

$$M[w] = \sum_{i,j=1}^{n} \frac{\partial^2 (a_{ij}w)}{\partial x_i \partial x_j} - \sum_{i=1}^{n} \frac{\partial (b_i w)}{\partial x_i} + cw.$$

With this definition of the adjoint, we find

$$\int_D \left(wL[u] - uM[w] \right) dx + \int_{\partial D} B[u, w] \, dx_1 \cdots \widehat{dx_i} \cdots dx_n = 0, \tag{43.6}$$

where $B[u, w]$ is defined by

$$B[u, w] = \sum_{i=1}^{n} (-1)^i \left\{ \left[b_i - \sum_{j=1}^{n} \frac{\partial a_{ij}}{\partial x_j} \right] uw + \sum_{j=1}^{n} a_{ij} \left[w \frac{\partial u}{\partial x_j} - u \frac{\partial w}{\partial x_j} \right] \right\}.$$

In (43.6), $dx_1 \cdots \widehat{dx_i} \cdots dx_n$ indicates the product $dx_1 \cdots dx_n$ with the factor dx_i removed. See Garabedian [1] or Zauderer [5] for details.

[3] If the elliptic operator $L[\cdot]$ is defined by $L[u] = -\nabla \cdot (p\nabla u) + qu$, then

$$wL[u] - uL[w] = \nabla \cdot (-pw\nabla u + pu\nabla w).$$

If the hyperbolic operator $\widetilde{L}[\cdot]$ is defined by $\widetilde{L}[u] = \rho u_{tt} + L[u]$, then

$$w\widetilde{L}[u] - u\widetilde{L}[w] = \widetilde{\nabla} \cdot [-pw\nabla u + pu\nabla w, \rho w u_t - \rho u w_t],$$

where $\widetilde{\nabla} = [\nabla, \partial/\partial t]$ is the space–time gradient operator. If the parabolic operator $\widehat{L}[\cdot]$ is defined by $\widehat{L}[u] = \rho u_t + L[u]$, then

$$w\widehat{L}[u] - u\widehat{L}^*[w] = \widetilde{\nabla} \cdot [-pw\nabla u + pu\nabla w, \rho uw],$$

where the operator $\widehat{L}^*[\cdot]$ is defined by $\widehat{L}^*[u] = -\rho u_t + L[u]$. Each of the last three equations can be integrated to obtain an expression similar to (43.6). See Zauderer [5] for details.

References

[1] P. R. Garabedian, *Partial Differential Equations*, Wiley, New York, 1964, pages 161–162.
[2] E. L. Ince, *Ordinary Differential Equations*, Dover Publications, Inc., New York, 1964, pages 123–125.
[3] W. Kaplan, *Operational Methods for Linear Systems*, Addison–Wesley Publishing Co., Reading, MA, 1962, pages 448–453.
[4] G. Valiron, *The Geometric Theory of Ordinary Differential Equations and Algebraic Functions*, Math Sci Press, Brookline, MA, 1950, pages 323–324.
[5] E. Zauderer, *Partial Differential Equations of Applied Mathematics*, John Wiley & Sons, New York, 1983, pages 483–486.

44.   Autonomous Equations

Applicable to   Ordinary differential equations of the form $F(y^{(n)}, y^{(n-1)}, \ldots, y'', y', y) = 0$.

Yields

An ordinary differential equation of lower order.

Idea

An autonomous equation is one left invariant under the transformation $x \to x + a$. Any ordinary differential equation in which the independent variable does not appear explicitly is an autonomous equation. Since we know something about the solution, we can reduce the order of the differential equation.

Procedure

Given the n-th order autonomous equation $F(y^{(n)}, y^{(n-1)}, \ldots, y'', y', y) = 0$, change the dependent variable from $y(x)$ to $u(y) = y'(x)$. The resulting ordinary differential equation for $u(y)$ will be of lower order. To find how the higher order derivatives transform, consult Table 44. After the ordinary differential equation of lower order has been solved for $u(y)$, $y(x)$ can be determined from integrating $u(y) = y'(x)$; i.e., $\displaystyle\int \frac{dy}{u(y)} = x$.

Example

Suppose we want to solve the nonlinear autonomous equation

$$\frac{d^2y}{dx^2} - \frac{dy}{dx} = 2y\frac{dy}{dx}. \tag{44.1}$$

Since there are no explicit occurrences of x in (44.1), we recognize the equation to be autonomous. Therefore, we change variables in (44.1) by $u(y) = \frac{dy}{dx}$. Using Table 44, (44.1) transforms into $u\frac{du}{dy} - u = 2yu$, or

$$u\left(\frac{du}{dy} - 1 - 2y\right) = 0. \tag{44.2}$$

From (44.2), either $u = 0$ or $\frac{du}{dy} - 1 - 2y = 0$. If $u(y) = 0$, then $\frac{dy}{dx} = 0$ and so one solution to (44.1) is

$$y(x) = A, \tag{44.3}$$

where A is a constant. Conversely, if $u(y) \neq 0$, then equation (44.2) requires that

$$\frac{du}{dy} - 1 - 2y = 0. \tag{44.4}$$

Equation (44.4) can be integrated to obtain

$$u(y) = y^2 + y + B, \tag{44.5}$$

where B is a constant. Using $u(y) = \frac{dy}{dx}$, equation (44.5) can be written as $\frac{dy}{dx} = y^2 + y + B$, so that $\displaystyle\int \frac{dy}{y^2 + y + B} = \int dx$, and therefore $\frac{2}{D}\tan^{-1}\left(\frac{2y+1}{D}\right) = x + C$, where $D^2 = 4B - 1$ and C is an additional constant. Inverting this last equation gives y explicitly as a function of x

$$y(x) = E\tan(Ex + F) - \tfrac{1}{2}, \tag{44.6}$$

where $E = D/2$ and $F = CE$. Hence, the two solutions to (44.1) are given by (44.3) and (44.6).

$$y_x = u,$$

$$y_{xx} = uu_y,$$

$$y_{xxx} = uu_y^2 + u^2 u_{yy},$$

$$y_{xxxx} = uu_y^3 + 4u_{yy}u_y u^2 + u^3 u_{yyy},$$

$$y_{xxxxx} = uu_y^4 + 7u_{yyy}u_y u^3 + 4u^3 u_{yyy}^2 + 11u^2 u_y^2 u_{yy} + u^4 u_{yyyy}.$$

To simplify notation, define $y_{x(n)}$ to be the n-th derivative of y with respect to x. Similarly for $u_{y(n)}$. The last equation is, therefore,

$$y_{x(5)} = uu_y^4 + 7u_{y(3)}u_y u^3 + 4u^3 u_{y(3)}^2 + 11u^2 u_y^2 u_{yy} + u^4 u_{y(4)},$$

$$y_{x(6)} = uu_y^5 + 11u_{y(4)}u_y u^4 + 15u^4 u_{y(3)}^2 u_{yy} + 32u^3 u_y^2 u_{y(3)}$$
$$\qquad + 34u^3 u_y u_{yy}^2 + 26u^2 u_y^3 u_{yy} + u^5 u_{y(5)},$$

$$y_{x(7)} = uu_y^6 + 57u^2 u_y^4 u_{yy} + 122u^3 u_y^3 u_{yyy} + 34u^4 u_{yy}^3 + 180u^3 u_y^2 u_{yy}^2$$
$$\qquad + 76u^4 u_{yy}^3 + 15u^5 u_{y(3)}2 + 192u^4 u_y u_{yy} u_{y(3)}$$
$$\qquad + 26u^5 u_{yy} u_{y(4)} + 16u^5 u_y u_{y(5)} + u^6 u_{y(6)}.$$

Table 44. How to transform derivatives under the change of independent variable: $u(y) = y_x(x)$.

Notes

[1] This method is derivable from Lie group methods (see page 314).

[2] In Schwarz's paper [2], there is a description of a REDUCE program that will automatically determine first integrals for an autonomous system of equations.

[3] The easiest way to make the necessary transformation in an autonomous differential equation is by replacing every occurrence of $\dfrac{d}{dx}$ with $u\dfrac{d}{dy}$. For instance, writing (44.1) in the form

$$\frac{d}{dx}\left(\frac{d}{dx}(y)\right) - \frac{d}{dx}(y) = 2y\frac{d}{dx}(y),$$

leads immediately to equation (44.2) via

$$u\frac{d}{dy}\left(u\frac{d}{dy}(y)\right) - u\frac{d}{dy}(y) = 2yu\frac{d}{dy}(y).$$

[4] Sometimes it is advantageous to write a pair of first order autonomous equations as a single first order equation, by diving the two equations. For example, the predator-prey equations

$$\frac{dx}{dt} = ax - bxy, \qquad \frac{dy}{dt} = -cy + dxy \qquad (44.7)$$

can be written in the form

$$\frac{dx}{dy} = \frac{ax - bxy}{-cy + dxy}. \qquad (44.8)$$

Although equation (44.7) cannot be solved explicitly in finite terms, from equation (44.8) we can show that $F(x, y) := dx + by - c \log x - a \log y$ is a constant on the solution curves $\{x(t), y(t)\}$.

[5] It is straightforward to create a MACSYMA program that will, for autonomous equations, perform the necessary change of variables. In the following copy of a terminal session, the input equation

$$\frac{1}{y}\frac{d^2y}{dx^2} - \frac{1}{y^2}\left(\frac{dy}{dx}\right)^2 - 1 + \frac{1}{y^3} = 0$$

is transformed into

$$y^3 - u\frac{du}{dy}y^2 + u^2y - 1 = 0.$$

```
AUTONOMOUS(EQN,Y,X):= BLOCK([NEW,A,U,MAX_DEGREE,J],
          DEPENDS(U,Y),
          MAX_DEGREE:DERIVDEGREE(EQN,Y,X),
          KILL(A),
          A[0]:Y,
          FOR J:1 THRU MAX_DEGREE DO (
                  A[J]:EXPAND( SUBST(U,DIFF(Y,X),DIFF(A[J-1],X)) ) ),
          FOR J:1 THRU MAX_DEGREE DO (
                  NEW: SUBST( A[J], DIFF(Y,X,J), NEW ) ),
          FACTOR(NEW) )$

DEPENDS(Y,X)$
EQN: DIFF( DIFF(Y,X)/Y, X) -1+1/Y**3;
```

$$\frac{y_{xx}}{y} - \frac{(y_x)^2}{y^2} - 1 + \frac{1}{y^3}$$

```
AUTONOMOUS(EQN,Y,X);
```

$$\frac{y^3 - u\,u_y\,y^2 + u^2\,y - 1}{y^3}$$

[6] Autonomous systems of ordinary differential equations can have *center manifolds*, which are a classification of the solution surface. As a simple example, consider the system

$$\mathbf{x}' = A\mathbf{x} + \mathbf{f}(\mathbf{x}, \mathbf{y}), \qquad \mathbf{y}' = B\mathbf{x} + \mathbf{g}(\mathbf{x}, \mathbf{y}), \qquad (44.9)$$

where A is a constant matrix all of whose eigenvalues are imaginary, B is a constant matrix all of whose eigenvalues have negative real part, and the functions $\mathbf{f}$ and $\mathbf{g}$, and their first derivatives, vanish as the point $(0,0)$. Then there is a function $\mathbf{h}$ such that

(A) **h** is an invariant manifold under (44.9),
(B) **h** and its first derivatives vanish at $(\mathbf{0}, \mathbf{0})$,
(C) the stability of the solution $(\mathbf{0}, \mathbf{0})$ is the same as that of the smaller system $\mathbf{x}' = A x + \mathbf{f}(\mathbf{x}, \mathbf{h}(\mathbf{x}))$.

References

[1] C. M. Bender and S. A. Orszag, *Advanced Mathematical Methods for Scientists and Engineers*, McGraw–Hill, New York, 1978, pages 24–25.
[2] F. Schwarz, "An Algorithm for Determining Polynomial First Integrals of Autonomous Systems of Ordinary Differential Equations," *J. Symbolic Comp.*, **1**, 1985, pages 229–233.
[3] E. D. Rainville and P. E. Bedient, *Elementary Differential Equations*, The MacMillan Company, New York, 1964, pages 268–269.

45.   Bernoulli Equation

Applicable to   Ordinary differential equations of the form: $y' + P(x)y = Q(x)y^n$.

Yields

An exact solution of the given equation.

Idea

By a change of dependent variable, a Bernoulli equation (which is a nonlinear equation of the form $y' + P(x)y = Q(x)y^n$, where n is not equal to 1) can be transformed to a first order linear equation. This linear equation can be solved by the use of integrating factors.

Procedure

Suppose we have the equation

$$y' + P(x)y = Q(x)y^n, \tag{45.1}$$

which we recognize to be a Bernoulli equation. To solve, we divide the equation by y^n and change the dependent variable from $y(x)$ to $u(x)$ by

$$u(x) = y(x)^{1-n}.$$

This changes (45.1) into the first order linear differential equation

$$\frac{1}{1-n} u' + P(x)u = Q(x). \tag{45.2}$$

An exact solution of (45.2) can be found by integrating factors (see page 305). The solution is given by

$$u(x) = \exp\left[(n-1)\int^x P(t)\,dt\right]\left\{\int^x \exp\left[(1-n)\int^s P(t)\,dt\right]Q(s)\,ds\right\}.$$

Example

Suppose we have the equation

$$y' + y = y^3 \sin x. \tag{45.3}$$

To solve this equation, divide it by y^3 and then define $u(x) = y(x)^{-2}$ so that (45.3) becomes

$$-\tfrac{1}{2}u' + u = \sin x. \tag{45.4}$$

The solution to (45.4) (obtained by the method of integrating factors) is

$$u(x) = Ae^{2x} + \tfrac{2}{5}(\cos x + 2\sin x),$$

where A is an arbitrary constant. Using $y(x) = u(x)^{-1/2}$, the final solution is found to be

$$y(x) = \left[Ae^{2x} + \tfrac{2}{5}(\cos x + 2\sin x)\right]^{-1/2}.$$

Notes

[1] If $n = 1$, then the original equation is in the form of (45.2) and it can be solved directly by the use of integrating factors.

References

[1] W. E. Boyce and R. C. DiPrima, *Elementary Differential Equations and Boundary Value Problems*, Fourth Edition, John Wiley & Sons, New York, 1986, page 28.

[2] E. L. Ince, *Ordinary Differential Equations*, Dover Publications, Inc., New York, 1964, page 22.

[3] E. D. Rainville and P. E. Bedient, *Elementary Differential Equations*, The MacMillan Company, New York, 1964, pages 69–71.

[4] G. F. Simmons, *Differential Equations with Applications and Historical Notes*, McGraw–Hill Book Company, New York, 1972, page 49.

46.　Clairaut's Equation

Applicable to　Differential equations of the form: $f(xy' - y) = g(y')$.

Yields

An exact implicit solution. Sometimes a singular solution may also be obtained.

Idea

A solution of the differential equation $f(xy' -$

Procedure

Given the equation

$$f(xy' - \qquad \tag{46.1}$$

a general solution (for which $y'' = 0$,) an implicitly by

$$f(xC - y, \quad (C)), \tag{46.2}$$

where C is an arbitrary constant. Equation (46.1) may also have a singular solution. If it does, it can be obtained by differentiating (46.1) with respect to x to obtain

$$y'' \left[f'(xy' - y)x - g'(y') \right] = 0. \tag{46.3}$$

If the first term in (46.3) is zero, then (46.2) is recovered. If the second term in (46.3) is zero, then (46.1) and (46.2) can be solved together to eliminate y'. The resulting equation for $y = y(x)$ will have no arbitrary constants and so will be a singular solution.

Example

Suppose we have the ordinary differential equation

$$(xy' - y)^2 - (y')^2 - 1 = 0. \tag{46.4}$$

Since (46.4) is of the same form as (46.1) (with $f(x) = x^2$, $g(x) = x^2 - 1$), a general solution can immediately be written down as $(xC - y)^2 = C^2 + 1$, or

$$y = Cx \pm \sqrt{C^2 - 1}, \tag{46.5}$$

where C is an arbitrary constant.

To find the singular solution, we differentiate (46.4) with respect to x to obtain

$$y''[2(xy' - 2)x - 2y'] = 0.$$

If the second term is set equal to zero, then we find

$$y' = \frac{xy}{x^2 - 1}. \tag{46.6}$$

Using (46.6) in (46.4) we determine the singular solution to be

$$x^2 + y^2 = 1. \tag{46.7}$$

Note that (46.7) is not derivable from (46.5), for any choice of C.

Notes

[1] The singular solution obtained by this method turns out to be the locus of the solutions in (46.2). That is, the envelope of the solutions in (46.2), for all possible values of the parameter C, will be the singular solution. See Ford [1] for details.

[2] A generalization of Clairaut's equation is Lagrange's equation (see page 311).

[3] Clairaut's partial differential equation

$$z = \sum_{i=1}^{n} x_i \frac{\partial z}{\partial x_i} + f\left(\frac{\partial z}{\partial x_1}, \dots, \frac{\partial z}{\partial x_n}\right)$$

has the solution $z = \sum_{i=1}^{n} a_i x_i + f(a_1, a_2, \dots, a_n)$. See Kamke [3].

References

[1] L. R. Ford, *Differential Equations*, McGraw–Hill Book Company, New York, 1955, pages 16–18

[2] E. L. Ince, *Ordinary Differential Equations*, Dover Publications, Inc., New York, 1964, pages 39–40.

[3] E. Kamke, *Differentialgleichungen Lösungsmethoden und Lösungen*, Volume II, Chelsea Publishing Company, New York, 1947, section 13.8, page 123.

[4] E. D. Rainville and P. E. Bedient, *Elementary Differential Equations*, The MacMillan Company, New York, 1964, pages 263–265.

47. Computer-Aided Solution

Applicable to Some classes of ordinary differential equations, most frequently first and second order equations.

Yields

An exact solution.

Idea

Some of the computer algebra languages that are available have a symbolic differential equation solver.

Procedure

Find a computer system that runs any of the following computer languages: DERIVE, FORMAC, MACSYMA, MAPLE, Mathematica, mu-Math, REDUCE, Scratchpad, SMP. Then learn how to use the language, and find the routine that solves differential equations automatically.

Example 1

The following is a copy of a terminal session using MACSYMA, which was run on a SUN3 workstation. Note that (c2), (c3), (c4), (c5) are input lines ("command" lines) and that (d2), (d3), (d4), (d5) are output lines ("display" lines).

In line (c2) a second order differential equation is input. In line (c3) the computer is asked for the solution. In line (c4) a nonlinear first order differential equation is entered. In line (c5) the solution is requested. The terms {%c, %k1, %k2} are arbitrary constants in the solution that MACSYMA found.

```
DOE MACSYMA Version 10.
Copyright 1982, Massachusetts Institute of Technology
Dumped by tony on Wed Jun 22 04:41:34 1988.
Loading global MACSYMA init file.
Welcome to DOE-MACSYMA

(c2) 'diff(y,x,2)+4*y=0;

                              2
                             d y
(d2)                         --- + 4 y = 0
                              2
                             dx

(c3) ode(%,y,x);

<< several lines deleted >>

(d3)                  y = %k1 sin(2 x) + %k2 cos(2 x)

(c4) 'diff(y,x)=y/(x+y*log(y));

                             dy        y
(d4)                         -- = ------------
                             dx    y log(y) + x

(c5) ode(%,y,x);

                                      2
                             2 x - y log (y)
(d5)                       - --------------- = %c
                                   2 y
```

Example 2

The following is a copy of a terminal session of SMP that was run on a SUN3 workstation. Note that #I[1], and #I[2] are input lines and that #0[1], and #0[2] are output lines. On line #I[1] the second order differential equation, $y'' + 4y = 0$ is input. On line #0[1] SMP returned the same differential equation in a slightly different format. On line #I[2] the solution is asked for. The terms #k1 and #k2 are arbitrary constants in the solution that SMP found.

```
#I[1]::  Dt[y,{x,2}] + 4 u = 0
#0[1]:   4u + Dt[y,x,{x,1}] = 0
#I[2]::  Odesol[%,y,x]

0.D.E. Solver

equation:     4u + Dt[y,x,{x,1}] = 0

1) order  =    2
2) type   = linear: a[x] y''+b[x] y'+c[x] y=f[x]
3) homogeneous: f[x] = 0

#0[2]:    {y -> #k1 Cos[2x] + #k2 Sin[2x]}
```

Example 3

The following copy of a terminal session of MAPLE is due to Keith Geddes. On the first line the differential equation $y'' + 4y = 0$ is input; on the second line a solution is requested. Note that C and C1 are arbitrary constants in the solution that MAPLE found.

```
      |\^/|
._|\|   |/|_. Watmum at Univ of Waterloo
 \ MAPLE / Version 4.2 --- Dec 1987
 <____ ____> For on-line help, type  help();
      |
> diff(y(x),x$2) + 4*y(x) = 0;
                    2
                    d
                  ----- y(x) + 4 y(x) = 0
                    2
                  dx

> dsolve(",y(x));
                                              2
              y(x) = 2 C sin(x) cos(x) + 2 C1 cos(x)  - C1
```

Example 4

The following is a modification of a terminal session of REDUCE which was originally created by Malcolm MacCallum. After the program is told that y depends on x, the differential equation $y'' + 4y = 0$ is input; on the next line a solution is requested.

```
Cambridge ORION/UNIX REDUCE System

REDUCE 3.3 (7 Dec 1987) :

1: in odesolve $

<< several lines deleted >>

2: depend y,x;

3: ode := df(y,x,2)+4*y;

ode := df(y,x,2)+4*y;

4: odesolve(ode,y,x);

This is a linear ODE of order 2 with constant coefficients

The solution is given by equating the following to zero

arbconst(2)*sin(2*x) + arbconst(1)*cos(2*x) - y
```

Example 5

The following is a copy of a terminal session using muMATH, which was run on an IBM-XT. Note that input lines begin with a question mark and output lines begin with an "at" sign.

Initially a package is loaded in that contains all of the routines for algebraic manipulation and for solving differential equations. On the next line the differential equation $y'' + 4y = 0$ is input; then a solution is requested. On the following line a nonlinear first order ordinary differential equation is input; its implicit solution is then found. The terms ARB(1), ARB(2), ARB(3) are arbitrary constants in the solution that muMATH found.

```
muSIMP-83 4.06 (12/18/83)
MS-DOS Version
Copyright (C) 1982 The SOFT WAREHOUSE
Licensed by MICROSOFT Corp.

<< several lines deleted >>

? DE: DIF(Y(X),X,2) + 4*Y(X)==0;
@: 4 Y (X) + DIF (Y (X), X, 2) == 0

? SOLVE(DE,Y(X));
@: {Y (X) == ARB(1) SIN (2 X) + ARB(2) COS (2 X)}
```

```
? DIF(Y(X),X)==Y(X)/(X+Y(X)*LN(Y(X)));
@: DIF (Y (X), X) == Y (X)/(X + Y (X) LN Y (X))

? SOLVE(@,Y(X));
@: {2 X - Y (X) LN Y(X)^2 + 2 Y (X) ARB(3) == 0}
```

Example 6

The following is a modified copy of a terminal session using Mathematica on a SUN4. Note that the nth input line is denoted $In[n]$ and the nth output line is denoted $Out[n]$. On the first input line the differential equation $y'' + 4y = 0$ is input; on the second input line a solution is requested. The terms $C[1]$ and $C[2]$ are arbitrary constants in the solution that Mathematica found.

```
Mathematica (sun4) 1.2 (November 6, 1989) [With pre-loaded data]
by S. Wolfram, D. Grayson, R. Maeder, H. Cejtin,
   S. Omohundro, D. Ballman and J. Keiper
with I. Rivin and D. Withoff
Copyright 1988,1989 Wolfram Research Inc.
 -- Terminal graphics initialized --

In[1]:= equation= y''[x]+4 y[x]==0
Out[1]= 4 y[x] + y''[x] == 0

In[2]:= DSolve[ equation, y[x], x]
Out[2]= {{y[x] -> C[2] Cos[-2 x] + C[1] Sin[-2 x]}}
```

Notes

[1] All of these symbolic manipulation languages can be used interactively or through a batch processor.

[2] Packages that can handle a wider variety of differential equations are constantly being created. See, for example, Schmidt [5], Kovacic [4], Watanabe [9], or Chan [1].

[3] Many of the programs illustrated above, and many others (such as the package by Hubbard and West [8]) can be run on a microcomputer (such as an IBM PC or a Macintosh).

[4] Given a homogeneous linear differential equation, whose coefficients are in a finite algebraic extension of $\mathbf{Q}[x]$, Singer's paper [14] has a decision procedure to determine a basis for the Liouvillian solutions. Liouvillian functions are, essentially, those functions that can be built up from rational functions by algebraic operations, taking exponentials and by integration.

[5] The theory underlying the computer solution of ordinary differential equations is quite complex. We need the following definitions:

(A) Let K be a field of functions. The function θ is a *Liouvillian generator* over K if it is:

(A) algebraic over K, that is if θ satisfies a polynomial equation with coefficients in K;

(B) exponential over K, that is if there is a ζ in K such that $\theta' = \zeta'\theta$, which is an algebraic way of saying that $\theta = \exp \zeta$; or

(C) an integral over K, that is if there is a ζ in K such that $\theta' = \zeta$, which is an algebraic way of saying that $\theta = \int \zeta$.

(B) Let K be a field of functions. An over-field $K(\theta_1, \ldots, \theta_n)$ of K is called a *field of Liouvillian functions over K* if each θ_i is a Liouvillian generator over K. A function is *Liouvillian over K* if it belongs to a Liouvillian field of functions over K.

Then, some of the important theorems in this area are:

Theorem: There is an algorithm which, given a second order linear differential equation, $y'' + ay' + by = 0$ with a and b rational functions of x, either finds two Liouvillian solutions such that every solution is a linear combination with constant coefficients of these two solutions or proves that there is no Liouvillian solution (except zero).

Theorem: There is an algorithm which, given a linear differential equation of any order, the coefficients of which are rational or algebraic functions: either finds a Liouvillian solution, or proves that there is none.

Theorem: Let A be a class of functions containing the coefficients of a linear differential operator L, let g be an element of A, and let us suppose that the equation $L[y] = g$ has an elementary solution over A. Then either $L[w] = 0$ has an algebraic solution over A, or y belongs to A.

Theorem: Let A be a class of functions, which contains the coefficients of a linear differential operator L, let g be an element of A, and let us suppose that the equation $L[y] = g$ has a Liouvillian solution over A. Then either $L[w] = 0$ has a solution $e^{\int z(x)\,dx}$ with z algebraic over A, or y belongs to A.

See Davenport, Siret, and Tournier [3] for details.

[6] An example of the use of FORMAC may be found in Hanson *et al.* [6].

[7] Shtokhamer [13] presents a MACSYMA program that implements the Prelle–Singer algorithm, and gives several examples.

References

[1] M. Bronstein, "The Transcendental Risch Differential Equation," *J. Symbolic Comp.*, **9**, No. 1, 1990, pages 49–60.

[2] W. C. Chan, "A Novel Symbolic Ordinary Differential Equation Solver," *SIGSAM Bulletin*, **15**, No. 3, August 1981, Issue Number 59, pages 9–14.

[3] J. H. Davenport, Y. Siret, and E. Tournier, *Systems and Algorithms for Algebraic Computation*, Academic Press, New York, 1988.

[4] DERIVE, SoftWarehouse, Inc., 3615 Harding Avenue, Honolulu, Hawaii 96816.

[5] K. O. Geddes, G. H. Gonnet, and B. W. Char, *MAPLE User's Manual*, University of Waterloo, Waterloo, Ontario, 1983.

[6] J. H. Hanson, A. C. Benander, and B. A. Benander, "The Computer Generated Symbolic Solution of a System of Linear First Order Differential Equations," *Comp. & Maths. with Appls.*, **19**, No. 7, 1990, pages 7–12.

[7] A. C. Hearn, *REDUCE 2 User's Manual*, Second Edition, University of Utah, Computational Physics Group Report UCP-19, Salt Lake City, Utah, 1973.

[8] J. Hubbard and B. West, *MacMath: A Dynamical Systems Software Package*, Springer–Verlag, New York, 1991.

[9] J. J. Kovacic, "An Algorithm for Solving Second Order Linear Homogeneous Differential Equations," *J. Symbolic Comp.*, 1986, **2**, pages 3–43.

[10] K. R. Meyer and D. Schmidt, *Computer Aided Proofs in Analysis*, Springer–Verlag, New York, 1990.

[11] M. J. Prelle and M. F. Singer, "Elementary First Integrals of Differential Equations," *Trans. Amer. Math. Soc.*, **279**, No. 1, 1983.

[12] P. Schmidt, "Substitution Methods for the Automatic Symbolic Solution of Differential Equations of First Order and First Degree," in E. Ng (ed.), *EUROSAM 79*, Springer–Verlag, New York, 1979, pages 164–176.

[13] R. Shtokhamer, *Solving First Order Differential Equations Using the Prelle–Singer Algorithm*, Technical Report No. 88-09, May 1988, University of Delaware, Newark, Delaware.

[14] M. F. Singer, "Liouvillian Solution of *n*-th Order Homogeneous Linear Differential Equations," *Am. J. Math.*, **103**, No. 4, 1981, pages 661–682.

[15] SMP, Inference Corp., Suite 203, 3916 S. Sepulveda Blvd., Culver City, Calif., 90230.

[16] E. Tournier, *Computer Algebra and Differential Equations*, Academic Press, New York, 1990.

[17] *VAX UNIX MACSYMA Reference Manual*, Symbolics Inc., Cambridge, MA, 1985.

[18] S. Watanabe, "An Experiment Towards a General Quadrature for Second Order Linear Ordinary Differential Equations by Symbolic Computation," in J. Fitch (ed.), *EUROSAM 1984*, Springer–Verlag, New York, 1984, pages 13–22.

[19] S. Watanabe, "A Technique for Solving Ordinary Differential Equations Using Riemann's *P*–functions," in P. S. Wang (ed.), *SYMSAC 81: Proceedings of the 1981 ACM Symposium on Symbolic and Algebraic Computation*, 1981, ACM, New York, pages 36–43.

[20] S. Wolfram, *Mathematica: A System of Doing Mathematics*, Addison–Wesley Publishing Co., Reading, MA, 1988.

[21] C. Wooff and D. Hodgkinson, *muMATH: A Microcomputer Algebra System*, Academic Press, New York, 1987.

48.   Constant Coefficient Linear Equations

Applicable to   Homogeneous linear ordinary differential equations with constant coefficients.

Yields

An exact solution.

Idea

Linear constant coefficient ordinary differential equations have exponential solutions. The method of undetermined coefficients can be used to solve this type of equation after a polynomial has been factored.

Procedure

Given the n-th order linear equation

$$y^{(n)} + a_{n-1}y^{(n-1)} + \cdots + a_1 y' + a_0 y = 0, \qquad (48.1)$$

where the $\{a_i\}$ are constants, look for a solution of the form

$$y(x) = Ce^{\lambda x}, \qquad (48.2)$$

where C is an arbitrary constant. Substituting (48.2) into (48.1) yields

$$e^{\lambda x}\left[\lambda^n + a_{n-1}\lambda^{(n-1)} + \cdots + a_1\lambda + a_0\right] = 0.$$

Hence, (48.2) is a solution of (48.1) if λ is a root of the *characteristic equation*, defined by

$$\lambda^n + a_{n-1}\lambda^{(n-1)} + \cdots + a_1\lambda + a_0 = 0. \qquad (48.3)$$

If (48.3) has n different roots $\{\lambda_i\}$, then the general solution to (48.1) is, by use of superposition,

$$y(x) = C_n e^{\lambda_n x} + C_{n-1} e^{\lambda_{n-1} x} + \cdots + C_1 e^{\lambda_1 x},$$

where the $\{C_i\}$ are arbitrary constants. If some of the roots of (48.3) are repeated (say $\lambda_1 = \lambda_2 = \ldots = \lambda_m$), then the solution corresponding to these $\{\lambda_i\}$ is

$$y(x) = (C_m x^{m-1} + C_{m-1} x^{m-2} + \cdots + C_2 x + C_1) e^{\lambda_1 x}.$$

Example
Given the linear differential equation

$$y^{(7)} - 14y^{(6)} + 80y^{(5)} - 242y^{(4)} + 419y^{(3)} - 416y'' + 220y' - 48y = 0, \quad (48.4)$$

we substitute $y(x) = e^{\lambda x}$ to find the characteristic equation

$$\lambda^7 - 14\lambda^6 + 80\lambda^5 - 242\lambda^4 + 419\lambda^3 - 416\lambda^2 + 220\lambda - 48 = 0,$$

which factors as

$$(\lambda - 1)^3 (\lambda - 2)^2 (\lambda - 3)(\lambda - 4) = 0. \qquad (48.5)$$

The roots of equation (48.5) are $\{1, 1, 1, 2, 2, 3, 4\}$. The general solution to (48.4) is therefore

$$y(x) = \{C_0 + C_1 x + C_2 x^2\} e^x + \{C_3 + C_4 x\} e^{2x} + C_5 e^{3x} + C_6 e^{4x},$$

where $\{C_0, \ldots, C_6\}$ are arbitrary constants.

Notes
[1] Using the transformation described on page 118, the system in (48.1) can be written in the form $\mathbf{y}' = A\mathbf{y}$, where A is an $n \times n$ constant matrix. Then the techniques described on page 360 may be used.

References
[1] W. E. Boyce and R. C. DiPrima, *Elementary Differential Equations and Boundary Value Problems*, Fourth Edition, John Wiley & Sons, New York, 1986, Section 5.3 (pages 263–268).
[2] G. F. Simmons, *Differential Equations with Applications and Historical Notes*, McGraw–Hill Book Company, New York, 1972, pages 83–86.

49.   Contact Transformation

Applicable to   First order and (occasionally) second order ordinary differential equations.

Yields
A reformulation, which may lead to an exact solution (sometimes in parametric form).

Idea
By changing variables a different, and sometimes easier, differential equation may be found.

Procedure
Given a relation between three variables

$$\phi(x, y, p) = 0, \tag{49.1}$$

it will be a first order ordinary differential equation if $dy - p\,dx = 0$. If the variables in (49.1) are changed by

$$\begin{aligned} x &= x(X, Y, P), \\ y &= y(X, Y, P), \\ p &= p(X, Y, P), \end{aligned} \tag{49.2}$$

then the transformed equation $\phi(X, Y, P) = 0$ will also be an ordinary differential equation if $dY - P\,dX = 0$. If this is true, then (49.2) is a *contact transformation*. For example, the change of variables

$$\begin{Bmatrix} x = P \\ y = PX - Y \\ p = X \end{Bmatrix} \Longleftrightarrow \begin{Bmatrix} X = p \\ Y = px - y \\ P = x \end{Bmatrix} \tag{49.3}$$

is a contact transformation. It is easy to show this:

$$\begin{aligned} 0 = dy - p\,dx \\ = d(PX - Y) - X\,dP \\ = P\,dX - dY. \end{aligned}$$

If the new differential equation, $\Phi(X, Y, P) = 0$, can be solved, then the solution to $\phi(x, y, p) = 0$ may be determined by eliminating X, Y, and P from the original equation, using the solution found and the transformation rules.

Example

Suppose we are given the nonlinear first order ordinary differential equation

$$2y \left(\frac{dy}{dx}\right)^2 - 2x \frac{dy}{dx} - y = 0, \qquad (49.4)$$

which we may write as

$$2yp^2 - 2xp - y = 0.$$

We utilize the contact transformation in (49.3) to obtain, after some algebra, the new first order ordinary differential equation

$$P + Y \left(\frac{1 - 2X^2}{2X^3 - 3X}\right) = 0. \qquad (49.5)$$

This differential equation can be solved by integrating factors to obtain

$$Y = C \left(2X^3 - 3X\right)^{1/3}, \qquad (49.6)$$

where C is an arbitrary constant. Now that we have the solution of the transformed equation, we can find the solution of the original differential equation.

Utilizing $Y = xX - y$ and $P = x$ from (49.3), equations (49.5) and (49.6) can be written as

$$x + (xX - y) \left(\frac{1 - 2X^2}{2X^3 - 3X}\right) = 0,$$

$$xX - y = \left(2X^3 - 3X\right)^{1/3}. \qquad (49.7)$$

Now X can be eliminated between these two equations by, say, the method of resultants (see page 46). This produces the solution to (49.4) in the form $f(x, y) = 0$ (there are 21 algebraic terms in this representation). Alternately, we can obtain a parametric representation of the solution by solving (49.7) for $x = x(X)$ and $y = y(X)$ and then treating X as a parameter.

Notes

[1] The condition $dy - pdx = 0$ states that, if the point (x, y) is on a curve, then p should be its tangent. The change of variables in this method gives a different parameterization of the same curve. In particular, if two curves touch in the old parameterization, then they also touch in the new parameterization; hence the name of the transformation.

[2] Some second order ordinary differential equations may also be solved by this method. If $R = \dfrac{dP}{dX} = \dfrac{d^2Y}{dX^2}$ and $\dfrac{1}{R} = \dfrac{dp}{dx} = \dfrac{d^2y}{dx^2}$ then we may use the relation $dP - RdX = dx - Rdp$.

[3] In more generality, a transformation of the $2n + 1$ variables $\{z, x_j, p_j \mid j = 1, \ldots, n\}$ to the $2n + 1$ variables $\{Z, X_j, P_j \mid j = 1, \ldots, n\}$ is a contact transformation if the total differential equation

$$dz - p_1 dx_1 - p_2 dx_2 - \ldots - p_n dx_n = 0$$

is invariant under the transformation; i.e., if the equality

$$dZ - P_1 dX_1 - P_2 dX_2 - \ldots - P_n dX_n = \rho\,[dz - p_1 dx_1 - p_2 dx_2 - \ldots - p_n dx_n]$$

holds identically for some nonzero function $\rho(\mathbf{x}, \mathbf{p}, z)$. See Iyanaga and Kawada [5] for details.

[4] A contact transformation is also a canonical transformation (see page 105). The generating function of the canonical transformation, Ω, satisfies the three relations: $\Omega(x, z, X, Z) = 0$, $\dfrac{\partial \Omega}{\partial X_j} + P_j \dfrac{\partial \Omega}{\partial Z} = 0$, and $\dfrac{\partial \Omega}{\partial x_j} + p_j \dfrac{\partial \Omega}{\partial z} = 0$.

[5] The Legendre transformation (see page 400) is an example of a contact transformation. For this transformation: $\Omega = Z + z + \sum x_j X_j$, $Z = \sum_j p_j x_j - z$, $X_j = -p_j$, $P_j = -x_j$, and $\rho = -1$.

[6] The Pedal transformation is given by: $\Omega = Z^2 - zZ - \sum x_j X_j + \sum X_j^2$, $X_j = -p_j Z$, $p_j = -\dfrac{2X_j - x_j}{2Z - z}$, and $\rho = \dfrac{Z}{2Z - z}$.

[7] The similarity transformation is given by: $\Omega = (Z - z)^2 - a^2 + \sum (X_j - x)j)^2$, $X_j = x_j - ap_j \left(1 + \sum p_j^2\right)^{-1/2}$, $P_j = p_j$, $Z = x_j + a\left(1 + \sum p_j^2\right)^{-1/2}$, and $\rho = 1$.

[8] Composing two contact transformations, or taking the inverse of a contact transformation, results in another contact transformation. Since the identity transformation is also a contact transformation, the set of all contact transformations forms an infinite dimensional topological group.

[9] This method is derivable from the method of Lie groups (see page 314), where it goes by the name of the *extended group of transformations*. See Ince [4] or Seshadri and Na [6] for details.

[10] Some other contact transformations are:

$$\left\{\begin{array}{l} x = X - YP \\ y = -Y\sqrt{P^2 - 1} \\ p = \dfrac{P}{\sqrt{P^2 - 1}} \end{array}\right\} \Longleftrightarrow \left\{\begin{array}{l} X = x - yp \\ Y = y\sqrt{p^2 - 1} \\ P = -\dfrac{p}{\sqrt{p^2 - 1}} \end{array}\right\}$$

$$\left\{\begin{array}{l} x = X - \dfrac{aP}{\sqrt{1 + P^2}} \\ y = Y + \dfrac{a}{\sqrt{1 + P^2}} \\ p = P \end{array}\right\} \Longleftrightarrow \left\{\begin{array}{l} X = x + \dfrac{ap}{\sqrt{1 + p^2}} \\ Y = y - \dfrac{a}{\sqrt{1 + p^2}} \\ P = p \end{array}\right\}$$

References

[1] H. Bateman, *Partial Differential Equations of Mathematical Physics*, Dover Publications, Inc., New York, 1944, pages 81–83.

[2] C. Carathéodory, *Calculus of Variations and Partial Differential Equations of the First Order*, Holden–Day, Inc., San Francisco, 1965, Chapter 7 (pages 102–120).

[3] C. R. Chester, *Techniques in Partial Differential Equations*, McGraw–Hill Book Company, New York, 1970, pages 206–207.

[4] E. L. Ince, *Ordinary Differential Equations*, Dover Publications, Inc., New York, 1964, pages 40–42.

[5] S. Iyanaga and Y. Kawada, *Encyclopedic Dictionary of Mathematics*, MIT Press, Cambridge, MA, 1980, pages 286 and 1448.

[6] R. Seshadi and T. Y. Na, *Group Invariance in Engineering Boundary Value Problems*, Springer–Verlag, New York, 1985, pages 18–20.

50. Delay Equations

Applicable to Ordinary differential delay equations.

Yields

In many cases, an exact analytical solution.

Idea

There are several standard techniques for delay equations.

Procedure

The standard methods for solving delay equations are by the use of

(A) Laplace transforms,
(B) Fourier transforms,
(C) Generating functions,
(D) General expansion theorems,
(E) The method of steps.

For the first two methods, the technique is the same as it is for ordinary differential equations(see page 295). That is, the transform is taken of the delay equation; by algebraic manipulations the transform is explicitly determined; and then an inverse transformation is taken. See Example 1.

For a delay equation with a single delay, the *method of steps* consists of solving the delay equation in successive intervals, whose length is the time delay. In each interval, only an ordinary differential equation needs to be solved. See Example 2.

The method of generating functions is frequently used when only integral values of the variables are of interest. The technique is similar to the technique for integral transforms described above. For generating functions the integration is replaced by a summation, and the "inverse transformation" is generally a differentiation (see page 265 for more details). See Example 3.

The general expansion theorems are all of the same form; given a delay equation, the solution can be expressed as a sum over the roots of a transcendental equation called the *characteristic equation*.

Example 1

Suppose we have the delay equation

$$y'(t) + ay(t-1) = 0, \qquad (50.1)$$

with the boundary conditions

$$y(t) = y_0 \qquad \text{when } -1 \le t \le 0, \qquad (50.2)$$

where a is a constant. We define the Laplace transform of $y(t)$ to be $Y(s)$ by $Y(s) = \int_0^\infty e^{-st} y(t)\, dt$. Multiplying equation (50.1) by e^{-st} and integrating with respect to t yields

$$\int_0^\infty e^{-st} y'(t)\, dt + a \int_0^\infty e^{-st} y(t-1)\, dt = 0. \qquad (50.3)$$

The first integral in (50.3) can be integrated by parts to yield

$$\int_0^\infty e^{-st} y'(t)\, dt = sY(s) - y_0. \qquad (50.4)$$

The second integral in (50.3) can be evaluated by changing the variable of integration from t to $u = t - 1$:

$$a \int_0^\infty e^{-st} y(t-1) \, dt = a \int_{-1}^\infty e^{-s(u+1)} y(u) \, du$$

$$= a \int_0^\infty e^{-s(u+1)} y(u) \, du + a \int_{-1}^0 e^{-s(u+1)} y(u) \, du$$

$$= ae^{-s} Y(s) + ay_0 \frac{1 - e^{-s}}{s}.$$

$$(50.5)$$

Utilizing both (50.4) and (50.5) in (50.3) results in the algebraic equation

$$sY(s) - y_0 + ae^{-s} Y(s) + ay_0 \frac{1 - e^{-s}}{s} = 0,$$

which can be solved for $Y(s)$:

$$Y(s) = \frac{y_0}{s} - \frac{ay_0}{s(s + ae^{-s})}. \qquad (50.6)$$

If this formula for $Y(s)$ is expanded as

$$Y(s) = \frac{y_0}{s} - y_0 \sum_{n=0}^\infty (-1)^n a^{n+1} e^{-ns} s^{-n-2},$$

then an inverse Laplace transform may be taken term by term to conclude that

$$y(t) = y_0 \sum_{n=0}^{\lfloor t \rfloor + 1} (-a)^n \frac{(t - n + 1)^n}{n!}, \qquad (50.7)$$

where the floor function, $\lfloor t \rfloor$, is the greatest integer less than or equal to t.

Another way of expressing the solution in (50.7) is by taking the inverse transform of $Y(s)$, as defined in (50.6), directly, and using Cauchy's theorem to evaluate the Bromwich contour integral. This results in

$$y(t) = -ay_0 \sum_r \frac{e^{s_r t}}{s_r(1 + s_r)}, \qquad (50.8)$$

where the summation is over all roots of the equation

$$s + ae^{-s} = 0. \qquad (50.9)$$

All the roots of (50.9) will be simple unless $a = e^{-1}$, when there is a double root at $s = -1$. The solution in (50.8) can be approximated (for large t) by just using the s_r that has the smallest real part. There exist theorems (see Pinney [10] for instance) that allow the solution of (50.1) to be written in the form of (50.8) immediately.

Example 2

In the method of steps, only a sequence of ordinary differential equations need to be solved. To illustrate this method, consider equations (50.1) and (50.2). In the interval $0 \leq y \leq 1$ the solution satisfies

$$y'(t) + ay_0 = 0,$$
$$y(0) = y_0. \tag{50.10}$$

The equation in (50.10) has the solution

$$y(t) = y_0(1 - at), \qquad \text{for } 0 \leq y \leq 1. \tag{50.11}$$

Now we solve for $y(t)$ in the next interval of length one. Using (50.11) we find that, in the interval $1 \leq y \leq 2$, the solution satisfies

$$y'(t) + ay_0[1 - a(t - 1)] = 0,$$
$$y(0) = y_0(1 - a). \tag{50.12}$$

The equation in (50.12) has the solution

$$y(t) = y_0 \left[1 - at + \tfrac{1}{2}a^2(t - 1)^2\right], \qquad \text{for } 1 \leq y \leq 2.$$

This process can be repeated indefinitely. The solution obtained is identical to the solution in (50.7).

Example 3

This example shows how generating functions may be used to solve delay equations. Consider equations (50.1) and (50.2). We define the generating function associated with $y(t)$, for $0 \leq t \leq 1$, by

$$Y(t, k) = \sum_{p=0}^{\infty} y(t + p)k^p. \tag{50.13}$$

Once this generating function is known, $y(t)$ may be obtained in either or the two ways

$$y(t + p) = \frac{1}{p!} \left(\frac{\partial^p}{\partial k^p} Y(t, k)\right)\Bigg|_{k=0}$$
$$= \frac{1}{2\pi i} \int_C Y(t, k)k^{-p-1} \, dk,$$

where C is a closed contour surrounding the origin in the k-plane and lying wholly within the region of analyticity in k of $Y(t, k)$.

By differentiating (50.13) with respect to t, multiplying k and redefining p we find that

$$
\begin{aligned}
Y_t(t,k) &= \sum_{p=0}^{\infty} y'(t+p)k^p, \\
kY(t,k) &= \sum_{p=1}^{\infty} y(t+p+1)k^p.
\end{aligned}
\tag{50.14}
$$

If we now evaluate (50.1) when t has the value $t+p$, multiply by k^p, and sum with respect to p from one to infinity we find (by using (50.14))

$$
Y_t(t,k) + a\left(kY(t,k) + y(t-1)\right) = 0
$$

or, since $0 \le t \le 1$,

$$
Y_t(t,k) + akY(t,k) = -ay_0.
$$

This equation is an ordinary differential equation and can be readily solved to yield

$$
Y(t,k) = e^{-akt}F(k) - \frac{y_0}{k}
\tag{50.15}
$$

where $F(k)$ is some unknown function. We can determine this function by a judicious use of the initial conditions. Evaluating (50.13) at $t = 1$ we find

$$
\begin{aligned}
kY(1,k) &= k\sum_{p=0}^{\infty} y(1+p)k^p \\
&= \sum_{p=0}^{\infty} y(1+p)k^{p+1} \\
&= y(0) + \sum_{p=0}^{\infty} y(p)k^p \\
&= y(0) + Y(0,k).
\end{aligned}
\tag{50.16}
$$

Evaluating (50.16) by use of (50.15) results in

$$
k\left(e^{-ak}F(k) - \frac{y_0}{k}\right) = y_0 + \left(F(k) - \frac{y_0}{k}\right),
$$

or

$$
F(k) = \frac{y_0}{k\left(1 - e^{-ak}\right)}.
$$

This leads to the complete determination of the generating function

$$Y(t, k) = \frac{y_0}{k} \left(\frac{e^{-akt}}{1 - ke^{-ak}} - 1 \right).$$

Via some algebraic manipulations, we can obtain

$$Y(t, k) = y_0 \sum_{p=0}^{\infty} k^p \sum_{q=0}^{p+1} \frac{(-a(p + t - q + 1))^q}{q!}, \qquad (50.17)$$

so that the solution can be read off (compare (50.17) with (50.13)):

$$y(t) = y_0 \sum_{q=0}^{\lfloor t \rfloor + 1} (-a)^q \frac{(t - q + 1)^q}{q!}$$

where the floor function indicates the least integer.

Notes

[1] In the literature, equations of the form $y_h'(t) = y_{h-1}(t)$ are often called *differential–difference equations*, while equations of the form $y'(t) = y(t - 1)$ are called *mixed differential–difference equations*. Delay equations are also known as *functional equations, differential–delay equations, differential equations with deviating argument*, and as *equations with retarded arguments*. *Neutral differential equations* are differential equations in which the highest order derivative of the unknown function is evaluated both at the present state t and at one of more past or future states.

[2] The sunflower equation is: $\ddot{x}(t) + a\dot{x}(t) + b\sin(t - r) = 0$.

[3] The Cherwell–Wright differential equation is: $\dot{x}(t) = (a - x(t - 1))x(t)$.

[4] Marsaglia *et al.* [8] numerically evaluate the following functions:
   (A) Renyi's function: $[(x - 1)y(x)]' = 2y(x - 1)$
   (B) Dickman's function: $xy'(x) = -y(x - 1)$
   (C) Buchstab's function: $[xy(x)]' = y(x - 1)$

[5] Several authors have tried to analyze delay equations by replacing $y(t - r)$ with the first few terms of a Taylor series, say

$$y(t - r) \simeq y(t) - ry'(t) + \frac{1}{2}r^2 y''(t) - \ldots + (-1)^m \frac{1}{m!} r^m y^{(m)}(t).$$

This is generally a bad technique, as the approximations that are obtained are often unrelated to the original equation. See Driver [6] (page 235) for more details.

[6] The book by Pinney [10] contains a large compilation of delay equations that have appeared in the literature. References are cited, and the (then) current knowledge of each of the equations is given.

[7] The system of linear delay equations

$$\mathbf{u}'(t) = A\mathbf{u}(t) + B\mathbf{u}(t - d), \qquad \text{for } t \geq t_0,$$
$$\mathbf{u}(t) = \mathbf{g}(t), \qquad \text{for } -d \leq t \leq t_0,$$

where $d \geq 0$ is the delay and A and B are constant square matrices has a solution of the form $\mathbf{u}(t) = \mathbf{c}e^{st}$ if and only if s is a zero of the transcendental equation: $\det\left(Is - A - Be^{-ds}\right) = 0$.

[8] As an example of the general expansion theorems, the equation

$$au'(t) + bu(t) + cu(t - d) = 0,$$

where a, b, c, d are all constant, and d is positive, is satisfied by

$$u(t) = \sum_r p_r(t)e^{ts_r}, \qquad (50.18)$$

where $\{s_r\}$ are complex numbers satisfying $as_r + b + ce^{-ds_r} = 0$, and $p_r(t)$ is a polynomial in t of degree less than the multiplicity s_r (see Bellman and Cooke [3], page 55). The sum in (50.18) is either finite or infinite, with suitable conditions to insure convergence. In actuality, finding all the solutions to (50.15) is very difficult. This technique generalizes to higher order ordinary differential equations and partial differential equations, but the work in obtaining a solution becomes prohibitive unless numerical methods are used.

[9] Delay equations are usually solved numerically. A survey of numerical techniques for solving delay equations may be found in Cryer []. In Nieves' paper [9] is the description of a computer algorithm that will numerically approximate the solution of functional equations with a minimal amount of user input.

References

[1] A. N. Al-Butib, "One-Step Implicit Methods for Solving Delay Differential Equations," *Int. J. Comp. Math.*, **16**, 1984, pages 157–168.

[2] V. L. Bakke and Z. Jackiewicz, "Stability Analysis of Linear Multistep Methods for Delay Differential Equations," *Int. J. Math. & Math. Sci.*, **9**, No. 3, 1986, pages 447–458.

[3] R. E. Bellman and K. L. Cooke, *Differential–Difference Equations*, Academic Press, New York, 1963.

[4] T. A. Burton, *Stability and Periodic Solutions of Ordinary and Functional Differential Equations*, Academic Press, New York, 1985.

[5] C. W. Cryer, "Numerical Methods for Functional Differential Equations," in K. Schmitt (ed.), *Delay and Functional Differential Equations and Their Applications*, Academic Press, New York, 1972, pages 17–101.

[6] R. D. Driver, *Introduction to Ordinary Differential Equations*, Harper & Row, Publishers, New York, 1978, Chapter 6 (pages 206–237).

[7] L. E. El'sgol'ts and S. B. Norkin, *Introduction to the Theory and Application of Differential Equations with Deviating Arguments*, Academic Press, New York, 1973.

[8] G. Marsaglia, A. Zaman, and J. C. W. Marsaglia, "Numerical Solution of Some Classicial Differential-Difference Equations," *Math. of Comp.*, **53**, No. 187, July 1989, pages 191–201.

[9] K. W. Nieves, "Automatic Integration of Functional Differential Equations: An Approach," *ACM Trans. Math. Software*, **1**, No. 4, December 1975, pages 357–368.

[10] E. Pinney, *Ordinary Difference–Differential Equations*, University of California Press, Berkeley, 1959.

[11] T. L. Saaty, *Modern Nonlinear Equations*, Dover Publications, Inc., New York, 1981, Chapter 5 (pages 213–261).

[12] L. Torelli, "Stability of Numerical Methods for Delay Differential Equations," *J. Comput. Appl. Math.*, **25**, 1989, pages 15–26.

[13] R. Weiner and K. Strehmel, "A Type Insensitive Code for Delay Differential Equations Basing on Adaptive and Explicit Runge–Kutta Interpolation Methods," *Computing*, **40**, No. 3, 1988, pages 255–265.

51.   Dependent Variable Missing

Applicable to   Ordinary differential equations of the form $G(y^{(n)}, y^{(n-1)}, \ldots, y'', y', x) = 0$

Yields

An ordinary differential equation of lower order.

Idea

If the dependent variable does not appear explicitly in an ordinary differential equation, then the order of the ordinary differential equation can be reduced by one.

Procedure

Suppose we have the n-th order ordinary differential equation

$$G(y^{(n)}, y^{(n-1)}, \ldots, y'', y', x) = 0. \tag{51.1}$$

Notice that the variable $y(x)$ does not appear explicitly in (51.1).

If we define $p(x) = y'(x)$, then equation (51.1) becomes

$$G(p^{(n-1)}, p^{(n-2)}, \ldots, p', p, x) = 0, \tag{51.2}$$

which is an ordinary differential equation of order $(n-1)$ for the dependent variable $p(x)$. After solving (51.2) for $p(x)$, $y(x)$ can be found by integrating $p(x)$.

Example

Suppose we have the second order equation

$$y'' + y' = x. \tag{51.3}$$

Using $y'(x) = p(x)$, equation (51.3) can be written as

$$p' + p = x. \tag{51.4}$$

Equation (51.4) can be solved by integrating factors (see page 305) to obtain

$$p(x) = Ae^{-x} + x - 1,$$

where A is an arbitrary constant. Then $p(x)$ can be integrated to obtain $y(x)$

$$y(x) = \int^x p(t)\, dt = B - Ae^{-x} + \frac{x^2}{2} - x,$$

where B is another arbitrary constant.

Notes
[1] This solution technique can be derived from Lie group methods (see page 314).

References
[1] W. E. Boyce and R. C. DiPrima, *Elementary Differential Equations and Boundary Value Problems*, Fourth Edition, John Wiley & Sons, New York, 1986, pages 111–112.
[2] M. E. Goldstein and W. H. Braun, *Advanced Methods for the Solution of Differential Equations*, NASA SP-316, U.S. Government Printing Office, Washington, D.C., 1973, pages 74–76.
[3] E. L. Ince, *Ordinary Differential Equations*, Dover Publications, Inc., New York, 1964, page 43.
[4] E. D. Rainville and P. E. Bedient, *Elementary Differential Equations*, The MacMillan Company, New York, 1964, pages 266–268.

52.   Differentiation Method

Applicable to   Nonlinear ordinary differential equations.

Yields
An explicit solution.

Idea
Sometimes differentiating an ordinary differential equation will result in an ordinary differential equation that is easier to solve.

Procedure
Given an ordinary differential equation, differentiate it with respect to the independent variable. This will yield a new equation that may sometimes factor (see page 245), or simplify in some other way. By considering each term in this new equation to be equal to zero, several possible solutions may be found.

The general solution of each term must then be used in the original equation, possibly to constrain some of the parameters.

Example
Suppose that we have the nonlinear ordinary differential equation

$$2yy'' - (y')^2 = \tfrac{1}{3}(y' - xy'')^2. \tag{52.1}$$

If this equation is differentiated with respect to x, the simplified result is

$$y''' \left(x^2 y'' - xy' - 3y\right) = 0,$$

from which we recognize that

$$y''' = 0 \quad \text{or} \quad x^2 y'' - xy' - 3y = 0. \tag{52.2}$$

In the first case, a candidate for the general solution is

$$y(x) = ax^2 + bx + c.$$

Using this form in the original equation, (52.1), we find after some simplification that $3ac = b^2$. Using this equation to determine c, a general solution to (52.1) is found to be

$$y(x) = ax^2 + bx + \frac{b^2}{3a}. \tag{52.3}$$

Another possibility is that the second expression in (52.2) is equal to zero. This second equation is an Euler equation (see page 235), and so the general solution is found to be

$$y(x) = \alpha x^3 + \frac{\beta}{x}.$$

Using this form in the original equation, (52.1), we find after some simplification that $\alpha\beta = 0$. Hence, two different solutions to (52.1) are given by

$$y(x) = \alpha x^3 \quad \text{and} \quad y(x) = \frac{\beta}{x}. \tag{52.4}$$

Between (52.3) and (52.4) are three different solutions to (52.1).

Notes
[1] The above example is from Bateman [1].
[2] This procedure is used to find the singular solutions to Clairaut's equation (see page 196).

References
[1] H. Bateman, *Partial Differential Equations of Mathematical Physics*, Dover Publications, Inc., New York, 1944, pages 66–67.

53. Differential Equations with Discontinuities*

Applicable to   Equations that contain discontinuous functions.

Yields
An exact solution.

Idea
Equations can be solved locally and then patched together at the points of discontinuity.

Procedure

The following discussion is limited to linear ordinary differential equations, but the general techniques apply to linear and nonlinear ordinary differential equations and partial differential equations.

Suppose we have the equation

$$a_n(x)y^{(n)} + a_{n-1}y^{(n-1)} + \ldots + a_1(x)y' + a_0(x)y = b(x), \qquad (53.1)$$

where the $\{a_i(x)\}$ and $b(x)$ may all be discontinuous. For example, $a_1(x)$ may look like

$$a_1(x) = \begin{cases} x & \text{if } 0 < x < 3, \\ \sin x & \text{if } 3 \leq x < 8. \end{cases}$$

We presume that the $\{a_i(x)\}$ and $b(x)$ are discontinuous at only a finite number of points, say $\{x_1, x_2, \ldots, x_m\}$, and that we wish to find the solution at the point x_f with $x_0 < x_1 < \ldots < x_m < x_f$. Assume further that the initial data $\{y(x_0), y'(x_0), y''(x_0), \ldots, y^{(n-1)}(x_0)\}$ are all given.

The general technique is to divide the interval from x_0 to x_f into m intervals and solve (53.1) separately on each interval. Since the equation is continuous on these intervals, we can use any technique known to us to find the solution. Define $y_j(x)$ to be the solution in the interval $[x_j, x_{j+1}]$.

To determine $y_j(x)$ completely, we need to specify the value of $\{y_j(x_j), y_j'(x_j), \ldots, y_j^{(n-1)}(x_j)\}$. These can be determined from $y_{j-1}(x)$. Since an equation of n-th order (which is what equation (53.1) is) must have continuous derivatives of all orders up to $n-1$, we simply match the values of $y_j(x)$ and its derivatives to the values of $y_{j-1}(x)$ and its derivatives, all at the point x_j.

To illustrate this technique on equation (53.1), we would solve

$$a_n(x)y_j^{(n)} + a_{n-1}y_j^{(n-1)} + \ldots + a_1(x)y_j' + a_0(x)y_j = b(x)$$

in the interval $[x_j, x_{j+1}]$, for $j = 0, 1, 2, \ldots, m$. To obtain the initial values for each equation we take

$$\begin{pmatrix} y_0(x_0) \\ y_0'(x_0) \\ \vdots \\ y_0^{(n-1)}(x_0) \end{pmatrix} = \begin{pmatrix} y(x_0) \\ y'(x_0) \\ \vdots \\ y^{(n-1)}(x_0) \end{pmatrix},$$

and then

$$\begin{pmatrix} y_j(x_j) \\ y_j'(x_j) \\ \vdots \\ y_j^{(n-1)}(x_j) \end{pmatrix} = \begin{pmatrix} y_{j-1}(x_j) \\ y_{j-1}'(x_j) \\ \vdots \\ y_{j-1}^{(n-1)}(x_j) \end{pmatrix}, \qquad \text{for } j = 1, 2, \ldots, m.$$

Finally, the solution at $x = x_f$ will be given by $y_m(x_f)$.

Example

Suppose we want to determine the value of $y(t)$ at $t = T$ when

$$y'' + f(t)y = 0,$$

and $f(t)$ is given by

$$f(t) = \begin{cases} -1 & \text{for } 0 \le t < \tau, \\ 1 & \text{for } \tau \le t \le T, \end{cases}$$

given that $y(0) = 1$, $y'(0) = 0$. (Here, τ and T are fixed constants.) To solve this problem, we break the interval from 0 to T into two intervals; interval I will be from 0 to τ while interval II will be from τ to T.

In interval I, $f(t)$ can be replaced by -1, so we solve

$$y_1'' - y_1 = 0, \qquad y_1(0) = 1, \quad y_1'(0) = 0.$$

This equation has the solution

$$y_1(t) = \cosh t.$$

In interval II, $f(t)$ can be replaced by 1, so we solve

$$y_2'' + y_2 = 0, \tag{53.2}$$

in the interval from τ to T. For the *initial* values of $y_2(t)$, we use the *final* values of $y_1(t)$, that is,

$$\begin{aligned} y_2(\tau) &= y_1(\tau) = \cosh \tau, \\ y_2'(\tau) &= y_1'(\tau) = \sinh \tau. \end{aligned} \tag{53.3}$$

The solution of (53.2) and (53.3) is

$$y_2(t) = (\sin \tau \cosh \tau + \cos \tau \sinh \tau) \sin t + (\cos \tau \cosh \tau - \sin \tau \sinh \tau) \cos t,$$

and hence, the value of $y(t)$ at $t = T$ is given by

$$y_2(T) = (\sin \tau \cosh \tau + \cos \tau \sinh \tau) \sin T + (\cos \tau \cosh \tau - \sin \tau \sinh \tau) \cos T.$$

Notes

[1] When the discontinuities involve the *dependent* variable, then the problem is generally a free boundary problem. See Elliot and Ockendon [2] or Fleishman [5] for a discussion.

[2] If the discontinuity appearing in a linear differential equation is a single delta function, which appears as a forcing function, then the solution will be a Green's function (see page 268).

[3] If the discontinuities include generalized functions (such as a delta function), then the solution may only exist in the weak sense. See Gear and Østerby [6] for details.

[4] There exist computer programs for numerically approximating differential equations with discontinuities. See Enright *et al.* [3] or Gear and Østerby [6].

[5] Fleishman [5] analyzes the equation $\dot{x} = A(t)x + \text{sgn}(x) + t(t)$, where "sgn" represents the signum function.

References

[1] W. E. Boyce and R. C. DiPrima, *Elementary Differential Equations and Boundary Value Problems*, Fourth Edition, John Wiley & Sons, New York, 1986, Section 6.3.1 (pages 304–309).

[2] C. M. Elliot and J. R. Ockendon, *Weak and Variational Methods for Moving Boundary Problems*, Pitman Publishing Co., Marshfield, MA, 1982.

[3] W. H. Enright, K. R. Jackson, S. P. NØorsett, and P. G. Thomsen, "Effective Solution of Discontinuous IVPs Using a Runge–Kutta Formula Pair with Interpolants," *Appl. Math. and Comp.*, **27**, 1988, pages 313–335.

[4] A. F. Filippov, *Differential Equations with Discontinuous Righthand Sides*, Kluwer Academic Publishers, Dordrecht, the Netherlands, 1988.

[5] B. A. Fleishman, "Convex Superposition in Piecewise-Linear Systems," *J. Math. Anal. Appl.*, **6**, No. 2, April 1963, pages 182–189.

[6] C. W. Gear and O. Østerby, "Solving Ordinary Differential Equations with Discontinuities," *ACM Trans. Math. Software*, **10**, No. 1, March 1984, pages 23–44.

[7] I. N. Hajj and S. Skelboe, "Steady-State Analysis of Piecewise-Linear Dynamic Systems," *IEEE Trans. Circ. & Syst.*, **CAS-28**, No. 3, March 1981, pages 234–241.

[8] H. H. Pan and R. M. Hohenstein, "A Method of Solution of an Ordinary Differential Equation Containing Symbolic Functions," *Quart. Appl. Math.*, April 1981, pages 131–136.

[9] T. S. Parker and L. O. Chua, "Efficient Solution of the Variational Equation for Piecewise-Linear Differential Equations," *Circuit Theory and Appl.*, **14**, No. 4, 1986, pages 305–314.

[10] D. Stewart, "A High Accuracy Method for Solving ODEs with Discontinuous Right-hand Side," *Numer. Math.*, **58**, 1990, pages 299–328.

[11] D. Westreich, "Numerical Solution of the Eigenvalue Problem for Discontinuous Linear Ordinary Differential Equations," *J. Inst. Maths. Applics*, **25**, 1980, pages 147–160.

54. Eigenfunction Expansions*

Applicable to Linear differential equations with linear boundary conditions.

Yields

An exact solution in terms of an infinite series.

Idea

Any "well-behaved" function can be expanded in a complete set of eigenfunctions. In this method, we expand the dependent variable in a differential equation as a sum of the eigenfunctions with unknown coefficients. From the given equation and boundary conditions, equations can then be determined for the unknown coefficients.

Procedure

We will describe the procedure for ordinary differential equations, but the same procedure can be used for partial differential equations (see Example 2). Assume that we want to solve the inhomogeneous linear ordinary differential equation

$$L[y] := \sum_{r=1}^{n} p_r(x) \frac{d^r y}{dx^r} = h(x),$$

$$B_i[y] := \sum_{r=1}^{n} \left(c_{ir} \frac{d^r y}{dx^r}(a) + d_{ir} \frac{d^r y}{dx^r}(b) \right) = 0, \qquad i = 1, 2, \ldots, n,$$

$$(54.1.a\text{--}b)$$

for $y(x)$, where $x \in [a, b]$ and $\{c_{ir}, d_{ir}, p_r(x), h(x)\}$ are all known.

Let us suppose that we know a complete set of eigenfunctions $\{u_k(x)\}$ that satisfy the boundary conditions in (54.1) and are orthogonal with respect to some weighting function $w(x)$. These could be obtained from a table (e.g., see page 292), or we might look for a set that is related to the differential equation in (54.1). A common approach is to choose a set of eigenfunctions $\{u_k\}$ that satisfy

$$H[u_k] = \lambda_k u_k,$$
$$R_i[u_k] = 0, \qquad i = 1, 2, \ldots, n,$$

$$(54.2.a\text{--}b)$$

where $H[\cdot]$ is a linear operator related to $L[\cdot]$ in some way, the $R_i[\cdot]$ are linear boundary conditions related to $B_i[\cdot]$ in some way, and λ_k is a constant (λ_k is an eigenvalue of the $(H, \{R_i\})$ system). The orthogonality condition requires that

$$(u_k, u_m) := \int_a^b u_k(x)u_m(x)w(x)\, dx = N_k\delta_{km} = \begin{cases} 0 & \text{for } m \neq k, \\ N_k & \text{for } m = k. \end{cases}$$
(54.3)

Frequently the operator $H[\cdot]$ is chosen to be same as the operator $L[\cdot]$, and the $\{R_i\}$ are chosen to be the same as the $\{B_i\}$. This is not required, nor must the degree of the differential equation in (54.2.a) be n (which is the degree of the differential equation in (54.1.a)).

Since the presumed eigenfunctions are complete, we can write any "sufficiently smooth" function as a linear combination of these functions. In particular, we choose to represent $y(x)$ and $h(x)$ as

$$y(x) := \sum_{k=1}^{\infty} y_k u_k(x), \qquad h(x) := \sum_{k=1}^{\infty} h_k u_k(x).$$
(54.4.a–b)

Once the $\{y_k\}$ are known, the problem is solved. The $\{h_k\}$ can be determined, given $h(x)$, by multiplying (54.4.b) by $w(x)u_m(x)$ and integrating with respect to x from a to b. This calculation can be written as

$$(h(x), u_m(x)) = \left(\sum_{k=1}^{\infty} h_k u_k(x), u_m(x) \right),$$

$$= \sum_{k=1}^{\infty} h_k \left(u_k(x), u_m(x) \right),$$

$$= \sum_{k=1}^{\infty} h_k \left(N_k \delta_{km} \right),$$

$$= N_m h_m,$$

where we have utilized (54.3). If we take the $\{R_i\}$ to be identical to the $\{B_i\}$ then, from (54.2.b), the boundary conditions for $y(x)$ (in (54.1.b)) are automatically satisfied. Hence, only equation (54.1.a) needs to be satisfied. Using (54.4.a) in (54.1.a) results in

$$L[y] = L\left[\sum_{k=1}^{\infty} y_k u_k(x) \right]$$

$$= \sum_{k=1}^{\infty} y_k L[u_k]$$
(54.5)

$$= h(x).$$

The $\{y_k\}$ can now be determined from (54.5) by multiplying (54.5) by $w(x)u_m(x)$ and integrating with respect to x from a to b. This produces

$$\sum_{k=1}^{\infty} y_k \left(L[u_k], u_m \right) = (h(x), u_m) = N_m h_m, \quad \text{for } m = 1, 2, \ldots, \qquad (54.6)$$

which is an infinite system of linear algebraic equations. In principle, all of the $\{y_k\}$ in (54.6) are coupled together.

In practice, if a good choice was made for the eigenfunctions, then (54.6) will simplify and y_m can be determined directly from (54.6). For instance, if $H[\cdot]$ is chosen to be equal to $L[\cdot]$ then $L[u_n] = \lambda_n u_n$ (from (54.2)) and (54.6) becomes $\sum_{k=1}^{\infty} y_k \lambda_k (u_k, u_m) = N_m h_m$ or, by orthogonality, $y_m = h_m / \lambda_m$.

Example 1

Suppose we have the fourth order differential equation and boundary conditions

$$\begin{aligned} L[y] &:= y'''' + \alpha y'' + \beta y = h(x), \\ y(0) &= 0, \qquad y(1) = 0, \\ y''(0) &= 0, \qquad y''(1) = 0, \end{aligned} \qquad (54.7)$$

to solve for $y(x)$ on the interval $x \in [0, 1]$.

For this case we choose to use the eigenfunctions corresponding to the Sturm–Liouville operator (see page 82)

$$\begin{aligned} H[u] &= u'', \\ u(0) &= 0, \\ u(1) &= 0. \end{aligned} \qquad (54.8)$$

For the operator in (54.8) it is easy to determine that the eigenfunctions are $u_k(x) = \sin k\pi x$, the eigenvalues are $\lambda_k = k\pi$, and the weighting function is $w(x) = 1$. Since this is a self-adjoint problem (see page 74), we know that these eigenfunctions are complete. Now that we have a set of eigenfunctions, we observe that they satisfy the four boundary conditions given in (54.7).

We write $y(x)$ in terms of these eigenfunctions as

$$y(x) = \sum_{k=1}^{\infty} y_k \sin k\pi x. \qquad (54.9)$$

Using (54.9) in (54.7) and then multiplying by $u_m(x)$ and integrating from $x = 0$ to $x = 1$ results in

$$\int_0^1 L[y(x)]u_m(x)\,dx = \int_0^1 L\left[\sum_{k=1}^\infty y_k \sin k\pi x\right]u_m(x)\,dx$$

$$= \sum_{k=1}^\infty y_k \int_0^1 L[\sin(k\pi x)]u_m(x)\,dx$$

$$= \int_0^1 h(x)u_m(x)\,dx.$$

Equating the last two expressions, using $u_m(x) = \sin m\pi x$ and simplifying gives

$$\sum_{k=1}^\infty y_k \int_0^1 \left(k^4\pi^4 - \alpha k^2\pi^2 + \beta\right)\sin k\pi x \sin m\pi x\,dx =$$

$$\int_0^1 h(x)\sin m\pi x\,dx,$$

or (since $\int_0^1 \sin k\pi x \sin m\pi x\,dx = \frac{1}{2}\delta_{km}$)

$$\frac{1}{2}y_k\left(k^4\pi^4 - \alpha k^2\pi^2 + \beta\right) = \int_0^1 h(x)\sin k\pi x\,dx. \tag{54.10}$$

Hence, solving (54.10) for y_k and using this value in (54.9) results in the explicit solution

$$y(x) = \sum_{k=1}^\infty \left(\frac{2\int_0^1 h(x)\sin k\pi x\,dx}{k^4\pi^4 - \alpha k^2\pi^2 + \beta}\right)\sin k\pi x.$$

If α and β are such that $k^4\pi^4 - \alpha k^2\pi^2 + \beta = 0$, for some value of k, then there will be no solution unless $\int_0^1 h(x)\sin k\pi x\,dx = 0$. Even then, the solution will not be unique; this is because the differential equation $L[u] = 0$, with the boundary conditions in (54.7), will have the solution $u(x) = C\sin k\pi x$, where C is arbitrary. See the section on alternative theorems (page 14).

Example 2

Suppose we want to solve the partial differential equation

$$\phi_t = \phi_{xx},$$
$$\phi(x,0) = f(x),$$
$$\phi(0,t) = 0, \qquad (54.11.a\text{-}d)$$
$$\phi(1,t) = 0,$$

for $\phi = \phi(x,t)$. We can use the eigenfunctions in (54.8) to solve this problem. In this case we expand $\phi(x,t)$ as

$$\phi(x,t) = \sum_{n=1}^{\infty} a_n(t) \sin n\pi x. \qquad (54.12)$$

By using this representation for $\phi(x,t)$, the boundary conditions in (54.11.b) and (54.11.c) are automatically satisfied. By multiplying (54.12) by $\sin(m\pi x)$ and integrating from $x = 0$ to $x = 1$, we find that

$$a_n(t) = 2 \int_0^1 \phi(z,t) \sin n\pi z \, dz. \qquad (54.13)$$

Using the boundary condition from (54.11.b) in (54.13) produces the initial values for the $\{a_n(t)\}$

$$a_n(0) = 2 \int_0^1 \phi(z,0) \sin n\pi z \, dz = 2 \int_0^1 f(z) \sin n\pi z \, dz. \qquad (54.14)$$

Now, the correct procedure is to multiply the original equation, (54.11.a), by one of the eigenfunctions, $\sin m\pi x$, and integrate from $x = 0$ to $x = 1$ to obtain

$$\int_0^1 \phi_t \sin m\pi x \, dx = \int_0^1 \phi_{xx} \sin m\pi x \, dx. \qquad (54.15)$$

After utilizing (54.12) for ϕ in (54.15), the resulting equation should be integrated by parts, using the information in (54.13). This results in

$$a_n'(t) = -n^2 \pi^2 a_n(t), \qquad (54.16)$$

where a prime denotes a derivative with respect to t. The solution of (54.16) is

$$a_n(t) = a_n(0) e^{-n^2 \pi^2 t},$$
$$= \left(2 \int_0^1 f(z) \sin n\pi z \, dz \right) e^{-n^2 \pi^2 t}, \qquad (54.17)$$

where we have used (54.14). Combining (54.12) and (54.17) we determine the final solution to (54.11) to be

$$\phi(x,t) = \sum_{n=1}^{\infty} \left(2 \int_0^1 f(z) \sin n\pi z \, dz \right) e^{-n^2\pi^2 t} \sin n\pi x.$$

Be aware that it would have been *incorrect*, when trying to obtain an ordinary differential equation for $a_n(t)$, to substitute (54.12) into (54.11.a) and then multiply by one of the eigenfunctions and perform the integration. While this would have resulted in the same differential equation and boundary conditions for a_n in this example, it might not work in other cases (see the next example). The proper technique is to multiply the original equation by one of the eigenfunctions, and then integrate by parts.

Example 3

Consider solving Laplace's equation in two dimensions in the unit square

$$u_{xx} + u_{yy} = 0,$$
$$u(x,1) = u(0,y) = u(1,y) = 0, \qquad\qquad (54.18.a\text{--}c)$$
$$u(x,0) = f(x).$$

Since the functions $\{\sin n\pi y\}$ are complete on the interval $[0,1]$, we choose to represent the solution to (54.18) in the form

$$u(x,y) = \sum_{n=0}^{\infty} c_n(x) \sin n\pi y, \qquad\qquad (54.19)$$

from which we can deduce that

$$c_n(x) = 2 \int_0^1 u(x,y) \sin n\pi y \, dy. \qquad\qquad (54.20)$$

From the boundary conditions on $u(x,y)$ at $x=0$ and at $x=1$ we also find that $c_n(0) = c_n(1) = 0$.

We will show that an incorrect answer is obtained if the $\{c_n\}$ are determined in a naive way. If we substituted the assumed form of the solution, e.g. (54.19), into the equation in (54.18.a) then we would find

$$u_{xx} + u_{yy} = \sum_{n=0}^{\infty} \left(c_n'' - n^2\pi^2 c_n \right) \sin n\pi y = 0.$$

Hence, by orthogonality, we would find that $c_n'' - n^2\pi^2 c_n = 0$. Solving this differential equation with the boundary conditions on c_n (e.g. $c_n(0) =$

$c_n(1) = 0$) we would be led to $c_n(x) = 0$ and so $u(x,y) = 0$. This is clearly *wrong*.

If, instead, the equation in (54.18.a) is multiplied by $2\sin n\pi y$ and then integrated with respect to y from 0 to 1, then we obtain

$$
\begin{aligned}
0 &= \int_0^1 2\sin n\pi y (u_{xx} + u_{yy})\, dy \\
&= \frac{d^2}{dx^2} \int_0^1 2u(x,y)\sin n\pi y\, dy + 2u_y(x,y)\sin n\pi y \Big|_0^1 \\
&\quad - 2n\pi u(x,y)\cos n\pi y \Big|_0^1 - n^2\pi^2 \int_0^1 2u(x,y)\sin n\pi y\, dy \\
&= c_n'' + 2n\pi f(x) - n^2\pi^2 c_n
\end{aligned}
$$

where we have integrated by parts twice, used (54.20) to substitute for the integral, and used the boundary conditions in (54.18.b-c). Solving this last equation for $c_n(x)$ we find

$$
c_n(x) = 2n\pi \int_0^1 G(x;t)f(t)\, dt
$$

where $G(x;t)$ is the Green's function $G(x;t) = \dfrac{\sinh n\pi x_< \sinh n\pi(1 - x_>)}{n\pi \sinh n\pi}$,

where $x_>$ ($x_<$) indicates the larger (smaller) of x and t.

This second approach gives the correct solution to this problem. The reason that the first approach would not work is that the series chosen to represent the solution does not have uniform convergence.

Notes
[1] Note that the solution in Example 2 would have been obtained in exactly the same form if separation of variables had been used (see page 419).
[2] If the chosen eigenfunctions do not come from a self-adjoint operator, then it will be necessary to know the eigenfunctions of the adjoint operator. This is because the orthogonality condition will utilize the eigenfunctions of the adjoint operator.
[3] Since the eigenfunctions we used in the examples were just sine functions, the expansions obtained here are identical to the results that would have been obtained from a Fourier sine series (see page 293).
[4] To determine that a set of functions is complete, it is not necessary that they be derived from a self-adjoint operator. See Minzoni [6] for an example of a set of functions proved complete by using theorems from analysis.

References

[1] G. Birkhoff and G.-C. Rota, *Ordinary Differential Equations*, John Wiley & Sons, New York, 1978, Chapter 11.

[2] E. Butkov, *Mathematical Physics*, Addison–Wesley Publishing Co., Reading, MA, 1968, pages 304–318.

[3] Z. Divis, "A Note on the Rate of Convergence of Sturm–Liouville Expansions," *J. Approx. Theory*, **50**, 1987, pages 200–207.

[4] S. J. Farlow, *Partial Differential Equations for Scientists and Engineers*, John Wiley & Sons, New York, 1982, Lesson 9 (pages 64–71).

[5] M. Kobayashi, "Eigenfunction Expansion: A Discontinuous Version," *SIAM J. Appl. Math.*, **50**, No. 3, June 1990, pages 910–917.

[6] A. A. Minzoni, "On the Completeness of the Functions $e^{-np(x)}\cos nx$, $e^{-np(x)}\sin nx$ for $n \geq 0$ and $p(x)$ a 2π Periodic Function," *Stud. Appl. Math.*, **75**, 1986, pages 265–269.

[7] I. Stakgold, *Green's Functions and Boundary Value Problems*, John Wiley & Sons, New York, 1979.

[8] E. C. Titchmarsh, *Eigenfunction Expansions Associated with Second-Order Differential Equations*, Clarendon Press, Oxford, 1946.

55.   Equidimensional-In-x Equations

Applicable to   Ordinary differential equations of a certain form.

Yields

An autonomous ordinary differential equation of the same order (which can then be reduced to an ordinary differential equation of lower order).

Idea

An equidimensional-in-x equation is one in which the scaling of the x variable does not change the equation. By a change of independent variable, we can change an equation of this type into an autonomous equation.

Procedure

An equidimensional-in-x equation is one that is left invariant under the transformation $x \to ax$, where a is a constant. That is, if the original equation is an equation for $y(x)$ and the x variable is replaced by the variable ax', then the new equation (in terms of y and x') will be identical to the original equation (which is in terms of y and x). An equation of this type can be converted to an autonomous equation of the same order by changing the independent variable from x to t by the transformation $x = e^t$.

Example

Suppose we have the nonlinear second order ordinary differential equation

$$x \frac{d^2y}{dx^2} = 2y \frac{dy}{dx}. \tag{55.1}$$

First we will show that this equation is equidimensional-in-x. Substituting ax' for x in (55.1) produces

$$(ax') \frac{d^2y}{d(ax')^2} = 2y \frac{dy}{d(ax')}, \tag{55.2}$$

or, multiplying (55.2) by the constant a

$$x' \frac{d^2y}{d(x')^2} = 2y \frac{dy}{dx'},$$

which is identical to equation (55.1).

Since we now know that (55.1) is equidimensional-in-x, we change variables from $y(x)$ to $y(t)$ by $x = e^t$. Using Table 55, we find that

$$e^t e^{-2t}(y_{tt} - y_t) = 2y(e^{-t}y_t),$$

or

$$y_{tt} - y_t = 2yy_t. \tag{55.3}$$

The equation in (55.3) is autonomous (there is no explicit t dependence). Hence, it can be reduced to an ordinary differential equation of order one by the transformation $u(y) = y_t(t)$ (see page 190 for more information).

Carrying out the details (equation (55.3) was the example in the section on autonomous equations), it is easy to derive that either $y(t)$ is a constant for all t, or $y(t)$ satisfies

$$y(t) = E \tan(F + Et) - \tfrac{1}{2},$$

where E and F are arbitrary constants. Changing the independent variable from t to x we have

$$y(x) = E \tan(F + E \log x) - \tfrac{1}{2}.$$

$$y_x = e^{-t}(y_t),$$

$$y_{xx} = e^{-2t}(y_{tt} - y_t),$$

$$y_{xxx} = e^{-3t}(y_{ttt} - 3y_{tt} + 2y_t),$$

$$y_{xxxx} = e^{-4t}(y_{tttt} - 6y_{ttt} + 11y_{tt} - 6y_t),$$

$$y_{xxxxx} = e^{-5t}(y_{ttttt} - 10y_{tttt} + 35y_{ttt} - 50y_{tt} + 24y_t),$$

To simplify notation, define $y_{x(n)}$ to be the n-th derivative of y with respect to x. Similarly for $u_{y(n)}$. The last equation is, therefore,

$$y_{x(5)} = e^{-5t}(y_{t(5)} - 10y_{t(4)} + 35y_{ttt} - 50y_{tt} + 24y_t),$$

$$y_{x(6)} = e^{-6t}(y_{t(6)} - 15y_{t(5)} + 85y_{t(4)} - 225y_{ttt} + 274y_{tt} - 120y_t),$$

$$y_{x(7)} = e^{-7t}(y_{t(7)} - 21y_{t(6)} + 175y_{t(5)} - 735y_{t(4)} + 1624y_{ttt} - 1764y_{tt} + 720y_t).$$

Table 55. How to transform derivatives under the change of dependent variable: $x = e^t$.

Notes

[1] This method is derivable from Lie group methods (see page 314).

[2] It is easy to write a MACSYMA program that will perform the necessary change of variables. In the terminal session shown, the second order equidimensional-in-x equation

$$\left(\frac{dy}{dx}\right)^2 - y\frac{d^2y}{dx^2} = 0$$

is converted into the second order autonomous equation

$$u\frac{d^2u}{dt^2} - \left(\frac{du}{dt}\right)^2 - u\frac{du}{dt} = 0.$$

This autonomous equation could then be reduced to a first order equation by the MACSYMA program given on page 193.

```
(c1) DEPENDS(Y,X)$

(c2) EQUIDIMENSIONAL_IN_X(EQN,Y,X):= BLOCK([NEW,HOLD,J],
            DEPENDS([U],[T]),
            GRADEF(T, X, %E**(-T) ),
            NEW:SUBST( U, Y, EQN ),
            NEW:EV(NEW, DIFF),
            NEW:SUBST( %E**T, X, NEW),
            NEW:FACTOR(NEW),
            NEW)$

(c3) EQN: DIFF(Y,X)**2-Y*DIFF(Y,X,2);
```

```
                                2
(d3)                         (y )  - y y
                               x        xx
```

```
(c4) EQUIDIMENSIONAL_IN_X(EQN,Y,X);
```

$$
\text{(d4)} \qquad - \%e^{-2t} \; (u_t u_{tt} - (u_t)^2 - u u_t)
$$

References

[1] C. M. Bender and S. A. Orszag, *Advanced Mathematical Methods for Scientists and Engineers*, McGraw–Hill, New York, 1978, page 25.

56. Equidimensional-In-y Equations

Applicable to   Ordinary differential equations of a certain form.

Yields

An ordinary differential equation of lower order.

Idea

An equidimensional-in-y equation is one in which the scaling of the y variable does not change the equation. This information can be used to lower the order of the equation by a change of the dependent variable.

Procedure

An equidimensional-in-y equation is one that is left invariant under the transformation $y \to ay$, where a is a constant. That is, if the original equation is an equation for $y(x)$ and the y variable is replaced by the variable ay', then the new equation (in terms of y' and x) will be identical to the original equation (which is in terms of y and x). An equation of this type can be converted to an equation of lower order by changing the dependent variable from $y(x)$ to $e^{u(x)}$.

Example

Suppose we have the equation

$$
(1-x)\left[y\frac{d^2y}{dx^2} - \left(\frac{dy}{dx}\right)^2 \right] + x^2 y^2 = 0, \tag{56.1}
$$

to solve. We can tell by inspection that this equation is equidimensional-in-y since all of the y terms in (56.1) all appear to the same power. That is, the y terms in (56.1) are all quadratic, the terms being of the form $\{y^2, y_x^2, y_{xx}^2, \ldots, yy_x, yy_{xx}, y_x y_{xx}, \ldots\}$.

To formally show that (56.1) is equidimensional-in-y, substitute ay' for y in (56.1) to find

$$(1-x)\left[(ay')\frac{d^2(ay')}{dx^2} - \left(\frac{d(ay')}{dx}\right)^2\right] + x^2(ay')^2 = 0.$$

Or, since a is a non-zero constant,

$$(1-x)\left[y\frac{d^2 y'}{dx^2} - \left(\frac{dy'}{dx}\right)^2\right] + x^2 y'^2 = 0, \tag{56.2}$$

which has the same form as (56.1). Now, substituting $e^{u(x)}$ for $y(x)$ in (56.1) produces

$$(1-x)\left[y^2\left(\frac{d^2 u}{dx^2} + \left(\frac{du}{dx}\right)^2\right) - \left(y\frac{du}{dx}\right)^2\right] + x^2 y^2 = 0, \tag{56.3}$$

where Table 56 has been used to determine how the derivatives transform under this change of variable. For $y \neq 0$, equation (56.3) becomes

$$(1-x)\frac{d^2 u}{dx^2} + x^2 = 0. \tag{56.4}$$

Note that equation (56.4) does not have any explicit y dependence. If it did have any such terms, then the original equation could not have been equidimensional-in-y. The solution to equation (56.3) is (see page 185)

$$u(x) = \int^x \left[\int^w \frac{z^2}{z-1}\, dz\right] dw,$$

$$= \frac{x^3}{6} + \frac{x^2}{2} + (x-1)\log(x-1) + Ax + B,$$

where A and B are arbitrary constants. Hence, the solution of the original equation is

$$y(x) = e^{u(x)} = (x-1)^{(x-1)}\exp\left(\frac{x^3}{6} + \frac{x^2}{2} + Ax + B\right).$$

$$y = e^u,$$
$$y_x = y u_x,$$
$$y_{xx} = y(u_{xx} + u_x^2],$$
$$y_{xxx} = y(u_{xxx} + 3u_x u_{xx} + u_x^3],$$
$$y_{xxxx} = y(u_{xxxx} + 4u_x u_{xxx} + 3u_{xx}^2 + 6u_x^2 u_{xx} + u_x^4].$$

To simplify notation, define $y_{x(n)}$ to be the n-th derivative of y with respect to x. Similarly for $u_{x(n)}$.

$$y_{x(4)} = y(u_{x(4)} + 4u_x u_{xxx} + 3u_{xx}^2 + 6u_x^2 u_{xx} + u_x^4),$$
$$y_{x(5)} = y(u_{x(5)} + 5u_x u_{x(4)} + 10u_{xx} u_{xxx} + 10u_x^2 u_{xxx} + 15u_x u_{xx}^2 + 10u_x^3 u_{xx} + u_x^5),$$
$$y_{x(6)} = y(u_{x(6)} + 6u_x u_{x(5)} + 15u_{xx} u_{x(4)} + 15u_x^2 u_{x(4)} + 10u_{xxx}^2 + 20u_x^3 u_{xxx}$$
$$+ 15u_{xx}^3 + 60u_x u_{xx} u_{xxx} + 45u_x^2 u_{xx}^2 + 15u_x^4 u_{xx} + u_x^6).$$

Table 56.   How to transform derivatives under the change of independent variable: $y(x) = e^{u(x)}$.

Notes

[1] This method is derivable from Lie group methods (see page 314).

[2] Equidimensional-in-y equations are also called *equations homogeneous in y*.

References

[1] C. M. Bender and S. A. Orszag, *Advanced Mathematical Methods for Scientists and Engineers*, McGraw–Hill, New York, 1978, page 27.

57.   Euler Equations

Applicable to   Linear ordinary differential equations of the form
$a_0 x^n y^{(n)} + a_1 x^{n-1} y^{(n-1)} + \ldots + a_{n-1} x y' + a_n y = 0$.

Yields

An exact solution.

Idea

An equation of the above type can be turned into a linear constant coefficient ordinary differential equation by a change of independent variable. This new equation can be solved exactly.

Procedure

An Euler equation has the form

$$a_0 x^n y^{(n)} + a_1 x^{n-1} y^{(n-1)} + \ldots + a_{n-1} xy' + a_n y = 0. \qquad (57.1)$$

If the independent variable is changed from x to t (via the transformation $x = e^t$) then the resulting equation becomes a linear constant coefficient ordinary differential equation. This type of equation can be solved exactly.

Use Table 57 to determine how the derivatives of y with respect to x become derivatives of y with respect to t.

Alternately, a solution of the form $y = x^k$ can be tried directly in (57.1).

Example 1

Given the Euler equation

$$x^2 y_{xx} - 2xy_x + 2y = 0,$$

we change variables by $x = e^t$ to obtain

$$y_{tt} - 3y_t + 2y = 0. \qquad (57.2)$$

The standard technique for solving a linear constant coefficient ordinary differential equation is to look for exponential solutions (see page 204). Using $y = e^{\lambda t}$ in (57.2) we find the characteristic equation to be $\lambda^2 - 3\lambda + 2 = 0$. The roots of this equation are $\lambda = 1$ and $\lambda = 2$. Therefore, the solution to (57.2) is

$$y(t) = C_1 e^t + C_2 e^{2t},$$

where C_1 and C_2 are arbitrary constants. Writing this solution in the original variables, we determine the final solution

$$y(x) = C_1 x + C_2 x^2.$$

Example 2

Given the Euler equation

$$x^3 y''' - x^2 y'' - 2xy' - 4y = 0 \qquad (57.3)$$

we use $y = x^k$ to find the characteristic equation:

$$\lambda(\lambda - 1)(\lambda - 2)x^k - \lambda(\lambda - 1)x^k - 2\lambda x^k - 4x^k = 0$$

or

$$\left(\lambda^2 + 1\right)(\lambda - 4) = 0.$$

This equation has the roots $\lambda = 4$ and $\lambda = \pm i$. Hence, the general solution to (57.3) is

$$y = C_1 x^4 + C_2 \cos(\log x) + C_3 \sin(\log x)$$

$$y_x = e^{-t}(y_t),$$

$$y_{xx} = e^{-2t}(y_{tt} - y_t),$$

$$y_{xxx} = e^{-3t}(y_{ttt} - 3y_{tt} + 2y_t),$$

$$y_{xxxx} = e^{-4t}(y_{tttt} - 6y_{ttt} + 11y_{tt} - 6y_t),$$

$$y_{xxxxx} = e^{-5t}(y_{ttttt} - 10y_{tttt} + 35y_{ttt} - 50y_{tt} + 24y_t).$$

To simplify notation, define $y_{x(n)}$ to be the n-th derivative of y with respect to x. Similarly for $y_{t(n)}$.

$$y_{x(5)} = e^{-5t}(y_{t(5)} - 10y_{t(4)} + 35y_{ttt} - 50y_{tt} + 24y_t),$$

$$y_{x(6)} = e^{-6t}(y_{t(6)} - 15y_{t(5)} + 85y_{t(4)} - 225y_{ttt} + 274y_{tt} - 120y_t),$$

$$y_{x(7)} = e^{-7t}(y_{t(7)} - 21y_{t(6)} + 175y_{t(5)} - 735y_{t(4)} + 1624y_{ttt} - 1764y_{tt} + 720y_t),$$

$$y_{x(8)} = e^{-8t}(y_{t(8)} - 28y_{t(7)} + 322y_{t(6)} - 1960y_{t(5)} + 6769y_{t(4)} - 13132y_{ttt}$$
$$+ 13068y_{tt} - 5040y_t).$$

Table 57. How to transform derivatives under the change of dependent variable $x = e^t$.

Notes

[1] This method is also applicable to the equation

$$a_0(Ax + B)^n y^{(n)} + a_1(Ax + B)^{n-1} y^{(n-1)} + \ldots + a_{n-1}(Ax + B)y' + a_n y = 0$$

which is only a trivial modification of an Euler equation.

[2] Equations of the form $\dfrac{dx}{\sqrt{P(x)}} = \pm \dfrac{dy}{\sqrt{P(y)}}$, where $P(x)$ is a polynomial of degree three or four have also been called Euler equations, see Valiron [5].

[3] Euler matrix differential equations (in which the $\{a_i\}$ in (57.1) are all matrices) are discussed in Jódar [3].

References

[1] W. E. Boyce and R. C. DiPrima, *Elementary Differential Equations and Boundary Value Problems*, Fourth Edition, John Wiley & Sons, New York, 1986, Section 4.4.

[2] N. Finizio and G. Ladas, *Ordinary Differential Equations with Modern Applications*, Wadsworth Publishing Company, Belmont, Calif, 1982, pages 103–105.

[3] L. Jódar, "Boundary Value Problems for Second Order Operator Differential Equations," *Linear Algebra and its Appls.*, **91**, 1987, pages 1–12.

[4] G. F. Simmons, *Differential Equations with Applications and Historical Notes*, McGraw–Hill Book Company, New York, 1972, page 86.

[5] G. Valiron, *The Geometric Theory of Ordinary Differential Equations and Algebraic Functions*, Math Sci Press, Brookline, MA, 1950, pages 201–202.

58.   Exact First Order Equations

Applicable to   First order ordinary differential equations.

Yields
   An exact solution (generally implicit).

Idea
   Some first order ordinary differential equations can be integrated directly.

Procedure
   If the given ordinary differential equation has the form

$$\frac{dy}{dx} = \frac{N(x,y)}{M(x,y)}, \tag{58.1}$$

and $N(x,y)$ and $M(x,y)$ are such that

$$\frac{\partial M}{\partial x} + \frac{\partial N}{\partial y} = 0, \tag{58.2}$$

then (58.1) is said to be an exact ordinary differential equation. Such an equation can be solved exactly, though the answer may be in terms of an integral. The (implicit) solution will be of the form

$$\phi(x,y) = C, \tag{58.3}$$

where C is an arbitrary constant. Motivating this is straightforward. Differentiating (58.3) with respect to x and rearranging terms gives

$$\frac{dy}{dx} = -\frac{\phi_x}{\phi_y}. \tag{58.4}$$

Comparing (58.4) to (58.1) we have

$$\phi_x = -N, \quad \phi_y = M, \tag{58.5.a–b}$$

and hence (58.2) is satisfied (since $\phi_{xy} = \phi_{yx}$). Conversely, if (58.2) is satisfied, then there is a ϕ such that (58.5) is satisfied. To solve (58.5) for ϕ, integrate (58.5.a) with respect to x and integrate (58.5.b) with respect to y for

$$\phi(x,y) = -\int N(x,y)\,dx + f(y),$$

$$\phi(x,y) = \int M(x,y)\,dy + g(x), \tag{58.6.a–b}$$

where $f(y)$ and $g(x)$ are unknown functions. Comparing (58.6.a) to (58.6.b) will determine $f(y)$ and $g(x)$. Knowing either of these, the full solution is then given by (58.6.a) or (58.6.b).

Example

Suppose we have the equation

$$\frac{dy}{dx} = \frac{3x^2 - y^2 - 7}{e^y + 2xy + 1}. \tag{58.7}$$

In (58.7) we identify

$$N(x, y) = 3x^2 - y^2 - 7,$$

$$M(x, y) = e^y + 2xy + 1.$$

Following our procedure, we find $M_x = -N_y = 2y$ and so we know that we can solve (58.7) exactly. Integrating N and M we find

$$\phi(x, y) = -\int N(x, y)\, dx + f(y) = -(x^3 + y^2 x - 7x) + f(y),$$
$$\phi(x, y) = \int M(x, y)\, dy + g(x) = (e^y + y^2 x + y) + g(x). \tag{58.8.a-b}$$

Comparing (58.8.a) to (58.8.b) we deduce that

$$-x^3 - y^2 x + 7x + f(y) = e^y + y^2 x + y + g(x),$$

or

$$f(y) - (e^y + y) = g(x) - (7x + x^3). \tag{58.9}$$

From (58.9) we conclude that

$$f(y) = e^y + y + A, \quad g(x) = 7x - x^3 + A, \tag{58.10.a-b}$$

where A is an arbitrary constant. Using either (58.10.a) in (58.8.a) or (58.10.b) in (58.8.b) we conclude

$$\phi(x, y) = -x^3 - y^2 + 7x + e^y + y + A.$$

The solution is then given by $\phi(x, y) = C$, where C is an arbitrary constant. Therefore

$$-x^3 - y^2 + 7x + e^y + y = B, \tag{58.11}$$

is the final solution, where $B := A - C$ is a final arbitrary constant. Note that the solution in (58.11) is implicit.

References

[1] W. E. Boyce and R. C. DiPrima, *Elementary Differential Equations and Boundary Value Problems*, Fourth Edition, John Wiley & Sons, New York, 1986, pages 79–84.

[2] E. D. Rainville and P. E. Bedient, *Elementary Differential Equations*, The MacMillan Company, New York, 1964, pages 29–33.

[3] G. F. Simmons, *Differential Equations with Applications and Historical Notes*, McGraw–Hill Book Company, New York, 1972, pages 38–41.

59. Exact Second Order Equations

Applicable to   Some nonlinear second order ordinary differential equations of the form $f(x, y, y')y'' + g(x, y, y') = 0$.

Yields

A first integral (which will be a first order ordinary differential equation).

Idea

Some second order ordinary differential equations can be integrated once.

Procedure

The second order differential equation

$$F(x, y, y', y'') = 0 \qquad (59.1)$$

is said to be exact if it is the total differential of some function; i.e., $F = d\phi/dx$ where $\phi = \phi(x, y, y')$. If (59.1) is exact, then $\phi = C$ is a solution to (59.1), with C an arbitrary constant. Differentiating $\phi = C$ with respect to x we find

$$\frac{d\phi}{dx} = \frac{\partial \phi}{\partial x} + \frac{\partial \phi}{\partial y}y' + \frac{\partial \phi}{\partial y'}y''. \qquad (59.2)$$

Comparing (59.2) to (59.1), we conclude that, for (59.1) to be exact, $F(x, y, y', y'')$ must have the form

$$F(x, y, y', y'') = f(x, y, y')y'' + g(x, y, y'), \qquad (59.3)$$

for some functions f and g with

$$f(x, y, y') = \frac{\partial \phi}{\partial y'}, \qquad g(x, y, y') = \frac{\partial \phi}{\partial x} + \frac{\partial \phi}{\partial y} y'. \qquad (59.4.a-b)$$

By differentiating (59.4.a–b) with respect to x, y, and p, (using $p := dy/dx$), all dependence on ϕ can be eliminated between the two equations in (59.4) to obtain

$$f_{xx} + 2pf_{xy} + p^2 f_{yy} = g_{xp} + pg_{yp} - g_y,$$
$$f_{xp} + pf_{yp} + 2f_y = g_{pp}. \qquad (59.5)$$

If the conditions in (59.5) hold, then (59.3) is exact. If (59.3) is exact, then we can integrate (59.4.a) (with respect to p) to determine $\phi(x, y, y')$ as

$$\phi = h(x, y) + \int f(x, y, p)\, dp, \qquad (59.6)$$

where $h(x, y)$ is, so far, an arbitrary function of integration. This function will be restricted when (59.6) is used in (59.4.b).

Example
Given the equation

$$xyy'' + x(y')^2 + yy' = 0, \qquad (59.7)$$

which has the form of (59.3), we identify: $f = xy$, $g = x(y')^2 + yy' = xp^2 + yp$. It is easy to verify that (59.5) holds. Hence, equation (59.7) is exact. Equation (59.6) now becomes

$$\phi = h(x, y) + \int xy\, dp$$
$$= h(x, y) + xyp. \qquad (59.8)$$

Using (59.8) in (59.4.b) yields

$$g = xp^2 + yp = \frac{\partial \phi}{\partial x} + \frac{\partial \phi}{\partial y} y'$$
$$= (h_x + yp) + (h_y + xp)p. \qquad (59.9)$$

Hence, if h is constant, say $h = D$, then (59.9) will be satisfied. Therefore a first integral of (59.7) is given by $\phi = C$, or

$$C = \phi(x, y, p)$$
$$= D + xyp \qquad (59.10)$$
$$= D + xy\frac{dy}{dx}.$$

In this example the first integral (59.10) can itself be integrated in closed form (this is often true). A solution to (59.7), obtained by solving the ordinary differential equation in (59.10), is thus given by

$$\frac{y^2}{2} = (C - D) \log x + E,$$

where E is another arbitrary constant.

Notes
[1] The most general solution for $h(x, y)$ in (59.9) is $h = h(y-x)$. With this form for h, however, the first integral cannot be integrated to yield an explicit solution.
[2] Exact second order linear ordinary differential equations have factorable operators (see page 246).
[3] Given the differential equation

$$f(x, y, \ldots, y^{(n)}) = 0, \tag{59.11}$$

define $f_i = \dfrac{\partial f}{\partial y^{(i)}}$. Then (59.11) will be exact if

$$f_0 - \frac{df_1}{dx} + \frac{d^2 f_2}{dx^2} - \cdots + (-1)^n \frac{d^n f_n}{dx^n} = 0. \tag{59.12}$$

If the differential equation in (59.11) is exact, then a first integral can be found by a repetitive sequence of steps: First integrate the highest order term in f and call this result F_1. Then integrate the highest order term in $f \, dx - dF_1$ and call this result F_2. Continue in this manner until $f \, dx - dF_1 - dF_2 - \ldots = 0$. Then a first integral is given by $F_1 + F_2 + \ldots = $ constant.
For example, given the nonlinear third order equation

$$f = yy''' - y'y'' + y^3 y' = 0, \tag{59.13}$$

we identify $f_3 = y$, $f_2 = -y'$, $f_1 = -y'' + y^3$, $f_0 = y''' + 3y^2 y'$; and verify that (59.12) is satisfied. We then calculate $F_1 = yy''$, since the highest order term in f is yy'''. Then $f \, dx - dF_1 = (-2y'y'' + y^3 y')dx$, and so we take $F_2 = -(y')^2$. Then $f \, dx - dF_1 - dF_2 = y^3 y' dx$, and so $F_3 = \frac{1}{4}y^4$. Finally then, $f \, dx - dF_1 - dF_2 - dF_3 = 0$, so that

$$yy'' - (y')^2 + \tfrac{1}{4}y^4 = \text{constant}$$

is a first integral of (59.13).

References

[1] M. E. Goldstein and W. H. Braun, *Advanced Methods for the Solution of Differential Equations*, NASA SP-316, U.S. Government Printing Office, Washington, D.C., 1973, page 93.

[2] G. Murphy, *Ordinary Differential Equations*, D. Van Nostrand Company, Inc., New York, 1960, pages 221–222.

60. Exact N-th Order Equations

Applicable to   Linear n-th order ordinary differential equations.

Yields

A first integral.

Idea

Some linear differential equations can be integrated exactly without modifying the equation in any way.

Procedure

The linear n-th order ordinary differential equation

$$P_n(x)\frac{d^n y}{dx^n} + P_{n-1}(x)\frac{d^{n-1}y}{dx^{n-1}} + \cdots + P_1(x)\frac{dy}{dx} + P_0(x)y = R(x), \quad (60.1)$$

is said to be *exact* if it can be integrated once to yield

$$Q_{n-1}(x)\frac{d^{n-1}y}{dx^{n-1}} + Q_{n-2}(x)\frac{d^{n-2}y}{dx^{n-2}} + \cdots + Q_1(x)\frac{dy}{dx} + Q_0(x)y = \int R(x)\,dx. \tag{60.2}$$

If (60.1) is exact, then the $\{Q_i(x)\}$ may be found from

$$\begin{aligned}
Q_{n-1} &= P_n, \\
Q_{n-2} &= P_{n-1} - P'_n, \\
Q_{n-3} &= P_{n-2} - P'_{n-1} + P''_n, \\
&\ \vdots \\
Q_0 &= P_1 - P'_2 + P''_3 - \cdots + (-1)^{n-1}P_n^{(n-1)}.
\end{aligned}$$

A necessary and sufficient condition for (60.1) to be exact can be found by differentiating (60.2) with respect to x and comparing terms with (60.1).

This condition is

$$\frac{d^n P_n}{dx^n} - \frac{d^{n-1} P_{n-2}}{dx^{n-1}} + \frac{d^{n-2} P_{n-3}}{dx^{n-2}} - \cdots + (-1)^{n-1}\frac{dP_1}{dx} + (-1)^n P_0 = 0.$$

Special Case

The second order linear ordinary differential equation

$$P(x)y'' + Q(x)y' + R(x)y = 0$$

will be exact if and only if $P''(x) - Q'(x) + R(x) = 0$.

Example

If we have the linear ordinary differential equation of third order

$$(1 + x + x^2)\frac{d^3 y}{dx^3} + (3 + 6x)\frac{d^2 y}{dx^2} + 6\frac{dy}{dx} = 6x, \qquad (60.3)$$

then we have: $R(x) = 6x$, $P_0 = 0$, $P_1 = 6$, $P_2 = 3 + 6x$, $P_3 = 1 + x + x^2$. It is easy to verify that

$$\frac{d^3 P_3}{dx^3} - \frac{d^2 P_2}{dx^2} + \frac{dP_1}{dx} - P_0 = 0,$$

and so (60.3) is exact. Integrating equation (60.3) directly we obtain

$$(1 + x + x^2)\frac{d^2 y}{dx^2} + (2 + 4x)\frac{dy}{dx} + 2y = 3x^2 + A, \qquad (60.4)$$

where A is an arbitrary constant. Now the equation in (60.4) is again exact, and so it can be integrated again to yield

$$(1 + x + x^2)\frac{dy}{dx} + (1 + 2x)y = x^3 + Ax + B, \qquad (60.5)$$

where B is an arbitrary constant.

Finally, the equation in (60.5) is exact once again. It can be integrated to yield the general solution of (60.3)

$$(1 + x + x^2)y = \frac{x^4}{4} + A\frac{x^2}{2} + Bx + C,$$

where C is an arbitrary constant.

References
[1] L. R. Ford, *Differential Equations*, McGraw–Hill Book Company, New York, 1955, pages 77–78.
[2] G. Murphy, *Ordinary Differential Equations*, D. Van Nostrand Company, Inc., New York, 1960, pages 221–222.

61.   Factoring Equations*

Applicable to   Ordinary differential equations and partial differential equations.

Yields

Equations of lower degree.

Idea

If a differential equation can be factored into simple terms, then the solution to each of the factors is a solution to the original equation.

Procedure

Given a differential equation, attempt to factor it. If this is possible, then solve each factor separately. Each of the solutions of the different factors will be a solution of the original differential equation.

Example

The nonlinear ordinary differential equation

$$y'(y' + y) = x(x + y) \tag{61.1}$$

for $y(x)$ may be factored into

$$(y' + y + x)(y' - x) = 0. \tag{61.2}$$

Solving each of the factors appearing in (61.2) separately, the solutions to (61.1) are given by

$$y(x) = \begin{cases} Ae^{-x} + 1 - x, \\ \\ B + \dfrac{x^2}{2}, \end{cases}$$

where A and B are constants.

Notes

[1] The complete solution to the original differential equation may switch from one solution branch to another.

References

[1] I. K. Argyros, "On the Cardinality of Solutions of Multilinear Differential Equations and Applications," *Int. J. Math. & Math. Sci.*, **9**, No. 4, 1986, pages 757–766.

[2] H. Bateman, *Partial Differential Equations of Mathematical Physics*, Dover Publications, Inc., New York, 1944, pages 97–98.

[3] M. Fogiel, *The Differential Equations Problem Solver*, Volume 1, Research and Education Association, New York, 1978, pages 1222–1229.

[4] M. S. Klamkin, "On Soluble nth Order Linear Differential Equations," *J. Math. Anal. Appl.*, **84**, 1981, pages 6–11.

62.   Factoring Operators*

Applicable to   Ordinary differential equations and partial differential equations.

Yields

A sequence of equations to solve that are of lower order.

Idea

If the operator representing a differential equation can be "factored" into two or more operators, it may be easier to find a solution.

Procedure

Suppose we wish to solve the differential equation $Q[u] = 0$ for the quantity $u(\mathbf{x})$, where $Q[\cdot]$ is a differential operator. When possible, "factor" the differential equation $Q[u] = 0$ as $L[H[u]] = 0$ where $L[\cdot]$ and $H[\cdot]$ are also differential operators. Then solve the two equations: $L[v] = 0$ for v, and then $H[u] = v$.

Example 1
   The fourth order partial differential equation

$$(\nabla^4 - a^2)u = 0, \tag{62.1}$$

where a is a constant and ∇^2 is the usual Laplacian, may be factored as

$$(\nabla^2 - a)(\nabla^2 + a)u = 0.$$

The general solution of (62.1), therefore, is given by the solution of the two successive second order differential equations

$$\begin{aligned} (\nabla^2 - a)v &= 0, \\ (\nabla^2 + a)u &= v. \end{aligned} \tag{62.2}$$

Alternatively, (62.1) could have factored equation as

$$(\nabla^2 + a)(\nabla^2 - a)u = 0$$

so that the general solution of (62.1) can also be written as the solution of

$$\begin{aligned} (\nabla^2 + a)w &= 0, \\ (\nabla^2 - a)u &= w. \end{aligned} \tag{62.3}$$

Solving (62.2) or (62.3) as a sequence of two second order differential equations may be easier than solving the fourth order equation (62.1) directly.

Example 2
   If we want to solve the nonlinear ordinary differential equation

$$\begin{aligned} Q[u] &= u_{xx}^2 - 2u_x u_{xx} + 2uu_x - u^2 = 0 \\ &= (u_{xx} - u_x)^2 - (u_x - u)^2 = 0, \end{aligned} \tag{62.4}$$

then we might factor the operator $Q[\cdot]$ as $Q[u] = L[H[u]]$, where $L[v] = v_x^2 - v^2$, and $H[u] = u_x - u$. Therefore, the equation $Q[u] = 0$ can be solved by solving the sequence of first order differential equations

$$L[v] = 0, \qquad H[u] = v.$$

The solution of $L[v] = 0$ is $v = Ce^{\pm x}$, where C is an arbitrary constant. The general solution of (62.4) can then be determined by solving

$$H[u] = u_x - u = v = Ce^{\pm x}. \tag{62.5}$$

Equation (62.5) can be solved by the use of integrating factors (see page 305) to obtain the two possible forms of the solution

$$u = \begin{cases} (A + Cx)e^x, \\ \\ Ce^{-x} + Be^x, \end{cases}$$

where A and B are also arbitrary constants.

Example 3

The relativistic wave equation

$$\frac{1}{c^2}\frac{\partial^2\psi}{\partial t^2} - \frac{\partial^2\psi}{\partial x^2} - \frac{\partial^2\psi}{\partial y^2} - \frac{\partial^2\psi}{\partial z^2} + \frac{m^2c^2}{\hbar^2}\psi = 0$$

was factored by Dirac using hypercomplex algebra. If $\{\alpha_1, \alpha_2, \alpha_3, \alpha_4\}$ represent four of the elements in this algebra that obey the relation $\alpha_\mu\alpha_\nu + \alpha_\nu\alpha_\mu = 2\delta_{\mu\nu}$, then the factored equation is

$$\left(\frac{1}{c}\frac{d}{dt} - \alpha_1\frac{d}{dx} - \alpha_2\frac{d}{dy} - \alpha_3\frac{d}{dz} - \alpha_4\frac{imc}{\hbar}\right)$$
$$\left(\frac{1}{c}\frac{d}{dt} + \alpha_1\frac{d}{dx} + \alpha_2\frac{d}{dy} + \alpha_3\frac{d}{dz} + \alpha_4\frac{imc}{\hbar}\right)\psi = 0.$$

The first factor led to the correct relativistic theory for the electron, while the second factor led to Dirac's prediction of the positron. See Dirac [4] for details.

Example 4

The formally self-adjoint homogeneous fourth order operator

$$\frac{d^2}{dx^2}\left(P(x)\frac{d^2y}{dx^2}\right)\frac{d}{dx}\left(Q(x)\frac{dy}{dx}\right) + R(x)y$$

may be factored into $L[\nu(x)L[y]]$, where $L[\cdot]$ is the second order operator

$$L[y] = \frac{d}{dx}\left[\lambda(x)\frac{dy}{dx}\right] + \mu(x)y,$$

where $\{\nu(x), \mu(x), \lambda(x)\}$ satisfy

$$\nu(x) = \frac{\beta'}{\alpha^2},$$
$$\lambda(x) = \alpha^2\beta',$$
$$\mu(x) = \frac{\alpha}{\beta'}\left(\alpha'' + \tfrac{1}{2}\gamma\alpha\right),$$

and $\{\alpha(x), \beta(x), \gamma(x), \delta(x)\}$ are any solution to

$$P(x) = \alpha^2\beta'^3,$$
$$Q(x) = \alpha^2\beta''' + 2\alpha\alpha'\beta'' + \left(4\alpha\alpha'' - 2\alpha'^2 + \gamma\alpha^2\right)\beta',$$
$$R(x) = \frac{\alpha}{\beta'}\left(\alpha'''' + \alpha\gamma'' + \alpha'\gamma' + \alpha\delta\right),$$

with $4\delta = 2\gamma'' + \gamma^2$. See Hill [9] for details.

Notes

[1] It is not true that the number of distinct factorizations is limited by the order of the differential equation. For example, the second order ordinary differential equation

$$(x^2 - x^3)u'' + (2x^2 - 4x)u' + (6 - 2x)u = 0,$$

has the three distinct factorizations

$$\left(x\frac{d}{dx} - 2\right)\left[(x - x^2)\frac{d}{dx} + 2x - 3\right]u = 0,$$

$$\left(x\frac{d}{dx} - 3\right)\left[(x - x^2)\frac{d}{dx} + x - 2\right]u = 0,$$

$$\left[(x - x^2)\frac{d}{dx} + x - 3\right]\left(x\frac{d}{dx} - 2\right)u = 0.$$

[2] The Laplacian in two dimensions admits the factorization:

$$\nabla^2 := \frac{\partial^2}{\partial x^2} + \frac{\partial^2}{\partial y^2} = \left(\frac{\partial}{\partial x} - i\frac{\partial}{\partial y}\right)\left(\frac{\partial}{\partial x} + i\frac{\partial}{\partial y}\right) := \left(\frac{\partial}{\partial z}\right)\left(\frac{\partial}{\partial \bar{z}}\right),$$

where $i = \sqrt{-1}$. Therefore, using $z = x + iy$, Laplace's equation may be written as $\nabla^2 u = \frac{\partial^2 u}{\partial z \partial \bar{z}} = 0$. This shows that the most general solution to Laplace's equation in two dimensions is $u = f(z) + g(\bar{z})$, where $f(z)$ and $g(\bar{z})$ are arbitrary functions. Also, since the biharmonic equation may be written as $\nabla^4 u = 16\frac{\partial^4 u}{\partial z^2 \partial \bar{z}^2} = 0$, the general solution of the biharmonic equation is seen to be $u = f(z) + g(\bar{z}) + zh(\bar{z}) + \bar{z}j(z)$.

The operators $\partial/\partial z$ and $\partial/\partial \bar{z}$ are known as *Wirtinger derivatives*. In two dimensions, solutions of Poisson's equation may sometimes be found by use of Wirtinger derivatives. See Henrici [8] for details.

[3] It is possible to write down an "explicit" factorization of any n-th order linear differential equation. To do so, however, requires explicit knowledge of the n linearly independent solutions. For example, if $L[\cdot]$ is the differential operator

$$L[u] = u'' + p(x)u' + q(x)u,$$

and u_1, u_2 are any two linearly independent solutions of $L[u] = 0$, then

$$L[u] = \frac{W(u_1, u_2)}{u_1}\frac{d}{dx}\left[\frac{u_1^2}{W(u_1, u_2)}\frac{d}{dt}\left(\frac{u}{u_1}\right)\right],$$

where $W(u_1, u_2)$ is the Wronskian of $u_1(x)$ and $u_2(x)$. In the n-th order case, consider the differential operator

$$H[u] = u^{(n)} + p_1(x)u^{(n-1)} + p_2(x)u^{(n-2)} + \cdots + p_n(x)u.$$

If $\{u_1, u_2, \ldots, u_n\}$ are n linearly independent solutions of $H[u] = 0$, then define W_k (for $k = 1, 2, \ldots, n$) to be the Wronskian of the first k linearly independent solutions; i.e., $W_k := W(u_1, u_2, \ldots, u_k)$. Using this definition, we can write $H[u]$ as

$$H[u] = \frac{W_n}{W_{n-1}} \frac{d}{dx} \left(\frac{W_{n-1}^2}{W_{n-1}W_n} \cdots \right.$$
$$\left. \cdots \frac{d}{dx} \left(\frac{W_2^2}{W_1 W_3} \left(\frac{d}{dx} \left(\frac{W_1^2}{W_0 W_2} \left(\frac{d}{dx} \left(\frac{u}{W_1} \right) \right) \right) \right) \right) \cdots \right).$$

See Rainville [11] for details.

[4] The factorization

$$\left\{ \frac{d}{dt} - q(t) \right\} \left\{ \frac{d}{dt} - q(t) \right\} w = \frac{d^2 w}{dt^2} + w \left\{ \frac{dq}{dt} - q^2 \right\}$$

leads to the technique for solving Riccati equations described on page 332.

[5] Differential resultants can be used to analyze the factoring of operators for linear differential equations. See Berkovich and Tsirulik [1] for details.

[6] Two differential operators P and Q are said to be permutable if $P(Q) = Q(P)$. From Ince [10] we have

> If P and Q are permutable operators of orders m and n respectively, they satisfy identically an algebraic relation of the form $F(P, Q) = 0$ of degree n in P and of degree m in Q.

For example, the operators

$$P = \frac{d^2}{dx^2} - \frac{2}{x^2},$$
$$Q = \frac{d^3}{dx^3} - \frac{3}{x^2} \frac{d}{dx} + \frac{3}{x^3},$$

are permutable since $PQ = QP$. We can also find the algebraic relation $P^3 - Q^2 = 0$, observe:

$$P(P(P(f))) = f'''''' - \frac{6}{x^2} f'''' + \frac{24}{x^3} f''' - \frac{72}{x^4} f'' + \frac{144}{x^5} f' - \frac{144}{x^6} f = Q(Q(f)).$$

This example is due to Ince [10]. See also Grünbaum [7].

References

[1] L. M. Berkovich and V. G. Tsirulik, "Differential Resultants and Some of Their Applications," *Differentsial'nye Uravneniya*, **22**, No. 5, May 1986, pages 750–757.

[2] W. D. Brownawell, "On The Factorization of Partial Differential Equations," *Can. J. Math.*, **39**, No. 4, 1987, pages 825–834.

[3] J. S. R. Chisholm and A. K. Common, "A Class of Second-Order Differential Equations and Related First-Order Systems," *J. Phys. A: Math. Gen.*, **20**, 1987, pages 5459–5472.

[4] P. A. M. Dirac, *The Principle of Quantum Mechanics*, Third Edition, Clarendon Press, Oxford, 1974, Chapter 11.

[5] G. J. Etgen, G. D. Jones, and W. E. Taylor, Jr., "On the Factorizations of Ordinary Linear Differential Operators," *Trans. Amer. Math. Soc.*, **297**, No. 2, 1986, pages 717–728.

[6] A. P. Fordy and J. Gibbons, "Factorization of Operators I. Miura Transformations," *J. Math. Physics*, **21**, No. 10, October 1980, pages 2508–2510.

[7] F. A. Grünbaum, "Commuting Pairs of Linear Ordinary Differential Operators of Orders Four and Six," *Physica D*, **31**, 1988, pages 424–433.

[8] P. Henrici, *Applied and Computational Complex Analysis*, Volume 3, Wiley, New York, 1986, pages 300–302.

[9] J. M. Hill, *Solution of Differential Equations by Means of One-Parameter Groups*, Pitman Publishing Co., Marshfield, MA, 1982.

[10] E. L. Ince, *Ordinary Differential Equations*, Dover Publications, Inc., New York, 1964, page 131.

[11] E. D. Rainville, *Intermediate Differential Equations*, The MacMillan Company, New York, 1964, pages 292–299.

[12] D. C. Sandell and F. M. Stein, "Factorization of Operators of Second Order Linear Homogeneous Ordinary Differential Equations," *Two Year College Mathematics Journal*, **8**, 1977, pages 132–141.

[13] V. H. Weston, "Factorization of the Wave Equation in Higher Dimensions," *J. Math. Physics*, **28**, No. 5, May 1987, pages 1061–1068.

63. Factorization Method

Applicable to   Eigenvalue/eigenfunction problems for homogeneous second order ordinary differential equations.

Yields

An equation from which a single eigenfunction can be used to calculate additional eigenfunctions.

Idea

By "factoring" an ordinary differential equation into a certain form, a ladder of eigenfunctions may be formed.

Procedure

Suppose we have the linear second order ordinary differential equation

$$\frac{d^2y}{dx^2} + r(x, m)y + \lambda y = 0, \tag{63.1}$$

where m is an integer, for which we would like to determine the eigenfunctions $\{y\}$ corresponding to a single value of the eigenvalue λ. We denote the eigenfunction by $y(\lambda, m)$ and suppress the x dependence. The equation in (63.1) is said to be *factorizable* if it is equivalent to each of

$$H_+^{m+1} H_-^{m+1} y(\lambda, m) = L(\lambda, m+1)y(\lambda, m),$$
$$H_-^m H_+^m y(\lambda, m) = L(\lambda, m)y(\lambda, m), \tag{63.2.a-b}$$

where $L(\lambda, m)$ is a function and the $H_\pm^m$ are differential operators. For a factorizable equation, finding $L(\lambda, m)$ and the $H_\pm^m$ is a difficult task. Also, not all equations in the form of (63.1) are factorizable.

If (63.1) is factorizable, and if $y(\lambda, m)$ is a solution of (63.1), then (see the notes)

$$y(\lambda, m+1) = H_-^{m+1} y(\lambda, m),$$
$$y(\lambda, m-1) = H_+^m y(\lambda, m), \tag{63.3.a-b}$$

are also solutions corresponding to the same value of λ, but different values of m. Hence, given one solution of (63.1) (for a specific value of λ), a *ladder of solutions* belonging to this value of λ may be formed by repeatedly iterating (63.3).

Example 1

The equation for the associated spherical harmonics may be put in the form

$$\frac{d^2y}{d\theta^2} - \frac{m^2 - \frac{1}{4}}{\sin^2\theta} + \left(\lambda + \frac{1}{4}\right)y = 0. \tag{63.4}$$

This equation is factorizable, and we find

$$H_\pm^m = \left(m - \tfrac{1}{2}\right)\cot\theta \pm \frac{d}{dx},$$
$$L(\lambda, m) = \lambda - \left(m - \tfrac{1}{2}\right)^2, \tag{63.5}$$

The eigenvalues of (63.4) are of the form $\lambda = l(l+1)$ for $l = m, m+1, \ldots$.
Some of the eigenfunctions of (63.4) are of the form

$$y_l^l(\theta) = \left[\frac{1 \cdot 3 \cdot 5 \cdots (2l+1)}{2 \cdot 2 \cdot 4 \cdots (2l)} \right]^{1/2} \sin^{l+1/2} \theta.$$

All of the remaining eigenfunctions may be found from (63.3) and (63.5)
to be given by

$$y_l^{m-1}(\theta) = \frac{1}{\sqrt{(l+m)(l+1-m)}} \left[\left(m - \frac{1}{2} \right) \cot \theta + \frac{d}{d\theta} \right] y_l^m(\theta),$$

$$y_l^{m+1}(\theta) = \frac{1}{\sqrt{(l+m+1)(l-m)}} \left[\left(m + \frac{1}{2} \right) \cot \theta - \frac{d}{d\theta} \right] y_l^m(\theta).$$

Example 2

As another example, Legendre's differential equation

$$(1 - x^2) \left[(1 - x^2) y_m' \right]' + m(m+1) y_m = 0$$

has the factorizations

$$H_-^m H_+^m y_m = -m^2 y_m,$$

$$H_+^{m+1} H_-^{m+1} y_m = -(m+1)^2 y_m,$$

where $H_\pm^m = (1 - x^2) \dfrac{d}{dx} \pm mx$. This factorization leads to the ladder of
solutions: $y_{m+1} = H_-^m y_m$.

Notes

[1] The results in (63.3) are straightforward to derive. For example, operating
on (63.2.b) with H_+^m results in

$$H_+^m H_-^m \left\{ H_+^m y(\lambda, m) \right\} = L(\lambda, m) \left\{ H_+^m y(\lambda, m) \right\}. \tag{63.6}$$

Since this has the same form as (63.2.a), which is by hypothesis equivalent
to (63.1), it must be that $y = H_+^m y(\lambda, m)$ is a solution of (63.1). In (63.3)
we called this $y(\lambda, m-1)$ since, when (63.6) is compared to (63.2.a), the
parameter m is replaced by $m-1$.

[2] The factorization method has been generalized to systems of equations in
Humi [4].

[3] The operators in (63.3) are sometimes called raising and lowering operators.
This method is sometimes called the ladder method.

[4] Infeld and Hull [5] have a large list of equations to which this method applies.

[5] The paper by Hermann [3] relates the technique in this section to Lie groups.
Sattinger and Weaver [8] also consider the relation to Lie groups.

References

[1] A. O. Barut, A. Inomata, and R. Wilson, "A New Realization of Dynamical Groups and Factorization Method," *J. Phys. A: Math. Gen.*, **20**, 1987, pages 4075–4083.

[2] N. Bessis and G. Bessis, "Algebraic Recursive Determination of Matrix Elements from Ladder Operator Considerations," *J. Phys. A: Math. Gen.*, **20**, 1987, pages 5745–5754.

[3] R. Hermann, "Infeld–Hull factorization, Galois–Picard–Vessiot Theory for Differential Operators," *J. Math. Physics*, **22**, No. 6, June 1981, pages 1163–1167.

[4] M. Humi, "Factorization of Systems of Differential Equations," *J. Math. Physics*, **27**, January 1986, pages 76–81.

[5] L. Infeld and T. E. Hull, "The Factorization Method," *Rev. Mod. Physics*, **23**, No. 1, January 1951, pages 21–68.

[6] G. L. Lamb, *Elements of Soliton Theory*, John Wiley & Sons, New York, 1980, pages 38–41.

[7] P. M. Morse and H. Feshback, *Methods of Theoretical Physics*, McGraw-Hill Book Company, New York, 1953, pages 788–789.

[8] D. H. Sattinger and O. L. Weaver, *Lie Groups and Algebras with Applications to Physics, Geometry, and Mechanics*, Springer–Verlag, New York, 1986, pages 49–54.

[9] E. Schrödinger, "A Method of Determining Quantum-Mechanical Eigenvalues and Eigenfunctions," *Proc. Roy. Irish Acad.*, **A46**, 1940, pages 9–16.

64.   Fokker–Planck Equation

Applicable to   Linear ordinary differential equations with linearly appearing "white Gaussian noise" terms (a single differential equation, or a system).

Yields

A Fokker–Planck equation (which is a parabolic partial differential equation) for the probability density of the solution.

Idea

If a differential equation contains random terms, then the solution to the differential equation can only be described statistically. The solution to the Fokker–Planck equation is the probability density of the solution to the original differential equation.

Procedure

Here we present the technique for constructing the Fokker–Planck equation for a linear system of ordinary differential equations depending on several white noise terms. Consider the linear system for the m component vector $\mathbf{x}(t)$

$$\frac{d}{dt}\mathbf{x}(t) = \mathbf{b}(t,\mathbf{x}) + \sigma(t,\mathbf{x})\,\mathbf{n}(t),$$

$$\mathbf{x}(t_0) = \mathbf{y},$$

(64.1a–b)

where $\sigma(t,\mathbf{x})$ is a real $m \times n$ matrix and $\mathbf{n}(t)$ is a vector of n independent white noise terms. That is

$$E[n_i(t)] = 0,$$

$$E[n_i(t)n_j(t+\tau)] = \delta_{ij}\delta(\tau),$$

(64.2)

where $E[\cdot]$ is the expectation operator, δ_{ij} is the Kronecker delta, and $\delta(\tau)$ is the delta function. The Fokker–Planck equation corresponding to (64.1.a) is given by

$$\frac{\partial P}{\partial t} = -\sum_{i=1}^{m}\frac{\partial}{\partial x_i}(b_i P) + \frac{1}{2}\sum_{i,j=1}^{m}\frac{\partial^2}{\partial x_i \partial x_j}(a_{ij}P),$$

(64.3)

where $P = P(t,\mathbf{x})$ is a probability density and the matrix $A = (a_{ij})$ is defined by $A(t,\mathbf{x}) = \sigma(t,\mathbf{x})\sigma^{\mathrm{T}}(t,\mathbf{x})$. The initial conditions for (64.3) come from (64.1.b), they are

$$P(t_0,\mathbf{x}) = \prod_{i=1}^{m}\delta(x_i - y_i).$$

(64.4)

The solution of (64.3) and (64.4) is the probability density of the solution to (64.1). Any statistical information about $\mathbf{x}(t)$ that could be ascertained from (64.1), can be derived from $P(t,\mathbf{x})$. For example, the expected value of some function of $\mathbf{x}$ and t, say $h(\mathbf{x},t)$, at a time t, can be calculated by

$$E[h(\mathbf{x}(t),t)] = \int_{-\infty}^{\infty} h(\mathbf{x}(t),t)P(t,\mathbf{x})\,d\mathbf{x}.$$

Special Case

In the special case of one dimension, the stochastic differential equation

$$\frac{dx}{dt} = f(x) + g(x)n(t),$$

(64.5)

with $x(0) = z$, corresponds to the Fokker–Planck equation

$$\frac{\partial P}{\partial t} = -\frac{\partial}{\partial x}(f(x)P) + \frac{1}{2}\frac{\partial^2}{\partial x^2}(g^2(x)P),$$

for $P(t,x)$ with $P(0,x) = \delta(x-z)$.

Example

Consider the *Langevin equation*

$$x'' + \beta x' = N(t), \tag{64.6}$$

with the initial conditions

$$x(0) = 0, \qquad x'(0) = u_0, \tag{64.7}$$

where $N(t)$ satisfies

$$\begin{aligned} E[N(t)] &= 0, \\ E[N(t)N(t+\tau)] &= \delta(\tau). \end{aligned} \tag{64.8}$$

From (64.8), we recognize that $N(t)$ is a white noise term. Therefore, we can use the Fokker–Planck equation to determine the probability density of $x(t)$. Since (64.6) has second derivative terms, we rewrite (64.6) and (64.7) as the vector system (see page 118)

$$\frac{d}{dt}\begin{pmatrix} x \\ u \end{pmatrix} = \begin{pmatrix} u \\ -\beta u \end{pmatrix} + \begin{pmatrix} 0 & 0 \\ 0 & 1 \end{pmatrix}\begin{pmatrix} n_1(t) \\ n_2(t) \end{pmatrix},$$

$$\begin{pmatrix} x \\ u \end{pmatrix}_{t=0} = \begin{pmatrix} 0 \\ u_0 \end{pmatrix}.$$

The Fokker–Planck equation for $P(t, x, u)$, the joint probability density of x and u at time t, is

$$\frac{\partial P}{\partial t} = -\frac{\partial}{\partial x}(uP) + \frac{\partial}{\partial u}(\beta u P) + \frac{1}{2}\frac{\partial^2 P}{\partial u^2}, \tag{64.9}$$

$$P(0, x, u) = \delta(x)\delta(u - u_0).$$

In this example, we can solve (64.9) exactly by taking a Fourier transform in x (see page 299), and then using the method of characteristics (see page 368). We eventually determine

$$P(t, x, u) = \frac{1}{\det D}\exp\left[-\begin{pmatrix} x - \mu_x \\ u - \mu_u \end{pmatrix} D \begin{pmatrix} x - \mu_x \\ u - \mu_u \end{pmatrix}^{\mathrm{T}}\right],$$

where $D = \begin{pmatrix} \sigma_{xx} & \sigma_{xu} \\ \sigma_{xu} & \sigma_{uu} \end{pmatrix}$, and the parameters $\{\mu_x, \mu_u, \sigma_{xx}, \sigma_{xu}, \sigma_{uu}\}$ are given by

$$\mu_x = \frac{u_0}{\beta} \left(1 - e^{-\beta t}\right),$$

$$\mu_u = u_0 e^{-\beta t},$$

$$\sigma_{xx}^2 = \frac{t}{\beta^2} - \frac{2}{\beta^3} \left(1 - e^{-\beta t}\right) + \frac{1}{2\beta^3} \left(1 - e^{-2\beta t}\right),$$

$$\sigma_{xu}^2 = \frac{1}{\beta^2} \left(1 - e^{-\beta t}\right) - \frac{1}{2\beta^2} \left(1 - e^{-2\beta t}\right),$$

$$\sigma_{uu}^2 = \frac{1}{2\beta} \left(1 - e^{-2\beta t}\right).$$

The details of this calculation are presented in Schuss [7].

Notes

[1] With a Fourier transform, the method of characteristics can often solve a Fokker–Planck equation in one dimension.

[2] Since a Fokker–Planck equation and the equation for a Green's function (see page 268) both have delta function forcing terms, the solution techniques are similar.

[3] Not all noise terms are white Gaussian noise (the requirements in (64.2) are very stringent). The book by Srinivasan and Vasudevan [8] has descriptions of several approximate techniques for other types of noise.

[4] When the coefficient of the noise term (i.e., $g(x)$ in (64.5)) is small, then a singular perturbation problem generally results.

[5] The solution of (64.1) is a Markov process; the density of its probability transition function is given by the solution to the Fokker–Planck equation and its initial conditions.

[6] Another name for the Fokker–Planck equation is the forward Kolmogorov equation.

[7] The solution of the Fokker–Planck equation in (64.3) (and its initial conditions in (64.4)) might be better represented by $P(t, \mathbf{x}; t_0, \mathbf{y})$. The function $P(t, \mathbf{x}; t_0, \mathbf{y})$ also satisfies the backwards Kolmogorov equation, which is the adjoint of (64.3). This equation:

$$\frac{\partial P}{\partial t_0} = -\sum_{i=1}^{m} b_i \frac{\partial P}{\partial y_i} - \frac{1}{2} \sum_{i,j=1}^{m} a_{ij} \frac{\partial^2 P}{\partial y_i \partial y_j}, \tag{64.10}$$

$$P(t_0, \mathbf{x}; t_0, \mathbf{y}) = \delta(\mathbf{x} - \mathbf{y}),$$

has as its independent variables the "backward variables" $\{t_0, \mathbf{y}\}$.

[8] When only moments of the probability density $P(t, \mathbf{x})$ are required, the method of moments (see page 491) may sometimes be used to calculate these moments without having to solve the Fokker–Planck equation.

[9] Another equivalent form of equation (64.1.a) that often appears is

$$dx(t) = b(t, x) \, dt + \sigma(t, x) \, dw(t), \qquad (64.11)$$

where $w(t)$ is a vector of independent standard Wiener processes.

References

[1] J. S. Chang and G. Cooper, "A Practical Difference Scheme for Fokker–Planck Equations," *J. Comput. Physics*, **6**, 1970, pages 1–16.

[2] P. Diţă, "The Fokker–Planck Equation with Absorbing Boundary," *J. Phys. A: Math. Gen.*, **18**, 1985, pages 2685–2690.

[3] C. W. Gardiner, "Handbook of Stochastic Methods," Springer–Verlag, New York, 1985.

[4] L. Garrido and J. Masoliver, "On a Class of Exact Solutions to the Fokker–Planck Equations," *J. Math. Physics*, **23**, No. 6, June 1982, pages 1155–1158.

[5] G. W. Harrison, "Numerical Solution of the Fokker Planck Equation Using Moving Finite Elements," *Num. Meth. Part. Diff. Eqs.*, **4**, 1988, pages 219–232.

[6] H. Risken, *The Fokker–Planck Equation*, Springer–Verlag, New York, 1984.

[7] Z. Schuss, *Theory and Applications of Stochastic Differential Equations*, John Wiley & Sons, New York, 1980.

[8] S. K. Srinivasan and R. Vasudevan, *Introduction to Random Differential Equations and Their Applications*, American Elsevier Publishing Company, New York, 1971, pages 45–47.

65. Fractional Differential Equations[*]

Applicable to Fractional differential equations.

Yields

An exact solution.

Idea

There are two common ways to solve fractional differential equations; using an integral transform, or transforming to an ordinary differential equation.

Procedure

There are two main methods for solving fractional differential equations

(A) transformation to an ordinary differential equation,
(B) using the Laplace transform.

To transform to an ordinary differential equation, care must be taken since the ordinary chain rule from calculus does not apply to fractional derivatives.

Example 1

This example will convert a fractional differential equation into an ordinary differential equation. Suppose we wish to solve the fractional differential equation

$$\frac{d^{1/2} f}{dx^{1/2}} + f = 0 \tag{65.1}$$

for $f(x)$. To convert this to an ordinary differential equation, we will differentiate with respect to x one-half time. This will produce a new differential equation that involves $\dfrac{d^{1/2} f}{dx^{1/2}}$. Eliminating this term between the new equation and equation (65.1), we will have determined an ordinary differential equation.

To differentiate (65.1) with respect to x one-half time, we have to use the differentiation rule (from Oldham and Spanier [4], page 155)

$$\frac{d^{1-Q}}{dx^{1-Q}} \frac{d^{Q}}{dx^{Q}} f = \frac{df}{dx} + C_1 x^{Q-2} + C_2 x^{Q-3} + \ldots + C_m x^{Q-m-1},$$

where $0 < Q \le m < Q + 1$, m is an integer and the $\{C_i\}$ are arbitrary constants. Hence, differentiating (65.1) one-half time results in

$$\frac{df}{dx} - C_1 x^{-3/2} + \frac{d^{1/2} f}{dx^{1/2}} = 0. \tag{65.2}$$

Eliminating the $d^{1/2}/dx^{1/2}$ term between (65.1) and (65.2) results in

$$\frac{df}{dx} - f = C_1 x^{-3/2}, \tag{65.3}$$

which is an ordinary differential equation for $f(x)$. Equation (65.3) has the solution (obtained by use of integrating factors)

$$f(x) = De^x - 2C_1 \left[\sqrt{\pi} e^x \operatorname{erf}(\sqrt{x}) + \frac{1}{\sqrt{x}} \right], \tag{65.4}$$

where D is another arbitrary constant. If we now utilize (65.4) in (65.1), it turns out that D and C_1 are related by $D = 2C_1\sqrt{\pi}$. This is because of the identities

$$\frac{d^{1/2}}{dx^{1/2}}e^x \operatorname{erf}(\sqrt{x}) = e^x, \qquad \frac{d^{1/2}}{dx^{1/2}}\frac{1}{\sqrt{x}} = 0,$$

$$\frac{d^{1/2}}{dx^{1/2}}e^x = \frac{1}{\sqrt{\pi x}} + e^x \operatorname{erf}(\sqrt{x}),$$

from Oldham and Spanier [4], pages 119 and 123. Therefore, the solution of (65.1) is

$$f(x) = D\left[e^x \operatorname{erfc}(\sqrt{x}) - \frac{1}{\sqrt{\pi x}}\right].$$

Example 2

This example will solve a fractional differential equation by use of Laplace transforms. Suppose we wish to solve the fractional differential equation

$$\frac{df}{dx} + \frac{d^{1/2}f}{dx^{1/2}} - 2f = 0. \tag{65.5}$$

The Laplace transform of (65.5) is

$$sF(s) - f(0) + \sqrt{s}F(s) - \frac{d^{-1/2}f(0)}{dx^{-1/2}} + F(s) = 0 \tag{65.6}$$

where $F(s)$ is defined to be the Laplace transform of $f(x)$; that is,

$$F(s) := \int_0^\infty f(x)e^{-xs}ds.$$

If we define the constant C by $C = f(0) + d^{-1/2}f(0)/dx^{-1/2}$, then the solution to (65.6) is given by

$$F(s) = \frac{C}{(\sqrt{s} - 1)(\sqrt{s} + 2)} = \frac{C}{3(\sqrt{s} - 1)} - \frac{C}{3(\sqrt{s} + 2)}, \tag{65.7}$$

and so the final solution to (65.5) can be obtained by finding the inverse Laplace transform to (65.7), which is

$$f(x) = \frac{C}{3}\left[2e^{4x} \operatorname{erfc}(2\sqrt{x}) + e^x \operatorname{erfc}(-\sqrt{x})\right].$$

Notes

[1] Fractional differential equations are also called *extraordinary differential equations*.

[2] One of many equivalent definitions for fractional derivatives is the following

$$\frac{d^q}{dx^q} f(x) = \frac{d^n}{dx^n} \left[\frac{1}{\Gamma(n-q)} \int_a^x \frac{f(y)}{(x-y)^{q-n+1}}\, dy \right],$$

for $n > q \geq 0$.

[3] Certain diffusion problems can be reduced to the solution of a semi-differential equation (one in which all the derivatives are either to an integer order, or a half integer order). See Chapter 11 of Oldham and Spanier [4] for details.

[4] A third technique for solving fractional differential equations is by the use of power series (see page 342). For fractional differential equations, a series of the form

$$f(x) = x^p \sum_{k=0}^{\infty} a_k x^{k/n}$$

is used, where $p > -1$, n is an integer, $a_0 \neq 0$, and the $\{a_i\}$ are unknowns.

[5] In Erdélyi's paper [2], there are several boundary value problems for ordinary differential equations that are solved by using fractional differential techniques.

References

[1] M. A. Al-Bassam, "On Generalized Power Series and Generalized Operational Calculus and its Applications," *Nonlinear Analysis*, World Sci. Publishing, Singapore, 1987.

[2] A. Erdélyi, "Axially Symmetric Potentials and Fractional Integration," *J. Soc. Indust. Appl. Math.*, **13**, No. 1, March 1965, pages 216–228.

[3] K. Nishimoto, "Applications to the Solutions of Linear Second Order Differential Equations of Fuchs Type," in A. C. McBride and G. F. Roach (eds.), *Fractional Calculus*, Pitman Publishing Co., Marshfield, MA, 1985, pages 140–153.

[4] K. B. Oldham and J. Spanier, *The Fractional Calculus*, Academic Press, New York, 1974.

[5] B. Ross, *Fractional Calculus and Its Applications*, Proceedings of the International Conference at the University of New Haven, June 1974, Springer–Verlag, New York, Lecture Notes in Mathematics #457, 1975.

[6] W. Wyss, "The Fractional Diffusion Equation," *J. Math. Physics*, **27**, No. 11, 1986, pages 2782–2785.

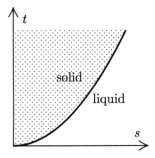

Figure 66. This diagram illustrates the location of the freezing boundary for the system given in (66.1).

66.   Free Boundary Problems[*]

Applicable to   Systems of differential equations in which the location of the boundary of the domain is one of the unknowns to be determined.

Idea

Sometimes a similarity solution may be used to determine the location of the free boundary. In more difficult problems, a numerical technique may be required.

Procedure

In free boundary problems, a differential equation must be solved in a domain whose size can vary. One of the unknowns to be determined is the size of the domain on which the equation is to be satisfied.

Differential equations of this type are most often solved numerically. In rare cases, an analytical solution may be obtained. These solutions are generally found by use of similarity methods (see page 424).

Example

Consider a mass of water in $x \geq 0$ at time $t = 0$. Initially the water has the constant temperature $T_H > 0$. If a constant temperature $T_C < 0$ is maintained at the surface $x = 0$, then the boundary of freezing, $x = s(t)$, will move into the fluid. The unknowns to solve for in this problem are the temperature of the water $w(x, t)$, the temperature of the ice $u(x, t)$, and the location of the unknown boundary, $x = s(t)$. See Figure 66.

The equations that describe the unknowns are:

$$u_t = u_{xx}, \quad \text{for} \quad 0 < x < s(t), \quad t \geq 0,$$
$$w_t = w_{xx}, \quad \text{for} \quad s(t) < x < \infty, \quad t \geq 0,$$
$$u(0, t) = T_C,$$
$$w(x, 0) = T_H, \quad\quad\quad\quad (66.1.a\text{-}g)$$
$$u(s(t), t) = 0,$$
$$w(s(t), t) = 0,$$
$$u_x(s(t), t) - w_x(s(t), t) = \lambda s'(t).$$

Here we have defined the freezing boundary to be the curve along which the temperature is zero, and equation (66.1.g) represents the transfer of latent heat necessary to create the ice. The parameter λ is the latent heat of fusion times the density divided by the coefficient of heat conduction.

Now we propose the similarity solution. Since diffusion equations often have time scaling as the square of a distance, we assume that a solution to (66.1) can be found with

$$u(x, t) = f(\eta) = f\left(\frac{x}{\sqrt{t}}\right), \quad w(x, t) = g(\eta) = g\left(\frac{x}{\sqrt{t}}\right), \quad (66.2)$$

for some unknown functions $f(\eta)$ and $g(\eta)$. Using these proposed forms in (66.1.g) shows that these forms are possible only if the freezing boundary is given by

$$s(t) = \alpha\sqrt{t}, \quad\quad\quad\quad (66.3)$$

for some value of α. Using (66.2) and (66.3) in (66.1) we find the equivalent system

$$f''(\eta) + \tfrac{1}{2}\eta f'(\eta) = 0, \quad \text{for } 0 < \eta < \alpha,$$
$$g''(\eta) + \tfrac{1}{2}\eta g'(\eta) = 0, \quad \text{for } \alpha < \eta < \infty,$$
$$f(0) = T_C, \quad f(\alpha) = 0, \quad\quad (66.4)$$
$$g(\infty) = T_H, \quad g(\alpha) = 0,$$
$$f'(\alpha) - g'(\alpha) = \frac{\lambda\alpha}{2}.$$

The ordinary differential equations in (66.4) may be solved to determine that

$$f(\eta) = T_C - T_H \frac{\text{erf}(\eta/2)}{\text{erf}(\alpha/2)},$$

$$g(\eta) = \frac{T_H}{\text{erfc}(\alpha/2)} \left[\text{erf}(\eta/2) - \text{erf}(\alpha/2)\right],$$

where α satisfies the transcendental equation

$$\frac{T_H}{\text{erf}(\alpha/2)} + \frac{T_C}{\text{erfc}(\alpha/2)} = -\lambda\alpha\frac{\sqrt{\pi}}{2}e^{\alpha^2/4}. \qquad (66.5)$$

Notes

[1] In writing (66.1.a) and (66.1.b) we have assumed that the thermo-physical parameters in both the ice and the water are the same (i.e., the Stefan number, which is a ratio of these parameters, is equal to one). In reality, these parameters are different and a constant which cannot be scaled out must be introduced into either (66.1.a) or (66.1.b).

[2] The example illustrated above is described in more detail in Chapter 3 of Crank [2].

[3] Melting problems for a pure material are also known as *Stefan problems.*

[4] Another technique often used in free boundary problems is changing coordinates so that the free boundaries become fixed in the new coordinate space. This is the idea behind the hodograph method (see page 390).

[5] Free boundary problems often arise in hydrodynamics, when the flow over an airfoil is being computed. When the flow becomes supersonic the type of governing equation changes from hyperbolic to elliptic and a different type of numerical scheme is required. Where the equation changes type is not known *a priori.*

[6] Some of the popular numerical techniques for solving free boundary problems go by the name of *front tracking methods* or *front fixing methods.* These techniques generally require that the location of the free boundary be approximately known before the computer code is run. A better approach is to use *enthalpy methods.* These methods do not need initial information about the interfaces, and multiple fronts can also occur.

[7] The paper by Hill and Dewynne [5] discusses several different approximation techniques applied to a single physical problem involving a free boundary.

References

[1] P. Charrier and B. Tessieras, "On Front-Tracking Methods Applied to Hyperbolic Systems of Nonlinear Conservation Laws," *SIAM J. Numer. Anal.,* **23**, No. 3, June 1986, pages 461–472.

[2] J. Crank, *Free and Moving Boundary Problems,* Clarendon Press, Oxford, 1984.

[3] D. B. Duncan, "A Simple and Effective Self-adaptive Moving Mesh for Enthalpy Formulations of Phase Change Problems," *IMA J. Num. Analysis,* **11**, 1991, pages 55–78.

[4] C. M. Elliot and J. R. Ockendon, *Weak and Variational Methods for Moving Boundary Problems,* Pitman, London, 1982.

[5] R. M. Furzeland, "A Comparative Study of Numerical Methods for Moving Boundary Problems," *J. Inst. Maths. Applics,* **26**, 1980, pages 411–429.

[6] J. M. Hill and J. N. Dewynne, "On the Inward Solidification of Cylinders," *Quart. Appl. Math.,* **44**, No. 1, April 1986, pages 59–70.

[7] G. Marshall, "A Front Tracking Method for One-Dimensional Moving Boundary Problems," *SIAM J. Sci. Stat. Comput.*, **7**, No. 1, January 1986, pages 252–263.

[8] L. I. Rubenšteĭn, *The Stefan Problem*, translated by A. D. Solomon, Amer. Math. Soc., Providence, Rhode Island, 1971.

[9] A. S. Wood, "An Efficient Finite-Dimensional Scheme for Multidimensional Stefan Problems," *Int. J. Num. Meth. Eng.*, **23**, 1986, pages 1757–1771.

[10] D. E. Womble, "A Front-Tracking Method for Multiphase Free Boundary Problems," *SIAM J. Numer. Anal.*, **26**, No. 2, April 1989, pages 380–396.

67.   Generating Functions*

Applicable to   Systems of differential equations, where each equation has a similar form.

Yields

An exact analytic solution.

Idea

Sometimes a single function can be used to contain the information in several equations.

Procedure

We illustrate the method as it applies to ordinary differential equations. Suppose we have a system of ordinary differential equations for $\{u_k(t)\}$, all of the form

$$\frac{d}{dt}u_N = f(u_{N-m}, \ldots, u_N, \ldots, u_{N+m}, t), \tag{67.1}$$

for $N = 1, 2, \ldots, \infty$ or $N = \pm 1, \pm 2, \ldots, \pm\infty$. We might introduce the ordinary generating function

$$G(s, t) = \sum_k u_k(t) s^k, \tag{67.2}$$

or the exponential generating function

$$H(s, t) = \sum_k u_k(t) \frac{s^k}{k!}. \tag{67.3}$$

Using (67.2) (or (67.3)) and (67.1), we can sometimes find a partial differential equation for $G(s,t)$ (or $H(s,t)$). After solving the partial differential equation we can determine the $\{u_k(t)\}$ from either

$$u_k(t) = \frac{1}{k!}\left(\frac{d}{ds}\right)^k G(s,t)\Big|_{s=0},$$

or

$$u_k(t) = \left(\frac{d}{ds}\right)^k H(s,t)\Big|_{s=0}.$$

After we have solved for the $\{u_k(t)\}$ we must then check that (67.2) (or (67.3)) converges for the values of t that are of interest.

Example

The classic equations relating to service times are called the *birth and death equations* (see Karlin and Taylor [2], page 135). For the special case of "constant death" and "linear birth," these equations have the form

$$\frac{d}{dt}P_0(t) = -\lambda P_0(t) + \mu P_1(t),$$
$$\frac{d}{dt}P_N(t) = \lambda P_{N-1}(t) - (\lambda + N\mu)P_N(t) + (N+1)\mu P_{N+1}(t), \tag{67.4}$$

where μ and λ are constants and $N = 1, 2, \ldots, \infty$. The initial conditions for (67.4) are

$$P_N(0) = \delta_{Nj}, \tag{67.5}$$

where δ_{Nj} is the Kronecker delta and j is a given positive integer. The ordinary generating function is defined in this case by

$$G(t,s) = \sum_{k=0}^{\infty} P_k(t)s^k. \tag{67.6}$$

Differentiating $G(t,s)$ with respect to t leads to

$$\begin{aligned}
\frac{\partial G}{\partial t} &= \sum_{k=0}^{\infty}\left[\frac{d}{dt}P_k(t)\right]s^k \\
&= [-\lambda P_0(t) + \mu P_1(t)]\,s^0 \\
&\quad + \sum_{k=1}^{\infty}(\lambda P_{k-1}(t) - (\lambda + k\mu)P_k(t) + (k+1)\mu P_{k+1}(t))\,s^k \\
&= \lambda(s-1)\left[P_0 + P_1 s + P_2 s^2 + \cdots\right] \\
&\quad + \mu(1-s)\left[P_1 + 2P_2 s + 3P_3 s^2 + \cdots\right] \\
&= (1-s)\left[-\lambda G + \mu\frac{\partial G}{\partial s}\right].
\end{aligned} \tag{67.7}$$

The initial condition for $G(t, s)$, from (67.5) and (67.6), becomes

$$G(0, s) = s^j. \tag{67.8}$$

The partial differential equation in (67.7), with the initial condition in (67.8), can be solved by the method of characteristics (see page 368). The solution is

$$G(t, s) = e^{-\lambda(1-s)\left(1-e^{-\mu t}\right)/\mu} \left[1 - (1 - s)e^{-\mu t}\right]^j.$$

Taking a Taylor series of (67.8) with respect to s (see (67.6)) results in

$$P_0(t) = e^{-\lambda(1-y)/\mu}(1 - y)^j,$$

$$P_1(t) = e^{-\lambda(1-y)/\mu}\frac{(1 - y)^{j-1}}{\mu} \left[\lambda y^2 + (j\mu - 2\lambda)\,y + \lambda\right],$$

$$P_2(t) = e^{-\lambda(1-y)/\mu}\frac{(1 - y)^{j-2}}{\mu^2} \left\{\lambda^2 y^4 + (2j\lambda\mu - 4\lambda^2)y^3\right.$$

$$\left. + [j(j - 1)\mu^2 - 2\lambda(2j\mu - 3\lambda)]y^2 + (2j\lambda\mu - 4\lambda^2)y + \lambda^2\right\},$$

where $y = e^{-\mu t}$.

Notes

[1] For the example given above, Laplace transforms (see page 300) could also have been used to solve (67.7) with (67.8).

[2] Nonlinear systems of differential equations can also be solved by this method. A classic application is to equations describing the aggregation of particles (see Feller [1]).

References

[1] W. Feller, *An Introduction to Probability Theory and Its Applications*, John Wiley & Sons, New York, 1968, Chapter 17 (pages 444–482).

[2] S. Karlin and H. M. Taylor, *A First Course in Stochastic Processes*, Second Edition, Academic Press, New York, 1975.

[3] J. Letessier, "The Numerical Resolution of Birth and Death Kolmogorov Equations," *Comp. & Maths. with Appls.*, **13**, No. 7, 1987, pages 595–600.

[4] H. M. Taylor and S. Karlin, *An Introduction to Stochastic Modeling*, Academic Press, New York, 1984, pages 310–316 and 337–338.

68.   Green's Functions*

Applicable to   Linear differential equations with linear boundary conditions.

Yields
An exact solution, in the form of an integral or an infinite series.

Idea
Initially, the solution of the linear differential equation with a "point source" is determined. Then, using superposition, the "forcing function" (appearing in either the differential equation or the boundary condition) is treated as a collection of point sources.

Procedure
Suppose we have the following linear differential equation for $u(\mathbf{x})$

$$L[u] = f(\mathbf{x}), \tag{68.1}$$

with the linear homogeneous boundary conditions

$$B_i[u] = 0, \tag{68.2}$$

for $i = 1, 2, \ldots, n$. Suppose we can solve for $G(\mathbf{x}; \mathbf{z})$, where $G(\mathbf{x}; \mathbf{z})$ satisfies

$$L[G(\mathbf{x}; \mathbf{z})] = \delta(\mathbf{x} - \mathbf{z}),$$
$$B_i[G(\mathbf{x}; \mathbf{z})] = 0,$$

and $\delta(\mathbf{x})$ is the usual delta function. Then the solution to (68.1) with (68.2) can be written as

$$u(\mathbf{x}) = \int G(\mathbf{x}; \mathbf{z}) f(\mathbf{z}) \, d\mathbf{z}, \tag{68.3}$$

integrated over some appropriate region.
Conversely, suppose we want to solve the linear homogeneous differential equation

$$\begin{aligned} L[v] &= 0, \\ B[v] &= h(\mathbf{x}). \end{aligned} \tag{68.4}$$

If we can solve

$$L[g(\mathbf{x}; \mathbf{z})] = 0,$$
$$B[g(\mathbf{x}; \mathbf{z})] = \delta(\mathbf{x} - \mathbf{z}),$$

for $g(\mathbf{x}; \mathbf{z})$, then the solution to (68.4) is given by

$$v(\mathbf{x}) = \int g(\mathbf{x}; \mathbf{z}) h(\mathbf{z}) \, d\mathbf{z}.$$

Both $G(\mathbf{x}; \mathbf{z})$ and $g(\mathbf{x}; \mathbf{z})$ are called Green's functions. The functions $f(\mathbf{x})$ and $h(\mathbf{x})$ are often referred to as "forcing functions." If, for example, $f(\mathbf{x}) \equiv 0$, then by (68.3) $u(\mathbf{x}) \equiv 0$.

Green's functions can be calculated once, then used repeatedly for different functions $f(\mathbf{x})$ and $h(\mathbf{x})$. Some Green's functions are tabulated in Table 68. To calculate the Green's function $G(\mathbf{x}; \mathbf{z})$, we require:

(A) $L[G(\mathbf{x}; \mathbf{z})] = 0,$ except at $\mathbf{x} = \mathbf{z}$.

(B) $B_i[G(\mathbf{x}; \mathbf{z})] = 0.$

(C) If $L[\cdot]$ is an n-th order ordinary differential equation,
    then $G(\mathbf{x}; \mathbf{z})$ must be continuous (with its derivatives up to order $n - 1$) at $\mathbf{x} = \mathbf{z}$.

(D) $\displaystyle\int_{\mathbf{z}-}^{\mathbf{z}+} L[G(\mathbf{x}; \mathbf{z})] \, d\mathbf{x} = 1.$

$$(68.5.a\text{--}d)$$

The conditions on $g(\mathbf{x}; \mathbf{z})$ are very similar:

(A) $L[g(\mathbf{x}; \mathbf{z})] = 0.$

(B) $B[g(\mathbf{x}; \mathbf{z})] = 0,$ except at $\mathbf{x} = \mathbf{z}$.

(C) If $L[\cdot]$ is an n-th order ordinary differential equation,
    then $g(\mathbf{x}; \mathbf{z})$ must be continuous (with its derivatives up to order $n - 1$) at $\mathbf{x} = \mathbf{z}$.

(D) $\displaystyle\int_{\mathbf{z}-}^{\mathbf{z}+} B[g(\mathbf{x}; \mathbf{z})] \, d\mathbf{x} = 1.$

$$(68.6.a\text{--}d)$$

Conditions (68.5.a,d) and (68.6.b,d) follow from the definition of the delta function. Conditions (68.5.c) and (68.6.c) follow from the definition of what a solution to an n-th order differential equation means, and (68.5.b) and (68.6.c) follow from the defining equations for $G(\mathbf{x}; \mathbf{z})$ and $g(\mathbf{x}; \mathbf{z})$.

Many methods can be used to construct a $G(\mathbf{x}; \mathbf{z})$ or a $g(\mathbf{x}; \mathbf{z})$ that satisfies the above four requirements. We will illustrate two methods for constructing $G(\mathbf{x}; \mathbf{z})$ for the special case of a second order linear ordinary differential equation. Then we illustrate the construction process for $g(\mathbf{x}; \mathbf{z})$ for a partial differential equation.

Method 1

Define the general linear second order ordinary differential equation with linear homogeneous boundary conditions by

$$L[u] := \frac{d}{dx}\left(p(x)\frac{du}{dx}\right) - s(x)u,$$

$$B_1[u] := \alpha_1 u(a) + \alpha_2 u'(a) = 0,$$

$$B_2[u] := \beta_1 u(a) + \beta_2 u'(b) = 0,$$

and suppose that we wish to solve $L[u] = f(x)$. If $y_1(x)$ and $y_2(x)$ are non-trivial (i.e., not identically equal to zero) and satisfy

$$L[y_1] = 0, \qquad\qquad B_1[y_1] = 0,$$
$$L[y_2] = 0, \qquad\qquad B_2[y_2] = 0,$$

then we can write $G(x;z)$ as

$$G(x;z) = \begin{cases} \dfrac{y_1(x)y_2(z)}{p(z)W(z)} & \text{for } a \leq x \leq z, \\[2ex] \dfrac{y_2(x)y_1(z)}{p(z)W(z)} & \text{for } z \leq x \leq b, \end{cases}$$

where $W(z) = \begin{vmatrix} y_1(z) & y_2(z) \\ y_1'(z) & y_2'(z) \end{vmatrix}$ is the Wronskian of $y_1(x)$ and $y_2(x)$ at the point $x = z$.

Method 2

Suppose that $L[\cdot]$ is a self-adjoint operator, so that it has a complete set of orthogonal eigenfunctions (see page 82). Suppose further that we know the eigenvalues $\{\lambda_n\}$ and the eigenfunctions $\{\phi_n\}$ for $\{L, B_1, B_2\}$ (see page 82). That is

$$L[\phi_n] = \lambda_n \phi_n,$$
$$B_1[\phi_n] = 0,$$
$$B_2[\phi_n] = 0,$$

then $G(x;z)$ is found to be

$$G(x;z) = \sum_{n=1}^{\infty} \frac{\phi_n(x)\phi_n(z)}{\lambda_n \int \phi_n^2(x)\, dx}.$$

Example 1

Suppose we wish to solve

$$y'' = f(x),$$
$$y(0) = 0, \quad y(L) = 0. \tag{68.7}$$

For the first method, we require the solutions $y_1(x)$ and $y_2(x)$ of

$$y_1'' = 0, \qquad y_1(0) = 0,$$
$$y_2'' = 0, \qquad y_2(L) = 0.$$

The solutions to these equations are

$$y_1(x) = Ax, \qquad y_2(x) = B(x - L),$$

where A and B are arbitrary constants. We compute the Wronskian to be $W(z) = ABL$. Therefore

$$G(x; z) = \begin{cases} \dfrac{x(z - L)}{L} & \text{for } 0 \le x \le z, \\[2mm] \dfrac{z(x - L)}{L} & \text{for } z \le x \le L. \end{cases} \tag{68.8}$$

For the second method, we find the eigenvalues and eigenfunctions to be

$$\lambda_n = \frac{n\pi}{L}, \quad \phi_n(x) = \sin \lambda_n x = \sin\left(\frac{n\pi x}{L}\right),$$

so that

$$G(x; z) = \frac{2L}{n\pi} \sum_{n=1}^{\infty} \sin\left(\frac{n\pi x}{L}\right) \sin\left(\frac{n\pi z}{L}\right). \tag{68.9}$$

Using either of (68.8) or (68.9) for $G(x; z)$, the solution to (68.7) can be written as

$$y(x) = \int_0^L G(x; z) \, f(z) \, dz. \tag{68.10}$$

For example, using (68.8) in (68.10), the solution to (68.7) can be written as

$$y(x) = \int_x^L \frac{x(z - L)}{L} f(z) \, dz + \int_0^x \frac{z(x - L)}{L} f(z) \, dz. \tag{68.11}$$

Note the similarity between (68.11) and the form of the solution shown in the section on variation of parameters (see page 356).

If, for example, $f(x) = x^3$, then evaluation of (68.11) results in

$$y(x) = \frac{x}{20}(x^4 - L^4).$$

The second method yields the same answer. For this example, the second method is equivalent to using finite Fourier series (see page 295).

Example 2

Suppose we are given the parabolic partial differential equation

$$\frac{\partial^2 u}{\partial x^2} = \frac{1}{a^2}\frac{\partial u}{\partial t} \tag{68.12}$$

for $u(x,t)$ with the initial and boundary conditions

$$u(x,0) = h(x), \quad u(\pm\infty, t) = 0. \tag{68.13}$$

We choose to write the solution as

$$u(x,t) = \int_{-\infty}^{\infty} g(x,t;z)h(z)\,dz, \tag{68.14}$$

where the Greens function $g(x,t;z)$ satisfies

$$\frac{\partial^2 g}{\partial x^2} = \frac{1}{a^2}\frac{\partial g}{\partial t},$$
$$g(x,0;z) = \delta(z-x), \quad g(\pm\infty,t;z) = 0.$$

Taking a Fourier transform (in x) of the equation for $g(x,t;z)$ results in

$$\frac{d\widehat{g}}{dt} = -a^2\omega^2\widehat{g},$$
$$\widehat{g}(\omega,0;z) = \frac{1}{\sqrt{2\pi}}e^{i\omega z}, \tag{68.15}$$

where $\widehat{g}(\omega,t;z)$ is defined to be the Fourier transform of $g(x,t;z)$; that is,

$$\widehat{g}(\omega,t;z) := \frac{1}{\sqrt{2\pi}}\int_{-\infty}^{\infty} g(x,t;z)e^{i\omega x}\,dx.$$

Solving the ordinary differential equation in (68.15) results in

$$\widehat{g}(\omega,t;z) = \frac{1}{\sqrt{2\pi}}e^{i\omega z}e^{-a^2\omega t}.$$

Using the inverse Fourier transform we then have our solution

$$g(x,t;z) = \frac{1}{\sqrt{2\pi}}\int_{-\infty}^{\infty} \widehat{g}(\omega,t;z)e^{-i\omega x}\,dx.$$

By using the convolution theorem for Fourier transforms, we can determine that

$$g(x,t;z) = \frac{1}{\sqrt{4\pi a^2 t}}e^{-(x-z)^2/4a^2 t}.$$

This should be used in (68.14) to determine the solution to (68.12) and (68.13).

Table 68:

Green's functions for various partial differential equations.
In the following, $\mathbf{r} = (x, y, z)$, $\mathbf{r}_0 = (x_0, y_0, z_0)$, $R^2 = (x - x_0)^2 + (y - y_0)^2 + (z - z_0)^2$, $P^2 = (x - x_0)^2 + (y - y_0)^2$, and $H(\cdot)$ is the Heaviside function (equal to one when the argument is positive, otherwise zero).

- For the potential equation

$$\nabla^2 G + k^2 G = -4\pi\delta(\mathbf{r} - \mathbf{r}_0),$$

with the radiation condition (outgoing waves only), the solution is

$$G = \begin{cases} \dfrac{2\pi i}{k} e^{ik|x - x_0|} & \text{in one dimension,} \\[2mm] i\pi H_0^{(1)}(kP) & \text{in two dimensions,} \\[2mm] \dfrac{e^{ikR}}{R} & \text{in three dimensions,} \end{cases}$$

where $H_0^{(1)}(\cdot)$ is a Hankel function (also called a Bessel function of the third kind).

- For the diffusion equation

$$\nabla^2 G - a^2 \frac{\partial G}{\partial t} = -4\pi\delta(\mathbf{r} - \mathbf{r}_0)\delta(t - t_0),$$

with the initial condition $G = 0$ for $t < t_0$, and the boundary condition $G = 0$ at $r = \infty$ in N dimensions, the solution is

$$G = \frac{4\pi}{a^2} \left(\frac{a}{2\sqrt{\pi(t - t_0)}} \right)^N e^{-a^2\|\mathbf{r} - \mathbf{r}_0\|^2 / 4(t - t_0)}.$$

- For the wave equation

$$\nabla^2 G - \frac{1}{c^2} \frac{\partial^2 G}{\partial t^2} = -4\pi\delta(\mathbf{r} - \mathbf{r}_0)\delta(t - t_0),$$

with the initial conditions $G = G_t = 0$ for $t < t_0$, and the boundary condition $G = 0$ at $r = \infty$ the solution is

$$G = \begin{cases} 2c\pi H\left[(t - t_0) - \dfrac{|x - x_0|}{c} \right] & \text{for one space dimension,} \\[3mm] \dfrac{2c}{\sqrt{c^2(t - t_0)^2 - P^2}} H\left[(t - t_0) - \dfrac{P}{c} \right] & \text{for two space dimensions,} \\[3mm] \dfrac{1}{R}\delta\left[\dfrac{R}{c} - (t - t_0) \right] & \text{for three space dimensions.} \end{cases}$$

Notes

[1] If $\mathbf{z}$ is in a n-dimensional space, then the integrals appearing in (68.5.d) and (68.6.d) are n single integrals, each one over one of the coordinate axes.

[2] Delta functions, in non-rectangular coordinate systems, are easily determined by a change of variables in the defining relation: $\int \delta(\mathbf{z})\, d\mathbf{z} = 1$. In changing variables, the Jacobian of the transformation will then divide the delta function terms. For example:

(A) In a spherical coordinate system, usually denoted by the coordinates r, θ, and ϕ, the delta function located at the point $\mathbf{x}' = (r', \theta', \phi')$ is given by

$$\delta(\mathbf{x} - \mathbf{x}') = \frac{1}{r^2 \sin\theta} \delta(r - r')\delta(\theta - \theta')\delta(\phi - \phi'),$$

for $r' \neq 0$ and $\theta' \neq 0, \pi$. For a point source at $r = r'$ and $\theta = 0$, the representation $\delta(r - r')\delta(\theta)/2\pi r^2 \sin\theta$ may be used, while a point source at the origin has the representation $\delta(r)/4\pi r^2$.

(B) In a cylindrical coordinate system, usually denoted by the coordinates ρ, θ, and z, the delta function located at the point $\mathbf{x}' = (\rho', \theta', z')$ is given by

$$\delta(\mathbf{x} - \mathbf{x}') = \frac{\delta(\rho - \rho')\delta(\theta - \theta')\delta(z - z')}{\rho},$$

for $\rho' > 0$. A point source at the origin has the representation $\dfrac{\delta(z)\delta(\rho)}{2\pi\rho}$.

[3] If $G^*(\mathbf{x}; \mathbf{z})$ satisfies the problem adjoint to $L[\cdot]$ (see page 74), then $G(\mathbf{x}; \mathbf{z}) = G^*(\mathbf{z}; \mathbf{x})$. Therefore, if $L[\cdot]$ and its associated boundary conditions are self-adjoint and $L[G(\mathbf{x}; \mathbf{z})] = \delta(\mathbf{x} - \mathbf{z})$, then $G(\mathbf{x}; \mathbf{z}) = G(\mathbf{z}; \mathbf{x})$. This is called the *reciprocity principle*. It can be observed in our example (see (68.9)).

[4] When the operator is self-adjoint, the Green's function is sometimes written in terms of the variables $x_<$ and $x_>$ instead of x and z. When this is done, $x_<$ ($x_>$) represents the smaller (larger) of x and z. For example, (68.11) could be written as $G(x; z) = \dfrac{x_<(x_> - L)}{L}$.

As another example, the differential equation with boundary conditions

$$y'' + k^2 y = f(x),$$
$$y(0) = 0, \quad y'(1) = 0,$$

has the Green's function $G(x; z) = -\dfrac{\cos k(1 - x_<)\sin kx_>}{k\cos k}$.

[5] Consider the self-adjoint second order operator $L[u] = (p(x)u'(x))' + q(x)u(x)$, and consider the boundary conditions

$$B_1[u] := a_1 u(a) + a_2 u'(a) = 0,$$
$$B_2[u] := b_1 u(b) + b_2 u'(b) = 0. \tag{68.16}$$

Define $\phi(x)$ and $\psi(x)$ to be the solutions to:

$$L[\phi] = \lambda r(x)\phi, \qquad B_1[\phi] = 0,$$
$$L[\psi] = \lambda r(x)\psi, \qquad B_2[\psi] = 0.$$

Then, the Green's function for the operator $L - \lambda r$, which satisfies the boundary conditions in (68.16) is given by $G_\lambda(x; z) = \dfrac{\phi(x_<)\psi(x_>)}{p(x)W(\phi, \psi)}$, where $W(\phi, \psi)$ represents the Wronskian.

[6] There will not exist a Green's function if the solution of the original problem is indeterminate. In this case, a *generalized* Green's function will exist. As an example, consider the system

$$y'' = f(x),$$
$$y(0) = y(1),$$
$$y'(0) = y'(1).$$

If $u(x)$ is any solution to the above system, then so is $u(x) + C$ where C is any constant. Since the solution of the original system is indeterminate an ordinary Green's function cannot be found. See the section on alternative theorems (page 14) or Farlow [4] for details.

   Sometimes, in such problems, the specific solution in which the Green's function is symmetric in both **x** and **z** is chosen. This results in the *modified Green's function*. See Stakgold [8] (pages 215–218) for details.

[7] Fokker–Planck equations have delta function initial conditions. The methods used for solving these equations are the same as the methods used for finding Green's functions.

[8] Butkovskiy's book [2] has a comprehensive listing of Green's functions. Any particular Green's function problem is partitioned into one of several separate disjoint groups labeled by a triple of integers: (r, m, n). In this partitioning, r represents the dimension of the spatial domain, m is the order of the highest derivative with respect to t, and n is the order of the highest derivative with respect to the space variables. Over 500 problems are catalogued and solved.

References

[1] E. Butkov, *Mathematical Physics*, Addison–Wesley Publishing Co., Reading, MA, 1968, Chapter 12 (pages 503–552).

[2] A. G. Butkovskiy, *Green's Functions and Transfer Functions Handbook*, Halstead Press, John Wiley & Sons, New York, 1982.

[3] R. Courant and D. Hilbert, *Methods of Mathematical Physics*, Interscience Publishers, New York, 1953.

[4] S. J. Farlow, *Partial Differential Equations for Scientists and Engineers*, John Wiley & Sons, New York, 1982, pages 290–298.

[5] M. D. Greenberg, *Application of Green's Functions in Science and Engineering*, Prentice–Hall Inc., Englewood Cliffs, NJ, 1971.

[6] K. E. Jordan, G. R. Richter, and P. Sheng, "An Efficient Numerical Evaluation of the Green's Function for the Helmholtz Operator on Periodic Structures," *J. Comput. Physics*, **63**, No. 1, 1986, pages 222–235.

[7] P. M. Morse and H. Feshback, *Methods of Theoretical Physics*, McGraw–Hill Book Company, New York, 1953.

[8] I. Stakgold, *Green's Functions and Boundary Value Problems*, John Wiley & Sons, New York, 1979, Chapter 1 (pages 42–85).

[9] E. Zauderer, *Partial Differential Equations of Applied Mathematics*, John Wiley & Sons, New York, 1983, pages 353–449.

69.   Homogeneous Equations

Applicable to   First order ordinary differential equations of a certain form.

Yields

An exact solution.

Idea

If $P(x, y)$ and $Q(x, y)$ are homogeneous functions of x and y of the same degree, then, by the change of variable $y = vx$, the differential equation $y' = P(x, y)/Q(x, y)$ can be made separable.

Procedure

A function $H(x, y)$ is called homogeneous of degree n if $H(tx, ty) = t^n H(x, y)$. In particular, a polynomial, $P(x, y)$, of two variables is said to be homogeneous of degree n if every term of $P(x, y)$ is of the form $x^j y^{n-j}$ for $j = 0, 1, \ldots, n$. A homogeneous function of degree n can be written as $H(x, y) = x^n H(1, y/x)$. Therefore, given an ordinary differential equation of the form

$$\frac{dy}{dx} = \frac{P(x, y)}{Q(x, y)}, \tag{69.1}$$

where $P(x, y)$ and $Q(x, y)$ are both homogeneous polynomials of degree n, we change variables by $y = vx$ to obtain

$$x\frac{dv}{dx} + v = \frac{P(1, v)}{Q(1, v)}.$$

Since this is a separable equation, it can be integrated to yield (see page 341)

$$\int \frac{dv}{\dfrac{P(1, v)}{Q(1, v)} - v} = \log x + C,$$

where C is an arbitrary constant.

Example

Suppose we have the ordinary differential equation

$$\frac{dy}{dx} = \frac{2x^3 y - y^4}{x^4 - 2xy^3}. \tag{69.2}$$

Since both the numerator and denominator of the right-hand side of (69.2) are homogeneous polynomials of degree four, we set $y = vx$ to obtain

$$x\frac{dv}{dx} + v = \frac{2v - v^4}{1 - 2v^3}.$$

or

$$x\frac{dv}{dx} = \frac{v + v^4}{1 - 2v^3}.$$

This last equation is separable, and the solution is given by

$$\int^x \frac{dx}{x} = \int^v \frac{1 - 2v^3}{v + v^4}\, dv,$$

$$\log x = \int^v \left(\frac{1}{v} - \frac{3v^2}{1 + v^3}\right) dv$$

$$= \log v - \log(1 + v^3) + \log C,$$

or

$$x(1 + v^3) = Cv, \tag{69.3}$$

where C is an arbitrary constant. Substituting $v = y/x$ in (69.3) yields the final solution $x^3 + y^3 = Cxy$.

Notes

[1] Equation (69.1) may be made exact (see page 238) by multiplying by the integrating factor $1/(Px - Qy)$.

[2] This method is derivable from Lie group methods (see page 314).

[3] This method is contained in the method for scale invariant equations (see page 338).

[4] Beware that the expression "homogeneous equation" has two entirely different meanings, see the definitions (page 1).

[5] It may be simpler to think of homogeneous equations as ordinary differential equations of the form $dy/dx = f(y/x)$. This is equivalent to (69.1).

[6] The equation

$$\frac{dy}{dx} = f\left(\frac{a_1 x + b_1 y + c_1}{a_2 x + b_2 y + c_2}\right) \tag{69.4}$$

can be made homogeneous if $a_1 b_2 \neq a_2 b_1$. The change of variables

$$x = X + h,$$
$$y = Y + k,$$

changes (69.4) into the homogeneous equation

$$\frac{dY}{dX} = f\left(\frac{a_1 X + b_1 Y}{a_2 X + b_2 Y}\right),$$

when h and k satisfy the equations: $\begin{pmatrix} a_1 & b_1 \\ a_2 & b_2 \end{pmatrix} \begin{pmatrix} h \\ k \end{pmatrix} = \begin{pmatrix} -c_1 \\ -c_2 \end{pmatrix}.$

[7] If $a_1 b_2 = a_2 b_1$, then (69.4) can be made separable (see page 341). Changing variables via $Y = x + \frac{b_1}{a_1} y = x + \frac{b_2}{a_2} y$ results in the equation

$$\frac{dY}{dx} = 1 + \frac{b_1}{a_1} f\left(\frac{a_1 Y + c_1}{a_2 Y + c_2}\right).$$

References

[1] W. E. Boyce and R. C. DiPrima, *Elementary Differential Equations and Boundary Value Problems*, Fourth Edition, John Wiley & Sons, New York, 1986, pages 87–91.

[2] L. R. Ford, *Differential Equations*, McGraw–Hill Book Company, New York, 1955, pages 40–45.

[3] M. E. Goldstein and W. H. Braun, *Advanced Methods for the Solution of Differential Equations*, NASA SP-316, U.S. Government Printing Office, Washington, D.C., 1973, pages 81–84.

[4] E. L. Ince, *Ordinary Differential Equations*, Dover Publications, Inc., New York, 1964, pages 18–20.

[5] G. F. Simmons, *Differential Equations with Applications and Historical Notes*, McGraw–Hill Book Company, New York, 1972, pages 35–37.

70.   Method of Images*

Applicable to   Differential equations with homogeneous boundary conditions and sources present.

Yields

An exact solution.

Idea

If we know the solution to a free space problem, then we can often use superposition to find a solution in a finite domain with homogeneous boundary conditions.

Procedure

Given a problem with a source present, solve the free space problem (that is, disregarding the boundary conditions). By superposition, determine the solution when there are sources at different points, of different strengths. Choose the position and strengths of these sources so as to obtain the desired boundary conditions.

The added sources cannot appear in the physical domain of the problem. Symmetry considerations tend to simplify the process of determining where the sources should go.

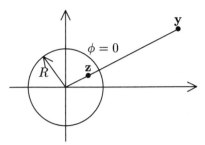

Figure 70.1 Equation (70.1) represents the potential outside of a grounded sphere of radius R, with a source point present.

Example 1

Suppose we wish to find the potential, $\phi(\mathbf{x})$, outside of a grounded sphere of radius R, when there is a point source at position $\mathbf{y}$ (with $||\mathbf{y}|| = \lambda > R$). The equations that represent this problem are:

$$\nabla^2 \phi = \delta(\mathbf{x} - \mathbf{y}),$$

$$\phi\Big|_{||\mathbf{x}||=R} = 0, \qquad \phi\Big|_{||\mathbf{x}||=\infty} = 0, \qquad (70.1.a\text{-}c)$$

in the region $R < ||\mathbf{x}|| < \infty$. See Figure 70.1. If the boundary condition at $||\mathbf{x}|| = R$ is ignored, then the problem

$$\nabla^2 \Psi = \delta(\mathbf{x} - \mathbf{y}),$$

$$\Psi\Big|_{||\mathbf{x}||=\infty} = 0,$$

has the solution (using Green's functions, see page 273)

$$\Psi = -\frac{1}{4\pi||\mathbf{x} - \mathbf{y}||}. \qquad (70.2)$$

If we place an additional source of strength S at the point $\mathbf{z}$ and solve

$$\nabla^2 \Phi = \delta(\mathbf{x} - \mathbf{y}) + S\delta(\mathbf{x} - \mathbf{z}),$$

$$\Phi\Big|_{||\mathbf{x}||=\infty} = 0, \qquad (70.3)$$

then we obtain (using (70.2) and superposition)

$$\Phi = -\frac{1}{4\pi||\mathbf{x} - \mathbf{y}||} - \frac{S}{4\pi||\mathbf{x} - \mathbf{z}||}. \qquad (70.4)$$

Note that the point $\mathbf{z}$ cannot be in the region $R < ||\mathbf{x}|| < \infty$, since then equation (70.3) (whose solution we want to be the solution to (70.1)) will not satisfy (70.1.a).

To determine the strength and location of the additional source (S and $\mathbf{z}$), we calculate the potential at $\mathbf{x} = \mathbf{p}$ where $||\mathbf{p}|| = R$ (i.e., on the surface of the sphere). We find

$$\Phi\Big|_{\mathbf{x}=\mathbf{p}} = -\frac{1}{4\pi}\left[\frac{1}{||\mathbf{p}-\mathbf{y}||} + \frac{S}{||\mathbf{p}-\mathbf{z}||}\right].$$

For this to be zero (and so $\Phi = \phi$) we require (after some vector algebra)

$$S = -\frac{R^4}{\lambda^4}, \qquad \mathbf{z} = \frac{R^2}{\lambda^2}\mathbf{y}.$$

Hence,

$$\Phi = -\frac{1}{4\pi}\left[\frac{1}{||\mathbf{x}-\mathbf{y}||} - \frac{R^4}{\lambda^4}\frac{1}{||\mathbf{x}-\mathbf{y}R^2/\lambda^2||}\right] \tag{70.5}$$

satisfies (70.3) and also (70.1.b). Since $||\mathbf{z}|| < R$ (by virtue of $||\mathbf{y}|| = \lambda > R$) the point source, we added is not in the physical domain of the problem. Therefore, the solution to (70.1) is given by (70.5).

Example 2

Suppose we wish to solve Laplace's equation in the half plane:

$$\nabla^2 u = 0, \qquad \text{for } y > 0,\ -\infty < x < \infty,$$
$$u(x,0) = f(x), \tag{70.6}$$
$$u \to 0, \qquad \text{as } \left|x^2 + y^2\right| \to \infty.$$

The solution to (70.6) can be obtained by Green's functions (see page 273):

$$u(\zeta,\eta) = -\int f(x)\frac{\partial G}{\partial y}(x,0;\zeta,\eta)\,dx, \tag{70.7}$$

where the Green's function $G(x,y;\zeta,\eta)$ satisfies:

$$\nabla^2 G = \frac{\partial^2 G}{\partial x^2} + \frac{\partial^2 G}{\partial y^2} = \delta(x-\zeta)\delta(y-\eta),$$
$$G(x,0;\zeta,\eta) = 0. \tag{70.8.a–b}$$

A solution to (70.8.a) is given by

$$G(x,y;\zeta,\eta) = \frac{1}{2\pi}\log\sqrt{(x-\zeta)^2 + (y-\eta)^2}. \tag{70.9}$$

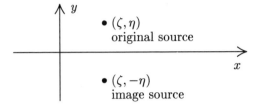

Figure 70.2 The original source and the image source for (70.8).

But (70.9) does not satisfy (70.8.b). If we place an image source at $(\zeta, -\eta)$, having the opposite sign of the source at (ζ, η) then $G(x, y; \zeta, \eta)$ will vanish along $y = 0$ by symmetry. See Figure 70.2.

Hence, the solution to (70.8) is

$$G(x, y; \zeta, \eta) = \frac{1}{2\pi} \log \sqrt{(x - \zeta)^2 + (y - \eta)^2} - \frac{1}{2\pi} \log \sqrt{(x - \zeta)^2 + (y + \eta)^2}.$$

Using this is in (70.7), we obtain the solution to (70.6):

$$u(\zeta, \eta) = \frac{1}{\pi} \int_{-\infty}^{\infty} f(x) \frac{\eta \, dx}{(x - \zeta)^2 + \eta^2}.$$

This solution is known as *Poisson's integral*.

Notes

[1] The method of images is often used to solve Laplace's equation in hydrodynamics and electrostatics.

[2] The method of images is also often used for diffusion problems and hyperbolic problems. See, for example, Butkov [1] or Stakgold [4].

References

[1] E. Butkov, *Mathematical Physics*, Addison–Wesley Publishing Co., Reading, MA, 1968, pages 529–530 and 595–599.

[2] A. K. Gautesen, "Oblique Derivative Boundary Conditions and the Image Method for Wedges," *SIAM J. Appl. Math.*, **48**, No. 6, December 1988, pages 1487–1492.

[3] J. D. Jackson, *Classical Electrodynamics*, John Wiley & Sons, New York, 1962, pages 26–29.

[4] O. D. Kellog, *Foundations of Potential Theory*, Dover Publications, Inc., New York, 1953, pages 228–230.

[5] I. Stakgold, *Green's Functions and Boundary Value Problems*, John Wiley & Sons, New York, 1979, pages 72–73 and 491–493.

[6] E. Zauderer, *Partial Differential Equations of Applied Mathematics*, John Wiley & Sons, New York, 1983, pages 420–432.

71. Integrable Combinations

Applicable to Systems of ordinary differential equations.

Yields

One or more ordinary differential equations that can be integrated exactly.

Idea

Sometimes, by combining pieces of a system of differential equations, a combination of the dependent variables can be determined explicitly in terms of the independent variable.

Procedure

Integration of the system of ordinary differential equations

$$\frac{dx_i}{dt} = f_i(t, x_1, x_2, \ldots, x_n), \quad \text{for } i = 1, 2, \ldots, n,$$

is often accomplished by choosing *integrable combinations*. An integrable combination is a differential equation which is derived from a system of differential equations and is readily integrable.

Example 1

Given the two equations

$$\frac{dx}{dt} = y, \qquad \frac{dy}{dt} = x, \tag{71.1}$$

an integrable combination can be obtained by adding the two equations to obtain

$$\frac{d(x+y)}{dt} = x + y.$$

This last equation can be integrated (treating $x + y$ as a single variable) to yield

$$x + y = Ae^t, \tag{71.2}$$

where A is an arbitrary constant. For the equations in (71.1), another integrable combination may be obtained by subtracting the equations. Integrating this new equation results in

$$x - y = Be^{-t}, \tag{71.3}$$

where B is another arbitrary constant. The explicit solution for $x(t)$ and $y(t)$ may be obtained by combining (71.2) and (71.3).

Example 2

Suppose we have the nonlinear system of ordinary differential equations

$$\frac{dx}{dt} = -3yz,$$

$$\frac{dy}{dt} = 3xz,$$

$$\frac{dz}{dt} = -xy.$$

Multiplying the first equation by x, the second by $2y$, and the third by $3z$ and adding, results in

$$x\frac{dx}{dt} + 2y\frac{dy}{dt} + 3z\frac{dz}{dt} = 0.$$

This last equation may be integrated to obtain $x^2 + 2y^2 + 3z^2 = C$, where C is an arbitrary constant. For this example, another integrable combination can be found by multiplying the first equation by x, multiplying the second by y, and adding. After solving this new differential equation, we determine the additional relation $x^2 + y^2 = D$, where D is another arbitrary constant.

Notes

[1] Each linearly independent integrable combination yields a first integral of the original system.

References

[1] L. E. Elsgolts, *Differential Equations and the Calculus of Variations*, MIR Publishers, Moscow, 1970, pages 186–189.

72.   Integral Representations: Laplace's Method*

Applicable to   Linear ordinary differential equations.

Yields

An integral representation of the solution.

Idea

Sometimes the solution of a linear ordinary differential equation can be written as a contour integral.

Procedure

Let $L_z[\cdot]$ be a linear differential operator with respect to z, and suppose that the ordinary differential equation we wish to solve has the form

$$L_z[u(z)] = 0. \tag{72.1}$$

We look for a solution of (72.1) in the form

$$u(z) = \int_C K(z,\xi)v(\xi)\,d\xi, \tag{72.2}$$

for some function $v(\xi)$ and some contour C in the complex ξ plane. The function $K(z,\xi)$ is called the *kernel*. Some common kernels for Laplace's method are:

$$\text{Laplace kernel:} \quad K(z,\xi) = e^{\xi z} \tag{72.3}$$
$$\text{Euler kernel:} \quad K(z,\xi) = (z - \xi)^N \tag{72.4}$$

We combine (72.2) and (72.1) for

$$\int_C L_z[K(z,\xi)]v(\xi)\,d\xi = 0. \tag{72.5}$$

Now we must find a linear differential operator $A_\xi[\cdot]$, operating with respect to ξ, such that

$$L_z[K(z,\xi)] = A_\xi[K(z,\xi)].$$

After $A_\xi[\cdot]$ has been found, then (72.5) can be rewritten as

$$\int_C A_\xi[K(z,\xi)]v(\xi)\,d\xi = 0. \tag{72.6}$$

Now we integrate (72.6) by parts. The resulting expression will be a differential equation for $v(\xi)$ with some boundary terms. The boundary terms determine the contour C, and the differential equation determines $v(\xi)$. Knowing both $v(\xi)$ and C, the solution to (72.1) is given by (72.2).

Special Case

For the case where $L_z[\cdot]$ is a linear operator with polynomial coefficients, the solution is easy to find using the Laplace kernel. Let $L_z[\cdot]$ have the form

$$L_z = \sum_{r=0}^{N} \left(\sum_{s=0}^{M} a_{rs} z^s \right) \frac{d^r}{dz^r}, \tag{72.7}$$

where the $\{a_{rs}\}$ are constants. Then define the linear differential operator $M_\xi[\cdot]$ by

$$M_\xi = \sum_{r=0}^{N} \left(\sum_{s=0}^{M} a_{rs} \frac{d^s}{d\xi^s} \right) \xi^r. \tag{72.8}$$

Now define $M_\xi^*[\cdot]$ to be the adjoint of $M_\xi[\cdot]$. Then $L_z[u(z)] = 0$ will have a solution of the form

$$u(z) = \int_C e^{z\xi} v(\xi) \, d\xi,$$

if $v(\xi)$ satisfies

$$M_\xi^*[v(\xi)] = 0, \tag{72.9}$$

and C is determined by

$$\left[P\{e^{z\xi}, v(\xi)\} \right]_C = 0, \tag{72.10}$$

where $P\{e^{z\xi}, v(\xi)\}$ is the bilinear concomitant of $e^{z\xi}$ and $v(\xi)$ (see page 187). Note the order of the original differential operator in (72.7) was N while the order of the differential operators in (72.8) and (72.9) is M.

Example

Consider Airy's equation

$$u'' - zu = 0. \tag{72.11}$$

We assume that the solution of (72.11) has the form

$$u(z) = \int_C e^{z\xi} v(\xi) \, d\xi, \tag{72.12}$$

for some $v(\xi)$ and some contour C. Substituting (72.12) into (72.11) we find

$$\int_C \xi^2 v(\xi) e^{z\xi} \, d\xi - z \int_C v(\xi) e^{z\xi} \, d\xi = 0. \tag{72.13}$$

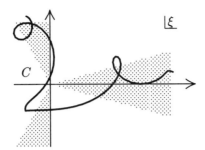

Figure 72. A solution to (72.11) is determined by any contour C that starts and ends in the shaded regions. All of the shaded regions extend to infinity. One possible contour is shown.

The second term in (72.13) can be integrated by parts to obtain

$$\int_C \xi^2 v(\xi) e^{z\xi}\, d\xi - \left[v(\xi) e^{z\xi} \right]\Big|_C + \int_C v'(\xi) e^{z\xi}\, d\xi = 0,$$

or

$$\left[v(\xi) e^{z\xi} \right]\Big|_C + \int_C e^{z\xi} \left[\xi^2 v(\xi) + v'(\xi) \right] d\xi = 0. \tag{72.14}$$

We choose

$$\xi^2 v(\xi) + v'(\xi) = 0, \tag{72.15}$$

and

$$\left[v(\xi) e^{z\xi} \right]\Big|_C = 0. \tag{72.16}$$

With these choices, equation (72.14) is satisfied. From (72.15) we can solve for $v(\xi)$

$$v(\xi) = \exp\left(-\frac{\xi^3}{3} \right). \tag{72.17}$$

Using (72.17) in (72.16) we must choose the contour C so that

$$\left[v(\xi) e^{z\xi} \right]\Big|_C = \left[\exp\left(-z\xi - \frac{\xi^3}{3} \right) \right]\Big|_C = 0, \tag{72.18}$$

for all real values of z. The only restriction that (72.18) places on C is that the contour start and end in one of the shaded regions in Figure 72. Finally, the solution to (72.11) can now be written

$$u(z) = \int_C e^{(\xi z - \xi^2/3)}\, d\xi. \tag{72.19}$$

Asymptotic methods can be applied to (72.19) to determine information about $u(z)$.

For this example, we also could have used the general results in (72.8)–(72.10). Identifying equation (72.11) with the operator in equation (72.7) we find

$$L_z = \frac{d^2}{dz^2} - z,$$

so that (from (72.8))

$$M_\xi = \xi^2 - \frac{d}{d\xi},$$

and also

$$M_\xi^* = \xi^2 + \frac{d}{d\xi}.$$

So we have to solve (from (72.9))

$$M_\xi^*[v(\xi)] = \xi^2 v + v' = 0.$$

Since this last equation is identical to (72.15), we find the same $v(\xi)$. We compute the bilinear concomitant to be

$$P\{e^{z\xi}, v(\xi)\} = v(\xi)\frac{d}{d\xi}e^{z\xi} - e^{z\xi}\frac{d}{d\xi}v(\xi),$$
$$= (z + \xi^2)\exp\left(-z\xi - \frac{\xi^3}{3}\right),$$

and we find the same contour C as before (see 18).

Notes

[1] Two linearly independent solutions of Airy's equation are often taken to be

$$\mathrm{Ai}(x) = \frac{1}{\pi}\int_0^\infty \cos\left(\frac{t^3}{3} + xt\right)\,dt,$$

$$\mathrm{Bi}(x) = \frac{1}{\pi}\int_0^\infty \left[\exp\left(-\frac{t^3}{3} + xt\right) + \cos\left(\frac{t^3}{3} + xt\right)\right]\,dt.$$

These solutions represent two different choices of the contour in (72.19).

[2] The Laplace equations

$$(a_0x + b_0)y^{(n)} + (a_1x + b_1)y^{(n-1)} + \ldots + (a_nx + b_n)y = 0$$

have solutions in the form of (72.2). Indeed, this was Laplace's original example. See Valiron [6] or Davies [3] for details.

[3] When the kernel of the transformation is some function of the product $z\xi$, then this method is sometimes called the Mellin transformation. See Ince [4] for details.

[4] Sometimes a double integral may be required to find an integral representation. In this case, a solution of the form $u(z) = \iint K(z; s, t) w(s, t)\, ds\, dt$ is proposed. Details may be found in Ince [4], page 197. As an example, the equation

$$(x^2 - 1)\frac{d^2 y}{dx^2} + (a + b + 1)x\frac{dy}{dx} + aby = 0$$

has the two linearly independent solutions

$$y_{\pm}(x) = \int_0^\infty \int_0^\infty \exp\left[\pm xst - \frac{1}{2}(s^2 + t^2)\right] s^{a-1} t^{b-1}\, ds\, dt.$$

[5] Equations of the form

$$\left[x^n F\left(x\frac{d}{dx}\right) + G\left(x\frac{d}{dx}\right)\right] y = 0,$$

which are sometimes called Pfaffian differential equations, can also be solved by this method. See Bateman [2] or Ince [4] (page 190) for details.

[6] An application of this method to partial differential equations may be found in Bateman [2], pages 268–275.

[7] The Mellin–Barnes integral representation for an ordinary differential equation has the form

$$u(z) = \int_C K(z, \xi) z^\xi \left[\frac{\displaystyle\prod_{j=1}^m \Gamma(b_j - \xi) \prod_{j=1}^n \Gamma(1 - a_j + \xi)}{\displaystyle\prod_{j=m+1}^q \Gamma(1 - b_j + \xi) \prod_{j=n+1}^r \Gamma(a_j - \xi)}\right] d\xi.$$

In this representation, only the contour C and the constants $\{a_i, b_j, m, n, q, r\}$ are to be determined (see Babister [1] for details).

References

[1] A. W. Babister, *Transcendental Functions Satisfying Nonhomogeneous Linear Differential Equations*, The MacMillan Company, New York, 1967, pages 24–26.

[2] H. Bateman, *Differential Equations*, Longmans, Green and Co., 1926, Chapter 10, (pages 260–264).

[3] B. Davies, *Integral Transforms and Their Applications*, Springer–Verlag, New York, 1978, pages 342–367.

[4] E. L. Ince, *Ordinary Differential Equations*, Dover Publications, Inc., New York, 1964, pages 186–203 and 438–468.

[5] F. W. J. Olver, *Asymptotics and Special Functions*, Academic Press, New York, 1974.

[6] G. Valiron, *The Geometric Theory of Ordinary Differential Equations and Algebraic Functions*, Math Sci Press, Brookline, MA, 1950, pages 306–319.

73.   Integral Transforms: Finite Intervals*

Applicable to   Linear differential equations.

Idea

In order to solve a linear differential equation, it is sometimes easier to transform the equation to some "space," solve the equation in that "space," and then transform the solution back.

Procedure

Given a linear differential equation, multiply the equation by a kernel and integrate over a specified region (see Table 73 for a listing of common kernels and limits of integration). Use integration by parts to obtain an equation for the transform of the dependent variable.

You will have used the "correct" transform (i.e., you have chosen the correct kernel and limits) if the boundary conditions given with the original equation have been utilized. Now solve the equation for the transform of the dependent variable. From this, obtain the solution by multiplying by the inverse kernel and performing another integration. Table 73 also lists the inverse kernel.

Example 1

Suppose we have the boundary value problem for $y = y(x)$

$$y_{xx} + y = 1,$$
$$y(0) = 0, \quad y(1) = 0. \qquad (73.1.a\text{--}c)$$

Since the solution vanishes at both of the endpoints, we suspect that a finite sine transform might be a useful transform to try. Define the finite sine transform of $y(x)$ to be $z(\xi)$, so that

$$z(\xi) := \int_0^1 y(x) \sin \xi x \, dx. \qquad (73.2)$$

(See "finite sine transform–2" in Table 73). Now multiply equation (73.1.a) by $\sin \xi x$ and integrate with respect to x from 0 to 1. This results in

$$\int_0^1 y_{xx}(x) \sin \xi x \, dx + \int_0^1 y(x) \sin \xi x \, dx = \int_0^1 \sin \xi x \, dx. \qquad (73.3)$$

If we integrate the first term in (73.3) by parts, twice, we obtain

$$\int_0^1 y_{xx}(x)\sin\xi x\,dx = y_x(x)\sin\xi x\Big|_{x=0}^{x=1} - \xi y(x)\cos\xi x\Big|_{x=0}^{x=1}$$

$$+ \xi^2\int_0^1 y(x)\sin\xi x\,dx. \tag{73.4}$$

Since we are only interested in $\xi = 0, \pi, 2\pi, \ldots$, (see Table 73) the first term on the right-hand side of (73.4) is identically zero. Because of the boundary conditions in (73.1.b-c), the second term on the right-hand side of (73.4) also vanishes. Since we have used the given boundary conditions to simplify certain terms appearing in the transformed equation, we suspect we have used an appropriate transform. If we had taken a finite cosine transform, instead of the one that we did, the boundary terms from the intergration by parts would not have vanished.

Using (73.4), simplified, in (73.3) results in

$$\xi^2\int_0^1 y(x)\sin\xi x\,dx + \int_0^1 y(x)\sin\xi x\,dx = \frac{1-\cos\xi}{\xi}.$$

Using the definition of $z(\xi)$ (from (73.2)) this becomes

$$\xi^2 z(\xi) + z(\xi) = \frac{1-\cos\xi}{\xi},$$

or

$$z(\xi) = \frac{1-\cos\xi}{(1+\xi^2)\xi}.$$

Now that we have found an explicit formula for the transformed function, we can use the summation formula (inverse transform) in Table 73 to determine that

$$y(x) = \sum_{\xi=0,\pi,2\pi,\ldots} 2z(\xi)\sin\xi x,$$

$$= \sum_{\xi=0,\pi,2\pi,\ldots} 2\frac{1-\cos\xi}{(1+\xi^2)\xi}\sin\xi x,$$

$$= \sum_{k=0}^\infty 2\frac{1-(-1)^k}{(1+\pi^2 k^2)\pi k}\sin k\pi x, \tag{73.5}$$

$$= \sum_{k=1,3,5,\ldots} \frac{4\sin k\pi x}{(1+\pi^2 k^2)\pi k},$$

where we have defined $k = \xi/\pi$.

The exact solution of (73.1) is $y(x) = 1 - \cos x + \dfrac{\cos 1 - 1}{\sin 1}\sin x$. If this solution is expanded in a finite Fourier series, we obtain the representation in (73.5).

Example 2

Suppose we have the following partial differential equation for $\phi(r,t)$ (this corresponds to the temperature of a long circular cylinder whose surface is at a constant temperature)

$$\frac{\partial^2 \phi}{\partial r^2} + \frac{1}{r}\frac{\partial \phi}{\partial r} = \frac{1}{\kappa}\frac{\partial \phi}{\partial t}, \qquad \text{for } 0 \le r < 1 \text{ and } t > 0,$$

$$\phi(1,t) = \phi_0, \qquad \text{for } t > 0, \qquad\qquad (73.6)$$

$$\phi(r,0) = 0, \qquad \text{for } 0 \le r < 1.$$

Multiplying this equation by $rJ_0(pr)$ (where p is positive and satisfies $J_0(p) = 0$, see "finite Hankel transform–1" in Table 73), and integrating with respect to r from 0 to 1, we find

$$p\phi_0 J_0'(p) - p^2 \Phi = \frac{1}{\kappa}\frac{d\Phi}{dt}, \qquad\qquad (73.7)$$

where we have defined $\Phi(p,t) = \int_0^1 \phi(r,t) r J_0(pr)\, dr$. This follows from the relation: $\int_0^1 \left(\frac{\partial^2 \phi}{\partial r^2} + \frac{1}{r}\frac{\partial \phi}{\partial r}\right) r J_0(pr)\, dr = p\phi_0 J_0'(p) - p^2 \Phi(p,t)$. The initial condition in (73.6) is transformed to $\Phi(p,0) = 0$. Using this, we can solve (73.7) to find $\Phi(p,t) = \frac{\phi_0}{p} J_0'(p)\left(e^{-\kappa p^2 t} - 1\right)$. Taking the inverse transform (and noting that $J_0'(p) = -J_1(p)$) we arrive at the final solution to (73.6):

$$\phi(r,t) = 2\phi_0 \sum_p \left(e^{-\kappa p^2 t} - 1\right)\frac{J_0(pr)}{pJ_1(p)},$$

where the summation is over all positive roots of $J_0(p) = 0$.

Table 73:

Different transform pairs of the form

$$v(\xi_k) = \int_\alpha^\beta u(x)K(x,\xi_k)\, dx, \qquad u(x) = \sum_{\xi_k} H(x,\xi_k)v(\xi_k).$$

Finite cosine transform $-$ 1, (see Miles [6], page 86) here l and h are arbitrary, and the $\{\xi_k\}$ satisfy $\xi_k \tan \xi_k l = h$.

$$v(\xi_k) = \int_0^1 u(x)\cos(x\xi_k)\, dx, \quad u(x) = \sum_{\xi_k} \frac{(2 - \delta_{\xi_k 0})(\xi_k^2 + h^2)\cos(\xi_k x)}{h + l(\xi_k^2 + h^2)}v(\xi_k).$$

Finite cosine transform − 2, (see Butkov [2], page 161) this is the last transform with $h = 0$, $l = 1$, so that $\xi_k = 0, \pi, 2\pi, \dots$.

$$v(\xi_k) = \int_0^1 u(x) \cos(x\xi_k) \, dx, \quad u(x) = \sum_{\xi_k} (2 - \delta_{\xi_k 0}) \cos(\xi_k x) \, v(\xi_k).$$

Finite sine transform − 1, (see Miles [6], page 86) here l and h are arbitrary, and the $\{\xi_k\}$ satisfy $\xi_k \cot(\xi_k l) = -h$.

$$v(\xi_k) = \int_0^1 u(x) \sin(x\xi_k) \, dx, \quad u(x) = \sum_{\xi_k} 2\frac{(\xi_k^2 + h^2) \sin(\xi_k x)}{h + l(\xi_k^2 + h^2)} v(\xi_k).$$

Finite sine transform − 2, (see Butkov [2], page 161) this is the last transform with $h = 0$, $l = 1$, so that $\xi_k = 0, \pi, 2\pi, \dots$.

$$v(\xi_k) = \int_0^1 u(x) \sin(x\xi_k) \, dx, \quad u(x) = \sum_{\xi_k} 2 \sin(\xi_k x) \, v(\xi_k).$$

Finite Hankel transform − 1, (see Tranter [9], page 88) here n is arbitrary and the $\{\xi_k\}$ are positive and satisfy $J_n(\xi_k) = 0$.

$$v(\xi_k) = \int_0^1 u(x) x J_n(x\xi_k) \, dx, \quad u(x) = \sum_{\xi_k} 2\frac{J_n(x\xi_k)}{J_{m+1}^2(\xi_k)} v(\xi_k).$$

Finite Hankel transform − 2, (see Miles [6], page 86) here n and h are arbitrary and the $\{\xi_k\}$ are positive and satisfy $\xi_k J_n'(a\xi_k) + h J_n(a\xi_k) = 0$.

$$v(\xi_k) = \int_0^a u(x) x J_n(x\xi_k) \, dx, \quad u(x) = \sum_{\xi_k} \frac{2\xi_k^2 J_n(x\xi_k)}{\left\{ \left(h^2 + \xi_k^2\right) a^2 - m^2 \right\} J_n^2(a\xi_k)} v(\xi_k).$$

Finite Hankel transform − 3, (see Miles [6], page 86) here $b > a$ and the $\{\xi_k\}$ satisfy $Z_n(b\xi_k) = 0$ where $Z_n(x\xi_k) := Y_n(a\xi_k) J_n(x\xi_k) - J_n(a\xi_k) Y_n(x\xi_k)$.

$$v(\xi_k) = \int_a^b u(x) x Z_n(x\xi_k) \, dx, \quad u(x) = \sum_{\xi_k} \frac{\pi^2}{2} \frac{\xi_k^2 J_n^2(b\xi_k) Z_n(x\xi_k)}{J_n^2(a\xi_k) - J_n^2(b\xi_k)} v(\xi_k).$$

Legendre transform, (see Miles [6], page 86) here $\xi_k = 0, 1, 2, \dots$.

$$v(\xi_k) = \int_{-1}^1 u(x) P_{\xi_k}(x) \, dx, \quad u(x) = \sum_{\xi_k} \frac{2\xi_k + 1}{2} P_{\xi_k}(x) v(\xi_k).$$

Notes

[1] There are many tables of transforms available (see Bateman [1] or Magnus, Oberhettinger, and Soni [5]). It is generally easier to look up a transform than to compute it.

[2] Transform techniques may also be used with systems of linear equations.

[3] The Legendre transform is useful for differential equations that contain the operator $L[u] = \dfrac{\partial}{\partial r}\left((1-r^2)\dfrac{\partial u}{\partial r}\right)$. For example, the transform of $L[u]$ is simply $-\xi_k(\xi_k+1)v(\xi_k)$.

[4] The finite Hankel transforms are useful for differential equations that contain the operator $L[u] = u_{rr} + \dfrac{u_r}{r} - \dfrac{n^2}{r^2}u$.

[5] Integral transforms can be constructed by integrating the Green's function for a Sturm–Liouville eigenvalue problem. This involves explicitly finding a representation of the delta function in terms of the eigenfunctions. For example, the relation

$$\frac{\delta(x-\xi_k)}{x} = \sum_\zeta \frac{2J_0(x\zeta)J_0(\xi_k\zeta)}{J_1^2(\zeta)}$$

where the sum is over all of the roots of $J_0(\zeta) = 0$, can be used to derive the *Fourier–Bessel series* (i.e., the "finite Hankel transform – 1," with $n = 0$ and $a = 1$). For more details see Stakgold [8].

[6] A transform pair that is continuous in each variable, on a finite interval, is the finite Hilbert transform

$$v(\xi) = \frac{1}{\pi}\int_{-1}^{1}\frac{u(x)}{x-\xi}\,dx, \quad u(x) = \frac{1}{\sqrt{1-x^2}}\left[C - \frac{1}{\pi}\int_{-1}^{1}\frac{\sqrt{1-\xi^2}}{\xi-x}v(\xi)\,d\xi\right],$$

where C is an arbitrary constant, and the integrals are to be evaluated in the principal value sense. See Sneddon [7], page 467, for details.

[7] Assuming that $g(u+1) = g(u)$, $f(t+1) = f(t)$, $\int_0^1 g(u)\,du = 0$, and $\int_0^1 f(t)\,dt = 0$, then we have the transform pair

$$f(t) = \int_0^1 g(u)\cot\pi(t-u)\,du, \quad g(u) = \int_0^1 f(t)\cot\pi(t-u)\,du.$$

See Courant and Hilbert [3], page 98.

References

[1] Staff of the Bateman Manuscript Project, A. Erdélyi (ed.), *Tables of Integral Transforms*, in 3 volumes, McGraw–Hill Book Company, New York, 1954.

[2] E. Butkov, *Mathematical Physics*, Addison–Wesley Publishing Co., Reading, MA, 1968, Chapter 5 (pages 179–220) and Section 8.5 (pages 299–304).

[3] R. Courant and D. Hilbert, *Methods of Mathematical Physics*, Interscience Publishers, New York, 1953.

[4] B. Davies, *Integral Transforms and Their Applications*, Springer–Verlag, New York, 1978.

[5] W. Magnus, F. Oberhettinger, and R. P. Soni, *Formulas and Theorems for the Special Functions of Mathematical Physics*, Springer–Verlag, New York, 1966.

[6] J. W. Miles, *Integral Transforms in Applied Mathematics*, Cambridge at the University Press, 1971.

[7] I. N. Sneddon, *The Use of Integral Transforms*, McGraw–Hill Book Company, New York, 1972.

[8] I. Stakgold, *Green's Functions and Boundary Value Problems*, John Wiley & Sons, New York, 1979.

[9] C. J. Tranter, *Integral Transforms in Mathematical Physics*, Methuen & Co. Ltd., London, 1966.

74. Integral Transforms: Infinite Intervals*

Applicable to   Linear differential equations.

Idea

In order to solve a linear differential equation, it is sometimes easier to transform the equation to some "space," solve the equation in that "space," and then transform the solution back.

Procedure

Given a linear differential equation, multiply the equation by a kernel and integrate over a specified region (see Table 74 for a listing of common kernels and limits of integration). Use integration by parts to obtain an equation for the transform of the dependent variable.

You will have used the "correct" transform (i.e., you have chosen the correct kernel and limits) if the boundary conditions given with the original equation have been utilized. Now solve the equation for the transform of the dependent variable. From this, obtain the solution by multiplying by the inverse kernel and performing another integration. Table 74 also lists the inverse kernel.

Warning

After a solution is obtained by a transform method, it must be checked that the solution satisfies the requirements of the transform. For example, for a function to have a Laplace transform, it must be a L_2 function (i.e., square integrable).

Example 1

Suppose we wish to find the solution to the parabolic partial differential equation

$$u_t = a^2 u_{xx} \tag{74.1}$$

with the initial condition and boundary conditions given by

$$
\begin{aligned}
u(x,0) &= 0, \\
u(0,t) &= u_0, && \text{for } t > 0, \\
u(\infty,t) &= 0, && \text{for } t > 0,
\end{aligned}
\tag{74.2.a–c}
$$

where a and u_0 are given constants.

Since this problem is in a semi-infinite domain (i.e., t varies from 0 to ∞), we suspect that a Laplace transform in t may be useful in finding the solution. Let $\mathcal{L}\{\cdot\}$ denote the Laplace transform operator, and define

$$v(x,s) := \mathcal{L}\{u(x,t)\} := \int_0^\infty e^{-st} u(x,t)\, dt \tag{74.3}$$

to be the Laplace transform of $u(x,t)$. We want to manipulate (74.1) into a form such that there are $v(x,s)$ terms present. To obtain this form, multiply (74.1) by e^{-st} and integrate wth respect to t from 0 to ∞ to obtain

$$\int_0^\infty e^{-st} u_t(x,t)\, dt = a^2 \int_0^\infty e^{-st} u_{xx}(x,t)\, dt. \tag{74.4}$$

The left-hand side of (74.4) can be integrated by parts while the x derivatives can be taken out of the integral in the right-hand side to obtain

$$-\frac{u(x,t)e^{-st}}{s}\Big|_0^\infty + \int_0^\infty s e^{-st} u(x,t)\, dt = a^2 \frac{\partial^2}{\partial x^2} \int_0^\infty e^{-st} u(x,t)\, dt.$$

If we assume that $\lim_{t \to \infty} e^{-st} u(x,t) = 0$ and use (74.2.a), then we obtain

$$\int_0^\infty s e^{-st} u(x,t)\, dt = a^2 \frac{\partial^2}{\partial x^2} \int_0^\infty e^{-st} u(x,t)\, dt.$$

Finally, using the definition of $v(x, s)$, from (74.3), we obtain

$$s\,v(x, s) = a^2 \frac{\partial^2}{\partial x^2} v(x, s), \tag{74.5}$$

which is essentially an ordinary differential equation in the independent variable x. The boundary conditions for this equation come from taking the Laplace transform of (74.2.b–c). We calculate

$$v(0, s) := \mathcal{L}\{u(0, t)\} = \mathcal{L}\{u_0\} = \int_0^\infty e^{-st} u_0 \, dt = \frac{u_0}{s},$$

$$v(\infty, s) := \mathcal{L}\{u(\infty, t)\} = \mathcal{L}\{0\} = 0. \tag{74.6}$$

Solving (74.5) with the boundary conditions in (74.6) results in

$$v(x, s) = \frac{u_0}{s} e^{-x\sqrt{s}/a}. \tag{74.7}$$

A table of inverse Laplace transforms, when applied to (74.7), results in

$$u(x, t) = \mathcal{L}^{-1}\{v(x, s)\}$$

$$= \frac{1}{2\pi i} \int_{\sigma-i\infty}^{\sigma+i\infty} e^{st} v(x, s) \, ds \tag{74.8}$$

$$= u_0 \left[1 - \operatorname{erf}\left(\frac{x}{2t\sqrt{a}}\right) \right],$$

which is the final solution.

Now that we have the solution we must either verify that it solves the differential equation and initial condition and boundary conditions that we started with (equation (74.1)), or we must verify that the steps we performed in obtaining the solution are valid. In this case it means verifying that $\lim_{t\to\infty} e^{-st} u(x, t) = 0$, and that $u(x, t)$ is square integrable. Since each of these are true, the solution found in (74.8) is correct.

Example 2

Suppose we have the ordinary differential equation

$$\frac{d^4 y}{dx^4} = y + p(x) \tag{74.9}$$

for $y(x)$, for $-\infty < x < \infty$, with the boundary conditions: $y(\pm\infty) = 0$, $y'(\pm\infty) = 0$. Since the equation is on a (doubly) infinite domain, we try to use a Fourier transform in x to find the solution.

Let $\mathcal{F}\{\cdot\}$ denote the Fourier transform operator, and define

$$z(\omega) = \mathcal{F}\{y(x)\} := \int_{-\infty}^{\infty} y(x) e^{i\omega x} \, dx$$

to be the Fourier transform of $y(x)$. If we apply the operator $\mathcal{F}\{\cdot\}$ to (74.9) (by multiplying by $e^{i\omega x}$ and integrating with respect to x), we find

$$\int_{-\infty}^{\infty} e^{i\omega x} \frac{d^4 y}{dx^4} \, dx = \int_{-\infty}^{\infty} e^{i\omega x} y \, dx + \int_{-\infty}^{\infty} e^{i\omega x} p(x) \, dx.$$

Integrating by parts and using the given boundary conditions, this can be simplified to

$$(i\omega)^4 z(\omega) = z(\omega) + \int_{-\infty}^{\infty} e^{i\omega x} p(x) \, dx.$$

This last expression can be solved to yield

$$z(\omega) = \frac{1}{\omega^4 - 1} \int_{-\infty}^{\infty} e^{i\omega x} p(x) \, dx \tag{74.10}$$

For any given $p(x)$, the integral in (74.10) can be evaluated, and then an inverse Fourier transform can be taken to determine $y(x) = \mathcal{F}^{-1}\{z(\omega)\}$.

Table 74:

Different integral transform pairs of the form

$$v(\xi) = \int_\alpha^\beta K(x,\xi)u(x)\,dx, \qquad u(x) = \int_a^b H(x,\xi)v(\xi)\,d\xi.$$

Fourier transform, (see Butkov [5], Chapter 7)

$$v(\xi) = \frac{1}{\sqrt{2\pi}}\int_{-\infty}^\infty e^{ix\xi}\,u(x)\,dx, \quad u(x) = \frac{1}{\sqrt{2\pi}}\int_{-\infty}^\infty e^{-ix\xi}\,v(\xi)\,d\xi.$$

Fourier cosine transform, (see Butkov [5], page 274)

$$v(\xi) = \sqrt{\frac{2}{\pi}}\int_0^\infty \cos(x\xi)\,u(x)\,dx, \quad u(x) = \sqrt{\frac{2}{\pi}}\int_0^\infty \cos(x\xi)\,v(\xi)\,d\xi.$$

Fourier sine transform, (see Butkov [5], page 274)

$$v(\xi) = \sqrt{\frac{2}{\pi}}\int_0^\infty \sin(x\xi)\,u(x)\,dx, \quad u(x) = \sqrt{\frac{2}{\pi}}\int_0^\infty \sin(x\xi)\,v(\xi)\,d\xi.$$

Hankel transform, (see Sneddon [23], Chapter 5)

$$v(\xi) = \int_0^\infty x J_\nu(x\xi)u(x)\,dx, \quad u(x) = \int_0^\infty \xi J_\nu(x\xi)v(\xi)\,d\xi.$$

Hartley transform, (see Bracewell [4])

$$v(\xi) = \frac{1}{\sqrt{2\pi}}\int_{-\infty}^\infty (\cos x\xi + \sin x\xi)\,u(x)\,dx,$$

$$u(x) = \frac{1}{\sqrt{2\pi}}\int_{-\infty}^\infty (\cos x\xi + \sin x\xi)\,v(\xi)\,d\xi.$$

Hilbert transform, (see Sneddon [23], pages 233–238)

$$v(\xi) = \int_{-\infty}^\infty \frac{1}{\pi(x-\xi)}u(x)\,dx, \quad u(x) = \int_{-\infty}^\infty \frac{1}{\pi(\xi-x)}v(\xi)\,d\xi.$$

K–transform, (see Bateman [2])

$$v(\xi) = \int_0^\infty K_\nu(x\xi)\sqrt{\xi x}\,u(x)\,dx, \quad u(x) = \frac{1}{\pi i}\int_{\sigma-i\infty}^{\sigma+i\infty} I_\nu(x\xi)\sqrt{\xi x}\,v(\xi)\,d\xi.$$

Kontorovich–Lebedev transform, (see Sneddon [23], Chapter 6)

$$v(\xi) = \int_0^\infty \frac{K_{i\xi}(x)}{x}\, u(x)\, dx, \quad u(x) = \frac{2}{\pi^2}\int_0^\infty \xi \sinh(\pi\xi) K_{i\xi}(x)\, v(\xi)\, d\xi.$$

Kontorovich–Lebedev transform (alternative form), (see Jones [13])

$$v(\xi) = \int_0^\infty H_\xi^{(2)}(x)\, u(x)\, dx, \quad u(x) = -\frac{1}{2x}\int_{-i\infty}^{i\infty} \xi J_\xi(x)\, v(\xi)\, d\xi.$$

Laplace transform, (see Sneddon [23], Chapter 3)

$$v(\xi) = \int_0^\infty e^{-x\xi}\, u(x)\, dx, \quad u(x) = \frac{1}{2\pi i}\int_{\sigma-i\infty}^{\sigma+i\infty} e^{x\xi}\, v(\xi)\, d\xi.$$

Mehler–Fock transform of order m, (see Sneddon [23], Chapter 7)

$$v(\xi) = \int_0^\infty \sinh(x) P_{i\xi-1/2}^m(\cosh x)\, u(x)\, dx,$$

$$u(x) = \int_0^\infty \xi \tanh(\pi\xi) P_{i\xi-1/2}^m(\cosh x)\, v(\xi)\, d\xi.$$

Mellin transform, (see Sneddon [23], Chapter 4)

$$v(\xi) = \int_0^\infty x^{\xi-1}\, u(x)\, dx, \quad u(x) = \frac{1}{2\pi i}\int_{\sigma-i\infty}^{\sigma+i\infty} x^{-\xi}\, v(\xi)\, d\xi.$$

Weber formula, (see Titchmarsh [25], page 75)

$$v(\xi) = \int_a^\infty \sqrt{x}\left[J_\nu(x\xi)Y_\nu(a\xi) - Y_\nu(x\xi)J_\nu(a\xi)\right] u(x)\, dx,$$

$$u(x) = \sqrt{x}\int_0^\infty \frac{J_\nu(x\xi)Y_\nu(a\xi) - Y_\nu(x\xi)J_\nu(a\xi)}{J_\nu^2(a\xi) + Y_\nu^2(a\xi)}\, v(\xi)\, d\xi.$$

Weierstrass transform, (see Hirschman and Widder [11], Chapter 8)

$$v(\xi) = \frac{1}{\sqrt{4\pi}}\int_{-\infty}^\infty e^{(\xi-x)^2/4}\, u(x)\, dx, \quad u(x) = \frac{1}{\sqrt{4\pi}}\lim_{T\to\infty}\int_{-T}^T e^{(x-i\xi)^2/4} v(i\xi)\, d\xi.$$

Unnamed transform, (see Naylor [18])

$$v(\xi) = \int_{-\infty}^\infty K_0(|\xi-x|)\, u(x)\, dx, \quad u(x) = -\frac{1}{\pi^2}\left(\frac{d^2}{dx^2} - 1\right)\int_{-\infty}^\infty K_0(|\xi-x|)\, v(\xi)\, d\xi.$$

Notes

[1] Note that many of the transforms in Table 74 do not have a standard form. In the Fourier transform, for example, the two $\sqrt{2\pi}$ terms might not be symmetrically placed as we have shown them. Also, a small variation of the K-transform is known as the Meijer transform (see Ditkin and Prudnikov [8], page 75).

[2] There are many tables of transforms available (see Bateman [2] or Magnus, Oberhettinger and Soni [16]). It is generally easier to look up a transform than to compute it.

[3] Transform techniques may also be used with systems of linear equations.

[4] If a function $f(x, y)$ has radial symmetry, then a Fourier transform in both x and y is equivalent to a Hankel transform of $f(r) = f(x, y)$, where $r^2 = x^2 + y^2$. See Sneddon [23], pages 79–83.

[5] Integral transforms can be constructed by integrating the Green's function for a Sturm–Liouville eigenvalue problem. This involves explicitly finding an integral representation of the delta function. For example, the relation

$$\delta(\eta) = \frac{1}{2\pi} \int_{-\infty}^{\infty} e^{i\eta\nu} \, d\nu \tag{74.11}$$

can be used to derive the Fourier transform. To see this, change η to $x - \xi$ in (74.11), multiply by $f(\xi)$ and integrate with respect to ξ to obtain

$$f(x) = \frac{1}{\sqrt{2\pi}} \int_{-\infty}^{\infty} e^{ix\nu} \left[\frac{1}{\sqrt{2\pi}} \int_{-\infty}^{\infty} f(\xi) e^{i\xi\nu} \, d\xi \right] d\nu$$

For more details, see Davies [7] (pages 267–287), or Stakgold [24].

[6] Many of the transforms in the table have a *convolution theorem*, which describes how the transform of the product of two functions is related to the transforms of the individual functions. For example, if $g(t)$ (respectively $h(t)$, $k(t)$) has the Laplace transform $G(s)$ (respectively $H(s)$, $K(s)$), and $G(s) = H(s)K(s)$, then

$$g(t) = \int_0^t h(t - \tau)k(\tau) \, d\tau.$$

This is called a *convolution product* and is often denoted by $g(t) = h(t) * k(t)$. See Miles [17], Table 2.3 (page 85).

[7] Most of the transforms in Table 74 have simple formulae relating the transform of the derivative of a function to the transform of the function. For example, if $G(s)$ is the Laplace transform of $g(t)$ then

$$\mathcal{L}\{g^{(n)}(t)\} = s^n G(s) - g^{(n-1)}(0) + sg^{(n-2)}(0) + \cdots + (-1)^n s^{n-1} g(0).$$

[8] Two transform pairs that are continuous in one variable and discrete in the other variable, on an infinite interval, are the Hermite transform

$$u(x) = \sum_{n=0}^{\infty} v_n H_n(x) e^{-x^2/2}, \qquad v_n = \frac{1}{(2^n)!\sqrt{\pi}} \int_{-\infty}^{\infty} u(x) H_n(x) e^{-x^2/2}\, dx,$$

where $H_n(x)$ is the n-th Hermite polynomial, and the Laguerre transform

$$u(x) = \sum_{n=0}^{\infty} v_n L_n^\alpha(x) \frac{n!}{\Gamma(n+\alpha+1)}, \qquad v_n = \int_0^\infty u(x) L_n^\alpha(x) x^\alpha e^{-x}\, dx,$$

where $L_n^\alpha(x)$ is the Laguerre polynomial of degree n, and $\alpha \geq 0$. See Haimo [10] for details.

[9] Integral transforms are generally created for solving a specific differential equation with a specific class of boundary conditions. The Mathieu integral transform (see Inayat-Hussain [12]) has been constructed for the two-dimensional Helmholtz equation in elliptic-cylinder coordinates.

[10] The papers by Namias [19], [20] on fractional order Fourier and Hankel transforms contain several examples of how the transforms may be used to solve differential equations.

[11] If we recognize that

$$\frac{d^r}{dx^r} = \prod_{i=1}^{r-1} \left(\frac{1}{x^{-i}} \frac{d}{x^{r-1}dx} x^{r-i} \right) \frac{d}{x^{r-1}dx} \tag{74.12}$$

then we see that the ν-transform, defined by

$$g(x;\nu) = Z[f(x);\nu] = \int_0^\infty \cdots \int_0^\infty f\left(x \prod t_i^{1/r}\right) e^{\sum t_i} \prod t_i^{\nu_i}\, dt_i,$$

$$f(x) = \frac{1}{(2\pi i)^{r-1}} \int_{-\infty}^{(0+)} \cdots \int_{-\infty}^{(0+)} g\left(x \prod t_i^{-1/r};\nu\right) e^{\sum t_i} \prod t_i^{-\nu_i - 1}\, dt_i,$$

where $\nu = (\nu_1, \ldots, \nu_{r-1})$ and i runs from 1 to $r-1$ in each sum and product, can be used with (74.12) to obtain:

$$Z\left[\frac{d^r u}{dx^r};\nu_r\right] = \left(\frac{r}{x}\right)^{r-1} \frac{dZ[u;\nu_r]}{dx} - \sum_{i=1}^{r-1} \frac{C_i}{x^{r-i}}$$

where $\nu_r = (-1/r, -2/r, \ldots, -(r-1)/r)$. This transform can be applied, for example, to the equations $y^{(r)} + axy' + by = f(x)$ and

$$\left(\frac{d^r}{dx^r} + \frac{b_1}{x} \frac{d^{r-1}}{dx^{r-1}} + \ldots + \frac{b_{r-1}}{x^{r-1}} \frac{d}{dx} \right) y + axy' + by = f(x).$$

See Klyuchantsev [14] for details.

[12] Classically, the Fourier transform of a function only exists if the function being transformed decays quickly enough at $\pm\infty$. The Fourier transform can be extended, though, to handle generalized functions. For example, the Fourier transform of the n-th derivative of the delta function is given by $\mathcal{F}\left(\delta^{(n)}(t)\right) = (i\omega)^n$.

Another way to approach the Fourier transform of functions that do not decay quickly enough at either ∞ or $-\infty$ is to use the *one-sided Fourier transforms*. See Chester [6] for details.

[13] Many of the transforms listed generalize naturally to n dimensions. For example, in n dimensions we have:

(A) Fourier transform: $v(\boldsymbol{\xi}) = (2\pi)^{-n/2} \int_{\mathrm{R}^n} e^{i\boldsymbol{\xi}\cdot\mathbf{x}} u(\mathbf{x})\, d\mathbf{x}$,

$u(\mathbf{x}) = (2\pi)^{-n/2} \int_{\mathrm{R}^n} e^{-i\boldsymbol{\xi}\cdot\mathbf{x}} v(\boldsymbol{\xi})\, d\boldsymbol{\xi}$.

(B) Hilbert transform (see Bitsadze [3]):

$$\frac{\partial f}{\partial x_i} = \frac{\Gamma(n/2)}{\pi^{n/2}} \int_{R^{n-1}} \frac{y_i - x_i}{|\mathbf{y} - \mathbf{x}|^n} \phi(y)\, dy, \quad i = 1, 2, \ldots, n-1,$$

$$\phi(y) = -\frac{\Gamma(n/2)}{\pi^{n/2}} \int_{R^{n-1}} \frac{(\mathbf{y} - \mathbf{x}) \cdot \nabla f}{|\mathbf{y} - \mathbf{x}|^n}\, d\mathbf{y},$$

[14] The name *Bessel transform* is given to an integral transform that involves a Bessel function. This class includes Hankel, K, Kontorovich–Lebedev, and many other transforms.

[15] Note that, for the Hilbert transform, the integrals in Table 74 are to be taken in the principal value sense.

References

[1] M. Abramowitz and I. A. Stegun, *Handbook of Mathematical Functions*, National Bureau of Standards, Washington, DC, 1964, pages 1019–1030.

[2] Staff of the Bateman Manuscript Project, A. Erdélyi (ed.), *Tables of Integral Transforms*, in 3 volumes, McGraw–Hill Book Company, New York, 1954.

[3] A. V. Bitsadze, "The Multidimensional Hilbert Transform," *Soviet Math. Dokl.*, **35**, No. 2, 1987, pages 390–392.

[4] R. N. Bracewell, *The Hartley Transform*, Oxford University Press, New York, 1986.

[5] E. Butkov, *Mathematical Physics*, Addison–Wesley Publishing Co., Reading, MA, 1968, Chapter 5 (pages 179–220), and Section 8.5 (pages 299–304).

[6] C. R. Chester, *Techniques in Partial Differential Equations*, McGraw–Hill Book Company, New York, 1970.

[7] B. Davies, *Integral Transforms and Their Applications*, Springer–Verlag, New York, 1978.

[8] V. A. Ditkin and A. P. Prudnikov, *Integral Transforms and Operational Calculus*, translated by D. E. Brown, English translation edited by I. N. Sneddon, Pergamon Press, New York, 1965.

[9] H.-J. Glaeske, "Operational Properties of a Generalized Hermite Transformation," *Aequationes Mathematicae*, **32**, 1987, pages 155–170.

[10] D. T. Haimo, "The Dual Weierstrass–Laguerre Transform," *Trans. AMS*, **290**, No. 2, August 1985, pages 597–613.

[11] I. I. Hirschman and D. V. Widder, *The Convolution Transform*, Princeton Univ Press, Princeton, NJ, 1955.

[12] A. A. Inayat-Hussain, "Mathieu Integral Transforms," *J. Math. Physics*, **32**, No. 3, March 1991, pages 669–675.

[13] D. S. Jones, "The Kontorovich–Lebedev Transform," *J. Inst. Maths. Applics*, **26**, 1980, pages 133–141.

[14] M. I. Klyuchantsev, "An Integral-Transformation Method of Solving some Types of Differential Equations," Translated from *Differentsial'nye Uravneniya*, **23**, No. 10, October 1987, pages 1668–1679.

[15] O. I. Marichev, *Handbook of Integral Transforms of Higher Transcendental Functions: Theory and Algorithmic Tables*, translated by L. W. Longdon, Halstead Press, John Wiley & Sons, New York, 1983.

[16] W. Magnus, F. Oberhettinger, and R. P. Soni, *Formulas and Theorems for the Special Functions of Mathematical Physics*, Springer–Verlag, New York, 1966.

[17] J. W. Miles, *Integral Transforms in Applied Mathematics*, Cambridge at the University Press, 1971.

[18] D. Naylor, "On an Integral Transform," *Int. J. Math. & Math. Sci.*, **9**, No. 2, 1986, pages 283–292.

[19] V. Namias, "The Fractional Order Fourier Transform and its Application to Quantum Mechanics," *J. Inst. Maths. Applics*, **25**, 1980, pages 241–265.

[20] V. Namias, "Fractionalization of Hankel Transforms," *J. Inst. Maths. Applics*, **26**, 1980, pages 187–197.

[21] C. Nasim, "The Mehler–Fock Transform of General Order and Arbitrary Index and its Inversion," *Int. J. Math. & Math. Sci.*, **7**, No. 1, 1984, pages 171–180.

[22] F. Oberhettinger and T. P. Higgins, *Tables of Lebedev, Mehler, and Generalized Mehler Transforms*, Mathematical Note No. 246, Boeing Scientific Research Laboratories, October 1961.

[23] I. N. Sneddon, *The Use of Integral Transforms*, McGraw–Hill Book Company, New York, 1972.

[24] I. Stakgold, *Green's Functions and Boundary Value Problems*, John Wiley & Sons, New York, 1979, Chapter 7 (pages 411–466).

[25] E. C. Titchmarsh, *Eigenfunction Expansions Associated with Second-Order Differential Equations*, Clarendon Press, Oxford, 1946.

[26] C. J. Tranter, *Integral Transforms in Mathematical Physics*, Methuen & Co. Ltd., London, 1966.

75.   Integrating Factors*

Applicable to   Linear first order ordinary differential equations.

Yields

An exact equation that can then be integrated.

Idea

When a given equation is not exact, it may be possible to multiply the equation by a certain term so that it does become exact. The term that is used is called an integrating factor.

Procedure

Let us suppose that the nonlinear ordinary differential equation

$$M(x, y)\, dx + N(x, y)\, dy = 0 \qquad (75.1)$$

is not exact (see pages 238–243). It may be, however, that if (75.1) is multiplied by an integrating factor $u(x, y)$ the resulting equation

$$uM\, dx + uN\, dy = 0$$

is exact. For this to be the case, we require $\partial(uM)/\partial y = \partial(uN)/\partial x$, or

$$u\left(\frac{\partial M}{\partial y} - \frac{\partial N}{\partial x}\right) = N\frac{\partial u}{\partial x} - M\frac{\partial u}{\partial y}. \qquad (75.2)$$

In general, solving the partial differential equation in (75.2), for $u(x, y)$, is more difficult than solving the ordinary differential equation in (75.1). But, in certain cases, it may be easier. For example,

(A) if $\dfrac{1}{N}\left(\dfrac{\partial M}{\partial y} - \dfrac{\partial N}{\partial x}\right) = f(x)$, a function of x alone, then

$u(x, y) = u(x) = \exp\left(\int^x f(z)\, dz\right)$ is an integrating factor for (75.1);

(B) if $\dfrac{1}{M}\left(\dfrac{\partial M}{\partial y} - \dfrac{\partial N}{\partial x}\right) = g(y)$, a function of y alone, then

$u(x, y) = u(y) = \exp\left(\int^y g(z)\, dz\right)$ is an integrating factor for (75.1).

Example

Suppose we have the general linear first order ordinary differential equation

$$y' + P(x)y = Q(x). \tag{75.3}$$

We recognize that the homogeneous equation corresponding to (75.3) $y' + P(x)y = 0$, or

$$dy + (P(x)y)dx = 0,$$

has the integrating factor $u(x) = \exp\left(\int^x P(z)\,dz\right)$, since $M \equiv yP(x)$, $N \equiv 1$ and case (A) applies with $f(x) := P(x)$. Hence equation (75.3) can be written as

$$(y' + P(x)y)\exp\left(\int^x P(z)\,dz\right) = Q(x)\exp\left(\int^x P(z)\,dz\right),$$

or

$$\frac{d}{dx}\left[y\exp\left(\int^x P(z)\,dz\right)\right] = Q(x)\exp\left(\int^x P(z)\,dz\right),$$

and therefore (by integrating) we find the solution to be

$$y(x) = \exp\left(-\int^x P(z)\,dz\right)\int^x Q(w)\exp\left(\int^w P(z)\,dz\right)dw.$$

Special Case

For a concrete illustration, the equation

$$y' + \frac{1}{x}y = x^2 \tag{75.4}$$

has $\{P(x) = 1/x, Q(x) = x^2\}$, so that

$$u(x) = \exp\left(\int^x \frac{1}{z}\,dz\right)$$
$$= \exp\left(\log x\right)$$
$$= x$$

is an integrating factor. When (75.4) is multiplied by $u(x) = x$, we obtain

$$xy' + y = x^3,$$
$$\frac{d(xy)}{dx} = x^3, \tag{75.10}$$
$$xy = \frac{x^4}{4} + C,$$

or

$$y = \frac{x^3}{4} + \frac{C}{x}, \tag{75.11}$$

where C is an arbitrary constant.

Notes

[1] If equation (75.1) admits a one parameter Lie group with generators $\{\xi, \eta\}$ (see page 314), then an integrating factor is given by $u(x,y) = \dfrac{1}{N\eta - M\xi}$.

For example, the differential equation $y(y^2 - x)\, dx + x^2\, dy = 0$ is invariant under the transformation $\{y' = e^{\varepsilon/2}y,\ x' = e^\varepsilon x\}$. Therefore the infinitesimal operator of the group is described by $\{\eta = \frac{1}{2}y,\ \xi = x\}$. This leads to the integrating factor $u = \dfrac{2}{3xy(x - 2y^2)}$, which leads to the solution $y = \dfrac{x}{\sqrt{2x + C}}$.

[2] If $Mx + Ny \neq 0$, and equation (75.1) is homogeneous (see page 276), then an integrating factor is given by $u(x,y) = \dfrac{1}{Mx + Ny}$.

For example, the differential equation $(xy - 2y^2)\, dx - (x^2 - 3xy)\, dy = 0$ is homogeneous and has the integrating factor $u = 1/xy^2$. This leads to the solution $\dfrac{x}{y} - \log(x^2 y^3) = C$.

[3] If $M = M_1(x)y - M_2(x)y^n$ and $N = 1$, then an integrating factor is given by $u(x,y) = y^{-n}\exp((1-n)/int M_1\, dx)$.

[4] The differential equation $M_1(x)M_2(y)\, dx + N_1(x)N_2(y)\, dy = 0$ has the integrating factor $u = (M_2 N_1)^{-1}$.

[5] The differential equation $yf(xy)\, dx + xg(xy)\, dy = 0$, when $f \neq g$, has the integrating factor $u = 1/[xy(f-g)]$. For example, the equation $y(1-xy)\, dx - x(1 + xy)\, dy = 0$ has $\{f(z) = 1 - z,\ g(z) = -1 - z\}$ so that an integrating factor is given by $u = 1/2xy$. This leads to the implicit solution $ye^{xy} = Cx$.

[6] Given equation (75.1), if $z = N - iM$ is an analytic function of x and y (i.e., the Cauchy–Riemann equations $\{N_x = -M_y,\ N_y = M_x\}$ are satisfied), then an integrating factor is given by $1/(N^2 + M^2)$.

For example, the homogeneous equation

$$\left(y^2 + 2xy - x^2\right) dy - \left(y^2 - 2xy - x^2\right) dx = 0$$

has the integrating factor $u = 1/\left[2\left(x^2 + y^2\right)^2\right]$ which leads to the solution $y + x = C(x^2 + y^2)$.

[7] Sometimes an integrating factor of the form $x^k y^n$ can be found (for specific values of k and n). This form of the integrating factor will always be adequate for differential equations of the form $x^a y^b(py\, dx + qx\, dy) + x^d y^e(ry\, dx + sx\, dy) = 0$, where $\{a, b, d, e, p, q, r, s\}$ are constants.

[8] The technique presented here also applies to linear ordinary differential equations of higher order. For example, the second order ordinary differential equation

$$\sqrt{x}\frac{d^2 y}{dx^2} + 2x\frac{dy}{dx} + 3y = 0 \tag{75.5}$$

can be made exact (see page 240) by use of the integrating factor $u(x) = \sqrt{x}$. Multiplying equation (75.5) by $\sqrt{x}$ results in

$$x\frac{d^2y}{dx^2} + 2x^{3/2}\frac{dy}{dx} + 3y\sqrt{x} = \frac{d}{dx}\left[x\frac{dy}{dx} + (2x^{3/2} - 1)y\right].$$

Murphy [2] has a discussion of how to make second order ordinary differential equations exact (see page 165).

[9] When the quasilinear partial differential equation in two independent variables, $M(x, y, u)u_x = N(x, y, u)u_y$, has $M_x = N_y$, then the solution is given implicitly by $\Phi(x, y, u) = 0$, where $M = \Phi_y$ and $N = \Phi_x$. If, alternately, $M_x \neq N_y$, then it may be possible to find an integrating factor $v(x, y)$ such that $(vM)_x = (vN)_y$. For example, if $(N_y - M_x)/M$ is a function of x alone, then $v(x) = \exp\left(\displaystyle\int \frac{N_y - M_x}{M}\, dx\right)$ will be an integrating factor.

As an illustration, the equation $u_x = yu_y$ has the integrating factor $v(x) = e^x$. The solution can then be found to be $u(x, y) = -Cy^3 e^{3x}$, where C is an arbitrary constant.

References

[1] W. E. Boyce and R. C. DiPrima, *Elementary Differential Equations and Boundary Value Problems*, Fourth Edition, John Wiley & Sons, New York, 1986, pages 84–87.

[2] G. Murphy, *Ordinary Differential Equations*, D. Van Nostrand Company, Inc., New York, 1960.

[3] J. D. Murray, *Asymptotic Analysis*, Springer–Verlag, New York, 1984, pages 22–27.

[4] M. J. Prelle and M. F. Singer, "Elementary First Integrals of Differential Equations," *Trans. Amer. Math. Soc.*, **279**, No. 1, September 1983, pages 215–229.

[5] E. D. Rainville and P. E. Bedient, *Elementary Differential Equations*, The MacMillan Company, New York, 1964, pages 35–37 and 59–66.

[6] G. F. Simmons, *Differential Equations with Applications and Historical Notes*, McGraw–Hill Book Company, New York, 1972, pages 42–46.

76.   Interchanging Dependent and Independent Variables

Applicable to   Ordinary differential equations.

Yields

A reformulation of the original equation.

Idea

Sometimes it is easier to solve an ordinary differential equation by interchanging the role of the dependent variable with the role of the independent variable. If this technique works, then the solution is given implicitly by $x = x(y)$ instead of the usual $y = y(x)$.

Procedure

Given the equation

$$\frac{dy}{dx} = f(x, y)$$

to solve, it might be easier to solve the equivalent equation

$$\frac{dx}{dy} = \frac{1}{f(x, y)}.$$

This method can also be used for ordinary differential equations with an order greater than 1. For these cases, Table 76 can be used to determine how the derivatives $\{y_x, y_{xx}, \ldots\}$ transform into the derivatives $\{x_y, x_{yy}, \ldots\}$.

Example 1

Suppose the solution is desired to the ordinary differential equation

$$\frac{dy}{dx} = \frac{x}{x^2 y^2 + y^5}.$$

Interchanging the dependent and independent variables in this equation produces

$$\frac{dx}{dy} = \frac{x^2 y^2 + y^5}{x} = y^2 x + \frac{y^5}{x}. \qquad (76.1)$$

Equation (76.1) is now a Bernoulli equation with $n = -1$ and can be solved exactly (see the section on Bernoulli equations, page 194). The solution is

$$x(y) = \left(A e^{2y^3/3} - \frac{y^3}{2} - \frac{3}{4} \right)^{1/2},$$

where A is an arbitrary constant.

$$y_x = x_y^{-1},$$

$$y_{xx} = -x_y^{-3}x_{yy},$$

$$y_{xxx} = 3x_y^{-5}x_{yy}^2 - x_y^{-4}x_{yyy},$$

$$y_{xxxx} = -15x_y^{-7}x_{yy}^3 + 10x_y^{-6}x_{yy}x_{yyy} - x_y^{-5}x_{yyyy}$$

$$y_{xxxxx} = 105x_y^{-9}x_{yy}^4 - 105x_y^{-8}x_{yy}^2x_{yyy} + 10x_y^{-7}x_{yyy}^2$$

$$+ 15x_y^{-7}x_{yy}x_{yyyy} - x_y^{-6}x_{yyyyy}$$

Table 76. How higher order derivatives transform when the dependent and independent variables are switched.

Example 2

As a second illustration, the following formidable nonlinear ordinary differential equation

$$y'' + xy(y')^3 = 0 \tag{76.2}$$

becomes, after interchanging the dependent and independent variables, Airy's equation

$$\frac{d^2x}{dy^2} = xy.$$

Hence, the solution to (76.2) is

$$x(y) = C_1 \operatorname{Ai}(y) + C_2 \operatorname{Bi}(y),$$

where C_1 and C_2 are arbitrary constants.

Notes

[1] When this method is applied to partial differential equations (and not ordinary differential equations), then the method is called the *hodograph transformation* (see page 390).

References

[1] C. M. Bender and S. A. Orszag, *Advanced Mathematical Methods for Scientists and Engineers*, McGraw–Hill, New York, 1978, Section 1.6.

[2] M. E. Goldstein and W. H. Braun, *Advanced Methods for the Solution of Differential Equations*, NASA SP-316, U.S. Government Printing Office, Washington, D.C., 1973, page 107.

[3] B. L. McAllister and C. J. Thorne, *Reverse Differential Equations and Others That Can Be Solved Exactly*, University of Utah, Salt Lake City, Technical Report Number 111, Project Number NR-056-239, 1952. (Available through NTIS).

77.   Lagrange's Equation

Applicable to   Equations of the form: $y = xF\left(\dfrac{dy}{dx}\right) + G\left(\dfrac{dy}{dx}\right)$.

Yields

An exact solution, sometimes given parametrically.

Idea

Equations of this form can be solved by quadratures.

Procedure

Given an equation of the form

$$y = xF\left(\frac{dy}{dx}\right) + G\left(\frac{dy}{dx}\right), \qquad (77.1)$$

use p to represent dy/dx so that (77.1) can be written as

$$y = xF(p) + G(p). \qquad (77.2)$$

Now differentiate (77.2) with respect to x to obtain

$$\frac{dy}{dx} \equiv p = F(p) + \frac{dp}{dx}\left[xF'(p) + G'(p)\right]. \qquad (77.3)$$

Equation (77.3) can be rewritten as

$$\frac{dx}{dp} = x\left(\frac{F'(p)}{p - F(p)}\right) + \left(\frac{G'(p)}{p - F(p)}\right), \qquad (77.4)$$

which is now a linear differential equation in x and p. It can be solved by the method of integrating factors (see page 305) to determine

$$x = \phi(p, C), \qquad (77.5)$$

where C is an arbitrary constant. Now there are two possibilities:

(A) Eliminate p between (77.2) and (77.5) to find $\Phi(y, x, C) = 0$

(B) Use (77.5) in (77.2) to obtain the parametric solution

$$x = \phi(P, C),$$
$$y = \phi(P, C)F(P) + G(P),$$

where P is a free parameter.

Example 1

Suppose we have the equation

$$y = 2x\frac{dy}{dx} - a\left(\frac{dy}{dx}\right)^3,$$ (77.6)

where a is a constant. Comparing (77.6) to (77.1), we identify $F(p) = 2p$, $G(p) = ap^3$. Hence, equation (77.5) becomes

$$\frac{dx}{dp} = -\frac{2x}{p} + 3ap.$$

This last equation has an integrating factor of p^2 and so

$$x = \frac{3a}{4}p^2 + \frac{C}{p^2},$$ (77.7)

where C is an arbitrary constant. Using (77.7) in (77.6) we can remove the x dependence to obtain

$$y = \frac{a}{2}p^3 + \frac{2C}{p}.$$

Hence, a parametric solution of (77.6) is given by

$$x = \frac{3a}{4}P^2 + \frac{C}{P^2},$$
$$y = \frac{a}{2}P^3 + \frac{2C}{P},$$ (77.8)

where P can have any value. By use of resultants (see page 46), the parameter P can be removed from (77.8) to determine the implicit solution (see page 46),

$$(27ay^2 - 16x^3)y^2 + 16a^2x(9ay^2 - 4x^3)C - 128a^3x^2C^2 - 64a^4C^3 = 0.$$

If C is taken to be zero, for instance, then the explicit solution $y = \frac{4}{3\sqrt{3a}}x^{3/2}$ is obtained.

Example 2

If we have the equation

$$y = 2x\frac{dy}{dx} - \left(\frac{dy}{dx}\right)^2,$$ (77.9)

then we make the identification $\{F(p) = 2p, G(p) = -p^2\}$ so that (77.4) becomes

$$\frac{dx}{dp} = x\left(-\frac{2}{p}\right) + 2,$$

or (using the integrating factor p^2)

$$x = \frac{2}{3}p^2 + \frac{C}{p^2},$$ (77.10)

where C is an arbitrary constant. Using (77.10) in (77.9) results in

$$y = \frac{C}{p} - \frac{p^2}{3}.$$

Hence, a parametric solution of (77.9) is given by

$$x = \frac{2}{3}P^2 + \frac{C}{P^2},$$
$$y = \frac{C}{P} + \frac{C}{P^2},$$ (77.11)

where P can have any value. By use of resultants the parameter P can be removed from (77.11) to determine the implicit solution

$$y^2(4y - 3x^2) + 6x(2x^2 - 3y)C + 9C^2 = 0.$$

Notes

[1] Equation (77.1) is known as d'Alembert's equation and also as an *equation linear in x and y*.

[2] If $F \equiv 1$, then (77.1) is the same as Clairaut's equation (see page 196).

[3] The technique presented in this section is only an application of the more general technique described on page 350.

References

[1] E. L. Ince, *Ordinary Differential Equations*, Dover Publications, Inc., New York, 1964, pages 38–39.
[2] G. Murphy, *Ordinary Differential Equations*, D. Van Nostrand Company, Inc., New York, 1960, pages 65–66.
[3] G. Valiron, *The Geometric Theory of Ordinary Differential Equations and Algebraic Functions*, Math Sci Press, Brookline, MA, 1950, pages 217–218.

78.   Lie Groups: ODEs

Applicable to   Linear and nonlinear ordinary differential equations.

Yields

Invariants and symmetries of a differential equation. Often these can be used to solve a differential equation.

Idea

By determining the *transformation group* under which a given differential equation is invariant, we can obtain information about the invariants and symmetries of a differential equation. Sometimes these can be used to solve a given differential equation.

Procedure

A one parameter Lie group of transformations is a family of coordinate transformations of the form

$$
\begin{aligned}
x_\varepsilon &= f(x, y; \varepsilon), \\
y_\varepsilon &= g(x, y; \varepsilon),
\end{aligned}
\tag{78.1}
$$

such that $\varepsilon = 0$ gives the identity transformation. It is also required (for the transformations to form a group) that $f(x, y; \varepsilon + \delta) = f(x_\varepsilon, y_\varepsilon; \delta)$, and $f^{-1}(x, y; \varepsilon) = f(x, y; -\varepsilon)$, with analogous formulae for $g(x, y; \varepsilon)$.

Equation (78.1) is called the *global transformation group*. Expanding (78.1) for small values of ε yields

$$
\begin{aligned}
x_\varepsilon &= x + \xi(x, y)\varepsilon + O(\varepsilon^2), \\
y_\varepsilon &= y + \eta(x, y)\varepsilon + O(\varepsilon^2),
\end{aligned}
$$

where

$$
\xi(x, y) = \left(\frac{\partial f}{\partial \varepsilon} \right)_{\varepsilon=0}, \qquad \eta(x, y) = \left(\frac{\partial g}{\partial \varepsilon} \right)_{\varepsilon=0}.
\tag{78.2}
$$

The quantities ξ and η are called the *infinitesimal transformations* of the group. Lie's first fundamental theorem states that knowing the infinitesimals $\{\xi(x,y), \eta(x,y)\}$ is equivalent to knowing the functions $\{f, g\}$ that appear in (78.1).

An n-th order differential equation

$$G\left(x, y, y', \ldots, y^{(n)}\right) = 0 \tag{78.3}$$

is said to be invariant under the group defined by (78.1) if the differential equation

$$G\left(x_\epsilon, y_\epsilon, y_\epsilon', \ldots, y_\epsilon^{(n)}\right) = 0$$

is equivalent to (78.3) under the change of variables in (78.1). The differential equation $G = 0$ will be invariant with respect to the one parameter group (defined by (78.1)) if

$$U^{(n)}G = 0, \tag{78.4}$$

on the manifold $G = 0$ in the space of the variables $\{x, y, y', \ldots, y^{(n)}\}$. The operator $U^{(n)}$ (sometimes called the *n-th prolongation*, see Olver [10]) is defined by

$$U^{(n)} = \xi \frac{\partial}{\partial x} + \eta \frac{\partial}{\partial y} + \sum_{l=1}^{n} \xi_l \frac{\partial}{\partial y^{(l)}}, \tag{78.5}$$

where $\xi_0 = \eta$ and $\xi_l = D(\xi_{l-1}) - y^{(l)} D(\xi)$, for $l = 1, 2, \ldots, n$, and the *total derivative operator* D is defined by $D := \dfrac{\partial}{\partial x} + y' \dfrac{\partial}{\partial y} + y'' \dfrac{\partial}{\partial y'} + \ldots$. Hence, (78.4) is a quasilinear equation, and the method of characteristics may be used to solve it.

If the differential equation $G = 0$ is invariant with respect to the group, then the subsidiary equations of (78.4) can be written as (see page 368)

$$\frac{dx}{\xi} = \frac{dy}{\eta} = \frac{d(y')}{\xi_1} = \cdots = \frac{d(y^{(n)})}{\xi_n}.$$

We can sometimes integrate two of these equations to obtain the two integrals: $u = u(x, y, y', \ldots)$ and $v = v(x, y, y', \ldots)$. If the original equation, $G = 0$, is written in terms of these new variables, then the resulting differential equation will only be of order $n-1$. Hence, we will have reduced the order of the given differential equation.

Example

Given the class of second order ordinary differential equations

$$G(x, y, y', y'') \equiv xy'' - F\left(\frac{y}{x}, y'\right) = 0, \tag{78.6}$$

we ask if this differential equation is invariant under the magnification group

$$\begin{aligned} x_\varepsilon &= xe^\varepsilon, \\ y_\varepsilon &= ye^\varepsilon. \end{aligned} \tag{78.7}$$

If it is, then we should be able to reduce (78.6) to a sequence of first order ordinary differential equations. Using (78.7) in the definitions in (78.2) and (78.5) we can sequentially calculate

$$\begin{aligned} &\xi(x, y) = x, \qquad \eta(x, y) = y, \\ &\xi_0 = \eta = y, \\ &\xi_1 = D(\xi_0) - y' D(\xi) = D(y) - y' D(x) = 0 \\ &\xi_2 = D(\xi_1) - y'' D(\xi) = D(0) - y'' D(x) = -y'', \\ &U^{(2)} = x\frac{\partial}{\partial x} + y\frac{\partial}{\partial y} - y''\frac{\partial}{\partial y''}. \end{aligned}$$

Applying $U^{(2)}$ to G we find

$$\begin{aligned} U^{(2)}G &= \left(x\frac{\partial}{\partial x} + y\frac{\partial}{\partial y} - y''\frac{\partial}{\partial y''}\right)\left[xy'' - F\left(\frac{y}{x}, y'\right)\right] \\ &= x\left(y'' + \frac{y}{x^2}F_1\right) + y\left(-\frac{1}{x}F_1\right) - y''(x) \\ &= 0, \end{aligned}$$

where F_1 denotes the derivative of F with respect to its first argument. We conclude, then, that $G = 0$ is invariant under the magnification group.

Now we form the subsidiary equations:

$$\frac{dx}{x} = \frac{dy}{y} = \frac{dy'}{0} = \frac{dy''}{-y''}.$$

From the first equality, $\dfrac{dx}{x} = \dfrac{dy}{y}$, we find that y/x is a constant; we write this as $y/x = u$. From the second equality, $\dfrac{dy}{y} = \dfrac{dy'}{0}$, we find that y' is a constant; we write this as $y' = v$.

Now we will write the equation $G = 0$ in terms of the "constants" which parameterize the solution space: $\{u, v\}$. To change variables we will need

$$y'' = \frac{dy'}{dx} = \frac{dv}{dx} = \frac{dv}{du}\frac{du}{dx} = \frac{dv}{du}\left(\frac{y'}{x} - \frac{y}{x^2}\right) = \frac{dv}{du}\frac{v - u}{x}.$$

Hence

$$G = xy'' - F\left(\frac{y}{x}, y'\right) = (v - u)\frac{dv}{du} - F(u, v) = 0. \qquad (78.8)$$

Finally, then, we have transformed the second order differential equation $G = 0$ into a first order differential equation in terms of u and v. After this equation is solved for $v = v(u)$, we then have a first order equation for $y(x)$ (using $u = y/x$ and $v = y'$).

Special Case 1

If we choose the special case $F(u, v) = v - u$ (for which (78.6) becomes the linear equation $x^2 y'' - xy' + y = 0$, with solutions $y = x$ and $y = x \log x$), equation (78.8) becomes $(v - u)\left(\frac{dv}{du} - 1\right) = 0$. The most general solution to this equation is $v = u + C$, where C is an arbitrary constant. Changing to our original variables, this becomes $\frac{dy}{dx} = \frac{y}{x} + C$. This equation has the solution $y = Cx \log x + Dx$, where D is another arbitrary constant.

Special Case 2

If we choose the special case $F(u, v) = u^2 - v^2$ (for which (78.6) becomes the nonlinear equation $x^3 y'' + x^2(y')^2 - y^2 = 0$), equation (78.8) becomes $\frac{dv}{du} = -v - u$. This first order equation can be integrated to yield $v = (u^2 - 2u + 2) + Ce^{-u}$, where C is an arbitrary constant. In this case we cannot integrate again to obtain $y = y(x)$ in closed form.

Notes

[1] Lie group analysis is the most useful and general of all the techniques presented in this book. Many of the other methods presented in this book can be derived from the method of Lie groups. For example:

(A) Equations with the dependent variable missing (see page 216) are invariant under the translation group $\{x_\varepsilon = x, y_\varepsilon = y + \varepsilon\}$.

(B) Equations with the independent variable explicitly missing (see page 190) are invariant under the translation group $\{x_\varepsilon = x + \varepsilon, y_\varepsilon = y\}$.

(C) Homogeneous equations (see page 276) are invariant under the affine group $\{x_\varepsilon = x, y_\varepsilon = ye^\varepsilon\}$.

(D) Scale invariant equations (see page 338) are invariant under the group $\{x_\varepsilon = xe^\varepsilon, y_\varepsilon = ye^{p\varepsilon}\}$.

(E) In Kumei and Bluman [9] it is shown that the hodograph transformation (see page 390) and the Legendre transformation (see page 400) are derivable from Lie group methods.

(F) Similarity solutions (see page 424) are all derivable from Lie group methods.

(G) Contact transformations (see page 206) and the Riccati transformation (see page 332) are derivable from Lie group methods.

[2] Easily readable books that explain Lie groups more fully are Bluman and Kumei [2] and Stephani [15].

[3] Using Lie groups to find symmetries of differential equations can be computationally intensive. Algorithms have been developed for computerized handling of the calculations, see Bocharov and Bronstein [3], Champagne and Winternitz [4], or Eliseev, Fedorova, and Kornyak [6].

[4] In the older literature, transformation groups were found and then classes of equations that were invariant under that group were determined. This was what was done in the example in this section. For example, it can be shown that the most general second order differential equation invariant under a group of the form

$$x_\varepsilon = f(x; \varepsilon) = x + \varepsilon \xi(x) + O(\varepsilon^2),$$
$$y_\varepsilon = g(x; \varepsilon)y = y + \varepsilon \eta(x)y + O(\varepsilon^2),$$

has the form

$$y'' + \left(\frac{\xi' - 2\eta}{\xi}\right) y' + \left(\frac{\eta^2 - \xi\eta'}{\xi^2}\right) y = \frac{\Phi(A, B)}{s\xi^2},$$

where Φ is an arbitrary function of its arguments, and $\{A, B, s\}$ are defined by

$$A(x, y) = sy,$$
$$B(x, y) = (\xi x - \eta y)s,$$
$$s(x) = \exp\left(-\int_{x_0}^{x} \frac{\eta(t)}{\xi(t)} \, dt\right).$$

See Hill [7] (page 84) for details.

[5] Recently the procedure in the last note has been reversed: given a differential equation, find a transformation group which leaves the equation invariant. To derive the transformation group, a set of partial differential equations arising from the equation $U^{(n)}G = 0$ must be solved. For example, for the second order ordinary differential equation $\ddot{x} = f(t, x, \dot{x})$ to be invariant under the group

$$x_\varepsilon = x + \varepsilon \psi(t, x) + O(\varepsilon^2),$$
$$t_\varepsilon = t + \varepsilon \phi(t, x) + O(\varepsilon^2),$$

requires that the following equation

$$\psi_{tt} + (2\psi_{xt} - \phi_{tt})\dot{x} + (\psi_{xx} - 2\phi_{xt})\dot{x}^2 - \phi_{xx}\dot{x}^3$$
$$+ [(\psi_x - 2\phi_t) - 3\phi_x\dot{x}]\, f(t,x,\dot{x}) - \phi f_t(t,x,\dot{x}) - \psi f_x(t,x,\dot{x})$$
$$- \left[\psi_t + (\psi_x - \phi_t)\dot{x} - \phi_x\dot{x}^2\right] f_{\dot{x}}(t,x,\dot{x}) = 0$$

hold for all $(t, x, \dot{x})$. See Aguirre and Krause [1] for details.

[6] The operator V, with $V = \xi(x,y)\dfrac{\partial}{\partial x} + \eta(x,y)\dfrac{\partial}{\partial y}$, is called the *infinitesimal operator* of the group. Observe that an arbitrary function of x_ε and y_ε, $F(x_\varepsilon, y_\varepsilon)$, can be formally expanded in terms of x and y as

$$F(x_\varepsilon, y_\varepsilon) = F(x,y) + \varepsilon \left(\frac{\partial f}{\partial \varepsilon}\frac{\partial}{\partial x} + \frac{\partial g}{\partial \varepsilon}\frac{\partial}{\partial y}\right)_{\varepsilon=0} F(x,y) + \cdots$$
$$= F(x,y) + \varepsilon V F(x,y) + \tfrac{1}{2}\varepsilon^2 V^2 F(x,y) + \cdots$$
$$= e^{\varepsilon V} F(x,y),$$

[7] If the parameter ε appearing in (78.1) had been an r-dimensional vector, then there would be r infinitesimal operators $\{V_1, V_2, \ldots, V_r\}$. Lie's second fundamental theorem states that these operators generate an r-dimensional Lie group under commutation: $[V_a, V_b] = K_{ab}^c V_c$, where the K's are called *structure constants* and summation occurs over repeated indices. Lie's third fundamental theorem relates the structure constants to one another.

If, in the above, $r = 1$ then the order of the original equation can be reduced by one. If $n \geq 2$ and $r = 2$, then the order of the original equation can be reduced by two. If $n \geq 3$ and $r \geq 3$, then it does not follow that the order of the original equation can be reduced by more than two. However, if the r-dimensional Lie algebra has a q-dimensional *solvable subalgebra*, then the order of the original equation can be reduced by q. See Bluman and Kumei [2] for details.

[8] The analysis in this section can be obtained from the general results of Lie algebras. For example, if $x(t)$ satisfies the equation $\ddot{x} = f(x, \dot{x})$, where f is in C^∞, and the solution is analytic for all t, then the solution may be obtained from $x_{t+\tau} = e^{t\Omega_\tau} x_\tau$, where

$$\Omega_\tau = v_\tau \left(\frac{\partial}{\partial x_\tau}\right) + f(x_\tau, v_\tau)\left(\frac{\partial}{\partial v_\tau}\right) + \left(\frac{\partial}{\partial \tau}\right),$$

and we have used x_τ to denote $x(\tau)$. For example, for the differential equation $\ddot{x} = 1$, we have $f = 1$ so that $\Omega_\tau = v_\tau \partial_{x_\tau} + \partial_{v_\tau} + \partial_\tau$, and we can calculate

$$\Omega_\tau x_\tau = v_\tau,$$
$$\Omega_\tau^2 x_\tau = \Omega_\tau v_\tau = 1,$$
$$\Omega_\tau^3 x_\tau = \Omega_\tau 1 = 0,$$
$$\Omega_\tau^k x_\tau = 0, \qquad \text{for } k \geq 3.$$

Using these calculations, we can then find

$$x_{t+\tau} = e^{t\Omega_\tau} x_\tau$$

$$= \sum_{k=0}^{\infty} \frac{t^k \Omega_\tau^k}{k!} x_\tau$$

$$= x_\tau + t v_\tau + \frac{t^2}{2},$$

or $x(t+\tau) = x(\tau) + t\dot{x}(\tau) + t^2/2$.

This also generalizes to higher dimensions. For example, the solution of the vector equation $\ddot{\mathbf{x}} = \mathbf{f}(\mathbf{x}, \dot{\mathbf{x}})$ may be written as $\mathbf{x}_{t+\tau} = e^{t\Omega_\tau}\mathbf{x}_\tau$, where

$$\Omega_\tau = \mathbf{v}_\tau \cdot \nabla_{\mathbf{x}_\tau} + \mathbf{f}(\mathbf{x}_\tau, \mathbf{v}_\tau) \cdot \nabla_{\mathbf{v}_\tau} + \frac{\partial}{\partial \tau}$$

[9] Technically, a Lie group is a topological group (i.e., a group that is also a topological space) which is also an analytic manifold on which the group operations are analytic. The tangent space to that manifold is a Lie algebra, which is a linear vector space. See Sattinger and Weaver [12] for an algebraic approach to Lie groups.

[10] It is also possible to find discrete groups that transform solutions of ordinary differential equations to other solutions, see Zaĭtsev [18]. For example, the generalized Emden–Fowler equation $y'' = Ax^n y^m (y')^l$ is described by the parameters $\mathbf{c} = (n, m, l)$. Under the discrete transformation $\{y = at, x = bu\}$ the solution $y = y(x; \mathbf{c})$ is mapped to the solution $u = u(y, \mathbf{c}')$, where $\mathbf{c}' = (n, m, 3 - l)$. Another such discrete transformation is given by $\{y = au^{-1/m}, x = bt^{1/(n+1)}\}$ for which $\mathbf{c}' = \left(-\dfrac{n}{n+1}, \dfrac{1}{1-l}, \dfrac{2m+1}{m} \right)$.

Zaĭtsev [1] illustrates this method by writing the solution of $y'' = x^{-15/8} y \sqrt{y'}$ in terms of the solutions to $u'' = 6u^2$ (which are elliptic functions).

References

[1] M. Aguirre and J. Krause, "Infinitesimal Symmetry Transformations of Some One-Dimensional Linear Systems," *J. Math. Physics*, **25**, No. 2, February 1984, pages 210–218.

[2] G. W. Bluman and S. Kumei, *Symmetries and Differential Equations*, Springer–Verlag, New York, 1989.

[3] A. V. Bocharov and M. L. Bronstein, "Efficiently Implementing Two Methods of the Geometrical Theory of Differential Equations: An Experience in Algorithm and Software Design," *Acta Appl. Math.*, **16**, 1989, pages 143–166.

[4] B. Champagne and P. Winternitz, "A MACSYMA Program for Calculating the Symmetry Group of a System of Differential Equations," CRM-1278, Université de Montréal, 1985.

[5] L. Dressner, *Similarity Solutions of Nonlinear Partial Differential Equations*, Pitman Publishing Co., Marshfield, MA, 1983.

[6] V. P. Eliseev, R. N. Fedorova, and V. V. Kornyak, "A REDUCE Program for Determining Point and Contact Lie Symmetries of Differential Equations," *Comput. Physics Comm.*, **36**, 1985, pages 383–389.

[7] J. M. Hill, *Solution of Differential Equations by Means of One-Parameter Groups*, Pitman Publishing Co., Marshfield, MA, 1982.

[8] E. L. Ince, *Ordinary Differential Equations*, Dover Publications, Inc., New York, 1964, Chapter 4 (pages 93–113).

[9] S. Kumei and G. W. Bluman, "When Nonlinear Differential Equations are Equivalent to Linear Differential Equations," *SIAM J. Appl. Math.*, **42**, No. 5, October 1982, pages 1157–1173.

[10] P. J. Olver, *Applications of Lie Groups to Differential Equations*, Graduate Texts in Mathematics #107, Springer–Verlag, New York, 1986.

[11] L. V. Ovsiannikov, *Group Analysis of Differential Equations*, translated by W. F. Ames, Academic Press, New York, 1982.

[12] D. H. Sattinger and O. L. Weaver, *Lie Groups and Algebras with Applications to Physics, Geometry, and Mechanics*, Springer–Verlag, New York, 1986.

[13] F. Schwarz, "A REDUCE Package for Determining Lie Symmetries of Ordinary and Partial Differential Equations," *Comput. Physics Comm.*, **27**, 1982, pages 179–186.

[14] R. Seshadi and T. Y. Na, *Group Invariance in Engineering Boundary Value Problems*, Springer–Verlag, New York, 1985.

[15] H. Stephani, *Differential Equations: Their Solution Using Symmetries*, M. MacCallum (ed.), Cambridge University Press, New York, 1989.

[16] S. Steinberg, "Lie Series and Nonlinear Differential Equations," *J. Math. Anal. Appl.*, **101**, 1984, pages 39–63.

[17] P. Winternitz, "Lie Groups and Solutions of Nonlinear Differential Equations," in K. B. Wolf (ed.), *Nonlinear Phenomena*, Lecture Notes in Physics #189, Springer–Verlag, New York, 1983, pages 263–331.

[18] V. F. Zaĭtsev, "On Discrete-Group Analysis of Ordinary Differential Equations," *Soviet Math. Dokl.*, **37**, No. 2, 1988, pages 403–406.

79.   Operational Calculus[*]

Applicable to   Ordinary differential equations and partial differential equations.

Yields
   A reformulation of the original differential equation.

Idea
   It may sometimes be easier to solve a differential equation in a transformed space.

Procedure
   Given an ordinary differential equation, transform it to a field of operators, solve the equation in that field, and then transform back. In this field, ordinary functions, generalized functions, and differential operators are all treated as objects in a single algebraic structure.
   The operator field that is used has, among other elements, an identity operator $(\mathcal{I})$, a differentiation operator (often denoted by D or s) and an integration operator (often denoted by D^{-1}). The operator D, when applied to the operator corresponding to a function $f(t)$, results in

$$D\{f\} = \{f'\} + \{f(0)\}, \qquad (79.1.)$$

The operator D^{-1}, when applied to the operator corresponding to a function $f(t)$ results in

$$D^{-1}\{f\} = \left\{ \int_0^t f(u)\, du \right\}.$$

The braces around the above expressions emphasize that they are operators in the field. In many applications, the operator D is formally treated as being a "large constant."
   There are tables of formulae describing how operators interact in their quotient field. For example, since

$$\frac{\mathcal{I}}{D - \alpha} = \{e^{\alpha t}\} \qquad (79.2)$$

we can calculate

$$\frac{\mathcal{I}}{(D - \alpha)^2} = \frac{\mathcal{I}}{(D - \alpha)} \frac{\mathcal{I}}{(D - \alpha)}$$
$$= \{e^{\alpha t}\} \{e^{\alpha t}\}$$
$$= \left\{ \int_0^t e^{\alpha u} e^{\alpha(t-u)}\, du \right\}$$
$$= \{t e^{\alpha t}\},$$

since the "product" of two operators is the operator corresponding to a convolution. The formula in (79.2) follows from (79.1) when $f(t) = e^{\alpha t}$, since

$$(D - \alpha)\left\{e^{\alpha t}\right\} = \left(\left\{\alpha e^{\alpha t}\right\}\right) + \{1\} - \alpha\left\{e^{\alpha t}\right\} = \mathcal{I}.$$

It is easy to represent generalized functions and non-continuous functions in the field. For example, a square wave of period $2c$ has the operator representation $\dfrac{\mathcal{I}}{D\left(\mathcal{I} + e^{-cD}\right)}$.

Example 1

The following ordinary differential equation for $y(t)$

$$y'' + y = 0,$$

has the operator representation

$$\left(D^2 + 1\right)\{y\} = 0, \tag{79.3}$$

or

$$D^2\left(1 + D^{-2}\right)\{y\} = 0.$$

By applying D^{-2} to the left of the above equation, we obtain

$$\left(1 + D^{-2}\right)\{y\} = D^{-2}\{0\}$$
$$= At + B,$$

where A and B are arbitrary constants. This equation may be formally solved by "dividing" by the operator on the left and expanding terms. We find

$$\begin{aligned}
\{y(t)\} &= \left\{\frac{1}{1 + D^{-2}}(At + B)\right\} \\
&= \left\{\left(1 - D^{-2} + D^{-4} - \cdots\right)(At + B)\right\} \\
&= \left\{(At + B) + \left(-\frac{At^3}{6} - \frac{Bt^2}{2}\right) + \left(-\frac{At^5}{120} - \frac{Bt^4}{24}\right) + \cdots\right\} \\
&= \{A\sin t + B\cos t\}.
\end{aligned}$$

$$\tag{79.4}$$

Hence, $y(t) = A\sin t + B\cos t$. Really, in this last calculation, there would be many more terms than those illustrated. For instance, when D^{-4} is applied to $(At + B)$, we obtain $\left(-\dfrac{At^5}{120} - \dfrac{Bt^4}{24}\right)$ plus some terms of the form $\left(C_1 t^3 + C_2 t^2 + C_3 t + C_4\right)$. When the form of the solution, with all these additional terms, is substituted into the defining equation (79.3), these additional constants turn out to be zero.

Example 2

Consider the constant coefficient linear ordinary differential equation for $z(t)$

$$z'' + 3z' + 2z = f(t),$$
$$z(0) = 1, \quad z'(0) = 0,$$

Because of the formula

$$z^{(n)} = D^n z - \left\{ z^{(n-1)}(0) + D z^{(n-2)}(0) + \cdots + D^{n-1} z(0) \right\}$$

(which parallels the rule for Laplace transforms) the equation for $z(t)$ has the operator representation

$$\left[D^2 \{z\} - D \right] + 3 \left[D \{z\} - I \right] + 2 \{z\} = \{f\},$$

This operator equation can be manipulated into

$$\{z\} = \frac{D + 3I}{D^2 + 3D + 2} + \frac{\{f\}}{D^2 + 3D + 2}$$

$$= \frac{2I}{D+1} - \frac{I}{D+2} + \left(\frac{I}{D+1} - \frac{I}{D+2} \right) \{f\}$$

$$= \left\{ 2e^{-t} \right\} - \left\{ e^{-2t} \right\} + \left\{ e^{-t} - e^{-2t} \right\} \{f\},$$

and hence

$$z(t) = 2e^{-t} - e^{-2t} + \int_0^t \left(e^{-u} - e^{-2u} \right) f(u) \, du,$$

which is the same result that would be obtained by use of Laplace transforms.

Notes

[1] The operational calculus is also called the Heaviside calculus.

[2] The operational calculus, at its simplest level, has a great similarity with Laplace transforms. One school of thought is that any integral transform creates an operational calculus.

[3] It is sometimes difficult to justify the formal steps that are employed in using the operation calculus. One solution (see Erdélyi [3]) is to use a more precisely defined operator, such as the primary operator

$$\widehat{D}_\lambda f(t) = f(t) + \lambda \int_0^t e^{\lambda(t-\theta)} f(\theta) \, d\theta$$

which has the inverse $\widehat{D}_\lambda^{-1} g(t) = g(t) - \lambda \int_0^t g(\theta) \, d\theta$.

[4] Infinite order differential equations are often solved by techniques similar to those described above. For example, the ordinary differential equations $\left(\alpha\dfrac{d}{dx}+1\right)^{-1}y+(\beta x-a)y=0$ and $\left[\cosh\left(i\dfrac{d}{dx}\right)+H(x)-a\right]y=0$ are infinite order differential equations for $y(x)$ (here $H(x)$ represents the step function). Recent results (as well as the solutions to the two above equations) may be found in Dimitrov [2].

[5] The extension of this technique to partial differential equations is straightforward. Using D for $\dfrac{\partial}{\partial x}$ and D' for $\dfrac{\partial}{\partial t}$, a partial differential equation can sometimes be written in the form $P(D,D')\{y\}=\{f\}$. The "inversion" process will then proceed in two steps. For example, a calculation analogous to the one in (79.4) might proceed as follows:

$$\{y\}=\frac{1}{P(D,D')}\{f\}$$

$$=\frac{1}{D^2-6DD'+9D'^2}\left(12x^2+36xt\right)$$

$$=\frac{1}{D^2}\left(1-\frac{3D'}{D}\right)^{-2}\left(12x^2+36xt\right)$$

$$=\frac{1}{D^2}\left(1+6\frac{D'}{D}+27\frac{D'^2}{D^2}+\ldots\right)\left(12x^2+36xt\right)$$

$$=\frac{1}{D^2}\left(12x^2+36xt\right)+\frac{6}{D^3}\left(36x\right)$$

$$=\left(x^4+6x^3t\right)+\left(9x^4\right)=10x^4+6x^3t.$$

References

[1] R. Courant and D. Hilbert, *Methods of Mathematical Physics*, Interscience Publishers, New York, 1953, Volume 2, pages 507–535.

[2] H. D. Dimitrov, "On the Solutions of Some Linear Operator Non-Polynomial Differential Equations," *J. Phys. A: Math. Gen.*, **15**, 1982, pages 367–379.

[3] A. Erdélyi, *Operational Calculus and Generalized Functions*, Holt, Rinehart and Winston, New York, 1962.

[4] H.-J. Glaeske, "Operational Properties of a Generalized Hermite Transformation," *Aequationes Mathematicae*, **32**, 1987, pages 155–170.

[5] W. Kaplan, *Operational Methods for Linear Systems*, Addison–Wesley Publishing Co., Reading, MA, 1962, pages 515–538.

[6] J. Mikusiński, *Operational Calculus*, Fifth Edition, Pergamon Press, New York, 1959.

[7] I. Z. Shtokalo, *Operational Calculus*, translated by V. Kumar, Pergamon Press, New York, 1976.

[8] K. Yosida, *Operational Calculus*, Springer–Verlag, New York, 1984.

80.   Pfaffian Differential Equations

Applicable to   Pfaffian differential equations.

Yields
  Knowledge of whether the equation is integrable.

Idea
  Pfaffian differential equations are partial differential equations of the form

$$\mathbf{f(x)} \cdot d\mathbf{x} = \sum_{i=1}^{n} F_i(x_1, x_2, \ldots, x_n)\, dx_i = 0. \tag{80.1}$$

For equations of this type:

(A) If $n = 3$, then a necessary and sufficient condition that (80.1) be integrable is that

$$\mathbf{f(x)} \cdot \operatorname{curl} \mathbf{f(x)} = 0.$$

(B) If $n \geq 4$, then a necessary and sufficient condition that (80.1) be integrable is that

$$F_p \left[\frac{\partial F_r}{\partial x_q} - \frac{\partial F_q}{\partial x_r} \right] + F_q \left[\frac{\partial F_p}{\partial x_r} - \frac{\partial F_r}{\partial x_p} \right] + F_r \left[\frac{\partial F_q}{\partial x_p} - \frac{\partial F_p}{\partial x_q} \right] = 0,$$

where p, q, and r are any three of the integers $1, 2, 3, \ldots, n$.

There exist a number of techniques for integrating Pfaffian equations.

Example
  If we have the equation

$$(y^2 + yz)\, dx + (xz + z^2)\, dy + (y^2 - xy)\, dz = 0, \tag{80.2}$$

then we identify

$$\mathbf{f(x)} = (y^2 + yz, xz + z^2, y^2 - xy),$$

so that

$$\operatorname{curl} \mathbf{f(x)} = \nabla \times \mathbf{f(x)} = 2(-x + y - z, y, -y).$$

Therefore $\mathbf{f(x)} \cdot \operatorname{curl} \mathbf{f(x)} = 0$, and there exists a solution to (80.2). The solution is, in fact, given by $y(x + z) = C(y + z)$, where C is any constant.

Procedure 1

If a Pfaffian equation is integrable, then there exists an integrating factor μ such that

$$d\phi = \sum_{i=1}^{n} \mu F_i \, dx_i.$$

By appropriate manipulations of (80.1), it may be shown that μ satisfies any of the equations

$$-\frac{d\mu}{\mu} = \sum_{j=1}^{n} \frac{1}{F_i} \left[\frac{\partial F_i}{\partial x_j} - \frac{\partial F_j}{\partial x_i} \right] dx_j, \qquad (80.3)$$

for $i = 1, 2, \ldots, n$. Any one of these equations may be solved to determine an integrating factor. Alternately, if two integrating factors can be found, say μ and ν, then a solution to (80.1) is given by $\mu/\nu = $ constant.

Example 1

The Pfaffian differential equation

$$y(x^2 - y^2 - xy) \, dx + x(y^2 - x^2 - xz) \, dy + xy(x + y) \, dz = 0 \qquad (80.4)$$

can be shown to pass the integrability requirements. Substituting into (80.3) results in

$$\begin{aligned}
-\frac{d\mu}{\mu} &= \frac{2(x-y)(2x+2y+z)}{y(x^2 - y^2 - yz)} \, dy - \frac{2(x+y)}{x^2 - y^2 - yz} \, dz, \\
&= -\frac{2(x-y)(2x+2y+z)}{x(y^2 - x^2 - xz)} \, dx - \frac{2(x+y)}{y^2 - x^2 - xz} \, dz, \qquad (80.5) \\
&= 2 \left(\frac{dx}{x} + \frac{dy}{y} \right),
\end{aligned}$$

for $j = 1, 2, 3$. The last equation in (80.5) can be integrated to determine $\mu = 1/(xy)^2$. Hence, multiplying (80.4) by $1/(xy)^2$ results in

$$d\phi = \left(\frac{x^2 - y^2 - yz}{x^2 y} \right) dx + \left(\frac{y^2 - x^2 - xz}{xy^2} \right) dy + \left(\frac{x+y}{xy} \right) dz,$$

which can be integrated to yield

$$\phi = \frac{x}{y} + \frac{y}{x} + \left(\frac{x+y}{xy} \right) z + C,$$

where C is an arbitrary constant.

Procedure 2

If an integrable Pfaffian differential equation is of the form $Pdx + Qdy + Rdz = 0$, where P, Q, and R are homogeneous functions of the same degree, then a solution may be found. First, define $Z = Px + Qy + Rz$. Then form

$$\frac{Pdx + Qdy + Rdz - dZ}{Z} + \frac{dZ}{Z} = 0, \tag{80.6}$$

and integrate (we have only addressed the case of $Z \neq 0$, although there are special techniques that can be used when $Z = 0$).

Example 2

Given the Pfaffian equation

$$(yz + z^2)\, dx - xz\, dy + xy\, dz = 0,$$

we define $Z = xz(y + z)$. Forming (80.6) we obtain

$$\frac{dZ}{Z} - \frac{2(dy + dz)}{y + z} = 0,$$

which can be immediately integrated to yield $Z = C(y + z)^2$ or $xz = C(y + z)$, where C is an arbitrary constant.

Procedure 3

The Pfaffian differential equation $Pdx + Qdy + Rdz = 0$ can sometimes be solved by taking one variable, say z, as a constant. Then the solution of $Pdx + Qdy = 0$ (since $z = \text{constant}$ means that $dz = 0$) will be given by $u(x, y) = \text{constant}$.

We take the "constant" in this last expression to be $f(z)$. Differentiating $u(x, y) = f(z)$, and comparing to the original equation, we may sometimes obtain an ordinary differential equation for $f(z)$.

Example 3

Given the Pfaffian equation

$$2x\, dx + dy + (1 + 2z^2 + 2yz + 2x^2z)\, dz = 0,$$

we treat z as a constant to obtain $2x\, dx + dy = 0$, which has the solution $x^2 + y = \text{constant} = f(z)$. This can be differentiated to obtain

$$2x\, dx + dy + f'(z)\, dz = 0.$$

Comparing this to the original equation, we find that $f(z)$ satisfies the ordinary differential equation: $f' = 1 + 2z^2 + 2zf$. Solving this equation to obtain $f(z) = Ce^{-z^2} - z$, where C is an arbitrary constant, we find the solution to the original equation to be

$$x^2 + y + z = Ce^{-z^2}.$$

Notes

[1] Another name for a Pfaffian differential equation is a *total differential equation*.

[2] One way to solve Pfaffian differential equations in three dimensions is by the observation: if $\operatorname{curl} \mathbf{f}(\mathbf{x}) = 0$, then $\mathbf{f}(\mathbf{x})$ must be the gradient of a scalar. Hence, the set of partial differential equations

$$f_i(\mathbf{x}) = \frac{\partial v(\mathbf{x})}{\partial x_i}, \quad \text{for } i = 1, \ldots, n,$$

may be solvable for $v(\mathbf{x})$. The solution to (80.1) would then be given implicitly by $v(\mathbf{x}) = \text{constant}$.

[3] If the Pfaffian differential equation is of the form $\sum_{i=1}^{n} f_i(x_i)\, dx_i = 0$, then the *integral surfaces* are defined by $\sum_{i=1}^{n} \int f_i(x_i)\, dx_i = C$, where C is an arbitrary constant.

[4] Sometimes a Pfaffian differential equation can be reduced to a system of ordinary differential equations. One such procedure is called Mayer's method. See Carathéodory [1] for details.

[5] Given the system of m Pfaffian differential equations in m dependent variables $\{z_j \mid j = 1, 2, \ldots, m\}$ and n independent variables $\{x_k \mid k = 1, 2, \ldots, n\}$

$$dz_j = \sum_{k=1}^{n} P_{jk}(\mathbf{x}, \mathbf{z})dx_k, \qquad j = 1, 2, \ldots, m,$$

the condition for complete integrability is given by

$$\frac{\partial P_{jk}}{\partial x_l} + \sum_{i=1}^{m} \frac{\partial P_{jk}}{\partial z_i} P_{il} = \frac{\partial P_{jl}}{\partial x_k} + \sum_{i=1}^{m} \frac{\partial P_{jl}}{\partial z_i} P_{ik},$$

for $j = 1, 2, \ldots, m$ and $k, l = 1, 2, \ldots, n$. See Iyanaga and Kawada [6] for details on how this system may be solved.

[6] Using the notation of exterior calculus, a total differential equation is an equation of the form $\omega = 0$, where ω is a differential 1-form $\sum_{i=1}^{n} a_i(\mathbf{x})\, dx_i$ on a manifold. A 1-form ω is called a Pfaffian form. See Iyanaga and Kawada [6] for details.

References

[1] C. Carathéodory, *Calculus of Variations and Partial Differential Equations of the First Order*, Holden–Day, Inc., San Francisco, 1965, pages 121–133.

[2] L. R. Ford, *Differential Equations*, McGraw–Hill Book Company, New York, 1955, pages 135–141.

[3] P. A. Griffiths and G. R. Jensen, *Differential Systems and Isometric Imbeddings*, Princeton University Press, Princeton N.J., 1987.

[4] E. R. Haack and W. N. Wendland, *Lectures on Partial and Pfaffian Differential Equations*, translated by E. R. Dawson and W. N. Everitt, Pergamon Press, New York, 1972.

[5] E. L. Ince, *Ordinary Differential Equations*, Dover Publications, Inc., New York, 1964, pages 52–59.

[6] S. Iyanaga and Y. Kawada, *Encyclopedic Dictionary of Mathematics*, MIT Press, Cambridge, MA, 1980.

[7] P. Moon and D. E. Spencer, *Partial Differential Equations*, D. C. Heath, Lexington, MA, 1969, pages 23–27.

[8] I. N. Sneddon, *Elements of Partial Differential Equations*, McGraw–Hill Book Company, New York, 1957, pages 18–33.

81.   Reduction of Order

Applicable to   Linear ordinary differential equations.

Yields

A lower order differential equation, if any non-trivial solution of the homogeneous equation is known.

Idea

For an n-th order linear ordinary differential equation, any non-trivial solution of the homogeneous equation can be used to reduce the order of the equation by one. For the special case of second order linear differential equations, knowing any solution of the homogeneous equation allows the general solution to be found.

Procedure

We choose to illustrate the method for second order equations. If we have the general second order linear ordinary differential equation

$$y'' + p(x)y' + q(x)y = r(x), \tag{81.1}$$

let $z(x)$ be any non-trivial solution to the corresponding homogeneous equation; i.e., $z(x)$ satisfies

$$z'' + p(x)z' + q(x)z = 0. \tag{81.2}$$

If we look for a solution of (81.1) in the form of $y(x) = z(x)v(x)$, then we can obtain a solvable equation for $v(x)$. Substituting $y(x) = z(x)v(x)$ into (81.1) yields

$$zv'' + (2z' + pz)v' + (z'' + pz' + qz)v = r. \tag{81.3}$$

Since $z(x)$ satisfies (81.2), equation (81.3) becomes

$$zv'' + (2z' + pz)v' = r. \tag{81.4}$$

If we now let $w(x) = v'(x)$, then (81.4) becomes a first order linear ordinary differential equation for $w(x)$. It can be solved by the use of integrating factors (see page 305).

Example

Given the second order linear differential equation

$$\frac{d^2y}{dx^2} - 2x\frac{dy}{dx} + 2y = 3, \tag{81.5}$$

we recognize that $z(x) = x$ is a solution of the homogeneous equation. Equation (81.4) becomes

$$x\frac{d^2v}{dx^2} + 2(1 - x^2)\frac{dv}{dx} = 3.$$

This equation may be solved by recognizing that it is a linear first order ordinary differential equation in the unknown dv/dx. Hence, integrating factors can be used to find dv/dx. After dv/dx is determined, it can be integrated directly to yield

$$v(x) = \frac{3}{2x} + \frac{A}{x}\int^x \frac{e^{t^2}}{t^2}\,dt + B,$$

where A and B are arbitrary constants. Using the relationship $y(x) = z(x)v(x)$, the general solution of (81.5) is

$$y(x) = \frac{3}{2} + A\int^x \frac{e^{t^2}}{t^2}\,dt + Bx.$$

Notes

[1] In both Rainville and Bedient [3] and in Finizio and Ladas [2] are an account of the general n-th order linear ordinary differential equation. The general result is that:

If $z(x)$ is a solution of the linear homogeneous equation

$$z^{(n)} + p_1(x)z^{(n-1)} + \ldots + p_n(x)z = 0 \qquad (81.6)$$

and if $y(x) = v(x)z(x)$, then the equation

$$y^{(n)} + p_1(x)y^{(n-1)} + \ldots + p_n(x)y = r(x) \qquad (81.7)$$

transforms into

$$v^{(n)} + q_1(x)v^{(n-1)} + \ldots + q_{n-1}v' = r(x).$$

This last equation may be reduced in order by defining $w(x) = v'(x)$.

[2] More generally, if $\{z_1(x), \ldots, z_p(x)\}$ are linearly independent solutions of (81.6), then the substitution

$$y(x) = \begin{vmatrix} z_1 & \cdots & z_p & v \\ z_1' & \cdots & z_p' & v' \\ \vdots & & \vdots & \vdots \\ z_1^{(p)} & \cdots & z_p^{(p)} & v^{(p)} \end{vmatrix}$$

reduces (81.7) to a linear ordinary differential equation of order $n-p$ for $v(x)$.

References

[1] W. E. Boyce and R. C. DiPrima, *Elementary Differential Equations and Boundary Value Problems*, Fourth Edition, John Wiley & Sons, New York, 1986, section 3.4 (pages 127–131).

[2] N. Finizio and G. Ladas, *Ordinary Differential Equations with Modern Applications*, Wadsworth Publishing Company, Belmont, Calif, 1982, pages 108–116.

[3] E. D. Rainville and P. E. Bedient, *Elementary Differential Equations*, The MacMillan Company, New York, 1964, pages 127–129.

82. Riccati Equation

Applicable to Ordinary differential equations of the form $y' = a(x)y^2 + b(x)y + c(x)$.

Yields

A reformulation as a linear second order ordinary differential equation, or a second solution if one solution is already known.

Idea

A change of dependent variable can transform a Riccati equation to a linear second order ordinary differential equation. Also, if one solution to a Riccati equation is known, then the other solution can be written down explicitly.

Procedure 1

Suppose we have the Riccati equation

$$y' = a(x)y^2 + b(x)y + c(x). \tag{82.1}$$

If the dependent variable in (82.1) is changed from $y(x)$ to $w(x)$ by

$$y(x) = \frac{w'(x)}{w(x)} \frac{1}{a(x)}, \tag{82.2}$$

then we obtain the equivalent second order linear ordinary differential equation

$$w'' - \left[\frac{a'(x)}{a(x)} + b(x) \right] w' + a(x)c(x)w = 0. \tag{82.3}$$

It might be easier to solve (82.3) than to solve (82.1) by other means.

Procedure 2

Suppose we have the Riccati equation

$$y' = a(x)y^2 + b(x)y + c(x), \tag{82.4}$$

and suppose further that one solution to this equation is already known to us, say, $y(x) = z(x)$. If $y(x) = z(x) + u(x)$ is substituted in (82.4), then the solvable Bernoulli equation

$$u' = (b + 2az)u + au^2$$

is obtained for $u(x)$. To solve this equation, the new dependent variable $v(x) = 1/u(x)$ should be introduced and then integrating factors used (see pages 194 and 305).

Example 1

Suppose we have the Riccati equation

$$y' = e^x y^2 - y + e^{-x} \tag{82.5}$$

to solve. By identifying $a(x) = e^x$, $b(x) = -1$, and $c(x) = e^{-x}$, the change of variables in (82.2) becomes

$$y(x) = \frac{w'(x)}{w(x)} e^{-x}, \tag{82.6}$$

so that (82.5) becomes $w'' + w = 0$, which could have been obtained directly from (82.3). The solution to this equation is $w(x) = A \sin x + B \cos x$, where A and B are arbitrary constants. Using this solution in (82.6) leads to the general solution of (82.5):

$$y(x) = -e^{-x} \left(\frac{A \cos x - B \sin x}{A \sin x + B \cos x} \right).$$

There should be only one arbitrary constant in the solution to (82.5), since it is a first order ordinary differential equation. In fact, this last equation may be written as

$$y(x) = -e^{-x} \left(\frac{\cos x - C \sin x}{\sin x + C \cos x} \right),$$

where we have defined $C = B/A$ (and assumed $A \neq 0$).

Example 2

Suppose we have the equation

$$y' = y^2 - xy + 1 \tag{82.7}$$

to solve. A solution to (82.7), obtained by inspection, is $y(x) = x$. We utilize this solution in forming

$$y(x) = x + u(x), \tag{82.8}$$

and then (using (82.8) in (82.7)) the equation $u' = u^2 + xu$ is obtained. This Bernoulli equation has the solution $u(x) = \dfrac{e^{x^2/2}}{A - \displaystyle\int_0^x e^{t^2/2} dt}$, where A is an arbitrary constant. Thus, the second solution to (82.7) is

$$y(x) = x + \frac{e^{x^2/2}}{A - \displaystyle\int_0^x e^{t^2/2} dt}.$$

Notes

[1] The transformation in (82.2) is known as the *Riccati transformation*.

[2] The following identity

$$\left[\frac{d}{dx} - q(x)\right]\left[\frac{d}{dx} + q(x)\right]u = u'' + \left(q' - q^2\right)u \qquad (82.9)$$

shows that the differential equation $u'' + p(x)u = 0$ can be factored into the form of (82.9) if $q' - q^2 = p$, which is a Riccati equation. This is specialized case of (82.1)–(82.3).

References

[1] C. M. Bender and S. A. Orszag, *Advanced Mathematical Methods for Scientists and Engineers*, McGraw–Hill, New York, 1978, Section 1.6.

[2] W. E. Boyce and R. C. DiPrima, *Elementary Differential Equations and Boundary Value Problems*, Fourth Edition, John Wiley & Sons, New York, 1986, pages 93–94 and 142–143.

[3] M. E. Goldstein and W. H. Braun, *Advanced Methods for the Solution of Differential Equations*, NASA SP-316, U.S. Government Printing Office, Washington, D.C., 1973, pages 45–46.

[4] E. L. Ince, *Ordinary Differential Equations*, Dover Publications, Inc., New York, 1964, pages 23–25 and 295.

[5] W. T. Reid, *Riccati Differential Equations*, Academic Press, New York, 1972.

[6] G. F. Simmons, *Differential Equations with Applications and Historical Notes*, McGraw–Hill Book Company, New York, 1972, pages 62–63.

83.   Matrix Riccati Equations

Applicable to   Systems of quadratic ordinary differential equations.

Yields

An exact solution.

Idea

There is an exact solution available for matrix Riccati equations. If the given ordinary differential equations can be put in the form of a matrix Riccati equation, then the solution can be found.

Procedure

If $Z(t)$, $A(t)$ and $K(t)$ are all $N \times N$ matrices, then we can use the following theorem:

If $Z(t)$ satisfies the following matrix Riccati equation

$$\frac{d}{dt}Z = ZAZ + KZ + ZK^{\mathrm{T}}, \qquad Z(t = 0) = Z_0, \tag{83.1}$$

then $Z(t)$ is explicitly given by

$$Z(t) = Q(t) \left[Z_0^{-1} - \int_0^t Q^{\mathrm{T}}(s)A(s)Q(s)\,ds \right]^{-1} Q^{\mathrm{T}}(t), \tag{83.2}$$

where $Q(t)$ is defined to be the solution of

$$\frac{d}{dt}Q(t) = K(t)Q(t), \qquad Q(t = 0) = I, \tag{83.3}$$

I is the $N \times N$ identity matrix, and the required matrix inverses are assumed to exist.

If a given system of ordinary differential equations can be placed in the form of (83.1), then the solution can be found from (83.2).

Example

Suppose we wish to solve the following system of equations for $x(t)$ and $y(t)$

$$\frac{dx}{dt} = a(t)(y^2 - x^2) + 2b(t)xy + 2cx,$$
$$\frac{dy}{dt} = b(t)(y^2 - x^2) - 2a(t)xy - 2cy, \tag{83.4}$$

with

$$x(0) = D, \quad y(0) = E.$$

If we form the matrices

$$Z = \begin{pmatrix} x & y \\ y & -x \end{pmatrix}, \quad A = \begin{pmatrix} -a(t) & b(t) \\ b(t) & a(t) \end{pmatrix},$$
$$K = \begin{pmatrix} c & 0 \\ 0 & c \end{pmatrix}, \quad Z_0 = \begin{pmatrix} D & E \\ E & -D \end{pmatrix},$$

then the equations in (83.4) are the same as those in (83.1). The solution for $Q(t)$ from (83.3) is

$$Q(t) = e^{ct}I.$$

Therefore, the solution for Z is

$$Z(t) = e^{2ct} \left[Z_0^{-1} - \int_0^t e^{2cs} A(s) \, ds \right]^{-1}.$$

If we define

$$\alpha(t) = \int_0^t e^{2cs} a(s) \, ds,$$

$$\beta(t) = \int_0^t e^{2cs} b(s) \, ds,$$

then by equating the corresponding entries of (83.2) we can find $\{x(t), y(t)\}$ in terms of $\{\alpha(t), \beta(t)\}$. We have

$$x(t) = e^{2ct} \left[\alpha(t)(E^2 + D^2) + D \right] / \Delta,$$
$$y(t) = e^{2ct} \left[\beta(t)(E^2 + D^2) + D \right] / \Delta,$$

where $\Delta = \Delta(x)$ is defined by

$$\Delta(x) = \left[\beta^2(t) + \alpha^2(t) \right] \left[E^2 + D^2 \right] - 2\beta(t)E + 2\alpha(t)D + 1.$$

Notes

[1] Matrix Riccati equations arise naturally in a number of physical settings. For example, the gains in a Kalman–Bucy filter satisfy a matrix Riccati equation (see Schuss [11], page 261). Also, the deflection of a beam can be described by such equations (see Distéfano [2]). They also appear quite often in the context of control theory (see Jodar and Abou-Kandil [3]) and invariant embedding solutions (see page 669).

[2] Kerner [5] shows that nonlinear differential systems of arbitrary order

$$\dot{\zeta}_i = X_i(\zeta_1, \zeta_2, \ldots, \zeta_k, t), \quad \text{for } i = 1, 2, \ldots, k,$$

may often be reduced to *Riccati systems*

$$\dot{x}_i = A_i + B_{i\alpha} x_\alpha + C_{i\alpha\beta} x_\alpha x_\beta,$$
$$\text{for } i = 1, 2, \ldots, n, \quad n \geq k, \quad \text{and } A, B, C \text{ constant,}$$

and then to *elemental Riccati systems*

$$\dot{z}_i = E_{i\alpha\beta} z_\alpha z_\beta, \quad \text{for } i = 1, 2, \ldots, p, \quad p(n) > n,$$

where $E_{i\alpha\beta}$ equals 0 or 1. His examples include ordinary differential equation systems that contain exponential functions and elliptic functions.

References

[1] S. Bittanti, A. J. Laub, and J. C. Willems (eds.), *The Riccati Equation*, Springer–Verlag, New York, 1991.

[2] N. Distéfano, *Nonlinear Processes in Engineering*, Academic Press, New York, 1974, pages 58–59.

[3] L. Jodar and H. Abou-Kandil, "A Resolution Method for Riccati Differential Systems Coupled in Their Quadratic Terms," *SIAM J. Appl. Math.*, **19**, No. 6, November 1988, pages 1425–1430.

[4] R. A. Jones, "Existence Theorems for the Matrix Riccati Equation $W' + WP(t)W + Q(t) = 0$," *Int. J. Math. & Math. Sci.*, **1**, 1978, pages 13–19.

[5] E. H. Kerner, "Universal Formats for Nonlinear Ordinary Differential Equations," *J. Math. Physics*, **22**, No. 7, July 1981, pages 1366–1371.

[6] K. N. Murty, K. R. Prasad, and M. A. S. Srinivas, "Upper and Lower Bounds for the Solution of the General Matrix Riccati Differential Equations," *J. Math. Anal. Appl.*, **147**, No. 1, 1990, pages 12–21.

[7] D. W. Rand and P. Winternitz, "Nonlinear Superposition Principles: A New Numerical Method for Solving Matrix Riccati Equations," *Comput. Physics Comm.*, **33**, 1984, pages 305–328.

[8] M. Razzaghi, "A Computational Solution for the Matrix Riccati Equation Using Laplace Transforms," *Int. J. Comp. Math.*, **11**, 1982, pages 297–304.

[9] W. T. Reid, *Riccati Differential Equations*, Academic Press, New York, 1972.

[10] W. T. Reid, "Solutions of a Riccati Matrix Differential Equation as Functions of Initial Values," *J. Math. Mech.*, **8**, 1959, pages 221–230.

[11] Z. Schuss, *Theory and Applications of Stochastic Differential Equations*, John Wiley & Sons, New York, 1980.

[12] R. M. Wilcox and L. P. Harten, "MACSYMA-Generated Closed-Form Solutions to Some Matrix Riccati Equations," *Appl. Math. and Comp.*, **14**, 1984, pages 149–166.

84.   Scale Invariant Equations

Applicable to   Ordinary differential equations of a certain form.

Yields

An equidimensional-in-x ordinary differential equation of the same order (which can then be reduced to an ordinary differential equation of lower order).

Idea

A scale invariant equation is one in which the equation is unchanged when x and y are scaled in a certain way. When an equation is scale invariant, we can convert the equation into an equidimensional-in-x ordinary differential equation of the same order by a change of the dependent variable. This equidimensional-in-x ordinary differential equation can then be changed into an autonomous equation of lower order.

Procedure

A scale invariant equation is one that is left invariant under the transformation $\{x \rightarrow ax, y \rightarrow a^p y\}$, where a and p are constants. That is, if the original equation is an equation for $y(x)$ and the x variable is replaced by the variable ax' and the y variable is replaced by the variable $a^p y'$, then the new equation (in terms of y' and x') will be identical to the original equation (which is in terms of y and x). The way to determine the value of p is to change variables and then see what value of p leaves the equation unchanged.

A scale invariant equation can be converted to an equidimensional-in-x equation by the substitution for y

$$y(x) = x^p u(x). \tag{84.1}$$

By the techniques on page 230, this equidimensional-in-x equation may then be made autonomous, and then (after another transformation) the order of the equation can be reduced.

Example

Suppose we have the nonlinear second order ordinary differential equation

$$x^2 \frac{d^2 y}{dx^2} + 3x \frac{dy}{dx} = \frac{1}{y^3 x^4}. \tag{84.2}$$

To determine if this equation is scale invariant, and if it is, what the value of p is, we substitute ax' for x and $a^p y'$ for y to obtain

$$(ax')^2 \frac{d^2(a^p y')}{d(ax')^2} + 3(ax') \frac{d(a^p y')}{d(ax')} = \frac{1}{(a^p y')^3 (ax')^4}$$

or

$$a^p x'^2 \frac{d^2 y'}{dx'^2} + 3a^p x' \frac{dy'}{dx'} = a^{(-3p-4)} \frac{1}{y'^3 x'^4}. \tag{84.3}$$

Hence, if we choose p so that $p = -3p - 4$, then the form of (84.3) will be the same as the form of (84.). So the equation is scale invariant, with the value $p = -1$. To make this equation equidimensional-in-x, we change variables by (84.1): $y(x) = u(x)/x$. Using this change of variables in (84.2) produces

$$x^2 \frac{d^2u}{dx^2} + x\frac{du}{dx} + u = \frac{1}{u^3}. \tag{84.4}$$

Equation (84.4) is equidimensional-in-x, so we use the substitution $x = e^t$ (see page 230) for

$$\frac{d^2u}{dt^2} + u = \frac{1}{u^3}. \tag{84.5}$$

Equation (84.5) is autonomous, so we change the independent variable by $v(u) = u'(t)$ (see page 190) for

$$v\frac{dv}{du} + u = \frac{1}{u^3}. \tag{84.6}$$

The solution of (84.6) can be found by separating variables (see page 419)

$$v(u) = \pm\sqrt{A - u^2 - \frac{1}{u^2}},$$

where A is an arbitrary constant. To find $u(t)$, we must now solve

$$\frac{du}{dt} = v(u) = \pm\sqrt{A - u^2 - \frac{1}{u^2}}. \tag{84.7}$$

Equation (84.7) is a separable equation whose solution is

$$u(t) = \pm\sqrt{\cosh B + \sinh B \sin(2t + C)},$$

where B and C are arbitrary constants. The last step is to recall that $y(x) = u(x)/x$ and that $x = e^t$. The final solution is therefore

$$y(x) = \pm\frac{1}{x}\sqrt{\cosh B + \sinh B \sin(2\log x + C)}.$$

Notes

[1] This method is derivable from Lie group methods (see page 314). The infinitesimal operator in this case is given by $U = x\frac{\partial}{\partial x} + py\frac{\partial}{\partial y}$.

[2] A special case of this method (when $p = 1$) is the method for homogeneous equations (see page 276).

[3] Euler equations (see page 235) are scale invariant equations for any value of the parameter p.

[4] Scale invariant equations are also called *isobaric equations*.

[5] In Rosen's paper [3] a change of variable is proposed, different from the one presented above, that often allows parametric solutions to be obtained.

References

[1] C. M. Bender and S. A. Orszag, *Advanced Mathematical Methods for Scientists and Engineers*, McGraw–Hill, New York, 1978, pages 25–26.

[2] M. E. Goldstein and W. H. Braun, *Advanced Methods for the Solution of Differential Equations*, NASA SP-316, U.S. Government Printing Office, Washington, D.C., 1973, pages 81–84.

[3] G. Rosen, "Alternative Integration Procedure for Scale-Invariant Ordinary Differential Equations," *Int. J. Math. & Math. Sci.*, **2**, 1979, pages 143–145.

85.   Separable Equations

Applicable to   First order ordinary differential equations.

Yields

An exact solution, often implicit.

Idea

First order ordinary differential equations can be solved directly if the forcing term factors into a term involving only the independent variable and a term involving only the dependent variable.

Procedure

Given an equation of the form

$$\frac{dy}{dx} = f(y)g(x), \tag{85.1}$$

both sides can be formally multiplied by $dx/f(y)$ and then integrated to obtain

$$\int \frac{dy}{f(y)} = \int g(x)\,dx. \tag{85.2}$$

The evaluation of (85.2) only requires that two integrals be evaluated. An arbitrary constant of integration must be included to obtain the most general solution of (85.1).

Example

Suppose we have the equation

$$\frac{dy}{dx} = \frac{9x^8 + 1}{y^2 + 1} \qquad (85.3)$$

to solve. Multiplying both sides of (85.3) by $(y^2+1)\,dx$ and then integrating results in

$$\int (y^2 + 1)\,dy = \int (9x^8 + 1)\,dx.$$

Evaluating the integrals yields

$$\frac{y^3}{3} + y = x^9 + x + C,$$

where C is an arbitrary constant.

Notes

[1] The solution obtained by this method will generally be implicit.

[2] The formal procedure of multiplying (85.1) by $dx/f(y)$ can be rigorously shown to give the correct answer.

References

[1] W. E. Boyce and R. C. DiPrima, *Elementary Differential Equations and Boundary Value Problems*, Fourth Edition, John Wiley & Sons, New York, 1986, Section 2.4 (pages 37–42).

[2] E. L. Ince, *Ordinary Differential Equations*, Dover Publications, Inc., New York, 1964, pages 17–18.

[3] G. F. Simmons, *Differential Equations with Applications and Historical Notes*, McGraw–Hill Book Company, New York, 1972, pages 35–36.

86. Series Solution[*]

Applicable to   Homogeneous linear ordinary differential equations. Most frequently second order differential equations.

Yields

An infinite series expansion of the two independent solutions.

Idea

If an infinite series is substituted into a linear equation, the different coefficients may be matched to obtain recurrences for the coefficients of the series. Solving these recurrences results in an explicit solution.

Procedure

Given a homogeneous linear second order ordinary differential equation in the form

$$y'' + P(x)y' + Q(x)y = 0, \tag{86.1}$$

we search for a series solution around the point $x = 0$. Clearly, an expansion about any other point, x_0, could be determined by changing the independent variable to $t = x - x_0$, and then analyzing the resulting equation near $t = 0$.

If $x = 0$ is an ordinary point of (86.1) (the definitions of ordinary points and singular points are given on page 11) then we may assume that $P(x)$ and $Q(x)$ have the known Taylor expansions

$$P(x) = \sum_{n=0}^{\infty} P_n x^n, \qquad Q(x) = \sum_{n=0}^{\infty} Q_n x^n, \tag{86.2}$$

in the region $|x| < \rho$, where ρ represents the minimum of the radii of convergence of the two series in (86.2). In this case, equation (86.1) will have two linearly independent solutions of the form

$$y(x) = \sum_{n=0}^{\infty} a_n x^n. \tag{86.3}$$

Alternately, if $x = 0$ is a regular singular point of (86.1) then we may assume that $P(x)$ and $Q(x)$ have the known expansions

$$P(x) = \sum_{n=-1}^{\infty} P_n x^n, \qquad Q(x) = \sum_{n=-2}^{\infty} Q_n x^n, \tag{86.4}$$

in the region $|x| < \rho$. After determining the expansions in (86.4), we need to determine the roots to the *indicial equation*

$$\alpha^2 + \alpha(P_{-1} - 1) + Q_{-2} = 0, \tag{86.5}$$

which is obtained by utilizing $y = x^\alpha$ in (86.1), along with the expansions in (86.4), and then determining the coefficient of the lowest order term. The two roots of this equation are called *the exponents of the singularity*.

There are now several cases, depending on the values of the exponents of the singularity:

[1] If $\alpha_1 \neq \alpha_2$ and $\alpha_1 - \alpha_2$ is not equal to an integer, then (86.1) will have two linearly independent solutions in the forms

$$y_1(x) = |x|^{\alpha_1} \left(1 + \sum_{n=1}^{\infty} b_n x^n\right),$$

$$y_2(x) = |x|^{\alpha_2} \left(1 + \sum_{n=1}^{\infty} c_n x^n\right). \tag{86.6}$$

[2] If $\alpha_1 = \alpha_2$, then (calling $\alpha = \alpha_1$) (86.1) will have two linearly independent solutions in the forms

$$y_1(x) = |x|^{\alpha} \left(1 + \sum_{n=1}^{\infty} d_n x^n\right),$$

$$y_2(x) = y_1(x) \log |x| + |x|^{\alpha} \sum_{n=0}^{\infty} e_n x^n. \tag{86.7}$$

[3] If $\alpha_1 = \alpha_2 + M$, where M is an integer greater than 0, then (86.1) will have two linearly independent solutions in the forms

$$y_1(x) = |x|^{\alpha_1} \left(1 + \sum_{n=1}^{\infty} f_n x^n\right),$$

$$y_2(x) = h y_1(x) \log |x| + |x|^{\alpha_2} \sum_{n=0}^{\infty} g_n x^n, \tag{86.8}$$

where the parameter h may be equal to zero.

The procedure in each of the four cases is the same: Substitute the given forms ((86.3), (86.6), (86.7), or (86.8)) into the original equation (86.1) and equate the coefficients of the x^j and $x^j \log x$ terms for different values of j. This will yield recurrence relations for the unknown coefficients. Solving these recurrence relations will determine the solution.

In the case of an ordinary point, there will be two unknown coefficients that parameterize the series solutions in (86.3). These two coefficients will generate the two linearly independent solutions of (86.1).

Example 1

Given the equation

$$y'' + y = 0, \tag{86.9}$$

we easily see that $x = 0$ is an ordinary point. Using (86.3) in (86.9) we find

$$(2a_2 + a_0) + (6a_3 + a_1)x + (12a_4 + a_2)x^2 + \dots$$
$$+ [(n+1)(n+2)a_{n+2} + a_n] x^n + \dots = 0.$$

Hence, we must have $a_{n+2} = -\dfrac{a_n}{(n+1)(n+2)}$. Iterating this relation we find

$$a_{2m} = (-1)^m \frac{1}{(2m)!}, \qquad a_{2m+1} = (-1)^m \frac{1}{(2m+1)!}. \tag{86.10}$$

Hence, using (86.10) in (86.3),

$$y(x) = a_0 \left(1 - \frac{x^2}{2!} + \frac{x^4}{4!} - \dots \right) + a_1 \left(x - \frac{x^3}{3!} + \frac{x^5}{5!} - \dots \right). \tag{86.11}$$

Of course, the exact solution to (86.9) is $y(x) = a_0 \cos x + a_1 \sin x$, which is what (86.11) has reproduced.

Example 2

Given the equation

$$y'' + \frac{1 + 2x}{2x} y' - \frac{1}{2x^2} y = 0, \tag{86.12}$$

we easily see that $x = 0$ is a regular singular point. In this case we have (see (86.4)) $P_{-1} = \frac{1}{2}$, $Q_{-2} = -\frac{1}{2}$. Therefore, the indicial equation (from (86.5)) becomes

$$\alpha^2 - \tfrac{1}{2}\alpha - \tfrac{1}{2} = (\alpha - 1)\left(\alpha - \tfrac{1}{2}\right) = 0.$$

Since the roots $\alpha_1 = 1$, $\alpha_2 = -\frac{1}{2}$ are unequal and do not differ by an integer, then we have case [1]. Using (86.6) in (86.12), for $\alpha_1 = 1$, and equating powers of x we readily find that

$$\sum_{n\geq 1}(n+1)(n)b_n x^{n-1} + \frac{1 + 2x}{2x}\left(1 + \sum_{n\geq 1} b_n x^n\right) - \frac{1}{2x^2}\left(x + \sum_{n\geq 1} b_n x^{n+1}\right) = 0.$$

Equating the coefficients for different powers of x we find that

$$b_1 = -\tfrac{2}{5}, \qquad b_{j+1} = -\frac{2(j+1)}{2j^2 + 7j + 5}b_j.$$

Hence, one solution of (86.12) is of the form

$$y_1(x) = x\left(1 - \tfrac{2}{5}x + \tfrac{4}{35}x^2 - \ldots\right).$$

The other solution can be obtained by using $\alpha_2 = -\tfrac{1}{2}$ in (86.6) and (86.12). For this solution we find

$$y_2(x) = x^{-1/2}\left(1 - x + \tfrac{1}{2}x^2 - \ldots\right).$$

The general solution of (86.12) is a linear combination of $y_1(x)$ and $y_2(x)$.

Notes

[1] This method is similar to the method of Taylor series (see page 548) but is different in that

    (A) it allows for logarithmic terms to be present, as well as fractional powers,

    (B) the recurrence relations are computed just once,

    (C) the method only applies to linear ordinary differential equations.

[2] The series solution found in (86.3), (86.6), (86.7) and (86.8) will always converge in the region $|x| < \rho$.

[3] The series in (86.6) are sometimes called Frobenius series. For regular singular points, this method is sometimes called the *method of Frobenius*.

[4] Della Dora and Tournier [4] describe a computer package that will symbolically produce the series for singular points.

[5] When the given linear ordinary differential equation has an irregular singular point, then series solutions are difficult to obtain and they may be slowly convergent. See Goldstein and Braun [6] or Bender and Orszag [1] for details. Often the WKB method (see page 558) is used to approximate the solution near an irregular singular point.

[6] Understanding the nature of the singular points in an ordinary differential equation leads to an understanding of the types of boundary conditions to be expected for that equation. For example, the ordinary differential equation $xy' = 1$ has the solution $y = C + \log x$, where C is an arbitrary constant. Only if $y(x)$ is specified at some point other than $x = 0$ will it be possible to determine the constant C. The point $x = 0$ is a regular singular point of this equation.

[7] This method extends easily to the general n-th order homogeneous linear ordinary differential equation at a regular singular point x_0. If the differential equation is given by

$$y^{(n)} + \frac{q_{n-1}(x)}{(x-x_0)} y^{(n-1)} + \frac{q_{n-2}(x)}{(x-x_0)^2} y^{(n-2)} + \cdots + \frac{q_0(x)}{(x-x_0)^n} y = 0,$$

where $\{q_0(x), \ldots, q_{n-1}(x)\}$ are analytic at x_0, then the indicial equation for α is given by

$$(\alpha)_n + q_{n-1}(x_0)(\alpha)_{n-1} + q_{n-2}(x_0)(\alpha)_{n-2} + \cdots + q_0(x_0)(\alpha)_0 = 0. \quad (86.13)$$

where $(\alpha)_n := (\alpha)(\alpha - 1) \cdots (\alpha - n + 1)$ and $(\alpha)_0 := 1$. If the n roots of (86.13) do not differ by integers, then there are n linearly independent solutions of the form of (86.6). Otherwise, the forms in (86.7) and (86.8) must be generalized. See Bender and Orszag [1] for details.

[8] Series solutions can also be used to find the solutions of partial differential equations (see Collatz [3] or Garabedian [5]), or to approximate the solution of nonlinear differential equations (see, for example, Leavitt [8]).

[9] The computer language MACSYMA has a function called SERIES that will compute the series expansion of a second order ordinary differential equation.

In the following terminal session with MACSYMA, Airy's equation $(y_{xx} + xy = 0)$ was input and the power series representation of the solution was obtained.

```
(c1) DERIVABBREV:TRUE$

(c2) LOAD(SERIES)$

(c3) DEPENDS(Y,X)$

(c4) DIFF(Y,X,2) + X*Y = 0;

(d4)                          y     + x y = 0
                               x x

(c5) NICEINDICES( SERIES(D4,Y,X) );

DIAGNOSIS: ORDINARY POINT
```

$$\text{(d5)} \quad y = \%k2\, x \sum_{i=0}^{\inf} \frac{(-1)^i\, x^{3i}}{\dfrac{4^i}{3}\, fff\,(-,\,i)\,9^i\, i!} + \%k1 \sum_{i=0}^{\inf} \frac{(-1)^i\, x^{3i}}{\dfrac{2^i}{3}\, fff(-,\,i)\,9^i\, i!}$$

The function $\texttt{fff}\,(\texttt{n},\texttt{i})$ is defined (in the MACSYMA manual) by $\texttt{fff}\,(\texttt{n},\texttt{i}) = (n)_i = n(n-1) \cdots (n-i+1)$. Note that $\%k1$ and $\%k1$ are arbitrary constants that appear in the general solution.

[10] When all of the singular points in an ordinary differential equation are regular, then the equation is said to be of Fuchs' type. A second order Fuchsian equation with 3 regular singular points can be transformed by a linear fractional transformation into the Riemann differential equation:

$$y'' + \left(\frac{A_1}{x} + \frac{A_2}{x-1}\right) + \left(\frac{A_3}{x^2} + \frac{A_4}{(x-1)^2} + \frac{A_5}{x(x-1)}\right) = 0$$

where the $\{A_i\}$ are constants. This equation can then be changed to a hypergeometric equation by a change of dependent variable.

[11] Morse and Feshback [9] discuss the canonical second order equations that have: 1, 2, and 3 regular singular points, 1 regular and 1 irregular singular points, 1 and 2 irregular singular points,

References

[1] C. M. Bender and S. A. Orszag, *Advanced Mathematical Methods for Scientists and Engineers*, McGraw–Hill, New York, 1978, Chapter 3 (pages 61–88).

[2] W. E. Boyce and R. C. DiPrima, *Elementary Differential Equations and Boundary Value Problems*, Fourth Edition, John Wiley & Sons, New York, 1986, Chapter 4 (pages 187–256).

[3] L. Collatz, *The Numerical Treatment of Differential Equations*, Springer–Verlag, New York, 1966, pages 222–226 and 419–422.

[4] J. Della Dora and E. Tournier, "Formal Solutions of Differential Equations in the Neighborhood of Singular Points," in P. S. Wang (ed.), *SYMSAC 81: Proceedings of the 1981 ACM Symposium on Symbolic and Algebraic Computation*, pages 25–29.

[5] P. R. Garabedian, *Partial Differential Equations*, Wiley, New York, 1964, Chapter 1 (pages 1–17).

[6] M. E. Goldstein and W. H. Braun, *Advanced Methods for the Solution of Differential Equations*, NASA SP-316, U.S. Government Printing Office, Washington, D.C., 1973, Chapter 9 (pages 251–279).

[7] E. L. Ince, *Ordinary Differential Equations*, Dover Publications, Inc., New York, 1964, Chapter 16 (pages 396–437).

[8] J. A. Leavitt, "A Power Series Method for Solving Nonlinear Boundary Value Problems," *Quart. Appl. Math.*, **27**, No. 1, 1969, pages 67–77.

[9] P. M. Morse and H. Feshback, *Methods of Theoretical Physics*, McGraw–Hill Book Company, New York, 1953, pages 667–674.

87.   Equations Solvable for x

Applicable to   First order ordinary differential equations that are of the first degree in x; i.e., equations of the form $x = f(y, y')$.

Yields

An exact solution, sometimes implicit.

Idea

Equations of the form $x = f(y, y')$ can be solved by finding a second equation involving x, y, and y', and then eliminating y' between the two equations.

Procedure

Given an equation of the form

$$x = f\left(y, \frac{dy}{dx}\right), \tag{87.1}$$

define, as usual, $p = \dfrac{dy}{dx}$, so that (87.1) may be written

$$x = f(y, p). \tag{87.2}$$

Now differentiate this with respect to y to obtain

$$\frac{dx}{dy} = \phi\left(y, p, \frac{dp}{dy}\right),$$

or

$$\frac{1}{p} = \phi\left(y, p, \frac{dp}{dy}\right), \tag{87.3}$$

for some function ϕ. Now the ordinary differential equation in (87.3), for $p = p(y)$, may sometimes be integrated to obtain

$$F(y, p; C) = 0, \tag{87.4}$$

for some function F, where C is an arbitrary constant. By elimination, the p may sometimes be removed from equations (87.2) and (87.4) to determine $y = y(x; C)$. In cases where it cannot be removed, we obtain a parametric solution.

Example

Suppose we wish to solve the nonlinear ordinary differential equation

$$y = 2x\frac{dy}{dx} + y\left(\frac{dy}{dx}\right)^2,$$ (87.5)

for $y(x)$. Solving (87.5) for x results in

$$x = -\frac{py}{2} + \frac{y}{2p},$$ (87.6)

where we have used $y' = p$. Differentiating (87.6) with respect to y, and factoring results in

$$\left(1 + \frac{1}{p^2}\right)\left(p + y\frac{dp}{dy}\right) = 0.$$

This equation may be integrated to yield

$$py = C.$$ (87.7)

Solving (87.7) for p, and using this is (87.5) results in the explicit solution

$$2xC - y^2 + C^2 = 0.$$

References

[1] H. T. H. Piaggio, *An Elementary Treatise on Differential Equations and Their Applications*, G. Bell & Sons, Ltd, London, 1926, page 64.

88.   Equations Solvable for y

Applicable to   First order ordinary differential equations that can be explicitly solved for y; i.e., equations of the form $y = f(x, y')$.

Yields

An exact solution, sometimes implicit.

Idea

Equations of the form $y = f(x, y')$ can be solved by finding a second equation involving x, y, and y', and then eliminating the y' term between the two equations.

Procedure

Given an equation of the form

$$y = f\left(x, \frac{dy}{dx}\right),\tag{88.1}$$

define, as usual, $p = \dfrac{dy}{dx}$, so that (88.1) may be written

$$y = f(x, p).\tag{88.22}$$

Now differentiate with respect to x to obtain

$$\frac{dy}{dx} = \phi\left(x, p, \frac{dp}{dx}\right),$$

or

$$p = \phi\left(x, p, \frac{dp}{dx}\right),\tag{88.3}$$

for some function ϕ. Now the ordinary differential equation in (88.3), for $p = p(x)$, may sometimes be integrated to obtain

$$F(x, p; C) = 0,\tag{88.4}$$

for some function F, where C is an arbitrary constant. By elimination, the p may sometimes be removed from equations (88.2) and (88.4) to determine $y = y(x; C)$. In cases where it cannot be removed, we obtain a parametric solution.

Example

Suppose we wish to solve the nonlinear ordinary differential equation

$$x = y\frac{dy}{dx} - x\left(\frac{dy}{dx}\right)^2 = yp - xp^2,\tag{88.5}$$

for $y(x)$. Differentiating (88.5) with respect to x, and using $p = y'$, results in

$$\frac{dp}{dx} = \frac{px}{p^2 - 1}.$$

This last equation may be integrated to determine

$$\tfrac{1}{2}x^2 = C + \tfrac{1}{2}p^2 - \log p,\tag{88.6}$$

where C is an arbitrary constant. Together, equations (88.5) and (88.6) constitute a parametric representation of the solution to (88.5). In this representation, p is treated as a running variable.

Notes

[1] The technique used for Lagrange's equation is a specialization of the present technique applied to a restricted class of equations (see page 311).

References

[1] H. T. H. Piaggio, *An Elementary Treatise on Differential Equations and Their Applications*, G. Bell & Sons, Ltd, London, 1926, page 63.

89.   Superposition*

Applicable to   Linear differential equations.

Yields

A set of linear differential equations with "easier" initial conditions or boundary conditions. The sum of the solutions to these new equations will produce the solution to the original equation.

Idea

By use of superposition, the solution to an inhomogeneous linear differential equation may be determined in terms of simpler systems.

Procedure

Given a linear differential equation with a forcing term, inhomogeneous initial conditions, or inhomogeneous boundary conditions, construct a set of equations with each equation having more homogeneous parts than the original system. Solve each of these parts separately, and then combine them for the final solution.

Example

Given the linear second order ordinary differential equation

$$L[y] = y'' + a(x)y' + b(x) = f(x), \tag{89.1}$$

we choose $y_1(x)$ and $y_2(x)$ to be any linearly independent solutions of $L[y_i] = 0$. If C_1 and C_2 are any constants then

$$y_c(x) = C_1 y_1(x) + C_2 y_2(x)$$

is called the *homogeneous solution* or the *complementary solution* of (89.1). We also define $y_p(x)$ to be any solution to $L[y_p] = f(x)$. The function $y_p(x)$ is called a *particular solution*.

Any solution of (89.1) (there will be different solutions, depending on what initial conditions or boundary conditions are chosen with (89.1)) may be written in the form

$$y(x) = y_c(x) + y_p(x),$$

for some choice of C_1 and C_2.

Notes

[1] In fluid dynamics, the influence of an obstacle in a flow can be simulated by a continuous superposition of sources. See, for instance, Homentcovschi [2].

[2] There also exist superposition principles for *nonlinear* equations. These are relations that allow new solutions, with arbitrary constants in them, to be calculated from other solutions. For instance, if y_1, y_2, and y_3 are solutions of the Riccati equation (see page 332), then y will also be solution if it satisfies

$$\frac{y - y_2}{y - y_1} = C \frac{y_3 - y_2}{y_3 - y_1},$$

where C is an arbitrary constant. See Ince [4] for details.

[3] More generally, Lie and Scheffers [6] showed that a necessary and sufficient condition for a system of n first order ordinary differential equations to have a (nonlinear) superposition formula is that the system of equations be of the

form $\dfrac{dy}{dt} = \displaystyle\sum_{k=1}^{r} f_k(t)\zeta_k(\mathbf{y})$ and that the vector fields $X_k := \displaystyle\sum_{m=1}^{n} \zeta_k^m(\mathbf{y})\dfrac{\partial}{\partial y_m}$

generate a finite dimensional Lie algebra.

Given a set of vector fields, $Z = \{X_1, \ldots, X_r\}$, and a Lie bracket $[\,,\,]$, a Lie algebra is generated by adding to Z all elements of the form $[X_i, X_j]$. This process is repeated with the new, potentially larger, set Z until no new elements enter Z. The resulting Z is closed under the $[\,,\,]$ operation and is a Lie algebra; it may contain a finite or an infinite number of elements.

References

[1] W. E. Boyce and R. C. DiPrima, *Elementary Differential Equations and Boundary Value Problems*, Fourth Edition, John Wiley & Sons, New York, 1986, Section 7.4 (pages 352–357).

[2] D. Homentcovschi, "Uniform Asymptotic Solutions of the Potential Field Around a Thin Oblate Body of Revolution," *SIAM J. Appl. Math.*, **42**, No. 1, February 1982, pages 44–65.

[3] J. Harnad, P. Winternitz, and R. L. Anderson, "Superposition Principles for Matrix Riccati Equations," *J. Math. Physics*, **24**, No. 5, May 1983, pages 1062–1072.

[4] E. L. Ince, *Ordinary Differential Equations*, Dover Publications, Inc., New York, 1964, pages 23–25.

[5] A. S. Jones, "Quasi-Additive Solutions of Nonlinear Differential Equations," *J. Austral. Math. Soc. (Series A)*, **42**, 1987, pages 92–116.

[6] S. Lie and G. Scheffers, *Vorlesungen über Continuierlichen Gruppen mit geometrischen und anderen Anwendungen*, Teubner, Leipzig, 1893.

[7] M. A. del Olmo, M. A. Rodriguez, and P. Winternitz, "Superposition Formulas for Rectangular Matrix Riccati Equations," *J. Math. Physics*, **28**, No. 3, March 1987, pages 530–535.

[8] S. Shnider and P. Winternitz, "Classification of Systems of Nonlinear Ordinary Differential Equations with Superposition Principles," *J. Math. Physics*, **25**, No. 11, November 1984, pages 3155–3165.

90. Method of Undetermined Coefficients*

Applicable to   Linear or nonlinear differential equations, a single equation or a system.

Yields

An exact homogeneous solution, an exact particular solution, or both.

Idea

If the general form of the solution of a given differential equation is known (or can be guessed), it can be substituted into the defining equations with unknown coefficients. Then the unknown coefficients can be determined.

Procedure

Very often we can guess the form of a solution to a differential equation. Or, we could just guess blindly. By having several unknown parameters in the assumed form of the solution, the solution should be able to fit the defining equation(s). By forcing the guessed solution to satisfy the equation, we may be able to determine these unknown quantities.

Example 1

Suppose we have the equation

$$y'' - \frac{2}{x^2}y = 7x^4 + 3x^3. \tag{90.1}$$

If we suspect that this equation has a power type solution for $y(x)$, we might search for a solution in the form

$$y(x) = ax^b, \tag{90.2}$$

90. Method of Undetermined Coefficients*

where a and b are unknowns to be determined. In this example, we p... that a and b are constants (in more complicated problems, the unkn... can be functions to be determined). We try to determine a and b... substituting our guess in the original equation for $y(x)$. Using (90.2) in (90.1) yields

$$ax^{b-2}(b^2 - b - 2) = 7x^4 + 3x^3. \tag{90.3}$$

This equation must be satisfied for all values of x. There is no single set of values for a and b for which this will be true. However, note the following:

(A) If $b = 6, a = 1/4$, then the left-hand side of (90.3) becomes $7x^4$.
(B) If $b = 5, a = 1/6$, then the left-hand side of (90.3) becomes $3x^3$.
(C) If $b = -1$, then the left-hand side of (90.3) becomes zero.
(D) If $b = 2$, then the left-hand side of (90.3) becomes zero.

The first two facts enable us to write the particular solution of (90.1) as

$$y_p(x) = \tfrac{1}{4}x^6 + \tfrac{1}{6}x^5.$$

The second two facts tell us that $y(x) = x^2$ and $y(x) = 1/x$ are both solutions to the homogeneous equation

$$y'' - \frac{2}{x^2}y = 0.$$

Therefore, the complete solution to (90.1) is

$$y(x) = \frac{1}{4}x^6 + \frac{1}{6}x^5 + Ax^2 + \frac{B}{x},$$

where A and B are arbitrary constants.

Example 2
Suppose we have the partial differential equation

$$\begin{aligned} u_{xx} &= u_t, \\ u(0,t) &= 0, \\ u(1,t) &= 0, \\ u(x,0) &= \sin \pi x. \end{aligned} \tag{90.4}$$

An appropriate guess for the form of the solution would be

$$u(x,t) = f(t)\sin \pi x,$$

for some unknown function $f(t)$. Using this guess in (90.4) results in the system

$$f' + \pi^2 f = 0, \qquad f(0) = 1.$$

Hence, $f(t) = e^{-\pi^2 t}$.

Example 3

A guess for the form of the solution of the equation

$$u_t = (uu_x)_x \tag{90.5}$$

might be

$$u(x,t) = f(t) + g(t)x^p, \tag{90.6}$$

for some functions $f(t)$ and $g(t)$, and some constant p. Using (90.6) in (90.5) leads to the choice $p = 2$. With this value, $f(t)$ and $g(t)$ can be determined so that

$$u(x,t) = (C - 6t)^{-1} x^2 + (C - 6t)^{-1/6}.$$

See Ames [1] for more details.

Notes

[1] In Table 3.1 of Boyce and DiPrima [2] is a description of general solution forms for a forced linear second order constant coefficient differential equation when the forcing function is a polynomial, a trigonometric function, an exponential function, or a combination of these terms. By utilizing this general form with unknown coefficients, a solution may be obtained.

[2] The reason that we suspected (90.1) to have a power type solution is that the homogeneous part of (90.1) is a Euler equation.

References

[1] W. F. Ames, "Ad Hoc Exact Techniques for Nonlinear Partial Differential Equations," in W. F. Ames (ed.), *Nonlinear Partial Differential Equations in Engineering*, Academic Press, New York, 1967.

[2] W. E. Boyce and R. C. DiPrima, *Elementary Differential Equations and Boundary Value Problems*, Fourth Edition, John Wiley & Sons, New York, 1986, Section 3.6.1 (pages 146–155).

[3] E. D. Rainville and P. E. Bedient, *Elementary Differential Equations*, The MacMillan Company, New York, 1964, pages 115–118.

[4] G. F. Simmons, *Differential Equations with Applications and Historical Notes*, McGraw–Hill Book Company, New York, 1972, pages 87–90.

91.   Variation of Parameters

Applicable to   Forced, linear ordinary differential equations.

Yields

An integral representation of the particular solution.

Idea

If we know the solution to the homogeneous equation, we can write down an expression for the particular solution.

Procedure

We illustrate the general technique for the linear ordinary differential equation of second order. Suppose we have the equation

$$y'' + P(x)y' + Q(x)y = R(x), \tag{91.1}$$

and suppose that we know that $\{y_1(x), y_2(x)\}$ are two linearly independent solutions to the homogeneous (unforced) equation

$$y'' + P(x)y' + Q(x)y = 0. \tag{91.2}$$

That is, every solution of (91.2) is a linear combination of $y_1(x)$ and $y_2(x)$. We look for the particular solution of (91.1) in the form

$$y(x) = v_1(x)y_1(x) + v_2(x)y_2(x), \tag{91.3}$$

where $v_1(x)$ and $v_2(x)$ are to be determined. Differentiating (91.3) with respect to x yields

$$y' = (v_1y_1' + v_2y_2') + (v_1'y_1 + v_2'y_2). \tag{91.4}$$

We choose the second term in (91.4) to vanish, so that

$$(v_1'y_1 + v_2'y_2) = 0. \tag{91.5}$$

If we now differentiate (91.4) with respect to x, and use this expression (with (91.3), (91.4) and (91.5)) in (91.2) then we obtain

$$v_1'y_1' + v_2'y_2' = R(x). \tag{91.6}$$

Equations (91.5) and (91.6) constitute two algebraic equations for the two unknowns $v_1'(x)$ and $v_2'(x)$. Solving these two algebraic equations yields

$$v_1' = -\frac{y_2(x)R(x)}{W(y_1, y_2)}, \qquad v_2' = \frac{y_1(x)R(x)}{W(y_1, y_2)}, \tag{91.7}$$

where $W(y_1, y_2) := y_1'y_2 - y_1y_2'$ is the usual Wronskian. The equations in (91.7) can be integrated, and the results used in (91.3) for

$$y(x) = -y_1(x) \int \frac{y_2(x)R(x)}{W(y_1, y_2)} \, dx + y_2(x) \int \frac{y_1(x)R(x)}{W(y_1, y_2)} \, dx.$$

Example

Suppose we have the equation

$$y'' + y = \csc x \tag{91.8}$$

to solve. The solutions to the homogeneous equation, $y'' + y = 0$, are clearly $y_1(x) = \sin x$ and $y_2(x) = \cos x$. Hence, we can compute the Wronskian to be $W(y_1, y_2) = -1$. Using this in (91.7) results in

$$v_1(x) = \int \frac{-\cos x \csc x}{-1} \, dx = \log(\sin x),$$

$$v_2(x) = \int \frac{-\sin x \csc x}{-1} \, dx = -x.$$

Hence, the particular solution to (91.8) is $y(x) = \sin x \log(\sin x) - x \cos x$.

Notes

[1] In Boyce and DiPrima [1] or Finizio and Ladas [4] may be found the generalization of the analysis presented above for differential equations of higher order. The result is:

> If $\{y_1, y_2, \ldots, y_n\}$ form a fundamental system of solutions for the equation
>
> $$y^{(n)} + a_{n-1}(x)y^{(n-1)} + \cdots + a_1(x)y' + a_0(x)y = 0,$$
>
> and if the functions $\{u_1, u_2, \ldots, u_n\}$ satisfy the system of equations
>
> $$y_1 u_1' + y_2 u_2' + \cdots + y_n u_n' = 0,$$
> $$y_1' u_1' + y_2' u_2' + \cdots + y_n' u_n' = 0,$$
> $$y_1'' u_1' + y_2'' u_2' + \cdots + y_n'' u_n' = 0,$$
> $$\vdots$$
> $$y_1^{(n-2)} u_1' + y_2^{(n-2)} u_2' + \cdots + y_n^{(n-2)} u_n' = 0,$$
> $$y_1^{(n-1)} u_1' + y_2^{(n-1)} u_2' + \cdots + y_n^{(n-1)} u_n' = f(x),$$
>
> then $y = u_1 y_1 + u_2 y_2 + \cdots + u_n y_n$ is a particular solution of
>
> $$y^{(n)} + a_{n-1}(x)y^{(n-1)} + \cdots + a_1(x)y' + a_0(x)y = f(x).$$

[2] This last result could also have been obtained by applying variation of parameters to a system of linear first order ordinary differential equations. Suppose we have the system

$$\mathbf{x}' = P(t)\mathbf{x} + \mathbf{g}(t),$$
$$\mathbf{x}(t_0) = \mathbf{x}_0, \tag{91.9}$$

where $\mathbf{g}(t)$ is a time dependent vector and $P(t)$ is a time dependent matrix. Then the solution can be written as

$$\mathbf{x}(t) = \Psi(t)\mathbf{x}_0 + \Psi(t) \int_{t_0}^{t} \Psi^{-1}(s)\mathbf{g}(s)\,ds,$$

where $\Psi(t)$ is a fundamental matrix of the system. This means that $\Psi(t)$ satisfies

$$\Psi' = P(t)\Psi, \quad \Psi(t_0) = I,$$

where I is an identity matrix of appropriate size. See Boyce and DiPrima [1] or Coddington and Levinson [3] for details.

[3] If (91.9) is stiff, that is $P(t)$ has eigenvalues with widely separated positive and negative real parts (see page 690), then $\Psi(t)$ may become numerically singular for $t \gg t_0$. For example, the problem $\mathbf{u}' = \begin{pmatrix} 0 & 1 \\ \lambda^2 & 0 \end{pmatrix} \mathbf{u}$ has the

fundamental matrix $\Psi(t) = \begin{pmatrix} \cosh\lambda(t-t_0) & \dfrac{1}{\lambda}\sinh\lambda(t-t_0) \\ \lambda\sinh\lambda(t-t_0) & \cosh\lambda(t-t_0) \end{pmatrix}$. For $\lambda(t-t_0) \geq 16$, this matrix is numerically singular even in 64-bit arithmetic.

References

[1] W. E. Boyce and R. C. DiPrima, *Elementary Differential Equations and Boundary Value Problems*, Fourth Edition, John Wiley & Sons, New York, 1986, pages 156–162, 275–277, and 391–393.

[2] E. A. Coddington and N. Levinson, *Theory of Ordinary Differential Equations*, McGraw–Hill Book Company, New York, 1955, pages 87–88.

[3] N. Finizio and G. Ladas, *Ordinary Differential Equations with Modern Applications*, Wadsworth Publishing Company, Belmont, Calif, 1982, page 136.

[4] E. L. Ince, *Ordinary Differential Equations*, Dover Publications, Inc., New York, 1964, pages 122–123.

[5] E. D. Rainville and P. E. Bedient, *Elementary Differential Equations*, The MacMillan Company, New York, 1964, pages 130–136.

[6] G. F. Simmons, *Differential Equations with Applications and Historical Notes*, McGraw–Hill Book Company, New York, 1972, pages 90–93.

92.　Vector Ordinary Differential Equations

Applicable to　A system of constant coefficient linear ordinary differential equations.

Yields

An exact solution is obtained.

Idea

Very often a system of coupled equations with constant coefficients can be transformed to a system of decoupled equations with constant coefficients.

Procedure

Given a system of n ordinary differential equations with constant coefficients, write the system as a vector ordinary differential equation in the following form

$$\mathbf{y}' = A\mathbf{y}, \qquad \mathbf{y}(t_0) = \mathbf{y}_0, \tag{92.1}$$

where $\mathbf{y}$ is a vector of the unknowns and A is a constant $n \times n$ matrix. Then determine the eigenvectors of A (i.e., those vectors $\mathbf{x}$ that satisfy $A\mathbf{x} = \lambda\mathbf{x}$ for some value of λ), and construct a diagonalizing matrix S whose columns are the eigenvectors of A. Then change variables by the transformation $\mathbf{y} = S\mathbf{u}$, so that (92.1) becomes

$$(S\mathbf{u})' = A(S\mathbf{u}),$$

or

$$\mathbf{u}' = S^{-1}AS\mathbf{u}. \tag{92.2}$$

By our choice of S, and assuming that A has n linearly independent eigenvectors, the matrix $S^{-1}AS$ will be diagonal. Hence the equations in (92.2) will decouple and each row of (92.2) will be an ordinary differential equation in one dependent variable (u_i). These equations can be solved by the method applicable to linear constant coefficient ordinary differential equations (see page 204). Once $\mathbf{u}$ is known, then $\mathbf{y}$ can be recovered from $\mathbf{y} = S\mathbf{u}$.

Example

Suppose we have the system of equations

$$\frac{dy_1}{dt} = 9y_1 + 2y_2,$$

$$\frac{dy_2}{dt} = y_1 + 8y_2.$$

This system of equations can be written as a vector ordinary differential equation as follows:

$$\frac{d}{dt}\begin{pmatrix} y_1 \\ y_2 \end{pmatrix} = \begin{pmatrix} 9 & 2 \\ 1 & 8 \end{pmatrix}\begin{pmatrix} y_1 \\ y_2 \end{pmatrix}, \tag{92.3}$$

or $\mathbf{y}' = A\mathbf{y}$, where $\mathbf{y} = (y_1, y_2)^{\mathrm{T}}$ and $A = \begin{pmatrix} 9 & 2 \\ 1 & 8 \end{pmatrix}$. The eigenvalues of A are $\lambda = 7$ and $\lambda = 10$ with the corresponding eigenvectors $(1, -1)^{\mathrm{T}}, (2, 1)^{\mathrm{T}}$. Therefore, the diagonalizing matrix, S, whose columns are the eigenvectors of A, is $S = \begin{pmatrix} 1 & 2 \\ -1 & 1 \end{pmatrix}$. We will also need the inverse of S, which is $S^{-1} = \begin{pmatrix} 1/3 & -2/3 \\ 1/3 & 1/3 \end{pmatrix}$. If we change variables by $\mathbf{y} = S\mathbf{u}$ then, (92.3) attains the form of (92.2). Specifically, we find

$$\frac{d}{dt}\begin{pmatrix} u_1 \\ u_2 \end{pmatrix} = \begin{pmatrix} 1/3 & -2/3 \\ 1/3 & 1/3 \end{pmatrix}\begin{pmatrix} 9 & 2 \\ 1 & 8 \end{pmatrix}\begin{pmatrix} 1 & 2 \\ -1 & 1 \end{pmatrix}\begin{pmatrix} u_1 \\ u_2 \end{pmatrix},$$

$$= \begin{pmatrix} 7 & 0 \\ 0 & 10 \end{pmatrix}\begin{pmatrix} u_1 \\ u_2 \end{pmatrix}. \tag{92.4}$$

Equation (92.4) can be expanded as

$$\frac{du_1}{dt} = 7u_1, \qquad \frac{du_1}{dt} = 10u_2.$$

Note that these last equations are decoupled and have constant coefficients. The solutions to these equations are given by

$$u_1 = Be^{7t}, \qquad u_2 = Ce^{10t},$$

where B and C are arbitrary constants. Therefore, using our original transformation we obtain $\mathbf{y} = S\mathbf{u}$, or

$$\begin{pmatrix} y_1 \\ y_2 \end{pmatrix} = \begin{pmatrix} 1 & 2 \\ -1 & 1 \end{pmatrix}\begin{pmatrix} u_1 \\ u_2 \end{pmatrix} = \begin{pmatrix} 1 & 2 \\ -1 & 1 \end{pmatrix}\begin{pmatrix} Be^{7t} \\ Ce^{10t} \end{pmatrix},$$

and therefore

$$y_1 = Be^{7t} + 2Ce^{10t},$$
$$y_2 = -Be^{7t} + Ce^{10t}. \tag{92.5}$$

The constants B and C may be found by evaluating (92.5) at $t = t_0$ and using (92.1):

$$\mathbf{y}_0 = B \begin{pmatrix} 1 \\ -1 \end{pmatrix} e^{7t_0} + C \begin{pmatrix} 2 \\ 1 \end{pmatrix} e^{10t_0},$$

$$= \begin{pmatrix} e^{7t_0} & 2e^{10t_0} \\ -e^{7t_0} & e^{10t_0} \end{pmatrix} \begin{pmatrix} B \\ C \end{pmatrix}.$$

Notes

[1] For a review of eigenvalues and eigenvectors see Strang [6].

[2] Of course, some systems of equations that are not of first order can also be reduced to the form of (92.). See page 118 for details.

[3] For a similar technique applied to partial differential equations, see page 384.

[4] This method is the same as "solving" the system in (92.1) by writing $\mathbf{y} = e^{At}\mathbf{y}_0$, where the exponential of a matrix is another matrix. See Coddington and Levinson [4] or Moler and Van Loan [5] for details.

[5] Similar results apply when A is a function of t. The equation $\mathbf{y}' = A(t)\mathbf{y}$, with $\mathbf{y}(t_0) = \mathbf{y}_0$, has the solution $\mathbf{y}(t) = e^{B(t)}\mathbf{y}(t_0)$, where $B(t) := \int_{t_0}^t A(t)\, dt$, whenever $BA = AB$.

[6] Given (92.1), a faster technique to find the solution (analogous to the method for constant coefficient linear equations on page 204) is to find the eigenvalues $\{\lambda_i\}$ and eigenvectors $\{\mathbf{x}_i\}$ of A and then write the most general solution in the form

$$\mathbf{y} = \sum_{i=1}^{n} C_i \mathbf{x}_i e^{\lambda_i t}, \tag{92.6}$$

where the $\{C_i\}$ are unknown constants.

For the example given, we can directly write the solution as

$$\mathbf{y} = C_1 \mathbf{x}_1 e^{\lambda_1 t} + C_2 \mathbf{x}_2 e^{\lambda_2 t}$$

$$= C_1 \begin{pmatrix} 1 \\ -1 \end{pmatrix} e^{7t} + C_2 \begin{pmatrix} 2 \\ 1 \end{pmatrix} e^{10t},$$

which is identical to (92.5).

[7] If the matrix A cannot be diagonalized (that is, if A does not have n linearly independent eigenvectors), then A has *generalized eigenvectors*. If the vector $\mathbf{z}_i^{(m)}$ satisfies $(A - \lambda_i I)^m \mathbf{z}_i^{(m)} = \mathbf{0}$ and $(A - \lambda_i I)^{m-1} \mathbf{z}_i^{(m)} \neq \mathbf{0}$, then $\mathbf{z}_i^{(m)}$ is called a generalized eigenvector of order m. (Note that a generalized eigenvector of order 1 is a usual eigenvector). Given $\mathbf{z}_i^{(m)}$, define $\mathbf{z}_i^{(n-1)} = (A - \lambda_i I) \mathbf{z}_i^{(n)}$ for $n = m, m-1, \ldots, 2$. Then define

$$\mathbf{y}_{ir} = e^{\lambda_i t}\left(\mathbf{z}_i^{(r)} + t\mathbf{z}_i^{(r-1)} + \frac{t^{r-1}}{(r-1)!}\mathbf{z}_i^{(1)}\right)$$

for $r = 1, 2, \ldots, m$. Then the $\{\mathbf{y}_{ir}\}$ will be a collection of linearly independent vectors and all solutions of (92.1) will be of the form $\sum_i \sum_r C_{ir} \mathbf{y}_{ir}$ (as in (92.6)). See Campbell [3] for details.

[8] Nonhomogeneous systems of linear equations, of the form

$$\mathbf{y}' = A(t)\mathbf{y} + \mathbf{g}(t)$$

may also be analyzed (see Boyce and DiPrima [2]. The easiest method is a generalization of the method of variation of parameters (see page 356). Alternately, if the nonhomogeneous system is of the form $\mathbf{y}' = A\mathbf{y} + t\mathbf{u}$, where A is a constant matrix and $\mathbf{u}$ is an arbitrary vector, then the system may be re-written as

$$\frac{d}{ds}\begin{pmatrix} \mathbf{y} \\ t \end{pmatrix} = \begin{pmatrix} A\mathbf{y} + t\mathbf{u} \\ 1 \end{pmatrix} = \begin{pmatrix} A & \mathbf{u} \\ 0 & 1 \end{pmatrix}\begin{pmatrix} \mathbf{y} \\ t \end{pmatrix},$$

which is now in the form of (92.1).

[9] The solution of

$$\frac{dX}{dt} = AX + XB, \qquad X(0) = C,$$

where A, B, C and X are *all* matrices is $X(t) = e^{At}Ce^{Bt}$. See Bellman [1] for details.

References

[1] R. Bellman, *Introduction to Matrix Analysis*, McGraw–Hill Book Company, New York, 1960, page 175.

[2] W. E. Boyce and R. C. DiPrima, *Elementary Differential Equations and Boundary Value Problems*, Fourth Edition, John Wiley & Sons, New York, 1986, Chapter 7 (pages 323–395).

[3] S. L. Campbell, *Singular Systems of Differential Equations*, Pitman Publishing Co., Marshfield, MA, 1980.

[4] E. A. Coddington and N. Levinson, *Theory of Ordinary Differential Equations*, McGraw–Hill Book Company, New York, 1955, pages 67–77.

[5] C. Moler and C. Van Loan, "Nineteen Dubious Ways to Compute the Exponential of a Matrix," *SIAM Review*, **20**, No. 4, October 1978, pages 801–836.

[6] G. Strang, *Linear Algebra and Its Applications*, Academic Press, New York, 1976, Chapter 5 (pages 171–230).

II.B

Exact Methods for PDEs

93.   Bäcklund Transformations

Applicable to   Nonlinear partial differential equations.

Yields

If a Bäcklund transformation can be found, then the solution of a nonlinear partial differential equation can be used to obtain either a different solution to the same partial differential equation, or to obtain a solution to a different nonlinear partial differential equation.

Idea

From a solution of a nonlinear partial differential equation, we can sometimes find a relationship that will generate the solution of

- (A) a different partial differential equation (this is a Bäcklund transformation),
- (B) the same partial differential equation (this is an auto-Bäcklund transformation).

Procedure

The first step (which is extremely difficult) is to determine a Bäcklund transformation between two partial differential equations. There are various methods described in the literature (see the references) that can be utilized for certain classes of equations. This transformation will utilize a solution of one of the partial differential equations to determine a solution to the other partial differential equation.

Example 1

Suppose we wish to find solutions to Burgers' equation

$$u_t + uu_x = \sigma u_{xx}. \tag{93.1}$$

Suppose that a solution of (93.1), $u(x,t)$, is already known. If $\phi(x,t)$ is defined to be any solution of the linear partial differential equation for $\phi(x,t)$

$$\phi_t + w\phi_x = \sigma\phi_{xx}, \tag{93.2}$$

(where $w(x,t)$ also satisfies (93.1)) and $v(x,t)$ is defined by

$$v(x,t) = -2\sigma\frac{\phi_x}{\phi} + w, \tag{93.3}$$

then $v(x,t)$ also satisfies Burgers' equation. Hence, two solutions of Burgers' equation (i.e., $u(x,t)$ and $w(x,t)$) can be used to generate another.

For example, a solution to (93.1) is clearly $u(x,t) = 0$. Using this and $w(x,t) = 0$ in (93.2) results in $\phi_t = \sigma\phi_{xx}$, which has as a solution $\phi(x,t) = \dfrac{e^{-x^2/4\sigma t}}{\sqrt{4\pi\sigma t}}$. Using this in (93.3) results in a different solution to Burgers' equation $v(x,t) = \dfrac{x}{t}$. This solution may be utilized to determine another solution, and the process can be repeated indefinitely.

Example 2

Suppose we wish to determine solutions to the sine–Gordon equation

$$u_{xt} = \sin u. \tag{93.4}$$

An auto-Bäcklund transformation is given by the pair of partial differential equations

$$v_x = u_x + 2\lambda \sin\left(\frac{v+u}{2}\right),$$
$$v_t = -u_t + \frac{2}{\lambda}\sin\left(\frac{v-u}{2}\right). \tag{93.5.a–b}$$

That is, given a solution to (93.4), $u(x,t)$, if $v(x,t)$ satisfies (93.5), then $v(x,t)$ will also be a solution of (93.4). This may be verified by determining v_{xt} both by differentiating (93.5.a) with respect to t and by differentiating (93.5.b) with respect to x. This results in

$$
\begin{aligned}
v_{xt} &= u_{xt} + 2\sin\left(\frac{v-u}{2}\right)\cos\left(\frac{v+u}{2}\right), \\
v_{xt} &= -u_{xt} + 2\sin\left(\frac{v+u}{2}\right)\cos\left(\frac{v-u}{2}\right).
\end{aligned}
\tag{93.6}
$$

Equating the two expressions in (93.6) results in (93.4), while adding them results in

$$v_{xt} = \sin v.$$

If we choose to utilize the solution $u(x,t) = 0$ of (93.4), then we can use the auto-Bäcklund transformation to determine another solution. Using $u(x,t) = 0$ in (93.5) results in

$$v_x = 2\lambda \sin\frac{v}{2}, \qquad v_t = \frac{2}{\lambda}\sin\frac{v}{2}.$$

This system of equations is easily solved to yield a new solution of the sine–Gordon equation

$$\tan\frac{v}{4} = C\exp\left(\lambda t + \frac{x}{\lambda}\right).$$

This solution may be used to determine another solution, and so on.

Notes
[1] The transformation in (93.2) and (93.3) with:
    (A) $w = 0$ (which we used in Example 1) is the Cole–Hopf transformation (see Whitham [11], pages 97–98).
    (B) $w = \phi$ was first found in Fokas [5].
        The transformation as we have presented it was found in Weiss, Tabor, and Carnevale [10].
[2] The Cole–Hopf transformation may also be written as the set of partial differential equations for the unknown $v(x,t)$

$$v_x = \frac{uv}{2\sigma}, \qquad v_t = \left(2\sigma u_x - u^2\right)\frac{v}{4\sigma}.$$

[3] Sometimes a Bäcklund transformation cannot be used to generate an infinite sequence of new solutions; the solutions repeat after some point. See Chan and Zheng [3] for some techniques to find new Bäcklund transformations when this occurs.

[4] Sakovich [9] determines all evolution equations ($w_t = f(w_x, w_{xx}, \ldots, w_{x \ldots x})$) and all Klein–Gordon equations ($w_{xy} = f(w)$) that admit a Bäcklund autotransformation (i.e., a mapping of the form $\phi = a[w]$, where $a[w]$ includes finite derivatives of w, that maps a solution of an equation to itself). Besides the linear equations, they include only the Liouville equation and the Burgers equation hierarchy.

References

[1] R. L. Anderson and N. H. Ibragimov, *Lie–Bäcklund Transformation in Applications*, SIAM, Philadelphia, 1979.

[2] G. W. Bluman and G. J. Reid, "Sequences of Related Linear PDEs," *J. Math. Anal. Appl.*, **144**, 1989, pages 565–585.

[3] W. L. Chan and Y.-K. Zheng, "Bäcklund Transformations for the Caudrey–Dodd–Gibbon–Sawada–Kotera Equation and its λ-Modified Equation," *J. Math. Physics*, **30**, No. 9, September 1989, pages 2065–2068.

[4] R. K. Dodd, J. C. Eilbeck, and H. C. Morris, *Solitons and Nonlinear Wave Equations*, Academic Press, London, 1982.

[5] A. Fokas, "Invariants, Lie–Bäcklund Operators, and Bcklund Transformations," PhD thesis, California Institute of Technology, Pasadena, California, 1979.

[6] G. L. Lamb, *Elements of Soliton Theory*, John Wiley & Sons, New York, 1980, Chapter 8 (pages 243–258).

[7] P. J. Olver, *Applications of Lie Groups to Differential Equations*, Graduate Texts in Mathematics #107, Springer–Verlag, New York, 1986.

[8] C. Rogers and W. F. Shadwick, *Backlund Transformations and Their Applications*, Academic Press, New York, 1982.

[9] S. Yu Sakovich, "On Special Bäcklund Autotransformations," *J. Phys. A: Math. Gen.*, **24**, 1991, pages 401–405.

[10] J. Weiss, M. Tabor, and G. Carnevale, "The Painlevé Property for Partial Differential Equations," *J. Math. Physics*, **24**, No. 3, March 1983, pages 522–526.

[11] G. B. Whitham, *Linear and Nonlinear Waves*, Wiley Interscience, New York, 1974, pages 609–611.

94.   Method of Characteristics

Applicable to   Systems of quasilinear partial differential equations (i.e., one or more partial differential equations linear in the first derivatives of the dependent variables, with no higher order derivatives present).

Yields

If the initial data is not given along a characteristic, then an exact solution can be obtained (generally implicit).

Idea

A quasilinear partial differential equation can be transformed into a set of ordinary differential equations that define the characteristics and a set of ordinary differential equations that describe how the solution changes along any specific characteristic.

Procedure

Suppose we have the quasilinear partial differential equation

$$a_1(\mathbf{x}, u)u_{x_1} + a_2(\mathbf{x}, u)u_{x_2} + \cdots + a_N(\mathbf{x}, u)u_{x_N} = b(\mathbf{x}, u), \qquad (94.1)$$

for the unknown $u(\mathbf{x}) = u(x_1, x_2, \ldots, x_N)$. If we were to differentiate $u(\mathbf{x})$ with respect to the variable s, then we obtain

$$\frac{du}{ds} = \left(\frac{\partial x_1}{\partial s}\right)u_{x_1} + \left(\frac{\partial x_2}{\partial s}\right)u_{x_2} + \cdots + \left(\frac{\partial x_N}{\partial s}\right)u_{x_N}. \qquad (94.2)$$

If we define

$$\frac{\partial x_k}{\partial s} = a_k(\mathbf{x}, u), \qquad (94.3)$$

for $k = 1, 2, \ldots, N$, then using (94.1) in (94.2) results in

$$\frac{du}{ds} = b(\mathbf{x}, u). \qquad (94.4)$$

To determine the solution of the partial differential equation in (94.1), we need to integrate the ordinary differential equations given in (94.3) and (94.4). (Equation (94.3) may look like a partial differential equation, but it is an ordinary differential equation with respect to s.) To perform this integration, initial conditions are needed in s for the $\{x_k\}$ and for u. Generally, the initial data for (94.1) will be given in the form

$$g(\mathbf{x}, u) = 0, \qquad (94.5)$$

on some manifold in $\mathbf{x}$ space. We identify this surface as corresponding to $s = 0$. If we think of $\mathbf{x}$ and u as depending on the variables $\{s, t_1, t_2, \ldots, t_{N-1}\}$, then the variables $\{t_1, t_2, \ldots, t_{N-1}\}$ can be used to parametrize the initial data in (94.5) (the examples will make this clear). That is,

$$\begin{aligned}
x_1(s=0) &= h_1(t_1, t_2, \ldots, t_{N-1}), \\
x_2(s=0) &= h_2(t_1, t_2, \ldots, t_{N-1}), \\
&\;\;\vdots \\
x_N(s=0) &= h_N(t_1, t_2, \ldots, t_{N-1}), \\
u(s=0) &= v(t_1, t_2, \ldots, t_{N-1}).
\end{aligned} \qquad (94.6)$$

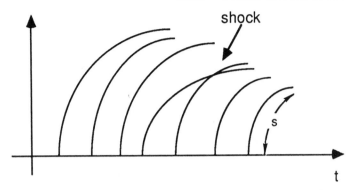

Figure 94. Depiction of the characteristics for a quasilinear equation.

Hence (94.6) supplies the initial conditions for the differential equations in (94.3) and (94.4).

After $\mathbf{x}$ and u are determined from (94.3), (94.4), and (94.6), then an implicit solution will have been obtained. If the $\{s, t_1, t_2, \dots, t_{N-1}\}$ can be analytically eliminated, then an explicit solution will be obtained. It is not always possible to perform this elimination analytically.

The physical picture of the construction of the solution is shown in Figure 94. The solution u is determined by the ordinary differential equation (94.4) along each characteristic. A characteristic is specified by the $\{t_i\}$ values. The parameter s represents scaled distance along a characteristic. When two characteristics cross, a *shock* is formed.

Note that a shock cannot form if the equation in (94.1) is linear; that is, each $\{a_i\}$ is only a function of $\mathbf{x}$, and not of u. At a shock, extra conditions are required. (See Landau and Lifshitz [2] for a discussion of the *Rankine–Hugoniot adiabatic*, which is used in fluid mechanics.)

Example 1

Suppose we want to solve the quasilinear partial differential equation

$$u_x + x^2 u_y = -yu,$$
$$u = f(y) \qquad \text{on } x = 0,$$

$$(94.7.a\text{--}b)$$

where $f(y)$ is a given function. Forming du/ds we have

$$\frac{du}{ds} = \left(\frac{\partial x}{\partial s}\right) u_x + \left(\frac{\partial y}{\partial s}\right) u_y. \qquad (94.8)$$

Comparing (94.8) to (94.7), we take

$$\frac{\partial x}{\partial s} = 1, \qquad \frac{\partial y}{\partial s} = x^2, \qquad \frac{du}{ds} = -yu. \qquad (94.9.a\text{--}c)$$

The initial data in (94.7.b) can be written parametrically as

$$x(s = 0) = 0,$$
$$y(s = 0) = t_1,$$
$$u(s = 0) = f(t_1).$$

(94.10.a–c)

The solution of (94.9.a) with (94.10.a) is

$$x(s, t_1) = s.$$

(94.11)

Therefore, (94.9.b) and (94.10.b) can be written as

$$\frac{\partial y}{\partial s} = s^2, \qquad y(s = 0) = t_1,$$

with the solution

$$y(s, t_1) = \frac{s^3}{3} + t_1.$$

(94.12)

Finally, the equation for u (from (94.9.c), (94.10)c, and (94.12)) becomes

$$\frac{du}{ds} = -\left(\frac{s^3}{3} + t_1\right) u, \qquad u(s = 0) = f(t_1),$$

with the solution

$$u(s, t_1) = f(t_1) \exp\left(-\frac{s^4}{12} - s t_1\right).$$

(94.13)

Equations (94.11), (94.12), and (94.13) constitute an implicit solution of (94.7).

In this case it is possible to analytically eliminate the s and t_1 variables to obtain an explicit solution. From (94.11) we obtain $s = x$. Using this in (94.12) results in $t_1 = y - \frac{x^3}{3}$. Using these two values in (94.13) results in the explicit solution

$$u(x, y) = f\left(y - \frac{x^3}{3}\right) \exp\left(\frac{x^4}{4} - xy\right).$$

Example 2

If we have the quasilinear partial differential equation in three dependent variables

$$u_x + u_y + xyu_z = u^2,$$

$$u = x^2 \qquad \text{on } y = z,$$

$$(94.14)$$

then we can write (94.3), (94.4), and (94.6) as

$$\frac{\partial x}{\partial s} = 1, \qquad \frac{\partial y}{\partial s} = 1, \qquad \frac{\partial z}{\partial s} = xy,$$

$$\frac{du}{ds} = u^2,$$

$$x(s = 0) = t_1, \quad y(s = 0) = t_2, \quad z(s = 0) = t_2, \quad u(s = 0) = t_1^2.$$

The equations for x and y can be integrated to yield

$$x = s + t_1, \qquad y = s + t_2. \tag{94.15}$$

Using these values for x and y, the equation for z becomes

$$\frac{\partial z}{\partial s} = (s + t_2)(s + t_1),$$

which can be integrated to yield

$$z = \frac{s^3}{3} + \frac{s^2}{2}(t_2 + t_1) + st_2t_1 + t_2. \tag{94.16}$$

The equation for u can also be integrated to obtain

$$u = -\frac{t_1^2}{1 + st_1^2}. \tag{94.17}$$

The equations in (94.15), (94.16), and (94.17) constitute an implicit solution to (94.14). The variables t_1 and t_2 can be eliminated to yield

$$u = -\frac{(x - s)^2}{1 + s(x - s)^2},$$

$$z = \frac{s^3}{3} + \frac{s^2}{2}(x + y) + s(xy - 1) + y. \tag{94.18.a–b}$$

To actually evaluate $u(x, y, z)$ at some given value of x, y, and z requires two steps. First the equation in (94.18.b) must be solved for s, and then this value is utilized in equation (94.18.a).

Alternately, the method of resultants (see page 46) could be used to obtain a single polynomial equation in terms of x, y, z, and u, alone. This results in an equation with 123 terms; the implicit solution given by (94.18) is more useful and more compact.

Notes

[1] This technique extends naturally to systems of partial differential equations, with virtually no increase in complexity. This allows a single partial differential equation of higher order (and hyperbolic type) to be analyzed. For example, the wave equation $u_{xx} = u_{tt}$ can be written, in the variables $\{v := u_x,\ w := u_t\}$, as the system of two quasilinear equations $\{v_t = w_x,\ w_t = v_x\}$.

[2] The general quasilinear system of N equations for the N unknowns $\mathbf{u} = (u_1, u_2, \ldots, u_n)$ in the two independent variables $\{x, t\}$ has the form

$$\sum_{j=1}^{N} A_{ij}(\mathbf{u}, x)\frac{\partial u_j}{\partial t} + \sum_{j=1}^{N} a_{ij}(\mathbf{u}, x)\frac{\partial u_j}{\partial x} + b_i = 0,$$

for $i = 1, 2, \ldots, N$. This equation will be hyperbolic (and hence solvable by the method of characteristics) if there exist N linearly independent real-valued N-dimensional vectors $\{\mathbf{v}^{(1)}, \mathbf{v}^{(2)}, \ldots, \mathbf{v}^{(N)}\}$ and N non-zero real-valued two-dimensional vectors $\{\boldsymbol{\alpha}^{(k)}, \boldsymbol{\beta}^{(k)}\}$ such that

$$\sum_{i,j=1}^{N} v_i^{(k)}\left[A_{ij}\boldsymbol{\alpha}^{(k)} - a_{ij}\boldsymbol{\beta}^{(k)}\right] = 0,$$

for $k = 1, \ldots, N$. See Whitham [4] for details and for several examples using this formalism.

[3] Referring to (94.1), it turns out that discontinuities in ∇u can propagate along characteristics, but discontinuities in u cannot. In fact, if u satisfies a second order linear hyperbolic partial differential equation in x and y, and if $\{u, u_x, u_y, u_{xx}, u_{xy}\}$ are all continuous across a curve C but u_{yy} suffers a jump upon crossing C, then C is necessarily a characteristic of the partial differential equation.

[4] Eliminating the $\{s, t\}$ variables at the end of the calculation will be possible, in principle, whenever the Jacobian of the transformation does not vanish; i.e., $\dfrac{\partial(u, x_1, x_2, \ldots)}{\partial(s, t_1, t_2, \ldots)} \neq 0$.

[5] An equivalent way of writing (94.3) is the form

$$\frac{dx_1}{a_1} = \frac{dx_2}{a_2} = \cdots = \frac{dx_N}{a_N},$$

which are called the *subsidiary equations*. When one or more of the a_k are zero, this equation looks peculiar, but is should be interpreted to be the same as (94.3). This form is used in place of (94.3) in many older texts. This formulation has been used occasionally in this book.

References

[1] S. J. Farlow, *Partial Differential Equations for Scientists and Engineers*, John Wiley & Sons, New York, 1982, Lesson 27 (pages 205–212).

[2] L. D. Landau and E. M. Lifshitz, *Fluid Mechanics*, Pergamon Press, New York, 1959, Chapter 9 (pages 310–346).

[3] P. Moon and D. E. Spencer, *Partial Differential Equations*, D. C. Heath, Lexington, MA, 1969, pages 27–29.

[4] G. B. Whitham, *Linear and Nonlinear Waves*, Wiley Interscience, New York, 1974, Chapter 5 (pages 113–142).

[5] E. Zauderer, *Partial Differential Equations of Applied Mathematics*, John Wiley & Sons, New York, 1983, Chapter 3 (pages 78–121).

95. Characteristic Strip Equations

Applicable to   Some partial differential equations in two independent variables.

Yields

When the technique is applicable, an implicit solution.

Idea

This method appears to be a generalization of the method of characteristics, but it can in fact be derived from that method. The formulae presented here are handy to use directly.

Procedure

Given the partial differential equation

$$F(x, y, u, p, q) = 0, \tag{95.1}$$

where $p = u_x$, $q = u_y$, we search for a solution $u = u(x, y)$. The technique is to solve the system of "strip equations" given by

$$
\begin{aligned}
\frac{\partial x}{\partial s} &= F_p, & \frac{\partial p}{\partial s} &= -F_x - pF_u, \\
\frac{\partial y}{\partial s} &= F_q, & \frac{\partial q}{\partial s} &= -F_y - qF_u, \\
\frac{\partial u}{\partial s} &= pF_p + qF_q,
\end{aligned}
\tag{95.2}
$$

where we now consider $\{x, y, p, q, u\}$ to all be functions of the two variables $\{s, t\}$. The equations in (95.2) are also called *Charpit's equations*.

The "initial" values for (95.2) (corresponding to $s = 0$) are given in terms of the other independent variable t. It will be possible to give initial values to all of the terms in (95.2) since the original equation (95.1) will have data with it that can be parameterized in terms of t.

After we have determined $\{x, y, u\}$ as functions of $\{s, t\}$, we must solve the equations implicitly to obtain the final solution in the form $u = u(x, y)$.

Example

Suppose we have the nonlinear partial differential equation

$$u_x u_y - u = 0, \tag{95.3}$$

with the initial data

$$u = y^2 \qquad \text{on } x = 0. \tag{95.4}$$

By comparing (95.3) with (95.1) we find that $F = pq - u$. Hence, the equations in (95.2) can be written as

$$\frac{\partial x}{\partial s} = q, \qquad \frac{\partial p}{\partial s} = p,$$

$$\frac{\partial y}{\partial s} = p, \qquad \frac{\partial q}{\partial s} = q, \tag{95.5.a–e}$$

$$\frac{\partial u}{\partial s} = 2pq.$$

The initial conditions for (95.5) are given by parameterizing (95.4) in terms of the dummy variable t. One such parameterization (there are always infinitely many) is

$$x = 0, \qquad y = t, \qquad u = t^2. \tag{95.6}$$

To determine the initial conditions for p and q, we utilize the chain rule

$$\frac{\partial u}{\partial t} = p\frac{dx}{dt} + q\frac{dy}{dt},$$

which can be evaluated at $s = 0$ (using (95.6)) to yield

$$2t = p(0, t) \cdot 0 + q(0, t) \cdot 1,$$

or $q(0, t) = 2t$. The original equation, (95.4), can be used (at $s = 0$) to determine that $p(0, t) = u(0, t)/q(0, t) = t/2$. Now that we have the initial conditions for all five variables appearing in (95.5), we can find the solution.

Equations (95.5.b) and (95.5.d) can be integrated directly to yield

$$p = \tfrac{1}{2}te^s, \qquad q = 2te^s.$$

Substituting these expressions in (95.5.a), (95.5.c), and (95.5.e), and integrating, results in

$$
\begin{aligned}
x &= 2t(e^s - 1), \\
y &= \tfrac{1}{2}t(e^s + 1), \\
u &= t^2 e^{2s}.
\end{aligned}
\qquad (95.7.a\text{–}c)
$$

Equations (95.7.a) and (95.7.b) can be inverted to produce s and t as functions of x and y:

$$e^s = \frac{4y + x}{4y - x}, \qquad t = \frac{4y - x}{4}.$$

Using these relations in (95.7.c) yields the final answer

$$u(x,t) = \frac{(x + 4y)^2}{16}.$$

Notes

[1] This method is sometimes called the Lagrange–Charpit method.

[2] Frequently, inverting the variables at the end (i.e., finding $s = s(x,y)$ and $t = t(x,y)$) is the step that cannot be carried out analytically.

[3] The variable s really specifies a characteristic, while t represents distance along any single characteristic.

[4] This technique works, as the example shows, even when the original equation is not quasi–linear. That is, the method of characteristics could not have been applied directly to (95.3).

References

[1] E. T. Copson, *Partial Differential Equations*, Cambridge University Press, New York, 1975, pages 5–9.

[2] P. R. Garabedian, *Partial Differential Equations*, Wiley, New York, 1964, pages 24–31.

[3] I. N. Sneddon, *Elements of Partial Differential Equations*, McGraw–Hill Book Company, New York, 1957, pages 61–66.

[4] E. Zauderer, *Partial Differential Equations of Applied Mathematics*, John Wiley & Sons, New York, 1983, pages 56–68.

96.   Conformal Mappings

Applicable to   Laplace's equation ($\nabla^2 u = 0$) in two dimensions.

Yields

A reformulation of the original problem.

Idea

Laplace's equation in two dimensions with a given boundary can be transformed to Laplace's equation with a different boundary by a conformal map. The idea is to choose the conformal map in such a way that the new boundary makes the problem easy to solve.

Procedure

Given Laplace's equation in the variables $\{x, y\}$ (i.e., $\nabla^2 u = u_{xx} + u_{yy} = 0$), we define the complex variable $z = x + iy$, where $i = \sqrt{-1}$. All of the boundaries of the original problem can now be described by values of z.

Any analytic transformation between two complex variables, say $\zeta = F(z)$, for which $d\zeta/dz$ is never zero, is said to be *conformal*. It turns out that Laplace's equation is invariant under a conformal map. That is, if $\zeta = \xi + i\eta = F(z)$, $u_{xx} + u_{yy} = 0$, and $F(z)$ is a conformal map, then $u_{\xi\xi} + u_{\eta\eta} = 0$.

In the new variables, $\{\xi, \eta\}$, the boundary might be very simple. If so, then Laplace's equation can be solved in this new domain. Then the solution of Laplace's equation in the original domain can be found by the change of variables induced by the conformal map.

A commonly used conformal map is the *Schwartz–Christoffel transformation*. This maps a closed polygonal figure (with n vertices) into a half plane. The mapping is given by the solution of

$$\frac{dz}{d\zeta} = C(\zeta - \zeta_1)^{\beta_1/\pi - 1}(\zeta - \zeta_2)^{\beta_2/\pi - 1} \cdots (\zeta - \zeta_n)^{\beta_n/\pi - 1} \qquad (96.1)$$

for appropriate $\{\beta_1, \beta_2, \ldots, \beta_n\}$ and $\{\zeta_1, \zeta_2, \ldots, \zeta_n\}$. The $\{\beta_i\}$ are the interior angles of the polygon, and the $\{\zeta_i\}$ are the (complex valued) positions of the polygon's vertices.

After the differential equation in (96.1) is formulated, it must be solved. The unknown constant C, as well as the arbitrary constant resulting from the integration, will be determined when the $\{\zeta_i\}$ are prescribed. The resulting function $\zeta = F(z)$ is the conformal map that maps the interior of the given polygonal figure into the half plane. See Trefethen [10] for a numerical implementation.

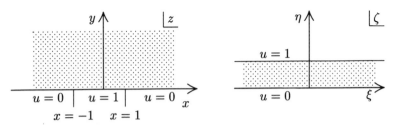

Figure 96.1 The original domain for Laplace's equation and the domain after a conformal mapping has been applied.

Example 1

Suppose we have Laplace's equation $(u_{xx} + u_{yy} = 0)$ to solve in the half plane $H = \{-\infty < x < \infty, 0 < y < \infty\}$, with the boundary conditions

$$u(x,0) = \begin{cases} 0 & \text{for } |x| > 1, \\ 1 & \text{for } |x| \le 1. \end{cases}$$

Under the mapping

$$\zeta = \xi + i\eta = F(z) = \log\left(\frac{z-1}{z+1}\right) = \log\left(\frac{x+iy-1}{x+iy+1}\right), \qquad (96.2)$$

the half plane H is mapped into a strip of height π in the (ξ, η) plane. See Figure 96.1 for pictures of the two geometrical regions involved.

In the (ξ, η) plane the boundary conditions become

$$u(\xi, 0) = 0,$$
$$u(\xi, \pi) = 1.$$

The solution to Laplace's equation in this domain is simply $u(\xi, \eta) = \eta/\pi$. To transform back to (x, y) coordinates, the transformation in (96.2) must be inverted. After some algebra it can be shown that

$$\eta = \arg\left(\frac{z-1}{z+1}\right) = \tan^{-1}\left(\frac{2y}{x^2+y^2-1}\right),$$

so that

$$u(x,y) = \frac{1}{\pi}\tan^{-1}\left(\frac{2y}{x^2+y^2-1}\right).$$

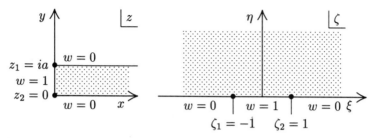

Figure 96.2 The original domain for Laplace's equation and the domain after the Schwartz–Christoffel transformation has been applied.

Example 2

Suppose we have Laplace's equation ($\nabla^2 w = 0$) in the channel open on the right (see Figure 96.2), with the boundary conditions

$$w(x,0) = 0 \qquad \text{for } 0 \le x < \infty$$
$$w(x,a) = 0 \qquad \text{for } 0 \le x < \infty$$
$$w(0,y) = 1 \qquad \text{for } 0 \le y \le 1.$$

The polygon in which this problem is being solved has vertices at $z_1 = ia$ and $z_2 = 0$, with the corresponding interior angles $\beta_1 = \beta_2 = \pi/2$. Using the Schwartz–Christoffel transformation we choose the vertices in the z plane to map to the vertices $\zeta_1 = -1$ and $\zeta_2 = 1$ in the ζ plane. The differential equation in (96.1) becomes:

$$\frac{dz}{d\zeta} = C \left(\zeta + 1\right)^{1/2} \left(\zeta - 1\right)^{1/2}$$

with the solution $z = C \cosh^{-1} \zeta + D$, where D is an arbitrary constant. To determine the constants C and D we must enforce that the vertices in the z plane mapped to the vertices in the ζ plane. We have the two simultaneous equations:

$$z_1 = ia = C \cosh^{-1}(\zeta_1) + D = C \cosh^{-1}(-1) + D = Ci\pi + D,$$
$$z_2 = 0 = C \cosh^{-1}(\zeta_2) + D = C \cosh^{-1}(1) + D = D,$$

with the solution $\{D = 0, \; C = a/\pi\}$. Hence, the desired conformal mapping is $\zeta = \cosh\left(\dfrac{\pi z}{a}\right)$. The problem in the ζ domain is now identical to the problem solved in Example 1.

Notes

[1] The *Joukowski transformation*, given by $\zeta = z + a^2/z$, maps an ellipse into a circle, or a circle into a strip.

[2] Algebraic mappings, given by $\zeta = z^{\beta/\pi}$, with $\beta > 0$, map a corner with angle α to a corner with angle $\alpha\beta/\pi$. For instance, if $\beta = 2\pi$, then a quarter plane ($\alpha = \pi/2$) is mapped to a half plane.

[3] Numerical implementation of the Schwartz–Christoffel transformation can fail on some seemingly very simple polygons. Mapping a rectangle with an aspect ration of 20 to 1, or most other regions with a similar degree of elongation, onto a half-plane may cause problems because the points in the transformed plane will be very close together. (This is known as the "crowding phenomenon.")

[4] The Schwartz–Christoffel transformation can also be used for doubly connected domains, see Iyanaga and Kawada [4].

[5] Conformal mappings are often used in hydrodynamics and electrostatics because, under a conformal mapping, lines of flow and equipotential lines are mapped into lines of flow and equipotential lines.

[6] Conformal mappings are often used to obtain an orthogonal coordinate system inside of a two-dimensional body. This may be used, for instance, when a grid is required on which the solution to a partial differential equation will be approximated numerically.

[7] Even when an analytic conformal map cannot be found, there are fast numerical techniques for finding an approximate conformal map. Riemann's mapping theorem states that all bounded simply connected plane regions can be conformally mapped onto the unit disk, and all bounded doubly connected plane regions can be conformally mapped onto an annulus. Using Poisson's formula (see page 411) exact solutions can be written down for these two geometries. See Trefethen [11] or Fornberg [2] for details.

[8] The mapping used in this method need not be conformal everywhere, it only needs to be conformal in the domain in which Laplace's equation is being solved. (Very few maps are conformal everywhere.)

[9] Kober [6] has a large collection of conformal mappings, with the geometric regions in both the (x, y) and (η, ζ) planes clearly illustrated.

[10] Seymour [9] describes a computer package that permits real time manipulation and display of conformal mappings of one complex plane onto another.

[11] If $\nabla_{x,y}^2$ represents the Laplacian in $\{x, y\}$ space, then under the conformal mapping $\zeta = F(z)$ the operator $\nabla_{x,y}^2$ is mapped to the operator $|F'(z)|^2 \nabla_{\eta,\zeta}^2$. Hence, the biharmonic equation $\nabla^4 u := \nabla_{x,y}^2 \nabla_{x,y}^2 u = 0$ becomes, under a conformal mapping, $|F'(z)|^2 \nabla_{\eta,\zeta}^2 \left(|F'(z)|^2 \nabla_{\eta,\zeta}^2 \right) u = 0$.

References

[1] T. K. DeLillo and A. R. Elcrat, "A Comparison of Some Numerical Conformal Mapping Methods for Exterior Regions," *SIAM J. Sci. Stat. Comput.*, **12**, No. 2, March 1991, pages 399–422.

[2] S. J. Farlow, *Partial Differential Equations for Scientists and Engineers*, John Wiley & Sons, New York, 1982, Lesson 47 (pages 379–388).

[3] J. M. Floryan and C. Zemach, "Schwarz–Christoffel Mappings: A General Approach," *J. Comput. Physics*, **72**, 1987, pages 347–371.

[4] B. Fornberg, "A Numerical Method for Conformal Mapping," *SIAM J. Sci. Stat. Comput.*, **1**, 1980, pages 386–400.

[5] P. Henrici, *Applied and Computational Complex Analysis*, Volume 3, Wiley, New York, 1986, Chapters 16 and 17 (pages 323–506).

[6] L. V. Kantorovich and V. I. Krylov, *Approximate Methods of Higher Analysis*, Interscience Publishers, New York, 1958, Chapters 5 and 6 (pages 358–615).

[7] S. Iyanaga and Y. Kawada, *Encyclopedic Dictionary of Mathematics*, MIT Press, Cambridge, MA, 1980, page 1156.

[8] H. Kober, *Dictionary of Conformal Representations*, Dover Publications, Inc., New York, 1952.

[9] N. Levinson and R. M. Redheffer, *Complex Variables*, Holden–Day, Inc., San Francisco, 1970, Chapter 5 (pages 259–332).

[10] H. R. Seymour, "Conform: A Conformal Mapping System," in B. W. Char (ed.), *SYMSAC '86*, ACM, New York, 1986, pages 163–168.

[11] L. N. Trefethen, "Numerical Computation of the Schwarz–Christoffel Transformation," *SIAM J. Sci. Stat. Comput.*, **1**, No. 1, March 1980, pages 82–102.

[12] L. N. Trefethen, *Numerical Conformal Mapping*, North–Holland Publishing Co., New York, 1986.

[13] M. Walker, *The Schwarz–Christoffel Transformation and Its Applications – A Simple Exposition*, Dover Publications, Inc., New York, 1964.

97.   Method of Descent

Applicable to   Partial differential equations (most often, wave equations).

Yields

An exact solution.

Idea

For some partial differential equations (in particular some wave equations) odd dimensional problems are easier than even dimensional problems. Hence it is reasonable, when given a $2n$ dimensional problem, to instead solve a $2n + 1$ dimensional problem and then "come down one dimension."

Procedure

Given a partial differential equation in n dimensions for the quantity $u(\mathbf{x}) = u(x_1, x_2, \ldots, x_n)$

$$L[u] = 0,$$

it might be easier to solve the $n + 1$ dimensional problem

$$L[v] + H[v] = 0,$$

for $v(\mathbf{x}, z) = v(x_1, x_2, \ldots, x_n, z)$, where $H[\cdot]$ is a differential operator with respect to z. Then, when $v(\mathbf{x}, z)$ is known, $u(\mathbf{x})$ can be obtained by either (1) an appropriate integral over z, or (2) taking v to be independent of z.

Example

Suppose we are given the two-dimensional wave equation

$$u_{tt} = c^2 \left(u_{xx} + u_{yy} \right), \qquad (97.1)$$

with the initial conditions

$$u(0, \mathbf{x}) = f(\mathbf{x}), \qquad u_t(0, \mathbf{x}) = g(\mathbf{x}), \qquad (97.2)$$

where $\mathbf{x} = (x, y)$. We might choose to instead solve the three-dimensional wave equation

$$v_{tt} = c^2 \left(v_{xx} + v_{yy} + v_{zz} \right),$$

with the initial conditions

$$v(0, \mathbf{x}, z) = f(\mathbf{x}), \qquad v_t(0, \mathbf{x}, z) = g(\mathbf{x}).$$

The three-dimensional wave equation has the well-known solution (see page 429)

$$v(t, \mathbf{x}, z) = ctM[g] + \frac{\partial}{\partial t}\Big(ctM[f]\Big), \qquad (97.3)$$

where $M[\cdot]$ is a functional defined to be the average value of its argument on a circle of radius ct; i.e.,

$$M[h(x, y, z)] := \frac{1}{4\pi c^2 t^2} \int_{S(t)} h \, dS$$

$$= \frac{1}{4\pi c^2 t^2} \int_0^\pi \int_0^{2\pi} h(x + ct\sin\theta\cos\phi, y + ct\sin\theta\sin\phi, z + ct\cos\theta)$$

$$\times \sin\theta \, d\theta \, d\phi,$$

(97.4)

where $S(t)$ is the surface of a sphere with origin at (x, y, z) and radius ct.

To solve the two-dimensional wave equation (97.1), we merely utilize the fact that f and g are independent of the variable z. Performing some algebraic manipulations, (97.4) becomes

$$M[h(x, y)] = \frac{1}{2\pi ct} \int\int_{\sigma(t)} \frac{h(\zeta, \eta) \, d\zeta \, d\eta}{\sqrt{c^2 t^2 - (x - \zeta)^2 - (y - \eta)^2}}, \qquad (97.5)$$

where $\sigma(t)$ is the interior of the circle: $(x - \zeta)^2 + (y - \eta)^2 = c^2 t^2$. Using (97.5) in (97.3) results in the solution to (97.1) and (97.2).

Notes

[1] This method is also called *Hadamard's method of descent*.

[2] If the descent step was applied once again, the solution of the one-dimensional wave equation, $w_{tt} = c^2 w_{xx}$, could be obtained from (97.3) and (97.5).

[3] Note that a line source, in three dimensions, might be viewed as a point source in two dimensions.

[4] One reason that odd space dimensional problems are sometimes easier than even dimensional problems is Huygen's principle. Huygen's principle (see Garabedian [4] or Chester [2]) states that the wave equation in an odd number of space dimensions only depends on the initial data (and its derivatives) on the *perimeter* of the domain of dependence. See the section on exact solutions of the wave equation (starting on page 429).

References

[1] E. T. Copson, *Partial Differential Equations*, Cambridge University Press, New York, 1975, pages 95–96.

[2] C. R. Chester, *Techniques in Partial Differential Equations*, McGraw–Hill Book Company, New York, 1970, pages 154–156.

[3] S. J. Farlow, *Partial Differential Equations for Scientists and Engineers*, John Wiley & Sons, New York, 1982, pages 187–188.

[4] P. R. Garabedian, *Partial Differential Equations*, Wiley, New York, 1964, Section 6.3 (pages 204–210).

[5] G. B. Whitham, *Linear and Nonlinear Waves*, Wiley Interscience, New York, 1974, pages 219–235.

[6] E. Zauderer, *Partial Differential Equations of Applied Mathematics*, John Wiley & Sons, New York, 1983, pages 226–232.

98.   Diagonalization of a Linear System of PDEs

Applicable to   A linear system of partial differential equations in two independent variables, of the form $\mathbf{u}_t + A\mathbf{u}_x = 0$, where A is a constant matrix.

Yields

A set of uncoupled equations.

Idea

By diagonalizing the coefficient matrix, the equations can be uncoupled and then solved.

Procedure

Given the linear system of differential equations

$$\mathbf{u}_t + A\mathbf{u}_x = 0, \tag{98.1}$$

we change the dependent variables to decouple the system. If the matrix A is $n \times n$ and has the eigenvectors $\{v_1, v_2, \ldots, v_n\}$ (which we assume to be linearly independent), then we define the matrix S by $S = (\,v_1 \quad v_2 \quad \ldots \quad v_n\,)$. Changing variables in (98.1) by $\mathbf{u} = S\mathbf{w}$ results in $S\mathbf{w}_t + AS\mathbf{w}_x = 0$, or

$$\mathbf{w}_t + \Lambda \mathbf{w}_x = 0, \tag{98.2}$$

where $\Lambda = S^{-1}AS$ is a diagonal matrix. The equations in (98.2) are now decoupled and can be solved separately for $\{w_1(x,t), w_2(x,t), \ldots, w_n(x,t)\}$. After they have been found, $\mathbf{u}$ may be determined from $\mathbf{u} = S\mathbf{w}$.

Example

Given the system of linear partial differential equations in two independent variables

$$\frac{\partial u_1}{\partial t} + 9\frac{\partial u_1}{\partial x} + 2\frac{\partial u_2}{\partial x} = 0,$$

$$\frac{\partial u_2}{\partial t} + \frac{\partial u_1}{\partial x} + 8\frac{\partial u_2}{\partial x} = 0,$$

(98.3)

we define the vector $\mathbf{u} = \begin{pmatrix} u_1 \\ u_2 \end{pmatrix}$ and the matrix $A = \begin{pmatrix} 9 & 9 \\ 1 & 8 \end{pmatrix}$ so that
(98.3) may be written in the form of (98.1).

The eigenvalues of A are $\lambda = 7$ and $\lambda = 10$ with the corresponding
eigenvectors: $v_1 = (1, -1)^T$ and $v_2 = (2, 1)^T$. Hence the matrix S
is given by $S = \begin{pmatrix} 1 & -1 \\ 2 & 1 \end{pmatrix}$, which has the inverse $S^{-1} = \begin{pmatrix} 1/3 & -2/3 \\ 1/3 & 1/3 \end{pmatrix}$.
Making the change of variables $\mathbf{u} = S\mathbf{w}$ turns (98.1) into (98.2) with Λ
defined by

$$\Lambda = S^{-1}AS,$$

$$= \begin{pmatrix} 1/3 & -2/3 \\ 1/3 & 1/3 \end{pmatrix} \begin{pmatrix} 9 & 9 \\ 1 & 8 \end{pmatrix} \begin{pmatrix} 1 & 2 \\ -1 & 1 \end{pmatrix},$$

$$= \begin{pmatrix} 10 & 0 \\ 0 & 7 \end{pmatrix}.$$

The equations in (98.2) can then be separated to obtain

$$\frac{\partial w_1}{\partial t} + 10\frac{\partial w_1}{\partial x} = 0,$$

$$\frac{\partial w_2}{\partial t} + 7\frac{\partial w_2}{\partial x} = 0.$$

These equations have the solution

$$w_1(x, t) = f(x - 10t),$$

$$w_2(x, t) = g(x - 7t),$$

where f and g are arbitrary functions of their arguments. Knowing $\mathbf{w}$ we
can determine $\mathbf{u}$ to be

$$u_1(x, t) = w_1(x, t) + 2w_2(x, t) = f(x - 10t) + 2g(x - 7t),$$
$$u_2(x, t) = -w_1(x, t) + w_2(x, t) = -f(x - 10t) + g(x - 7t).$$

(98.4)

Knowing the general form of the solution, any initial conditions for $u_1(x, t)$
and $u_2(x, t)$ could be utilized. For example, if we had

$$u_1(x, 0) = 3\sin 2x,$$

$$u_2(x, 0) = 0,$$

(98.5)

then utilizing (98.4) in (98.5) produces

$$\begin{aligned} f(x) + 2g(x) &= 3\sin 2x, \\ -f(x) + g(x) &= 0, \end{aligned}$$

and so $f(z) = g(z) = \sin 2z$ and the final solution can be written

$$\begin{aligned} u_1(x,t) &= \sin(2x - 20t) + 2\sin(2x - 14t), \\ u_2(x,t) &= -\sin(2x - 20t) + \sin(2x - 14t). \end{aligned}$$

References

[1] S. J. Farlow, *Partial Differential Equations for Scientists and Engineers*, John Wiley & Sons, New York, 1982, Lesson 29 (pages 223–231).

99. Duhamel's Principle

Applicable to   Linear parabolic and hyperbolic partial differential equations.

Yields
    An integral representation in terms of the solution of a more tractable partial differential equation.

Idea
    To solve a parabolic partial differential equation with a time varying source function and time varying boundary conditions, only a parabolic partial differential equation with a constant source term and constant boundary conditions needs to be solved.

Procedure
    Suppose we have the parabolic partial differential equation for $u(\mathbf{x}, t)$

$$\begin{aligned} \frac{\partial}{\partial t} u(\mathbf{x}, t) &= L[u(\mathbf{x}, t)] + F(\mathbf{x}, t), \\ u(\mathbf{y}, t) &= G(\mathbf{y}, t), \qquad \text{for } t > 0, \\ u(\mathbf{x}, 0) &= H(\mathbf{x}), \end{aligned} \tag{99.1}$$

where $L[\cdot]$ is an elliptic operator in $\mathbf{x}$ and $\mathbf{y}$ denotes a point on the boundary. Note that (99.1) has a time dependent source function $F(\mathbf{x}, t)$ and time dependent surface conditions $G(\mathbf{y}, t)$. Instead of solving (99.1) for $u(\mathbf{x}, t)$, we choose to solve the parabolic partial differential equation

$$\frac{\partial}{\partial t} v(\mathbf{x}, t, \tau) = L[v(\mathbf{x}, t, \tau)] + F(\mathbf{x}, \tau),$$
$$v(\mathbf{y}, t, \tau) = G(\mathbf{y}, \tau), \qquad \text{for } t > 0, \tag{99.2}$$
$$v(\mathbf{x}, 0, \tau) = H(\mathbf{x}),$$

for $v(\mathbf{x}, t, \tau)$. Note that the variable of integration in (99.2) is t, while the source term and the surface conditions depend upon the parameter τ. Hence, the equation for $v(\mathbf{x}, t, \tau)$ has (effectively) a *constant* source term and *constant* surface conditions. Thus, it should be easier to determine $v(\mathbf{x}, t, \tau)$ than it was to determine $u(\mathbf{x}, t)$.

Knowing the solution of (99.2), the solution to (99.1) can be written as

$$u(\mathbf{x}, t) = \frac{\partial}{\partial t} \int_0^t v(\mathbf{x}, t - \tau, \tau) \, d\tau. \tag{99.3}$$

This is easily derived from manipulations of the Laplace transforms of equation (99.1) and (99.2). See any of the references for details.

Example

Suppose we want to solve the equations describing the temperature of an initially cool, insulated rod with a temperature $f(t)$ specified at one end

$$\begin{aligned}
u_t &= u_{xx}, & \text{for } 0 < x < 1, \, 0 < t < \infty, \\
u(0, t) &= 0, & \text{for } 0 < t < \infty, \\
u(1, t) &= f(t), & \text{for } 0 < t < \infty, \\
u(x, 0) &= 0, & \text{for } 0 \le x \le 1.
\end{aligned} \tag{99.4}$$

Instead of solving (99.4) for $u(x, t)$ we solve

$$\begin{aligned}
v_t &= v_{xx}, & \text{for } 0 < x < 1, \, 0 < t < \infty, \\
v(0, t, \tau) &= 0, & \text{for } 0 < t < \infty, \\
v(1, t, \tau) &= f(\tau), & \text{for } 0 < t < \infty, \\
v(x, 0, \tau) &= 0, & \text{for } 0 \le x \le 1,
\end{aligned} \tag{99.5}$$

for $v(x, t, \tau)$. By separation of variables (see page 419) the solution of (99.5) is found to be

$$v(x, t, \tau) = f(\tau) \left[x + \frac{2}{\pi} \sum_{n=1}^{\infty} \frac{(-1)^n}{n} e^{-n^2 \pi^2 t} \sin n\pi x \right],$$

which, for notational convenience, we choose to write as $v(x, t, \tau) = f(\tau)g(x, t)$. Using (99.3), the solution for $u(x, t)$ can then be written as

$$
\begin{aligned}
u(x, t) &= \frac{\partial}{\partial t} \int_0^t v(x, t - \tau, \tau) \, d\tau \\
&= \frac{\partial}{\partial t} \int_0^t f(\tau)g(x, t - \tau) \, d\tau \\
&= \frac{\partial}{\partial t} \int_0^t f(t - T)g(x, T) \, dT \\
&= f(0)g(x, t) + \int_0^t f'(t - T)g(x, T) \, dT,
\end{aligned} \tag{99.6}
$$

where we defined $T = t - \tau$ in the above. If, for example, $f(t) = e^{-t}$, then (99.6) may be simplified to yield

$$
u(x, t) = x - e^{-t} - 1 - \frac{2}{\pi} \sum_{n=1}^{\infty} \left(\frac{(-1)^n \sin n\pi x}{n(1 - n^2 \pi^2)} \left\{ n^2 \pi^2 e^{-n^2 \pi^2 t} - e^{-t} \right\} \right).
$$

Notes

[1] The procedure for hyperbolic partial differential equations is analogous to the procedure for parabolic partial differential equations. Consider, for example, the hyperbolic equation

$$
u_{tt} + L[u] = b(x, t),
$$

(where $L[\cdot]$ is uniformly elliptic) with the boundary conditions

$$
u(x, 0) = u_t(x, 0) = 0.
$$

If $v(x, t, \tau)$ is defined to be the solution of

$$
\begin{aligned}
v_{tt} + L[v] &= 0, \qquad \text{for } t > \tau, \\
v(x, \tau, \tau) &= 0, \\
v_t(x, \tau, \tau) &= b(x, \tau),
\end{aligned}
$$

then we have $u(x, t) = \int_0^t v(x, t, \tau) \, d\tau$.

Using this formulation, it can be shown that the solution to $u_{tt} - c^2 \nabla^2 u = F(x, y, z, t)$, is given by

$$
u(x, y, z, t) = \frac{1}{4\pi c} \int \int_{\xi^2 + \eta^2 + \zeta^2 \le c^2 t^2} \int \frac{F(\xi, \eta, \zeta, t - r/c)}{r} \, d\xi \, d\eta \, d\zeta, \tag{99.7}
$$

where $r^2 = (x - \xi)^2 + (y - \eta)^2 + (z - \zeta)^2$. The integral in (99.7) is called the *retarded potential*.

References
[1] C. R. Chester, *Techniques in Partial Differential Equations*, McGraw–Hill Book Company, New York, 1970, pages 156–158.
[2] R. Courant and D. Hilbert, *Methods of Mathematical Physics*, Interscience Publishers, New York, 1953, Volume 2, pages 202–204.
[3] S. J. Farlow, *Partial Differential Equations for Scientists and Engineers*, John Wiley & Sons, New York, 1982, Lesson 14 (pages 106–111).
[4] I. N. Sneddon, *Elements of Partial Differential Equations*, McGraw–Hill Book Company, New York, 1957, pages 278–282.
[5] E. Zauderer, *Partial Differential Equations of Applied Mathematics*, John Wiley & Sons, New York, 1983, pages 159–165.

100. Exact Equations

Applicable to   Quasilinear partial differential equations.

Yields

An exact solution.

Idea

Some quasilinear partial differential equations can be integrated directly.

Procedure

Consider the quasilinear partial differential equation

$$M(x, y, u)u_x = N(x, y, u)u_y. \qquad (100.1)$$

If this equation satisfies the exactness condition $M_x = N_y$, then an implicit solution to (100.1) will be given by $\phi(x, y, u) = 0$, where

$$M = \phi_y, \qquad N = \phi_x. \qquad (100.2.a\text{–}b)$$

To determine the function ϕ, integrate (100.2.a) to obtain

$$\phi = \int M \, dy + g(x, u). \qquad (100.3)$$

Then, using (100.2.b) we have

$$\int M_x \, dy + g_x(x, u) = N$$

or

$$g(x, u) = \int \left(N - \int M_x \, dy \right) dx + h(u), \qquad (100.4)$$

where $h(u)$ is an arbitrary function. Using (100.4) in (100.3) results in the final solution.

Example

Consider the equation

$$y u_x = x u u_y,$$

for which $M = y$ and $N = xu$. This equation is exact since $M_x = 0 = N_y$. From (100.3) we have

$$\phi = \int M \, dy + g(x, u) = \tfrac{1}{2} y^2 + g(x, u).$$

From (100.2.b) we have $\phi_x = g_x = N = xu$, or $g = \tfrac{1}{2} x^2 u + h(u)$. This leads to the general implicit solution:

$$\phi = \tfrac{1}{2} \left(y^2 + x^2 u \right) + h(u) = 0.$$

Choosing, for example, $h(u) = \tfrac{1}{2} au + b$ results in the explicit solution

$$u(x, y) = -\frac{b + y^2}{a + x^2}.$$

Notes

[1] The above example is from Benton [1].

References

[1] S. H. Benton, Jr., *The Hamilton–Jacobi Equation*, Academic Press, New York, 1977.

101. Hodograph Transformation

Applicable to   Quasilinear partial differential equations, a single equation or a system of equations.

Yields

A new formulation of the original equations.

Idea

In a partial differential equation it may be easier to solve the equation with the dependent and independent variables switched.

Procedure

This procedure works on a quasilinear equation or a system of such equations. That is, every term of each equation must have one and only one first derivative term, and there can be no higher order derivative terms in the equations.

Consider the case of two dependent variables (u, v) in two independent variables (x, y). Suppose $L[u, v] = 0$ represents the equation(s) to be solved for $u(x, y)$ and $v(x, y)$. This equation is transformed to the "hodograph" plane by writing $x = x(u, v)$ and $y = y(u, v)$ and transforming $L[u, v] = 0$ into a new equation $H[x, y] = 0$. In this new equation, x and y are treated as the dependent variables.

The solution obtained will, in general, be implicit. After the solution is obtained in the hodograph plane, the transformation must be checked to ensure that it is not singular.

Example 1

Suppose we have a pair of nonlinear equations arising from gas dynamics (from Whitham [9], page 182)

$$v_y + uv_x + bvu_x = 0,$$
$$u_y + uu_x + \frac{1}{b}vv_x = 0, \tag{101.1}$$

where b is a constant. Because the equations in (101.1) are quasilinear, the method of characteristics can be used to solve them. However, it is difficult to use that method directly.

The hodograph transformation can be used on (101.1) by inverting $u(x, y)$, $v(x, y)$ to find (see, for example, Kaplan [6], on how to change variables in this manner)

$$x_u = -v_y/J, \qquad x_v = u_y/J,$$
$$y_u = v_x/J, \qquad y_v = -u_x/J. \tag{101.2}$$

In (101.2), J is the Jacobian of the transformation, $J = u_y v_x - v_y u_x$. Using (101.2) in (101.1) results in the equations

$$x_u - uy_u + bvy_v = 0,$$
$$x_v - uy_v + \frac{1}{b}vy_u = 0. \tag{101.3}$$

Because the original equations were quasilinear, the Jacobian factors out of the equations (assuming it never vanishes) and does not appear in (101.3).

The equations in (101.3) are now quasilinear in the dependent variables (x, y). They may easily be solved by the method of characteristics, the details may be found in Whitham [9].

Example 2

An equation that arises in transonic small disturbance theory is

$$\phi_x \phi_{xx} - \phi_{yy} = 0. \tag{101.4}$$

Using $a := \phi_x$ and $b := \phi_y$, equation (101.4) can be written as the system of quasilinear equations:

$$a_y - b_x = 0, \qquad -a a_x + b_y = 0.$$

Using the Hodograph transformation, these equations simplify to:

$$x_b - y_a = 0, \qquad a y_b - x_a = 0,$$

with $J = x_b y_a - y_b x_a$. Combining these equations results in the familiar Tricomi equation: $a y_{bb} - y_{aa} = 0$.

Notes

[1] The hodograph transformation is frequently used in fluid mechanics for problems with unknown boundaries. In many situations, the boundaries become fixed in the hodograph plane.

[2] The transformation will be non-singular if the Jacobian of the transformation, J, does not vanish in the region of interest.

[3] Ames [1] shows that the nonlinear equations describing a vibration problem

$$u_t - v_x = 0, \qquad v_t - F^2(u) u_x = 0,$$

become, after applying the Hodograph transformation, the linear equations:

$$x_v - y_u = 0, \qquad x_u - F^2(u) y_v = 0.$$

[4] Whitham [9] (page 617) shows how the Born–Infeld equation

$$\left(1 - u_t^2\right) u_{xx} + 2 u_x u_t u_{xt} - \left(1 + u_x^2\right) u_{tt} = 0$$

may be linearized with the Hodograph transformation.

[5] This technique can also be applied to ordinary differential equations, a differential equation for $y(x)$ is inverted to become a differential equation for $x(y)$. See page 308.

References

[1] W. F. Ames, "Ad Hoc Exact Techniques for Nonlinear Partial Differential Equations," in W. F. Ames (ed.), *Nonlinear Partial Differential Equations in Engineering*, Academic Press, New York, 1967, pages 35–37.

[2] S. Bergman, *The Hodograph Method in the Theory of Compressible Fluid*, Supplement to *Fluid Dynamics* by von Mises and Friedrichs, Brown University, 1942.

[3] P. A. Clarkson, A. S. Fokas and M. J. Ablowitz, "Hodograph Transformations of Linearizable Partial Differential Equations," *SIAM J. Appl. Math.*, **49**, No. 4, August 1989, pages 1188–1209.

[4] J. Crank and T. Ozis, "Numerical Solution of a Free Boundary Problem by Interchanging Dependent and Independent Variables," *J. Inst. Maths. Applics*, **26**, 1980, pages 77–85.

[5] J. Crank, *Free and Moving Boundary Problems*, Clarendon Press, Oxford, 1984.

[6] W. Kaplan, *Advanced Calculus*, Addison–Wesley Publishing Co., Reading, MA, 1952, pages 132–135.

[7] A. R. Manwell, *The Hodograph Equations: An Introduction to the Mathematical Theory of Plane Transonic Flow*, Hafner, Darien, Conn., 1971.

[8] A. M. Siddiqui, P. N. Kaloni, and O. P. Chandna, "Hodograph Transformation Methods in Non-Newtonian Fluids," *J. Eng. Math.*, **19**, 1985, pages 203–216.

[9] G. B. Whitham, *Linear and Nonlinear Waves*, Wiley Interscience, New York, 1974.

102. Inverse Scattering

Applicable to   Nonlinear evolution equations, a single equation or a system.

Yields

A reformulation into an inverse problem, which can sometimes result in an exact solution.

Idea

By rewriting the evolution equation, some natural eigenfunction problems emerge.

Procedure

An evolution equation for $u(t, \mathbf{x}) = u(t, x_1, \ldots, x_m)$ may be written in the form

$$u_t = K(u), \tag{102.1}$$

where $K(\ )$ denotes a nonlinear differential operator in x. For a system of equations, the u in (102.1) represents a vector of unknowns $(u_1, \ldots, u_n)$.

The procedure is to write (102.1) in the Lax pair form (this is often the hardest part of the procedure)

$$L_t = i[L, A] = i(LA - AL), \tag{102.2}$$

where L and A are linear differential operators in $\mathbf{x}$ whose coefficients are polynomials in u and its $\mathbf{x}$ derivatives. Here, L_t refers to differentiation of u (and its derivatives) with respect to t in the expression for L. See Example 1 for how (102.2) is to be interpreted. Note that, if A were a Hamiltonian, then (102.2) would be a Heisenberg equation.

A straightforward calculation now shows that

$$i\phi\lambda_t = (L - \lambda)(A\phi - i\phi_t),$$

for arbitrary $\phi(t, \mathbf{x})$ and λ. If we assume that $\phi(t = 0, \mathbf{x})$ and $\lambda(t)$ are an eigenfunction–eigenvalue pair for L, that is

$$L\phi = \lambda\phi, \tag{102.3}$$

and if the eigenfunctions $\{\phi_j(t, \mathbf{x})\}$ evolve in time as

$$i\phi_t = A\phi, \tag{102.4}$$

then the eigenvalues will be independent of time (i.e., $\lambda_t = 0$).

Hence, the time evolution of the eigenfunctions can be determined from (102.4). Using the eigenfunctions $\{\phi_j(t, \mathbf{x})\}$, an inverse problem must be solved; the operator L must be determined from knowledge of its eigenfunctions. Since L depends on u, this might lead to a solution for u. For some problems, the time evolution of the eigenfunctions can be used in the Gelfand–Levitan linear integral equation (see Faddeyev [7]), which may (sometimes) be solved to determine $u(t, \mathbf{x})$.

Given equation (102.1) and the initial conditions $u(t = 0, \mathbf{x})$, the procedure can be summarized as

(A) Find the Lax pair representation of the evolution equation(s).
(B) Using $u(t = 0, \mathbf{x})$, evaluate L at $t = 0$ and then determine the eigenvalues $\{\lambda_j\}$ and the initial values of the eigenfunctions $\{\phi_j(0, \mathbf{x})\}$. These are the solutions to (102.3).

(C) Find the time evolution of the eigenfunctions by solving (102.4).

(D) Determine $u(t, \mathbf{x})$ by solving an inverse problem; i.e., using $\{\phi_j(t, \mathbf{x})\}$ as the solutions to (102.3), determine L for $t > 0$.

Note that $\{\phi_j(t), \mathbf{x}\}$ is called the *scattering data*. Even if step [4] cannot be carried out, useful information may be obtained from the scattering data.

Example 1

For the KdV equation

$$u_t + u_{xxx} - 6uu_x = 0,$$

a Lax pair is given by

$$L = \frac{\partial^2}{\partial x^2} - u,$$

$$A = -i \left(4\frac{\partial^3}{\partial x^3} - 6u\frac{\partial}{\partial x} - 3\frac{\partial u}{\partial x} \right).$$

$(102.5.a\text{--}b)$

This may be verified by calculating

$$L(A(\psi)) = i(3\psi u_{xxx} + 12\psi_x u_{xx} - 3\psi uu_x + 15\psi_{xx} u_x - 6\psi_x u^2$$
$$+ 10\psi_{xxx} u - 4\psi_{xxxxx}),$$
$$A(L(\psi)) = i(4\psi u_{xxx} + 12\psi_x u_{xx} - 9\psi uu_x + 15\psi_{xx} u_x - 6\psi_x u^2$$
$$+ 10\psi_{xxx} u - 4\psi_{xxxxx}),$$
$$(LA - AL)\psi = -i(u_{xxx} - 6uu_x)\psi$$

(102.6)

where $\psi = \psi(x)$ is an arbitrary function. Using (102.5.a) and (102.6), we then determine

$$L_t = -u_t,$$
$$[L, A] = [L(A(\cdot)) - A(L(\cdot))] = -i(u_{xxx} - 6uu_x).$$

(102.7)

When (102.7) is used in (102.2), the KdV equation is the result.

Example 2

For the sine–Gordon equation

$$u_{xt} = \sin u,$$

the scattering equations (which determine the initial values of the eigen-functions) for the vector eigenfunction $(\phi, \psi)^{\mathrm{T}}$ may be written as

$$L \begin{pmatrix} \phi \\ \psi \end{pmatrix} = i \begin{pmatrix} \dfrac{\partial}{\partial x} & \dfrac{1}{2}\dfrac{\partial u}{\partial x} \\ \dfrac{1}{2}\dfrac{\partial u}{\partial x} & -\dfrac{\partial}{\partial x} \end{pmatrix} \begin{pmatrix} \phi \\ \psi \end{pmatrix} = \lambda \begin{pmatrix} \phi \\ \psi \end{pmatrix},$$

while the evolution equations for the vector eigenfunction may be written as

$$i\frac{\partial}{\partial t} \begin{pmatrix} \phi \\ \psi \end{pmatrix} = A \begin{pmatrix} \phi \\ \psi \end{pmatrix} = -\frac{1}{4\lambda} \begin{pmatrix} \cos u & \sin u \\ \sin u & -\cos u \end{pmatrix} \begin{pmatrix} \phi \\ \psi \end{pmatrix}.$$

Notes

[1] The formulation of inverse scattering presented here is not the only possible formulation. There are other formulations, which may be easier to carry out on specific problems.

[2] The paper by Case and Kac [5] discusses a discrete inverse scattering problem; their problem illustrates many of the ideas from scattering theory without all of the mathematical difficulties.

References

[1] M. J. Ablowitz, D. J. Kaup, A. C. Newell, and H. Segur, "The Inverse Scattering Transform — Fourier Analysis for Nonlinear Problems," *Stud. Appl. Math.*, **53**, No. 4, December 1974, pages 249–315.

[2] M. J. Ablowitz and H. Segur, *Solitons and the Inverse Scattering Transform*, SIAM, Philadelphia, 1981.

[3] F. Calogero and A. Degasperis, *Spectral Transform and Solitons: Tools to Solve and Investigate Nonlinear Evolution Equations*, North–Holland Publishing Co., New York, 1982.

[4] F. Calogero and M. C. Nucci, "Lax Pairs Galore," *J. Math. Physics*, **32**, No. 1, January 1991, pages 72–74.

[5] K. M. Case and M. Kac, "A Discrete Version of the Inverse Scattering Problem," *J. Math. Physics*, **14**, No. 5, 1973, pages 594–603.

[6] W. Eckhaus and A. V. Harten, *The Inverse Scattering Transformation and the Theory of Solitons*, North–Holland Publishing Co., New York, 1981.

[7] L. D. Faddeyev, "The Inverse Problem in the Quantum Theory of Scattering," *J. Math. Physics*, **4**, 1963, pages 72–104.

[8] M. Ito, "A REDUCE program for Evaluating a Lax Pair Form," *Comput. Physics Comm.*, **34**, 1985, pages 325–331.

[9] J. R. McLaughlin, "Analytical Methods for Recovering Coefficients in Differential Equations from Spectral Data," *SIAM Review*, **28**, No. 1, March 1986, pages 53–72.

[10] M. Musetts and R. Conte, "Algorithmic Method for Deriving Lax Pairs from the Invariant Painlevé Analysis of Nonlinear Partial Differential Equations," *J. Math. Physics*, **32**, No. 6, June 1991, pages 1450–1457.

[11] M. C. Nucci, "Pseudopotentials, Lax Equations, and Bäcklund Transformations for Non-Linear Evolution Equations," *J. Phys. A: Math. Gen.*, **21**, 1988, pages 73–79.

[12] M. Tabor, *Chaos and Integrability in Nonlinear Dynamics*, John Wiley & Sons, New York, 1989, Chapter 7.

[13] G. B. Whitham, *Linear and Nonlinear Waves*, Wiley Interscience, New York, 1974.

103. Jacobi's Method

Applicable to   First order partial differential equations with three or more dependent variables. In the special case that the dependent variable appears explicitly in the equation, then it also applies to equations with two dependent variables.

Yields

An explicit solution if a certain step can be carried out.

Idea

Given a partial differential equation for $z(\mathbf{x}) = z(x_1, x_2, \ldots, x_n)$, if the set of n first derivatives $\{p_i = \partial z/\partial x_i \mid i = 1, 2, \ldots, n\}$ are explicitly known, then $z(\mathbf{x})$ may be found by integrating the Pfaffian differential equation: $dz = p_1 dx_1 + \cdots + p_n dx_n$. Jacobi's method determines the $\{p_i\}$ from a given partial differential equation.

Procedure

Let us presume that the given partial differential equation for $z = z(\mathbf{x}) = z(x_1, \ldots, x_n)$, with $n = 3$, is of the form

$$F(\mathbf{x}, \mathbf{p}) = 0, \qquad (103.1)$$

where $p_i = \partial z/\partial x_i$. If we could find two other equations, that have the same solution as (103.1), of the form $\{F_2(\mathbf{x}, \mathbf{p}) = 0, F_3(\mathbf{x}, \mathbf{p}) = 0\}$, then we might be able to determine $\{p_1 = p_1(\mathbf{x}), \ldots, p_n = p_n(\mathbf{x})\}$ by combining these three equations. Then we could find $z(\mathbf{x})$ by solving the Pfaffian differential equation (see page 326)

$$dz = p_1\, dx_1 + \cdots + p_n\, dx_n. \tag{103.2}$$

So, we need to determine $\{F_2, F_3\}$ in such a way that their solutions are the same as the solution to (103.1). This requirement results in (see the section on compatible systems, page 39)

$$[F, F_2] := \sum_{i=1}^{n} \left(\frac{\partial F}{\partial x_i} \frac{\partial F_2}{\partial p_i} - \frac{\partial F}{\partial p_i} \frac{\partial F_2}{\partial x_i} \right) = 0,$$
$$[F, F_3] = 0, \tag{103.3}$$
$$[F_2, F_3] = 0.$$

where $[\,,\,]$ is the usual Poisson bracket. The characteristic equations for F_2 (or F_3), from (103.3), can be written as (see page 368)

$$\frac{dx_1}{-\dfrac{\partial F}{\partial p_1}} = \frac{dp_1}{\dfrac{\partial F}{\partial x_1}} = \ldots = \frac{dx_n}{-\dfrac{\partial F}{\partial p_n}} = \frac{dp_n}{\dfrac{\partial F}{\partial x_n}}. \tag{103.4}$$

(These are also known as the subsidiary equations.) Hence, the procedure is to solve (103.4) for $F_2(\mathbf{x}, \mathbf{p}) = 0$ and $F_3(\mathbf{x}, \mathbf{p}) = 0$. It must be then be verified that $[F_2, F_3] = 0$. Then solving $\{F = 0, F_1 = 0, F_2 = 0\}$ for $p_i = p_i(\mathbf{x})$ and integrating (103.2) results in a solution to (103.1).

Example

This example is from Piaggio [3]. Suppose we have the following nonlinear partial differential equation in three independent variables

$$0 = F(\mathbf{x}, \mathbf{p}) = 2x_1 x_3 \frac{\partial z}{\partial x_1} + 3x_3^2 \frac{\partial z}{\partial x_2} + \left(\frac{\partial z}{\partial x_2} \right)^2 \frac{\partial z}{\partial x_3}, \tag{103.5}$$
$$= 2x_1 x_3 p_1 + 3x_3^2 p_2 + p_2^2 p_3,$$

The subsidiary equations in (103.4) can be written as

$$\frac{dx_1}{-2x_1 x_3} = \frac{dp_1}{2x_3 p_1} = \frac{dx_2}{-3x_3^2 - 2p_2 p_3} = \frac{dp_2}{0} = \frac{dx_3}{-p_2^2} = \frac{dp_3}{2x_1 p_1 + 6x_3 p_2}. \tag{103.6}$$

From the first equality in (103.6) we have

$$F_2(\mathbf{x}, \mathbf{p}) = p_1 x_1 - A_1 = 0, \tag{103.7}$$

where A_1 is an arbitrary constant. From the fourth term in (103.6) we have

$$F_3(\mathbf{x}, \mathbf{p}) = p_2 - A_2 = 0, \tag{103.8}$$

where A_2 is another arbitrary constant. Clearly, $[F_2, F_3] = 0$ for our chosen F_2 and F_3. Combingin (103.7) and (103.8) with the original equation, (103.5), we find that

$$p_3 = -\frac{1}{A_2^2}(2A_1 x_3 + 3A_2 x_3^2). \tag{103.9}$$

In equations (103.7)–(103.9) we have found expressions for the $\{p_i\}$. Hence

$$dz = p_1\, dx_1 + p_2\, dx_2 + p_3\, dx_3$$
$$= \frac{A_1}{x_1}\, dx_1 + A_2\, dx_2 - \frac{1}{A_2^2}(2A_1 x_3 + 3A_2 x_3^2)\, dx_3,$$

which can be integrated to yield the solution

$$z = A_1 \log x_1 + A_2 x_2 - \frac{1}{A_2^2}(A_1 x_3^2 + A_2 x_3^3) + A_3,$$

where A_3 is another arbitrary constant.

Notes

[1] If the given partial differential equation has only two independent variables and if the dependent variable z is explicit in the partial differential equation, then we can transform the partial differential equation into the form of (103.1). For example, if we have $F(x, y, z, p, q) = 0$ (where, as usual, $p = \partial z/\partial x$, $q = \partial z/\partial y$), suppose that $u(x, y, z) = 0$ is an integral of this equation. If we define $u_1 = \partial u/\partial x$, $u_2 = \partial u/\partial y$, $u_3 = \partial u/\partial z$, then we can write $p = -u_1/u_3, q = -u_2/u_3$. Using these definitions for p and q in the original equation yields an equation of the form $f(x, y, z, u_1, u_2, u_3) = f(\mathbf{x}, \mathbf{p}) = 0$.

[2] When this method is specialized to two independent variables, it is often called *Charpit's method*. See Chester [2] or Piaggio [3] for details.

[3] When $n > 3$, then the only change in the procedure is that we must now determine $\{F_2, F_3, \ldots, F_n\}$ and use these (with F) to solve for the $\{p_i\}$.

References

[1] W. F. Ames, "Ad Hoc Exact Techniques for Nonlinear Partial Differential Equations," in W. F. Ames (ed.), *Nonlinear Partial Differential Equations in Engineering*, Academic Press, New York, 1967, pages 54–57.

[2] C. R. Chester, *Techniques in Partial Differential Equations*, McGraw–Hill Book Company, New York, 1970, page 212 and Chapter 15 (pages 315–337).

[3] H. T. H. Piaggio, *An Elementary Treatise on Differential Equations and Their Applications*, G. Bell & Sons, Ltd, London, 1926, pages 162–170.

[4] I. N. Sneddon, *Elements of Partial Differential Equations*, McGraw–Hill Book Company, New York, 1957, pages 69–73 and 78–80.

104. Legendre Transformation

Applicable to   Partial differential equations in one dependent variable that are *not* of the form $F\left(u_{x_1}, u_{x_2}, \ldots, u_{x_n}\right) = 0$.

Yields

An alternative formulation of the original problem.

Idea

A surface in space may be described by a point or as an envelope of tangent planes. Changing variables from one representation to the other may facilitate finding a solution. After a solution is obtained, it can be transformed back to the original variables.

Procedure

We illustrate the technique for two independent variables, the notes show how the technique may be extended to n independent variables. Given a function $u(x, y)$, we change to the new variables $w(\zeta, \eta)$ by the transformation

$$w(\zeta, \eta) + u(x, y) = x\zeta + y\eta, \qquad (104.1)$$

with the following definitions

$$u_x = \zeta, \quad w_\zeta = x, \quad u_y = \eta, \quad w_\eta = y. \qquad (104.2)$$

From (104.1) and (104.2) it is easy to derive that

$$u_{xx} = J w_{\eta\eta},$$
$$u_{xy} = u_{yx} = -J w_{\zeta\eta},$$
$$u_{yy} = J w_{\zeta\zeta},$$

where J is the Jacobian of the transformation. The Jacobian may be expressed as

$$J = u_{xx}u_{yy} - (u_{xy})^2 = \frac{1}{w_{\eta\eta}w_{\zeta\zeta} - (w_{\eta\zeta})^2}.$$

To be able to transform from the $\{u, x, y\}$ variables to the $\{w, \zeta, \eta\}$ variables, the Jacobian must not vanish. If $J \neq 0$, then the surface is said to be *developable*. The solutions with $J = 0$ are said to be *non-developable* solutions. The non-developable solutions are not obtainable by the Legendre transformation.

Summary

For the partial differential equation of at most second order in the variables $\{u, x, y\}$,

$$F\left(x, y, u, u_x, u_y, u_{xx}, u_{xy}, u_{yy}\right) = 0, \tag{104.3}$$

we make the Legendre transformation to obtain the new equation

$$F\left(w_\zeta, w_\eta, \zeta w_\zeta + \eta w_\eta - w, \zeta, \eta, J w_{\eta\eta}, -J w_{\zeta\eta}, J w_{\zeta\zeta}\right) = 0, \tag{104.4}$$

in the new variables $\{w, \eta, \zeta\}$.

Sometimes equation (104.4) is easier to solve than equation (104.3). After equation (104.4) is solved to determine $w(\eta, \zeta)$, we must change back to the original variables. Changing from the $\{w, \eta, \zeta\}$ variables to the $\{u, x, y\}$ variables can be done (due to the implicit function theorem) but may be difficult.

Example

Consider the nonlinear partial differential equation

$$u_x u_y = x, \tag{104.5}$$

which we want to solve for $u(x, y)$. The Legendre transformation of (104.5) is (using the transformations in (104.1) and (104.2), or using (104.4) directly)

$$w_\zeta = \zeta\eta. \tag{104.6}$$

This has the solution

$$w(\zeta, \eta) = \tfrac{1}{2}\eta\zeta^2 + f(\eta), \tag{104.7}$$

where $f(\eta)$ is an arbitrary function of η. We have now finished solving the differential equation. Since we have the solution in terms of the new variables, all that remains is to transform to the old variables. This change of variables will utilize the $w(\zeta, \eta)$ that was found.

Using (104.6) and $w_\zeta = x$ (from (104.2)) we have

$$x = \eta\zeta. \qquad (104.8)$$

Differentiating (104.7) with respect to η and using $y = w_\eta$ (from (104.2)) yields

$$y = \tfrac{1}{2}\zeta^2 + f'(\eta). \qquad (104.9)$$

Using (104.7)–(104.9) in (104.1) produces the equation

$$u = x\zeta + y\eta - w(\zeta, \eta) = \eta\zeta^2 + \eta f'(\eta) - f(\eta). \qquad (104.10)$$

Solving (104.8) for ζ, and then substituting that result in (104.9) and (104.10) produces

$$y = \frac{x^2}{2\eta^2} + f'(\eta),$$

$$u = \frac{x^2}{\eta} + \eta f'(\eta) - f(\eta). \qquad (104.11.a\text{–}b)$$

This is a parametric representation of the solution $u(x, y)$. All of the developable solutions of (104.5) are completely characterized by (104.11). Given any $f(\eta)$ we can, in principle, find $\eta = \eta(x, y)$ from (104.11.a). Using this value for η in (104.11.b) then gives u as a function of x and y.

To illustrate this, if we choose

$$f(\eta) = \frac{A}{\eta},$$

where A is an arbitrary constant, then (104.11) becomes

$$y = \left(\frac{1}{2}x^2 - A\right)\frac{1}{\eta^2}, \qquad u = \frac{x^2 - 2A}{\eta}. \qquad (104.12.a\text{–}b)$$

Solving (104.12.a) for η and using this expression in (104.12.b) produces

$$u(x, y) = \sqrt{2y(x^2 - 2A)}.$$

Now that we have an explicit solution, we must check that the Jacobian does not vanish. In this example, $J \neq 0$.

Notes

[1] Observe that

$$u = Dy + \frac{1}{2D}x^2 + C, \qquad (104.13)$$

where C and D are constants, is also a solution to equation, (104.5), but this solution is not contained in (104.11) for any $f(\eta)$. This is because the solution in (104.13) is non-developable $(J = 0)$.

[2] The Legendre transformation may be naturally extended to partial differential equations in n variables. The transformation (from $u(x_1, x_2, \ldots, x_n)$ to $w(\zeta_1, \zeta_1, \ldots, \zeta_n)$) and its inverse is given by

$$u(x_1, x_2, \ldots, x_n) = w(\zeta_1, \zeta_1, \ldots, \zeta_n) + x_1\zeta_1 + x_2\zeta_2 + \cdots + x_n\zeta_n,$$

$$u_{x_1} = \zeta_1, \qquad u_{x_2} = \zeta_2, \qquad \cdots, \qquad u_{x_n} = \zeta_n,$$
$$w_{\zeta_1} = x_1, \qquad w_{\zeta_2} = x_2, \qquad \cdots, \qquad w_{\zeta_n} = x_n.$$

See Courant and Hilbert [3] for more details.

[3] Clairaut's equation, $u = xu_x + yu_y + f(u_x, u_y)$, under the Legendre transformation, becomes the simple equation $w = -f(\zeta, \eta)$.

[4] The Legendre transformation is an example of a contact transformation (see page 206).

[5] The Legendre transformation is an involutory transformation; that is, the Legendre transformation applied twice results in the original equation.

[6] The Legendre transformation is used in mechanics when transforming from the Lagrangian formulation to the Hamiltonian formulation (or vice-versa). See Goldstein [5] for details.

[7] The Legendre transformation is used in thermodynamics when transforming the fundamental equation from internal energy (canonical variables are specific volume and specific entropy) to the Gibbs function (canonical variables are pressure and temperature), or to enthalpy (canonical variables are pressure and specific entropy), or to the Helmholtz function (canonical variables are specific volume and temperature). For more details of this application, see Kestin [6].

[8] If the Legendre transformation is applied to a partial differential equation of the form $F(u_x, u_y) = 0$, then the algebraic relation $F(\zeta, \eta) = 0$ results. Since $w(\zeta, \eta)$ cannot be determined from this equation, this class of equations cannot be solved by the use of the Legendre transformation.

References

[1] W. F. Ames, "Ad Hoc Exact Techniques for Nonlinear Partial Differential Equations," in W. F. Ames (ed.), *Nonlinear Partial Differential Equations in Engineering*, Academic Press, New York, 1967, pages 37–40.

[2] C. R. Chester, *Techniques in Partial Differential Equations*, McGraw–Hill Book Company, New York, 1970, pages 209–210.

[3] R. Courant and D. Hilbert, *Methods of Mathematical Physics*, Interscience Publishers, New York, 1953, Volume 2, pages 32–39.

[4] B. Epstein, *Partial Differential Equations An Introduction*, McGraw–Hill Book Company, New York, 1962, pages 65–68.

[5] H. Goldstein, *Classical Mechanics*, Addison–Wesley Publishing Co., Reading, MA, 1950, Section 7.1 (pages 215–218).

[6] J. Kestin, *A Course in Thermodynamics*, Blaisdell Publishing Co., Waltham, MA, 1966, Chapter 12.

105.   Lie Groups: PDEs

Applicable to   Linear and nonlinear partial differential equations.

Yields

Similarity variables that may be used to decrease the number of independent variables in a partial differential equation.

Idea

By determining the transformation group under which a given partial differential equation is invariant, we can obtain information about the invariants and symmetries of that equation. This information, in turn, can be used to determine similarity variables that will reduce the number of independent variables in the system.

Procedure

Some background material about Lie Groups may be found in the section "Lie Groups: ODEs" (starting on page 314). We will utilize terms that have been defined in that section.

We illustrate the general technique on one partial differential equation in two independent variables. Suppose we would like to solve the partial differential equation

$$N(u, x, y) = 0, \tag{105.1}$$

for $u(x, y)$. We first determine a one parameter Lie group of transformations, under which (105.1) is invariant; then we use this group to determine similarity variables. We suppose that the group has the form

$$\bar{u} = u + \varepsilon U(u, x, y) + O(\varepsilon^2),$$
$$\bar{x} = x + \varepsilon X(u, x, y) + O(\varepsilon^2), \tag{105.2}$$
$$\bar{y} = y + \varepsilon Y(u, x, y) + O(\varepsilon^2).$$

We want this group to leave equation (105.1) invariant; that is,

$$N(\bar{x}, \bar{y}, \bar{u}) = 0, \tag{105.3}$$

or, equivalently,

$$u(\bar{x}, \bar{y}) = \bar{u}(u, x, y; \varepsilon). \tag{105.4}$$

If we utilize the transformations in (105.2), then the chain rule produces

$$\frac{\partial x}{\partial \bar{x}} = 1 - \varepsilon \left(X_x + X_u u_x \right) + O\left(\varepsilon^2\right),$$
$$\frac{\partial x}{\partial \bar{y}} = -\varepsilon \left(X_y + X_u u_y \right) + O\left(\varepsilon^2\right),$$
$$\frac{\partial y}{\partial \bar{x}} = -\varepsilon \left(Y_x + Y_u u_x \right) + O\left(\varepsilon^2\right), \tag{105.5}$$
$$\frac{\partial y}{\partial \bar{y}} = 1 - \varepsilon \left(Y_y + Y_u u_y \right) + O\left(\varepsilon^2\right).$$

From (105.5), it is conceptually easy (though algebraically intensive) to determine how derivatives in the $\{\bar{u}, \bar{x}, \bar{y}\}$ system transform to derivatives in the $\{u, x, y\}$ system. For instance,

$$\frac{\partial \bar{u}}{\partial \bar{x}} = u_x + \varepsilon \left(U_x + (U_u - X_x)u_x - Y_x u_y - X_u u_x^2 - Y_u u_x u_y \right) + O\left(\varepsilon^2\right),$$
$$\frac{\partial \bar{u}}{\partial \bar{y}} = u_y + \varepsilon \left(U_y + (U_u - Y_y)u_y - X_y u_x - Y_u u_y^2 - X_u u_y u_x \right) + O\left(\varepsilon^2\right),$$
$$\frac{\partial^2 \bar{u}}{\partial \bar{x}^2} = u_{xx} + \varepsilon \Big(-Y_{uu} u_x^2 u_y - X_{uu} u_x^3 - 2Y_u u_x u_{xy} - (3X_u + 2Y_{yu})u_x u_y$$
$$- Y_u u_y^2 + (X_{uu} - 2Y_{ux})U_x^2 - 2Y_y u_{xy} + (U_u - 2X_x Y_{xx})u_y$$
$$+ U_{xx} + (2U_{xu} - Y_{xx})u_x \Big) + O\left(\varepsilon^2\right). \tag{105.6}$$

Now the group is determined (i.e., $\{U, X, Y\}$ are determined) by requiring (105.3) to be satisfied.

After the group has been determined, a solution to (105.1) may be found from the *invariant surface condition*

$$U(u, x, y) = X(u, x, y)\frac{\partial u}{\partial x} + Y(u, x, y)\frac{\partial u}{\partial y}, \tag{105.7}$$

which is just the first order term of (105.4) when that equation is expanded for small values of ε. The solution of (105.7) leads to similarity variables that reduce the number of independent variables in the system. Note that (105.7) is quasilinear and that the subsidiary equations may be written as

$$\frac{du}{U(u,x,y)} = \frac{dx}{X(u,x,y)} = \frac{dy}{Y(u,x,y)}. \tag{105.8}$$

Example 1

Suppose we wish to analyze the heat equation

$$u_y = u_{xx}. \tag{105.9}$$

We take $\overline{u}_{\overline{y}} = \overline{u}_{\overline{x}\overline{x}}$ and substitute for the derivatives from (105.6). We also substitute u_y for u_{xx} (from (105.9)). This leads to a large expression that must equal zero (for an idea of how the expression looks, see note number 5 on page 318).

Equating to zero the coefficients of $\{u, u_x, u_y, u_x^2, u_y^2, u_{xy}, u_x u_y, u_x u_{xy}\}$ in this expression leads to eight simultaneous equations involving $\{U, X, Y\}$. The solution to these equations will determine the transformation group. Three of these equations are

$$u_x^2: \quad Y_u = 0,$$
$$u_x u_y: \quad X_u = 0,$$
$$u_x u_{xy}: \quad U_{uu} = 0.$$

These equations produce: $X(u,x,y) = X(x,y)$, $Y(u,x,y) = Y(x,y)$ and $U(u,x,y) = f(x,y)u + g(x,y)$, where f and g are functions to be determined. Using this simplification for $\{U, X, Y\}$, the other five equations become

$$Y_x = 0, \qquad\qquad f_{xx} - f_y = 0,$$
$$2X_x - Y_y = 0, \qquad\qquad g_{xx} - g_y = 0, \tag{105.10}$$
$$X_y - X_{xx} + 2f_x = 0.$$

If we take $g \equiv 0$ (just to simplify the algebra), then the equations in (105.10) may be solved to determine the transformation group

$$X = 2c_1 y + 4c_2 xy + c_4 + c_5 x,$$
$$Y = 4c_2 y^2 + 2c_5 y + +c_6, \tag{105.11}$$
$$U = -\left(c_1 x + c_2(x^2 + 2y) + c_3\right)u.$$

where $\{c_1, \ldots, c_6\}$ are arbitrary constants. Now that we have found a transformation group, similarity variables may be found.

Special Case 1

If we take $c_1 = c_2 = c_4 = c_6 = 0$ in (105.11), the subsidiary equations (from (105.8)) become

$$\frac{du}{-c_3 u} = \frac{dx}{c_5 x} = \frac{dy}{2c_5 y}.$$

Two solutions to these equations are

$$\text{constant} = \frac{x}{\sqrt{y}}, \qquad \text{constant} = \frac{u}{y^\alpha},$$

where $\alpha = -c_3/2c_5$. From these similarity variables we propose a solution of the form

$$\eta = \frac{x}{\sqrt{y}}, \qquad h(\eta) = \frac{u}{y^\alpha}. \qquad (105.12)$$

Using this form in equation (105.9), we find that h satisfies the ordinary differential equation $h'' = \alpha h - \frac{1}{2}\eta h'$. Every solution to this equation will generate a solution to the equation (105.9), in the form $u(x, y) = y^\alpha h\left(\frac{x}{\sqrt{y}}\right)$.

Special Case 2

If we take $c_1 = c_2 = c_4 = c_5 = 0$ in (105.11), then the subsidiary equations (from (105.8)) become

$$\frac{du}{-c_3 u} = \frac{dx}{0} = \frac{dy}{c_6}.$$

Two solutions to these equations are

$$\text{constant} = x, \qquad \text{constant} = \frac{u}{e^{\beta y}},$$

where $\beta = -c_3/c_6$. From these similarity variables we propose a solution of the form

$$\eta = x, \qquad k(\eta) = \frac{u}{e^{\beta y}}.$$

Using this form in equation (105.9), we find that k satisfies the ordinary differential equation $k'' - \beta k = 0$. Every solution to this equation will generate a solution to the equation (105.9), in the form $u(x, y) = e^{\beta y} k(x)$.

Example 2

Consider similarity solutions of Laplace's equation in two dimensions: $\nabla^2 u = u_{xx} + u_{yy} = 0$. To find the Lie group of transformations that leaves this equation invariant, we consider the group defined in (105.2). After extensive algebra we find that, to lowest order, $\{X, Y, U\}$ may be expressed as

$$X = d_1 + d_3x - d_4y + d_5(x^2 - y^2) + 2d_6xy + \text{(cubic terms)},$$
$$Y = d_2 + d_3y + d_4x + 2d_5xy + d_6(y^2 - x^2) + \text{(cubic terms)}, \qquad (105.13)$$
$$U = d_7u + V(x, y),$$

where $V(x, y)$ is any solution to $\nabla^2 V = 0$ and $\{d_1, d_2, \ldots, d_7\}$ are arbitrary (complex) constants. The similarity solutions to $\nabla^2 u = 0$ may now be determined from the subsidiary equations in (105.8). We will take $V = 0$, and investigate two possibilities for the other parameters in (105.13).

Special Case 1

If we presume that the only nonzero parameters in (105.1) are d_1, d_2, and d_7, then the subsidiary equations become

$$\frac{du}{d_7u} = \frac{dx}{d_1} = \frac{dy}{d_2}.$$

Using the equation specified by the second equality sign, we determine that $d_2x - d_1y = $ constant. Using the equation specified by the first equality sign, we determine that $ue^{-d_7x/d_1} = $ constant. Hypothezing a solution of the form $u(x, y) = e^{d_7x/d_1}f(d_2x - d_1y)$, and then requiring that $\nabla^2 u = 0$, leads to a constant coefficient ordinary differential equation for f:

$$d_1^2\left(d_1^2 + d_2^2\right)f'' - 2d_1d_2d_7f' + d_7^2f = 0.$$

Special Case 2

If we presume that the only nonzero parameters in (105.1) are d_3 and d_7, then the subsidiary equations become

$$\frac{dx}{d_3x} = \frac{dy}{d_3y} = \frac{du}{d_7u}.$$

These equations can be solved to determine that: $u(x, y) = y^m g(\eta)$, $\eta = y/x$, where $m = d_7/d_3$. By requiring $\nabla^2 u = 0$ to hold, we find the following ordinary differential equation for $g(\eta)$:

$$\left(\eta^2 + \eta^4\right)g'' + 2\eta\left(m + \eta^2\right)g' + m(m-1)g = 0.$$

Notes

[1] Lie group analysis is the most useful and general of all the techniques presented in this book.

[2] There are other techniques for determining the group under which a given partial differential equation is invariant. A list of techniques is given in Seshadri and Na [12].

[3] If $u(x, y)$ is a solution of (105.9), then the following transformations also represent solutions:

$$T_1 : \begin{cases} x \to x + 2cy \\ u \to ue^{-c(x+y^2)} \end{cases}$$

$$T_2 : \begin{cases} x \to x/(1 - 4cy) \\ y \to y/(1 - 4cy) \\ u \to u\sqrt{1 - 4cy}\, \exp\left(-\dfrac{cx^2}{1 - 4cy}\right) \end{cases}$$

$$T_3 : \quad u \to e^c u$$

$$T_4 : \quad x \to x + c \qquad\qquad\qquad\qquad\qquad (105.14)$$

$$T_5 : \begin{cases} x \to e^c x \\ y \to e^{2c} y \end{cases}$$

$$T_6 : \quad y \to y + c$$

These transformations were all obtained from the group in (105.11). For example, the similarity variable $\eta = \dfrac{x}{\sqrt{y}}$ in (105.12) is equivalent to transformation T_5.

If $u = m(x, y)$ is a solution of the equation in (105.9), then another solution is given by (using all of the transformations listed in (105.14))

$$u = \frac{1}{\sqrt{1 + 4c_2 y}} \exp\left(c_3 - \frac{c_1 x + c_2 x^2 - c_1^2 y}{1 + 4c_2 y}\right)$$

$$\times m\left(\frac{e^{-c_5}(x - 2c_1 y)}{1 + 4c_2 y} - c_4, \frac{e^{-2c_5} y}{1 + 4c_2 y} - c_6\right)$$

See Olver [7] (pages 120–123) for details.

[4] Using Lie groups to find symmetries of partial differential equations can be computationally intensive. Algorithms have been developed for computerized handling of the calculations. A computer package in REDUCE is described in Schwarz [11], a package in FORMAC is in Fedorova and Kornyak [5], and a package in MACSYMA is in Rosencrans [10].

[5] The general equation of nonlinear heat conduction takes the form $u_t = (K(u)u_x)_x$. For this equation:

(A) If $K(u)$ is constant, then the symmetry group is infinite dimensional;

(B) If $K(u) = (au + b)^{-4/3}$, with $a \neq 0$, then there is a five-parameter symmetry group;

(C) If $K(u) = (au + b)^m$, for $m \neq -\frac{4}{3}$ and $a \neq 0$, then there is a four-parameter symmetry group;

(D) If $K(u) = ce^{au}$, then there is a four-parameter symmetry group;

(E) If $K(u)$ does not have one of the forms mentioned above, then there is a three-parameter symmetry group.

[6] A new technique for finding symmetries of partial differential equations, that are not point symmetries nor Lie–Bäcklund symmetries, may be found in Bluman, Reid, and Kumei [3].

[7] Olver [7] derives the complete symmetry group for many partial differential equations, including the: heat equation, wave equation, Euler equations, and Korteweg-de Vries equation. Ames and Nucci [1] studied the: Burgers' equation, Korteweg-de Vries equation (1 and 2 dimensions), Hopf equation, and Lin–Tsien equation.

[8] The section on similarity methods (beginning on page 424) shows how to find similarity variables of a specific form. The techniques in this section are, of course, much more general and will determine all possible similarity variables.

References

[1] W. F. Ames and M. C. Nucci, "Analysis of Fluid Equations by Group Methods," *J. Eng. Math*, **20**, 1985, pages 181–187.

[2] G. W. Bluman and S. Kumei, *Symmetries and Differential Equations*, Springer–Verlag, New York, 1989.

[3] G. W. Bluman, G. J. Reid, and S. Kumei, "New Classes of Symmetries for Partial Differential Equations," *J. Math. Physics*, **29**, No. 4, April 1988, pages 806–811.

[4] L. Dressner, *Similarity Solutions of Nonlinear Partial Differential Equations*, Pitman Publishing Co., Marshfield, MA, 1983.

[5] R. N. Fedorova and V. V. Kornyak, "Determination of Lie–Bäcklund Symmetries of Differential Equations Using FORMAC," *Comput. Physics Comm.*, **39**, 1986, pages 93–103.

[6] J. M. Hill, *Solution of Differential Equations by Means of One-Parameter Groups*, Pitman Publishing Co., Marshfield, MA, 1982.

[7] P. J. Olver, *Applications of Lie Groups to Differential Equations*, Graduate Texts in Mathematics #107, Springer–Verlag, New York, 1986.

[8] L. V. Ovsiannikov, *Group Analysis of Differential Equations*, translated by W. F. Ames, Academic Press, New York, 1982.

[9] A. Reiman, "Computer-Aided Closure of the Lie Algebra Associated with a Nonlinear Partial Differential Equation," *Comp. & Maths. with Appls.*, **7**, No. 5, 1981, pages 387–393.

[10] S. I. Rosencrans, "Computation of Higher–Order Fluid Symmetries Using MACSYMA," *Comput. Physics Comm.*, **38**, 1985, pages 347–356.

[11] F. Schwarz, "Automatically Determining Symmetries of Partial Differential Equations," *Computing*, **34**, 1985, pages 91–106.

[12] R. Seshadi and T. Y. Na, *Group Invariance in Engineering Boundary Value Problems*, Springer–Verlag, New York, 1985.

[13] S. Steinberg, "Applications of the Lie Algebraic Formulas of Baker, Campbell, Hausdorff, and Zassenhaus to the Calculation of Explicit Solutions of Partial Differential Equations," *J. Differential Equations*, **26**, 1977, pages 404–434.

106. Poisson Formula

Applicable to   Laplace's equation ($\nabla^2 u = 0$) in two dimensions with $u(\mathbf{x})$ prescribed on a circle; i.e., the Dirichlet problem in a disk.

Yields

An exact solution, given by an integral.

Idea

A simple extension of the Cauchy integral formula (from complex variable theory) allows the solution for Laplace's equation in a circle to be written down analytically.

Procedure

If $u(r, \theta)$ satisfies

$$\nabla^2 u = u_{rr} + \frac{1}{r}u_r + \frac{1}{r^2}u_{\theta\theta} = 0, \quad \text{for } 0 < r < R, \tag{106.1}$$

and

$$u(R, \theta) = f(\theta), \quad \text{for } 0 \leq \theta < 2\pi, \tag{106.2}$$

then $u(r, \theta)$ for $0 < r < R$ is given by

$$u(r, \theta) = \frac{1}{2\pi} \int_0^{2\pi} \frac{R^2 - r^2}{R^2 - 2Rr\cos(\theta - \phi) + r^2} f(\phi)\, d\phi. \tag{106.3}$$

This is known as the Poisson formula for a circle.

Example

If we have

$$\nabla^2 u = 0, \qquad u(R, \theta) = \sin\theta,$$

then

$$u(r, \theta) = \frac{1}{2\pi} \int_0^{2\pi} \frac{R^2 - r^2}{R^2 - 2Rr\cos(\theta - \phi) + r^2} \sin\phi \, d\phi$$

$$= \frac{r}{R} \sin\theta,$$

where the integral was carried out by using the method of residues.

Notes

[1] By use of conformal mappings, Laplace's equation in two dimensions for a non-circular region can often be changed to solving Laplace's equation in a circular region. Poisson's formula can be used for this new problem, and then the mapping can be used to find the solution for the original geometry. See the section on conformal mappings (page 376) for more details.

[2] The solution of (106.1) and (106.2) could also have been obtained by the use of Fourier series (see page 293). By this technique the solution to (106.1) and (106.2) becomes

$$u(r, \theta) = \frac{a_0}{2} + \sum_{n=1}^{\infty} \left(\frac{r}{a}\right)^n (a_n \cos n\theta + b_n \sin n\theta), \qquad (106.4)$$

where $\{a_n, b_n\}$ are defined by:

$$a_n = \frac{1}{\pi} \int_{-\pi}^{\pi} f(\theta) \cos n\theta \, d\theta, \qquad b_n = \frac{1}{\pi} \int_{-\pi}^{\pi} f(\theta) \sin n\theta \, d\theta. \qquad (106.5)$$

Note that this same solution would have been obtained by utilizing separation of variables. In Farlow [2] and Young [6] it is shown that the Poisson formula in (106.3) may be derived from the solution in (106.4) and (106.5).

[3] The Neumann problem for a disk

$$\nabla^2 v = 0, \qquad \frac{\partial v}{\partial n}(R, \theta) = g(\theta), \qquad (106.6)$$

may be converted to the Dirichlet problem (equations (106.1) and (106.2)) if we define

$$f(\theta) = \int_0^\theta g(\phi) \, d\phi,$$

$$v(x, y) = \int^{(x,y)} (u_y \, dx - u_x \, dy), \qquad (106.7)$$

see Young [6] for details. Note that the periodicity requirement of $f(\theta)$ requires that $g(\theta)$ satisfy (from (106.7)) $\int_0^{2\pi} g(\phi) \, d\phi = 0$. This must be satisfied if there is to exist any solution to (106.6). This requirement is related to the alternative theorems on page 14. (Note that the solution to (106.6) is indeterminate with respect to a constant.)

[4] The solution to the *exterior* problem

$$\nabla^2 w = 0,$$
$$w(R, \theta) = f(\theta), \quad w \text{ bounded at } r = \infty, \tag{106.8}$$

is given by

$$w(r, \theta) = -\frac{1}{2\pi} \int_0^{2\pi} \frac{R^2 - r^2}{R^2 - 2Rr\cos(\theta - \phi) + r^2} f(\phi)\, d\phi, \tag{106.9}$$

which is valid for $r \geq R$. See Kantorovich and Krylov [4] for details.

[5] Other exact solutions to Laplace's equation are also known. For example:

(A) If $\nabla^2 u = 0$ in a sphere of radius one and $u(1, \theta, \phi) = f(\theta, \phi)$, then

$$u(r, \theta, \phi) = \frac{1}{4\pi} \int_0^{\pi} \int_0^{2\pi} f(\Theta, \Phi) \frac{1 - r^2}{(1 - 2r\cos\gamma + r^2)^{3/2}} \sin\Theta\, d\Theta\, d\Phi, \tag{106.10}$$

where $\cos\gamma := \cos\theta\cos\Theta + \sin\theta\sin\Theta\cos(\phi - \Phi)$.

(B) If $\nabla^2 u = 0$ in the half plane, $y \geq 0$, and $u(x, 0) = f(x)$, then

$$u(x, y) = \frac{1}{\pi} \int_{-\infty}^{\infty} \frac{f(t)y}{(x - t)^2 + y^2}\, dt. \tag{106.11}$$

(C) If $\nabla^2 u = 0$ in the half space, $z \geq 0$, and $u(x, y, 0) = f(x, y)$, then

$$u(x, y, z) = \frac{z}{2\pi} \int_{-\infty}^{\infty} \int_{-\infty}^{\infty} \frac{f(\zeta, \eta)}{\left[(x - \zeta)^2 + (y - \eta)^2 + z^2\right]^{3/2}}\, d\zeta\, d\eta. \tag{106.12}$$

(D) If $\nabla^2 u = 0$ in the annulus, $0 < a \leq r \leq 1$, and $u(1, \theta)$ and $u(a, \theta)$ are given, then an explicit solution is given by Villat's integration formula. See Iyanaga and Kawada [3] for details.

References

[1] R. V. Churchill, *Complex Variables and Applications*, McGraw–Hill Book Company, New York, 1960, Chapter 11 (pages 242–258).

[2] S. J. Farlow, *Partial Differential Equations for Scientists and Engineers*, John Wiley & Sons, New York, 1982, Lesson 33 (pages 262–269).

[3] S. Iyanaga and Y. Kawada, *Encyclopedic Dictionary of Mathematics*, MIT Press, Cambridge, MA, 1980, page 1450.

[4] L. V. Kantorovich and V. I. Krylov, *Approximate Methods of Higher Analysis*, Interscience Publishers, New York, 1958, pages 572–575.

[5] N. Levinson and R. M. Redheffer, *Complex Variables*, Holden–Day, Inc., San Francisco, 1970, page 360.

[6] E. C. Young, *Partial Differential Equations*, Allyn and Bacon, Inc., Boston, MA, 1972, pages 273–285.

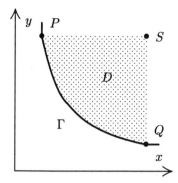

Figure 107.1 Domain in which equation (107.1) is solved.

107. Riemann's Method

Applicable to   Linear hyperbolic equations of the second order in two independent variables.

Yields

An exact solution in terms of the solution to the adjoint equation.

Idea

The solution of a non-characteristic initial value problem in two dimensions can be found if the adjoint equation with specified boundary conditions can be solved.

Procedure

Suppose we have the hyperbolic partial differential equation

$$L[u] = u_{xy} + a(x,y)u_x + b(x,y)u_y + c(x,y)u = f(x,y), \qquad (107.1)$$

where $u(x,y)$ is specified on the boundary Γ, which is not a characteristic. (See Figure 107.1.) Note that any linear hyperbolic equations of second order in two independent variables can be written in the form of (107.1).

We wish to find $u(S) = u(\zeta, \eta)$, where S represents an arbitrary point and is indicated in Figure 107.1. If we assume that the initial curve Γ is monotonically decreasing, then we can write the solution as

$$\begin{aligned}
u(\zeta, \eta) = {}& \tfrac{1}{2} R(P; \zeta, \eta) u(P) + \tfrac{1}{2} R(Q; \zeta, \eta) u(Q) \\
& - \int_P^Q B[u(x,y), R(x,y; \zeta, \eta)] \\
& + \iint_D f(x,y) R(x,y; \zeta, \eta)\, dx\, dy,
\end{aligned} \qquad (107.2)$$

where

$$B[u, v] = \left(avu + \tfrac{1}{2}vu_y - \tfrac{1}{2}v_yu\right) dy + \left(-bvu + \tfrac{1}{2}vu_x - \tfrac{1}{2}v_xu\right) dx,$$

(note that $B[u, v]$ includes the differential terms dx and dy) and $R(x, y; \zeta, \eta)$ is the Riemann function defined by

$$R_{xy} - aR_x - bR_y + (c - a_x - b_y)R = 0,$$

$$R(\zeta, y; \zeta, \eta) = \exp\left[\int_\eta^y a(\zeta, \sigma)\, d\sigma\right],$$

$$R(x, \eta; \zeta, \eta) = \exp\left[\int_\zeta^x b(\sigma, \eta)\, d\sigma\right],$$

$$R(\zeta, \eta; \zeta, \eta) = 1.$$

(107.3)

In this formulation, PS is a horizontal segment and QS is a vertical segment that contain the domain the dependence D. The derivation of this formula is more detailed than the format of this book allows. See Garabedian [6] for a full description. A simple motivation for the Riemann function is given in Kreith [8].

Example 1
   Suppose we have the partial differential equation

$$\alpha^2 w_{\beta\beta} - \beta^2 w_{\alpha\alpha} = 0,$$

$$w(\alpha, 1) = f(\alpha),$$

$$w_\beta(\alpha, 1) = g(\alpha),$$

(107.4.a–c)

where $-\infty < \alpha < \infty$, $1 < \beta < \infty$, and $f(\alpha)$ and $g(\alpha)$ are given functions. If we change variables in (107.1) from $\{w, \alpha, \beta\}$ to $\{u, x, y\}$ by

$$u(x, y) = w(\alpha, \beta),$$

$$x = \alpha\beta, \quad y = \frac{\beta}{\alpha},$$

(see the transformation on page 139), then equation (107.4.a) becomes

$$u_{xy} - \frac{1}{2x}u_y = 0.$$

(107.5)

The boundary conditions in (107.4) transform to

$$u\left(s, \frac{1}{s}\right) = f(s),$$

$$su_x\left(s, \frac{1}{s}\right) + \frac{1}{s}u_y\left(s, \frac{1}{s}\right) = g(s),$$

(107.6)

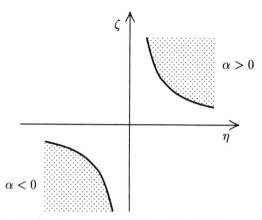

Figure 107.2 Domain in which equation (107.4) is solved.

where $-\infty < s < \infty$. By manipulations of (107.6), we can derive

$$
u\left(s, \frac{1}{s}\right) = f(s),
$$

$$
u_x\left(s, \frac{1}{s}\right) = \frac{1}{2}\left[f'(s) + \frac{1}{s}g(s)\right], \qquad (107.7)
$$

$$
u_y\left(s, \frac{1}{s}\right) = \frac{1}{2}\left[sg(s) - s^2 f(s)\right].
$$

The domain in which (107.5) and (107.7) are to be solved is shown in Figure 107.2.

To solve (107.5) and (107.7) we use Riemann's method. Comparing (107.5) to (107.1) we determine: $a = 0$, $b = -1/2x$, $c = 0$, $f = 0$. Hence the solution (from (107.2)) becomes

$$
u(\zeta, \eta) = \tfrac{1}{2}R(P; \zeta, \eta)u(P) + \tfrac{1}{2}R(Q; \zeta, \eta)u(Q)
$$
$$
- \int_P^Q \left[\left(\frac{1}{2}Ru_y - \frac{1}{2}R_y u\right)dy - \left(-\frac{1}{2x}Ru + \frac{1}{2}Ru_x - \frac{1}{2}R_x u\right)dx\right].
$$
$$
(107.8)
$$

All that remains is to find the Riemann function. From (107.3), $R(x, y; \zeta, \eta)$ satisfies

$$
R_{xy} + \frac{1}{2x}R_y = 0,
$$
$$
R(\zeta, y; \zeta, \eta) = 1,
$$
$$
R(x, \eta; \zeta, \eta) = \sqrt{\frac{\zeta}{x}}, \qquad (107.9.a\text{--}d)
$$
$$
R(\zeta, \eta; \zeta, \eta) = 1.
$$

Since (107.9.a) can be integrated directly with respect to x and then with respect to y, the general solution to (107.9) is easily seen to be of the form

$$R(x, y; \zeta, \eta) = M(x; \zeta, \eta) + \frac{K(y; \zeta, \eta)}{\sqrt{x}}, \tag{107.10}$$

for some $M(x; \zeta, \eta)$ and some $K(y; \zeta, \eta)$. Using (107.10) in the boundary conditions in (107.9), the solution is found to be

$$R(x, y; \zeta, \eta) = \sqrt{\frac{\zeta}{x}}. \tag{107.11}$$

Using (107.11) in (107.8), we can find $u(\zeta, \eta)$ and hence $w(\alpha, \beta)$ for any values of α and β.

Example 2

The Riemann function for the partial differential equation

$$u_{xy} = \tfrac{1}{4} k^2 u, \tag{107.12}$$

(when k is a constant) is

$$R(x, y; \zeta, \eta) = I_0 \left(k \sqrt{(x - \zeta)(y - \eta)} \right),$$

where I_0 is the usual modified Bessel function of order zero. Hence, the solution to (107.12) with the boundary conditions

$$u_x = \psi(x) \qquad \text{when } y = 0,$$
$$u_y = \phi(x) \qquad \text{when } x = 0,$$

is given by

$$u(x, y) = \int_0^y I_0 \left(k \sqrt{x(y - \eta)} \right) \phi(\eta) \, d\eta + \int_0^x I_0 \left(k \sqrt{y(x - \zeta)} \right) \psi(\zeta) \, d\zeta.$$

Notes

[1] Essentially, the Riemann function is a type of Green's function, the connection is made in Zauderer [11]. What we have called the Riemann function is sometimes called a Green's function or a Riemann–Green function.

[2] If the operator $L[u]$ in (107.1) is self-adjoint then we have the reciprocity principle: $R(x, y; \zeta, \eta) = R(\zeta, \eta; x, y)$.

[3] Numerical methods for solving hyperbolic equations that use the Riemann function are generally referred to as Godunov-type methods. A comparison of some Godunov-type methods with more classicial methods may be found in Woodward and Colella [10].

[4] Copson [3] suggests that the Riemann function may often have the form

$$R(x, y; \zeta, \eta) = \sum_{k=0}^{\infty} \frac{G_k \Upsilon^k}{(k!)^2}$$

where $\Upsilon = (x - \zeta)(y - \eta)$. When this is the case, then only the coefficients $\{G_k\}$ must be found. Copson [3] gives several examples of this approach.

[5] The technique presented here may be extended to higher order equations, for which the Riemann tensor must be determined. See Courant and Hilbert [4].

References

[1] H. Bateman, *Differential Equations*, Longmans, Green and Co., 1926, pages 280–285.

[2] C. R. Chester, *Techniques in Partial Differential Equations*, McGraw–Hill Book Company, New York, 1970, pages 222–231.

[3] E. T. Copson, *Partial Differential Equations*, Cambridge University Press, New York, 1975, pages 77–88.

[4] R. Courant and D. Hilbert, *Methods of Mathematical Physics*, Interscience Publishers, New York, 1953, Volume II, pages 450–461.

[5] J. L. Davis, *Finite Difference Methods in Dynamics of Continuous Media*, The MacMillan Company, New York, 1986, pages 75–79.

[6] P. R. Garabedian, *Partial Differential Equations*, Wiley, New York, 1964, pages 127–135.

[7] N. Iraniparast, "Green–Riemann Functions for a Class of Hyperbolic Focal Point Problems," *SIAM J. Math. Anal.*, **20**, No. 2, March 1989, pages 408–414.

[8] K. Kreith, "Establishing Hyperbolic Green's Functions via Leibniz's Rule," *SIAM Review*, **33**, No. 1, March 1991, pages 101–105.

[9] I. N. Sneddon, *Elements of Partial Differential Equations*, McGraw–Hill Book Company, New York, 1957, pages 119–122.

[10] P. Woodward and P. Colella, "The Numerical SImulation of Two-Dimensional Dluid FLow with Strong Shocks," *J. Comput. Physics*, **54**, 1984, pages 115–173.

[11] E. Zauderer, *Partial Differential Equations of Applied Mathematics*, John Wiley & Sons, New York, 1983, pages 485–492.

108. Separation of Variables

Applicable to   Most often, linear homogeneous partial differential equations.

Yields

An exact solution, generally in the form of an infinite series.

Idea

We look for a solution to a partial differential equation by separating the solution into pieces, where each piece deals with a single dependent variable.

Procedure

For linear homogeneous partial differential equations, try to represent the solution as a sum of terms where each term factors into a product of expressions, each expression dealing with a single independent variable. For nonlinear equations, try to represent the solution as a sum of such expressions. In all cases, not only must the equation admit a solution of the proposed form, but the boundary conditions must also have the right form.

In more detail, suppose that $L[u] = 0$ is a linear partial differential equation for $u(\mathbf{x})$ that has the form $L[u] = \sum_i L_i[u]$, where the $L_i[u]$ are differential operators. We look for a solution of this partial differential equation in the form

$$u(\mathbf{x}) = u(x_1, x_2, \cdots, x_n) = X_1(x_1)X_2(x_2)\ldots X_n(x_n),$$

where the functions $\{X_1, X_2, \ldots, X_n\}$ are to be determined. By using the above form in the original equation and reasoning about which terms depend upon which variables, we can often reduce the original partial differential equation into an ordinary differential equation for each of the X_i. In carrying this out, arbitrary constants will be introduced. After the resulting ordinary differential equations are solved, the arbitrary constants can generally be found by physical reasoning.

Since superposition can be used in linear equations, any number of terms (of the form shown above) will also be a solution of the original equation. Also, if each of these terms is multiplied by some constant, and then added together, the resulting expression will also be a solution. Hence, the final solution will frequently be a sum or an integral.

This sum will have unknown constants in it, due to the constants allowed in the superposition. These constants will be determined from the initial conditions and/or the boundary conditions.

The only time that we can be sure that we have found the most general solution to a given ordinary differential equation by this technique is when there exists a "completeness theorem" for each of the ordinary differential equations that we have found.

Example 1

Suppose we wish to solve the heat equation in a circle

$$\frac{\partial u}{\partial t} = \nabla^2 u \equiv \frac{1}{r}\frac{\partial}{\partial r}\left(r\frac{\partial u}{\partial r}\right) + \frac{1}{r^2}\frac{\partial^2 u}{\partial \theta^2}, \tag{108.1}$$

for $u(t, r, \theta)$. We try to separate variables in (108.1) by proposing a solution of the form

$$u(t, r, \theta) = T(t)R(r)\Theta(\theta). \tag{108.2}$$

Substituting (108.2) into (108.1) (and simplifying) yields

$$\frac{1}{rR}\frac{d}{dr}\left(r\frac{dR}{dr}\right) + \frac{1}{r^2\Theta}\frac{d^2\Theta}{d\theta^2} - \frac{1}{T}\frac{dT}{dt} = 0. \tag{108.3}$$

By the assumption in (108.2), only the third term in (108.3) has any dependence on the variable t. Since the other terms cannot have any t dependence, it must be that the third term also has no t dependence. Therefore, this term must be equal to some (unknown) constant; i.e.,

$$\frac{1}{T}\frac{dT}{dt} = -\lambda = \text{ some unknown constant.} \tag{108.4}$$

The minus sign in (108.4) is taken for convenience later on. Using (108.4) in (108.3) and simplifying we find

$$\frac{r}{R}\frac{d}{dr}\left(r\frac{dR}{dr}\right) + r^2\lambda + \frac{1}{\Theta}\frac{d^2\Theta}{d\theta^2} = 0. \tag{108.5}$$

The third term in (108.3) is the only one that could depend on θ, but we easily see that it cannot depend on θ because the first two terms in (108.5) could not cancel out any θ dependence. Therefore we must conclude that

$$\frac{1}{\Theta}\frac{d^2\Theta}{d\theta^2} = -\rho = \text{ another unknown constant.} \tag{108.6}$$

Using (108.6) in (108.5), we find

$$r\frac{d}{dr}\left(r\frac{dR}{dr}\right) + (-\rho + r^2\lambda)R = 0. \tag{108.7}$$

Note that we have, at this point, found ordinary differential equations that describe each of the terms in the solution proposed in (108.2).

But, in doing so, we have introduced two arbitrary constants; λ and ρ. Solving the ordinary differential equations in (108.4), (108.6), and (108.7) yields

$$T(t) = Ae^{-\lambda t},$$
$$\Theta(\theta) = B\sin(\sqrt{\rho}\theta) + C\cos(\sqrt{\rho}\theta), \tag{108.8}$$
$$R(r) = DJ_{\sqrt{\rho}}(\sqrt{\lambda}r) + EY_{\sqrt{\rho}}(\sqrt{\lambda}r),$$

where $\{A, B, C, D, E\}$ are arbitrary constants, and $\{J_*, Y_*\}$ are Bessel functions. By superposition, the most general solution to (108.2) can now be written as

$$u(t, r, \theta) = \int_{-\infty}^{\infty} d\lambda \int_{-\infty}^{\infty} d\rho\, e^{-\lambda t}\left[B(\lambda, \rho)\sin(\sqrt{\rho}\theta) + C(\lambda, \rho)\cos(\sqrt{\rho}\theta)\right]$$
$$\times \left[D(\lambda, \rho)J_{\sqrt{\rho}}(\sqrt{\lambda}r) + E(\lambda, \rho)Y_{\sqrt{\rho}}(\sqrt{\lambda}r)\right],$$
$$\tag{108.9}$$

where $\{B, C, D, E\}$ may now depend on λ and ρ. Now physical reasoning must be used to evaluate $\{B, C, D, E\}$.

For example, if the heat equation, (108.2), is being solved in the entire circle, then it must be that the solution is periodic in θ with period 2π. That is, $u(t, r, \theta) = u(t, r, \theta + 2\pi)$. This constraint (which is equivalent to $\Theta(\theta) = \Theta(\theta + 2\pi)$), placed on (108.8), restricts $\sqrt{\rho}$ to be an integer. Hence, in this case, the most general solution has the form (using $n^2 = \rho$)

$$u(t, r, \theta) = \int_{-\infty}^{\infty} d\lambda \sum_{n=0}^{\infty} e^{-\lambda t}\left[B(\lambda, n^2)\sin n\theta + C(\lambda, n^2)\cos n\theta\right]$$
$$\times \left[D(\lambda, n^2)J_n(\sqrt{\lambda}r) + E(\lambda, n^2)Y_n(\sqrt{\lambda}r)\right].$$

If the point $r = 0$ was included in the domain of the original problem, then we would require $E(\lambda, n^2) \equiv 0$ since $Y_n(r)$ is unbounded at $r = 0$. Likewise, only those values of $\lambda \geq 0$ will be physically realistic. Hence, in this case we find

$$u(t, r, \theta) = \int_{0}^{\infty} d\lambda \sum_{n=0}^{\infty} e^{-\lambda t}\left[B(\lambda, n^2)\sin n\theta + C(\lambda, n^2)\cos n\theta\right]J_n(\sqrt{\lambda}r).$$
$$\tag{108.10}$$

More conditions could be placed on the coefficients depending on the exact form of the initial conditions and boundary conditions.

Example 2

Suppose we have the equation

$$f(x)u_x^2 + g(y)u_y^2 = a(x) + b(y) \tag{108.11}$$

to solve. We might propose a solution of the form

$$u(x, y) = \phi(x) + \psi(y). \tag{108.12}$$

Using (108.12) in (108.11) results in the equation

$$f(x)[\phi'(x)]^2 - a(x) = g(y)[\psi'(y)]^2 - b(y). \tag{108.13}$$

The left-hand side of (108.13) must be independent of x (since the right-hand side is), hence we can set

$$f(x)[\phi'(x)]^2 - a(x) = \alpha = \text{ some constant}, \tag{108.14}$$

and then

$$g(y)[\psi'(y)]^2 - b(y) = \alpha. \tag{108.15}$$

Solving (108.14) and (108.15) we have determined that a solution to (108.11) is given by

$$v(x, y) = \int_{x_0}^x \sqrt{\frac{a(\xi) + \alpha}{f(\xi)}} \, d\xi + \int_{y_0}^y \sqrt{\frac{b(\eta) - \alpha}{g(\eta)}} \, d\eta + \beta, \tag{108.16}$$

where β is another arbitrary constant. The solution in (108.16) may not be the most general solution to (108.11). For nonlinear equations, it is very difficult to determine if the most general solution has been found.

Notes

[1] Note that the solution in (108.10) could also have been obtained by use of Fourier series (see page 293). The form of the solution in (108.10) (i.e., the $e^{-\lambda t}$ term) suggests that a Laplace transform might be an appropriate way to analyze (108.1).

[2] Carslaw and Jaeger [4] have the decompositions (similar to (108.9)) for many heat conduction problems.

[3] If the equation $L[u] = 0$ can be separated into ordinary differential equations when $u(\mathbf{x}) = \dfrac{u_1(x_1)u_2(x_2)\cdots u_n(x_n)}{R(\mathbf{x})}$, and $R \neq 1$, then the equation is said to be R separable.

[4] Moon and Spencer [10] list 11 common orthogonal coordinate systems in which both Laplace's equation and Helmholtz's equation separate. These coordinate systems are rectangular, circular cylinder, elliptic cylinder, parabolic cylinder, spherical, prolate spheroidal, oblate spheroidal, parabolic, conical, ellipsoidal, and paraboloidal. Also included are the exact decompositions that are obtained (similar to (108.9)).

The above analysis is repeated for 21 different cylindrical coordinate systems that are obtained by translating an orthogonal map in a direction perpendicular to the plane of the map. The above analysis is again carried out for 10 different rotational coordinate systems that are obtained by twirling an orthogonal map in a plane, about an axis. In each of these 31 coordinate systems, Laplace's equation or Helmholtz's equation separates (or is R separable).

[5] A necessary and sufficient condition for a system with two degrees of freedom, with the Hamiltonian $H = \frac{1}{2}(p_x^2 + p_y^2) + V(x, y)$, to be separable in elliptic, polar, parabolic, or cartesian coordinates is that the expression

$$(V_{yy} - V_{xx})(-2axy - b'y - bx + d)$$
$$+ 2V_{xy}(ay^2 - ax^2 + by - b'x + c')$$
$$+ V_x(6ay + 3b) + V_y(-6ax - 3b')$$

vanishes for some constants $(a, b, b', c, c', d) \neq (0, 0, 0, c, c, 0)$. The values of these constants determine in which of the above four coordinate systems the differential equations separate.

For three degrees of freedom, a similar expression has been devised that determines in which of 11 different coordinate systems the equations separate. For more details, see Marshall and Wojciechowski [8].

[6] The Hartree–Fock approximation is a technique for approximating the eigenfunctions $u(\mathbf{x})$ and eigenvalues λ of the partial differential equation

$$-\nabla^2 u + f(\mathbf{x})u = \lambda u, \tag{108.17}$$

when $f(\mathbf{x})$ is a prescribed function. The technique consists of approximating $f(\mathbf{x})$ by

$$f(\mathbf{x}) \simeq f_1(x_1)f_2(x_2)\cdots f_n(x_n).$$

If $f(\mathbf{x})$ has the form shown above, then equation (108.17) can be solved by separation of variables. The solution will be of the form

$$u(\mathbf{x}) = u_1(x_1)u_2(x_2)\cdots u_n(x_n),$$
$$\lambda = \lambda_1 + \lambda_2 + \cdots + \lambda_n.$$

In the Hartree–Fock approximation, a variational principle is used to determine what the "best" $\{f_j(x_j)\}$ are. See Fischer [5] for details.

[7] Miller [11] contains a group theoretical approach to the method of separation of variables. For many linear differential equations, the separated solutions are easily related to the Lie algebra generated by the equation.

References

[1] F. M. Arscott and A. Darai, "Curvilinear Co-ordinate Systems in which the Helmholtz Equation Separates," *IMA J. Appl. Mathematics*, **27**, 1981, pages 33–70.

[2] E. K. Blum and G. J. Reid, "On the Numerical Solution of Three-Dimensional Boundary Value Problems by Separation of Variables," *SIAM J. Numer. Anal.*, **25**, No. 1, February 1988, pages 75–90.

[3] W. E. Boyce and R. C. DiPrima, *Elementary Differential Equations and Boundary Value Problems*, Fourth Edition, John Wiley & Sons, New York, 1986, Chapter 10 (pages 513–580).

[4] H. S. Carslaw and J. C. Jaeger, *Conduction of Heat in Solids*, Clarendon Press, Oxford, 1984.

[5] C. F. Fischer, "Approximate Solution of Schrödinger's Equation for Atoms," in J. Hinze (ed.), *Numerical Integration of Differential Equations and Large Linear Systems*, Springer–Verlag, New York, 1982, pages 71–81.

[6] J. Hainzl, "On a General Concept for Separation of Variables," *SIAM J. Math. Anal.*, **13**, No. 2, March 1982, pages 208–225.

[7] L. Kaufman and D. D. Warner, "Algorithm 685: A Program for Solving Separable Elliptic Equations," *ACM Trans. Math. Software*, **16**, No. 4, December 1990, pages 325–351.

[8] I. Marshall and S. Wojciechowski, "When is a Hamiltonian System Separable?," *J. Math. Physics*, **29**, No. 6, June 1988, pages 1338–1346.

[9] P. Moon and D. E. Spencer, *Field Theory For Engineers*, D. Van Nostrand Company, Inc., New York, 1961.

[10] P. Moon and D. E. Spencer, *Field Theory Handbook*, Springer–Verlag, New York, 1961.

[11] W. Miller, Jr., *Symmetry and Separation of Variables*, Addison–Wesley Publishing Co., Reading, MA, 1977.

109.   Similarity Methods

Applicable to   Linear or nonlinear partial differential equations. Also systems of differential equations.

Yields

An equation with one fewer independent variables.

Idea

Sometimes the number of independent variables in a partial differential equation can be reduced by taking algebraic combinations of the independent variables.

Procedure

The idea of this method is to find new independent variables (called similarity variables) that are combinations of the old independent variables. The differential equation, when written in the new variables, will not depend on all of the new variables.

One technique for discovering the correct new variables is to choose temporary variables to be a parameter to some (unknown) power times the old variables. After writing the equation in terms of the temporary variables, the powers can be found by requiring homogeneity in the parameter. New variables are then constructed from the old variables in such a way that the parameter does not enter.

Example 1

Suppose the following linear partial differential equation

$$\frac{\partial u}{\partial t} + \frac{u}{2t} = \nu \frac{\partial^2 u}{\partial z^2}, \tag{109.1}$$

for $u(t, z)$ is to be simplified from being a function of the two independent variables $\{t, z\}$ to being a function of only one independent variable. We define the temporary variables u', z', t', and the parameter λ by

$$\begin{aligned} u &= u' \lambda, \\ t &= t' \lambda^m, \\ z &= z' \lambda^n, \end{aligned} \tag{109.2}$$

for some unknown values of n and m. In these temporary variables, equation (109.1) becomes

$$\frac{\partial u'}{\partial t'} \lambda^{1-m} + \frac{u'}{2t'} \lambda^{1-m} = \nu \frac{\partial^2 u'}{\partial (z')^2} \lambda^{1-2n}. \tag{109.3}$$

For the parameter λ to be eliminated from (109.3), we require that the exponents of λ in each term of (109.3) all be the same. That is, $1 - m = 1 - 2n$. This equation has the solution $m = 2n$. At this point we know that there are similarity solutions of (109.1), but still must determine what they are. Using $m = 2n$ in (109.2), the change of variables becomes

$$\begin{aligned} u &= u' \lambda, \\ t &= t' \lambda^{2n}, \\ z &= z' \lambda^n. \end{aligned} \tag{109.4}$$

Combining the original independent variables $\{t, z\}$, we form a new independent variable $\{\eta\}$ whose transformation from the old variables to the temporary variables does not depend on λ (we use (109.4) here):

$$\eta := \frac{z}{\sqrt{t}} = \frac{z'}{\sqrt{t'}}.$$

Now we have to propose the similarity solution. We look for a solution of the form

$$u(t, z) = v\left(\frac{z}{\sqrt{t}}\right) = v(\eta).\tag{109.5}$$

When the form of (109.5) is used in equation (109.1), the equation becomes

$$2\nu\frac{d^2v}{d\eta^2} + \eta\frac{dv}{d\eta} - v = 0,\tag{109.6}$$

which is now an ordinary differential equation. Every solution of (109.6) will generate a solution of (109.1).

Example 2

Consider the following nonlinear partial differential equation:

$$\frac{\partial u}{\partial t} + \frac{u}{2t} + \beta u\frac{\partial u}{\partial z} = \nu\frac{\partial^2 u}{\partial z^2},\tag{109.7}$$

for $u(t, z)$. This equation differs from (109.1) by the $\beta u u_z$ term. We wish to simplify this equation from being a function of the two independent variables $\{t, z\}$ to being a function of only one independent variable. After we do this, we will find a solution for the $\beta = 0$ case. We define the temporary variables u', z', t', and the parameter λ by (109.2). In these temporary variables, equation (109.7) becomes

$$\frac{\partial u'}{\partial t'}\lambda^{1-m} + \frac{u'}{2t'}\lambda^{1-m} + \beta u'\frac{\partial u'}{\partial z'}\lambda^{2-n} = \nu\frac{\partial^2 u'}{\partial(z')^2}\lambda^{1-2n}.\tag{109.8}$$

For the parameter λ to be eliminated from (109.8), we require that the exponents of λ in each term of (109.8) all be the same. That is,

$$1 - m = 2 - n = 1 - 2n.\tag{109.9}$$

The equations in (109.9) have the unique solution: $n = -1$, $m = -2$. At this point we know that there is a similarity solution of (109.7). Using $n = -1$, $m = -2$ in (109.2) changes the variables to $\{u = u'\lambda,\ t = t'\lambda^{-2},$

$z = z'\lambda^{-1}\}$. Combining the original independent variables $\{t, z\}$, we form a new independent variable $\{\eta\}$ whose transformation from the old variables to the temporary variables does not depend on λ:

$$\eta := \frac{z}{\sqrt{t}} = \frac{z'}{\sqrt{t'}}.$$

Combining the original dependent variable $\{u\}$ with the original independent variables $\{t, z\}$, we can also form a new dependent variable $\{w\}$ whose transformation from the old variables to the temporary variables does not depend on λ:

$$w = \frac{t}{z}u = \frac{t'}{z'}u'. \tag{109.10}$$

Now we have to propose the similarity solution. By solving (109.10) for u, we are led to the assumption

$$u(t, z) = \frac{z}{t}w\left(\frac{z}{\sqrt{t}}\right) = \frac{z}{t}w(\eta). \tag{109.11}$$

When the form in (109.11) is used in equation (109.7), the equation becomes

$$2\nu\eta\frac{d^2w}{d\eta^2} + \left(4\nu + \eta^2 - 2\beta\eta^2 w\right)\frac{dw}{d\eta} + (1 - 2\beta w)\,w = 0. \tag{109.12}$$

If we define $g(\eta)$ by $g(\eta) = \eta w(\eta)$, then equation (109.12) becomes

$$2\nu\frac{d^2g}{d\eta^2} + (\eta - 2\beta g)\frac{dg}{d\eta} = 0. \tag{109.13}$$

Every solution of this ordinary differential equation will lead to similarity solutions of (109.7). In the special case of $\beta = 0$ (when equation (109.7) becomes the identical to (109.1)), the general solution to (109.13) is given by

$$g(\eta) = A + B\,\mathrm{erf}\left(\frac{\eta}{\sqrt{4\nu}}\right),$$

where A and B are arbitrary constants. This results in the solution

$$u(t, z) = \frac{1}{\sqrt{t}}\left[A + B\,\mathrm{erf}\left(\frac{z}{\sqrt{4\nu t}}\right)\right]$$

to (109.1). Note that this similarity solution could *not* have been obtained from (109.6), since the scalings in (109.5) and (109.11) are different.

Notes

[1] In general, a partial differential equation may have some similarity solutions and some solutions that are not similarity solutions.

[2] This method is sometimes called the *method of one parameter groups*, due to the single parameter λ that was used in (109.2).

[3] This method is derivable from Lie group methods (see page 404).

[4] To solve a differential system (differential equation(s) with boundary condition(s)), the boundary conditions as well as the equation(s) must admit the similarity variable.

[5] This method also applies to systems of ordinary differential equations. If $\dfrac{d\mathbf{u}}{dx} = \mathbf{f}(x, \mathbf{u})$ is a system of first order ordinary differential equations for $\mathbf{u} = (u_1, \ldots, u_n)$, and if there exists a one parameter group of symmetries of the system, then there is a change of variables $(y, \mathbf{w}) = \Xi(x, \mathbf{u})$ which takes the system into $\dfrac{d\mathbf{w}}{dy} = \mathbf{g}(y, w_1, \ldots, w_{n-1})$. Hence, the original system reduces to a system of $n-1$ ordinary differential equations for $(w_1, \ldots, w_{n-1})$ together with the quadrature $w_n(y) = \int g_n(y, w_1(y), \ldots, w_{n-1}(y)) \, dy$.

[6] For some systems there are natural similarity variables. For instance, in a two-dimensional problem with radial symmetry, the variable r (where $r^2 = x^2 + y^2$) should be a similarity variable if the original equations were written in terms of x and y. Similarly, in a radially symmetric three-dimensional problem, the variable ρ (where $\rho^2 = x^2 + y^2 + z^2$) should be a similarity variable.

[7] For diffusion equations, similarity solutions are often of the form $f(x/\sqrt{t})$ or $t^\alpha f(x/\sqrt{t})$.

[8] The partial differential equation $F\left(tx, u, \dfrac{u_t}{x}, \dfrac{u_x}{t}\right) = 0$, for $u(x, t)$, has the similarity variable $w = tx$. Considering $u = u(w)$, we find the equivalent ordinary differential equation $F(w, u, u_w, u_w) = 0$.

References

[1] W. F. Ames, *Nonlinear Partial Differential Equations*, Volume 1, Academic Press, New York, 1967, pages 135–141.

[2] L. Dressner, *Similarity Solutions of Nonlinear Partial Differential Equations*, Pitman Publishing Co., Marshfield, MA, 1983.

[3] J. Kevorkian and J. D. Cole, *Perturbation Methods in Applied Mathematics*, Springer–Verlag, New York, 1981.

[4] J. R. King, "Exact Similarity Solutions to Some Nonlinear Diffusion Equations," *J. Phys. A: Math. Gen.*, **23**, 1990, pages 3681–3697.

[5] P. Roseneau and J. L. Schwarzmeier, "Similarity Solutions of Systems of Partial Differential Equations," *Comput. Physics Comm.*, **27**, 1982, pages 179–186.

[6] R. Seshadi and T. Y. Na, *Group Invariance in Engineering Boundary Value Problems*, Springer–Verlag, New York, 1985, pages 39–42.

110. Exact Solutions to the Wave Equation

Applicable to   The n-dimensional wave equation.

Yields

An explicit solution in terms of an integral.

Idea

An exact formula is available for the n-dimensional wave equation $u_{tt} = \nabla^2 u$.

Procedure

The n-dimensional wave equation

$$\frac{\partial^2 u}{\partial t^2} = \nabla^2 u = \frac{\partial^2 u}{\partial x_1{}^2} + \cdots + \frac{\partial^2 u}{\partial x_n{}^2}, \tag{110.1}$$

with the initial data (we use $\mathbf{x} = (x_1, \ldots, x_n)$)

$$u(0, \mathbf{x}) = f(\mathbf{x}), \qquad u_t(0, \mathbf{x}) = g(\mathbf{x}), \tag{110.2}$$

has two different (but similar) forms of the solution, depending on whether n is even or odd. When n is odd the solution is given by

$$
\begin{aligned}
u(t, \mathbf{x}) = \frac{1}{1 \cdot 3 \cdots (n-2)} \bigg\{ & \frac{\partial}{\partial t} \left(\frac{\partial}{t \, \partial t} \right)^{(n-3)/2} t^{n-2} \omega[f; \mathbf{x}, t] \\
& + \left(\frac{\partial}{t \, \partial t} \right)^{(n-3)/2} t^{n-2} \omega[g; \mathbf{x}, t] \bigg\},
\end{aligned}
\tag{110.3}
$$

where $\omega[h; \mathbf{x}, t]$ is defined to be the average of the function $h(\mathbf{x})$ over the surface of an n-dimensional sphere of radius t centered at $\mathbf{x}$. That is

$$\omega[h; \mathbf{x}, t] = \frac{1}{\sigma_n(t)} \int h(0, \boldsymbol{\zeta}) \, d\Omega,$$

where $|\boldsymbol{\zeta} - \mathbf{x}|^2 = t^2$, $\sigma_n(t)$ is the surface area of the n-dimensional sphere of radius t, and $d\Omega$ is an element of area. (Note that $\sigma_n(t) = 2\pi^{n/2} t^{n-1} / \Gamma\left(\frac{n}{2}\right)$.)

When n is even the solution to (110.1) and (110.2) is given by

$$
\begin{aligned}
u(t, \mathbf{x}) = \frac{1}{2 \cdot 4 \cdots (n-2)} \bigg\{ & \frac{\partial}{\partial t} \left(\frac{\partial}{t \, \partial t} \right)^{(n-2)/2} \int_0^t \omega[f; \mathbf{x}, \rho] \frac{\rho^{n-1} \, d\rho}{\sqrt{t^2 - \rho^2}} \\
& + \left(\frac{\partial}{t \, \partial t} \right)^{(n-2)/2} \int_0^t \omega[g; \mathbf{x}, \rho] \frac{\rho^{n-1} \, d\rho}{\sqrt{t^2 - \rho^2}} \bigg\},
\end{aligned}
\tag{110.4}
$$

where $w[h; \mathbf{x}, t]$ is defined as above. Since the expression in (110.4) is integrated over ρ, the values of f and g must be known everywhere in the *interior* of the n-dimensional sphere.

Special Case 1

When $n = 1$, the above formulae produce the D'Alembert solution (see Chester [1]) of the equation $u_{tt} = c^2 u_{xx}$:

$$u(x, t) = \frac{1}{2}\left[f(x - ct) + f(x + ct)\right] + \frac{1}{2c} \int_{x-ct}^{x+ct} g(\zeta)\, d\zeta. \qquad (110.5)$$

Special Case 2

When $n = 2$, the above formulae produce the Parseval solution

$$u(x, t) = \frac{1}{2\pi} \frac{\partial}{\partial t} \int\int_{R(t)} \frac{f(x_1 + \zeta_1, x_2 + \zeta_2)}{\sqrt{t^2 - \zeta_1^2 - \zeta_2^2}}\, d\zeta_1\, d\zeta_2$$

$$+ \frac{1}{2\pi} \int\int_{R(t)} \frac{g(x_1 + \zeta_1, x_2 + \zeta_2)}{\sqrt{t^2 - \zeta_1^2 - \zeta_2^2}}\, d\zeta_1\, d\zeta_2,$$

where $R(t)$ is the region $\{(\zeta_1, \zeta_2)\,|\, \zeta_1^2 + \zeta_2^2 \le t^2\}$.

Special Case 3

When $n = 3$, the above formulae produce the Poisson solution (also known as the Kirchoff solution)

$$u(x, t) = \frac{\partial}{\partial t}\left(tw[f; \mathbf{x}, t]\right) + tw[g; \mathbf{x}, t],$$

where

$$w[h; \mathbf{x}, t] = \frac{1}{4\pi} \int_0^{2\pi} \int_0^{\pi} h(x_1 + t\sin\theta\cos\phi, x_2 + t\sin\theta\sin\phi, x_3 + t\cos\theta)$$

$$\times \sin\theta\, d\theta\, d\phi.$$

Example

A string stretched in the shape of a sine wave and then released from rest will have the amplitude $u(x, t)$, where

$$u_{tt} = u_{xx},$$
$$u(x, 0) = \sin x,$$
$$u_t(x, 0) = 0.$$

By virtue of (110.5), this has the solution $u(x, t) = \frac{1}{2}\Big(\sin(x-t) + \sin(x+t)\Big)$.

Notes

[1] The name "D'Alembert solution" is also applied to the solution of the wave equation in a *semi-infinite* domain

$$v_{tt} = c^2 v_{xx},$$
$$v(0,t) = 0, \qquad \text{for } 0 < t < \infty,$$
$$v(x,0) = f(x), \qquad \text{for } 0 \le x < \infty,$$
$$v_t(x,0) = g(x), \qquad \text{for } 0 \le x < \infty.$$

This equation has the solution (see Farlow [2], page 143)

$$v(x,t) = \begin{cases} \dfrac{1}{2}[f(x+ct) + f(x-ct)] + \dfrac{1}{2c}\int_{x-ct}^{x+ct} g(\zeta)\,d\zeta, & \text{for } x \ge ct, \\[2mm] \dfrac{1}{2}[f(x+ct) - f(ct-x)] + \dfrac{1}{2c}\int_{ct-x}^{x+ct} g(\zeta)\,d\zeta, & \text{for } x < ct. \end{cases}$$

[2] Another useful formula for the solution of the inhomogeneous wave equation

$$\frac{\partial^2 u}{\partial t^2} - \frac{\partial^2 u}{\partial x^2} - \frac{\partial^2 u}{\partial y^2} - \frac{\partial^2 u}{\partial z^2} = F(t,x,y,z),$$

with the homogeneous initial conditions:

$$u(0,x,y,z) = 0, \qquad u_t(0,x,y,z) = 0.$$

The solution is given by

$$u(t,x,y,z) = \frac{1}{4\pi} \iiint\limits_{\rho \le t} \frac{F(t-\rho,\zeta,\eta,\xi)}{\rho}\,d\zeta\,d\eta\,d\xi,$$

with $\rho = \sqrt{(x-\zeta)^2 + (y-\eta)^2 + (z-\xi)^2}$.

[3] Another useful formula is for the solution of

$$\frac{\partial^2 u}{\partial t^2} = \frac{\partial^2 u}{\partial x^2} + \frac{\partial^2 u}{\partial y^2} + \frac{\partial^2 u}{\partial z^2} + \lambda u,$$
$$u(0,x,y,z) = f(x,y,z),$$
$$u_t(0,x,y,z) = g(x,y,z),$$

where λ is an arbitrary constant. The solution is given by

$$u(t,x,y,z) = \frac{\partial}{\partial t}\left[t w[f;\mathbf{x},t] + \lambda \int_0^t \rho^2 w[f;\mathbf{x},\rho] I(\lambda t^2 - \lambda \rho^2)\,d\rho \right]$$
$$+ t w[g;\mathbf{x},t] + \lambda \int_0^t \rho^2 w[g;\mathbf{x},\rho] I(\lambda t^2 - \lambda \rho^2)\,d\rho,$$

where $I(a) := I_0'(\sqrt{a})/\sqrt{a}$ and I_0 is the usual modified Bessel function.

[4] The solutions given in (110.3) and (110.4) may be derived from one another by the method of descent (see page 382).

References
[1] C. R. Chester, *Techniques in Partial Differential Equations*, McGraw–Hill Book Company, New York, 1970, pages 17–23.
[2] S. J. Farlow, *Partial Differential Equations for Scientists and Engineers*, John Wiley & Sons, New York, 1982, Lessons 17 and 18 (pages 129–145).
[3] P. R. Garabedian, *Partial Differential Equations*, Wiley, New York, 1964, pages 191–210.

111.   Wiener–Hopf Technique

Applicable to   Linear partial differential equations on an infinite interval that have different types of boundary data on different parts of the interval.

Yields

An exact solution.

Idea

In some linear partial differential equations, we would like to take a Fourier transform, but cannot because the type of boundary data changes along the boundary. The Wiener–Hopf technique is to take a Fourier transform anyway and allow part of the data to be "missing." Solving the problem (using Liouville's theorem) we determine the "missing" data and the solution simultaneously.

Procedure

Sometimes a linear partial differential equation has a form amenable to a Fourier transform, but the boundary conditions would seem to preclude it. For example, the reduced wave equation

$$\nabla^2 \phi + k^2 \phi = 0, \qquad (111.1)$$

in two dimensions may suggest the use of a Fourier transform in x. But, if the boundary conditions are given by, say,

$$\frac{\partial \phi(x, 0)}{\partial y} = 0 \quad \text{for } x \geq 0,$$
$$\phi \qquad \text{is continuous for } x < 0, \qquad (111.2)$$

then it is not clear how to take such a transform. Generally, we would require $\partial \phi / \partial y$ to be known for all x, before we could take a Fourier transform. The solution technique is to *assume* that $\partial\phi/\partial y$ is known for all x and then take a Fourier transform. The quantity $\partial\phi/\partial y$ for $x < 0$ will be determined when the final solution is determined.

The solution procedure uses Liouville's theorem, one form of which is

If $E(z)$ is an entire function (i.e., $E(z)$ is analytic in the finite $|z|$ plane) and if $E(z)$ is bounded by a constant as $|z| \to \infty$, then $E(z)$ is identically constant.

(See, for example, Levinson and Redheffer [4].)

The difficult part of the solution procedure will turn out to be the "factorization" step. That is, given the functions $A(\omega), B(\omega), C(\omega)$ (all analytic in the strip $\alpha < \operatorname{Im}\omega < \beta$), find functions $\Phi_+(\omega), \Psi_-(\omega)$ satisfying

$$A(\omega)\Phi_+(\omega) + B(\omega)\Psi_-(\omega) + C(\omega) = 0, \qquad (111.3)$$

where

(A) Equation (111.3) holds in the strip: $\alpha < \operatorname{Im}\omega < \beta$.
(B) $\Phi_+(\omega)$ is analytic in the upper half plane: $\alpha < \operatorname{Im}\omega$.
(C) $\Psi_-(\omega)$ is analytic in the lower half plane: $\operatorname{Im}\omega < \beta$.

We will continue to use the following standard notation: a subscript of "+" ("−") indicates a function that is analytic in the upper (lower) half plane $\alpha < \operatorname{Im}\omega$ ($\operatorname{Im}\omega < \beta$).

Example

Suppose we have the linear partial differential equation exterior to the half line ($y = 0, x \geq 0$)

$$\phi_{xx} + \phi_{yy} - \phi_x = 0, \qquad (111.4)$$

with the boundary conditions

$$\begin{aligned} \phi &\to 0 && \text{as } r = \sqrt{x^2 + y^2} \to \infty, \\ \phi &= e^{-x} && \text{on } y = 0, \ x \geq 0. \end{aligned} \qquad (111.5.a\text{–}b)$$

We define the Fourier transform of $\phi(x, y)$ by $\Phi(\omega, y) = \dfrac{1}{\sqrt{2\pi}} \int_{-\infty}^{\infty} \phi(x, y)e^{i\omega x}\, dx$. If we assume that $\phi_x \to 0$ as $r \to \infty$, then (111.4) can be Fourier transformed (by multiplying by $e^{i\omega x}$ and integrating with respect to x) to yield

$$\frac{d^2\Phi}{dy^2} - (\omega^2 - i\omega)\Phi = 0. \qquad (111.6)$$

If we extend the definition of $\phi(x, 0)$ in (111.5.b) to be

$$\phi(x, 0) = \begin{cases} e^{-x} & x \geq 0, \\ u(x) & x < 0, \end{cases} \tag{111.7}$$

where $u(x)$ is unknown, then we can transform (111.7) to find

$$\Phi(\omega, 0) = U(\omega) + \frac{1}{\sqrt{2\pi}} \frac{1}{1 - i\omega}, \tag{111.8}$$

where $U(\omega)$ is the Fourier transform of $u(x)$, i.e.,

$$U(\omega) = \frac{1}{\sqrt{2\pi}} \int_{-\infty}^{0} u(x) e^{i\omega x} \, dx.$$

The solution of (111.6) (which is an ordinary differential equation in y) and (111.8), which vanishes as $|y| \to \infty$, is

$$\Phi(\omega, y) = \left[U(\omega) + \frac{1}{\sqrt{2\pi}} \frac{1}{1 - i\omega} \right] \exp\left(-|y| \sqrt{\omega^2 - i\omega} \right), \tag{111.9}$$

where the square root branch is specified by $\mathrm{Re} \sqrt{\omega^2 - i\omega} \geq 0$.

Once we determine $U(\omega)$, we can (in principle) invert (111.9) by taking an inverse Fourier transform. This would yield $\phi(x, y)$. Finding $U(\omega)$ is the hard part of the calculation.

Since the solution of the original problem (and its derivatives) must be continuous across $y = 0$ (for $x < 0$) we define a function $f(x)$ by

$$f(x) := \phi_y(x, 0^+) - \phi_y(x, 0^-),$$

$$= \begin{cases} 0 & \text{for } x < 0, \\ v(x) & \text{for } x > 0, \end{cases} \tag{111.10.a–b}$$

where 0^+ (0^-) indicates a vanishingly small quantity that is greater (less) than zero and $v(x)$ is an unknown function. Taking the Fourier transform of (111.10.b) produces

$$F(\omega) := \frac{1}{\sqrt{2\pi}} \int_{-\infty}^{\infty} f(x) e^{i\omega x} \, dx$$

$$= \frac{1}{\sqrt{2\pi}} \int_{0}^{\infty} v(x) e^{i\omega x} \, dx, \tag{111.11}$$

while the Fourier transform of (111.10.a) produces

$$F(\omega) = \Phi_y(\omega, 0^+) - \Phi_y(\omega, 0^-)$$
$$= -2\left[U(\omega) + \frac{1}{\sqrt{2\pi}}\frac{1}{1 - i\omega}\right]\sqrt{\omega^2 - i\omega}, \qquad (111.12)$$

where the solution in (111.9) has been used.

Using our subscript convention (and the definition in (111.11)), we note that $F(\omega) = F_+(\omega)$, where, for instance, we could take $\alpha = 1/3$. We now *assume* that $U(\omega) = U_-(\omega)$, for, say, $\beta = 2/3$. This places a constraint on $u(x)$ that has to be verified at the end of the calculation.

By algebraic manipulations of (111.12) we can obtain (this step should not be trivialized, it is the hardest step in the calculation)

$$-\frac{F_+(\omega)}{2\sqrt{\omega}} - \left[\frac{\sqrt{-i}}{\sqrt{\pi}(1 - i\omega)}\right]_+ = U_-(\omega)\sqrt{\omega - i} + \left[\frac{\sqrt{\omega - i} - \sqrt{-2i}}{\sqrt{2\pi}(1 - i\omega)}\right]_-.$$
$$(111.13)$$

If we define $E(\omega)$ to be the left-hand side of (111.13), then $E(\omega)$ is entire. This is because the left-hand side and the right-hand side of (111.13) overlap in the strip $\alpha < \mathrm{Im}\,\omega < \beta$, and these two functions are analytic in their respective half planes. Hence, one side of (111.13) supplies the analytic continuation of the other side.

If we now assume that

(A) $F_+(\omega) \to 0$ as $|\omega| \to \infty$ in $\mathrm{Im}\,\omega > \beta$,
(B) $\omega U_-(\omega) \to 0$ as $|\omega| \to \infty$ in $\mathrm{Im}\,\omega < \alpha$,

then $E(\omega) \to 0$ as $|\omega| \to \infty$. By Liouville's theorem we can conclude that $E(\omega) \equiv 0$ and so from (111.13)

$$U(\omega) = U_-(\omega) = -\frac{1}{\sqrt{\omega - i}}\left[\frac{\sqrt{\omega - i} - \sqrt{-2i}}{\sqrt{2\pi}\sqrt{1 - i\omega}}\right].$$

Using this in (111.9) and taking an inverse Fourier transform yields $\phi(x, y)$.

Notes

[1] The Wiener–Hopf method was originally formulated for the solution of integral equations.
[2] The problem in (111.1) and (111.2) is analyzed in more detail in Carrier, Krook, and Pearson [1]. The same problem, with an incident oblique wave, is solved in Davies [2].

References

[1] G. F. Carrier, M. Krook, and C. E. Pearson, *Functions of a Complex Variable*, McGraw–Hill Book Company, New York, 1966, pages 376–386.

[2] B. Davies, *Integral Transforms and Their Applications*, Springer–Verlag, New York, 1978, pages 288–307.

[3] A. E. Heins, "The Scope and Limitations of the Method of Wiener and Hopf," *Communications on Pure and Applied Mathematics*, **9**, 1956, pages 447–466.

[4] N. Levinson and R. M. Redheffer, *Complex Variables*, Holden–Day, Inc., San Francisco, 1970.

[5] B. Noble, *Methods Based on the Wiener–Hopf Technique*, Pergamon Press, New York, 1958.

III

Approximate Analytical Methods

112. Introduction to Approximate Analysis

Sometimes an exact solution cannot be obtained for a differential equation and an approximate solution must be found. Other times, an approximate solution may convey more information than an exact solution.

There are essentially two types of approximations:

[1] those that give an approximation over a range of the independent variable, and

[2] those that give an approximation only near a single point.

Approximations of the second type are more common.

This section of the book is not broken up into methods for ordinary differential equations and methods for partial differential equations since most of the methods can be used for either type of differential equation.

Listed below are, in the author's opinion, those methods which are the most useful when approximating the solution to ordinary differential equations and partial differential equations. These are the methods that might be tried first.

Most Useful Methods for Differential Equations
- · Collocation
- · Dominant Balance
- · Graphical Analysis: The Phase Plane
- · Least Squares Method
- · Lyapunov Functions
- · Newton's Method
- · Perturbation Method: Method of Averaging
- · Perturbation Method: Boundary Layer Method
- · Perturbation Method: Regular Perturbation
- · WKB Method

113. Chaplygin's Method

Applicable to An initial value problem for a single first order ordinary differential equation.

Yields

Improved upper and lower bounds on the solution.

Idea

Using an upper and lower bound on the solution, a set of tighter bounds can be constructed.

Procedure

For an equation of the form $y' = f(x, y)$, $y(x_0) = y_0$ the method is derived from the following theorem (due to Chaplygin):

If the differential inequalities

$$u'(x) - f(x, u(x)) < 0,$$
$$v'(x) - f(x, v(x)) > 0,$$

$$(113.1)$$

hold for $x > x_0$, with $u(x_0) = y_0$ and $v(x_0) = y_0$, then

$$u(x) < y(x) < v(x) \qquad (113.2)$$

holds for all $x > x_0$.

The procedure is to determine (or "guess") a $u(x)$ and a $v(x)$ that satisfy (113.1). Then there are two different techniques available for computing $\{u_1(x), v_1(x)\}$ such that

$$u(x) < u_1(x) \le y(x) \le v_1(x) < v(x). \qquad (113.3)$$

For each of the two techniques, the functions $\{u_1(x), v_1(x)\}$ will be different. The functions obtained, $\{u_1(x), v_1(x)\}$, will also satisfy (113.1), and the process may be iterated.

Technique One

Let K be the Lipschitz constant of the function $f(x, y)$. Then if $\{u_1(x), v_1(x)\}$ are defined by

$$u_1(x) = u(x) + \int_{x_0}^{x} e^{-K(x-t)} \left[f(t, u(t)) - u'(t) \right] dt,$$

$$v_1(x) = v(x) - \int_{x_0}^{x} e^{-K(x-t)} \left[v'(t) - f(t, v(t)) \right] dt,$$

then (113.3) will be satisfied.

Technique Two

For this technique, it must be true that $\partial^2 f / \partial y^2$ is of constant sign in the region of interest. Once this has been established, define $\{M(x), N(x), \widehat{M}(x), \widehat{N}(x)\}$ by

$$M(x)y + N(x) = f(x, u(x)) + \frac{f(x, v(x)) - f(x, u(x))}{v(x) - u(x)}(y - u(x)),$$

$$\widehat{M}(x)y + \widehat{N}(x) = f(x, u(x)) + f_y(x, u(x))(y - u(x)).$$

$$(113.4)$$

(Note that both sides of each equation are linear in the indeterminate y.) Then define $u_1(x)$ to be the solution of

$$y' = M(x)y + N(x), \qquad y(x_0) = y_0. \tag{113.5}$$

Finally, define $v_1(x)$ to be the solution of

$$y' = \widehat{M}(x)y + \widehat{N}(x), \qquad y(x_0) = y_0. \tag{113.6}$$

With these definitions for $u_1(x)$ and $v_1(x)$, (113.3) will be satisfied. Note that the equations in (113.5) and (113.6) can be solved by the use of integrating factors (see page 305).

Example

Suppose we wish to bound the solution to the equation

$$y' = y^2 + x^2, \qquad y(0) = 0,$$

when x is in the range $[0, 1/\sqrt{2}]$.

First observe that $u(x) = x^3/3$ and $v(x) = 11x^3/30$ satisfy the conditions of Chaplygin's theorem, so that (113.2) holds. Using the first technique, we recognize that $K = \sqrt{2}$ in the region of interest, so that the functions

$$u_1(x) = \frac{x^3}{3} + \frac{1}{9} \int_0^x t^6 e^{-\sqrt{2}(x-t)} \, dt,$$

$$v_1(x) = \frac{11}{3} x^3 - \int_0^x \left(\frac{t^2}{10} - \frac{121}{900} t^6 \right) e^{-\sqrt{2}(x-t)} \, dt,$$

satisfy the constraint in (113.3). Using the second technique, we note that $\partial^2 f / \partial y^2$ and so we can use the results in (113.4), (113.5), and (113.6). It is straightforward to calculate

$$M(x) = \tfrac{7}{10} x^3, \qquad\qquad \widehat{M}(x) = \tfrac{2}{3} x^3,$$

$$N(x) = x^2 - \tfrac{11}{90} x^6, \qquad \widehat{N}(x) = x^2 - \tfrac{1}{9} x^6.$$

Solving the equations in (113.5) and (113.6) we find

$$u_1(x) = e^{x^4/6} \int_0^x \left(z^2 - \tfrac{1}{9} z^6 \right) e^{-z^4/6} \, dz,$$

$$v_1(x) = e^{7x^4/40} \int_0^x \left(z^2 - \tfrac{11}{90} z^6 \right) e^{-7z^4/40} \, dz. \tag{113.7}$$

Notes

[1] The above example is from Mikhlin and Smolitskiy [4].

[2] The solutions in (113.7) may be Taylor expanded about $x = 0$ to find that

$$u_1(x) = \tfrac{1}{3} x^3 + \tfrac{1}{63} x^7 + \frac{2}{2079} x^{11} + O\left(x^{12}\right),$$

$$v_1(x) = \tfrac{1}{3} x^3 + \tfrac{1}{63} x^7 + \frac{1}{990} x^{11} + O\left(x^{12}\right), \tag{113.8}$$

If only the terms shown in (113.8) are kept, then these approximations satisfy (113.1). Hence, they also satisfy (113.3). It can be shown that ,

[3] Another useful inequality (see McNabb [3]) is a small generalization of the following:

If $u(t)$, $v(t)$, and $f(t, w)$ satisfy sufficient smoothness conditions on $[a, b]$, if $u(a) < v(a)$, and if $u' - f(t, u) < v' - f(t, v)$ for $a < t \leq b$, then $u(t) < v(t)$ on $[a, b]$.

[4] This procedure can be implemented numerically.

References

[1] C. Fabry and P. Habets, "Upper and Lower Solutions for Second-Order Boundary Value Problems with Nonlinear Boundary Conditions," *Nonlinear Analysis*, **10**, No. 10, 1986, pages 985–1007.

[2] V. Lakshmikantham and S. Leela, *Differential and Integral Inequalities*, Academic Press, New York, 1969, pages 64–69.

[3] A. McNabb, "Comparison Theorems for Differential Equations," *J. Math. Anal. Appl.*, **119**, 1986, pages 417–428.

[4] S. G. Mikhlin and K. L. Smolitskiy, *Approximate Methods for Solutions of Differential and Integral Equations*, American Elsevier Publishing Company, New York, 1967, pages 9–12.

114. Collocation

Applicable to   Ordinary differential equations and partial differential equations.

Yields

An approximation to the solution, valid over an interval.

Idea

An approximation to the solution with some free parameters is proposed. The free parameters are determined by forcing the approximation to exactly satisfy the given equation at some set of points.

Procedure

Suppose we are given the differential equation

$$N[y] = 0, \tag{114.1}$$

for $y(\mathbf{x})$ in some region R, with the boundary conditions

$$B[y] = 0, \tag{114.2}$$

on some portion of the boundary of R. We choose an approximation to $y(\mathbf{x})$ that has several parameters in it, say $y(\mathbf{x}) \simeq w(\mathbf{x}; \boldsymbol{\alpha})$, where $\boldsymbol{\alpha}$ is a vector of parameters. This approximation is chosen in such a way that it satisfies the boundary conditions in (114.2). The unknown parameters are determined by requiring the approximation to satisfy (114.1) at some collection of points.

Example

Suppose we wish to approximate the solution to the ordinary differential equation

$$N[y] = y'' + y + x = 0,$$
$$y(0) = 0, \quad y(1) = 0,$$

(114.3)

by the method of collocation. We choose to approximate the exact solution by

$$y(x) \simeq w(x) = \alpha_1 x(1 - x) + \alpha_2 x(1 - x^2).$$

(114.)

Note that $w(x)$ satisfies the boundary conditions for $y(x)$. Using this approximation we find

$$N[w(x)] = -\alpha_1(2 - x + x^2) - \alpha_2(5x + x^3) + x.$$

Now we must choose the collocation points. We choose the two points $x = 1/3$ and $x = 2/3$. Requiring $N[w(x)]$ to be zero at these two points results in the simultaneous equations

$$-\frac{48}{27}\alpha_1 - \frac{46}{27}\alpha_2 - \frac{1}{3} = 0,$$
$$-\frac{48}{27}\alpha_1 - \frac{98}{27}\alpha_2 - \frac{2}{3} = 0.$$

The solution to these equations is $\alpha_1 = 9/416$, $\alpha_2 = 9/52$. Hence, our approximation of the solution of (114.3) is

$$y(x) \simeq \frac{9}{416}x(1 - x) + \frac{9}{52}x(1 - x^2).$$

(114.4)

Note that the exact solution to (114.3) is $y(x) = \dfrac{\sin x}{\sin 1} - x$. The maximum difference between the approximate solution in (114.4) and the exact solution in the range $0 < x < 1$, occurs at $x \simeq .7916$ where the error is approximately .00081.

Notes

[1] This method is an example of a *weighted residual method*.
[2] This method is often implemented numerically.
[3] The COLSYS computer program (in FORTRAN) utilizes collocation to solve boundary value problems. See Ascher, Christiansen, and Russell [1] for details.

References

[1] U. Ascher, J. Christiansen, and R. D. Russell, "Collocation Software for Boundary-Value ODEs," *ACM Trans. Math. Software*, **7**, No. 2, June 1981, pages 209–222.

[2] L. Collatz, *The Numerical Treatment of Differential Equations*, Springer-Verlag, New York, 1966.

[3] D. Gottlieb and S. A. Orszag, *Numerical Analysis of Spectral Methods: Theory and Applications*, SIAM, Philadelphia, 1977.

[4] M. Hanke, "On a Least-Squares Collocation Method for Linear Differential-Algebraic Equations," *Numer. Math.*, **54**, No. 1, 1988, pages 79–90.

[5] E. N. Houstis, C. C. Christara, and J. R. Rice, "Quadratic-Spline Collocation Methods for Two-Point Boundary Value Problems," *Int. J. Num. Methods Eng.*, **26**, No. 4, 1988, pages 935–952.

[6] E. N. Houstis, W. F. Mitchell, and J. R. Rice, "Collocation Software for Second-Order Elliptical Partial Differential Equations," *ACM Trans. Math. Software*, **11**, No. 4, December 1985, pages 379–412.

[7] I. Lie, "The Stability Function for Multistep Collocation Methods," *Numer. Math.*, **57**, No. 8, 1990, pages 779–787.

[8] M. Y. Hussaini, D. A. Kopriva, and A. T. Patera, "Spectral Collocation Methods," *Appl. Num. Math.*, **5**, 1989, pages 177–208.

[9] H. B. Keller, *Numerical Solutions of Two Point Boundary Value Problems*, SIAM, Philadelphia, 1976, pages 175–176.

[10] N. K. Madsen and R. F. Sincovec, "Algorithm 540: PDECOL, General Collocation Software for Partial Differential Equations," *ACM Trans. Math. Software*, **5**, No. 3, September 1979, pages 326–351.

[11] M. Sakai, "A Collocation Method for a Singular Boundary Value Problem," *Congr. Numer.*, **62**, 1988, pages 171–179.

[12] K. Wright, A. H. A. Ahmed, and A. H. Seleman, "Mesh Selection in Collocation for Boundary Value Problems," *IMA J. Num. Analysis*, **11**, No. 1, January 1991, pages 7–20.

115. Dominant Balance

Applicable to   Linear and nonlinear differential equations.

Yields

An approximation to the solution valid in a region.

Idea

A differential equation with many terms in it might be well determined by only a few of those terms.

Procedure

If there are M terms in a differential equation, try solving the differential equation in a region by only considering 2 (or 3, or 4, ..., or $M-1$) terms to be important in that region. Discard all the other terms and solve this differential equation with fewer terms. After a solution is obtained, check that the discarded terms are actually smaller than the terms that were retained.

Example

Suppose we have the equation

$$y'' - \frac{2}{x^{3/2}}y' = \frac{3}{16x^2}, \tag{115.1}$$

and we would like to find an approximate solution as $x \to 0$. To determine the solution uniquely in this region, we must specify some information about $y(x)$ as $x \to 0$. In this example we choose the condition: $y \to 0$ as $x \to 0$.

There are three different two-term balances of (115.1) that we can take; that is, the first two terms in (115.1) can be taken approximately equal, the first and third terms can be taken approximately equal, or the second and third terms can be taken approximately equal. These possibilities yield the following two term balances:

$$y'' - \frac{2}{x^{3/2}}y' \simeq 0, \quad \text{which requires that} \quad |y''| \gg \left|\frac{3}{16x^2}\right|, \tag{115.2}$$

or

$$y'' \simeq \frac{3}{16x^2}, \quad \text{which requires that} \quad |y''| \gg \left|\frac{2y'}{x^{3/2}}\right|, \tag{115.3}$$

or

$$-\frac{2}{x^{3/2}}y' \simeq \frac{3}{16x^2}, \quad \text{which requires that} \quad |y''| \ll \left|\frac{3}{16x^2}\right|. \tag{115.4}$$

We will investigate each of these in turn. The solution to (115.2) is

$$y_1(x) = A + B\int \exp\left(-\frac{4}{x^{1/2}}\right)dx,$$

where A and B are arbitrary constants. Note that this solution violates the condition in (115.2) since

$$|y_1''| = \frac{2|B|}{x^{3/2}}\exp\left(-\frac{4}{x^{1/2}}\right) \ll \frac{3}{16x^2} \quad \text{as} \quad x \to 0.$$

Therefore (115.2) is an *inconsistent balance.*

The solution to (115.3) is

$$y_2(x) = -\frac{3}{16} \log x + Cx + D,$$

where C and D are arbitrary constants. But this solution cannot satisfy $y \to 0$ as $x \to 0$, so it must also be discarded.

The solution to (115.4) is

$$y_3(x) = -\frac{3}{16} \sqrt{x},$$

where we have already used the fact that $y \to 0$ as $x \to 0$. For this solution, the condition in (115.4) is satisfied, since

$$|y''| = \frac{3}{32x^{3/2}} \ll \frac{3}{16x^2} \quad \text{as} \quad x \to 0.$$

Hence, we have found a *consistent balance.* We conclude that

$$y(x) \sim -\frac{3}{16} \sqrt{x} \quad \text{as} \quad x \to 0.$$

Notes

[1] Even if a consistent balance has been found, the solution associated with that balance may be unrelated to the true solution of the differential equation(s). This is because a consistent balance has *apparent consistency*, but not necessarily *genuine consistency*. Another set of words that express the same ideas are *honest methods* and *dishonest methods*. See Keller [2] or Lin and Segel [4] for more details.

References

[1] C. M. Bender and S. A. Orszag, *Advanced Mathematical Methods for Scientists and Engineers*, McGraw–Hill, New York, 1978, pages 83–88.

[2] J. B. Keller, "Wave Propagation in Random Media," *Proc. Sympos. Appl. Math.*, **13**, Amer. Math. Soc., Providence, Rhode Island, 1960.

[3] N. Levinson, "Asymptotic Behavior of Solutions of Non-Linear Differential Equations," *Stud. Appl. Math.*, **48**, 1969, pages 285–297.

[4] C. C. Lin and L. A. Segel, *Mathematics Applied to Deterministic Problems in the Natural Sciences*, Macmillan, New York, 1974, pages 188–189.

116. Equation Splitting

Applicable to   Differential equations.

Yields

An exact solution, but usually not the most general form of the solution.

Idea

By equating two parts of a differential equation to a common term, we may be able to find a fairly general solution to the given differential equation.

Procedure

Separate a differential equation into two (or more) terms such that a general solution is available for one of the terms. Use the other term(s) to restrict this general solution.

Example

Suppose we have, from fluid dynamics, the stream function form of the boundary layer equations to solve for $\Phi(x, y)$:

$$\Phi_y \Phi_{xy} - \Phi_x \Phi_{yy} = \nu \Phi_{yyy}. \tag{116.1}$$

We split this equation by choosing both the right and the left-hand sides of this equation to be identically equal to zero. That is, we break (116.1) into the two simultaneous equations

$$\Phi_y \Phi_{xy} - \Phi_x \Phi_{yy} = 0,$$
$$\nu \Phi_{yyy} = 0. \tag{116.2.a–b}$$

Any solution of (116.2) is also a solution of (116.1). Note that the converse is *not* true: a solution to (116.1) may not satisfy (116.2.a) or (116.2.b). Hence, the solution that is obtained from (116.2) will not be the most general solution.

The general solution to (116.2.b) can be easily found since it is essentially an ordinary differential equation in the independent variable y. The solution of (116.2.b) is

$$\Phi(x, y) = a(x)y^2 + b(x)y + c(x), \tag{116.3}$$

for arbitrary coefficient functions $a(x)$, $b(x)$ and $c(x)$. Using (116.3) in (116.2.a) we conclude that

$$(2ay + b)(2ya' + b') - (a'y^2 + b'y + c')(2a) = 0 \tag{116.4}$$

must hold for all values of x and y. Hence, $a(x)$, $b(x)$ and $c(x)$ can be restricted by equating the coefficients of y^2, y^1, and y^0 in (116.4) to zero. This results in

$$\text{coefficient of } y^2: \quad 4aa' - 2aa' = 0, \tag{116.5}$$

$$\text{coefficient of } y^1: \quad (2ab' + ba') - 2ab' = 0, \tag{116.6}$$

$$\text{coefficient of } y^0: \quad bb' - 2ac' = 0. \tag{116.7}$$

Now we solve the equations appearing in (116.5), (116.6), and (116.7). Equation (116.5) can only be valid if $a(x)$ is a constant, say A. Then equation (116.6) is valid for any $b(x)$ and equation (116.7) can be rewritten as

$$(b^2)' - 4Ac' = 0. \tag{116.8}$$

Equation (116.8) can be integrated to determine $c(x) = \dfrac{b^2(x)}{4A} + D$, where D is an arbitrary constant of integration. Now, using what we have found, the solution in (116.3) becomes

$$\Phi(x, y) = Ay^2 + b(x)y + \left(\frac{b(x)^2}{4A} + D\right), \tag{116.9}$$

for arbitrary A, D, and $b(x)$.

Notes

[1] The above example is from Ames [1].

[2] Note that, for the equations in (116.2) we could have found the general solution of (116.2.a) and then used (116.2.b) to restrict it. The general solution of (116.2.a) is $\Phi(x, y) = F(y + G(x))$, where F and G are arbitrary functions. Using this solution in (116.2.b), and determining conditions on F and G, results in the solution in (116.9).

References

[1] W. F. Ames, "Ad Hoc Exact Techniques for Nonlinear Partial Differential Equations," in W. F. Ames (ed.), *Nonlinear Partial Differential Equations in Engineering*, Academic Press, New York, 1967, pages 59 and 65–69.

[2] M. E. Goldstein and W. H. Braun, *Advanced Methods for the Solution of Differential Equations*, NASA SP-316, U.S. Government Printing Office, Washington, D.C., 1973, page 109.

[3] G. B. Whitham, *Linear and Nonlinear Waves*, Wiley Interscience, New York, 1974, page 421.

117. Floquet Theory

Applicable to   Linear ordinary differential equations with periodic coefficients and periodic boundary conditions.

Yields
   Knowledge of whether all solutions are stable.

Idea
   If a linear differential equation has periodic coefficients and periodic boundary conditions, then the solutions will generally be a periodic function times an exponentially increasing or an exponentially decreasing function. Floquet theory will determine if the solution is exponentially increasing (and so "unstable") or exponentially decreasing (and so "stable").

Procedure
   Suppose we have an n-th order linear ordinary differential equation whose coefficients are periodic with common period T. The general technique is to write the ordinary differential equation as a first order vector system, of dimension n (see page 118), and then solve this vector ordinary differential equation for any set of n linearly independent conditions, for $0 \leq t \leq T$.
   This yields a propagator matrix B, such that $\mathbf{y}(t + mT) = B^m \mathbf{y}(t)$, where $m = 1, 2, \ldots$. Hence, to determine the stability of the original problem, we need only determine the eigenvalues of B. If any of the eigenvalues are larger than one in magnitude, then the solution is "unstable."
   As an example of the general theory, we consider second order linear ordinary differential equations of the form

$$y'' + q(t)y = 0, \tag{117.1}$$

where $q(t)$ is periodic with period T, i.e., $q(t + T) = q(t)$. We can write (117.1) as a vector ordinary differential equation in the form

$$\mathbf{y}(t) = \begin{pmatrix} y(t) \\ y'(t) \end{pmatrix}, \qquad \mathbf{y}' = \begin{pmatrix} 0 & 1 \\ -q(t) & 0 \end{pmatrix} \mathbf{y},$$

where $\mathbf{y}(0) = \begin{pmatrix} y(0) \\ y'(0) \end{pmatrix}$ is known in principle. We now define $u(t)$ and $v(t)$ to be the solutions of

$$\begin{pmatrix} u(t) \\ u'(t) \end{pmatrix}' = \begin{pmatrix} 0 & 1 \\ -q(t) & 0 \end{pmatrix} \begin{pmatrix} u(t) \\ u'(t) \end{pmatrix}, \qquad \begin{pmatrix} u(0) \\ u'(0) \end{pmatrix} = \begin{pmatrix} 1 \\ 0 \end{pmatrix}, \tag{117.2}$$

and

$$\begin{pmatrix} v(t) \\ v'(t) \end{pmatrix}' = \begin{pmatrix} 0 & 1 \\ -q(t) & 0 \end{pmatrix} \begin{pmatrix} v(t) \\ v'(t) \end{pmatrix}, \qquad \begin{pmatrix} v(0) \\ v'(0) \end{pmatrix} = \begin{pmatrix} 0 \\ 1 \end{pmatrix}. \qquad (117.3)$$

Then, by superposition, $\mathbf{y}(t) = A(t)\mathbf{y}(0) = \begin{pmatrix} u(t) & v(t) \\ u'(t) & v'(t) \end{pmatrix} \mathbf{y}(0)$. Equivalently, $\mathbf{y}(T) = B\mathbf{y}(0)$, where $B = A(T)$. Hence, $\mathbf{y}(2T) = B\mathbf{y}(T) = B^2\mathbf{y}(0)$, $\mathbf{y}(3T) = B^3\mathbf{y}(0)$, etc. The eigenvalues of B are needed to determine stability. By the usual calculation, λ will be an eigenvalue of B if and only if $|B - \lambda I| = 0$. We calculate,

$$\begin{aligned} |B - \lambda I| &= \begin{vmatrix} u(T) - \lambda & v(T) \\ u'(T) & v'(T) - \lambda \end{vmatrix} \\ &= \lambda^2 - \lambda[u(T) + v'(T)] + [u(T)v'(T) - u'(T)v(T)] \\ &= \lambda^2 - \lambda\Delta + 1, \end{aligned} \qquad (117.4)$$

where we have defined $\Delta = u(T) + v'(T)$, and we set $u(T)v'(T) - u'(T)v(T)$ equal to one since the Wronskian of equation (117.1) is identically equal to one. Solving (117.4) for λ, we determine that $\lambda = \frac{1}{2}\Delta \pm \sqrt{\frac{1}{4}\Delta^2 - 1}$, and so we conclude

(A) If $|\Delta| < 2$, then, for both values of λ, we have $|\lambda| \leq 1$ and so all of the solutions to (117.1) are stable.

(B) If $|\Delta| > 2$, then there is least one value of λ with $|\lambda| > 1$ and so the solutions to (117.1) are unstable.

Example

Suppose we have the equation

$$y'' + f(t)y = 0, \qquad (117.5)$$

where $f(t)$ is a square wave function of period T

$$f(t + T) = f(t) = \begin{cases} -1 & \text{for} \quad 0 \leq t < T/2, \\ 1 & \text{for} \quad T/2 \leq t \leq T. \end{cases} \qquad (117.6)$$

Note that f(t) is *not* continuous. This does not change any of the analysis. We can solve (117.5) and (117.6) by using $f(t) = -1$ and solving for $\{u(t), v(t)\}$ in the interval $0 \leq t < T/2$. Then we set $f(t) = 1$ and solve for $\{u(t), v(t)\}$ in the interval $T/2 < t \leq T$, using as initial conditions the values calculated when we took $f(t) = -1$. See the section on solving equations with discontinuities (page 219).

The solutions of (117.2) and (117.3) are found to be (for $T/2 < t \leq T$)

$$v(t) = (\cosh \tau \sin \tau + \sinh \tau \cos \tau) \sin t + (\cosh \tau \cos \tau - \sinh \tau \sin \tau) \cos t,$$
(117.7)

and

$$u(t) = (\sinh \tau \sin \tau + \cosh \tau \cos \tau) \sin t + (\sinh \tau \cos \tau + \cosh \tau \sin \tau) \cos t,$$
(117.8)

where $\tau = T/2$. With (117.7) and (117.8), we determine Δ to be

$$\Delta = u(T) + v'(T) = 2 \cosh \tau \cos \tau.$$
(117.9)

The conclusion is that the solutions to (117.6) will be stable or unstable depending on whether the magnitude of Δ, as given by (117.9), is greater than or smaller than two. Different values of T will give different conclusions. For example

(A) If T=17 or $T = e^2$, then $|\Delta| > 2$ and some unstable solutions to (117.5) exist.

(B) If T=1 or $T = \pi$, then $|\Delta| < 2$ and all to the solutions to (117.5) are stable.

Notes

[1] Mathematicians call this technique Floquet theory, while physicists call it Bloch wave theory. Solid state physicists use this technique for determining band gap energies.

[2] Note that the periodicity of $f(t)$ in (117.5) does *not*, by itself, insure that $y(t)$ has a periodic solution. If, however, $f(t)$ is periodic and has mean zero, then equation (117.5) will have a periodic solution of the same period.

[3] The linear system $\mathbf{y}' = B(t)\mathbf{y}$ is said to be *noncritical* with respect to T if it has no periodic solution of period T except the trivial solution $\mathbf{y} = \mathbf{0}$. Otherwise, the system is said to be *critical*.

References

[1] M. Abramowitz and I. A. Stegun, *Handbook of Mathematical Functions*, National Bureau of Standards, Washington, DC, 1964, Section 20.3 (pages 727–730).

[2] G. Birkhoff and G.-C. Rota, *Ordinary Differential Equations*, John Wiley & Sons, New York, 1978, pages 325–326.

[3] E. A. Coddington and N. Levinson, *Theory of Ordinary Differential Equations*, McGraw–Hill Book Company, New York, 1955, pages 78–81.

[4] H. S. Hassan, "Floquet Solutions of Nonlinear Ordinary Differential Equations," *Proc. Roy. Soc. Edin. Sect. A*, **106**, 1987, No. 3–4, pages 267–275.

[5] W. Kaplan, *Operational Methods for Linear Systems*, Addison–Wesley Publishing Co., Reading, MA, 1962, pages 472–490.

[6]   D. L. Lukes, *Differential Equations: Classical to Controlled*, Academic Press, New York, 1982, Chapter 8, pages 162–179.

[7]   W. Magnus and S. Winkler, *Hill's Equation*, Dover Publications, Inc., New York, 1966, pages 3–10.

[8]   B. D. Sleeman and P. D. Smith, "Double Periodic Floquet Theory for a Second Order System of Ordinary Differential Equations," *Quart. J. Math. Oxford. Ser.*, **37**, No. 147, 1986, pages 347–356.

118.   Graphical Analysis: The Phase Plane

Applicable to   Two coupled autonomous first order ordinary differential equations or an autonomous second order ordinary differential equation.

Yields

A graphical representation of the solution.

Idea

The qualitative features of the solution of two coupled autonomous first order ordinary differential equations may be ascertained from the phase plane.

Procedure

Suppose we have the set of two coupled autonomous first order ordinary differential equations

$$\frac{dx}{dt} = f(x,y), \quad \frac{dy}{dt} = g(x,y). \tag{118.1}$$

As t increases, $x(t)$ and $y(t)$ will describe a path in (x,y) space. This will not be the case at those points (x_0, y_0) where

$$f(x_0, y_0) = 0, \quad g(x_0, y_0) = 0.$$

At these points the value does not change with t: $x(t) = x_0$ and $y(t) = y_0$. These points are called *critical points*. (They are also called equilibrium points or singular points).

To analyze the motion near a single critical point, we linearize (118.1) about that point. By a linear change of variables, we can place the critical point at the origin $(x, y) = (0, 0)$. Near a critical point at the origin, (118.1) can be written as

$$
\frac{dx}{dt} = ax + by + \widehat{f}(x, y),
$$
$$
\frac{dy}{dt} = cx + dy + \widehat{g}(x, y),
$$

(118.2)

where $\widehat{f}(x, y) = o(|x| + |y|)$ and $\widehat{g}(x, y) = o(|x| + |y|)$ as $x \to 0$, $y \to 0$. We assume that a, b, c, d are real numbers and they are not all equal to zero. If we discard the $\widehat{f}$ and $\widehat{g}$ terms in (118.2) and look for solutions of the form

$$
x(t) = Ae^{\lambda t}, \quad y(t) = Be^{\lambda t},
$$

then we find that λ must be an eigenvalue of the matrix $\begin{pmatrix} a & b \\ c & d \end{pmatrix}$. That is, λ must satisfy

$$
\lambda^2 - (a + d)\lambda + (ad - bc) = 0.
$$

(118.3)

There are five different types of behavior that can be observed near the critical point $(0, 0)$, based on the roots of (118.3). If the roots of (118.3) are:

(A) Real, distinct, and of the same sign, then the critical point is called a *node*. (See Figure 118.1.a for a typical picture.) Note that the symmetry axes are determined by the eigenvectors of the 2×2 matrix shown above.

(B) Real, distinct, and of opposite signs, then the critical point is called a *saddle point*. (See Figure 118.1.b for a typical picture.)

(C) Real and equal, then the critical point is again a node. (See Figure 118.1.c for a typical picture.)

(D) Pure imaginary, then the critical point is called a *center*. (See Figure 118.1.d for a typical picture.)

(E) Conjugate complex numbers, but not pure imaginary, then the critical point is called a *spiral* or a *focus*. (See Figure 118.1.e for a typical picture.)

In each of the figures, an arrow points in the direction of increasing t. For each case illustrated, there exist systems in which the arrows are pointing in the opposite direction than what we have illustrated. Each solution of (118.2) (corresponding to different initial conditions) describes a single trajectory. Every trajectory must either

(A) go to infinity, or

(B) approach a limit cycle (see page 63), or

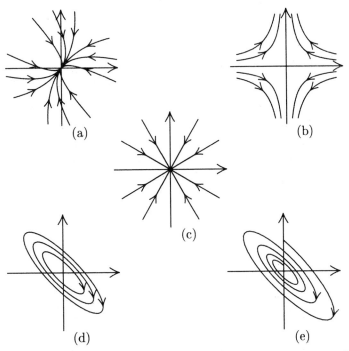

Figure 118.1 The different types of behavior in the phase plane: (a) and (c) are nodes, (b) is a saddle point, (d) is a center, and (e) is a spiral.

(C) tend to a critical point.

If the solution goes to infinity, then the solution is said to be unstable, otherwise it is said to be stable.

Example 1

Consider the simple linear differential equation system

$$\frac{d\mathbf{x}}{dt} = \begin{pmatrix} a & b \\ c & d \end{pmatrix} \mathbf{x}.$$

For this equation, the eigenvalues satisfy (118.3), which we write in the form $\lambda^2 - T\lambda + \Delta = 0$, where T is the trace of the matrix ($T = a + d$) and Δ is the determinant ($\Delta = ad - bc$). The eigenvalues, and the qualitative picture of the phase plane, can be deduced from T and Δ. Figure 118.2 shows the type of behavior to expect for different values of T and Δ. The curve Figure 118.2 is given by determinant= (trace)^2; only centers can occur along this curve.

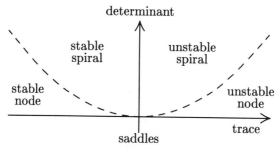

Figure 118.2 The different types of behavior in the phase plane, as a function of the trace and determinant of the 2×2 matrix.

Example 2

Consider the nonlinear autonomous second order ordinary differential equation

$$\frac{d^2x}{dt^2} + \beta\frac{dx}{dt} + \omega^2 \sin x = 0, \tag{118.4}$$

which can be written as the coupled system

$$\frac{dx}{dt} = y,$$
$$\frac{dy}{dt} = -\beta y - \omega^2 \sin x. \tag{118.5}$$

For the equations in (118.5) there are infinitely many critical points at the locations $\{x = n\pi,\, y = 0 \mid n = 0, 1, 2, \ldots\}$. To analyze the behavior near the point $(k\pi, 0)$ the new variables $\widetilde{y} = y$, $\widetilde{x} = x - k\pi$, are introduced. In these new variables, the system in (118.5) can be approximated by

$$\frac{d\widetilde{x}}{dt} = \widetilde{y},$$
$$\frac{d\widetilde{y}}{dt} = -\beta\widetilde{y} + (-1)^{k+1}\omega^2\widetilde{x}, \tag{118.6}$$

when $\widetilde{x}$ and $\widetilde{y}$ are both small. From (118.3) the characteristic equation for (118.6) becomes

$$\lambda^2 + \beta\lambda + \omega^2(-1)^k = 0,$$

with the roots

$$\lambda_1 = \frac{-\beta + \sqrt{\beta^2 + (-1)^{k+1}4\omega^2}}{2}, \quad \lambda_2 = \frac{-\beta - \sqrt{\beta^2 + (-1)^{k+1}4\omega^2}}{2}.$$

If we now assume that $\beta > 0$ and $\beta^2 > 4\omega^2$, then

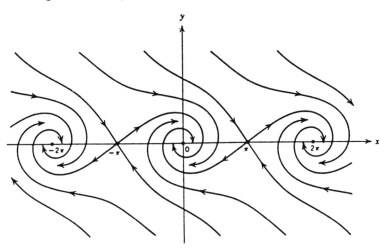

Figure 118.3. Phase plane for the equation in (118.4).

(A) For k even, $\lambda_1 < 0$ and $\lambda_2 < 0$. Hence, the point is a node.

(B) For k odd, $\lambda_1 > 0$ and $\lambda_2 < 0$. Hence, the point is a saddle point.

With this information, we can draw the phase plane for the system in (118.5) (see Figure 118.3). Since the system in (118.4) is dissipative (i.e., the total "energy" decays), all of the different possible solutions approach one of the nodes in infinite time. The trajectories in the phase plane clearly show this.

Notes

[1] In the above, we have presumed that the critical points are *isolated*; that is, each critical point has a neighborhood around it in which no other critical points are present.

[2] If, in (118.2), $ad - bc$ were equal to zero, then second degree (or higher) terms in the Taylor series of f and g would be required to determine the behavior near that critical point. See Boyce and DiPrima [2] for details.

If $ad - bc \neq 0$, then the solution curves of the nonlinear system in (118.1) will be qualitatively similar to the solution curves of the linear system in (118.2), with the single exception that a center for (118.2) may be either a center or a spiral for system (118.1).

[3] A second order autonomous ordinary differential equation can always be written as a first order system, (see page 118). Also, the general equation of first order $M(x,y)\,dx + N(x,y)\,dy = 0$ may be written as a system in the form of (118.1); i.e.,

$$\frac{dx}{dt} = N(x,y), \qquad \frac{dy}{dt} = -M(x,y).$$

[4] The point at infinity may be analyzed by changing variables by

$$x_1 = \frac{x}{x^2 + y^2}, \qquad y_1 = \frac{-y}{x^2 + y^2},$$

and then analyzing the point $(0,0)$ in the x_1, y_1-plane. This corresponds to the substitution $z_1 = 1/z$, when $z = x + iy$ is treated as a complex variable.

[5] Kath [8] describes a method that combines phase plane techniques with matched asymptotic expansions. This method can be used to analyze second order, nonlinear, non-autonomous, singular boundary value problems.

[6] Two different graphing programs for showing phase planes on a Macintosh computer are *DEGraph* and *Phase Portraits*. A review of these programs is in Hartz [4]. A program that runs on IBM personal computers (and compatibles) is *Phaser*; see Margolis [9] for a review.

References

[1] C. M. Bender and S. A. Orszag, *Advanced Mathematical Methods for Scientists and Engineers*, McGraw–Hill, New York, 1978, pages 171–197.

[2] W. E. Boyce and R. C. DiPrima, *Elementary Differential Equations and Boundary Value Problems*, Fourth Edition, John Wiley & Sons, New York, 1986, pages 456–486.

[3] E. A. Coddington and N. Levinson, *Theory of Ordinary Differential Equations*, McGraw–Hill Book Company, New York, 1955, Chapter 15 (pages 371–388).

[4] D. Hartz, "DEGraph and Phase Portraits," *Notices of the American Mathematical Society*, **36**, No. 5, May/June 1989, pages 559–561.

[5] J. Hubbard and B. West, *MacMath: A Dynamical Systems Software Package*, Springer–Verlag, New York, 1991.

[6] I. Huntley and R. M. Johnson, *Linear and Nonlinear Differential Equations*, Halstead Press, New York, 1983, Chapter 8 (pages 114–133).

[7] D. W. Jordan and P. Smith, *Nonlinear Ordinary Differential Equations*, Clarendon Press, Oxford, 1987, Chapters 1–3.

[8] W. L. Kath, "Slowly Varying Phase Planes and Boundary-Layer Theory," *Stud. Appl. Math.*, **72**, 1985, pages 221–239.

[9] M. S. Margolis, "Phaser," *Notices of the American Mathematical Society*, **37**, No. 4, April 1990, pages 430–434.

[10] D. Wang, "Computer Algebraic Methods for Investigating Plane Differential Systems of Center and Focus Type," in E. Kaltofen and S. M. Watt (eds.), *Computers and Mathematics*, Springer–Verlag, New York, 1990, pages 91–99.

119. Graphical Analysis:
The Tangent Field

Applicable to   First order ordinary differential equations.

Yields

A graphical representation the solutions corresponding to different initial conditions.

Idea

The qualitative features of the solution of a first order ordinary differential equation may be ascertained from the tangent field.

Procedure

Given a first order ordinary differential equation in the form

$$\frac{dy}{dx} = f(x, y), \tag{119.1}$$

the procedure is to draw small line segments in the (x, y) plane, such that the line segment that goes through the point (x_0, y_0) has the slope $f(x_0, y_0)$. Note that a slope of m corresponds to an angle of $\tan^{-1} m$. After a region of (x, y) space has been covered with these small line segments, it should be apparent how the solution curves of (119.1) behave. An approximate solution may then be drawn by "connecting up" the line segments that originate from a given point.

Constructing the tangent field by hand is often facilitated by the *method of isoclines*. In this method, a few curves of the form $f(x, y) = C$, with C being a constant, are constructed. Along each one of these curves, dy/dx is equal to the constant C. Hence, at every point on these curves, the small line segments all have the same slope.

Example 1

Suppose we have the nonlinear ordinary differential equation

$$\frac{dy}{dx} = 1 - xy^2. \tag{119.2}$$

It is straightforward to construct the tangent field, which is shown in Figure 119.1.

Every solution of (119.2) must be tangent to whatever line segments it passes near. For example, if (119.2) had the initial condition $y(0) = 1$, then

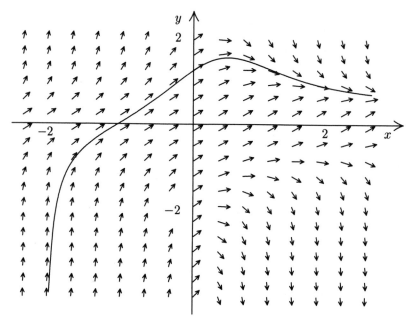

Figure 119.1 Tangent field for the equation in (119.2).

the solution can be approximately traced by starting at the point $(0, 1)$ and drawing a line that remains tangent to the line segments. For this equation and initial condition, y tends to zero as x tends to infinity. This behavior can be seen in Figure 119.1.

Example 2

Given the differential equation

$$\frac{dy}{dx} = 2x + y, \tag{119.3}$$

we find that the isoclines are the straight lines $2x + y = C$. Figure 119.2 shows the isoclines, with small line segments superposed, as well as three solutions to (119.3).

The exact solution to (119.3) is

$$y = 2(1 - x) + Ae^{-x},$$

where A is an arbitrary constant. The linear behavior for $x \gg 0$ and the exponential behavior for $x < 0$ can be identified in this figure.

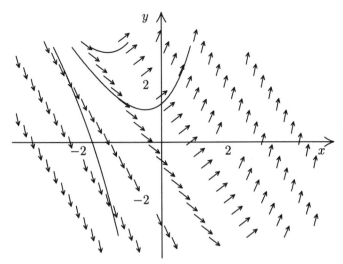

Figure 119.2 Tangent field for the equation in (119.3).

Notes

[1] Consider drawing a small circle Γ in the (x, y) plane that surrounds the point (x_0, y_0). Traversing the circle counter-clockwise, the direction field will change. In every case, the change in angle must be a multiple of 2π: $[\text{angle}]_\Gamma = 2\pi I_\Gamma$, where I_Γ is an integer called the *index of the vector field*. The index may be positive, negative or zero.

   If Γ surrounds no critical points, then the index is zero. If Γ surrounds a saddle point, then the index is -1. If Γ surrounds a center, spiral, or node, then the index is -1. If Γ surrounds more than one critical point, then the index is the sum of the indices for each critical point.

[2] Equation (119.1) sometimes arises from the autonomous system $\{\dot{x} = F(x, y),$ $\dot{y} = G(x, y)\}$, via $\dfrac{dy}{dx} = \dfrac{G(x, y)}{F(x, y)}$. In this case, we have $I_\Gamma = \dfrac{1}{2\pi} \displaystyle\oint_\Gamma \dfrac{F\, dG - G\, dF}{F^2 + G^2}$. See Jordan and Smith [3] for details.

[3] Hand construction of the tangent field produces only qualitative information. More information can be obtained when a computer is used to generate the tangent field,

References

[1] C. M. Bender and S. A. Orszag, *Advanced Mathematical Methods for Scientists and Engineers*, McGraw–Hill, New York, 1978, pages 148–149.

[2] W. E. Boyce and R. C. DiPrima, *Elementary Differential Equations and Boundary Value Problems*, Fourth Edition, John Wiley & Sons, New York, 1986, pages 34–35.

[3] D. W. Jordan and P. Smith, *Nonlinear Ordinary Differential Equations*, Clarendon Press, Oxford, Second Edition, 1987, Chapter 3.

120. Harmonic Balance

Applicable to   Nonlinear ordinary differential equations with periodic solutions.

Yields

An approximate solution valid over the entire period. There is a specified procedure for increasing the number of terms and, hence, for increasing the accuracy.

Idea

Harmonic balance is a way of looking for periodic solutions in non-linear systems by trying to fit a truncated Fourier series and choosing the frequency, amplitude, and phases so that any error occurs only in the discarded harmonics.

Procedure

Suppose we have a differential equation of the form

$$f(x, x_t, x_{tt}, t) = 0, \tag{120.1}$$

and we wish to find a periodic solution of period T. We look for an approximation to (120.1) in the form of a truncated Fourier series

$$x(t) \simeq y(t) := a_0 + \sum_{j=1}^{N} a_j \cos j\omega t + b_j \sin j\omega t,$$

where $\omega = 2\pi/T$. The unknowns to be determined are $\{a_0, a_j, b_j \mid j = 1, \ldots, N\}$ and possibly T.

If T is known, then we require the $2N + 1$ unknowns to satisfy the $2N + 1$ algebraic equations

$$\int_0^T f(y, y_t, y_{tt}, t) \sin k\omega t \, dt = 0,$$
$$\int_0^T f(y, y_t, y_{tt}, t) \cos k\omega t \, dt = 0, \tag{120.2.a–b}$$

for $k = 0, 1, \ldots, N$.

If the period T is unknown, then there are $2N + 2$ unknowns to be determined. To find algebraic equations for these unknowns, we require (120.2) to hold for $k = 0, 1, \ldots, N$ and, say, (120.2.a) for $k = N + 1$.

Example 1

Given the equation

$$\frac{d^2x}{dt^2} + x + \alpha \left(\frac{dx}{dt}\right)^2 = \sin t, \tag{120.3}$$

where α is a given constant, we search for a 2π periodic solution. If we take $T = 2\pi$ and $N = 2$, then we are assuming that

$$x(t) \simeq y(t) = a_0 + a_1 \cos t + a_2 \cos 2t + + b_1 \sin t + b_2 \sin 2t. \tag{120.4}$$

Using (120.3) and (120.4) in (120.2) produces the set of simultaneous algebraic equations

$$\alpha \left(4b_2^2 + b_1^2 + 4a_2^2 + a_1^2\right) + 2a_0 = 0,$$
$$\alpha \left(b_1 b_2 + a_1 a_2\right) = 0,$$
$$\alpha \left(b_1^2 - a_1^2\right) - 6a_2 = 0,$$
$$2\alpha \left(a_1 b_1 - a_2 b_1\right) - 1 = 0,$$
$$3b_2 + \alpha a_1 b_1 = 0.$$

These equations have the unique solution $\{a_0 = -(\alpha^{2/3} + 3^{4/3})/2(9\alpha)^{1/3}$, $a_1 = 0$, $a_2 = 1/2(3\alpha)^{1/3}$, $b_1 = -3^{1/3}/\alpha^{2/3}$, $b_2 = 0\}$. Hence, the approximation (for $N = 2$) becomes

$$x(t) \simeq -\left(\frac{3}{\alpha^2}\right)^{1/3} \sin t + \frac{1}{2(3\alpha)^{1/3}} \left(\cos 2t - 3\right) - \frac{(3\alpha)^{1/3}}{6}. \tag{120.5}$$

Note that this approximation indicates the qualitatively correct behavior, at least for small values of α. When α is small, equation (120.3) is a harmonic oscillator being forced near resonance. This would lead to a large magnitude solution, which is what (120.5) indicates.

Example 2

Given the equation

$$\frac{d^2x}{dt^2} + x = c(x^2 + \cos t),$$

we choose $N = 1$ and look for solutions of period $T = 2\pi$. Using the approximation

$$x(t) \simeq y(t) = a_0 + a_1 \cos t + b_1 \sin t,$$

we find that $b_1 = 0$, $a_1 = -1/2a_0$, and $a_0 = c^{1/3}z/2$, where z satisfies the cubic equation $c^{4/3}z^4 - 2z^3 + 2 = 0$. Here, the analytical solution for a_0 is available (implicitly), but is not very informative. However, if we assume that $|c| \ll 1$, then it can be shown that $a_0 = \dfrac{c^{1/3}}{2}\left[1 + \dfrac{c^{1/3}}{6} + O(c^{8/3})\right]$.

Example 3

The requirements in (120.2) are not the only way in which to obtain useful approximations. Consider the Duffing equation, $\ddot{x} + x = \varepsilon x^3$, with $\dot{x}(0) = 0$. If we presume that $x = A \cos \omega t$, then

$$x - \varepsilon x^3 = A \cos \omega t \left(1 - \tfrac{3}{4}\varepsilon A^2\right) - \tfrac{1}{4}\varepsilon A^3 \cos 3\omega t.$$

If we disregard the last, higher order, term, then we may write $x - \varepsilon x^3 \approx x \left(1 - \tfrac{3}{4}\varepsilon A^2\right)$. With this approximation, the original equation becomes $\ddot{x} - \left(1 - \tfrac{3}{4}\varepsilon A^2\right) x \approx 0$. Since we have presumed that $x = A \cos \omega t$, we can immediately identify the frequency: $\omega^2 \approx 1 - \tfrac{3}{4}\varepsilon A^2$. Hence, to leading order, our approximate solution becomes $x \approx A \cos \left(1 - \tfrac{3}{8}\varepsilon A^2\right) t$.

Notes

[1] This technique is known in the engineering literature as the *describing function method*.

[2] Strictly speaking, this method may also be used to obtain approximations to differential equations that do not have periodic solutions.

[3] This technique applies, in principle, to equations in which there is no small parameter. However, it may prove that the algebraic equations generated by (120.2) are not solvable in closed form unless a perturbation expansion is used (as in Example 2).

[4] Mees [6] has a very extensive bibliography, separated into categories (applications, theory, background theory, Hopf bifurcation, and harmonic balance).

[5] When this method is implemented numerically, it is known as the *spectral method*. See the section beginning on page 759 or see Gottlieb and Orszag [2] for details.

References

[1] A. A. Ferri, "On the Equivalence of the Incremental Harmonic Balance Method and the Harmonic Balance–Newton Raphson Method," *J. Appl. Mech.*, **53**, June 1986, pages 455–457.

[2] D. Gottlieb and S. A. Orszag, *Numerical Analysis of Spectral Methods: Theory and Applications*, SIAM, Philadelphia, 1977.

[3] F. R. Groves, Jr., "Numerical Solution of Nonlinear Differential Equations Using Computer Algebra," *Int. J. Comp. Math.*, **13**, 1983, pages 301–309.

[4] I. Huntley and R. M. Johnson, *Linear and Nonlinear Differential Equations*, Halstead Press, New York, 1983, Chapter 12 (pages 166–168).

[5] K. S. Kundert, G. B. Sorkin, and A. Sangiovanni-Vincentelli, "Applying Harmonic Balance to Almost-Periodic Circuits," *IEEE Trans. Microwave Theory and Tech.*, **36**, No. 2, February 1988, pages 366–378.

[6] A. I. Mees, "Describing Functions: Ten Years On," *IMA J. Appl. Mathematics*, **32**, 1984, pages 221–233.

121. Homogenization

Applicable to   "Microscopic" differential equations.

Yields
    "Macroscopic" differential equations.

Idea
    By averaging microscopic differential equations, differential equations for macroscopic quantities may be determined.

Procedure
    In many fields, the ("microscopic") equations of motion contain more information than is needed by a practitioner solving a specific problem. For instance, in a fluid flow problem, it may be that only the mass flow is required, not a detailed analysis of the flow field. Consequently, it is of interest to take an "average" of the "microscopic" differential equations to obtain a set of differential equations that describe the "macroscopic" quantities of interest.
    The average taken could be a time average, a space average, an ensemble average, or an average of some other type.
    In the homogenization method, it is usually assumed that there is a fast time (or a short length) scale, on which the "microscopic" differential equations vary. The dependence on this fast scale is usually assumed to be either periodic or random. In mechanics problems, the small length scale is often the length scale of the inclusions or heterogeneities.
    Often, a formal procedure for analyzing problems via homogenization is by a multi-scaling procedure (see page 524).

Example 1
    As an example of the general procedure, consider the elliptic problem

$$-\sum_{i,j} \frac{\partial}{\partial x_i}\left[a_{ij}^{\varepsilon}(\mathbf{x})\frac{\partial u}{\partial x_j}\right] = f(\mathbf{x}), \tag{121.1}$$

in some domain Ω. The equation in (121.1) probably came from a system of the form

$$-\frac{\partial p_i}{\partial x_i} = f(\mathbf{x}),$$

$$p_i = a_{ij}^{\varepsilon}(\mathbf{x})\frac{\partial u}{\partial x_j}, \tag{121.2}$$

via Hamilton's equations. In (121.1) and (121.2), it is now assumed that $a_{ij}^{\varepsilon}(\mathbf{x})$ is of the form $a_{ij}(\mathbf{x}/\varepsilon)$ and that $a_{ij}^{\varepsilon}(\mathbf{x})$ is periodic in $\mathbf{x}$, with the period in the x_i variable being L_i. A formal two scale procedure can be defined by (see page 524)

$$y_i = \frac{x_i}{\varepsilon},$$

$$u(\mathbf{x}) = u_0(\mathbf{x}, \mathbf{y}) + \varepsilon u_1(\mathbf{x}, \mathbf{y}) + \varepsilon^2 u_2(\mathbf{x}, \mathbf{y}) + \cdots,$$

$$p(\mathbf{x}) = \mathbf{p}^0(\mathbf{x}, \mathbf{y}) + \varepsilon \mathbf{p}^1(\mathbf{x}, \mathbf{y}) + \varepsilon^2 \mathbf{p}^2(\mathbf{x}, \mathbf{y}) + \cdots,$$

where $\mathbf{p} = (p_1, p_2, \ldots)$. In this case, we choose to define the average of some arbitrary function of $\mathbf{x}$ and $\mathbf{y}$ to be

$$\bar{A}(\mathbf{x}) := \frac{1}{L_1 L_2 \cdots L_n} \int A(\mathbf{x}, \mathbf{y}) \, d\mathbf{y}. \tag{121.3}$$

We integrate over $\mathbf{y}$ in (121.3) to average over the high frequency component of a function that depends on both $\mathbf{x}$ and $\mathbf{y}$. For our example, it is straightforward to show that

$$-\frac{\partial \bar{p}_i^0}{\partial x_i} = f(\mathbf{x}). \tag{121.4}$$

where $\bar{\mathbf{p}}^0 = (\bar{p}_1^0, \bar{p}_2^0, \ldots)$. Now, if an $a_{ij}^h(\mathbf{x})$ can be found such that

$$\bar{p}_i^0 = a_{ij}^h(\mathbf{x}) \frac{\partial u_0}{\partial x_j}, \tag{121.5}$$

then $a_{ij}^h(\mathbf{x})$ is said to be the homogenized coefficient, and equations (121.4) and (121.5) are the homogenized equations.

Example 2

For a more detailed example, consider the equation

$$A_\varepsilon u_\varepsilon := -\sum_{i,j} \frac{\partial}{\partial x_i} \left(a_{ij} \left(\frac{\mathbf{x}}{\varepsilon} \right) \frac{\partial u_\varepsilon}{\partial x_j} + a_0 \left(\frac{\mathbf{x}}{\varepsilon} \right) u_\varepsilon \right) = f(\mathbf{x}) \tag{121.6}$$

where $a_{ij}(\mathbf{y})$ and $a_0(\mathbf{y})$, with $\mathbf{y} := \mathbf{x}/\varepsilon$, are periodic on the unit cube Y. We assume that the solution can be expanded in the form

$$u_\varepsilon = u_0 \left(\mathbf{x}, \frac{\mathbf{x}}{\varepsilon} \right) + \varepsilon u_1 \left(\mathbf{x}, \frac{\mathbf{x}}{\varepsilon} \right) + \ldots$$

$$= u_0(\mathbf{x}, \mathbf{y}) + \varepsilon u_1(\mathbf{x}, \mathbf{y}) + \ldots. \tag{121.7}$$

Using the chain rule (i.e., ∂_{x_i} becomes $\partial_{x_i} + \dfrac{1}{\varepsilon}\partial_{y_i}$), using (121.7) into (121.6), and equating powers of ε results in

$$A_1 u_0 = 0,$$
$$A_1 u_1 = A_2 u_0, \qquad\qquad (121.8.a\text{--}c)$$
$$A_1 u_2 = A_2 u_1 + A_3 u_0 + f,$$

where

$$A_1 = -\sum_{i,j} \frac{\partial}{\partial y_i}\left(a_{ij}(\mathbf{y})\frac{\partial}{\partial y_j}\right),$$

$$A_2 = \sum_{i,j} \frac{\partial}{\partial y_i}\left(a_{ij}(\mathbf{y})\frac{\partial}{\partial x_j}\right) + \frac{\partial}{\partial x_i}\left(a_{ij}(\mathbf{y})\frac{\partial}{\partial y_j}\right),$$

$$A_3 = \sum_{i,j} \frac{\partial}{\partial x_i}\left(a_{ij}(\mathbf{y})\frac{\partial}{\partial x_j}\right) + a_0(\mathbf{y}).$$

If we define an averaging operator by

$$M\left[v\right] = \frac{1}{|Y|}\int_Y v(\mathbf{y})\,d\mathbf{y},$$

then it can be shown that the equation $A_1 v = h$ will have a unique solution only if $M\left[h\right] = 0$ (see the section on alternative theorems, page 14). This condition, applied to (121.8.b), indicates that $u_0 = u_0(\mathbf{x})$. This fact simplifies (121.8.b) to

$$A_1 u_1 = -\sum_{i,j}\frac{\partial}{\partial y_i}\left(a_{ij}(\mathbf{y})\frac{\partial}{\partial y_j}\right)u_1 = \sum_{i,j}\left(\frac{\partial a_{ij}(\mathbf{y})}{\partial y_i}\right)\frac{\partial u_0(\mathbf{x})}{\partial x_j} = A_2 u_0$$

Using separation of variables on this results in $u_1(\mathbf{x},\mathbf{y}) = \sum_k z_k(\mathbf{y})\dfrac{\partial u_0(\mathbf{x})}{\partial x_k}$,

where $z_k(\mathbf{y})$ is the unique periodic solution of

$$A_1 z_k = -\sum_{i,j}\frac{\partial}{\partial y_i}\left(a_{ij}(\mathbf{y})\frac{\partial}{\partial y_j}\right)z_k = \sum_i \frac{\partial a_{ik}(\mathbf{y})}{\partial y_i}. \qquad (121.9)$$

Equation (121.9) is known as the *cell problem*.

To finally obtain a solution, we require from (121.8.c) that $M[A_2 u_1 + A_3 u_0 + f] = 0$. This results in

$$-\sum_{i,j}p_{ij}(\mathbf{x})\frac{\partial^2 u_0(\mathbf{x})}{\partial x_i \partial x_j} + M\left[a_0\right]u_0(x) = f(\mathbf{x})$$

where $p_{ij}(\mathbf{x}) := M\left[a_{ij}\right] - M\left[\sum_k a_{ik}\frac{\partial z_j}{\partial y_k}\right].$

Notes

[1] Homogenization techniques are often used in fluid mechanics (two phase flow in particular), electric field theory, and solid mechanics.

[2] Homogenization is often the method used in ad hoc "mean field" theories, "effective media" theories, and "averaged equations."

[3] Homogenization seems to be related to renormalization group theory. Renormalization group methods study the asymptotic behavior of a system (i.e., the macroscopic behavior) when the scale of observation is much larger than the scale of microscopic description, see Goldenfeld, Martin, and Oono [5].

[4] In Persson and Wyller [7] it is shown that, for a sample problem, homogenization is equivalent to Whitham's averaged Lagrangian method.

[5] Averages, denoted by $\langle \cdot \rangle$, are generally required to satisfy "Reynold's rules"

$$\langle f + g \rangle = \langle f \rangle + \langle g \rangle,$$
$$\langle \langle f \rangle g \rangle = \langle f \rangle \langle g \rangle,$$
$$\langle c \rangle = c,$$

when f and g are random or periodic functions and c is a constant. It is also often required that

$$\left\langle \frac{\partial f}{\partial t} \right\rangle = \frac{\partial \langle f \rangle}{\partial t}$$

be satisfied for functions f that are "well behaved."

References

[1] M. Avellaneda, "Iterated Homogenization, Differential Effective Medium Theory and Applications," *Comm. Pure Appl. Math*, **60**, No. 5, September 1987, pages 527–554.

[2] A. Bensoussan, J. L. Lions, and G. Papanicolaou, *Asymptotic Analysis for Periodic Structures*, North–Holland Publishing Co., New York, 1978.

[3] J. M. Burgers, "On Some Problems of Homogenization," *Quart. Appl. Math.*, **35**, No. 4, January 1978, pages 421–434.

[4] J. Ericksen, D. Kinderlehrer, R. Kohn, and J.-L. Lions (eds.), *Homogenization and Effective Moduli of Materials and Media*, Springer–Verlag, New York, 1986.

[5] N. Goldenfeld, O. Martin, and Y. Oono, "Intermediate Asymptotics and Renormalization Group Theory," *J. Scientific Comput.*, **4**, No. 4, 1989.

[6] E. W. Larsen, "Two Types of Homogenization," *SIAM J. Appl. Math.*, **36**, No. 1, February, 1979, pages 26–33.

[7] L. Persson and J. Wyller, "A Note on Whithams Method and the Homogenization Procedure," *Physica Scripta.*, **38**, 1988, pages 774–776.

[8] E. Sanchez-Palencia, "Homogenization Method For the Study of Composite Media," in J. D. Murray (ed.), *Asymptotic Analysis*, Springer–Verlag, New York, 1984, pages 192–214.

[9] M. Vogelius, "A Variational Method to Find Effective Coefficients for Periodic Media. A Comparison with Standard Homogenization," in R. Burridge, S. Childress, and G. Papanicolaou (eds.), *Macroscopic Properties of Disordered Media*, Springer–Verlag, New York, 1982.

122. Integral Methods

Applicable to   Linear and nonlinear partial differential equations.

Yields

An approximation of the solution.

Idea

A sequence of physical approximations may lead to an approximate solution.

Procedure

There are generally three separate steps in using the common integral approximation techniques:

(A) A physical boundary (either natural or imposed mathematically) is assumed to be at some finite distance.

(B) A weak form of the equations is assumed to hold, up to the boundary described above.

(C) The form of the solution is guessed by the method of undetermined coefficients.

These concepts will be made clear in the following example.

Example

Suppose we want to approximate the solution of the linear parabolic partial differential equation

$$u_t = \alpha u_{xx}, \quad \text{for } x > 0, \, t > 0,$$
$$u(0, x) = u_0,$$
$$\frac{\partial u}{\partial x}(t, 0) = f(t),$$

$$(122.1.a\text{--}c)$$

where $f(t)$ is some prescribed function. Note that the value of $u(t, x)$ (which physically might represent a temperature) is initially u_0. For the first approximation, we suppose that there is a finite distance $\beta(t)$ which varies with time, beyond which the temperature is still u_0.

This assumption is contrary to fact; we know that the diffusion equation has an infinite propagation speed, and the value of u at *all* points is immediately changed from u_0. But the change from u_0 will be exponentially small at large distances, so we assume it is zero for $x \geq \beta(t)$. This adds the boundary conditions

$$u(t, \beta(t)) = u_0,$$
$$u_x(t, \beta(t)) = 0. \tag{122.2.a–b}$$

Equation (122.2.a) states that the temperature at the boundary $x = \beta(t)$ is always equal to u_0. Equation (122.2.b) states that there is no heat flux across $x = \beta(t)$; if there was such a flux, then the region beyond $x = \beta(t)$ would not maintain the temperature $u = u_0$.

The second approximation is to assume that a weak form of the differential equation will hold. To obtain this weak form we integrate (122.1.a) with respect to x from $x = 0$ to $x = \beta(t)$ to obtain

$$\int_0^{\beta(t)} u_t \, dx = \alpha \int_0^{\beta(t)} u_{xx} \, dx.$$

This expression can be integrated by parts to obtain

$$\frac{d}{dt} \int_0^{\beta(t)} u \, dx - u(t, \beta(t)) \frac{d\beta(t)}{dt} = \alpha \left[u_x(t, \beta(t)) - u_x(t, 0) \right]$$
$$= -\alpha f(t), \tag{122.3}$$

where we have used (122.2.b) and (122.1.c). If we define

$$w(t) = \int_0^{\beta(t)} u \, dx, \tag{122.4}$$

then (122.3) can be written as the ordinary differential equation

$$\frac{d}{dt} \left(w - u_0 \beta \right) = -\alpha f(t). \tag{122.5}$$

Note that, from (122.4), the average value of $u(t, x)$ in the region $0 \leq x \leq \beta(t)$ is given by $w(t)/\beta$.

Now we must determine $\beta(t)$ from (122.5). Before we can solve for $\beta(t)$, however, we need to determine $w(t)$. To determine $w(t)$, we presume some form of the general solution for $u(t, x)$. By the use of undetermined coefficients we suppose that $u(t, x)$ has the form

$$u(t, x) = \begin{cases} a(t) + b(t)x + c(t)x^2, & \text{for } 0 < x < \beta(t), \\ u_0, & \text{for } x > \beta(t), \end{cases}$$

where $a(t)$, $b(t)$ and $c(t)$ are all unknowns. If this form is to satisfy (122.1.c) and (122.2) then it must be restricted to be of the form

$$u(t,x) = \begin{cases} u_0 - \dfrac{f(t)}{2\beta(t)}[\beta(t) - x]^2, & \text{for } 0 \le x < \beta(t), \\ u_0, & \text{for } x \ge \beta(t). \end{cases} \tag{122.6}$$

Using this form in (122.4) results in $w(t) = u_0\beta(t) - \dfrac{\beta^2(t)f(t)}{6}$. Using this value for $w(t)$ in (122.5) results in $\dfrac{d}{dt}\left[\dfrac{\beta^2(t)f(t)}{6}\right] = \alpha f(t)$. The solution of this ordinary differential equation is

$$\beta(t) = \sqrt{\frac{6\alpha}{f(t)}\int_0^t f(s)\,ds}. \tag{122.7}$$

Using this form for $\beta(t)$ in (122.6) completes the determination of the approximate solution.

For comparison purposes: if $f(t)$ is the constant F, then the temperature at $x = 0$ is given by (using (122.7) and (122.6))

$$u(t,0) \simeq u_0 - \sqrt{\tfrac{3}{2}\alpha tF}. \tag{122.8}$$

Conversely, the exact solution of (122.1) can be found by the use of Laplace transforms to be $u(t,x) = u_0 - \sqrt{\dfrac{\alpha}{\pi}}\displaystyle\int_0^t \dfrac{f(t-\tau)}{\sqrt{\tau}}e^{-x^2/4\alpha\tau}\,d\tau$, and so, when $f(t)$ is the constant F, the exact solution becomes $u(t,0) = u_0 - \sqrt{\dfrac{4}{\pi}\alpha tF}$. The difference between this exact solution and the approximation in (122.8) is about nine percent.

Notes

[1] This method, in fluid mechanics, is known as the Kármán–Pohlausen technique. The distance $\beta(t)$ then represents the thickness of a boundary layer.

[2] When this technique is used, it is most often with partial differential equations that have only a single space variable.

[3] This technique is often used in free boundary problems (see page 262).

References

[1] W. F. Ames, *Nonlinear Partial Differential Equations*, Academic Press, New York, 1967, pages 271–278.

[2] T. R. Goodman, "Application of Integral Methods to Transient Nonlinear Heat Transfer," in T. F. Irvine, Jr. and J. P. Hartnett (eds.), *Advances in Heat Transfer*, Academic Press, New York, 1964, pages 51–122.

[3] D. S. Riley and P. W. Duck, "Application of the Heat-Balance Integral Method to the Freezing of a Cuboid," *Int. J. Heat Mass. Transfer*, **20**, 1977, pages 294–296.

123.   Interval Analysis

Applicable to   Ordinary differential equations and partial differential equations.

Yields

An analytical approximation with an *exact* bound on the error.

Idea

Initially, we bound the solution between an upper and lower bound. Then, iterating a contraction mapping we generate a sequence of approximations in which the upper bound decreases and the lower bound increases.

Procedure

We use the interval notation $[a, b]$ to indicate some number between the values of a and b. We allow the coefficients of polynomials to be intervals. For example, the interval polynomial

$$Q(x) = 1 + [2, 3]x^2 + [-1, 4]x^3,$$

evaluated at the point $x = y$ means that

$$\min_{\substack{2 \leq \eta \leq 3 \\ -1 \leq \zeta \leq 4}} \left(1 + \eta y^2 + \zeta y^3\right) \leq Q(y) \leq \max_{\substack{2 \leq \eta \leq 3 \\ -1 \leq \zeta \leq 4}} \left(1 + \eta y^2 + \zeta y^3\right).$$

There exists an algebra of interval polynomials. For example

$$\left(x + [2, 3]x^3\right) + \left([1, 2]x + [1, 4]x^3\right) = [2, 3]x + [3, 7]x^3,$$

$$([1, 3] + [-1, 2]x)^2 = [1, 9] + [-6, 12]x + [0, 4]x^2.$$

If $P(x)$ and $Q(x)$ are interval polynomials, then at any point y we can write $P(y) \in [P_L, P_U], Q(y) \in [Q_L, Q_U]$. We say that $P(x)$ contains $Q(x)$ on some interval $[c, d]$ if $P_L \leq Q_L$ and $Q_U \leq P_U$ for all $y \in [c, d]$. This is denoted by: $Q(x) \subset P(x)$.

To approximate the solution of an ordinary differential equation, we search for a contraction mapping (see page 54) that has the form: $P_{k+1} = F[P_k]$, where $F[\cdot]$ is a functional, $P_{k+1} \subset P_k$, and P_k tends to the solution of the differential equation as $k \to \infty$.

Example

Suppose we want to approximate the solution of

$$y' = y^2, \qquad y(0) = 1, \qquad (123.1)$$

for values of x in the interval $[0, 1/4]$. Equation (123.1) can be written as the equivalent integral equation

$$y(x) = 1 + \int_0^x y^2(z) \, dz. \qquad (123.2)$$

It is easy to see that the solution of (123.2) must lie in the interval $[1, 2]$ when $x \in [0, 1/4]$. This is because y' is always positive, so y cannot be smaller than 1 (which is what $y(0)$ is) and if it is assumed that $y(z_0) = 2$ for some $z_0 \in (0, 1/4)$, then a contradiction can be reached by using (123.2). We now define the iteration sequence (the contraction mapping) by

$$P_{k+1}(x) = 1 + \int_0^x P_k^2(z) \, dz,$$

for $k = 0, 1, 2, \ldots$, which is just Picard's integral formula (see page 535). We start the sequence off by $P_0(x) = [1, 2]$ and then calculate

$$P_1(x) = 1 + \int_0^x [1, 2]^2 \, dz,$$

$$= 1 + [1, 4]x,$$

$$P_2(x) = 1 + \int_0^x (1 + [1, 4]z)^2 \, dz,$$

$$= 1 + \int_0^x (1 + [2, 8]z + [1, 16]z^2) \, dz,$$

$$= 1 + x + [1, 4]x^2 + \left[\tfrac{1}{3}, \tfrac{16}{3}\right] x^3,$$

$$P_3(x) = 1 + x + x^2 + x^3 + [1, 2]x^4 + \cdots,$$

$$P_4(x) = 1 + x + x^2 + x^3 + x^4 + \left[1, \tfrac{7}{5}\right] x^5 + \cdots.$$

It is easy to show that $P_{k+1}(x) \subset P_k(x)$ and that $\{P_k(x)\}$ converges to the exact solution $y(x)$ of (123.1). Note that, from the $P_k(x)$, *exact* estimates of the solution are available. For example, from $P_2(x)$ we find, $1.141 < y(1/8) < 1.198$.

Notes

[1] The exact solution to the system in (123.1) is $y(x) = 1/(1 - x)$, which has the Taylor series: $y(x) = 1 + x + x^2 + x^3 + x^4 + \cdots$.

[2] The techniques presented in this section can be implemented numerically. Interval arithmetic packages are available in Algol (see Guenther and Marquardt [6]), FORTRAN (see Yohe [13]), and PASCAL (see Rall [10]).

[3] The book by Eijgenraam [4] contains some worked examples.

[4] To avoid dealing with polynomials of large degree, as in the example, we could observe that

$$x^n \subset \left[0, \left(\frac{1}{4}\right)^{n-m}\right] x^m,$$

for x in the interval $[0, \frac{1}{4}]$. This allows us to replace x^n by x^m, with a coarsening of the bounds.

[5] The real power of this method is that it can be applied to differential equations whose coefficients are given by intervals. For example, this would be the case in a problem where a parameter appearing in a differential equation is known only approximately.

[6] The paper by Ames and Nicklas [2] describes the solution of elliptic partial differential equations, using interval analysis to solve the finite difference equations produced by a numerical approximation. Schwandt [10] addresses the same issue, but with the use of a vector computer.

[7] When solving ordinary differential equations numerically, using interval techniques, the error bounds often exhibit spurious exponential growth due to the differential equation solver used. Numerical methods have been developed that prevent spurious exponential growth of the intervals for linear systems, see Gambill and Skeel [5] for details.

[8] The journal *Interval Computations* is a useful reference.

References

[1] G. Alefeld and J. Herzberger, *Introduction to Interval Computations*, Academic Press, New York, 1983.

[2] W. F. Ames and R. C. Nicklas, "Accurate Elliptic Differential Equation Solver," in W. L. Miranker and R. A. Toupin (eds.), *Accurate Scientific Computations*, Springer–Verlag, New York, 1986, pages 70–85.

[3] G. F. Corliss, "Survey of Interval Algorithms for Ordinary Differential Equations," *Appl. Math. and Comp.*, **31**, 1989, pages 112–120.

[4] P. Eijgenraam, *The Solution of Initial Value Problems Using Interval Arithmetic*, Mathematisch Centrum, Amsterdam, 1981.

[5] T. N. Gambill and R. D. Skeel, "Logarithmic Reduction of the Wrapping Effect with Application to Ordinary Differential Equations," *SIAM J. Numer. Anal.*, **25**, No. 1, February 1988, pages 153–162.

[6] G. Guenther and G. Marquardt, "A Programming System for Interval Arithmetic," in K. Nickel (ed.), *Interval Mathematics 1980*, Academic Press, New York, 1980, pages 355–366.

[7] S. Markov "Interval Differential Equations," in K. Nickel (ed.), *Interval Mathematics 1980*, Academic Press, New York, 1980, pages 145–164.

[8] R. E. Moore and S. Zuhe, "An Interval Version of Chebyshev's Method for Nonlinear Operator Equations," *Nonlinear Analysis*, **7**, No. 1, 1983, pages 21–34.

[9] E. P. Oppenheimer and A. N. Michel, "Application of Interval Analysis Techniques to Linear Systems: Part III—Initial Value Problems," *IEEE Trans. Circuits and Systems*, **35**, No. 10, October 1988, pages 1243–1256.

[10] L. B. Rall, "An Introduction to the Scientific Computing Language Pascal-SC," *Comp. & Maths. with Appls.*, **14**, No. 1, 1987, pages 53–69.

[11] H. Schwandt, "Newton-Like Interval Methods for Large Nonlinear Systems of Equations on Vector Computers," *Comput. Physics Comm.*, **37**, 1985, pages 223–232.

[12] H. Schwandt, "Interval Arithmetic Methods for Systems of Nonlinear Equations Arising from Discretizations of Quasilinear Elliptic and Parabolic Partial Differential Equations," *Applied Numerical Math.*, **3**, 1987, pages 257–287.

[13] J. M. Yohe, "Software for Interval Arithmetic: A Reasonable Portable Package," *ACM Trans. Math. Software*, **5**, No. 1, March 1979, pages 50–63.

124. Least Squares Method

Applicable to   Ordinary differential equations and partial differential equations.

Yields

An approximation to the solution.

Idea

A variational principle is created for a given differential equation, and then an approximation to the solution with some free parameters is proposed. By use of the variational principle, the free parameters are determined.

Procedure

Given the differential equation

$$N[u] = 0, \tag{124.1}$$

for $u(\mathbf{x})$ in some region of space R, with the homogeneous boundary conditions

$$B[u] = 0, \tag{124.2}$$

on some portion of the boundary of R, we define the functional

$$J[v(\mathbf{x})] = \int_R \Big(N[v(\mathbf{x})]\Big)^2 d\mathbf{x}. \tag{124.3}$$

Notice that $J[v(\mathbf{x})] \geq 0$, for all functions $v(\mathbf{x})$.

The solution to (124.1) and (124.2) clearly satisfies $J[u] = 0$ since the integrand is identically equal to zero in this case. Hence, the solution to (124.1) and (124.2) represents a minimum of the functional $J[\cdot]$.

Now we choose an approximation to $u(\mathbf{x})$ that has several parameters in it, say $u(\mathbf{x}) \simeq w(\mathbf{x}; \boldsymbol{\alpha})$, where $\boldsymbol{\alpha}$ is a vector of parameters. This approximation is chosen in such a way that it satisfies the conditions in (124.2). The parameters in $w(\mathbf{x}; \boldsymbol{\alpha})$ are determined by minimizing $J[w(\mathbf{x}, \boldsymbol{\alpha})]$; i.e., by solving the simultaneous system of equations

$$\frac{\partial}{\partial \alpha_k} J[w(\mathbf{x}, \boldsymbol{\alpha})] = 0, \qquad \text{for } k = 1, 2, \dots. \tag{124.4}$$

Example

Suppose we wish to approximate the solution of the two point boundary value problem

$$u'' + u + x = 0,$$
$$u(0) = 0, \quad u(1) = 0. \tag{124.5}$$

(Note that the exact solution of (124.5) is $y(x) = \dfrac{\sin x}{\sin 1} - x$.) In this case we may define $J[v(x)]$ to be

$$J[v(x)] = \int_0^1 (v'' + v + x)^2 \, dx.$$

We choose to approximate the solution of (124.5) by

$$u(x) \simeq w(x) = \alpha_1(x - x^2) + \alpha_2(x - x^3).$$

This approximation has been chosen in such a way that the boundary conditions for $u(x)$ are satisfied. Using $w(x)$ in the functional results in

$$J[w(x)] = \frac{1}{210} \left[707\alpha_1^2 + 2121\alpha_1\alpha_2 + 2200\alpha_2^2 - 385\alpha_1 - 784\alpha_2 + 70 \right].$$

Forming (124.4) for $k = 1, 2$, we determine that α_1 and α_2 must satisfy the simultaneous algebraic equations

$$\frac{\partial J[w(x)]}{\partial \alpha_1} = \frac{1}{210} [1414\alpha_1 + 2121\alpha_2 - 385] = 0,$$

$$\frac{\partial J[w(x)]}{\partial \alpha_2} = \frac{1}{210} [2121\alpha_2 + 4400\alpha_2 - 784] = 0.$$

These equations have the solution: $\{\alpha_1 = \frac{4448}{246137} \simeq .0181, \alpha_2 = \frac{413}{2437} \simeq .1694\}$. The function $w(x)$, with these values, becomes our approximation. The greatest difference between the exact solution and the approximate solution, in the range $0 < x < 1$, is at $x \simeq .5215$ where the difference is approximately .0016.

Notes
[1] Note that for the functional in (124.3) there may exist, in general, several different functions $\{v_k(\mathbf{x})\}$ that satisfy $J[v_k(\mathbf{x})] = 0$.
[2] This method is similar to the Rayleigh–Ritz method (see page 554) in that an approximation is utilized in a variational equation.
[3] This method is an example of a *weighted residual method*, see page 699.
[4] This technique is often implemented numerically.

References
[1] C. L. Chang and M. D. Gunzburger, "A Subdomain-Galerkin/Least Squares Method for First-order Elliptic Systems in the plane," *SIAM J. Numer. Anal.*, **27**, No. 5, 1990, pages 1197–1211.
[2] L. Collatz, *The Numerical Treatment of Differential Equations*, Springer–Verlag, New York, 1966, pages 184 and 220–221.
[3] M. Hanke, "On a Least Squares Collocation Method for Linear Differential-Algebraic Equations," *Numer. Math.*, **54**, No. 1, 1988, pages 79–90.

125. Lyapunov Functions

Applicable to   Ordinary differential equations and partial differential equations.

Yields
    Bounds on the solution in phase space.

Idea
    Even without solving a given differential equation, sometimes we can restrict the solution to be in a certain portion of phase space.

Procedure
    Given a differential equation, find a non-negative functional of the solution, which has a non-positive derivative. Then the solution of the differential equation will remain in a region described by the functional and the initial conditions. Most often, the functional will involve the dependent variable and some of its derivatives.

Example 1
    Suppose we wish to bound the solution of a damped harmonic oscillator

$$x_{tt} + \beta x_t + \omega^2 x = 0,$$
$$x(0) = A, \qquad x_t(0) = B, \tag{125.1}$$

with $\beta > 0$. In this case, we define the Lyapunov functional to be

$$L[x(t), x_t(t), x_{tt}(t), t] = \omega^2 x^2(t) + x_t^2(t).$$

Since $L[\cdot]$ is a sum of squares, it cannot be negative. Differentiating $L[\cdot]$ with respect to t produces

$$\begin{aligned} L_t[x, x_t, x_{tt}, t] &= (\omega^2 x^2 + x_t^2)_t \\ &= 2\omega^2 x x_t + 2 x_t x_{tt} \\ &= -2\beta x_t^2, \end{aligned} \tag{125.2.a–c}$$

where we have used the original differential equation (125.1) to replace the x_{tt} term in (125.2.b). Since β is positive, L_t is non-positive. Therefore, $L[\cdot]$ is a non-increasing function of t. Hence,

$$\begin{aligned} \omega^2 x^2(t) + x_t^2(t) &= L[x(t), x_t(t), x_{tt}(t), t] \\ &\leq L[x(0), x_t(0), x_{tt}(0), 0] \\ &\leq \omega^2 x^2(0) + x_t^2(0) \\ &\leq \omega^2 A^2 + B^2 \\ &\leq \text{ a prescribed constant.} \end{aligned} \tag{125.3}$$

Therefore, we have found an upper bound for $\omega^2 x^2(t) + x_t^2(t)$, without solving the original equation.

Example 2

Suppose we have the wave equation on a finite domain $(0 \leq x \leq L)$

$$u_{tt} = c^2 u_{xx},$$
$$u_x(0, t) = u_x(L, t) = 0, \quad u(x, 0) = g(x), \tag{125.4}$$

where c is a given constant and $g(x)$ is given. In this case, we choose the Lyapunov functional to be

$$V(t) = \frac{1}{2} \int_0^L [u_t^2 + c^2 u_x^2]\, dx. \tag{125.5}$$

Since $V(t)$ is the integral of a non-negative quantity, $V(t)$ is also non-negative. Differentiating $V(t)$ with respect to t produces

$$V_t = \int_0^L [u_t u_{tt} + c^2 u_x u_{xt}]\, dx$$
$$= \int_0^L [u_t(c^2 u_{xx}) + c^2 u_x u_{xt}]\, dx$$
$$= c^2 \int_0^L [u_t u_{xx} + u_x u_{xt}]\, dx. \tag{125.5.a-c}$$

Integration of the second term in (125.5.c) by parts yields

$$V_t = c^2 \int_0^L [u_t u_{xx} - u_{xx} u_t]\, dx + c^2 u_x u_t \Big|_{x=0}^{x=L}$$
$$= c^2 [u_x(L, t)u_t(L, t) - u_x(0, t)u_t(0, t)],$$

or, using the initial conditions in (125.4),

$$V_t = 0.$$

We conclude that $V(t) = V(0)$, for all values of t. This statement is essentially an "energy" statement: the energy (described by (125.5)) carried by a wave (described by (125.4)) remains constant.

Notes

[1] Lyapunov functionals are often devised from physical considerations. The Lyapunov functionals in both of these examples represent the "energy" of the system in a mathematical way.

[2] Finding Lyapunov functionals is, in general, a difficult task. It is often made easier by considering conservation laws: energy, momentum, etc. The "energy" in example one is *not* held constant, because of the dissipation due to the β term. If $\beta = 0$, then $L_t = 0$ and so (125.3) becomes

$$\omega^2 x^2(t) + x_t^2(t) = \omega^2 A^2 + B^2.$$

In this case, the energy is constant.

[3] There is a constructive method, due to Zubov [9], for obtaining Lyapunov functionals for systems of ordinary differential equations. The procedure requires the solution of a partial differential equation, which is derived from the given system of ordinary differential equations. See Hahn [3] (pages 78–82) or Willems [8] (pages 42–43) for details. Hahn [3] gives an example: A Lyapunov function for the system $\{\dot{x}+ = -x + 2x^2 y, \ \dot{y} = -y\}$ is $L = -1 + \exp\left(-\dfrac{y^2}{2} - \dfrac{x^2}{2(1-xy)}\right)$.

[4] A different constructive method is described in Oğuztöreli *et al.* [6]. A detailed algorithm is given for systems of ordinary differential equations of the form: $\{\dot{x} = f(t,x,y), \ \dot{y} = g(t,x,y)\}$. The Lyapunov function for a modification of the Mathieu differential equation, $\ddot{x} = (\alpha + 2\beta\cos 2t)x + \varepsilon x^2$, is derived for the region $\left(x^2 + y^2 < \rho^2\right)$, where ρ is a sufficiently small number.

[5] Suppose we have the nonlinear system $\mathbf{x}' = \mathbf{f}(\mathbf{x})$, with $\mathbf{f}(0) = \mathbf{0}$ and the Jacobian matrix $J(\mathbf{x}) = \dfrac{\partial \mathbf{f}}{\partial \mathbf{x}}$. If a constant, symmetric, positive definite matrix P can be found, such that $PJ(\mathbf{x})+J^{\mathrm{T}}(\mathbf{x})P$ is negative definite, then $V = \mathbf{x}^{\mathrm{T}}P\mathbf{x}$ is a Lyapunov function (with $V' = \mathbf{x}^{\mathrm{T}} \left\{\int \left[J^{\mathrm{T}}(z\mathbf{x})P + PJ(z\mathbf{x})\right] dz\right\} \mathbf{x}$).

If P is chosen to be the identity matrix then $V = \mathbf{x}^{\mathrm{T}}\mathbf{x}$ will be a Lyapunov function if all of the eigenvalues of the matrix $J(\mathbf{x}) + J^{\mathrm{T}}(\mathbf{x})$ are negative. This is known as Krasovskii's theorem.

[6] Burton [2] describes how Lyapunov functions may be constructed for delay differential equations.

[7] "Lyapunov" is sometimes written "Liapunov."

References

[1] W. E. Boyce and R. C. DiPrima, *Elementary Differential Equations and Boundary Value Problems*, Fourth Edition, John Wiley & Sons, New York, 1986, pages 502–512.

[2] T. A. Burton, "Perturbation and Delays in Differential Equations," *SIAM J. Appl. Math.*, **29**, No. 3, November 1975, pages 422–438.

[3] W. Hahn, *Theory and Application of Liapunov's Direct Method*, Prentice–Hall Inc., Englewood Cliffs, NJ, 1963.

[4] D. W. Jordan and P. Smith, *Nonlinear Ordinary Differential Equations*, Clarendon Press, Oxford, Second Edition, 1987, Chapter 10 (pages 283–319).

[5] R. E. Kalman and J. E. Bertram, "Control System Analysis and Design Via the 'Second Method' of Lyapunov, I: Continuous-Time Systems," *J. Basic Engrg. Trans. ASME*, **82**, No. 2, 1960, pages 371–393.

[6] J. Lasalle and S. Lefschetz, *Stability by Liapunov's Direct Method with Applications*, Academic Press, New York, 1961.

[7] M. N. Oğuztöreli, V. Lakshmikantham, and S. Leela, "An Algorithm for the Construction of Liapunov Functions," *Nonlinear Analysis*, **5**, No. 11, 1981, pages 1195–1212.

[8] G. F. Simmons, *Differential Equations with Applications and Historical Notes*, McGraw–Hill Book Company, New York, 1972, pages 316–322.

[9] J. L. Willems, *Stability Theory of Dynamical Systems*, John Wiley & Sons, New York, 1970.

[10] V. I. Zubov, *Methods of A. M. Lyapunov and Their Application*, P. Noordhoff, The Netherlands, 1964.

126. Equivalent Linearization and Nonlinearization

Applicable to   Nonlinear ordinary differential equations. This technique is most frequently used for ordinary differential equations with periodic solutions.

Yields

An approximate periodic solution.

Idea

We model the given equation by a linear or nonlinear equation for which the exact solution can be found.

Procedure

Suppose we want to approximate the solution to the nonlinear ordinary differential equation

$$D[x(t), t] = 0, \tag{126.1}$$

where $D[\cdot]$ is a differential operator. We represent the initial conditions and boundary conditions for $x(t)$ as $B[x(t)] = 0$, and assume that $x(t)$ is periodic on some interval, say for t from 0 to T. We do not need to know T *a priori*.

We model (126.1) by choosing a $D^*[\cdot]$ that has properties that are "similar" to the properties of $D[\cdot]$. This can be done by any technique. To allow some generality, we assume that $D^*[\cdot]$ depends on a set of parameters $\alpha = (\alpha_1, \alpha_2, \ldots, \alpha_n)$. Now we look for a solution $y(t; \alpha)$ of

$$D^*[y(t; \alpha), t; \alpha] = 0, \quad B[y(t; \alpha)] = 0 \qquad (126.2)$$

that is periodic on the interval $[0, T]$. We will approximate the solution to (126.1), $x(t)$, by the solution to (126.2), $y(t; \alpha)$. For this to be a good approximation, the error made must be small. We define the error made in using $y(t; \alpha)$ for $x(t)$ to be

$$\mathcal{E}(t, \alpha) := D[y(t; \alpha), t].$$

The claim is that $x(t) \simeq y(t; \alpha)$ if the "total error" is "small" in some sense. The "total error" could be measured as

$$\frac{1}{T} \int_0^T |\mathcal{E}(t, \alpha)|^2 \, dt \qquad \text{"mean square error,"}$$

or

$$\frac{1}{T} \int_0^T |\mathcal{E}(t, \alpha)| \, dt \qquad \text{"mean modulus,"}$$

or

$$\max |\mathcal{E}(t, \alpha)| \qquad \text{"extremum."}$$

The "total error" can be minimized by choosing the α. This is accomplished by differentiating the total error with respect to α_i and setting the resulting expression to zero (for $i = 1, 2, \ldots, n$). Solving these simultaneous algebraic equations yields the desired values of the α_i.

Example 1

Suppose we wish to approximate the periodic solution of the nonlinear ordinary differential equation

$$D[x(t), t)] = x'' + ax + bx^3 + cx^5 = 0,$$
$$x(0) = A, \qquad x'(0) = 0. \qquad (126.3)$$

Here $\{a, b, c, A\}$ are all known, fixed constants. We choose to approximate the solution of (126.3) by the solution of the linear ordinary differential equation

$$D^*[y(t), t; \omega] = y'' + \omega^2 y = 0,$$
$$y(0) = A, \quad y'(0) = 0, \qquad (126.4)$$

for some (unknown) value of ω. In this example, the vector of unknown parameters $\boldsymbol{\alpha}$ is the single variable ω.

The solution to (126.4) is

$$y(t) = A \cos \omega t. \tag{126.5}$$

The error in using (126.5) for the solution of (126.3) is

$$\begin{aligned}
\mathcal{E}(t, \omega) &= D[y(t), t], \\
&= y'' + ay + by^3 + cy^5, \\
&= (a - \omega^2) \cos \omega t + b \cos^3 \omega t + c \cos^5 \omega t.
\end{aligned}$$

We choose, in this example, to minimize the mean square error. Hence, we define the total error, $E(\omega)$, by

$$\begin{aligned}
E(\omega) &= \frac{1}{T} \int_0^T |\mathcal{E}(t, \omega)|^2 \, dt \\
&= \frac{1}{T} \int_0^T [(a - \omega^2) \cos \omega t + b \cos^3 \omega t + c \cos^5 \omega t]^2 \, dt.
\end{aligned} \tag{126.6}$$

Now, what is T? For equation (126.3), we do not know the true period of the solution. But, we are using the solution of (126.4) to approximate the solution of (126.3). And, for equation (126.4), the solution has period $T = 2\pi/\omega$ (see (126.5)). Hence, to evaluate (126.6), we use $T = 2\pi/\omega$ to obtain

$$\begin{aligned}
E(\omega) =\, &\big[128\omega^4 - (160c + 192b + 256a)\omega^2 + 63c^2 + (140b + 160a)c \\
&+ 80b^2 + 192ab + 128a^2\big]/256
\end{aligned} \tag{126.7}$$

Now, the goal is to minimize the total error. If (126.7) is differentiated with respect to ω, and the resulting equation is solved for ω, then

$$\omega^2 = a + \tfrac{3}{4}bA^2 + \tfrac{5}{8}cA^4 \qquad \text{or} \qquad \omega = 0. \tag{126.8}$$

Therefore, an approximation to the solution of (126.3) is found by using (126.8) in (126.5):

$$x(t) \simeq A \cos \left[t\sqrt{a + \tfrac{3}{4}bA^2 + \tfrac{5}{8}cA^4} \right].$$

Example 2

Suppose we wish to approximate the periodic solution of the undamped Duffing equation

$$D[x(t), t] = x'' + ax + bx^3 = B \cos \omega t, \tag{126.9}$$

where $\{a, b, B, \omega\}$ are all known constants. We choose to model the equation in (126.9) by the nonlinear equation

$$D^*[y(t), t] = y'' + ay + by^3 = \gamma \operatorname{cn}(\eta t, k), \tag{126.10}$$

where $\operatorname{cn}(\eta t, k)$ is the Jacobian elliptic cosine function with modulus k. The a and b in (126.10) are the same as the a and b in (126.9). The three remaining parameters in (126.10) that are at our control are $\{\gamma, \eta, k\}$. The solution to (126.10) is known to be (see the lookup solution technique, on page 148)

$$y(t) = \beta \operatorname{cn}(\eta t, k), \tag{126.11}$$

where β, γ, η, and k are related by

$$b\beta^3 + (a - \eta^2)\beta = \gamma, \qquad k^2 = \frac{b\beta^2}{2\eta^2}. \tag{126.12}$$

These equations determine β and k (in principle) in terms of γ and η. For the period of the forcing function in (126.10) to match the period of the forcing function in (126.9) (which is $2\pi/\omega$) we also require

$$\eta = \frac{2K(k)\omega}{\pi}, \tag{126.13}$$

where $K(k)$ is the complete elliptic integral of the first kind with modulus k. We will use (126.13) to determine η. This leaves us with one adjustable parameter, γ, with which to effect the minimization of the total error.

Now we calculate

$$\mathcal{E}(t, \gamma) = D[y(t)] = B \cos \omega t - \gamma \operatorname{cn}(\eta t, k). \tag{126.14}$$

If we choose to use the mean square error, with $T = 2\pi/\omega$, we find that the total error is minimized for

$$\gamma = \frac{B\pi K(k)}{2 \left[E(k) - k'^2 K(k) \right]} \operatorname{sech}\left(\frac{\pi K(k')}{2K(k)} \right), \tag{126.15}$$

where $E(k)$ is the complete elliptic integral of the second kind and k', given by $k'^2 = 1 - k^2$, is the complementary modulus.

Using (126.13), (126.14), and (126.15) in (126.11) results in the final approximation to the steady state periodic solution of (126.9).

Notes

[1] Note that in example 1 the effective frequency of the approximate solution depends on the initial conditions. This is generally expected in nonlinear problems.

[2] For example 2, the approximate solution $y(t)$ correctly tracks the frequency change of the solution when the magnitude of the forcing function is changed. More details on this example may be found in Iwan and Patula [4].

[3] This technique also works well for stochastic equations. In this application, the definition of the total error should include expectations taken over all of the random variables. This is sometimes called "statistical linearization." See Beaman [1] for details.

[4] This technique extends naturally to systems of equations. In this case there will be an error associated with each equation $\{E_i(t, \boldsymbol{\alpha})\}$, and we can define the total error by $E(t, \boldsymbol{\alpha}) = \sum_i |E_i(t, \boldsymbol{\alpha})|^2$.

[5] This technique can also be used for problems that do not have periodic solutions. The technique often used in this case is to minimize the integral of $|\mathcal{E}|^2$ from 0 to ∞.

[6] Differential operators representing differential equations may also be linearized directly, without minimizing some error functional. We have the definition:

> The operator $A[\cdot]$ is linearizable at u_0 if there exists a bounded linear operator $L[\cdot]$ such that $A[u] - A[u_0] = L[h] + r$, with $\lim\limits_{h \to 0} \dfrac{||r||}{||h||} = 0$, when $h = u - u_0$.

See Stakgold [10] for details.

References

[1] J. J. Beaman, "Accuracy of Statistical Linearization," in P. J. Holmes (ed.), *New Approaches to Nonlinear Problems in Dynamics*, SIAM, Philadelphia, 1980, pages 195–207.

[2] T. Caughey, "Equivalent Linearization Techniques," *J. Acoust. Soc. of America*, **35**, No. 11, November 1963, pages 1706–1711.

[3] P. Hagedorn, *Non-Linear Oscillations*, Clarendon Press, Oxford, 1982, pages 14–16.

[4] W. D. Iwan and E. J. Patula, "The Merit of Different Error Minimization Criteria in Approximate Analysis," *J. Appl. Mech.*, 1972, pages 257–262.

[5] W. D. Iwan and I.-M. Yang, "Application of Statistical Linearization Techniques to Nonlinear Multidegree-of-Freedom Systems," *J. Appl. Mech.*, June 1972, pages 545–550.

[6] H. Kröger, "Linearization of Nonlinear Differential Equations by Means of Cauchy's Integral," *J. Math. Physics*, **26**, No. 5, May 1985, pages 929–940.

[7] N. W. McLachlan, *Ordinary Non-Linear Differential Equations in Engineering and Physical Sciences*, Oxford University Press, New York, 1950, Chapter 6 (pages 103–112).

[8] P.-T. D. Spanos, *Linearization Techniques for Non-Linear Dynamical Systems*, Ph.D. thesis, Department of Applied Mechanics, California Institute of Technology, Pasadena, Calif., 1976.

[9] S. K. Srinivasan and R. Vasudevan, *Introduction to Random Differential Equations and Their Applications*, American Elsevier Publishing Company, New York, 1971, pages 45–47.

[10] I. Stakgold, *Green's Functions and Boundary Value Problems*, John Wiley & Sons, New York, 1979, pages 578–581.

127. Maximum Principles

Applicable to   Linear ordinary differential equations and linear partial differential equations.

Yields

Upper or lower bounds on the solution.

Idea

By the use of a maximum theorem, we can find bounds on certain types of equations.

Procedure

There are many theorems applicable to specialized equations and boundary conditions, that lead to bounds on the solutions. Maximum principles exist for all types of partial differential equations (hyperbolic, elliptic, and parabolic) as well as for ordinary differential equations. We choose to illustrate two theorems.

Example 1

A theorem from advanced calculus is:

> A continuous real-valued function on a bounded closed interval attains its maximum and minimum on the interval.

We will use this theorem to bound the solution to an ordinary differential equation. Consider the equation

$$e^x y'' + x(1 - x)y' = (1 + x^2)y, \qquad (127.1)$$

when $|y(a)| \leq M$ and $|y(b)| \leq M$. We claim that, for all x in the finite interval $[a, b]$, $y(x)$ is bounded in magnitude by M.

Suppose that $y(x)$ exceeded M in some region within the interval $[a, b]$. Then there would be maximum value of y on the interval; say it occurs at the point $x = c$. Since y is a maximum at $x = c$, we require $y'(c) = 0$ and $y''(c) \leq 0$. But this, with (127.1), implies that $e^c y''(c) = (1 + c^2) y(c)$. This cannot be correct, the right side is positive but the left side cannot be. Hence, y does not exceed M in the interval. It can similarly be shown that y cannot be less than $-M$. Hence $|y(x)| \leq M$ for x in the interval.

Example 2

In Ames [1], is the theorem:

Let $u(x)$ be a solution of the ordinary differential equation

$$
\begin{aligned}
L[u] &= u'' + H(x, u, u') = 0, & \text{for } a < x < b, \\
B_1[u] &= -u'(a) \cos \theta + u(a) \sin \theta = \gamma_1, \\
B_2[u] &= -u'(b) \cos \phi + u(b) \sin \phi = \gamma_2,
\end{aligned} \tag{127.2}
$$

where $0 \leq \theta \leq \pi/2$, $0 \leq \phi \leq \pi/2$, θ and ϕ are not both zero, H, H_u, $H_{u'}$ are all continuous, and $H_u \leq 0$.

If z_1 and z_2 satisfy

$$
\begin{aligned}
L[z_1] &\leq 0, & \text{for } a < x < b, \\
B_1[z_1] &\geq \gamma_1, \\
B_2[z_1] &\geq \gamma_2,
\end{aligned} \tag{127.3}
$$

$$
\begin{aligned}
L[z_2] &\geq 0, & \text{for } a < x < b, \\
B_1[z_2] &\leq \gamma_1, \\
B_2[z_2] &\leq \gamma_2,
\end{aligned} \tag{127.4}
$$

then we can conclude

$$
z_2(x) \leq u(x) \leq z_1(x), \tag{127.5}
$$

for $a < x < b$.

Hence, the solutions to (127.3) and (127.4) form bounds on the solution of (127.2).

As an illustration of this theorem, suppose we want to approximate the solution of the ordinary differential equation

$$
\begin{aligned}
u'' - u^3 &= 0, & \text{for } 0 < x < 1, \\
u(0) &= 0, \\
u(1) &= 1.
\end{aligned}
$$

This is in the form of (127.2) with $a = 0$, $b = 1$, $\theta = \phi = \pi/2$, $\gamma_1 = 0$, $\gamma_2 = 1$. We note that $z_1(x) = x$ satisfies (127.3) because

$$z_1'' - z_1^3 = -x^3 \le 0,$$
$$z_1(0) = 0,$$
$$z_1(1) = 1.$$

We now search for a $z_2(x)$ of the form x^α. Using $z_2(x) = x^\alpha$ in (127.4) yields

$$\alpha(\alpha - 1)x^{\alpha-2} - x^{3\alpha} \ge 0,$$
$$B_1[z_2] = 0 \text{ if } \alpha > 0, \qquad\qquad (127.6.a\text{–}c)$$
$$B_2[z_2] = 1.$$

Because of (127.6.b), we restrict our search to $\alpha > 0$. With this assumption, $x^{2(\alpha+1)} \le 1$ for x between 0 and 1. Hence, (127.6.a) will be satisfied if

$$\alpha(\alpha - 1) \ge 1. \qquad\qquad (127.7)$$

We choose $\alpha = (1 + \sqrt{5})/2$, so that (127.7) is satisfied. Hence, we can conclude, from (127.5)

$$x^{(1+\sqrt{5})/2} \le u(x) \le x, \qquad \text{for } 0 < x < 1. \qquad\qquad (127.8)$$

Notes

[1] Any value of α larger than $(1 + \sqrt{5})/2$ would also have yielded a bound for $u(x)$ in (127.8). The best bound corresponds to the minimal value of α, which was the one used.

[2] Some of the "classical" maximum principles are (see Sperb [7], pages 12–21 or Protter and Weinberger [5])

   (A) If $u(x)$ is non-constant and satisfies $u'' + b(x)u' \ge 0$ in an interval, and $b(x)$ is bounded, then $u(x)$ attains its maximum on the boundaries of the interval.

   (B) If $u(x)$ is non-constant and satisfies $u'' + b(x)u' + h(x) > 0$ in an interval, and $b(x)$ and $h(x)$ are bounded, and $h \le 0$, then a non-negative minimum of $u(x)$ can occur only on the boundaries of the interval.

   (C) If the elliptic operator $L[\cdot]$ has bounded coefficients and $u(\mathbf{x})$ satisfies the inequality

$$L[u] = \sum_{i,j} a_{ij}(\mathbf{x})\frac{\partial^2 u}{\partial x_i \partial x_j} + \sum_i b_i(\mathbf{x})\frac{\partial u}{\partial x_i} \ge 0$$

   in some bounded domain D, then $u(\mathbf{x})$ cannot assume its maximum at an interior point of D unless $u(\mathbf{x})$ is identically constant.

(D) If $L[\cdot]$ is a uniformly elliptic operator with bounded coefficients and $u(\mathbf{x}, t)$ satisfies the inequality

$$L[u] - \frac{\partial u}{\partial t} = \sum_{i,j} a_{ij}(\mathbf{x}) \frac{\partial^2 u}{\partial x_i \partial x_j} + \sum_i b_i(\mathbf{x}) \frac{\partial u}{\partial x_i} - \frac{\partial u}{\partial t} \geq 0$$

in $D \times (0, T)$, where D is a bounded domain and $T < \infty$, then $u(\mathbf{x})$ can attain its maximum only for $t = 0$ or on ∂D.

[3] A theorem in Durstine and Shaffer [3], applicable to ordinary differential equations and partial differential equations, states:

Let $L[\cdot]$ and $B[\cdot]$ be linear differential operators such that the equation

$$L[u] + \phi(\mathbf{x}) = 0, \qquad \text{in a domain } D,$$
$$B_j[u] = \lambda_j, \qquad \text{for } j = 1, 2, \ldots, q \text{ on } \partial D,$$

has a unique solution $u(\mathbf{x})$, and the Green's function does not change sign in D. If $w_1(\mathbf{x})$ and $w_2(\mathbf{x})$ satisfy

$$L[w_k] + \phi(\mathbf{x}) = \varepsilon_k(\mathbf{x}), \qquad \text{in } D,$$
$$B_j[w_k] = \lambda_j, \qquad \text{for } j = 1, 2, \ldots, q \text{ on } \partial D,$$

and

(A) $\varepsilon_1/\varepsilon_2$ is continuous,
(B) ε_1 does not change sign in D,
(C) either $1 > M \geq \dfrac{\varepsilon_2}{\varepsilon_1} \geq m$ or $M \geq \dfrac{\varepsilon_2}{\varepsilon_1} \geq m > 1$,

then

$$w_1 + \frac{w_1 - w_2}{M - 1} < u(\mathbf{x}) < w_1 + \frac{w_1 - w_2}{m - 1}.$$

[4] A theorem in Hille [4], applicable to first order ordinary differential equations, states:

Let $F(x, y)$ and $G(x, y)$ be continuous in a region D (which contains the initial data) and suppose that $F(x, y) < G(x, y)$ everywhere in D. Let $y(x)$ and $z(x)$ be the solutions of

$$y' = F(x, y), \qquad y(x_0) = y_0,$$
$$z' = G(x, y), \qquad z(x_0) = y_0.$$

Then, in the region where $y(x)$ and $z(x)$ are defined and continuous

$$z(x) < y(x), \qquad \text{for } x < x_0,$$
$$y(x) < z(x), \qquad \text{for } x_0 < x.$$

[5] A theorem in Ding [2] states:

> Consider the equation $\ddot{x} + g(x) = p(t)$ with
>   (A) $p(t)$ is continuous and 2π periodic,
>   (B) $g(x)$ is continuously differentiable and satisfies:
> $$\lim_{|x| \to \infty} \frac{g(x)}{x} = \infty.$$
> If $p(t)$ is an even function, or if $p(t)$ is odd and $g(x)$ is an even function, then all solutions of this equation are bounded.

References

[1] W. F. Ames, *Nonlinear Partial Differential Equations*, Academic Press, New York, 1967, page 181.

[2] T. Ding, *Boundedness of Solutions of Duffing's Equation*, IMA Preprint Series #58, University of Minnesota, Minneapolis, Minnesota, 1984.

[3] R. M. Durstine and D. H. Shaffer, "Determination of Upper and Lower Bounds for Solutions to Linear Differential Equations," *Quart. Appl. Math*, **16**, No. 3, 1958, pages 315–317.

[4] E. Hille, *Lectures on Ordinary Differential Equations*, Addison–Wesley Publishing Co., Reading, MA, 1969, pages 87–88.

[5] M. H. Protter and H. F. Weinberger, *Maximum Principles in Differential Equations*, Springer–Verlag, New York, 1984.

[6] M. J. Sewell, *Maximum and Minimum Principles*, Cambridge University Press, New York, 1987.

[7] R. Sperb, *Maximum Principles and Their Applications*, Academic Press, New York, 1981.

[8] A. Varma and W. Strieder, "Approximate Solutions of Non-linear Boundary-Value Problems," *IMA J. Appl. Mathematics*, **34**, 1985, pages 165–171.

128. McGarvey Iteration Technique

Applicable to   First order ordinary differential equations.

Yields

A sequence of approximations to the solution.

Idea

The method consists of generating a sequence of functions by a recurrence relation. The initial function used is arbitrary.

Procedure

Given the first order ordinary differential equation

$$\frac{dy}{dx} = f(x, y), \tag{128.1}$$

we choose $T_0(x, y) = T_0(y)$ to be an arbitrary function of y. Then we define the sequence of functions $\{T_n(x, y)\}$ by the recurrence relation

$$T_n(x, y) = -\int f(x, y) \left[\frac{\partial}{\partial y} T_{n-1}(x, y)\right] dx. \tag{128.2}$$

If we form $S_n(x, y) = \sum_{k=0}^{n} T_k(x, y)$, then $S_n(x, t) = $ constant is an approximate implicit solution to (128.1). As n increases, $S_n(x, y)$ will converge to the true solution of (128.1) if

$$\lim_{n \to \infty} \frac{\dfrac{\partial}{\partial y} T_n(x, y)}{\dfrac{\partial}{\partial y} S_n(x, y)} = 0.$$

Example

Suppose we wish to approximate the solution of the nonlinear ordinary differential equation

$$\frac{dy}{dx} = x + \frac{1}{y},$$

$$y(0) = 4.$$

In this case, (128.2) becomes

$$T_n(x, y) = -\int^x \left(x + \frac{1}{y}\right) \frac{\partial}{\partial y} T_{n-1}(x, y) \, dx. \tag{128.3}$$

We choose $T_0(y) = y$ (recall that T_0 is only a function of y). From (128.3) we can calculate

$$T_1(x, y) = -\frac{1}{2}x^2 - \frac{x}{y},$$

$$T_2(x, y) = -\frac{1}{2}\frac{x^2}{y^3} - \frac{1}{3}\frac{x^3}{y^2},$$

$$T_3(x, y) = -\frac{1}{2}\frac{x^3}{y^5} - \frac{13}{24}\frac{x^4}{y^4} - \frac{2}{15}\frac{x^5}{y^3}.$$

Note that we have not used any constants of integration in evaluating the $\{T_n\}$. This part of the analysis is independent of whether we choose such constants, or not. We can now calculate $S_3(x, y)$ as

$$S_3(x, y)$$

$$= \sum_{k=0}^{3} T_k(x, y)$$

$$= y + \left(-\frac{1}{2}x^2 - \frac{x}{y}\right) + \left(-\frac{1}{2}\frac{x^2}{y^3} - \frac{1}{3}\frac{x^3}{y^2}\right) + \left(-\frac{1}{2}\frac{x^3}{y^5} - \frac{13}{24}\frac{x^4}{y^4} - \frac{2}{15}\frac{x^5}{y^3}\right),$$

$$= \frac{120y^6 - 60x^2y^5 - 120xy^4 - 40x^3y^3 - 4x^2(4x^3 + 15)y^2 - 65x^4y - 60x^3}{120y^5}.$$

Now, for the first time, we use the initial condition: $y(0) = 4$. The implicit approximation to the solution of (128.1) is then given by

$$S_3(x, y) = S_3(x_0, y_0) = S_3(0, 4),$$

or

$$\frac{120y^6 - 60x^2y^5 - 120xy^4 - 40x^3y^3 - 4x^2(4x^3 + 15)y^2 - 65x^4y - 60x^3}{120y^5} = 4.$$

$$(128.4)$$

For any value of x, equation (128.4) is a polynomial in y. Thus, for any x we can solve for y. For this example, it turns out that the difference between the implicit solution given by (128.4) and the numerical solution is less than 2% for $0 \le x \le 20$.

Notes

[1] The above example is from Mcgarvey [1].

[2] This approximation technique may converge in cases where Picard approximations (see page 535) diverge.

[3] For certain classes of equations, error estimates can be obtained for this technique.

References

[1] J. F. McGarvey, "Approximating the General Solution of a Differential Equation," *SIAM Review*, **24**, No. 3, July 1982, pages 333–337.

129. Moment Equations: Closure

Applicable to A stochastic differential equation or a Fokker–Planck equation (which is a second order parabolic partial differential equation).

Yields

A system of ordinary differential equations from which different moments may be determined.

Idea

Interpreting the solution of the Fokker–Planck equation as a probability density, ordinary differential equations may sometimes be found for the moments of the random process.

Procedure

The solution of a Fokker–Planck equation is the probability density $P(\mathbf{x}, t)$ of a random process (see page 254). For an N-dimensional random process $\mathbf{x} = (x_1, x_2, \ldots, x_N)$, the Fokker–Planck equation has the form

$$\frac{\partial P}{\partial t} = -\sum_{i=1}^{N} \frac{\partial}{\partial x_i}(c_i P) + \sum_{i,j=1}^{N} \frac{\partial^2}{\partial x_i \partial x_j}(a_{ij} P), \qquad (129.1)$$

where the coefficients $\{c_i\}$ and $\{a_{ij}\}$ are, in general, functions of t and $\mathbf{x}$. All of the coefficients are determined by the stochastic differential equation that created (129.1).

The expectation of a function of $\mathbf{x}$, say $f(\mathbf{x})$, is defined to be the integral of $f(\mathbf{x})$ times $P(\mathbf{x}, t)$, integrated over all values of $\mathbf{x}$. That is,

$$E[f(\mathbf{x}(t))] = \int f(\mathbf{x}) P(\mathbf{x}, t)\, d\mathbf{x}.$$

Note that this expectation is a function of t. If equation (129.1) is multiplied by $f(\mathbf{x})$, and integrated over all values of $\mathbf{x}$, there results

$$\frac{d}{dt} E[f(\mathbf{x}(t))] = -\sum_{i=1}^{N} \int f(\mathbf{x}) \frac{\partial}{\partial x_i}(c_i P)\, d\mathbf{x} + \sum_{i,j=1}^{N} \int f(\mathbf{x}) \frac{\partial^2}{\partial x_i \partial x_j}(a_{ij} P)\, d\mathbf{x}.$$

$$(129.2)$$

Often, we may be able to integrate the right-hand side of (129.2) by parts to obtain an ordinary differential equation for $E[f(\mathbf{x})]$.

Example

The system of stochastic differential equations

$$\frac{dx}{dt} + x = z, \qquad\qquad x(0) = 0,$$
$$\frac{dz}{dt} + 2z = N(t), \qquad z(0) = 1,$$

$$(129.3.a\text{--}d)$$

where $N(t)$ is "white Gaussian noise" corresponds to the Fokker–Planck equation and initial condition

$$\frac{\partial P}{\partial t} = \frac{\partial}{\partial x}[(x - z)P] + 2\frac{\partial}{\partial z}[zP] + \frac{\partial^2}{\partial z^2}[P],$$
$$P(0, x, z) = \delta(x)\delta(z - 1),$$

$$(129.4)$$

for the probability density $P(t, x, z)$ (see page 254). Suppose we desire the expected value of $x(t)$:

$$E[x(t)] = \int_{-\infty}^{\infty} \int_{-\infty}^{\infty} x P(t, x, z)\, dx\, dz.$$

Multiplying (129.4) by x, and integrating from $-\infty$ to ∞ with respect to both x and z, produces

$$\frac{d}{dt}E[x(t)] = -E[x(t)] + E[z(t)], \qquad\qquad (129.5)$$

where we have made the physically reasonable assumptions that $|x|P(t, x, z) \rightarrow 0$ as $|x| \rightarrow \infty$, and both $|z|P(t, x, z) \rightarrow 0$ and $|z|P_z(t, x, z) \rightarrow 0$ as $|z| \rightarrow \infty$. These assumptions were required to carry out the integrations by parts in the right-hand side of (129.2).

Note that (129.5) involves the expected value of z. To obtain an equation for $E[z]$, (129.4) can be multiplied by z and then integrated to obtain

$$\frac{d}{dt}E[z(t)] = -2E[z(t)]. \qquad\qquad (129.6)$$

From (129.3.b) and (129.3.d), the initial conditions for (129.5) and (129.6) are

$$E[x(0)] = 0, \qquad E[z(0)] = 1. \qquad\qquad (129.7.a\text{--}b)$$

Alternately, these initial conditions can be obtained directly from the initial conditions in (129.4) by taking expectations.

If (129.6) is solved with (129.7.b), then (129.5) can be solved with (129.7.a) to determine the expectation of both $x(t)$ and $z(t)$

$$E[z(t)] = e^{-2t},$$
$$E[x(t)] = \tfrac{1}{3}\left(e^{-t} - e^{-2t}\right).$$

If the second order moments (i.e., $\{E[x^2(t)], E[x(t)z(t)], E[z^2(t)]\}$) are desired, the equations comparable to (129.5) and (129.6) are

$$\frac{d}{dt}\begin{pmatrix} E[x^2] \\ E[xz] \\ E[z^2] \end{pmatrix} = \begin{pmatrix} -1 & 2 & 0 \\ 0 & -3 & 1 \\ 0 & 0 & 4 \end{pmatrix}\begin{pmatrix} E[x^2] \\ E[xz] \\ E[z^2] \end{pmatrix} + \begin{pmatrix} 0 \\ 0 \\ 2 \end{pmatrix},$$

$$\begin{pmatrix} E[x^2] \\ E[xz] \\ E[z^2] \end{pmatrix}_{t=0} = \begin{pmatrix} 0 \\ 0 \\ 1 \end{pmatrix},$$

where we have dropped the explicit dependence on t for clarity. These equations were obtained by multiplying (129.4) by each of x^2, xz, and z^2, and then integrating with respect to x and z.

Notes
[1] Another procedure for determining ordinary differential equations for the moments is described on page 494.
[2] It is not always the case that the system of ordinary differential equations for the moments will close (that is, there will be m equations for the m unknowns). For example, the stochastic differential equation

$$\frac{d^2 x}{dt^2} + \frac{dx}{dt} + x + \varepsilon x^3 = N(t),$$

where $N(t)$ is white noise, corresponds to the Fokker–Planck equation

$$\frac{\partial P}{\partial t} = -\dot{x}\frac{\partial P}{\partial x} + \frac{\partial}{\partial \dot{x}}[(\dot{x} + x + \varepsilon x^3)P] + \frac{\partial^2 P}{\partial \dot{x}^2}.$$

In this case, the equations for the first moments become

$$\frac{d}{dt}E[x] = E[\dot{x}],$$
$$\frac{d}{dt}E[\dot{x}] = -E[\dot{x}] + E[x] - \varepsilon E[x^3].$$

Therefore, knowledge of $E[x]$ requires knowledge of $E[x^3]$. In this example, the system of ordinary differential equations that determine $E[x^3]$ involves the quantity $E[x^5]$, etc. However, if ε is small, then perturbation techniques may be used to approximately solve the moment equations.

[3] For systems that do not close, two "closing" approximations that are commonly used are (see Boyce [3]):

(A) Gaussian closure (also called "cumulant discard"),

(B) correlation discard.

In the Gaussian closure technique, a high odd cumulant of the probability density is set to zero. This procedure yields an equation for $E[x^k]$ in terms of $\{E[x^j] \mid 0 < j < k\}$. Correlation discard is generally used for equations that have "colored noise" forcing terms. In this approximation technique, some high power of the dependent variable in the stochastic differential equation and the "colored noise" is assumed to be uncorrelated.

Crandall [4] contains a review of non-Gaussian closure techniques. See also Ibrahim, Soundararajan and Heo [5].

[4] For determining the moments of random functions defined by partial differential equations, see, for instance, Wan's paper [6].

References

[1] S. A. Assaf and L. D. Zirkle, "Approximate Analysis of Non-Linear Stochastic Systems," *Int. J. Control*, **23**, No. 4, 1976, pages 477–492.

[2] R. V. Bobrik, "Hierarchies of Moment Equations for the Solution of the Schrödinger Equation with Random Potential and Their Closure," *Teoreticheskaya i Matematicheskaya Fizika*, **68**, No. 2, August 1986, pages 301–311.

[3] D. C. C. Bover, "Moment Equation Methods for Nonlinear Stochastic Systems," *J. Math. Anal. Appl.*, **65**, 1978, pages 306–320.

[4] W. E. Boyce, "Random Eigenvalue Problems," in A. T. Bharucha-Reid (ed.), *Probabilistic Methods in Applied Mathematics*, Academic Press, New York, 1968, pages 1–73.

[5] S. H. Crandall, "Non-Gaussian Closure Techniques for Stationary Random Vibration," *Int. J. Non-Linear Mechanics*, **20**, No. 1, 1985, pages 1–8.

[6] R. A. Ibrahim, A. Soundararajan, and H. Heo, "Stochastic Response of Nonlinear Dynamic Systems Based on a Non-Gaussian Closure," *J. Appl. Mech.*, **52**, December 1985, pages 965–970.

[7] F. Y. M. Wan, "Linear Partial Differential Equations with Random Forcing," *Stud. Appl. Math.*, **51**, No. 2, June 1972, pages 163–178.

130. Moment Equations: Itô Calculus

Applicable to   A set of stochastic differential equations.

Yields

A system of ordinary differential equations from which different moments may be determined.

Idea

Using Itô calculus, a set of ordinary differential equations may be determined that will describe the moments of a random process.

Procedure

In the Itô calculus, there are two different types of differential elements. There are dt terms, which are small; and there are $d\beta$ terms (*Brownian motion* terms), which are random. Brownian motion is the integral of *white noise*; i.e., $\beta(t) = \int_0^t n(s)\,ds$, when $n(s)$ is white noise.

We assume the standard scaling: $\mathrm{E}\left[(d\beta)^2\right] = dt$, where the $\mathrm{E}[\cdot]$ operator represents an expectation taken over the random variables in the system. The Brownian motion terms also have mean zero: $\mathrm{E}[d\beta] = 0$.

Suppose that $x_1(t)$ and $x_2(t)$ are random processes described by the two stochastic differential equations

$$\begin{aligned}
\frac{dx_1}{dt} &= a_1(t) + b_1(t)n(t), \\
\frac{dx_2}{dt} &= a_2(t) + b_2(t)n(t),
\end{aligned} \tag{130.1}$$

or

$$\begin{aligned}
dx_1 &= a_1(t)\,dt + b_1(t)\,d\beta, \\
dx_2 &= a_2(t)\,dt + b_2(t)\,d\beta.
\end{aligned}$$

Itô's lemma states that

$$d\left(x_1 x_2\right) = x_1\,dx_2 + x_2\,dx_1 + b_1 b_2\,dt. \tag{130.2}$$

This relation is different from the result in the classical calculus by the inclusion of the last term. This relationship may be used to determine moment equations for a random process.

Example

Given the stochastic differential equation

$$\frac{d^2y}{dt^2} - n(t)y = 0,$$

$$y(0) = 1, \quad y'(0) = 0,$$

where $n(t)$ is white noise, we can define $z = \dfrac{dy}{dt}$ and so obtain the coupled system of stochastic differential equations

$$dy = z\,dt, \quad y(0) = 1,$$
$$dz = y\,d\beta, \quad z(0) = 0. \tag{130.3}$$

Using Itô's lemma repeatedly on (130.3), we can derive the following relations

$$d\left(y^L\right) = Ly^{L-1}z\,dt,$$
$$d\left(z^K\right) = \frac{K(K-1)}{2}y^2z^{K-2}\,dt + Kz^{K-1}y\,d\beta. \tag{130.4}$$

If we define the $N+1$ different N-th order moments by

$$G_N^M(t) = \mathrm{E}\left[y^{N-M}(t)z^M(t)\right], \qquad M = 0, 1, \ldots, N,$$

then, from (130.4), we obtain the set of coupled ordinary differential equations

$$\frac{dG_N^M}{dt} = (N-M)G_N^{M+1} + \frac{M(M-1)}{2}G_N^{M-2}$$

for $\left\{G_N^M(t)\right\}$. For example, if we choose $N = 2$, then we obtain the system

$$\frac{d}{dt}\begin{pmatrix} G_2^0 \\ G_2^1 \\ G_2^2 \end{pmatrix} = \begin{pmatrix} 0 & 2 & 0 \\ 0 & 0 & 1 \\ 1 & 0 & 0 \end{pmatrix}\begin{pmatrix} G_2^0 \\ G_2^1 \\ G_2^2 \end{pmatrix}, \tag{130.5}$$

with the initial conditions

$$\begin{pmatrix} G_2^0 \\ G_2^1 \\ G_2^2 \end{pmatrix}_{t=0} = \begin{pmatrix} 1 \\ 0 \\ 0 \end{pmatrix}.$$

The eigenvalues of the matrix in (130.5) are the three cube roots of two. Hence, each of $\{G_2^0, G_2^1, G_2^2\}$ grows exponentially in time.

Notes

[1] Note that the Fokker–Planck equation corresponding to (130.3) is

$$\tfrac{1}{2}y^2 P_{zz} - z P_y = P_t,$$
$$P(y,z,0) = \delta(y-1)\delta(z),$$

where $P(y,z,t)$ represents the joint probability density of y and z at time t.

[2] The coupled ordinary differential equations that are derived for the moments in this section are identical to the equations obtained by the method described on page 491.

References

[1] Z. Schuss, *Theory and Applications of Stochastic Differential Equations*, John Wiley & Sons, New York, 1980.

[2] V. A. Kulkarny and B. S. White, "Focussing of Waves in Turbulent Inhomogeneous Media," *Phys. Fluids*, **25**, No. 10, 1982, pages 1770–1784.

131. Monge's Method

Applicable to Some nonlinear second order partial differential equations with two independent variables.

Yields

An exact solution.

Idea

Application of some algebraic identities and then the use of equation splitting permits some nonlinear partial differential equations to be solved.

Procedure

Monge's method works for some differential equations of the form

$$R\frac{\partial^2 z}{\partial x^2} + S\frac{\partial^2 z}{\partial x \partial y} + T\frac{\partial^2 z}{\partial y^2} = V,$$

or

$$Rr + Ss + Tt = V, \tag{131.1}$$

for $z = z(x,y)$ where, as usual, $r = z_{xx}$, $s = z_{xy}$, $t = z_{yy}$, $p = z_x$, $q = z_y$, and $\{R, S, T, V\}$ may be functions of $\{p, q, x, y, z\}$.

First, note that we can write

$$dp = p_x \, dx + p_y \, dy = r \, dx + s \, dy,$$
$$dq = q_x \, dx + q_y \, dy = s \, dx + t \, dy. \qquad (131.2.a\text{–}b)$$

Solving (131.2.a) for r and (131.2.b) for t, and then using these values in (131.1), we obtain

$$[R \, dp \, dy + T \, dq \, dx - V \, dy \, dx] - s \left[R \, (dy)^2 - S \, dy \, dx + T \, (dx)^2 \right] = 0. \qquad (131.3)$$

By use of equation splitting (see page 446) we look for the simultaneous solutions to

$$R \, dp \, dy + T \, dq \, dx - V \, dy \, dx = 0,$$
$$R \, (dy)^2 - S \, dy \, dx + T \, (dx)^2 = 0. \qquad (131.4.a\text{–}b)$$

Any solution of (131.4) is also a solution of (131.3).

Such a solution is called an intermediate integral and will depend on an arbitrary constant or function. If we can find two such integrals, say

$$f(x, y, z, p, q) = A, \qquad g(x, y, z, p, q) = B,$$

where A and B are arbitrary constants, then we may be able to solve for $\{p = p(x, y, z), \ q = q(x, y, z)\}$. If we could, then we might be able to integrate the Pfaffian differential equation (see page 326) $dz = p \, dx + q \, dy$ to determine $z = z(x, y)$.

Example

Suppose we have the partial differential equation

$$y^2 \frac{\partial^2 z}{\partial x^2} - 2y \frac{\partial^2 z}{\partial x \partial y} + \frac{\partial^2 z}{\partial y^2} = \frac{\partial z}{\partial y} + 6y, \qquad (131.5)$$

which can be written as: $y^2 r - 2ys + t = p + 6y$. Therefore, we have $\{R = y^2, \ S = -2y, \ T = 1, \ V = p + 6y\}$. The two equations in (131.4) then become

$$y^2 \, dp \, dy + dq \, dx - (p + 6y) \, dy \, dx = 0,$$
$$(y \, dy + dx)^2 = 0. \qquad (131.6.a\text{–}b)$$

Equation (131.6.b) can be integrated to obtain

$$2x + y^2 = A, \qquad (131.7)$$

where A is an arbitrary constant. Dividing (131.6.a) by dx (or, equivalently, by $(-y\,dy)$ from (131.6.b)) we obtain

$$-y\,dp + dq - (p + 6y)\,dy = 0,$$

which can be integrated to yield $-py + q - 3y^2 = -\phi(2x + y^2)$, or

$$y\frac{\partial z}{\partial x} - \frac{\partial z}{\partial y} + 3y^2 = \phi(2x + y^2), \tag{131.8}$$

where ϕ is an arbitrary function. Equation (131.8) is an intermediate integral, and the only one that equation (131.6) has (due to the double root appearing in (131.6.b)).

Since we do not have two intermediate integrals, we can not proceed with the derivation in the Procedure. However, we can solve (131.8) directly to obtain a solution of (131.5).

Since (131.8) is quasilinear, the method of characteristics (see page 368) may be used. The subsidiary equations are

$$\frac{dx}{y} = \frac{dy}{-1} = \frac{dz}{-3y^2 + \phi(2x + y^2)}. \tag{131.9}$$

From the first equality in (131.9) we recover the integral in (131.7).

Using (131.7) in the second equality in (131.9) yields

$$\frac{dy}{-1} = \frac{dz}{-3y^2 + \phi(A)},$$

with the solution $z - y^3 + y\phi(2x + y^2) = B$, where B is another arbitrary constant.

Hence, a general integral of (131.5) is

$$\Phi\left(z - y^3 + y\phi(2x + y^2), 2x + y^2\right) = 0.$$

This leads to a general solution of (131.5)

$$z = y^3 - y\phi(2x + y^2) + \psi(2z + y^2),$$

where ϕ and ψ are arbitrary functions of their arguments.

Notes

[1] Because equation splitting was used in going from (131.3) to (131.4) the solution obtained in (131.13) is not the most general solution.

References

[1] W. F. Ames, "Ad Hoc Exact Techniques for Nonlinear Partial Differential Equations," in W. F. Ames (ed.), *Nonlinear Partial Differential Equations in Engineering*, Academic Press, New York, 1967, pages 60–65.

[2] A. R. Forsyth, *Theory of Differential Equations*, Dover Publications, Inc., New York, 1959, Volume 6, pages 202–208.

[3] H. T. H. Piaggio, *An Elementary Treatise on Differential Equations and Their Applications*, G. Bell & Sons, Ltd, London, 1926, pages 181–187.

[4] I. N. Sneddon, *Elements of Partial Differential Equations*, McGraw–Hill Book Company, New York, 1957, pages 131–135.

132. Newton's Method

Applicable to   Partial differential equations and ODEs.

Yields

A sequence of approximations to the solution.

Idea

When a Newton iteration is applied to a nonlinear differential equation, each step of the iteration requires that a linear differential equation be solved.

Procedure

We illustrate the general procedure on an ordinary differential equation. Suppose we wish to approximate the solution to the first order ordinary differential equation

$$G(y', y, x) = 0,$$
$$y(0) = y_0, \tag{132.1}$$

for $y(x)$ when $G(y', y, x)$ is a nonlinear function.

If an approximate solution of (132.1), say $y_k(x)$, is known, then $G(y', y, x)$ could be expanded about $y_k(x)$ to obtain

$$G(y', y, x) \simeq G(y_k', y_k, x) + G_y(y_k', y_k, x)(y - y_k) + G_{y'}(y_k', y_k, x)(y' - y_k'), \tag{132.2}$$

to leading order. For the solution to (132.1), $G(y', y, x) = 0$ and so (132.2) becomes

$$(y' - y_k')G_{y'}(y_k', y_k, x) + (y - y_k)G_y(y_k', y_k, x) \simeq -G(y_k', y_k, x).$$

Therefore, if the linear ordinary differential equation

$$e'_k G_{y'} + e_k G_y = -G,$$
$$e_k(0) = 0,$$
(132.3)

is solved for the "correction term" $e_k(x)$, then defining

$$y_{k+1}(x) = y_k(x) + e_k(x)$$

should yield a better approximation, $y_{k+1}(x)$, to $y(x)$.

Equation (132.3) can be solved exactly by the use of integrating factors (see page 305). However, it only needs to be solved approximately because the higher order approximations (i.e., y_{k+2}, y_{k+3}, ...) will correct errors made in solving (132.3).

Special Case

In the special case that the original equation is linear in y' and hence of the form

$$G(y', y, x) = y' - f(x, y) = 0,$$

then the definition of y_{k+1} may be succinctly represented as

$$y'_{k+1} - f_y(x, y_k(x))y_{k+1} = f(x, y_k(x)) - f_y(x, y_k(x))y_k(x),$$
$$y_{k+1}(0) = y_0.$$
(132.4)

Example

Suppose we are looking for an approximation, near $x = 0$, of the solution to the nonlinear ordinary differential equation

$$y' + y^3 = 0,$$
$$y(0) = 1,$$

which has the known exact solution

$$y(x) = (1 + 2x)^{-1/2}$$
$$= 1 - x + \frac{3}{2}x^2 - \frac{5}{2}x^3 + \frac{35}{8}x^4 - \frac{63}{8}x^5 + \cdots.$$

For this problem we recognize that $f(x, y) = -y^3$ and so (132.4) becomes

$$y'_{k+1} + 3y_k^2 y_{k+1} = 2y_k^3,$$
$$y_{k+1}(0) = 1.$$
(132.5)

If we start with $y_0 = y(0) = 1$, then, from (132.5)

$$y_1' + 3y_1 = 2,$$
$$y_1(0) = 1,$$

with the solution

$$y_1(x) = \tfrac{1}{3}\left(2 + e^{-3x}\right)$$
$$= 1 - x + \tfrac{3}{2}x^2 - \tfrac{3}{2}x^3 + \cdots.$$

If we use the approximation $y_1 \simeq 1 - x$, then the equation for y_2 (from (132.5), with $k = 1$) is

$$y_2' + 3(1 - x)^2 y_2 = 2(1 - x)^3,$$
$$y_2(0) = 0,$$

with the solution

$$y_2(x) = -2e^{-x^3 + 3x^2 - 3x}\left(\int_0^x e^{x^3 - 3x^2 + 3x}(x - 1)^3\, dx + 1\right)$$
$$= 1 - x + \tfrac{3}{2}x^2 - \tfrac{5}{2}x^3 + \tfrac{35}{8}x^4 - \tfrac{261}{40}x^5 + \cdots.$$

We see then that $y_1(x)$ has the first 3 terms correct, while $y_2(x)$ (which only used the first order information in $y_1(x)$) has the first 5 terms correct.

Notes

[1] For symbolic manipulation of the formulae appearing above, Geddes [4] discusses the number of correct terms at each step.

[2] Most often, this iterative method will be implemented numerically and not performed analytically. This is because, by hand, it is often easier to find a Taylor series solution directly (see page 548) than to use Newton iterates. Rice and Boisvert [7] have a numerical example of using Newton's method to solve an elliptic equation.

[3] Error estimates for Newton's method (applied to first order equations) can be found in Mikhlin and Smolitskiy [6].

[4] When Newton's method is numerically applied to nonlinear boundary value problems, the method is often called *quasi-linearization*. This is the same algorithm that is obtained when multiple shooting is used (see page 631), and the number of rays becomes very large. See Stoer and Bulirsch [8] or Bellman and Kalaba [3] for details.

[5] Geddes [4] showed that the number of correct coefficients in a power series solution obtained by this method, when applied to an explicit first order nonlinear ordinary differential equation, more than doubles at each step.

References

[1] U. M. Ascher, R. M. M. Mattheij, and R. D. Russel, *Numerical Solution of Boundary Value Problems for Ordinary Differential Equations*, Prentice–Hall Inc., Englewood Cliffs, NJ, 1988, pages 52–55.

[2] M. A. Barkatou, "Rational Newton Algorithm for Computing Formal Solution of Linear Differential Equations," in *Lecture Notes in Computer Science # 358*, Springer–Verlag, New York, 1989, pages 183–195.

[3] R. E. Bellman and R. E. Kalaba, *Quasilinearization and Nonlinear Boundary-Value Problems*, American Elsevier Publishing Company, New York, 1965.

[4] K. O. Geddes "Convergence Behavior of the Newton Iteration for First Order Differential Equations," in E. W. Ng (ed.), *Symbolic and Algebraic Computation*, Springer–Verlag, New York, 1979, pages 189–199.

[5] R. B. Guenther abd J. W. Lee, "Convergence of the Newton–Raphson Method for Boundary Value Problems of Ordinary Differential Equations," in *Computation Solutions of Nonlinear Systems of Equations*, Amer. Math. Soc., Providence, Rhode Island, 1990, pages 257–264.

[6] S. G. Mikhlin and K. L. Smolitskiy, *Approximate Methods for Solutions of Differential and Integral Equations*, American Elsevier Publishing Company, New York, 1967, pages 12–16.

[7] J. R. Rice and R. F. Boisvert, *Solving Elliptic Problems Using ELLPACK*, Springer–Verlag, New York, 1985, pages 101–111.

[8] J. Stoer and R. Bulirsch, *Introduction to Numerical Analysis*, translated by R. Bartels, W. Gautschi, and C. Witzgall, Springer–Verlag, New York, 1976, pages 498–502.

133. Padé Approximants

Applicable to   Any type of function (whether it comes from a differential equation or not).

Yields

An approximation formula generally valid over an interval, and, often, information about whether singularities exist.

Idea

A Taylor series can be manipulated to produce information about the existence of singularities.

Procedure

When a power series representation of a function diverges, it indicates the inability of the power series to approximate the function in a certain region. A theorem of complex analysis states that if a Taylor series of a function diverges, then that function has singularities in the complex plane. A Padé approximant is a ratio of polynomials that contains the same information that a truncated power series does. Since the polynomial in the denominator may have roots in the region of interest, the Padé approximant may accurately indicate the presence of singularities.

Suppose we have found the k-th order Taylor series solution to a differential equation (see page 548)

$$y(x) = a_0 + a_1 x + a_2 x^2 + \cdots + a_k x^k. \tag{133.1}$$

The (N, M) Padé approximant, $P_M^N(x)$, is a ratio of polynomials, with the polynomial in the numerator having degree N and the polynomial in the denominator having degree M:

$$P_M^N(x) = \frac{B_0 + B_1 x + \cdots + B_N x^N}{A_0 + A_1 x + \cdots + A_M x^M}, \tag{133.2}$$

with $N + M + 1 = k$. Without loss of generality, we take $A_0 = 1$. The remaining $N + M + 1$ coefficients $\{A_1, A_2, \ldots, A_N, B_0, B_1, \ldots, B_M\}$ are chosen so that the first $N + M + 1$ terms in the Taylor series expansion of $P_M^N(x)$ match the first $N + M + 1$ terms of the Taylor series in (133.1).

It often happens that $P_M^N(x)$ converges (as $N, M \to \infty$) to the true solution of the differential equation, even when the Taylor series solution diverges!

Usually we only consider the convergence of the Padé sequence $\{P_0^J(x), P_1^{J+1}(x), P_2^{J+2}(x), \ldots\}$ having $N = M + J$ and J held constant while $M \to \infty$. The special sequence with $J = 0$ is called the *diagonal sequence*.

Example 1

Suppose we wish to approximate the solution of the ordinary differential equation

$$y' = y^2, \qquad y(0) = 1. \tag{133.3}$$

Since (133.3) is separable (see page 341), the solution to (133.3) can be found to be $y(x) = 1/(1 - x)$. If we tried to find the Taylor series of $y(x)$, directly from (133.3), we would obtain

$$y(x) = 1 + x + x^2 + x^3 + x^4 + \cdots. \tag{133.4}$$

This geometric series is convergent, of course, only for $|x| < 1$. The solution has a singularity at $x = 1$, but this fact is not readily apparent from the expansion in (133.4).

The diagonal sequence of Padé approximants corresponding to (133.4) is

$$P_1^1(x) = \frac{1}{1-x},$$

$$P_2^2(x) = \frac{1}{1-x},$$

$$P_3^3(x) = \frac{1}{1-x}.$$

Therefore, the diagonal sequence of Padé approximants recovers the *exact* solution to the differential equation, from only a few terms in the Taylor series. Of course, this is an exceptional example.

Example 2

Suppose we wish to approximate the solution of the ordinary differential equation

$$y' = 1 + y^2, \qquad y(0) = 0. \tag{133.5}$$

Since (133.5) is separable, the solution to (133.5) can be found to be $y(x) = \tan x$. If we tried to find the Taylor series of $y(x)$, directly from (133.5), we would find

$$y(x) = x + \frac{x^3}{3} + \frac{2x^5}{15} + \frac{17x^7}{315} + \cdots. \tag{133.6}$$

Note that the exact solution has singularities at $x = \pm(2n+1)\pi/2$, while the Taylor series approximation does not appear to show this behavior. Using (133.6) we can compute the first few elements of the diagonal sequence

$$P_2^2(x) = \frac{3x}{3 - x^2},$$

$$P_3^3(x) = \frac{x(x^2 - 15)}{3(2x^2 - 5)},$$

$$P_4^4(x) = \frac{5x(21 - 2x^2)}{x^4 - 45x^2 + 105}.$$

Note that these Padé approximants have singularities where the denominator vanishes:

(A) For $P_2^2(x)$, these singularities are at $x \simeq \pm 1.7$.
(B) For $P_3^3(x)$, these singularities are at $x \simeq \pm 1.58$.

(C) For $P_4^4(x)$, these singularities are at $x \simeq \pm 1.5712$, and $x \simeq \pm 6.5$.

We observe that these Padé approximants are attempting to recover the singularities of the exact solution at $x = \pm \pi/2$ and $x = \pm 3\pi/2$.

Because the Padé approximants have these singularities, they produce an accurate numerical approximation of the exact solution, over a wide range of values.

Notes

[1] Padé approximants are not always better than a Taylor series representation. In fact, it may happen that the Padé approximants diverge while the Taylor series converges.

[2] Padé approximants are also called *rational function approximations*.

[3] Prendergast [8] proposes a technique to find the Padé approximants for the solution of a nonlinear differential equation, without first finding the Taylor series. Martin and Zamudio–Cristi [6] address the same issue, but for a smaller class of equations.

[4] In Bender and Orszag [1] is a discussion of computational techniques for computing Padé approximants numerically.

[5] A *two-point Padé approximant* is one that utilizes Taylor series information about two different points. Often these points are chosen to be zero and infinity. For two-point Padé approximants the coefficients in equation (133.2) are chosen so that both Taylor series will be matched. See Bender and Orszag [1] or Magnus [7] for details.

[6] Many symbolic computer languages have a function that finds Padé approximants analytically when a Taylor series is input. See Czapor and Geddes [3].

[7] The Bulirsch–Stoer method is a numerical method for solving first order ordinary differential equations using Padé approximants, Richardson extrapolation, and the modified midpoint rule. See Press *et al.* [9] for details.

References

[1] C. M. Bender and S. A. Orszag, *Advanced Mathematical Methods for Scientists and Engineers*, McGraw–Hill, New York, 1978, pages 383–410.

[2] A. Cuyt, *Padé Approximants for Operators: Theory and Applications*, Springer–Verlag, Berlin, Lecture Notes in Mathematics #1065, 1984.

[3] S. R. Czapor and K. O. Geddes, "A Comparison of Algorithms for the Symbolic Computation of Padé Approximants," in J. Fitch (ed.), *EUROSAM '84*, Springer–Verlag, New York, 1984, pages 248–259.

[4] C. W. Gear, *Numerical Initial Value Problems in Ordinary Differential Equations*, Prentice–Hall Inc., Englewood Cliffs, NJ, 1971, pages 93–101.

[5] K. O. Geddes, "Symbolic Computation of Padé Approximants," *ACM Trans. Math. Software*, **5**, No. 2, June 1979, pages 218–233.

[6] P. Martin and J. Zamudio–Cristi, "Fractional Approximation For First-Order Differential Equations with Polynomial Coefficients—Application to $E_1(x)$," *J. Math. Physics*, **23**, No. 12, December 1982, pages 2276–2280.

[7] A. Magnus, "On the Structure of the Two-Point Padé Table," in W. B. Jones, W. J. Thron, and H. Waadeland (eds.), *Analytic Theory of Continued Fractions*, Springer–Verlag, New York, 1982, pages 176–193.

[8] K. H. Prendergast, *Rational Approximation for Non-Linear Ordinary Differential Equations*, in D. Chudnovsky and G. Chudnovsky (eds.), *The Riemann Problem, Complete Integrability and Arithmetic Applications*, Lecture Notes in Mathematics #925, Springer–Verlag, New York, 1982.

[9] W. H. Press, B. P. Flannery, S. Teukolsky, and W. T. Vetterling, *Numerical Recipes*, Cambridge University Press, New York, 1986, pages 563–568.

[10] M. F. Reusch, L. Ratzan, N. Pomphrey, and W. Park, "Diagonal Padé Approximations for Initial Value Problems," *SIAM J. Sci. Stat. Comput.*, **9**, No. 5, September 1988, pages 829–838.

[11] R. A. Williamson, "Padé Approximations in the Numerical Solution of Hyperbolic Differential Equations," in H. Werner and H. J. Bünger (eds.), *Padé Approximation and Its Applications, Bad Honnef 1983*, Springer–Verlag, New York, 1984, pages 252–264.

134. Perturbation Method: Method of Averaging

Applicable to   Nonlinear differential equations that have a periodic solution and a small parameter.

Yields

An approximation to the solution, valid over an entire period.

Idea

Write the solution of a given differential equation as a function with slowly varying parts. Then average those slowly varying parts over a complete cycle.

Procedure

We illustrate the method on a perturbed harmonic oscillator. Suppose we have the equation

$$\frac{d^2y}{dt^2} + y + \varepsilon f\left(y, \frac{dy}{dt}\right) = 0. \tag{134.1}$$

Note that, when $\varepsilon = 0$, equation (134.1) is a harmonic oscillator. The solution to (134.1), when $\varepsilon = 0$, is therefore: $y(t) = A\cos(t + \Theta)$ (where

A and Θ are constants). If ε is very small, we might expect a similar "looking" solution, so we assume that the solution to (134.1) is given by

$$y(t) = a\cos(t + \theta),\qquad (134.2)$$

where $a(t)$ and $\theta(t)$ are "slowly varying" (another expression often used is "nearly constant"). Differentiating (134.2) with respect to t yields

$$\frac{dy}{dt} = -a\sin(t + \theta) + \frac{da}{dt}\cos(t + \theta) - a\frac{d\theta}{dt}\cos(t + \theta).\qquad (134.3)$$

If a and θ are "slowly varying," then da/dt and $d\theta/dt$ will be "small" compared to a. Hence, we set the derivative of $y(t)$ to be

$$\frac{dy}{dt} = -a\sin(t + \theta).\qquad (134.4)$$

Comparing (134.3) to (134.4), it is clear that we have made the assumption

$$\frac{da}{dt}\cos(t + \theta) - a\frac{d\theta}{dt}\cos(t + \theta) = 0.\qquad (134.5)$$

This gives one equation relating the two unknowns, $a(t)$ and $\theta(t)$. Differentiating (134.4) and using (134.4) and (134.1) results in the expression

$$-\frac{da}{dt}\sin(t + \theta) - a\frac{d\theta}{dt}\cos(t + \theta) = \varepsilon f\left(a\cos(t + \theta), -a\sin(t + \theta)\right).\qquad (134.6)$$

The two equations in (134.5) and (134.6) can be solved to yield the relations

$$\frac{da}{dt} = \varepsilon f\left(a\cos(t + \theta), -a\sin(t + \theta)\right)\sin(t + \theta),$$
$$\frac{d\theta}{dt} = \frac{\varepsilon}{a}f\left(a\cos(t + \theta), -a\sin(t + \theta)\right)\cos(t + \theta).\qquad (134.7)$$

The equations in (134.2) and (134.7) are still exact. The change of variables from $\{y, y'\}$ to $\{a, \theta\}$ (by use of (134.2) and (134.5)) has been carried out without any approximation being made. The assumptions made have been motivated by the smallness of ε, but the system is still exact.

Now we use the "slowly varying" feature of a and θ to make the required approximation. If a and θ are "slowly varying," then the values of da/dt and $d\theta/dt$ should not change much over a single period of the solution. Hence, if we replace the right-hand sides of (134.7) by their averages over one period, then the solutions for $a(t)$ and $\theta(t)$ should not be changed very much. Therefore, we approximate the solution of (134.7) by the solution of

$$\frac{da}{dt} = \varepsilon F(a), \qquad \frac{d\theta}{dt} = \frac{\varepsilon}{a} G(a), \qquad (134.8)$$

where

$$F(a) = \frac{1}{2\pi} \int_0^{2\pi} f\left(a\cos(t+\theta), -a\sin(t+\theta)\right)\sin(t+\theta)\, d\theta,$$

$$G(a) = \frac{1}{2\pi} \int_0^{2\pi} f\left(a\cos(t+\theta), -a\sin(t+\theta)\right)\cos(t+\theta)\, d\theta. \qquad (134.9)$$

The prescription is to evaluate (134.9), then to solve (134.8). Knowing $a(t)$ and $\theta(t)$ we can evaluate (134.2) and so recover an approximation to $y(t)$.

Example 1

For the Van de Pol oscillator

$$\frac{d^2y}{dt^2} + y + \varepsilon(y^2 - 1)\frac{dy}{dt} = 0,$$

we identify $f(y, y') = (y^2 - 1)y'$. Evaluating (134.9) with this f results in: $F(a) = \frac{a}{2} - \frac{a^3}{8}$ and $G(a) = 0$. This, in turn, allows us to solve (134.8). We find:

$$a^2(t) = \frac{4}{1 + \left(\dfrac{4}{a_0^2} - 1\right)e^{-\varepsilon t}}, \qquad \theta(t) = \theta_0,$$

where $a_0 = a(0), \theta_0 = \theta(0)$. Note that as $t \to \infty$ the approximation in (134.2) tends to a sinusoidally varying function of magnitude two.

Example 2

For Duffing's equation

$$\frac{d^2y}{dt^2} + y + \varepsilon y^3 = 0,$$

we identify $f(y, y') = y^3$. Evaluating (134.9) with this f results in: $F(a) = 0$ and $G(a) = \frac{3}{4}a^3$. This, in turn, allows us to solve (134.8). We find:

$$a(t) = a_0, \qquad \theta(t) = \theta_0 + \frac{3}{8}a_0^2\varepsilon t.$$

Notes

[1] This method is also called the method of Krylov–Bogoliubov–Mitropolski.

[2] There are many ways in which averaging techniques can be applied to differential equations; we have only illustrated one technique. Another useful technique is the method of averaged Lagrangians, see Whitham [8]. This technique is applied by finding the Lagrangian corresponding to a given differential equation (see page 57), assuming an expansion of the Lagrangian that contains slowly-varying functions and a small parameter, and, at each order of the small parameter, solving the differential equation corresponding to that term of the Lagrangian.

References

[1] J. Golec and G. Ladde, "Averaging Principle and Systems of Singularly Perturbed Stochastic Differential Equations," *J. Math. Physics*, **31**, No. 5, May 1990, pages 1116–1123.

[2] M. I. Gromyak, "Justification of a Scheme for Averaging of Hyperbolic Systems with Fast and Slow Variables. A Mixed Problem," *Ukrain. Mat. Zh.*, **38**, No. 5, 1986, pages 575–582. In Russian.

[3] J. Kevorkian and J. D. Cole, *Perturbation Methods in Applied Mathematics*, Springer–Verlag, New York, 1981, pages 279–287.

[4] U. Kirchgraber, "An ODE-Solver Based on the Method of Averaging," *Numer. Math.*, **53**, 1988, pages 621–652.

[5] A. H. Nayfeh, *Perturbation Methods*, John Wiley, New York, 1973, Chapter 5 (pages 159–227).

[6] R. H. Rand and D. Armbruster, *Perturbation Methods, Bifurcation Theory and Computer Algebra*, Springer–Verlag, New York, 1987, Chapter 5 (pages 107–131).

[7] J. A. Sanders and F. Verhulst, *Averaging Methods in Nonlinear Dynamical Systems*, Springer–Verlag, New York, 1985.

[8] G. B. Whitham, *Linear and Nonlinear Waves*, Wiley Interscience, New York, 1974.

135.   Perturbation Method: Boundary Layer Method

Applicable to   Differential equations with a small parameter present, for which regular perturbation series are inadequate.

Yields

This singular perturbation technique yields an expansion of the solution in terms of the small parameter.

Idea

If a regular perturbation series cannot match all the boundary conditions in a differential equation, there may be one or more regions where the solution is rapidly varying.

Procedure

Given a differential equation with a small parameter ε attempt to find a solution in the form of a regular perturbation series (see page 528). Call this the "outer" solution. If the "outer" solution cannot match all of the initial conditions or boundary conditions, then attempt to place "boundary layers" (regions of rapid variation) near one or more of the boundaries.

Inside of each boundary layer, the solution will vary smoothly (in a stretched variable) from the value of a "outer" solution to the value on the boundary. If multiple "outer" solutions exist, then there may be internal boundary layers (called "shocks"). These internal boundary layers will change the solution smoothly from one "outer" solution to another.

Example

Consider the constant coefficient ordinary differential equation

$$\varepsilon \frac{d^2 y}{dx^2} + \frac{dy}{dx} + y = 0,$$
$$y(0) = 4, \quad y(1) = 5, \tag{135.1}$$

where ε is a number much smaller than one. Initially we look for an "outer" solution in the form of a regular perturbation series (see page 528)

$$y = y_{\text{outer}} = y_0 + \varepsilon y_1 + \varepsilon^2 y_2 + \cdots. \tag{135.2}$$

Using (135.2) in (135.1), and setting the coefficients of different powers of ε to zero, produces the sequence of equations

$$\frac{dy_0}{dx} + y_0 = 0,$$
$$\frac{dy_1}{dx} + y_1 = -\frac{d^2 y_0}{dx^2}, \tag{135.3.a–b}$$
$$\vdots$$

with the boundary conditions

$$y_0(0) = 4, \quad y_0(1) = 5,$$
$$y_i(0) = 0, \quad y_i(1) = 0, \quad \text{for } i = 1, 2, 3, \ldots. \tag{135.4.a–b}$$

The most general solution of (135.3.a) is

$$y_0(x) = Ce^{-x} \tag{135.5}$$

for some constant C. This solution cannot satisfy both of the boundary conditions in (135.4.a); so we suspect the existence of a boundary layer.

First, we search for a boundary layer near $x = 0$. If it is not possible to place one there, then we would attempt to place one near the other boundary, at $x = 1$. Since a change of order one is expected to take place in a thin x region we scale x so that the width of the thin region becomes of order one (in the new variable $\tilde{x}$)

$$\tilde{x} = \frac{x}{\varepsilon}. \tag{135.6}$$

(In other problems, the scaling may be different; it may be that $\tilde{x} = x/\varepsilon^\beta$, where β is an integer or a fraction.) Using the new independent variable in (135.6), the equation in (135.1) may be written as

$$\frac{d^2y}{d\tilde{x}^2} + \frac{dy}{d\tilde{x}} + \varepsilon y = 0. \tag{135.7}$$

The solution of this equation is called the "inner" solution. If we search for a regular perturbation series solution to (135.7), in the form of (135.2), then the sequence of equations begins

$$\frac{d^2y_0}{d\tilde{x}^2} + \frac{dy_0}{d\tilde{x}} = 0,$$
$$\frac{d^2y_1}{d\tilde{x}^2} + \frac{dy_1}{d\tilde{x}} = -y_0, \tag{135.8.a–b}$$
$$\vdots$$

Using the general solution to (135.8.a) we have

$$y_{\text{inner}}(\tilde{x}) = y_0(\tilde{x}) + O(\varepsilon) = D + Ee^{-\tilde{x}} + O(\varepsilon), \tag{135.9}$$

where D and E are constants.

Since we have assumed that the boundary layer is at $x = 0$, the "inner" solution in (135.9) must satisfy the boundary condition at $x = 0$, i.e., $y_{\text{inner}}(0) = 4$. The "outer" solution does not extend to $x = 0$ (since the boundary layer is present) but does extend to $x = 1$. Hence the solution in (135.5) must satisfy the boundary condition at $x = 1$; i.e., $y_{\text{outer}}(1) = 5$. Evaluating (135.5) and (135.9) at their respective boundaries results in

$$y_{\text{outer}}(x) = 5e^{1-x} + O(\varepsilon),$$

$$y_{\text{inner}}(\widetilde{x}) = (4 - E) + Ee^{-\widetilde{x}} + O(\varepsilon). \tag{135.10}$$

To determine the constant E, we need a "matching principle." The "matching principle" is needed to ensure continuity of the solution as it changes from y_{inner} to y_{outer}. Since the transition occurs for x just larger than zero, we require

$$\lim_{x \to 0+} y_{\text{inner}}(x) = \lim_{x \to 0+} y_{\text{outer}}(x).$$

Writing y_{inner} in terms of $\widetilde{x}$, and assuming that ε is arbitrarily small, this statement can be written as

$$\lim_{\widetilde{x} \to \infty} y_{\text{inner}}(\widetilde{x}) = \lim_{x \to 0+} y_{\text{outer}}(x). \tag{135.11}$$

Sometimes this is called an *intermediate expansion* since the matching occurs on an intermediate scale. Using the solutions from (135.10) in (135.11), we determine that $E = 4 - 5e$.

Finally we need to combine y_{inner} and y_{outer} together to obtain a uniformly valid approximation, y_{uniform}, over the entire interval: $x \in [0, 1]$. The uniform approximation is defined to be the sum of y_{inner} plus y_{outer}, minus the overlap value. That is

$$
\begin{aligned}
y_{\text{uniform}} &= y_{\text{inner}} + y_{\text{outer}} - \text{value in (135.11)} \\
&= \left[5e + (4 - 5e)e^{-\widetilde{x}} \right] + \left[5e^{1-x} \right] - [5e] + O(\varepsilon) \\
&= (4 - 5e)e^{-\widetilde{x}} + 5e^{1-x} + O(\varepsilon) \\
&= (4 - 5e)e^{-x/\varepsilon} + 5e^{1-x} + O(\varepsilon).
\end{aligned} \tag{135.12}
$$

Figure 135.1 has graphs of the exact solution of (135.1) and the approximate solution given by (135.12) for $\varepsilon = .1$.

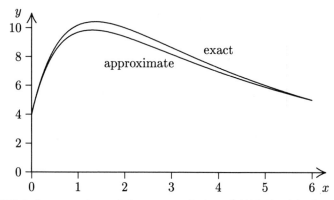

Figure 135.1 A comparison of the exact solution of (135.1) with the approximation in (135.12) for $\varepsilon = .1$.

Notes

[1] The exact solution to (135.1) is given by

$$y(x) = \frac{1}{e^{r_2} - e^{r_1}} \left[(4e^{r_2} - 5) e^{r_1 x} + (5 - 4e^{r_1}) e^{r_2 x} \right], \qquad (135.13)$$

where $r_1 = \left(-1 - \sqrt{1 - 4\varepsilon} \right)/2\varepsilon$ and $r_2 = \left(-1 + \sqrt{1 - 4\varepsilon} \right)/2\varepsilon$. For small values of ε, $r_1 \approx -1/\varepsilon$ and $r_2 \approx -1$. Using these approximations in (135.13), and expanding everything to leading order, results in (135.12).

[2] If the example were carried to second order in ε, then we would have found

$$y_{\text{outer}} = \left[5e^{1-x} \right] + \varepsilon \left[5(1-x)e^{1-x} \right] + O(\varepsilon^2),$$

$$y_{\text{inner}} = \left[5e + (4 - 5e)e^{-\tilde{x}} \right] + \varepsilon \left[5e \left(1 - e^{-\tilde{x}} \right) - 5e\tilde{x} + (4 - 5e)\tilde{x}e^{-\tilde{x}} \right]$$
$$+ O(\varepsilon^2),$$

$$y_{\text{uniform}} = \left[5e^{1-x} + (4 - 5e)(1 + x)e^{-x/\varepsilon} \right] + \varepsilon \left[5e^{1-x}(1 - x) - e^{1-x/\varepsilon} \right]$$
$$+ O(\varepsilon^2).$$

[3] In the example, we could have expected trouble initially. The original equation is of second order, and so needs two boundary conditions. But the first order term in the regular perturbation series, (135.3.a), is a differential equation of first order, so it would be unlikely to match the two boundary conditions.

[4] If it were not possible to match the "inner" and "outer" solutions in (135.11), then we would have tried to put a boundary layer at $x = 1$. To do this, we scale x so that it has a large variation near $x = 1$, say $\widehat{x} = (1 - x)/\varepsilon$. Using this new distance scale, the leading order terms in the "outer" and "inner" solutions would have the form of (135.5) and (135.9).

Now, however, the outer solution would extend to $x = 0$ (so that $y_{\text{outer}} = 4e^{-x}$), while the inner solution would extend to $x = 1$ (so that $y_{\text{inner}} = (5 - E) + Ee^{-\widehat{x}}$). At this point, we find that we cannot perform the necessary matching. We have: $\lim_{x \to 1^-} y_{\text{outer}}(x) = 4e^{-1}$ but

$$\lim_{x \to 1^-} y_{\text{inner}}(x) = \lim_{\widehat{x} \to \infty} y_{\text{inner}}(\widehat{x}) = \begin{cases} 5, & \text{if } E = 0, \\ \infty, & \text{if } E > 0, \\ -\infty, & \text{if } E < 0. \end{cases}$$

We conclude that there is no boundary layer near $x = 1$, at least with the scaling $\widehat{x} = (1 - x)/\varepsilon$.

[5] Sometimes a boundary layer can appear in the middle of the region of interest. As an example of a "shock" or an "interior transition layer," consider the problem $\varepsilon y'' + xy' = 0$ with the boundary values $y(-1) = 1$ and $y(1) = 2$. The solution to this problem is $y(t; \varepsilon) = \dfrac{1}{2}\left(3 + \dfrac{\text{erf}\left(x/2\sqrt{\varepsilon}\right)}{\text{erf}\left(1/2\sqrt{\varepsilon}\right)}\right)$.

Note the following limits which indicate the non-uniformity of convergence:

$$\lim_{x \to 0^+} \lim_{\varepsilon \to 0^+} y(t; \varepsilon) = 2,$$

$$\lim_{x \to 0^-} \lim_{\varepsilon \to 0^+} y(t; \varepsilon) = 1,$$

$$\lim_{x \to 0} y(t; \varepsilon) = \frac{3}{2}.$$

[6] For certain forms of simple equations, it is possible to predict the existence of boundary layers and other phenomena for generic boundary conditions. Table 135 shows the behavior that can be expected from the equation

$$\varepsilon y'' - p(x)y' - q(x)y = g(x), \qquad a \le x \le b$$
$$y(a) = \alpha, \quad y(b) = \beta$$

(135.14)

when ε is small and positive. For each case there are simple examples which exhibit the predicted behavior.

For non-generic boundary conditions, other solutions are possible. For example, (135.1) fits case (a) in Table 135, which predicts the existence of a boundary layer near $x = 0$. However, if the boundary conditions for (135.1) had been $y(0) = y(1) = 4$, then the solution would have been $y(x) = 4$, with no boundary layers present.

conditions on $p(x)$	type of solution
$p(x) \neq 0$ on $a \leq x \leq b$:	
(a) $p(x) < 0$	Boundary layer at $x = a$
(b) $p(x) > 0$	Boundary layer at $x = b$
$p(x) = 0$:	
(c) $q(x) > 0$	Boundary layers at $x = a$ and $x = b$
(d) $q(x) < 0$	Rapidly oscillating solution
(e) $q(x)$ changes sign	Classical turning point
$p' \neq q$, $p(0) = 0$ only at $x = 0$:	
(f) $p'(0) < 0$	No boundary layers,
	interior layer at $x = 0$
(g) $p'(0) > 0$	Boundary layers at $x = a$ and $x = b$,
	no interior layer at $x = 0$

Table 135. Possible behaviors for equation (135.14).

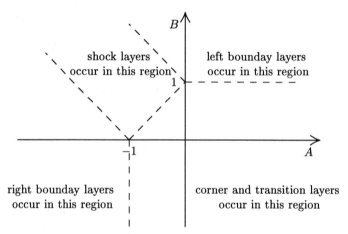

Figure 135.2 Different possible solutions to (135.15) for varying boundary conditions.

[7] A classic example showing the dependence of the solution on the boundary conditions is in Kevorkian and Cole [3], Section 2.5. This nonlinear equation,

$$\varepsilon y'' + yy' - y = 0,$$
$$y(0) = A, \qquad y(0) = B, \tag{135.15}$$

has the solution behaviors shown in Figure 135.2.

[8] There are many matching principles that can be used to determine the unknown constants in the "inner" and "outer" solutions. One that is used in Van Dyke [9] (page 64) is

> The n-term expansion of the inner solution (written in the outer variables) to m-terms is equal to the the m-term expansion of the outer solution (written in the inner variables) to n-terms.

[9] Sometimes there can be multiple boundary layers at a single boundary. That is, there are several layers of boundary layers (each with a different scaling) before the "outer" solution is matched to the value at the boundary.

[10] There exist special numerical procedures that can be used for equations that have boundary layers. See, for instance, Miranker [5].

[11] Lo [2] presents a technique for calculating many terms in an asymptotic expansion. The computer language MACSYMA is used to perform the asymptotic matching at each stage.

[12] This method is sometimes called the *method of matched asymptotic expansions*.

References

[1] C. M. Bender and S. A. Orszag, *Advanced Mathematical Methods for Scientists and Engineers*, McGraw–Hill, New York, 1978, Chapter 9 (pages 417–483).

[2] L. L. Lo, "Asymptotic Matching by the Symbolic Manipulator MACSYMA," *J. Comput. Physics*, **61**, 1985, pages 38–50.

[3] J. Kevorkian and J. D. Cole, *Perturbation Methods in Applied Mathematics*, Springer–Verlag, New York, 1981, pages 20–50 and 370–387.

[4] P. A. Lagerstrom, *Matched Asymptotic Expansions*, Springer–Verlag, New York, 1988.

[5] W. L. Miranker, *Numerical Methods for Stiff Equations*, D. Reidel Publishing Co., Boston, 1981, Chapter 5 (pages 88–108).

[6] A. H. Nayfeh, *Perturbation Methods*, John Wiley, New York, 1973, Chapter 4 (pages 110–158).

[7] C. E. Pearson, "On a Differential Equation of Boundary Layer Type," *J. of Math. and Physics*, **47**, 1968, pages 134–154.

[8] S. M. Roberts, "Further Examples of The Boundary Value Technique in Singular Perturbation Problems," *J. Math. Anal. Appl.*, **133**, 1988, pages 411–436.

[9] M. Van Dyke, *Perturbation Methods in Fluid Mechanics*, Parabolic Press, Stanford, Calif., 1975, Chapter 5 (pages 77–98).

136.   Perturbation Method: Functional Iteration

Applicable to   Differential equations with a "small" term and homogeneous initial conditions or boundary conditions. Without the "small" term, the differential equation must be a linear and have a known Green's function.

Yields

A sequence of approximations.

Idea

If the given equation is only a "small" perturbation from a linear equation (with a known Green's function), then we may obtain an equivalent integral equation. This integral equation may be expanded methodically. Diagrams are often used to keep track of the terms.

Procedure

We will illustrate the general technique on a specific class of partial differential equations. Suppose we have the differential equation

$$\frac{\partial \phi}{\partial t} = H(t, x, \partial_x)\phi + V(x, \partial_x)\phi + A(x),$$
$$\phi(0, x) = 0, \quad \phi(t, 0) = \phi(t, 1) = 0,$$

$$(136.1)$$

for the unknown $\phi(t, x)$, where H and V are functionals. Let us presume that, in some sense, $\|V\phi\| \ll \|H\phi\|$. If the solution $G(t, x; y)$ of

$$\frac{\partial G}{\partial t} = H(t, x, \partial_x)G + \delta(x - y),$$
$$G(0, x; y) = 0, \quad G(t, 0; y) = G(t, 1; y) = 0,$$

$$(136.2)$$

is known, then the solution to (136.1) can be written as the equivalent integral equation

$$\phi(t, x) = \int_0^1 G(t, x; y)[A(x) + V(x, \partial_x)\phi(t, x)]_{x=y}\, dy$$
$$= \phi_0(t, x) + \int_0^1 G(t, x; y)V(y, \partial_y)\phi(t, y),\, dy$$

$$(136.3.a\text{--}b)$$

where $\phi_0(t, x) := \int_0^1 G(t, x; y)A(y)\, dy$. This is because $G(t, x; y)$ is a Green's function (see page 268) and superposition can be used (note that the boundary conditions in (136.1) and (136.2) are homogeneous). If $\phi(t, y)$, as determined by the right-hand side of (136.3.b), is utilized in the integral in (136.3.b), then we obtain

$$\phi(t, x) = \phi_0(t, x) + \int_0^1 G(t, x; y)\phi_0(t, y)\, dy$$

$$+ \int_0^1 dy \int_0^1 du\, [G(t, x; y)V(y, \partial_y)]\, [G(t, y; u)V(u, \partial_u)]\, \phi(t, u).$$

$$(136.4)$$

If $\phi(t, u)$, as determined by the right-hand side of (136.3.b), is utilized in the double integral in (136.4), and the process repeated, then we find

$$\phi(t, x) = \phi_0(t, x) + \int_0^1 G(t, x; y)\phi_0(t, y)\, dy$$

$$+ \int_0^1 dy \int_0^1 du\, [G(t, x; y)V(y, \partial_y)]\, [G(t, y; u)V(u, \partial_u)]\, \phi_0(t, u)$$

$$+ \int_0^1 dy \int_0^1 du \int_0^1 dv\, [G(t, x; y)V(y, \partial_y)]\, [G(t, y; u)V(u, \partial_u)]$$

$$[G(t, u; v)V(v, \partial_v)]\, \phi_0(t, v) + \cdots.$$

$$(136.5)$$

Hence, we have produced a "natural" expansion of the solution to (136.1). Since writing the integrals in (136.5) becomes tedious, diagrams are often utilized. In a fairly obvious notation we may write (136.5) as

$$\phi(t, x) = \phi_0(t, x) + F_1 + F_2 + F_3 + \cdots, \tag{136.6}$$

where each F_i is represented by a diagram in Figure 136.1. The diagrams used in this method are never anything more than a shorthand notation for mathematical expressions. For each specific problem in which diagrammatic techniques are used, the diagrams must be appropriately defined. In this example, a node on a diagram corresponds to the operation $[G(t, \bullet; -)V(-, \partial_-)]$ and each line indicates an integral.

Example 1

We now show how functional iteration method can be used to approximate the solution of an ordinary differential equation, with a small parameter present. Given the differential equation with boundary conditions for $\phi(x)$

$$\frac{d^2\phi}{dx^2} = \varepsilon\, [1 - \phi],$$

$$\phi(0) = \phi(1) = 0, \tag{136.7}$$

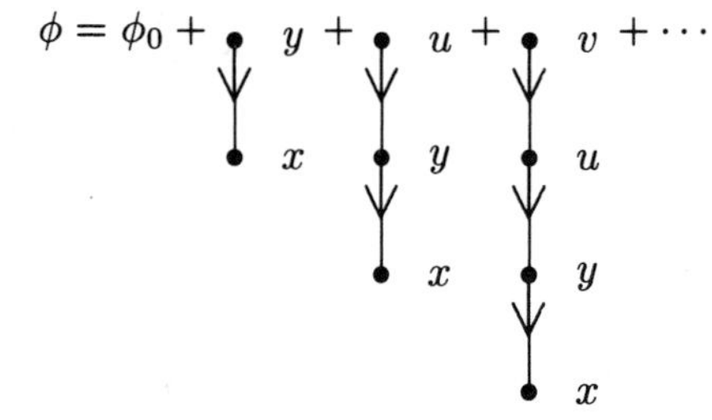

$$\phi = \phi_0 + \bullet \quad y + \bullet \quad u + \bullet \quad v + \cdots$$

Figure 136.1 Diagrammatic representation of the solution in (136.6).

we first note that the exact solution is given by

$$\phi(x) = 1 - \cos\sqrt{\varepsilon}x + \frac{\cos\sqrt{\varepsilon} - 1}{\sin\sqrt{\varepsilon}}\sin\sqrt{\varepsilon}x,$$

$$= \varepsilon\left(\frac{x^2 - x}{2}\right) - \varepsilon^2\left(\frac{x^4 - 2x^3 + x}{24}\right) + O(\varepsilon^3). \qquad (136.8.a\text{--}b)$$

The Green's function that we need, $G(x; y)$, will satisfy the equation

$$\frac{d^2G}{dx^2} = \delta(x - y),$$
$$G(0) = G(1) = 0,$$

and is given by (see the example for the Green's function method, page 271)

$$G(x; y) = \begin{cases} x(y - 1) & \text{for } 0 \le x \le y, \\ y(x - 1) & \text{for } y < x \le 1. \end{cases}$$

The differential equation in (136.7) can then be written as an integral equation, using this Green's function, as

$$\phi(x) = \varepsilon\int_0^1 G(x; y)\left[1 - \phi(y)\right] dy$$

$$= \varepsilon\phi_0(x) - \varepsilon\int_0^1 G(x; y)\phi(y)\, dy, \qquad (136.9.a\text{--}b)$$

where

$$\phi_0(x) := \int_0^1 G(x; y)\, dy$$

$$= \int_x^1 x(y - 1)\, dy + \int_0^x y(x - 1)\, dy$$

$$= \frac{x^2 - x}{2}.$$

If the value of $\phi(x)$ (as defined by the right-hand side of (136.9.b)) is inserted for the function $\phi(y)$ in (136.9.b), the natural expansion arises

$$
\phi(x) = \varepsilon\phi_0(x) - \varepsilon^2 \int_0^1 G(x; y)\phi_0(y)\, dy
$$

$$
+ \varepsilon^3 \int_0^1 G(x; y) \int_0^1 G(y; z)\phi_0(z)\, dz\, dy - O(\varepsilon^4),
$$

(136.10)

which can be represented by

$$
\phi(x) = \phi_0(x) + F_1 + F_2 + F_3 + \cdots,
$$

where the $\{F_i\}$ are given in Figure 136.1. In this example, a node on a diagram corresponds to multiplying by $G(\alpha; \beta)$ (for some specific α and β) and each line segment indicates an integration. It is easy to evaluate the first few diagrams, that is, to evaluate the first few terms in (136.10). The approximation obtained from (136.10) is identical to the expansion in (136.8.b).

Example 2

The reason for finding a Green's function is so the solution of the original differential equation may be written in terms of an integral (as in (136.3.b) or (136.9.b)). For a first order equation, though, an integral representation can be found immediately. We now analyze a nonlinear first order differential equation to indicate more fully how the diagrams may be used. Consider the nonlinear ordinary differential equation

$$
\frac{dz}{dt} = f(t) + g(t)z^2,
$$

$$
z(0) = 0.
$$

in which the nonlinear term (i.e., the $g(t)$ function) is "small." This equation may be integrated directly to obtain

$$
z(t) = \int_0^t f(\tau)\, d\tau + \int_0^t g(\tau)z^2(\tau)\, d\tau. \tag{136.11}
$$

If the value of $z(t)$ from the left-hand side of (136.11) is used in the right-hand side, then

$$
z(t) = \int_0^t f(\tau)\, d\tau + \int_0^t g(\tau) \left[\int_0^\tau f(\tau_1)\, d\tau_1 \right]^2 d\tau
$$

$$
+ 2 \int_0^t g(\tau) \left[\int_0^\tau f(\tau_1)\, d\tau_1 \right] \left[\int_0^\tau g(\tau_2)z^2(\tau_2)\, d\tau_2 \right] d\tau \tag{136.12}
$$

$$
+ \int_0^t g(\tau) \left[\int_0^\tau g(\tau_2)z^2(\tau_2)\, d\tau_2 \right]^2 d\tau.
$$

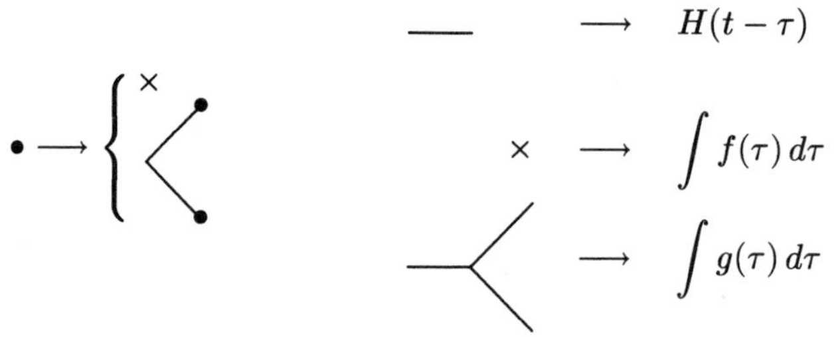

Figure 136.2 Rules for creating diagrams and interpreting them for example 2.

A "natural" perturbation expansion would be to keep the first two terms in the right-hand side of (136.12), and assume that the last two terms are "small." If $|z(t)| \ll 1$ then this may well be the case since the last two terms involve $|z|^2$ while the first two terms involve $|z|$.

The functional iteration technique can be used to derive (136.12) and the higher order extensions from diagrams. We need two sets of rules: one set of rules describes how the diagrams may be computed; the other set of rules describes how the diagrams are to be turned into mathematical expressions. If we use the rules in Figure 136.2, (where $H(\ )$ denotes the Heaviside function), then the first two steps in the diagrammatic solution to $z(t)$ (from (136.11)) are given by the diagrams in Figure 136.3.

Note that the third and fourth diagrams in Figure 136.3 represent the same mathematical expression since they are topologically equivalent. The purpose of the Heaviside function is to restrict the range of integration. By careful inspection, the mathematical expressions associated with the last set of diagrams will be seen to be identical to (136.12).

Notes

[1] In the physics literature, the Green's function is sometimes call the *propagator*. This is usually written in terms of a *path integral*, $G = \int e^{iS/\hbar}$, where S is the action, defined to be the integral of the Lagrangian. The diagrams produced in this context are sometimes called Feynman diagrams.

[2] When nonlinear equations are approximated by this technique, as in example 2, keeping track of the terms in the expansion that are of the some order is greatly facilitated by some shorthand notation. The diagrams presented above perform such a task.

[3] In more complicated problems, the diagrams will have several different types of line segments and several different types of nodes.

[4] Often an "algebra of diagrams" is created, so that diagrams can be added, subtracted and multiplied, without recourse to the mathematical expression that each diagram represents. This would require amplification of the rules that were used in example 2.

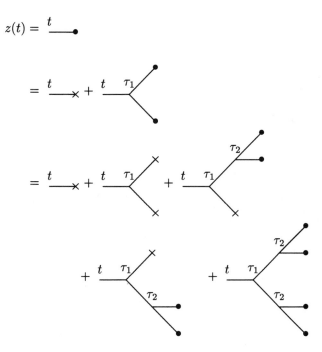

Figure 136.3 Two steps in the diagrammatic expansion of (136.11).

[5] Presented in this section has been just one type of functional iteration, there are many others. For example, Picard iteration (see page 535) is a functional iteration method. Another method is a decomposition method frequently used by Adomian [2].

[6] This technique is particularly important in problems in which there is no "small" parameter. In these cases, the *formally correct* diagrammatic expansion may be algebraically approximated by exactly summing certain classes of diagrams. See Mattuck [6] for details.

References

[1] A. A. Abrikosov, L. P. Gorkov, and I. E. Dzyaloshinski, *Methods of Quantum Field Theory in Statistical Physics*, Dover Publications, Inc., New York, 1963, pages 68–84.

[2] G. Adomian, *Stochastic Systems*, Academic Press, New York, 1983.

[3] L. Fishman and J. J. McCoy, "Factorization and Path Integration of the Helmholtz Equation: Numerical Algorithms," *J. Acoust. Soc. Am.*, **81**, No. 5, May 1987, pages 1355–1376.

[4] J.-M. Drouffe and C. Saclay, "Computer Algebra as a Research Tool in Physics," in B. Buchberger and B. F. Caviness (eds.), *EUROCAL '85*, Springer–Verlag, New York, 1985, pages 58–67.

[5] J. C. Houard and M. Irac-Astaud, "A New Approach to Perturbation Theory: Star Diagrams," *J. Math. Physics*, **24**, No. 8, August 1983, pages 1997–2005.

[6] R. D. Mattuck, *A Guide to Feynman Diagrams in the Many-Body Problem*, Academic Press, New York, 1976.

[7] A. H. Nayfeh, *Perturbation Methods*, John Wiley, New York, 1973, pages 360–372.

[8] P. Pascual and R. Tarrach, *QCD: Renormalization for the Practitioner*, Springer–Verlag, New York, 1984, pages 34–37.

[9] L. S. Schulman, *Techniques and Applications of Path Integration*, John Wiley & Sons, New York, 1981.

[10] S. K. Srinivasan and R. Vasudevan, *Introduction to Random Differential Equations and Their Applications*, American Elsevier Publishing Company, New York, 1971, Chapter 8 (pages 147–160).

137. Perturbation Method: Multiple Scales

Applicable to   Nonlinear differential equations that have a small parameter present.

Yields

An approximation to the solution.

Idea

This is a singular perturbation technique, applicable to problems for which regular perturbation techniques fail. The assumption in this technique is that the solution depends on more than one "length" (or "time") scale.

Procedure

We presume that the solution depends on two (or more) different length (or time) scales. By trying different possibilities, we determine what these length scales are. These different length scales are treated as dependent variables when transforming the given ordinary differential equation into a partial differential equation, but then the length scales are treated as independent variables when solving the equations.

The dependent variable is then expanded in a regular perturbation series (see page 528), where each functions in the series depends on all of the different length scales. The different orders of ε are collected, and the sequential set of partial differential equations are solved.

As these equation are solved, the requirement is that each successive term must vanish no slower (as ε tends to zero) than the previous term.

Example

Suppose we have the ordinary differential equation

$$\varepsilon y'' + y' = 2,$$
$$y(0) = 0, \quad y(1) = 1, \tag{137.1}$$

for $y(x; \varepsilon)$. We immediately recognize that (137.1) is likely to be a singular perturbation problem. This is because, when we set ε equal to zero, the equation becomes a first order differential equation, and it is very unlikely that the solution of this equation (which depends on a single constant) will match both boundary conditions.

We first need to determine what the proper length scales are for this problem. We guess that, for this problem, the proper length scales are $u := x$ and $v := x/\varepsilon$. If we had guessed incorrectly, then we would not be able to carry out all of the calculations. First, equation (137.1) must be written in terms of these new variables. Writing $\dfrac{d}{dx}$ as

$$\frac{d}{dx} = \left(\frac{du}{dx} \frac{\partial}{\partial u} + \frac{dv}{dx} \frac{\partial}{\partial v} \right) = \left(\frac{\partial}{\partial u} + \frac{1}{\varepsilon} \frac{\partial}{\partial v} \right),$$

the equation in (137.1) becomes

$$\varepsilon \left(\frac{\partial}{\partial u} + \frac{1}{\varepsilon} \frac{\partial}{\partial v} \right)^2 y + \left(\frac{\partial}{\partial u} + \frac{1}{\varepsilon} \frac{\partial}{\partial v} \right) y = 2. \tag{137.2}$$

We now propose the expansion of $y(x; \varepsilon)$ as a regular perturbation series in the dependent variables u and v

$$y(x; \varepsilon) = y_0(u, v) + \varepsilon y_1(u, v) + \varepsilon^2 y_2(u, v) + \cdots. \tag{137.3}$$

Using (137.3) in (137.2) and equating the different powers of ε results in an infinite sequence of equations, of which the first three are:

$$O(\varepsilon^{-1}): \quad \frac{\partial^2 y_0}{\partial v^2} + \frac{\partial y_0}{\partial v} = 0,$$

$$O(\varepsilon^0): \quad \frac{\partial^2 y_1}{\partial v^2} + \frac{\partial y_1}{\partial v} = 2 - 2\frac{\partial^2 y_0}{\partial u \partial v} - \frac{\partial y_0}{\partial u}, \quad (137.4.a\text{--}c)$$

$$O(\varepsilon^1): \quad \frac{\partial^2 y_2}{\partial v^2} + \frac{\partial y_2}{\partial v} = -2\frac{\partial^2 y_1}{\partial u \partial v} - \frac{\partial y_1}{\partial u} - \frac{\partial^2 y_0}{\partial u^2}.$$

The first partial differential equation can be solved to determine

$$y_0(u, v) = A(u) + B(u)e^{-v}, \quad (137.5)$$

where $A(u)$ and $B(u)$ are arbitrary functions of u. The second equation, (137.4.b), then becomes

$$\frac{\partial^2 y_1}{\partial v^2} + \frac{\partial y_1}{\partial v} = 2 - A'(u) + B'(u)e^{-v}, \quad (137.6)$$

which has the solution

$$y_1(u, v) = [2 - A'(u)]v + vB'(u)e^{-v} + D(u) + E(u)e^{-v}, \quad (137.7)$$

where $D(u)$ and $E(u)$ are arbitrary functions. Now we use our solvability condition, which states that the higher order terms will vanish no slower than the lower order terms. For $y_1(u, v)$ (as given in (137.7)) to vanish no slower than $y_0(u, v)$ (as given in (137.5)) we require that $2 - A'(u) = 0$ and $B'(u) = 0$. Otherwise, for $x \neq 0$ and $\varepsilon \ll 1$, the terms in y_1 would be larger than the terms in y_0 (since, in this case, $v \gg 1$).

Using these two constraints, we determine that $A(u) = 2u + A_0$ and $B(u) = B_0$, where A_0 and B_0 are constants. Hence, the first order solution becomes

$$y_0(u, v) = (2u + A_0) + B_0 e^{-v}. \quad (137.8)$$

Going back to the original variable (i.e., x), the leading term in the solution for $y(x; \varepsilon)$ is (from (137.3) and (137.8))

$$y(x; \varepsilon) \approx y_0(x) \approx (2x + A_0) + B_0 e^{-x/\varepsilon}.$$

This expression can be matched to both of the boundary conditions in (137.1) to determine that

$$y(x; \varepsilon) \approx 2x - \left(1 - e^{-x/\varepsilon}\right). \quad (137.9)$$

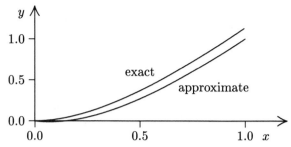

Figure 137. A comparison of the exact solution to (137.1) (given by (137.10)) and the approximate solution in (137.9), when $\varepsilon = .5$.

The exact solution to (137.1) is given by

$$y(x; \varepsilon) = 2x - \frac{1 - e^{-x/\varepsilon}}{1 - e^{-1/\varepsilon}}. \tag{137.10}$$

Hence we see that the approximate analysis has correctly obtained the first term in the expansion as ε tends to zero. Figure 137 has a comparison of (137.9) and (137.10) when $\varepsilon = .5$.

Notes

[1] It was not really necessary to solve equation (137.6) for y_1 to obtain the constraints on $A(u)$ and $B(u)$. By analysis of the equation for y_1, with an eye towards obtaining solutions that do not grow with v, the same conditions could have been obtained. This is an important procedure in more complicated problems where explicit solutions are not easy to find. See the section on alternative theorems, beginning on page 14.

[2] Any problem that can be solved by matched asymptotic expansions can also be solved by multiple scales, although the procedure may require more work.

[3] The method of multiple scales does not result in an answer that is valid over an indefinitely long range. If, for instance, the two scales are x and εx then the solution is valid, generally, for $x = O(\varepsilon^{-1})$.

[4] In Rubenfeld [7], there is an account of why the method of multiple scales sometimes gives *incorrect* results.

[5] In Fateman [3], there is a description of a MACSYMA program that will automatically utilize the method of multiple scales to approximate the solution of differential equations.

[6] The method of multiple scales is often called "two timing."

[7] The choice of length scales depends on the particular problem. For some problems three (or more) length scales may be appropriate. Each length scale may have a complicated dependence on the parameter ε.

References

[1] C. M. Bender and S. A. Orszag, *Advanced Mathematical Methods for Scientists and Engineers*, McGraw–Hill, New York, 1978, Chapter 11 (pages 544–568).

[2] J. U. Brackbill and B. I. Cohen, *Multiple Time Scales*, Academic Press, New York, 1985.

[3] R. J. Fateman, "An Approach to Automatic Asymptotic Expansions," in R. D. Jenks (ed.), *SYMSAC '76*, ACM, New York, 1976, pages 365–371.

[4] J. Kevorkian and J. D. Cole, *Perturbation Methods in Applied Mathematics*, Springer–Verlag, New York, 1981, pages 115–151.

[5] C. C. Lin and L. A. Segel, *Mathematics Applied to Deterministic Problems in the Natural Sciences*, Macmillan, New York, 1974, pages 325–331.

[6] A. H. Nayfeh, *Perturbation Methods*, John Wiley, New York, 1973, Chapter 6 (pages 228–307).

[7] L. A. Rubenfeld, "On a Derivative-Expansion Technique and Some Comments on Multiple Scaling in the Asymptotic Approximation of Solutions of Certain Differential Equations," *SIAM Review*, **20**, No. 1, January 1978, pages 79–105.

[8] J. A. Sanders and F. Verhulst, *Averaging Methods in Nonlinear Dynamical Systems*, Springer–Verlag, New York, 1985.

[9] M. Van Dyke, *Perturbation Methods in Fluid Mechanics*, Parabolic Press, Stanford, Calif., 1975, Section 10.4 (pages 198–200).

138. Perturbation Method: Regular Perturbation

Applicable to   Differential equations with a small parameter.

Yields

A series of terms of decreasing magnitude that approximate the solution of the original differential equation.

Idea

When an equation is changed by only a small amount, the solution will often only change by a small amount.

Procedure

Expand the dependent variables in a power series depending on the small parameter in the problem. Substitute this series into the original equation, the boundary conditions and the initial conditions. Expand all of the equations, equate the terms corresponding to different powers of the small parameter, and solve the equations sequentially.

Example

Suppose we have the equation

$$y'' + \varepsilon y' + y = 0,$$
$$y(0) = 1, \quad y'(0) = 0, \tag{138.1}$$

where ε is a number whose magnitude is much smaller than one. We suppose that the solution to (138.1), $y(x; \varepsilon)$, can be expanded in a power series in ε as follows

$$y(x; \varepsilon) = y_0(x) + \varepsilon y_1(x) + \varepsilon^2 y_2(x) + \cdots. \tag{138.2}$$

Then, using (138.2) in (138.1) we obtain

$$(y_0'' + \varepsilon y_1'' + \cdots) + \varepsilon(y_0' + \varepsilon y_1' + \cdots) + (y_0 + \varepsilon y_1 + \cdots) = 0, \tag{138.3}$$

$$y_0(0) + \varepsilon y_1(0) + \varepsilon^2 y_2(0) + \cdots = 1,$$
$$y_0'(0) + \varepsilon y_1'(0) + \varepsilon^2 y_2'(0) + \cdots = 0.$$

Equating powers of ε in (138.3) to zero produces the sequence of equations

$$O(\varepsilon^0): \qquad y_0'' + y_0 = 0,$$
$$y_0(0) = 1, \tag{138.4}$$
$$y_0'(0) = 0.$$

$$O(\varepsilon^1): \qquad y_1'' + y_1 = -y_0',$$
$$y_1(0) = 1, \tag{138.5}$$
$$y_1'(0) = 0.$$

The solution to (138.4) is

$$u_0(x) = \cos x. \tag{138.6}$$

Using (138.6) in (138.5) we must now solve the equation

$$y_1'' + y_1 = \sin x,$$
$$y_1(0) = 1, \tag{138.7}$$
$$y_1'(0) = 0.$$

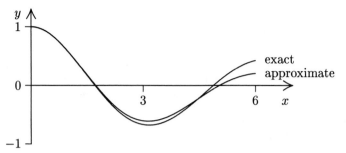

Figure 138. Comparison of the exact solution and the two term approximation to equation (138.1), when $\varepsilon = .25$.

The solution to (138.7) is

$$y_1(x) = \tfrac{1}{2}(\sin x - x \cos x).\tag{138.8}$$

Therefore, the solution for $y(x; \varepsilon)$ is approximately (using (138.6) and (138.8) in (138.2))

$$y(x; \varepsilon) = \cos x + \frac{\varepsilon}{2}(\sin x - x \cos x) + O(\varepsilon^2).\tag{138.9}$$

We could continue this process indefinitely and calculate as many terms as were needed to obtain a desired accuracy. Figure 138 is a comparison of the first two terms of (138.9), when $\varepsilon = .25$, with the exact solution.

Notes

[1] The exact solution to (138.1) is given by

$$y(x; \varepsilon) = \frac{\varepsilon}{\sqrt{4 - \varepsilon^2}} e^{-\varepsilon x/2} \sin\left(x\sqrt{1 - \frac{\varepsilon^2}{4}}\right) + e^{-\varepsilon x/2} \cos\left(x\sqrt{1 - \frac{\varepsilon^2}{4}}\right),$$

which can be expanded for small ε to yield

$$y(x; \varepsilon) = \cos x + \frac{\varepsilon}{2}(\sin x - x \cos x) + O(\varepsilon^2).$$

[2] This method will *not* work on all equations that have a small parameter. As a simple example, consider

$$\varepsilon y'' + y = 0, \quad y(0) = 1, \quad y(1) = 2.\tag{138.10}$$

In this example, the first order equation (corresponding to equation (138.4)) is

$$y_0 = 0, \quad y(0) = 1, \quad y(1) = 2.$$

Clearly, this equation has no solution. Hence the original expansion, (138.3), must not be adequate to represent the solution of (138.10).

[3] In deriving equations (138.4) and (138.5) from (138.3), it was implicitly assumed that each of $|y_1(x)|$, $|y_1'(x)|$, and $|y_1''(x)|$ are $O(1)$. Observe that this will *not* be the case when $x = O(1/\varepsilon)$ (see (138.8)). Hence, we conclude that only when $x \ll 1/\varepsilon$ can (138.9) be a good approximation to the solution of (138.1). If an approximation to the solution is desired over a larger range of x values, then the method of multiple scales might be used (see page 524). *Secular terms* is the name given to terms that become large and prevent a perturbation expansion from being valid.

[4] If the solution to a differential equation is not analytic at $\varepsilon = 0$, then the solution can *not* be expanded in the form of (138.2). Often, the best procedure is to utilize an expansion of the form

$$y(x;\varepsilon) = y_0(x) + \mu_1(\varepsilon)y_1(x) + \mu_2(\varepsilon)y_2(x) + \ldots,$$

and then determine the scaling functions $\{\mu_i\}$ as the $\{y_i\}$ are determined. It is frequently the case that the $\{\mu_i\}$ are given by terms of the form $\{\varepsilon^n \log^m \varepsilon\}$. Terms with $m \neq 0$ are sometimes called *switchback terms* (see Lagerstrom and Reinelt [4] or Van Dyke [7]).

[5] The functional iteration method described on page 518 produces the same terms that would be obtained by a regular perturbation expansion. The benefit of the diagrammatic method is that it allows easier manipulation of the terms.

References

[1] C. M. Bender and S. A. Orszag, *Advanced Mathematical Methods for Scientists and Engineers*, McGraw–Hill, New York, 1978, pages 319–335.

[2] S. J. Farlow, *Partial Differential Equations for Scientists and Engineers*, John Wiley & Sons, New York, 1982, Lesson 46, (pages 370–378).

[3] J. Kevorkian and J. D. Cole, *Perturbation Methods in Applied Mathematics*, Springer–Verlag, New York, 1981, pages 17–20.

[4] P. A. Lagerstrom and D. A. Reinelt, "Note on Logarithmic Switchback Terms in Regular and Singular Perturbation Expansion," *SIAM J. Appl. Math.*, **44**, No. 3, June 1984, pages 451–462.

[5] C. C. Lin and L. A. Segel, *Mathematics Applied to Deterministic Problems in the Natural Sciences*, Macmillan, New York, 1974, pages 45–55 and 225–241.

[6] A. H. Nayfeh, *Perturbation Methods*, John Wiley, New York, 1973.

[7] M. Van Dyke, *Perturbation Methods in Fluid Mechanics*, Parabolic Press, Stanford, Calif., 1975, pages 9–20 and 200–202.

139. Perturbation Method: Strained Coordinates

Applicable to   Differential equations that have a small parameter present.

Yields

An approximation to the solution, valid on a long time scale.

Idea

A regular perturbation expansion may give the correct answer, but at the wrong location. By scaling the dependent variable and one or more of the independent variables by the small parameter, the solution may be approximated at the correct location.

Procedure

If the regular perturbation solution to a differential equation has secular terms, but the original equation has bounded solutions, then the regular perturbation approximation is not valid for large values of the independent variables. One way to obtain a solution that is valid for longer scales is by "straining the coordinates"; that is, expanding the dependent variable *and* one or more of the independent variables in terms of the small parameter.

To completely specify the arbitrary functions and constants that arise, use the maxim: "Higher order approximation shall be no more singular than the first."

Example

Suppose we wish to approximate the solution to the nonlinear differential equation

$$\frac{d^2y}{dt^2} + \omega^2 y = \varepsilon y^3,$$

$$y(0) = 1, \quad y'(0) = 0.$$

$$(139.1)$$

This equation can be integrated once by first multiplying by y'. The resulting first order differential equation can be integrated in terms of elliptic functions. The explicit solution indicates that the solution is periodic.

If a regular perturbation technique is attempted, then the resulting equations can be solved in the usual manner (see page 528) to determine that

$$y(t;\varepsilon) = \cos\omega t + \varepsilon \left[\frac{3}{8}\frac{t}{\omega}\sin\omega t - \frac{1}{32\omega^2}\left(\cos 3\omega t - \cos\omega t\right)\right] + O(\varepsilon^2).$$

Note that the second term in this solution becomes unbounded as t increases. Hence, secular terms are present.

In the method of straining, both the dependent variable and the independent variable are expanded in terms of ε. For this example, we presume the expansion has the form

$$t = t(\tau; \varepsilon) = \tau + \varepsilon t_1(\tau) + O(\varepsilon^2),$$
$$y = y(\tau; \varepsilon) = y_0(\tau) + \varepsilon y_1(\tau) + O(\varepsilon^2).$$

$$(139.2.a\text{--}b)$$

Noting that the derivative with respect to t can be replaced with a derivative with respect to τ by

$$\frac{d}{dt} = (1 - \varepsilon t_1' + \cdots) \frac{d}{d\tau},$$

(where a prime ($'$) denotes differentiation with respect to τ), we find that (139.1) can be turned into a sequence of equations, with each equation involving the next $y_k(\tau)$ term. The first two equations are

$$\frac{d^2 y_0}{d\tau^2} + \omega^2 y_0 = 0,$$
$$\frac{d^2 y_1}{d\tau^2} + \omega^2 y_1 = y_0^3 + 2t_1' \frac{d^2 y_0}{d\tau^2} + t_1'' \frac{dt_0}{d\tau}.$$

$$(139.3)$$

The boundary conditions are similarly expanded. We find

$$y_0(0) = 1, \qquad \frac{dy_0}{d\tau}(0) = 0,$$
$$y_1(0) + t_1(0) \frac{dy_0}{d\tau}(0) = 0,$$
$$\frac{dy_1}{d\tau}(0) - t_1'(0) \frac{dy_0}{d\tau}(0) + t_1(0) + \frac{dy_0}{d\tau}(0) = 0.$$

$$(139.4)$$

Now we proceed to solve the equations sequentially, just as in the regular perturbation method. The first equation in (139.3) with the first pair of boundary conditions in (139.4) yields

$$y_0(\tau) = \cos \omega \tau. \tag{139.5}$$

Using this value for $y_0(\tau)$, the next equation in (139.3) (which is for $y_1(\tau)$) becomes

$$\frac{d^2 y_1}{d\tau^2} + \omega^2 y_1 = \tfrac{1}{4} \cos 3\omega\tau + \left(\tfrac{3}{4} - 2\omega^2 t_1' \right) \cos \omega\tau - \omega t_1'' \sin \omega\tau. \tag{139.6}$$

To prevent $y_1(\tau)$ from having any secular terms, this equation cannot be forced at resonance. This means that the right-hand side of (139.6) cannot have any terms that are solutions of the homogeneous equation. To keep the right-hand side of (139.6) from having any $\cos \omega\tau$ or $\sin \omega\tau$ terms, we choose

$$\left(\frac{3}{4} - 2\omega^2 t_1'\right) = 0 \quad \text{or} \quad t_1 = \frac{3\tau}{8\omega^2}. \tag{139.7}$$

If we now solve (139.6), there will be no secular terms. Utilizing (139.7) in equation (139.2.a) results in

$$t = \tau + \frac{3\varepsilon}{8\omega^2}\tau + \cdots,$$

or

$$\tau = \left(1 - \frac{3\varepsilon}{8\omega^2}\right)t + \cdots.$$

Using this last expression for τ in (139.5) results in our final form of the first order approximation

$$y_0(t) = \cos\left[\left(\omega - \frac{3\varepsilon}{8\omega}\right)t\right].$$

Notes

[1] Another common application of this method is to differential equations whose solutions are well behaved, but approximations by a regular perturbation scheme produce singular terms. For example, the differential equation

$$(x + \varepsilon u)\frac{du}{dx} + u = 0, \qquad u(1) = 1, \tag{139.8}$$

has a solution that is well behaved at $x = 0$, but the regular perturbation series $u(x;\varepsilon) = u_0(x) + \varepsilon u_1(x) + \ldots$ yields $u_0 = x^{-1}$, $u_1 = \frac{1}{2}\left(x^{-1} - x^{-3}\right)$, and higher order terms which are even more singular at $x = 0$. Applying strained coordinate techniques results in the exact solution of (139.8): $u(x) = \left(-x + \sqrt{x^2 + 2\varepsilon + 2\varepsilon^2}\right)/\varepsilon$. This solution shows that $u(x)$ cannot be expanded in a power series in ε near $x = 0$.

[2] The paper by Roberts and Shipman [6] concerns itself with equations of the form

$$[f(x) + \varepsilon y]\frac{dy}{dx} + q(x)y = r(x)$$

on the interval $0 < x < 1$, with $y(1) = c$, when the method of straining does *not* work.

[3] This technique is a useful tool in many areas including the theory of boundary layers and the structure and propagation of shock waves.

[4] This technique is also known as the Lighthill method, the Lindstedt method, and the Poincaré–Lighthill method.

[5] The computer language MACSYMA has a package for automatically implementing this technique, see Len [3] for details.

References

[1] C. Comstock, "The Poincaré–Lighthill Perturbation Technique and Its Generalizations," *SIAM Review*, **14**, No. 3, July 1972, pages 433–446.

[2] M. E. Goldstein and W. H. Braun, *Advanced Methods for the Solution of Differential Equations*, NASA SP-316, U.S. Government Printing Office, Washington, D.C., 1973, pages 306–311.

[3] J. L. Len, "Perturbation Solution of ODEs in MACSYMA: Lindstedt's Method," *MACSYMA Newsletter*, **5**, No. 2, April 1988, pages 6–9.

[4] M. J. Lighthill, "A Technique for Rendering Approximate Solutions to Physical Problems Uniformly Valid," *Z. Flugwiss.*, **9**, 1961, pages 267–275.

[5] A. H. Nayfeh, *Perturbation Methods*, John Wiley, New York, 1973, Chapter 3 (pages 56–109).

[6] S. M. Roberts and J. S. Shipman, "An Iteration Perturbation Technique," *J. Comput. Physics*, **16**, 1974, pages 285–297.

[7] G. Whitham, "The Flow Pattern of a Supersonic Projectile," *Comm. Pure Appl. Math*, **5**, 1952, pages 301–348.

[8] M. Van Dyke, *Perturbation Methods in Fluid Mechanics*, Parabolic Press, Stanford, Calif., 1975, Chapter 6 (pages 99–120).

140. Picard Iteration

Applicable to Differential equations, a single equation or a system.

Yields

A sequence of approximations to the solution.

Idea

We can write an ordinary differential equation as a fixed point formula. If we have a starting guess, we can iterate the equation to find an approximate solution to the original equation.

Procedure

Suppose we have the first order ordinary differential equation

$$\frac{dy}{dx} = f(y, x),$$

with the initial condition $y(x_0) = y_0$. This equation can be written as an integral equation in the form

$$y(x) = y_0 + \int_{x_0}^{x} f\left(y(z), z\right) \, dz. \tag{140.1}$$

Note that (140.1) already incorporates the initial conditions. If we had a guess of $y(x)$, say $y_1(x)$, then we might be able to improve our guess by forming $y_2(x)$ as follows

$$y_2(x) = y_0 + \int_{x_0}^{x} f\left(y_1(z), z\right)\, dz.$$

Then, knowing $y_2(x)$, we could form $y_3(x)$ by the same technique. We can continue this process indefinitely, each time using the formula

$$y_{n+1}(x) = y_0 + \int_{x_0}^{x} f\left(y_n(z), z\right)\, dz. \tag{140.2}$$

What we take for $y_1(x)$ is arbitrary; it is often easiest to take $y_1(x) = y_0$.

Example

Suppose we have the following ordinary differential equation

$$\frac{dy}{dx} = x^2 + y^2,$$

with $y(0) = 1$. We can write the iteration formula, (140.2), in this case, as

$$y_{n+1}(x) = 1 + \int_{0}^{x} [z^2 + y_n(z)^2]\, dz.$$

If we take $y_1(x) = 1$, then we find

$$
\begin{aligned}
y_2(x) &= 1 + x + \tfrac{1}{3}x^3, \\
y_3(x) &= 1 + x + x^2 + \tfrac{2}{3}x^3 + \cdots, \\
y_4(x) &= 1 + x + x^2 + \tfrac{4}{3}x^3 + \tfrac{5}{6}x^4 + \cdots, \\
y_5(x) &= 1 + x + x^2 + \tfrac{4}{3}x^3 + \tfrac{7}{6}x^4 + \tfrac{16}{15}x^8 + \cdots.
\end{aligned}
\tag{140.3}
$$

The Taylor series solution of this problem (see page 548) begins

$$y(x) = 1 + x + x^2 + \tfrac{4}{3}x^3 + \tfrac{7}{6}x^4 + \tfrac{6}{5}x^5 + \cdots.$$

Hence, each successive approximation in (140.3) appears to have one more correct term.

Notes

[1] The successive approximations found by this method are not guaranteed to converge.

[2] This method can also be used on systems of first order ordinary differential equations. For example, the scheme corresponding to the system

$$\frac{dy}{dt} = f(y, z, t), \qquad y(0) = y_0,$$

$$\frac{dz}{dt} = g(y, z, t), \qquad z(0) = z_0,$$

is

$$y_{n+1}(t) = y_0 + \int_0^t f(y_n(t), z_n(t), t) \, dt,$$

$$z_{n+1}(t) = z_0 + \int_0^t g(y_n(t), z_n(t), t) \, dt.$$

[3] Picard iteration can be applied to ordinary differential equations of n-th order, without first writing the equation as a first order system. For example, the second order ordinary differential equation

$$y'' = f\left(t, y(t), y'(t)\right),$$

$$y(a) = A, \qquad y(b) = B,$$

has the convenient iteration scheme

$$y_{n+1}(x) = A + (x - a)y_n'(a) + \int_a^x (x - t)f\left(t, y_n(t), y_n'(t)\right) \, dt$$

where $y_0(x) = A + (x - a)(B - A)/(b - a)$.

[4] It is also possible to approximate some partial differential equations by this technique. For example, the elliptic equation $\nabla^2 u = f\left(x, y, u, \dfrac{\partial u}{\partial x}, \dfrac{\partial u}{\partial y}\right)$ has the natural iteration formula $\nabla^2 u_n = f\left(x, y, u_{n-1}, \dfrac{\partial u_{n-1}}{\partial x}, \dfrac{\partial u_{n-1}}{\partial y}\right)$.
See Iyanaga and Kawada [2] for the technical conditions on when this iteration scheme will converge to the solution of the original equation. Rice and Boisvert [4] illustrate this technique with the use of ELLPACK.

References

[1] W. E. Boyce and R. C. DiPrima, *Elementary Differential Equations and Boundary Value Problems*, Fourth Edition, John Wiley & Sons, New York, 1986, pages 97–103.

[2] S. Iyanaga and Y. Kawada, *Encyclopedic Dictionary of Mathematics*, MIT Press, Cambridge, MA, 1980, page 998.

[3] M. Lal and D. Moffatt, "Picard's Successive Approximation for Non-Linear Two-Point Boundary Value Problems," *J. Comput. Appl. Math.*, **8**, No. 4, 1982, pages 233–236.

[4] J. R. Rice and R. F. Boisvert, *Solving Elliptic Problems Using ELLPACK*, Springer–Verlag, New York, 1985, pages 79–82.

[5] G. F. Simmons, *Differential Equations with Applications and Historical Notes*, McGraw–Hill Book Company, New York, 1972, pages 418–422.

141. Reversion Method

Applicable to   Forced nonlinear ordinary differential equations.

Yields

A local approximation.

Idea

To derive the method, we assume a certain parameter is small and develop a perturbation expansion in that parameter. In practice, we use the formulae obtained by this method when the parameter is equal to one.

Procedure

Suppose that the general nonlinear differential equation whose solution we wish to approximate near the initial value is given by

$$D_1 y + D_2 y^2 + \cdots + D_5 y^5 + \cdots = k\phi(x), \qquad (141.1)$$

where the $\{D_i\}$ represent differential operators. We seek $y = y(x)$, where k is a constant and $\phi(x)$ is a known forcing function. For this method to work we require that $D_1 \neq 0$.

We assume that $y(x)$ is analytic and k is sufficiently small so that the solution to (141.1) can be expanded in a power series in k. That is, we take

$$y(x) = a_1(x)k + a_2(x)k^2 + a_3(x)k^3 + \cdots. \qquad (141.2)$$

Using (141.2) in (141.1) and equating powers of k results in an infinite sequence of equations for the $\{a_i(x)\}$. This sequence of equations begins

$$D_1 a_1 = \phi(x),$$
$$D_1 a_2 = -D_2 a_1^2,$$
$$D_1 a_3 = -[2D_2 a_1 a_2 + D_3 a_1^3], \quad\quad (141.3.a\text{-}d)$$
$$D_1 a_4 = -[D_2(a_2^2 + 2a_1 a_3) + 3D_3 a_1^2 a_2 + D_4 a_1^4].$$

The reversion method is to assume the solution to (141.1) can be represented in the form of (141.2) when $k = 1$ and the coefficients are given by (141.3).

Example

Suppose we have the following nonlinear ordinary differential equation

$$\frac{dv}{dx} + \alpha v^2 = x, \quad\quad v(0) = v_0,$$

and we seek an approximation near $x = 0$. Changing variables to $y = v - v_0$ changes the equation into

$$\frac{dy}{dx} + 2v_0 \alpha y + \alpha y^2 = x - \alpha v_0^2, \quad\quad y(0) = 0, \quad\quad (141.4)$$

which simplifies the initial condition. Comparing (141.4) to (141.1), we make the identifications

$$D_1 = \frac{d}{dx} + 2v_0 \alpha, \quad D_2 = \alpha,$$
$$k = 1, \quad \phi(x) = x - \alpha v_0^2.$$

From (141.3.a), we obtain the following equation for a_1: $D_1 a_1 = \phi(x)$, or

$$\left(\frac{d}{dx} + 2v_0 \alpha \right) a_1 = x - \alpha v_0^2. \quad\quad (141.5)$$

Since $v(0) = 0$, we will take $a_1(0) = a_2(0) = \cdots = 0$. The solution to (141.5), with $a_1(0) = 0$, is

$$a_1 = \frac{x}{2v_0 \alpha} + \frac{e^{-2v_0 \alpha x} - 1}{4v_0^2 \alpha^2} + \alpha v_0^2 \frac{e^{-2v_0 \alpha x} - 1}{2v_0 \alpha},$$

which was obtained by using a Laplace transform (see page 300).

The function a_2 can be determined from equation (141.3.b)

$$\left(\frac{d}{dx} + 2v_0 \alpha \right) a_2 = \alpha a_1^2 = \alpha \left(\frac{x}{2v_0 \alpha} + \frac{e^{-2v_0 \alpha x} - 1}{4v_0^2 \alpha^2} + \alpha v_0^2 \frac{e^{-2v_0 \alpha x} - 1}{2v_0 \alpha} \right)^2,$$

with $a_2(0) = 0$. This can also be solved by using Laplace transforms. Proceeding in this way, many terms in the series (141.2) can be evaluated.

Notes

[1] The above example is from Pipes [2].

[2] The extension of equation (141.3) can be found in Orstrand [1], which lists formulae for the first 13 terms.

References

[1] C. E. Van Orstrand, *Philosophical Magazine*, **19**, 1910, page 366.

[2] L. A. Pipes and L. R. Harvill, *Applied Mathematics for Engineers and Physicists*, McGraw–Hill Book Company, New York, 1970, pages 653–665.

142. Singular Solutions

Applicable to   Nonlinear ordinary differential equations.

Yields

A singular solution.

Idea

Singular solutions may exist where the implicit function theorem does not hold in differential algebraic equations.

Procedure

The algebraic ordinary differential equation

$$F(x, y, y', \ldots, y^{(n)}) = 0, \tag{142.1}$$

can often be solved for the $y^{(n)}$ term to determine that

$$y^{(n)} = G_1(x, y, \ldots, y^{(n-1)})$$

or

$$y^{(n)} = G_2(x, y, \ldots, y^{(n-1)}) \tag{142.2}$$

or

$$\vdots$$

$$y^{(n)} \doteq G_m(x, y, \ldots, y^{(n-1)}).$$

By the implicit function theorem, if $\dfrac{\partial F}{\partial y^{(n)}}(x, y, y', \ldots, y^{(n)}) \neq 0$, then the solutions in equation (142.2) are the only solutions possible. However, at those points where $\dfrac{\partial F}{\partial y^{(n)}}(x, y, y', \ldots, y^{(n)}) = 0$, there exists the possibility of singular solutions.

If the $y^{(n)}$ term is eliminated from the two equations

$$F(x, y, y', \ldots, y^{(n)}) = 0,$$
$$\frac{\partial F}{\partial y^{(n)}}(x, y, y', \ldots, y^{(n)}) = 0,$$

then an equation of the form

$$H(x, y, y', \ldots, y^{(n-1)}) = 0 \tag{142.3}$$

results. This is called the *p-discriminant equation*. Its solution(s) describe the *singular loci*.

After equation (142.3) is solved to determine possible singular solutions, it must be verified that they are, in fact, actual solutions to the original equation (142.1). Typically, the solution to (142.3), being a differential equation of $(n-1)$-st order, will only involve $n-1$ arbitrary constants.

Example
Given the nonlinear first order ordinary differential equation

$$F(x, y, y') = xy'^2 - 3yy' + 9x^2 = 0, \tag{142.4}$$

it is straightforward to compute

$$\frac{\partial F}{\partial y'} = 2xy' - 3y = 0. \tag{142.5}$$

Eliminating the y' term between (142.4) and (142.5) results in

$$y = \pm 2x^{3/2}. \tag{142.6}$$

In this case, both of the solutions in (142.6) satisfy (142.4). Note that the singular solutions in (142.6) do not depend on any constants, even though (142.4) was a first order differential equation.

Notes

[1] The general n-th order ordinary differential equation, linear in the n-th derivative term,

$$U(x,y,y',\ldots,y^{(n-1)})y^{(n)} + V(x,y,y',\ldots,y^{(n-1)}) = 0,$$

has the singular solution $y = z(x)$ if $z(x)$ satisfies both of

$$U(x,z,z',\ldots,z^{(n-1)}) = 0,$$
$$V(x,z,z',\ldots,z^{(n-1)}) = 0.$$

[2] Another way to determine singular solutions of the differential equation $f(x,y,y') = 0$ is to obtain the general solution $\phi(x,y,C) = 0$ (where C is an arbitrary constant) and then formally eliminate C between the two equations

$$\phi(x,y,C) = 0,$$
$$\frac{\partial \phi}{\partial x}(x,y,C) = 0.$$

The resulting equation, which only involves x and y, is called the *c-discriminant equation*.

For example, the differential equation $y'^2 + 4 - 4y = 0$ has the general solution $y(x) = 1 + (x - C)^2$; hence $\phi(x,y,C) = y - 1 - (x + C)^2$. Forming the c-discriminant results in the singular solution $y = 1$.

[3] In general (see Piaggio [7]), the p-discriminant equation will contain the envelope of the solutions, the cusp-locus and the tac-locus squared. The c-discriminant equation will contain the envelope of the solutions, the cusp-locus cubed and the node-locus squared. Of these, only the envelope is a solution to the original differential equation.

[4] Some envelope solutions of differential equations may be found by use of Lie groups, see Bluman [1].

[5] For polynomial functions, the algebraic elimination in the computation of the c-discriminant (or the p-discriminant) can be done by the use of resultants (see page 46).

References

[1] G. Bluman, "Invariant Solution for Ordinary Differential Equations," *SIAM J. Appl. Math.*, **50**, No. 6, December 1990, pages 1706–1715.

[2] L. E. Elsgolts, *Differential Equations and the Calculus of Variations*, MIR Publishers, Moscow, 1970, pages 81–88.

[3] M. E. Goldstein and W. H. Braun, *Advanced Methods for the Solution of Differential Equations*, NASA SP-316, U.S. Government Printing Office, Washington, D.C., 1973, pages 18–24.

[4] E. L. Ince, *Ordinary Differential Equations*, Dover Publications, Inc., New York, 1964, pages 83–91.

[5] W. Kaplan, *Ordinary Differential Equations*, Addison–Wesley Publishing Co., Reading, MA, 1958, pages 330–337.

[6] G. Murphy, *Ordinary Differential Equations*, D. Van Nostrand Company, Inc., New York, 1960, pages 74–80.

[7] H. T. H. Piaggio, *An Elementary Treatise on Differential Equations and Their Applications*, G. Bell & Sons, Ltd, London, 1926, Chapter 6 (pages 65–79) and pages 192–201.

143. Soliton Type Solutions

Applicable to   Partial differential equations with wave-like solutions, often partial differential equations with only two independent variables.

Yields

Knowledge of whether solitons can be present.

Idea

See if there is a solitary wave solution to the partial differential equation. This indicates the possibility that the equation has solitons for solutions.

Procedure

A solitary wave is a localized, traveling wave and many nonlinear partial differential equations have solutions of this type. A soliton is a solitary wave that exhibits particle-like behavior. The particle-like properties include stability, localizability, and finite energy. A soliton is best described, however, in terms of its interaction with other solitary waves. We say that an equation possesses solitons when two or more colliding solitary waves do not break up and disperse but, instead, become more solitary waves.

In this technique we change variables in such a way as to make such a solitary wave more apparent. If the original partial differential equation were in the independent variables x and t, we search for a solution of the form $u(x - ct)$. Here c represents the wave speed; if $c > 0$ ($c < 0$), then $u(x - ct)$ represents a wave traveling to the right (left). Note that many partial differential equations have solitary waves as solutions; most of these partial differential equations do *not* exhibit soliton behavior.

Example

One representation of the Korteweg–de Vries (kdV) equation is given by

$$u_t + \sigma u u_x + u_{xxx} = 0. \tag{143.1}$$

We change the independent variables from $\{x, t\}$ to $\{\eta, \zeta\}$ via (see page 139) $\{\eta = t,\ \zeta = x - ct.\}$. This change of variable turns (143.1) into

$$u_\eta - cu_\zeta + \sigma u u_\zeta + u_{\zeta\zeta\zeta} = 0. \tag{143.2}$$

If we now *presume* that (143.1) admits a wave-like solution, we can then take $u(\eta, \zeta) = v(\zeta) = v(x - ct)$. By assuming this functional form for $u(\eta, \zeta)$, equation (143.2) becomes

$$cv_\zeta + \sigma v v_\zeta + v_{\zeta\zeta\zeta} = 0. \tag{143.3}$$

Equation (143.3) is an autonomous ordinary differential equation. Hence, the order can be reduced by one (see page 190). In fact, for the equation in (143.3), the exact solution can be obtained.

Equation (143.3) can be integrated with respect to ζ to obtain

$$-cv + \tfrac{1}{2}\sigma v^2 + v_{\zeta\zeta} = A,$$

where A is an arbitrary constant. This last equation, when multiplied by v_ζ, can be integrated again to obtain

$$-\tfrac{1}{2}cv^2 + \tfrac{1}{6}\sigma v^3 + (v_\zeta)^2 = Av + B, \tag{143.4}$$

where B is another arbitrary constant. Equation (143.4) can be solved algebraically for v_ζ and then this first order ordinary differential equation can be integrated in terms of elliptic functions (see Abramowitz and Stegun [1]).

Hence, we have shown that the KdV equation has solitary waves as solution. For a soliton type solution to exist for (143.1), it must be determined that a solution of (143.4) exists that is localized (i.e., differs appreciably from zero only in a bounded region). Finally, to actually show that the KdV has solitons, the interaction of these solitary waves must be investigated. From a much deeper analysis (see, for example, Whitham [9]) it is possible to show that the Korteweg–de Vries equation possesses solitons as solutions. In fact, the KdV equation can have, as its solutions, any number of solitons.

Notes

[1] The technique that we have presented is no more than using similarity variables (see page 424) to obtain a solution of a specific form. Of course, the boundary conditions must admit a traveling wave solution, as well as the equations.

[2] The wave speed (c in the Example) often must be determined as part of the solution. In the above example, it would be determined by the boundary conditions (as would A and B). Typically, in nonlinear problems, the velocity is amplitude dependent.

References

[1] M. Abramowitz and I. A. Stegun, *Handbook of Mathematical Functions*, National Bureau of Standards, Washington, DC, 1964.

[2] M. J. Ablowitz and H. Segur, *Solitons and the Inverse Scattering Transform*, SIAM, Philadelphia, 1981, Chapter 17 (pages 587–607).

[3] F. Calogero and A. Degasperis, *Spectral Transform and Solitons: Tools to Solve and Investigate Nonlinear Evolution Equations*, North–Holland Publishing Co., New York, 1982.

[4] R. K. Dodd, J. C. Eilbeck, and H. C. Morris, *Solitons and Nonlinear Wave Equations*, Academic Press, London, 1982.

[5] P. G. Drazin and R. S. Johnson, *Solitons: An Introduction*, Cambridge University Press, New York, 1989.

[6] W. Eckhaus and A. V. Harten, *The Inverse Scattering Transformation and the Theory of Solitons*, North–Holland Publishing Co., New York, 1981.

[7] G. L. Lamb, *Elements of Soliton Theory*, John Wiley & Sons, New York, 1980.

[8] A. C. Newell, *Solitons in Mathematics and Physics* SIAM, Philadelphia, 1985.

[9] G. B. Whitham, *Linear and Nonlinear Waves*, Wiley Interscience, New York, 1974, Chapter 17 (pages 577–620).

144.   Stochastic Limit Theorems

Applicable to   Linear differential equations that contain a small parameter and a random forcing term of a certain form.

Yields

A Fokker–Planck equation.

Idea

Some equations do not have a "white noise" forcing term and so a Fokker–Planck equation cannot be directly constructed (see page 254). However, it is often true that random forcing terms behave like "white noise" in some asymptotic limit. Hence, in this limit, a Fokker–Planck equation can be constructed.

Procedure

If $F(\mathbf{x}, t, \tau)$ is a "sufficiently random" mean zero function then, as ε tends to zero, the form

$$\frac{1}{\varepsilon} F\left(\mathbf{x}, t, \frac{t}{\varepsilon^2}\right), \tag{144.1}$$

behaves, in a certain sense, like a "white noise" term (see Papanicolaou and Kohler [4]). Using the "white noise" equivalent of (144.1), a Fokker–Planck equation can be obtained in the variables $\{\mathbf{x}, t\}$.

Hence, the prescription is to change a given equation into the form of (144.1) and then obtain and analyze the corresponding Fokker–Planck equation.

Example

Using the geometric optics approximation to the wave equation, the scaled position and velocity of a ray in a weakly random medium satisfy

$$\frac{dx}{dt} = v,$$

$$\frac{dv}{dt} = \frac{1}{\varepsilon} F\left(x, \frac{t}{\varepsilon^2}\right),$$

after a ray has traveled a long distance in the random medium. Here $F(\ )$ is a random function with mean zero (it represents the wave speed perturbation at any point). Assuming a "mixing condition" on F, which is a statement about how random $F(\ )$ is, the theorem in Papanicolaou and Kohler [4] can be used in the limit of ε going to zero.

Using this theorem, it can be shown that the probability density of the solution to (144.1) converges weakly to the solution of the following Fokker–Planck equation

$$\gamma \frac{\partial^2 P}{\partial v^2} - \frac{\partial P}{\partial x} = \frac{\partial P}{\partial t},$$

where the number γ is defined by $\gamma^2 = -\int_0^\infty E[F(0, y)(F(0, 0)]\, dy$, and $E[\cdot]$ is the expectation operator. The details of the derivation are beyond the scope of this book. More details of this example may be found in Kulkarny and White [3].

Notes

[1] There are many different limit theorems that yield a "white noise" limit. For example, Keston and Papanicolaou's paper [1] is concerned with random differential equations of the form

$$\frac{dx}{dt} = \frac{1}{\varepsilon^2} v,$$
$$\frac{dv}{dt} = \frac{1}{\varepsilon} F(x, v).$$

[2] The theorems in Papanicolaou and Kohler [4] and in Keston and Papanicolaou [1] have many technical requirements that must be satisfied. The "mixing condition" requirement has been verified for only a few physical process.

[3] For some limit theorems, the Fokker–Planck formalism can be eliminated completely. For example, in Khas'minskii [2] it is shown that the solution to the problem

$$\frac{dx}{dt} = \varepsilon F(x, t, \omega, \varepsilon), \qquad x(0) = x_0,$$

in an interval of order $O(1/\varepsilon)$, can be uniformly approximated by the solution to the problem $\dfrac{d\bar{x}}{dt} = \varepsilon \overline{F}(\bar{x})$, $\bar{x}(0) = x_0$, where

$$\overline{F}(x) := \lim_{T \to \infty} \frac{1}{T} \int_0^T E[F(x, t, \omega, \varepsilon)]\, dt,$$

if the stochastic process $F(x, t, \omega, \varepsilon)$ satisfies the law of large numbers for fixed x.

[4] Pardoux [5] finds a white noise limit of a partial differential equation.

References

[1] H. Keston and G. Papanicolaou, "A Limit Theorem for Stochastic Acceleration," *Comm. Math. Phys.*, **78**, 1980, pages 19–63.

[2] R. Z. Khas'minskii, "A Limit Theorem for the Solutions of Differential Equations with Random Right-Hand Sides," *Theory Prob. Appl.*, **11**, No. 3, 1966, pages 390–405.

[3] V. A. Kulkarny and B. S. White, "Focussing of Waves in Turbulent Inhomogeneous Media," *Phys. Fluids*, **25**, No. 10, 1982, pages 1770–1784.

[4] G. Papanicolaou and W. Kohler, "Asymptotic Theory of Mixing Stochastic Ordinary Differential Equations," *Comm. Pure Appl. Math*, **27**, 1974, pages 641–668.

[5] E. Pardoux, "Asymptotic Analysis of a Semi-Linear PDE with Wide-Band Noise Disturbances," in L. Arnold and P. Kotelenz (eds.), *Stochastic Space–Time Models and Limit Theorems*, D. Reidel Publishing Co., Boston, 1985, pages 227–242.

[6] C. Van Den Broeck, "Stochastic Limit Theorems: Some Examples from Nonequilibrium Physics," in L. Arnold and P. Kotelenz (eds.), *Stochastic Space-Time Models and Limit Theorems*, D. Reidel Publishing Co., Boston, 1985, pages 179–189.

[7] B. White and J. Franklin, "A Limit Theorem for Stochastic Two-Point Boundary Value Problems of Ordinary Differential Equations," *Comm. Pure Appl. Math*, **32**, 1979, pages 253–276.

145. Taylor Series Solutions

Applicable to   Initial value problems, both ordinary differential equations and partial differential equations.

Yields

An approximation to the solution near a point.

Idea

For an initial value problem, a Taylor series expansion can give an approximate solution.

Procedure

We will illustrate the general procedure on a first order linear ordinary differential equation. Suppose we have the differential equation

$$y'(x) = F(x, y), \tag{145.1}$$

(where $'$ indicates differentiation with respect to x) with the initial condition $y(a) = y_0$, where $F(x, y)$ is a known function. Evaluating (145.1) at $x = a$, we can determine $y'(a) = F(a, y_0)$. Differentiating (145.1) with respect to x, and using the chain rule, results in

$$y''(x) = F_x(x, y) + F_y(x, y)y_x. \tag{145.2}$$

Now equation (145.2) can be evaluated at $x = a$ to explicitly determine

$$y''(a) = F_x(a, y(a)) + F_y(a, y(a))y_x(a)$$
$$= F_x(a, y_0) + F_y(a, y_0)F(a, y_0),$$

where we have used $y'(a) = F(a, y_0)$.

We can continue this process of differentiating (145.1) and evaluating the result to determine the n-th derivative of $y(x)$ at the point $x = a$. The result will only involve the partial derivatives of $F(x, y)$ and the numerical values a and y_0. Knowing these values allows us to construct the Taylor series expansion of $y(x)$ about $x = a$ by use of

$$y(x) = y(a) + \frac{y'(a)}{1!}(x-a)^1 + \frac{y''(a)}{2!}(x-a)^2 + \frac{y'''(a)}{3!}(x-a)^3 + \cdots. \quad (145.3)$$

Example

Suppose we wish to approximate the solution of the nonlinear initial value problem

$$y' = x^2 - y^2, \qquad\qquad (145.4.a\text{--}b)$$
$$y(0) = 1.$$

From (145.4) it is straightforward to compute

$$\begin{aligned}
y'' &= 2x - 2yy', \\
y''' &= 2 - 2(y')^2 - 2yy', \\
y'''' &= -6y'y'' - 2yy''', \qquad\qquad (145.5) \\
&\;\;\vdots
\end{aligned}$$

Using (145.4.b), we evaluate (145.4.a) and then (145.5) sequentially, at $x = 0$, to determine

$$\begin{aligned}
y'(0) &= -1, \\
y''(0) &= 2, \\
y'''(0) &= -4, \qquad\qquad (145.6) \\
y''''(0) &= 20, \\
&\;\;\vdots
\end{aligned}$$

Using the values from (145.6) in (145.3), with $a = 0$, the solution of (145.4) for $y(x)$ near $x = 0$ is given by

$$y = 1 - x + \frac{2}{2!}x^2 - \frac{4}{3!}x^3 + \frac{20}{4!}x^4 + \cdots$$

$$= 1 - x + x^2 - \frac{2}{3}x^3 + \frac{5}{6}x^4 + \cdots.$$

Notes

[1] This method may be applied to higher order equations, and systems of equations.

[2] The method of series solution (see page 342), when used at an ordinary point, also yields a Taylor series solution.

[3] The Taylor series worked out by this method can be used to compute Padé approximates to the solution. These Padé approximates may give information about singularities of the exact solution (see the section on Padé approximants, page 503). Fernández, Arteca, and Castro [3] have developed a different technique for determining the location of singular points by postulating a form of the singularity.

[4] A direct representation of the Taylor series may be obtained by implicit differentiation. We find that the solution to the differential equation $y' = f(t, y)$, with $y(0) = 0$ has the Lie-series representation

$$y(t) = \sum_{n=1}^{\infty} \frac{t^n}{n!} \left[\left(\frac{\partial}{\partial t} + f(t, z) \frac{\partial}{\partial z} \right)^n z \right] \Bigg|_{z=0} \qquad (145.7)$$

See Igumnov [6] for a computationally efficient way to determine $y(t)$ from (145.7) when $f(t, y)$ has a known Taylor series. Finizio and Ladas [4] also have a numerical scheme on pages 293–298.

[5] The numerical technique of analytical continuation (see page 623) combines Taylor series at several different points to approximate the solution of a differential equation in a large region.

[6] A FORTRAN program for solving ordinary differential equations by the use of Taylor series may be found in Chang and Corliss [1].

[7] Taylor's theorem has been generalized in a way in which the general term is a fractional derivative (see Osler [8] for details).

References

[1] Y. F. Chang and G. F. Corliss, "Solving Ordinary Differential Equations Using Taylor Series," *ACM Trans. Math. Software*, **8**, 1982, pages 114–144.

[2] G. Corliss and D. Lowery, "Choosing a Stepsize for Taylor Series Methods for Solving ODE's," *J. Comput. Appl. Math.*, **3**, No. 4, 1977, pages 251–256.

[3] F. M. Fernández, G. A. Arteca, and E. A. Castro, "Singular Points from Taylor Series," *J. Math. Physics*, **28**, No. 2, February 1987, pages 323–329.

[4] N. Finizio and G. Ladas, *Ordinary Differential Equations with Modern Applications*, Wadsworth Publishing Company, Belmont, Calif, 1982, pages 116–120.

[5] C. Hunter and B. Guerrieri, "Deducing the Properties of Singularities of Functions from Their Taylor Series Coefficients," *SIAM J. Appl. Math.*, **39**, No. 2, October 1980, pages 248–263.

[6] V. P. Igumnov, "Representation of Solutions of Differential Equations by Modified Lie Series," *Differential Equations*, **20**, 1984, pages 683–688.

[7] E. Kochavi and R. Segev, "Numerical Solution of Field Problems by Nonconforming Taylor Discretization," *Appl. Math. Modeling*, **15**, March 1991, pages 152–157.

[8] T. J. Osler, "Taylor's Series Generalized for Fractional Derivatives and Applications," *SIAM Review*, **2**, No. 1, February 1971, pages 37–48.

[9] M. Razzaghi and M. Razzaghi, "Solution of Linear Two-Point Boundary Value Problems via Taylor Series," *J. Franklin Inst.*, **326**, No. 4, 1989, pages 511–521.

146. Variational Method: Eigenvalue Approximation

Applicable to   Differential equations with eigenvalues to be determined.

Yields

Estimates for the eigenvalues.

Idea

If we guess approximate eigenfunctions, then we will obtain approximations to the eigenvalues. The "better" we guess the eigenfunctions, the better the estimates of the eigenvalues will be.

Procedure

While the procedure is quite general, we will discuss it in the specific context of a Sturm–Liouville equation. Suppose we have the Sturm–Liouville equation on the interval $[a, b]$

$$L[y] = \frac{d}{dx}\left[p(x)\frac{dy}{dx}\right] - s(x)y = -\lambda r(x)y, \qquad (146.1)$$

with $p(x) > 0$, $s(x) \geq 0$, and $y(a) = y(b) = 0$. If we expand $y(x)$ as

$$y(x) = \sum_{n=1}^{\infty} c_n \phi_n(x), \qquad (146.2)$$

where the $\{\phi_n(x)\}$ are an arbitrary set of complete functions that vanish at $x = a$ and $x = b$, and the $\{c_n\}$ are constants, then the $\{c_n\}$ must satisfy

$$\sum_{n=1}^{\infty} (A_{mn} - \lambda R_{mn}) c_n = 0, \qquad (146.3)$$

for $m = 1, 2, \ldots$, where

$$A_{mn} = \int_a^b [p(x)\phi_m'(x)\phi_n'(x) + s(x)\phi_m(x)\phi_n(x)] \, dx,$$
$$R_{mn} = \int_a^b r(x)\phi_m(x)\phi_n(x) \, dx. \qquad (146.4)$$

Equation (146.3) is obtained by substituting (146.2) into (146.1), multiplying the result by $\phi_m(x)$, integrating with respect to x from a to b, and using integration by parts. If the $\{\phi_n(x)\}$ are the eigenfunctions of the operator in (146.1), then the matrices A and R are diagonal matrices and the eigenvalues $\{\lambda_i\}$ are easily obtained.

If, instead of (146.2), we use the finite sum

$$y(x) = \sum_{n=1}^{N} c_n \psi_n(x),$$

where the $\{\psi_n(x)\}$ are chosen to satisfy the boundary conditions, then (146.3) becomes

$$\sum_{n=1}^{N} \left(\overline{A}_{mn} - \bar{\lambda}\overline{R}_{mn} \right) c_n = 0, \qquad (146.5)$$

for $m = 1, 2, \ldots, N$. In this equation, $\overline{A}$ and $\overline{R}$ are given by (146.4) with $\phi_k(x)$ replaced by $\psi_k(x)$. For (146.5) to have a non-trivial solution, $\bar{\lambda}$ must satisfy

$$|\mathcal{A} - \bar{\lambda}\mathcal{R}| = 0 \qquad (146.6)$$

where $\mathcal{A}$ is the matrix formed out of the $\overline{A}_{mn}$ and $\mathcal{R}$ is the matrix formed out of the $\overline{R}_{mn}$. If the $\{\psi_k(x)\}$ that we have have chosen are "close" to the actual eigenfunctions of (146.1), then the $\{\bar{\lambda}_k\}$ obtained from (146.6) will be "close" to the eigenvalues $\{\lambda_k\}$ of (146.1).

It is always true that the smallest $\bar{\lambda}$ from (146.6) is larger than the smallest λ of (146.1).

Example

Suppose an approximation to the smallest eigenvalues of the Sturm–Liouville system

$$y'' = -\lambda y,$$
$$y(-1) = y(1) = 0,$$

$$(146.7)$$

is desired. Equation (146.7) has the same form as (146.1), with $p(x) = 1$, $s(x) = 0$, $r(x) = 1$, $a = -1$, and $b = 1$. We guess that $y(x)$ can be well approximated by

$$y(x) = c_1(1 - x^2),$$

which is (146.3) with $N = 1$ and $\psi_1(x) = (1 - x^2)$. Using (146.4) we calculate

$$\overline{A}_{11} = \int_{-1}^{1} (-2x)(-2x)\, dx = \frac{8}{3},$$

$$(146.8)$$

$$\overline{R}_{11} = \int_{-1}^{1} (1 - x^2)(1 - x^2)\, dx = \frac{16}{15}.$$

$$(146.9)$$

Using (146.8) and (146.9) in (146.5) yields the eigenvalue equation for $\overline{\lambda}$, $\frac{8}{3} - \frac{16}{15}\overline{\lambda} = 0$, and therefore, $\overline{\lambda} = 2.5$. For this example, it turns out that the smallest eigenvalue is exactly $\lambda = \pi^2/4 \simeq 2.467$, which corresponds to the eigenfunction $\phi(x) = \cos(\pi x/2)$.

Notes

[1] The above example is from Butkov [1].

[2] For the Sturm–Liouville equation in (146.1), it can be shown that

$$\lambda = \frac{\left. (-p y y_x) \right|_a^b + \int_a^b \left(p(y')^2 + sy^2 \right) dx}{\int_a^b ry^2\, dx}.$$

This is known as the *Rayleigh quotient*. This can be used to estimate the lowest eigenvalue since

$$\lambda_1 \leq \min_{u(x)} \left[\frac{\left. (-p u u_x) \right|_a^b + \int_a^b \left\{ p(u')^2 + su^2 \right\} dx}{\int_a^b ru^2\, dx} \right],$$

where λ_1 represents the smallest eigenvalue, and the minimization is taken over all continuous functions that satisfy the boundary conditions associated with (146.1) (but not necessarily the differential equation itself). See Haberman [2] for details.

[3] There are similar relations for for the eigenvalues of partial differential equations, which are also called the Rayleigh quotient. (See Butkov [1] for details.)

For example (see Haberman [2]), for the Helmholtz equation in a bounded region, $\nabla^2 u + \lambda u = 0$ there is the relation

$$\lambda = \frac{-\oint u\nabla u \cdot \mathbf{n}\, ds + \iint\limits_R |\nabla u|^2\, dx\, dy}{\iint\limits_R u^2\, dx\, dy}.$$

References

[1] E. Butkov, *Mathematical Physics*, Addison–Wesley Publishing Co., Reading, MA, 1968, pages 573–586.

[2] R. Haberman, *Elementary Applied Partial Differential Equations*, Prentice–Hall Inc., Englewood Cliffs, NJ, 1983, pages 172–176 and 224–226.

[3] H. F. Weinberger, *Variational Methods for Eigenvalue Approximation*, SIAM, Philadelphia, 1974.

[4] E. Zauderer, *Partial Differential Equations of Applied Mathematics*, John Wiley & Sons, New York, 1983, pages 450–483.

147. Variational Method: Rayleigh–Ritz

Applicable to   Differential equations that come from a variational principle.

Yields

An approximation valid over an interval.

Idea

The variational expression from which a differential equation is derived can be used to approximate the solution.

Procedure

Most equations of mathematical physics and engineering arise from a variational principle (see the section on variational equations, page 88). For example, the first variation of

$$J[u] = \iint_D \left(u_x^2 + u_y^2 + 2uf \right) \, dx \, dy, \tag{147.1}$$

(also known as the Euler–Lagrange equation associated with (147.1)) is given by

$$\delta J = u_{xx} + u_{yy} - f = 0.$$

Hence, the solution to

$$u_{xx} + u_{yy} = f, \qquad \text{in the region } D,$$
$$u = g, \qquad \text{on the boundary of } D,$$

is given by that function $u(x, y)$ that equals g on the boundary and minimizes (147.1).

The Rayleigh–Ritz method is to determine the functional that a differential equation comes from, and then to find an approximate minimum. This is done by choosing a sequence of functions $\{\phi_1, \phi_2, \ldots, \phi_n\}$ and then forming

$$u_N(x, y) = a_1 \phi_1(x, y) + a_2 \phi_2(x, y) + \cdots + a_n \phi_n(x, y), \tag{147.2}$$

where the $\{a_i\}$ are unknown. Of course, the $\{\phi_k\}$ must be chosen in such a way that the boundary conditions are satisfied. Now, the $\{a_i\}$ are chosen in such a way that the functional will be minimized. Specifically, using (147.2) in (147.1) (or the appropriate variational principal), the $\{a_i\}$ are chosen by solving the simultaneous system of equations given by

$$\frac{\partial}{\partial a_i} J[u_N] = 0, \qquad \text{for } i = 1, \ldots, N.$$

This will often be a simultaneous system of polynomial equations.

If the $\{\phi_i\}$ in (147.2) are chosen "well," then u_N will tend to u as $n \to \infty$.

Example 1

Suppose we wish to approximate the solution to the following Poisson equation in the unit square

$$u_{xx} + u_{yy} = \sin \pi x, \qquad \text{for } 0 < x < 1, \quad 0 < y < 1,$$
$$u = 0, \qquad \text{on } x = 0,\ x = 1,\ y = 0,\ y = 1. \qquad (147.3.a\text{--}b)$$

The above equation comes from the variational principle $\delta J = 0$, where

$$J[u] = \int_0^1 \int_0^1 \left(u_x^2 + u_y^2 + 2u \sin \pi x \right) dx\, dy. \qquad (147.4)$$

We choose to approximate $u(x, y)$ by a linear combination of

$$\phi_1(x, y) = x(1 - x)y(1 - y),$$
$$\phi_2(x, y) = x^2(1 - x)y(1 - y),$$
$$\phi_3(x, y) = x(1 - x)y^2(1 - y).$$

Note that each of the $\{\phi_i\}$ vanish on the boundary of the square, and so u_3 will also (as (147.3.b) requires).

Using (147.2) (with $N = 3$) in (147.4) results in the minimization of the function

$$\left[24\pi^3 a_3^2 + \left(35\pi^3 a_2 + 70\pi^3 a_1 + 2100\right) + 24\pi^3 a_2^2 \right.$$
$$\left. + \left(70\pi^3 a_1 + 2100\right) a_2 + 70\pi^3 a_1^2 + 4200a_1 \right]/3150\pi^3. \qquad (147.5)$$

Differentiating (147.5) with respect to each of a_1, a_2, and a_3 results in the linear system of equations

$$\begin{pmatrix} 140\pi^3 & 70\pi^3 & 70\pi^3 \\ 70\pi^3 & 48\pi^3 & 35\pi^3 \\ 70\pi^3 & 35\pi^3 & 48\pi^3 \end{pmatrix} \begin{pmatrix} a_1 \\ a_2 \\ a_3 \end{pmatrix} = \begin{pmatrix} -4200 \\ -2100 \\ -2100 \end{pmatrix},$$

with the solution: $\{a_1 = -\dfrac{30}{\pi^3},\ a_2 = 0,\ a_3 = 0\}$. Using these values in (147.2) yields an approximation to the solution of (147.3).

Note that the exact solution to the problem in (147.3) can be found by finite Fourier transforms (see page 293) to be

$$u(x, y) = \frac{\sin \pi x}{\pi^2 \sinh \pi} \left[\sinh \pi y + \sinh(\pi(1 - y)) - \sinh \pi \right]. \qquad (147.6)$$

Figure 147 has a comparison of the exact solution (147.6) and the approximate solution found above. This figure compares the values of $u(.1, y)$ and $u_3(.1, y)$ as y varies from 0 to 1.

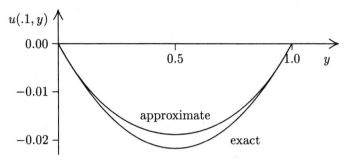

Figure 147. A comparison of the exact solution in (147.6) and the approximate solution in (147.2), when $x = .1$.

Example 2

A variation of this method, due to Kantorovich, is to choose the $\{\phi_k\}$ to depend only on y and to allow the $\{a_k\}$ to depend on x. For example, to approximate the solution of the Poisson equation

$$u_{xx} + u_{yy} = -2, \quad \text{for } 0 < x < 1, \quad 0 < y < 1,$$
$$u = 0, \quad \text{on } x = 0, \ x = 1, \ y = -1, \ y = 1, \quad (147.7.a\text{–}b)$$

which corresponds to the first variation of

$$J[u] = \int_0^1 \int_{-1}^1 \left(u_x^2 + u_y^2 - 4u\right) \, dx \, dy, \qquad (147.8)$$

we choose

$$u(x,y) \approx v(x,y) = f(x)(y^2 - 1). \qquad (147.9)$$

where $f(x)$ is unknown. Using (147.9) in (147.8) results in

$$J[v] = \int_0^1 \left(\frac{16}{15}f'^2 + \frac{8}{3}f^2 + \frac{16}{3}\right) \, dx, \qquad (147.10)$$

which must now be minimized. The first variation of (147.10) yields the following differential equation for $f(x)$

$$f'' - \tfrac{5}{2}f = \tfrac{5}{2}. \qquad (147.11)$$

The function $f(x)$ must satisfy $f(0) = f(1) = 0$ for (147.7.b) to be satisfied. Solving (147.11) with these boundary conditions results in

$$f(x) = -1 + \cosh \alpha x + \left(\frac{1 - \cosh \alpha}{\sinh \alpha}\right) \sinh \alpha x, \qquad (147.12)$$

where $\alpha = \sqrt{10}/2$. Combining (147.12) with (147.9) results in the final approximation to (147.7).

Notes

[1] Example 2 is from Casti and Kalaba [2].

[2] The Rayleigh–Ritz method also works for ordinary differential equations. For example, the variational principle corresponding to $J[u] = \int_0^1 [(y')^2 + y^2] \, dx$ is $\delta J = y'' + y = 0$.

[3] This method is an example of a *weighted residual method*, see page 699.

[4] This technique is often implemented numerically.

References

[1] E. Butkov, *Mathematical Physics*, Addison–Wesley Publishing Co., Reading, MA, 1968, pages 573–586.

[2] J. Casti and R. Kalaba, *Imbedding Methods in Applied Mathematics*, Addison–Wesley Publishing Co., Reading, MA, 1973, pages 68–69.

[3] S. J. Farlow, *Partial Differential Equations for Scientists and Engineers*, John Wiley & Sons, New York, 1982, Lesson 45 (pages 362–369).

[4] L. V. Kantorovich and V. I. Krylov, *Approximate Methods of Higher Analysis*, Interscience Publishers, New York, 1958, Chapter 4 (pages 241–357).

[5] S. G. Mikhlin and K. L. Smolitskiy, *Approximate Methods for Solutions of Differential and Integral Equations*, American Elsevier Publishing Company, New York, 1967, Chapter 3 (pages 147–269).

[6] I. Stakgold, *Green's Functions and Boundary Value Problems*, John Wiley & Sons, New York, 1979, pages 539–544.

[7] E. Zauderer, *Partial Differential Equations of Applied Mathematics*, John Wiley & Sons, New York, 1983, pages 470–483.

148. WKB Method

Applicable to   Linear differential equations.

Yields

A global approximation.

Idea

The solution of an ordinary differential equation near an irregular singular point is often in the form of an exponential. Conversely, an exponential will often be a good approximation to an ordinary differential equation (even one without an irregular singular point.)

Procedure

If a given ordinary differential equation does not have a small parameter in it, multiply the highest order derivative term by a "small" parameter ε^2. This turns the equation into a singularly perturbed differential equation. Later, we will set ε equal to one, and recover the original equation.

Given a singularly perturbed linear ordinary differential equation (of any order), look for a solution of the form

$$y(x) \sim \exp\left[\frac{1}{\delta} \sum_{n=0}^{\infty} \delta^n S_n(x)\right], \tag{148.1}$$

where we consider $\delta = \delta(\varepsilon)$ to be a small number.

The technique is to use (148.1) in the original equation and then apply dominant balance (see page 443) to determine a differential equation for $S_0(x)$. Solve this equation for $S_0(x)$. Then, using this solution for $S_0(x)$, apply dominate balance again to determine the next largest term. This will be a differential equation for the unknown $S_1(x)$. Solve this equation, and then iterate this procedure to determine several of the $\{S_i(x)\}$.

In order for the WKB approximation to be valid on an interval, we require that $\delta^n S_{n+1} \ll 1$ as $\delta \to 0$ and that $S_{n+1}(x)/S_n(x)$ be a bounded function of x on the given interval (for $n = 1, 2, \ldots$). If these do not hold, the expansion procedure is not valid. Note that if we have $\delta = 1$, the constraints on $\{S_i\}$ become constraints on the interval where the approximation is valid.

Special Case

For the singularly perturbed linear second order ordinary differential equation

$$\varepsilon^2 y'' = Q(x)y, \tag{148.2}$$

with $Q(x) \neq 0$, we use (148.1) in (148.2) to determine

$$\frac{\varepsilon^2}{\delta^2}(S_0')^2 + \frac{2\varepsilon^2}{\delta} S_0' S_1' + \frac{\varepsilon^2}{\delta} S_0'' + \cdots = Q(x), \tag{148.3}$$

where the exponential term common to both sides has been factored out. The largest terms in (148.3) are $(S_0')^2 \varepsilon^2/\delta^2$ and $Q(x)$. Since $Q(x)$ is presumed to be of order one, we must have $\delta = \varepsilon$ and $(S_0')^2 = Q(x)$, or

$$S_0(x) = \pm \int^x \sqrt{Q(t)}\, dt. \tag{148.4}$$

Using $\delta = \varepsilon$ and (148.4) in (148.3), and applying dominant balance again, yields a first order differential equation for $S_1(x)$

$$2S_0'S_1' + S_0'' = 0,$$

which can be integrated directly to yield

$$S_1(x) = -\tfrac{1}{4}\log Q(x). \tag{148.5}$$

Using (148.4) and (148.5) in (148.1), we determine the leading order approximation to the solution of (148.1) to be

$$y(x) \sim C_1\,[Q(x)]^{-1/4}\,\exp\!\left(\frac{1}{\varepsilon}\int^x \sqrt{Q(t)}\,dt\right)$$
$$+ C_2\,[Q(x)]^{-1/4}\,\exp\!\left(-\frac{1}{\varepsilon}\int^x \sqrt{Q(t)}\,dt\right), \tag{148.6}$$

for some constants C_1 and C_2. If a higher order approximation was desired, it is easy to derive that

$$S_2(x) = \pm\int^x \left[\frac{Q''}{8Q^{3/2}} - \frac{5(Q')^2}{32Q^{5/2}}\right] dt,$$
$$S_3(x) = \frac{Q''}{16Q^2} + \frac{5(Q')^2}{64Q^3},$$

since all of the equations for the higher order $\{S_i(x)\}$ are of first order.

In Marić and M. Tomić [11] it is shown that (148.6) is the correct asymptotic result if $\int^\infty \sqrt{Q}\,dt = \infty$ and $\int^\infty Q'^2 Q^{-5/2}\,dt < \infty$.

Example

Given the Airy equation

$$y'' = xy, \tag{148.7}$$

we introduce a small parameter ε^2 and write (148.7) as $\varepsilon^2 y'' = xy$. This is now an equation of the same form as (148.2), with $Q(x) = x$. Hence, the approximation in (148.6) (with $\varepsilon = 1$) yields

$$y(x) \sim C_1 x^{-1/4}\exp\!\left(\tfrac{2}{3}x^{3/2}\right) + C_2 x^{-1/4}\exp\!\left(-\tfrac{2}{3}x^{3/2}\right). \tag{148.8}$$

If we had included the $S_2(x)$ term, the approximation would be

$$y(x) \sim C_1 x^{-1/4}\exp\!\left(\tfrac{2}{3}x^{3/2}\right)\left(1 + \tfrac{5}{48}x^{-3/2}\right)$$
$$+ C_2 x^{-1/4}\exp\!\left(-\tfrac{2}{3}x^{3/2}\right)\left(1 - \tfrac{5}{48}x^{-3/2}\right). \tag{148.9}$$

In both (148.8) and (148.9) the approximations are valid only as $x \to \infty$.

Notes

[1] WKB stands for G. Wentzel, H. Kramers, and L. Brillouin. This method is also sometimes called the WKBJ method, or the Jeffreys method.

[2] The eigenvalue problem $z'' + \lambda^2 V(x)z = 0$ with $y(0) = y(l) = 0$ can also be analyzed by the WKB method. Using (148.6) we can write the approximate solution as $z(x) = A(x)\sin\left(-\frac{1}{\lambda}\int^x \sqrt{V(t)}\,dt + \phi(x)\right)$. The eigenvalues $\{\lambda_i\}$ are determined by where the oscillatory function vanishes. To leading order it can be shown that the eigenvalues satisfy $\lambda_n = \frac{n\pi}{L}$, as $n \to \infty$, where $L = \int_0^l \sqrt{V(t)}\,dt$. A new correction to this formula is in Lindblom and Robiscoe [8].

[3] Ludwig [9] illustrates how the WKB method may be applied to partial differential equations.

[4] The WKB approximation results in an asymptotic series. Hence, as more terms are taken in (148.1), the result may diverge.

[5] WKB is a singular perturbation technique, and boundary layer theory (see page 510) may be derived from it.

[6] The approximation $y(x) \simeq \exp\left[\dfrac{S_0(x)}{\delta}\right]$ is often called the geometrical optics approximation. The approximation $y(x) \simeq \exp\left[\dfrac{S_0(x)}{\delta} + S_1(x)\right]$ is often called the physical optics approximation.

[7] For the linear ordinary differential equation of degree n $\varepsilon\dfrac{d^n y}{dx^n} = Q(x)y$, the physical optics approximation is given by $y(x) \simeq \exp\left[\dfrac{S_0(x)}{\delta} + S_1(x)\right]$ with $\delta = \varepsilon^{1/n}$ and

$$S_0 = \omega \int^x [Q(x)]^{1/n}\,dt, \qquad S_1 = \frac{1-n}{2n}\log Q(x),$$

where ω is any of the n-th roots of unity (i.e., $\omega^n = 1$).

[8] In regions where $Q(x)$ does not vanish, the classical WKB solutions of (148.2) in (148.6) are valid. Points where $Q(x)$ is equal to zero are called turning points or transition points, the solutions in (148.6) are not valid at these points. However, the *Langer connection formula* shows how the solution on each side of a turning point may be connected.

Consider (148.2) when $Q(x)$ has a single, simple zero at $x = 0$, and is monotonically increasing everywhere. We presume the boundary condition $y(\infty) = 0$, to avoid the exponentially growing solution in (148.6) when $x \to \infty$. Consider a region that contains a turning point. Dividing this region into three smaller regions (with the turning point in the center region), asymptotic approximation may be obtained in each region. (Use WKB in the two outer regions, linearize $Q(x)$ in the center region and write the answer in terms of Airy functions). By appropriate matching (see page

510), the arbitrary constants in these three solutions can be related. Hence, a uniformly valid approximation is given by:

$$y_{\text{unif}}(x) = C S_0^{1/6} Q(x)^{-1/4} \, \text{Ai} \left[\left(\frac{3}{2\varepsilon} S_0(x) \right)^{2/3} \right]$$

where $S_0(x) = \int_0^x \sqrt{Q(t)} \, dt$ and C is an arbitrary constant.

Many extensions to this simple formula have been found. The ordinary differential equations considered can be of higher order, there can be multiple turning points, and the turning point need not be simple. Wazwaz [17] considers a singular perturbation problem for a second order ordinary differential equation with two interior points of second order.

[9] Note that WKB approximations to the two linearly independent solutions to $\varepsilon y'' + a(x)y' + b(x)y = 0$ have the form

$$y_1(x) \simeq c_1 \exp \left[-\int^x \frac{b(t)}{a(t)} \, dt \right],$$

$$y_2(x) \simeq \frac{c_2}{a(x)} \exp \left[\int^x \frac{b(t)}{a(t)} \, dt - \frac{1}{\varepsilon} \int^x a(t) \, dt \right],$$

as $\varepsilon \to 0^+$. See Example 4 in section 10.1 of Bender and Orszag.

[10] Fedoryuk [4] considers the equation $\varepsilon y'' + f(x, y) = 0$.

References

[1] C. M. Bender and S. A. Orszag, *Advanced Mathematical Methods for Scientists and Engineers*, McGraw–Hill, New York, 1978, Chapter 10 (pages 484–543).

[2] G. F. Carrier, M. Krook, and C. E. Pearson, *Functions of a Complex Variable*, McGraw–Hill Book Company, New York, 1966.

[3] P. A. Farrell, "Sufficient Conditions for the Uniform Convergence of a Difference Scheme for a Singularly Perturbed Turning Point Problem," *SIAM J. Numer. Anal.*, **25**, No. 3, June 1988, pages 618–643.

[4] M. V. Fedoryuk, "The WKB-Method for a Non-Linear Equation of the Second Order," *U.S.S.R. Comput. Maths. Math. Phys.*, **26**, No. 1, 1986, pages 121–128.

[5] S. Giler, "Generalised WKBJ Formulae," *J. Phys. A: Math. Gen.*, **21**, 1988, pages 909–930.

[6] R. N. Kesarwani and Y. P. Varshni, "Five-Term WKBJ Approximation," *J. Math. Physics*, **21**, 1980, pages 90–92.

[7] R. Langer, "The Asymptotic Solutions of Certain Linear Differential Equations of the Second Order," *Trans. Amer. Math. Soc.*, **36**, 1934, pages 90–106.

[8] L. Lindblom and R. T. Robiscoe, "Improving the Accuracy of WKB Eigenvalues," *J. Math. Physics*, **32**, No. 5, May 1991, pages 1254–1258.

[9] D. Ludwig, "Persistence of Dynamical Systems Under Random Perturbations," *SIAM Review*, **17**, No. 4, October 1975, pages 605–640.

[10] R. Lynn and J. B. Keller, "Uniform Asymptotic Solutions of Second-Order Linear Ordinary Differential Equations with Turning Points," *Comm. Pure Appl. Math*, **23**, 1970, pages 379–408.

[11] V. Marić and M. Tomić, "On Liouville–Green (WKB) Approximation for Second Order Linear Differential Equations,;; *Differential Integral Equations*, **1**, No. 3, 1988, pages 299–304.

[12] J. McHugh, "An Historical Survey of Ordinary Linear Differential Equations with a Large Parameter and Turning Points," *Arch. Hist. Exact. Sci.*, **7**, 1971, pages 277–324.

[13] M. El Sawi, "On the WKBJ Approximation," *J. Math. Physics*, **28**, No. 3, March 1987, pages 556–558.

[14] C. R. Steele, "Applications of the WKB method in Solid Mechanics," *Mechanics Today*, **3**, 1976, pages 243–295.

[15] J. G. Taylor, "Improved Error Bounds for the Liouville–Green (or WKB) Approximation," *J. Math. Anal. Appl.*, **85**, 1982, pages 79–89.

[16] W. Wasow, *Linear Turning Point Theory*, Springer–Verlag, New York, 1985.

[17] A.-M. Wazwaz, "Two Turning Points of Second Order," *SIAM J. Appl. Math.*, **50**, No. 3, June 1990, pages 883–892.

[18] B. Willner and L. A. Rubenfeld, "Uniform Asymptotic Solutions for a Linear Ordinary Differential Equation with one μ-th Order Turning Point: Analytic Theory," *Comm. Pure Appl. Math*, **26**, 1976, pages 343–367.

IV.A

Numerical Methods: Concepts

149. Introduction to Numerical Methods

Numerical analysis is a rapidly growing field, with new techniques being developed constantly. Presented in the last section of this book are some of the more commonly used methods.

The section has been separated into three parts

[1] Introductory material about numerical methods.

[2] Methods which can be used for ordinary differential equations and, sometimes, also partial differential equations. When a method in this part can be used for a partial differential equation, there is a star (∗) alongside the method number.

[3] Methods which can only be used for partial differential equations.

For some of the numerical methods presented in this section, a FORTRAN computer program has been given, when a short program could be written. None of the codes have been optimized for speed. To economize on space, many of the comments that would normally appear in a well-documented computer code have been removed. When a FORTRAN

computer code is given, the output is also indicated. These codes were
executed using FORTRAN 77.

Below are some useful comments when solving differential equations
numerically.

[1] Use prepared software packages whenever possible. Numerical codes
are available for solving nearly any type of non–stiff ordinary differential equation. See page 570.

[2] When writing a computer program, always test it on problems for
which you know the solution, either analytically or from a different,
reliable computer code.

[3] When choosing a numerical scheme to approximate the solution to a
differential equation, it is useful to balance the roundoff error with the
truncation error of the machine being used. A higher order method
will not give more accurate answers if the major component of the
error is due to roundoff. Likewise, performing calculations in "double
precision" will not give more accurate answers if the major component
of the error is due to the discretization scheme.

[4] Perform numerical calculations with as many digits of precision as is
reasonable for efficient execution. Single precision arithmetic on the
CYBER, which uses 64 bits, is usually sufficient. On some of the IBM
computers, however, double precision is required to obtain the same
accuracy.

[5] The standard way to determine if a numerical scheme is implemented
correctly and the mesh sizes are small enough to justify the *a priori*
error estimates is to reduce the size of the mesh and re-run the calculation. The resulting *a posteriori* error estimates should agree with
the *a priori* error estimates.

[6] As a rule of thumb, to calculate a first derivative by forward differences,
the roundoff error and the truncation error will be approximately
equal (and so accuracy will be high) if the difference in values used
is the square root of the number of significant digits. For example,
if your computer is working with 20 decimal digits of precision, then
an accurate numerical approximation to the derivative of $y(t)$ will be
obtained by $[y(t) - y(t + \Delta t)]/\Delta t$ for $\Delta t \simeq 10^{-10}$.

[7] Note that several of the methods described in earlier parts of this book
may be readily implemented numerically. For some of those methods,
references have been given that refer to numerical implementations.
No mention of those methods is made in this section.

[8] Listed below are, in the author's opinion, the most useful methods
appearing in this last section. These are the methods that might be
tried first, when a numerical approximation is required.

Most Useful Methods for ODEs
- · Boundary Value Problems: Box Method
- · Boundary Value Problems: Shooting Method*
- · Continuation Method*
- · Euler's Forward Method
- · Finite Element Method*
- · Predictor–Corrector Methods
- · Runge–Kutta Methods
- · Stiff Equations*
- · Weighted Residual Methods*

Most Useful Methods for PDEs
- · Continuation Method*
- · Finite Element Method*
- · Weighted Residual Methods*
- · Elliptic Equations: Finite Differences
- · Elliptic Equations: Relaxation
- · Hyperbolic Equations: Method of Characteristics
- · Hyperbolic Equations: Finite Differences
- · Method of Lines
- · Parabolic Equations: Implicit Method
- · Pseudo-Spectral Method

150. Definition of Terms for Numerical Methods

A-stable   A linear multistep method is A-stable if all solutions of the difference equation generated by the application of this method to the scalar test equation, $y' = \lambda y$, tend to zero as $x \to \infty$ for all complex λ with Re $\lambda < 0$ and for all fixed step sizes h with $h > 0$. Note that an explicit multistep method cannot be A-stable.

Computational molecule   A computational molecule is a pictorial representation of a finite difference scheme for a partial differential equation in two independent variables. In such a figure, the circles indicate which points are related by a difference scheme; the value being determined by the difference scheme is often shown shaded. For example, the computational molecule for the so-called "five-point star" approximation to the Laplacian,

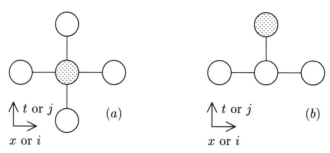

Figure 150. Computational molecules for two different approximations.

$\nabla^2 u_{i,j} \simeq \frac{1}{4}\left(u_{i+1,j} + u_{i,j+1} + u_{i-1,j} + u_{i,j-1}\right)$, is shown in Figure 150.a. The computational molecule for the following explicit finite difference approximation to $u_t = u_{xx}$

$$\frac{u_{i+1,j} - u_{i,j}}{\Delta t} = \frac{u_{i,j+1} - 2u_{i,j} + u_{i,j-1}}{(\Delta x)^2},$$

is shown in Figure 150.b.

Consistency of a finite difference scheme   A method is consistent if the truncation errors tend to zero as the mesh is refined (i.e., as the characteristic scales in the mesh $\{\Delta x,\ \Delta t,\ \ldots\}$ tend to zero). There are two types of consistency:

> **Conditionally consistent** If the truncation errors only tend to zero if $\{\Delta x,\ \Delta t,\ \ldots\}$ tend to zero in a certain way. For example, it may be required that $(\Delta x)^2 < \Delta t$.

> **Unconditionally consistent** If the truncation errors tend to zero no matter how $\{\Delta x,\ \Delta t,\ \ldots\}$, tend to zero.

Conservative scheme   A conservative numerical scheme is one in which the "total energy" described by the differential system is conserved during the integration of the system.

Difference scheme   A difference scheme is an approximation of a derivative term at a point by a collection of values near the point.

> **Centered scheme** A centered scheme is symmetric about the point at which the derivative is being approximated. For example, $y'(x) \simeq \dfrac{y(x+h) - y(x-h)}{2h}$, when $h \ll 1$.

> **One-sided scheme** A one-sided scheme only uses values from one side of the point at which a derivative is being approximated. Examples are forward and backward difference schemes.

Forward difference scheme A forward difference scheme is a one-sided difference scheme that uses points "ahead" of the point that is being approximated. For example, $y'(x) \simeq \dfrac{y(x+h) - y(x)}{h}$, when $h \ll 1$.

Backward difference scheme A backward difference scheme is a one-sided difference scheme that uses points "behind" the point that is being approximated. For example, $y'(x) \simeq \dfrac{y(x) - y(x-h)}{h}$, when $h \ll 1$.

Explicit method   An explicit method is one for which there is an explicit formula, at a point, for the value of the unknown terms appearing in the differential equation.

Grid   A grid is a set of points, called *mesh points*, on which the solution of a differential equation is approximated. If the points are uniformly spaced, then we have a *uniform grid*; otherwise we have a *non-uniform grid*. See page 606.

Implicit method   An implicit method is one for which there is not an explicit formula, at a point, for the value of the unknown terms appearing in the differential equation. Generally a nonlinear algebraic equation must be solved to determine the value at a given point.

Mesh   See Grid.

Order of a numerical method   See page 573.

Step size   See page 573.

Stiff equations   Stiff equations are differential equations that are ill-posed in a computational sense. There are many different definitions of stiffness, two common ones are

(A) A system of differential equations is said to be stiff on the interval $[0, T]$ if there exists a component of a solution of the system that has a variation on $[0, T]$ that is large compared with $1/T$.

(B) A system is stiff if there exists more than one scale, with a great difference in size, on which the solution evolves. For instance, the system of differential equations $\mathbf{y}' = \mathbf{A}\mathbf{y}$ (where $\mathbf{A}$ is a constant matrix with eigenvalues $\lambda_i(\mathbf{A})$) is stiff if $\max_i |\lambda_i(\mathbf{A})| \gg \min_i |\lambda_i(\mathbf{A})|$.

Truncation error   See page 573.

151. Available Software

Applicable to   Differential equations that are to be approximated numerically.

Idea

When numerically approximating the solution to a differential equation, it is best to use commercially available software whenever possible. The routines commonly available for ordinary differential equations are adequate for nearly all types of problems. The routines commonly available for partial differential equations are not as well developed. For linear problems with no singularities, however, the available software is very good.

There are a multitude of commercially available computer libraries and isolated computer routines available. A taxonomy for differential equation software has been developed as part of the GAMS project at the National Institute of Standards and Technology [6]. GAMS [5] has both a taxonomy of computer routines and a listing of some available software. Excerpts from GAMS may be found starting on page 586.

Since good software is readily available, we paraphrase the admonition that Byrne and Hindmarsh [8] give:

> ...if you are using a 10-line solver for differential equations ...you should consider using one of the programs referenced in this section. There is now commercially available "software" for differential equations with no error control, a user-specified step size, and no warning messages. We advise against using such programs, even on a small computer. The reasons are straightforward. For all but trivial problems, such programs cannot be sufficiently reliable for accurate computational results.

When using a prepared software package, it is always useful to test the package on problems similar to the one that you will use the package for. There are many collections of test problems for this purpose.

For example, Rice and Boisvert [20] describe 56 elliptic partial differential equation test problems defined on rectangular regions. Most of these problems involve some parameters; by selecting certain values for those parameters, 189 specific problems are identified. These problems are classified with respect to operator type (Poisson, Helmholtz, self–adjoint, constant coefficients, general), boundary conditions (Dirichlet, Neumann, mixed), and features of the solution (entire, analytic, singular, peak, oscillatory, boundary layer, wave front, singularities, irregular, discontinuities, computationally complex).

Notes

[1] Given a new problem to solve numerically, it is often attractive to design new software for this class of problem. However, it is usually more efficient to transform the problem and use well-tested codes. See, for example, Shampine and Zhang [22].

[2] Addison *et al.* [2] present a decision tree to assist in the process of selecting an appropriate algorithm for the numerical solution of initial value ordinary differential equations. The decision tree can be used in an interactive manner. Where possible, the recommended software routines are in maintained libraries that have been extensively tested. Addison *et al.* [3] contains a decision tree for boundary value problems.

[3] Periodically there are reviews in the literature of software applicable to a specific type of differential equation. See the references.

[4] The books by Press *et al.* [19], contain collections of FORTRAN, PASCAL, and C codes for both ordinary differential equations and partial differential equations.

[5] Some common FORTRAN routines (other than those listed in GAMS [5]) include: DASSL, DISPL1, LSODE, ODEPACK. These routines are available through the National Energy Software Center, Argonne Laboratories, Argonne, Illinois 60429.

[6] Many scientific software routines, including those for differential equations, may be obtained for free (via electronic mail) from a variety of computer networks. To obtain instructions on how to obtain this software, send the mail message "send index" to the following Internet or uucp addresses:

> netlib@research.att.com
> uunet!research!netlib

See the article by Dongarra and Grosse [12] for details.

[7] Even though it is possible, using spreadsheet programs to numerically approximate differential equations is *not* recommended; see Enloe [14].

[8] Software for small computers is summarized in Penn [18] and Teles *et al.* [23].

References

[1] J. C. Adams, "Mudpack — Multigrid Portable Fortran Software for the Efficient Solution of Linear Partial Differential Equations," *Appl. Math. and Comp.*, **34**, No. 2, 1989, pages 113–146.

[2] C. A. Addison, W. H. Enright, P. W. Gaffney, I. Gladwell, and P. M. Hanson, "A Decision Tree for the Numerical Solution of Initial Value Ordinary Differential Equations," *ACM Trans. Math. Software*, **17**, No. 1, March 1991, pages 1–10.

[3] C. A. Addison, W. H. Enright, P. W. Gaffney, I. Gladwell, and P. M. Hanson, "A Decision Tree for the Numerical Solution of Boundary Value Ordinary Differential Equations," SMU Math Report 89-7, Southern Methodist University.

[4] R. E. Bank, *PLTMG: A Software Package for Solving Elliptic Partial Differential Equations*, SIAM, Philadelphia, 1990.

[5] R. F. Boisvert, S. E. Howe, D. K. Kahaner, and J. L. Springmann, *Guide to Available Mathematical Software*, NISTIR 90-4237, Center for Computing and Applied Mathematics, National Institute of Standards and Technology, Gaithersburg, MD 20899, March 1990.

[6] R. F. Boisvert, S. E. Howe, and D. K. Kahaner, "GAMS: A Framework for the Management of Scientific Software," *ACM Trans. Math. Software*, **11**, No. 4, December 1985, pages 313–355.

[7] R. F. Boisvert and R. A. Sweet, "Mathematical Software for Elliptic Boundary Value Problems," in W. R. Cowell (ed.), *Sources and Development of Mathematical Software*, Prentice–Hall Inc., Englewood Cliffs, NJ, 1984, Chapter 9 (pages 200–263).

[8] G. D. Byrne and A. C. Hindmarsh, "Stiff ODE Solvers: A Review of Current and Coming Attractions," *J. Comput. Physics*, **70**, 1987, pages 1–62.

[9] B. Childs, M. Scott, J. W. Daniel, E. Denman, and P. Nelson (eds.), *Codes for Boundary-Value Problems in Ordinary Differential Equations*, Springer–Verlag, New York, 1979.

[10] L. M. Delves, A. McKerrell, and S. A. Peters, "Performance of GEM2 on the ELLPACK Problem Population," *Int. J. Num. Methods Eng.*, **23**, 1986, pages 229–238.

[11] P. M. Dew and J. E. Walsh, "A Set of Library Routines for Solving Parabolic Equations in One Space Variable," *ACM Trans. Math. Software*, **7**, No. 3, September 1981, pages 295–314.

[12] J. J. Dongarra and E. Grosse, "Distribution of Mathematical Software Via Electronic Mail," *Comm. of the ACM*, **30**, No. 5, 1987, pages 403–407.

[13] W. R. Dyksen and C. J. Ribbens, "Interactive ELLPACK: An Interactive Problem-Solving Environment for Elliptic Partial Differential Equations," *ACM Trans. Math. Software*, **13**, No. 2, June 1987, pages 113–132.

[14] C. L. Enloe, "Solving Coupled, Nonlinear Differential Equations with Commercial Spreadsheets," *Computers in Physics*, Jan/Feb 1989, pages 75–76.

[15] P. W. Gaffney, "A Performance Evaluation of Some FORTRAN Subroutines for the Solution of Stiff Oscillatory Ordinary Differential Equations," *ACM Trans. Math. Software*, **10**, No. 1, March 1984, pages 58–72.

[16] M. Machura and R. A. Sweet, "A Survey of Software for Partial Differential Equations," *ACM Trans. Math. Software*, **6**, No. 4, 1980, pages 461–488.

[17] D. K. Melgaard and R. F. Sincovec, "General Software for Two-Dimensional Nonlinear Partial Differential Equations," *ACM Trans. Math. Software*, **7**, No. 1, March 1981, pages 106–125.

[18] H. L. Penn, "A Review of Differential Equations Software," *Collegiate Microcomputer*, **6**, 1988, pages 33–42.

[19] W. H. Press, B. P. Flannery, S. Teukolsky, and W. T. Vetterling, *Numerical Recipes*, Cambridge University Press, New York, 1986.

[20] J. R. Rice and R. F. Boisvert, *Solving Elliptic Problems Using ELLPACK*, Springer–Verlag, New York, 1985.

[21] L. F. Shampine and H. A. Watts, "Software for Ordinary Differential Equations," in W. R. Cowell (ed.), *Sources and Development of Mathematical Software*, Prentice–Hall Inc., Englewood Cliffs, NJ, 1984, Chapter 6 (pages 113–133).

[22] L. F. Shampine and W. Zhang, "Efficient Integration of Ordinary Differential Equations by Transformations," *Comp. & Maths. with Appls.*, **15**, No. 3, 1988, pages 213–220.

[23] E. Teles, H. L. Penn, and J. Wilkin, "ODE Software for the IBM PC", *College Math. J.*, **21**, No. 3, 1990, pages 242–245, and "ODE Software for the Macintosh", *College Math. J.*, **21**, No. 4, 1990, pages 330–332.

152.   Finite Difference Methodology

Applicable to   Differential equations.

Yields

A finite difference scheme that can be used to numerically approximate a given differential equation.

Procedure

For the first order ordinary differential equation $y' = f(x, y)$ consider the general multistep (or k-step) method

$$N[v_n, v_{n+1}, \ldots, v_{n+k}] := \sum_{j=0}^{k} \alpha_j v_{n+j} - h \sum_{j=0}^{k} \beta_j f(x_{n+j}, v_{n+j}) = 0, \quad (152.1)$$

where $\alpha_0 \neq 0$, $n = k, k+1, \ldots$, and v_n is an approximation to $y(x_n)$ (where $x_n = nh$ and h is a small number called the *step size*). We presume the constants $\{\alpha_i\}$ and $\{\beta_i\}$ are known.

If $\beta_0 \neq 0$, then the scheme is an *implicit* difference method. If $\beta_0 = 0$, then the scheme is an *explicit* difference method. For explicit methods, equation (152.1) can be solved for v_n in terms of the other quantities in equation (152.1).

The exact solution to the equation $y' = f(x, y)$ will *not*, in general, satisfy $N[y_n, y_{n+1}, \ldots, y_{n+k}] = 0$ (here, $y_n = y(x_n)$). If $h \ll 1$, then a Taylor series can be employed to show that

$$y_{n+j} = y_n + jhy_n' + \frac{(jh)^2}{2} y_n'' + \cdots.$$

Using this expansion, a Taylor series can be taken of $N[y_n, y_{n+1}, \ldots, y_{n+k}]$ to obtain

$$
\begin{aligned}
N[y_n, y_{n+1}, \ldots, y_{n+k}] &= \sum_{j=0}^{k} \alpha_j y_{n-j} - h \sum_{j=0}^{k} \beta_j f(x_{n-j}, y_{n-j}) \\
&= h^{p+1} R_n + O(h^{p+2}),
\end{aligned}
\tag{152.2}
$$

for some numbers p and R_n.

If $p \geq 1$, then the method is said to be *consistent*. If a method is consistent, then p is called the *order of the method*. We say that "the method is p-th order accurate." The term $h^{p+1} R_n$ is called the *truncation error*. A theorem of numerical analysis states that there exist methods of order $p = 2k$.

The first and second *characteristic polynomials* of the method in (152.1) are defined as $\rho(x)$ and $\sigma(x)$, where

$$
\rho(x) = \sum_{j=0}^{k} \alpha_j x^j, \qquad \sigma(x) = \sum_{j=0}^{k} \beta_j x^j.
$$

If (152.1) is consistent, then it follows that $\rho(1) = 0$ and $\rho'(1) = \sigma(1)$.

If $p > k + 2$, then the method will always be unstable (stability for ordinary differential equations is defined on page 613). Specifically, if k is odd, then $p = k + 1$ is the largest p such that there is a stable method. Also, if k is even, then $p = k + 2$ is the largest p such that there is a stable method. If a difference method is stable and is of p-th order accuracy, then $|v_n - y_n| = o(h^p)$ in any finite interval, $0 \leq x \leq L$.

Many finite difference formulas are tabulated on page 578. For example, for Euler's method and the trapezoidal rule, $k = 1$. For Simpson's rule, $k = 2$ and $p = 4$. To obtain a discretization for a differential equation, it is possible to obtain a finite difference formula for every term in the differential equation and then combine these formulas in the obvious manner. (Just replace each term in the differential equation with its finite difference approximation.) However, combining formulas in this way for partial differential equations—without understanding the underlying physics of the problem and the approximations—can quickly produce results that are unrelated to the true problem (see also page 25).

Example

There are many procedures for generating finite difference formulas for the terms appearing in differential equations; we illustrate one straightforward method. Suppose we want to find an approximation to $f'(x_0)$, given the values $f(x_0 - h)$ and $f(x_0 + h)$. We write

$$f'(x_0) = \alpha f(x_0 - h) + \beta f(x_0 + h) + e(x_0; h), \qquad (152.3)$$

where α and β are constants to be determined, and $e(x_0; h)$ represents the error term. Taking a Taylor series of the right-hand side of (152.3) (and using f_0 to represent $f(x_0)$, f_0' for $f'(x_0)$, etc.), we find

$$f_0' = \alpha \left[f_0 - h f_0' + \frac{h^2}{2} f_0'' - \frac{h^3}{6} f_0''' + O(h^4) \right]$$
$$+ \beta \left[f_0 + h f_0' + \frac{h^2}{2} f_0'' + \frac{h^3}{6} f_0''' + O(h^4) \right] + e(x_0; h).$$

If we choose $\alpha = -\beta$, then this simplifies to

$$f_0' = \beta \left[2 h f_0' + \frac{h^3}{3} f_0''' + O(h^4) \right] + e(x_0; h).$$

Finally, if we choose $\beta = 1/2h$, then we obtain $f_0' = f_0' + \dfrac{h^2}{6} f''' + O(h^3) + e(x_0; h)$. Hence, $e(x_0; h) = O(h^2)$. Putting all of this together, we have the finite difference approximation

$$f'(x_0) = \frac{f(x_0 + h) - f(x_0 - h)}{2h} + O(h^2).$$

This formula could be used to approximate the ordinary differential equation $y' = y^2$, on a uniform mesh, by

$$\frac{u(x_0 + h) - u(x_0 - h)}{2h} = u^2(x_0),$$

where $u(x) \approx y(x)$. Using $x_0 := nh$ and $u_n := u(nh)$ in this formula, we find $\dfrac{u_{n+1} - u_{n-1}}{2h} = u_n^2$. This can be manipulated into the explicit formula: $u_{n+1} = u_{n-1} + 2h u_n^2$.

Notes

[1] Observe that a difference scheme can be stable and still not be consistent. Stability and accuracy are two entirely different concerns.

[2] The Dahlquist relations are

$$\sum_{j=0}^{p} \alpha_j j^k = -k \sum_{j=0}^{p} \beta_j j^{k-1}. \qquad (152.4)$$

If they hold for $k = 0, 1, \ldots, p$, then we have (compare with (152.1))

$$\sum_{j=0}^{p} \alpha_j y(t - jh) = \sum_{j=0}^{p} \beta_j y'(t - jh) + O\left(h^{p+1}\right).$$

[3] Finite difference schemes can be looked up (see page 578 or Isaacson and Keller [5]) or they can be constructed as needed (see Lapidus and Pinder [9] or Ganzha, Mazurik, and Shapeev [2]).

[4] When approximating a differential equation on a bounded interval, the limit $h \to 0$, $n \to \infty$, nh fixed, is of interest. If the *local error* of a discretization scheme (as determined by (152.2)) is $O(h^{p+1})$, then the *global error* (the error at the end of the integration) will be $O(h^p)$.

[5] Obrechkoff methods utilize derivatives of y in forming the finite difference scheme. The k-step Obrechkoff method using the first m derivatives of y may be written

$$\sum_{j=0}^{k} \alpha_j y_{n+j} = \sum_{i=1}^{m} h^i \sum_{j=0}^{k} \beta_{ij} y_{n+j}^{(i)}.$$

See Lambert [8] for details.

[6] Often, a differential equation will have invariants that remain constant during the evolution of the differential equation. For example, in a conservative system the energy should remain constant. A numerical scheme should be used that insures that these invariants remain constant; see Gear [3].

[7] State-of-the-art software packages for ordinary differential equations do not use a single discretization scheme with a fixed step size. Rather, they vary their order (i.e., they choose from a collection of discretization formulas) and they vary the step size. Ideally, the optimal step size and order are determined at each step; this is an important aspect of the code's efficiency (see page 690).

[8] A detailed derivation and example of Euler's method are given on page 653.

[9] To determine if a finite difference scheme for a partial differential equation is stable, see either the Courant consistency criterion (page 618) or the Von Neumann stability test (page 621).

[10] There are other types of finite difference approximations that are not in the form of (152.1). See, for example, the cosine method (see page 640), the predictor–corrector method (see page 679), or the method of Runge–Kutta (see page 684).

[11] There are many useful theorems in numerical analysis concerning methods for specific equations. For example: a method for $u_t = u_x$ with non-negative coefficients cannot have an accuracy of $p > 1$. See Iserles and Strang [6].

[12] Energy propagation under dispersive partial differential equations travels with the *group velocity*. Even if an equation is non-dispersive, any finite difference approximation to it will be dispersive. Hence, study of the group velocity is an important part of the analysis of a finite difference scheme. See Trefethen [11] for details.

References

[1] M. Abramowitz and I. A. Stegun, *Handbook of Mathematical Functions*, National Bureau of Standards, Washington, DC, 1964, pages 882–887.

[2] V. G. Ganzha, S. I. Mazurik, and V. P. Shapeev, "Symbolic Manipulations on a Computer and Their Application to Generation and Investigation of Difference Schemes," in B. Buchberger and B. F. Caviness (eds.), *EURO-CAL '85*, Springer–Verlag, New York, 1985, pages 335–347.

[3] C. W. Gear, "Maintaining Solution Invariants in the Numerical Solution of ODEs," *SIAM J. Sci. Stat. Comput.*, **7**, No. 3, July 1986, pages 734–743.

[4] S. K. Godunov and V. S. Ryabenkii, *Difference Schemes: An Introduction to the Underlying Theory*, (trans. by E. M. Gelbard), North-Holland, New York, 1987.

[5] E. Isaacson and H. B. Keller, *Analysis of Numerical Methods*, John Wiley & Sons, New York, 1966, Chapter 8 (pages 364–434).

[6] A. Iserles and G. Strang, "The Optimal Accuracy of Difference Schemes," *Trans. Amer. Math. Soc.*, **277**, No. 2, June 1983, pages 779–803.

[7] K. R. Jackson, "The Convergence of Integrand–Approximation Formulas for the Numerical Solution of IVPs for ODEs," *SIAM J. Numer. Anal.*, **25**, No. 1, February 1988, pages 163–188.

[8] J. D. Lambert, *Computational Methods in Ordinary Differential Equations*, Cambridge University Press, New York, 1973.

[9] L. Lapidus and G. F. Pinder, *Numerical Solution of Partial Differential Equations in Science and Engineering*, Wiley, New York, 1982, pages 153–162.

[10] L. F. Shampine, "Implementation of Implicit Formulas for the Solution of ODEs," *SIAM J. Sci. Stat. Comput.*, **1**, No. 1, March 1980, pages 103–118.

[11] L. N. Trefethen, "Group Velocity in Finite Difference Schemes," *SIAM Review*, **24**, No. 2, April 1982, pages 113–136.

[12] F. D. Van Niekerk, "Non-Linear One Step Methods for Initial Value Problems," *Comp. & Maths. with Appls.*, **13**, No. 4, 1987, pages 367–371.

153.   Finite Difference Formulas

Applicable to   Differential equations that will be solved by the method of finite differences.

Idea

A table of finite difference formulas for some common grids and common equations can be useful.

Procedure

Given a differential equation to be approximated by finite differences, and a grid (see page 606) on which the solution is desired, replace every derivative by a finite difference approximation to that derivative. Standard finite difference formulas presume that there is an underlying uniform grid with a grid spacing of h. (In two dimensions, the uniform grid spacing is commonly taken to be h in one direction, and k in another direction).

In the standard formulas for ordinary differential equations for the system $\mathbf{y}' = \mathbf{f}(x, \mathbf{y})$, we use the shorthand notation $x_n := x_0 + nh$, $\mathbf{y}_n := \mathbf{y}(x_n)$, $\mathbf{f}_n := \mathbf{f}(x_n, \mathbf{y}_n)$, and $\mathbf{v}_n \approx \mathbf{y}_n$.

In the standard formulas for partial differential equations for the system $L[\mathbf{z}] = \mathbf{f}(x, y, \mathbf{z})$ (where $L[]$ is a two-dimensional differential operator) we use the shorthand notation $x_n := x_0 + nh$, $y_n := y_0 + nk$, $\mathbf{x}_{n,m} := (x_n, y_m)$, $\mathbf{z}_{n,m} := \mathbf{z}(x_n, y_m)$, $\mathbf{f}_{n,m} := \mathbf{f}(x_n, y_m, \mathbf{z}_{n,m})$, and $\mathbf{v}_{n,m} \approx \mathbf{z}_{n,m}$.

In this section we include tables of formulas for the following cases:

[1] One Dimension: Rectilinear Grid

[2] Two Dimensions: Rectilinear Grid

[3] Two Dimensions: Irregular Grid

[4] Two Dimensions: Triangular Grid

[5] Formulas for the ODE: $y' = f(x, y)$

[6] Explicit formulas for the PDE: $au_x + u_t = 0$

[7] Implicit formulas for the PDE: $au_x + u_t = S(x, t)$

[8] Formulas for the PDE: $F(u)_x + u_t = 0$

[9] Formulas for the PDE: $u_x = u_{tt}$

One Dimension: Rectilinear Grid

The following is a list of finite difference formulas, of different accuracies, for a grid with uniform spacing.

Formulas for the first derivative:

$$f'(x_0) = \frac{f_1 - f_0}{h} + O(h)$$

$$f'(x_0) = \frac{f_1 - f_{-1}}{2h} + O(h^2)$$

$$f'(x_0) = \frac{-f_2 + 4f_1 - 3f_0}{2h} + O(h^2)$$

$$f'(x_0) = \frac{-f_2 + 8f_1 - 8f_{-1} + f_{-2}}{12h} + O(h^4)$$

Formulas for the second derivative:

$$f''(x_0) = \frac{f_2 - 2f_1 + f_0}{h^2} + O(h)$$

$$f''(x_0) = \frac{f_1 - 2f_0 + f_{-1}}{h^2} + O(h^2)$$

$$f''(x_0) = \frac{-f_3 + 4f_2 - 5f_1 + 2f_0}{h^2} + O(h^3)$$

$$f''(x_0) = \frac{-f_2 + 16f_1 - 30f_0 + 16f_{-1} - f_{-2}}{12h^2} + O(h^4)$$

Formulas for the third derivative:

$$f'''(x_0) = \frac{f_3 - 3f_2 + 3f_1 - f_0}{h^3} + O(h)$$

$$f'''(x_0) = \frac{f_2 - 2f_1 + 2f_{-1} - f_{-2}}{2h^3} + O(h^2)$$

Formulas for the fourth derivative:

$$f^{(4)}(x_0) = \frac{f_4 - 4f_3 + 6f_2 - 4f_1 + f_0}{h^4} + O(h)$$

$$f^{(4)}(x_0) = \frac{f_2 - 4f_1 + 6f_0 - 4f_{-1} + f_{-2}}{h^4} + O(h^2)$$

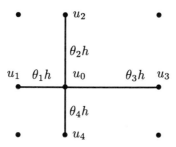

Figure 153.1 Spacing on an irregular domain.

Two Dimensions: Rectilinear Grid

The following is a list of finite difference formulas, of different accuracies, for rectangular grids with uniform spacing. Other formulas can be obtained from the last list by simply holding one variable constant.

Formulas for first order partial derivatives:

$$f_x(\mathbf{x}_{0,0}) = \frac{1}{2h}\left(f_{1,0} - f_{-1,0}\right) + O\!\left(h^2\right)$$

$$f_x(\mathbf{x}_{0,0}) = \frac{1}{4h}\left(f_{1,1} - f_{-1,1} + f_{1,-1} - f_{-1,-1}\right) + O\!\left(h^2\right)$$

Formulas for second order partial derivatives:

$$f_{xx}(\mathbf{x}_{0,0}) = \frac{1}{3h^2}\left(f_{1,1} - 2f_{0,1} + f_{-1,1} + f_{1,0} - 2f_{0,0} + f_{-1,0}\right.$$

$$\left. + f_{1,-1} - 2f_{0,-1} + f_{-1,-1}\right) + O\!\left(h^2\right)$$

$$f_{xy}(\mathbf{x}_{0,0}) = \frac{1}{4h^2}\left(f_{1,1} - f_{1,-1} - f_{-1,1} + f_{-1,-1}\right) + O\!\left(h^2\right)$$

Formulas for the Laplacian:

$$\nabla^2 f(\mathbf{x}_{0,0}) = \frac{1}{h^2}\left(f_{1,0} + f_{0,1} + f_{-1,0} + f_{0,-1} - 4f_{0,0}\right) + O\!\left(h^2\right)$$

$$\nabla^2 f(\mathbf{x}_{0,0}) = \frac{1}{12h^2}\left(-60f_{0,0} + 16(f_{1,0} + f_{0,1} + f_{-1,0} + f_{0,-1})\right.$$

$$\left. - (f_{2,0} + f_{0,2} + f_{-2,0} + f_{0,-2})\right) + O\!\left(h^4\right)$$

Two Dimensions: Irregular Grid

Nonuniform grids may be the only way to numerically solve some practical problems involving partial differential equations. For example, a non-uniform grid may be required near the boundaries of a domain. Also, adaptive grids and moving grids are sometimes more useful than a fixed grid (see page 606). The following finite difference formulas refer to the parameters defined in Figure 153.1.

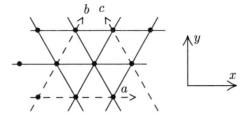

Figure 153.2 Definition of the coordinate system for a triangular domain.

Formulas for first order partial derivatives:

$$\frac{\partial u}{\partial x}\bigg|_{\mathbf{x}_{0,0}} = \frac{u_2 - u_4}{h(\theta_2 + \theta_4)} + O(h)$$

$$\frac{\partial u}{\partial y}\bigg|_{\mathbf{x}_{0,0}} = \frac{u_3 - u_1}{h(\theta_1 + \theta_3)} + O(h)$$

Formulas for second order partial derivatives:

$$\frac{\partial^2 u}{\partial x^2}\bigg|_{\mathbf{x}_{0,0}} = \frac{2}{h^2}\left[\frac{u_1 - u_0}{\theta_1(\theta_1 + \theta_3)} + \frac{u_3 - u_0}{\theta_3(\theta_1 + \theta_3)}\right] + O(h)$$

$$\frac{\partial^2 u}{\partial y^2}\bigg|_{\mathbf{x}_{0,0}} = \frac{2}{h^2}\left[\frac{u_2 - u_0}{\theta_2(\theta_2 + \theta_4)} + \frac{u_4 - u_0}{\theta_2(\theta_2 + \theta_4)}\right] + O(h)$$

$$\nabla^2 u\bigg|_{\mathbf{x}_{0,0}} = \frac{\partial^2 u}{\partial x^2} + \frac{\partial^2 u}{\partial y^2}\bigg|_{\mathbf{x}_{0,0}}$$

$$= \frac{2}{h^2}\left[\frac{u_1}{\theta_1(\theta_1 + \theta_3)} + \frac{u_2}{\theta_2(\theta_2 + \theta_4)} + \frac{u_3}{\theta_1(\theta_1 + \theta_3)} + \frac{u_4}{\theta_2(\theta_2 + \theta_4)}\right.$$
$$\left. - \left(\frac{1}{\theta_1\theta_3} + \frac{1}{\theta_2\theta_4}\right)u_0\right] + O(h)$$

Two Dimensions: Triangular Grid

Sometimes it is easier to perform computations on a uniform triangular grid (see Figure 153.2). If we represent the three directions on the triangular grid as $\{a, b, c\}$, then we can compute the partial derivatives:

$$\frac{\partial u}{\partial a} = u_x, \qquad\qquad \frac{\partial^2 u}{\partial a^2} = u_{xx},$$

$$\frac{\partial u}{\partial b} = \frac{1}{2}u_x + \frac{\sqrt{3}}{2}u_y, \qquad \frac{\partial^2 u}{\partial b^2} = \frac{1}{4}u_{xx} + \frac{\sqrt{3}}{2}u_{xy} + \frac{3}{4}u_{yy},$$

$$\frac{\partial u}{\partial c} = -\frac{1}{2}u_x + \frac{\sqrt{3}}{2}u_y, \qquad \frac{\partial^2 u}{\partial c^2} = \frac{1}{4}u_{xx} - \frac{\sqrt{3}}{2}u_{xy} + \frac{3}{4}u_{yy}.$$

These relations may be inverted to yield

$$u_x = \frac{\partial u}{\partial a}, \qquad\qquad u_{yy} = \frac{1}{3}\left(2\frac{\partial^2 u}{\partial b^2} + 2\frac{\partial^2 u}{\partial c^2} - \frac{\partial^2 u}{\partial a^2}\right),$$

$$u_y = \frac{1}{\sqrt{3}}\left(\frac{\partial u}{\partial b} + \frac{\partial u}{\partial c}\right), \qquad u_{xy} = \frac{1}{\sqrt{3}}\left(\frac{\partial^2 u}{\partial b^2} - \frac{\partial^2 u}{\partial c^2}\right),$$

$$u_{xx} = \frac{\partial^2 u}{\partial a^2}, \qquad\qquad \nabla^2 u = u_{xx} + u_{yy} = \frac{2}{3}\left(\frac{\partial^2 u}{\partial a^2} + \frac{\partial^2 u}{\partial b^2} + \frac{\partial^2 u}{\partial c^2}\right).$$

See Gerald and Wheatley [6] (section 7.9) for a worked example using triangular coordinates.

Schemes for the ODE: $y' = f(x, y)$

Some common difference formulas for the ordinary differential equation $y' = f(x, y)$ are:

Adams–Bashforth, order 2: $v_n - v_{n-1} = \frac{1}{2}h\left[3f_{n-1} - f_{n-2}\right]$

Adams–Bashforth, order 4: $v_n - v_{n-1} = \frac{1}{24}h\left[55f_{n-1} - 59f_{n-2} + 37f_{n-3} - 9f_{n-4}\right]$

Adams–Moulton, order 4: $v_n - v_{n-1} = \frac{1}{24}h\left[9f_n + 19f_{n-1} - 5f_{n-2} + f_{n-3}\right]$

backward Euler: $v_n - v_{n-1} = hf_n$

Euler's method: $v_n - v_{n-1} = hf_{n-1}$

leapfrog: $v_{n+1} - v_{n-1} = hf_n$

midpoint rule: $v_n - v_{n-1} = \frac{1}{2}h(f_n + f_{n-1})$

Simpson's rule*: $v_n - v_{n-2} = \frac{1}{3}h(f_n + 4f_{n-1} + f_{n-2})$

trapezoidal rule**: $v_n - v_{n-1} = \frac{1}{2}h(f_n + f_{n-1})$.

Of these methods, Euler's method and the leapfrog method are explicit; all the others are implicit methods.

---

* Also known as Milne's method.

** Also known as Heun's method and as the Adams–Moulton method of order 2.

Explicit formulas for the PDE: $au_x + u_t = 0$

Below we list named explicit difference formulas for the partial differential equation $au_x + u_t = 0$. DuChateau and Zachmann [3] (page 450) also list the local truncation error for each of these methods. In this listing, h is the uniform x spacing, and k is the uniform t spacing. The approximation to $u(x_n, t_j) = u(x_0 + nh, t_0 + jk)$ is represented by $u_{n,j}$.

Forward in time, forward in space (FTFS):
$$a\frac{u_{n+1,j} - u_{n,j}}{h} + \frac{u_{n,j+1} - u_{n,j}}{k} = 0$$

Forward in time, centered in space (FTCS) (unstable):
$$a\frac{u_{n+1,j} - u_{n-1,j}}{2h} + \frac{u_{n,j+1} - u_{n,j}}{k} = 0$$

Forward in time, backward in space (FTBS):
$$a\frac{u_{n,j} - u_{n-1,j}}{h} + \frac{u_{n,j+1} - u_{n,j}}{k} = 0$$

Lax–Friedrichs method:
$$a\frac{u_{n+1,j} - u_{n-1,j}}{2h} + \frac{u_{n,j+1} - \frac{1}{2}\left(u_{n-1,j} - u_{n+1,j}\right)}{k} = 0$$

Lax–Wendroff method:
$$u_{n,j+1} = u_{n,j} - \frac{ak}{2h}\left(u_{n+1,j} - u_{n-1,j}\right)$$
$$+ \frac{a^2 k^2}{2h^2}\left(u_{n-1,j} - 2u_{n,j} + u_{n+1,j}\right)$$

Implicit formulas for the PDE: $au_x + u_t = S(x,t)$

Below we list named implicit difference formulas for the partial differential equation $au_x + u_t = S(x,t)$. DuChateau and Zachmann [3] (page 460) also list the local truncation error for each of these methods. In this listing, h is the uniform x spacing, and k is the uniform t spacing. The approximation to $u(x_n, t_j) = u(x_0 + nh, t_0 + jk)$ is represented by $u_{n,j}$, and $S_{n,j}$ is used to represent $S(x_n, t_j)$.

Backward in time, backward in space (BTBS):
$$a\frac{u_{n+1,j+1} - u_{n,j+1}}{h} + \frac{u_{n+1,j+1} - u_{n+1,j}}{k} = S_{n+1,j+1}$$

Backward in time, centered in space (BTCS):
$$a\frac{u_{n+1,j+1} - u_{n-1,j+1}}{2h} + \frac{u_{n,j+1} - u_{n,j}}{k} = S_{n,j+1}$$

Crank–Nicolson:
$$\frac{1}{2}\left(a\frac{u_{n+1,j+1} - u_{n-1,j+1}}{2h} + a\frac{u_{n+1,j} - u_{n-1,j}}{2h}\right) + \frac{u_{n,j+1} - u_{n,j}}{k} = S_{n,j+1/2}$$

Wendroff method:
$$\frac{1}{2}\left(a\frac{u_{n+1,j+1} - u_{n,j+1}}{h} + a\frac{u_{n+1,j} - u_{n,j}}{h}\right)$$
$$+ \frac{1}{2}\left(\frac{u_{n+1,j+1} - u_{n+1,j}}{k} + \frac{u_{n,j+1} - u_{n,j}}{k}\right) = S_{n+1/2,j+1/2}$$

Formulas for the PDE: $F(u)_x + u_t = 0$

Below we list named difference formulas for the partial differential equation $F(u)_x + u_t = 0$ (see DuChateau and Zachmann [3], page 475, for more details). In this listing, h is the uniform x spacing, k is the uniform t spacing, and the ratio of these is $s = k/h$. The approximation to $u(x_n, t_j) = u(x_0 + nh, t_0 + jk)$ is represented by $u_{n,j}$ and $F_{m,n} := F(u_{m,n})$. A star superscript indicates an intermediate result (and $F_n^* := F(u_n^*)$). Finally, $a_n := F_n' = F'(u_n)$.

Note that some of the left-hand sides of the last listing can be obtained from this listing by taking $F(u) = au$.

Centered in time–centered in space (unstable):
$$u_{n,j+1} = u_{n,j} - \tfrac{1}{2}s\left(F_{n+1,j} - F_{n-1,j}\right)$$

Lax–Friedrichs method:
$$u_{n,j+1} = \tfrac{1}{2}\left(u_{n+1,j} + u_{n-1,j}\right) - \tfrac{1}{2}s\left(F_{n+1,j} + F_{n-1,j}\right)$$

Lax–Wendroff method:
$$u_{n,j+1} = u_{n,j} - \tfrac{1}{2}s\left(F_{n+1,j} - F_{n-1,j}\right)$$
$$+ \tfrac{1}{2}s^2\left[a_{n+1/2,j}\left(F_{n+1,j} - F_{n,j}\right) - a_{n-1/2,j}\left(F_{n,j} - F_{n-1,j}\right)\right]$$

Richtmeyer method:
$$u_{n+1/2}^* = \tfrac{1}{2}\left(u_{n+1,j} + u_{n,j}\right) - \tfrac{1}{2}\left(F_{n+1,j} - F_{n,j}\right)$$
$$u_{n,j+1} = u_{n,j} - s\left(F_{n+1/2}^* - F_{n-1}^*\right)$$

MacCormack method:
$$u_n^* = u_{n,j} - s\left(F_{n+1,j} - F_{n,j}\right)$$
$$u_{n,j+1} = \tfrac{1}{2}\left[u_{n,j} + u_n^* - s\left(F_n^* - F_{n-1}^*\right)\right]$$

FTBS upwind method (use when $F'(u) > 0$):
$$u_{n,j+1} = u_{n,j} + s\left(F_{n-1,j} - F_{n,j}\right)$$

FTFS upwind method (use when $F'(u) < 0$):
$$u_{n,j+1} = u_{n,j} - s\left(F_{n+1,j} - F_{n,j}\right)$$

Formulas for the PDE: $u_x = u_{tt}$

Below we list named difference formulas for the partial differential equation $u_x = u_{tt}$. Lapidus and Pinder [9] discuss each of these methods in some detail. In this listing, h is the uniform x spacing, k is the uniform t spacing, and ρ is defined to be $\rho = h/k^2$. The approximation to $u(x_n, t_j) = u(x_0 + nh, t_0 + jk)$ is represented by $u_{n,j}$.

Classic explicit approximation:
$$u_{n+1,j} = (1 - 2\rho)u_{n,j} + \rho\left(u_{n,j+1} + u_{n,j-1}\right)$$

DuFort–Frankel explicit approximation:
$$(1 + 2\rho)u_{n+1,j} = 2\rho\left(u_{n,j+1} + u_{n,j-1}\right) + (1 - 2\rho)u_{n-1,j}$$

Richardson explicit approximation:
$$u_{n+1,j} - u_{n-1,j} - 2\rho\left(u_{n,j+1} + u_{n,j-1}\right) + 4\rho u_{n,j} = 0$$

Backward implicit approximation:
$$(1 + 2\rho)u_{n+1,j} - \rho\left(u_{n+1,j+1} + u_{n+1,j-1}\right) = u_{n,j}$$
Crank–Nicolson implicit approximation:
$$2(\rho + 1)u_{n+1,j} - \rho\left(u_{n+1,j+1} + u_{n+1,j-1}\right) = 2(1 - \rho)u_{n,j} + \rho\left(u_{n,j+1} + u_{n,j-1}\right)$$
Variable weighted implicit approximation (with $0 \le \theta \le 1$):
$$(1 + 2\rho\theta)u_{n+1,j} = \rho(1 - \theta)\left(u_{n,j+1} + u_{n,j-1}\right) + \rho\theta\left(u_{n+1,j+1} + u_{n+1,j-1}\right)$$
$$+ \left[1 - 2\rho(1 - \theta)\right]u_{n,j}$$

Notes

[1] Fornberg [4] has a simple recursive technique for determining finite difference formula of high order.

[2] For problems with periodic boundary conditions, it is possible to obtain finite differential formulas that are of infinite order; see page 759.

[3] All of the discretization methods used should be of comparable order. That is, if one term in an equation has a discretization error of $O(h^2)$, then there is no reason for another term to have a discretization error of $O(h^4)$.

[4] Note that nonuniform grids may give rise to a number of consistency/stability phenomena that have no counterpart on uniform grids.

References

[1] M. Abramowitz and I. A. Stegun, *Handbook of Mathematical Functions*, National Bureau of Standards, Washington, DC, 1964, pages 883–885.

[2] I. Altas and J. W. Stephenson, "Finite Difference Schemes on Irregular Meshes," *Congr. Numer.*, **69**, 1989, pages 21–32.

[3] P. DuChateau and D. Zachmann, *Applied Partial Differential Equations*, Harper & Row, Publishers, New York, 1989.

[4] B. Fornberg, "Generation of Finite Difference Formulas on Arbitrarily Spaced Grids," *Math. of Comp.*, **51**, No. 184, October 1988, pages 699–706.

[5] V. G. Ganzha, S. I. Mazurik, and V. P. Shapeev, "Symbolic Manipulations on a Computer and Their Application to Generation and Investigation of Difference Schemes," in B. Buchberger and B. F. Caviness (eds.), *EURO-CAL '85*, Springer–Verlag, New York, 1985, pages 335–347.

[6] C. F. Gerald and P. O. Wheatley, *Applied Numerical Analysis*, Addison–Wesley Publishing Co., Reading, MA, 1984.

[7] B. Heinrich, *Finite Difference Methods on Irregular Networks*, Birkhäuser Basil, 1987.

[8] H. B. Keller and V. Pereyra, "Symbolic Generation of Finite Difference Formulas," *Math. of Comp.*, **32**, 1978, pages 955–971.

[9] L. Lapidus and G. F. Pinder, *Numerical Solution of Partial Differential Equations in Science and Engineering*, Wiley, New York, 1982, section 4.3, pages 153–162.

[10] D. Voss, "A Fifth-Order Exponentially Fitted Formula," *SIAM J. Math. Anal.*, **25**, No. 3, June 1988, pages 670–678.

154. Excerpts from GAMS

Applicable to   Differential equations.

Yields

Software appropriate for a specific problem.

Procedure

When numerically approximating the solution to a differential equation, it is best to use prepared software whenever possible. As described on page 570, there are a multitude of commercially available computer libraries and isolated computer routines for this purpose. A taxonomy for differential equation software has been developed as part of the GAMS project at the National Institute of Standards and Technology (NIST) [1]. GAMS [2] has both a taxonomy of computer routines and a listing of some available software. In this section we print part of that manual.

In the "Taxonomy" section we indicate the subject headings under which software has been classified. In the "Excerpts" section we give the text that appeared in GAMS [2], in a format similar to the original. For each topic the applicable routines in a specific library are described. (The author thanks Dr. Ronald Boisvert of NIST for making this text available electronically.)

The following computer libraries are referred to in the "Excerpts" section†. Their inclusion does not constitute an endorsement. Nor does it necessarily imply that unnamed packages are not worth trying.

- CMLIB
- Collected Algorithms of the ACM
- ELLPACK
- IMSL
- NAG
- NMS
- PLOD
- PORT
- Scientific Desk
- SCRUNCH

---

† Identification of commercial products does not imply recommendation or endorsement by NIST.

Notes

[1] In the excerpts section, ACM TOMS stands for *ACM Trans. Math. Software*.

[2] Software is not listed for all the taxonomy classes that have been established.

[3] The GAMS manual refers to the libraries PDELIB and the "IMSL Subprogram Library." PDELIB is an internal name at NIST that is not known elsewhere, and no source code is available. The "IMSL Subprogram Library" consists of old routines that IMSL no longer supports. Hence, all references to each of these libraries have been deleted in the excerpts section.

References

[1] R. F. Boisvert, S. E. Howe, and D. K. Kahaner, "GAMS: A Framework for the Management of Scientific Software," *ACM Trans. Math. Software*, **11**, No. 4, December 1985, pages 313–355.

[2] R. F. Boisvert, S. E. Howe, D. K. Kahaner, and J. L. Springmann, *Guide to Available Mathematical Software*, NISTIR 90-4237, Center for Computing and Applied Mathematics, National Institute of Standards and Technology, Gaithersburg, MD 20899, March 1990.

[3] CMLIB—this is a collection of codes from many sources that NIST has combined into a single library. The relevant sublibraries are:

  (A) BVSUP, see M. R. Scott and H. A. Watts, "Computational Solutions of Linear Two-Point Boundary Value Problems via Orthonormalization," *SIAM J. Numer. Anal.*, **14**, 1977, pages 40–70.

  (B) CDRIV and SDRIV, see D. Kahaner, C. Moler, and S. Nash, *Numerical Methods and Software*, Prentice–Hall Inc., Englewood Cliffs, NJ, 1989.

  (C) DEPAC: Code developed by L. Shampine and H. A. Watts.

  (D) FISHPAK: Code developed by P. N. Swarztrauber and R. A. Sweet.

  (E) SDASSL, see L. R. Petzold, "Differential/Algebraic Equations Are Not ODE's," *SIAM J. Sci. Stat. Comput.*, **3**, No. 3, 1982, pages 367–384.

  (F) VHS3: Code developed by R. A. Sweet.

[4] ELLPACK, see J. R. Rice and R. F. Boisvert, *Solving Elliptic Problems Using ELLPACK*, Springer–Verlag, New York, 1985.

[5] IMSL Inc., 2500 Park West Tower One, 2500 City West Blvd., Houston, TX 77042.

[6] NAG, Numerical Algorithms Group Limited, Wilkinson House, Jordan Hill, Oxford OX2 8DR, UK.

[7] NMS—this is an internal name at NIST. The code is from D. Kahaner, C. Moler, and S. Nash, *Numerical Methods and Software*, Prentice–Hall Inc., Englewood Cliffs, NJ, 1989.

[8] PLOD, see E. Agron, I. Change, G. Gunaratna, D. Kahaner, and M. Reed, "Mathematical Software: PLOD," *IEEE Micro*, **8**, No. 4, 1988, pages 56–61.

[9] PORT, see P. Fox *et al.*, *The PORT Mathematical Subroutine Library Manual*, Bell Laboratories, Murray Hill, NJ, 1977.

[10] Scientific Desk is distributed by M. McClain, NIST, Bldg 225 Room A151, Gaithersburg, MD 20899.

[11] SCRUNCH—these are old, unsupported codes in BASIC. The codes are translations of FORTRAN algorithms from G. Forsythe, M. Malcom, and C. Moler, *Computer Methods for Mathematical Computations*, Prentice–Hall Inc., Englewood Cliffs, NJ, 1977.

Taxonomy

I1.	Ordinary Differential Equations (ODE's)
I1a.	Initial value problems
I1a1.	General, nonstiff or mildly stiff
I1a1a.	One-step methods (e.g., Runge-Kutta)
I1a1b.	Multistep methods (e.g., Adams predictor-corrector)
I1a1c.	Extrapolation methods (e.g., Bulirsch-Stoer)
I1a2.	Stiff and mixed algebraic-differential equations
I1b.	Multipoint boundary value problems
I1b1.	Linear
I1b2.	Nonlinear
I1b3.	Eigenvalue (e.g., Sturm-Liouville)
I1c.	Service routines (e.g., interpolation of solutions, error handling, test programs)
I2.	Partial differential equations
I2a.	Initial boundary value problems
I2a1.	Parabolic
I2a1a.	One spatial dimension
I2a1b.	Two or more spatial dimensions
I2a2.	Hyperbolic
I2b.	Elliptic boundary value problems
I2b1.	Linear
I2b1a.	Second Order
I2b1a1.	Poisson (Laplace) or Helmholtz equation
I2b1a1a.	Rectangular domain (or topologically rectangular in the coordinate system)
I2b1a1b.	Nonrectangular domain
I2b1a2.	Other separable problems
I2b1a3.	Nonseparable problems
I2b1c.	Higher order equations (e.g., biharmonic)
I2b2.	Nonlinear
I2b3.	Eigenvalue
I2b4.	Service routines
I2b4a.	Domain triangulation *(search also class P)*
I2b4b.	Solution of discretized elliptic equations

Excerpts

I1a:	**Initial value problems for ordinary differential equations**

PLOD Interactive Program

PLOD   An easy to use interactive system on a personal computer for the solution of initial value problems for ordinary differential equations. The user can change initial conditions, intervals, parameters, etc., and examine various plots on the terminal. Little programming needed.

I1a1a:	One-step methods (e.g., Runge–Kutta) for general, nonstiff or mildly stiff initial value problems for ordinary differential equations

Collected Algorithms of the ACM

A497   DMRODE: a subprogram for the automatic integration of functional differential equations, such as retarded ordinary differential equations, Volterra integro-differential equations, and difference differential equations. (See K. W. Neves, ACM TOMS 1 (1975) pp. 369–371.)

A504   GERK: A FORTRAN subprogram to solve nonlinear systems of ordinary differential equations when it is important to have a global error estimate. Integrations are performed on different mesh spacings, and global extrapolation is applied to provide an estimate of the global error in the more accurate solution. The integrations are done using Runge–Kutta–Fehlberg methods of 4th and 5th order. (See L. F. Shampine and H. A. Watts, ACM TOMS 2 (1976) pp. 200–203.)

A553   M3RK: A FORTRAN subroutine for solving initial value problems for nonlinear first order systems of ordinary differential equations which originate from semi-discretization of parabolic partial differential equations. M3RK is based on stabilized, explicit three-step Runge–Kutta formulas of order one and two, and degree 2 through 12. (See J. G. Verwer, ACM TOMS 6 (1980) pp. 236–239.)

CMLIB Library (DEPAC Sublibrary)

DERKF   Solves a system of first order ordinary differential equations with arbitrary initial conditions by a Runge–Kutta method.

IMSL MATH/LIBRARY Subprogram Library

IVPRK   Solves an initial value problem for ordinary differential equations using the Runge–Kutta–Verner fifth- and sixth-order method.

NAG Subprogram Library

D02BAF   Integrates a system of first order ordinary differential equations over a range with suitable initial conditions, using a Runge–Kutta–Merson method.

D02BBF   Integrates a system of first order ordinary differential equations over a range with suitable initial conditions, using a Runge–Kutta–Merson method, and returns the solution at points specified by the user.

D02BDF   Integrates a system of first order ordinary differential equations over a range with suitable initial conditions, using a Runge–Kutta–Merson method, and computes a global error estimate check. A stiffness check is also available.

D02BGF   Integrates a system of first order ordinary differential equations over a range with suitable initial conditions, using a Runge–Kutta–Merson method, until a specified component attains a given value.

D02BHF   Integrates a system of first order ordinary differential equations over a range with suitable initial conditions, using a Runge–Kutta–Merson method, until a user-specified function of the solution is zero.

D02PAF   Integrates a system of first order ordinary differential equations over a range with suitable initial conditions, using a Runge–Kutta–Merson method. A variety of facilities for interrupting the calculation is provided. This routine is relatively complicated and is recommended only to experienced users.

D02YAF   Integrates a system of first order ordinary differential equations over one step, using Merson's Runge–Kutta method.

Scientific Desk PC Subprogram Library

I1A1A   Integrates a system of neqn first order ordinary differential equations of the form $dy(i)/dt = f(t,y(1),y(2),...,y(neqn))$, where the $y(i)$ are given at t (Runge–Kutta–Fehlberg method).

SCRUNCH Subprogram Library

RKF45   Runge–Kutta–Fehlberg method for the integration of a first order system of ordinary differential equations. In BASIC.

I1a1b: **Multistep methods (e.g., Adams predictor-corrector) for general, nonstiff or mildly stiff initial value problems for ordinary differential equations**

Collected Algorithms of the ACM

A658   ODESSA: A FORTRAN ordinary differential equation solver (a modification of LSODE) with explicit simultaneous sensitivity analysis. (See J. R. Leis and M. A. Kramer, ACM TOMS 14 (1988) pp. 61–67.)

CMLIB Library (CDRIV Sublibrary)

CDRIV1   Numerical integration of complex initial value problems for ordinary differential equations, Gear stiff formulas. Easy to use.

CDRIV2 Numerical integration of complex initial value problems for ordinary differential equations, Gear stiff and Adams formulas, root finding.

CDRIV3 Numerical integration of complex initial value problems for ODEs, Gear and Adams formulas, implicit equations, sparse Jacobians, root finding.

CMLIB Library (DEPAC Sublibrary)

DEABM Solves a system of first order ordinary differential equations with arbitrary initial conditions by a predictor-corrector method.

CMLIB Library (SDASSL Sublibrary)

SDASSL Solves the system of differential/algebraic equations of the form $g(t,y,y')=0$, with given initial values.

CMLIB Library (SDRIV Sublibrary)

SDRIV1 Numerical integration, initial value problems, ordinary differential equations, Gear stiff formulas. Easy to use.

SDRIV2 Numerical integration, initial value problems, ordinary differential equations, Gear/Adams formulas.

SDRIV3 Numerical integration, initial value problems, ordinary differential equations, implicit equations, sparse Jacobians.

IMSL MATH/LIBRARY Subprogram Library

IVPAG Solves an initial value problem for ordinary differential equations using an Adams-Moulton or Gear method.

NAG Subprogram Library

D02CAF Integrates a system of first order ordinary differential equations over a range with suitable initial conditions, using a variable order, variable step Adams method.

D02CBF Integrates a system of first order ordinary differential equations over a range with suitable initial conditions, using a variable order, variable step Adams method, and returns the solution at points specified by the user.

D02CGF Integrates a system of first order ordinary differential equations over a range with suitable initial conditions, using a variable order, variable step Adams method, until a specified component attains a given value.

D02CHF Integrates a system of first order ordinary differential equations over a range with suitable initial conditions, using a variable order, variable step Adams method, until a user-specified function of the solution is zero.

D02QAF Integrates a system of first order ordinary differential equations over a range with suitable initial conditions, using a variable order, variable step Adams method. A variety of facilities for interrupting the calculation are provided. This routine is relatively complicated and is recommended only to experienced users.

NMS Subprogram Library

SDRIV2 Numerical integration, initial value problems, ordinary differential equations, Gear/Adams formulas.

Scientific Desk PC Subprogram Library

I1A2 Numerical integration of initial value problems for ordinary differential equations, implicit equations, sparse Jacobians, root finding.

I1A2E Numerical integration of initial value problems for ordinary differential equations, Gear stiff formulas; easy to use.

I1A2F Numerical integration of initial value problems for ordinary differential equations, Gear/Adams formulas, root finding.

I1a1c: **Extrapolation methods (e.g., Bulirsch–Stoer) for general, nonstiff or mildly stiff initial value problems for ordinary differential equations**

IMSL MATH/LIBRARY Subprogram Library

IVPBS Solves an initial value problem for ordinary differential equations using the Bulirsch–Stoer extrapolation method.

PORT Subprogram Library

ODES Solves an initial value problem for a system of ordinary differential equations. Easy to use.

ODES1 Solves an initial value problem for a system of ordinary differential equations. Allows great flexibility and user control.

I1a2: **Stiff and mixed algebraic-ordinary differential equations**

Collected Algorithms of the ACM

A534 STINT: A FORTRAN subprogram for integrating a set of first order ordinary differential equations using stiffly stable, cyclic composite linear multistep methods. (See J. M. Tendler, T. A. Bickart, and Z. Picel, ACM TOMS 4 (1978) pp. 399–403.)

A658 ODESSA: A FORTRAN ordinary differential equation solver (a modification of LSODE) with explicit simultaneous sensitivity analysis. (See J. R. Leis and M. A. Kramer, ACM TOMS 14 (1988) pp. 61–67.)

CMLIB Library (CDRIV Sublibrary)

CDRIV1 Numerical integration of complex initial value problems for ordinary differential equations, Gear stiff formulas. Easy to use.

CDRIV2 Numerical integration of complex initial value problems for ordinary differential equations, Gear stiff and Adams formulas, root finding.

CDRIV3   Numerical integration of complex initial value problems for ODEs, Gear and Adams formulas, implicit equations, sparse Jacobians, root finding.

CMLIB Library (DEPAC Sublibrary)

DEBDF   Solves a system of first order stiff ordinary differential equations with arbitrary initial conditions by Gear's method.

CMLIB Library (SDRIV Sublibrary)

SDRIV1   Numerical integration, initial value problems, ordinary differential equations, Gear stiff formulas. Easy to use.

SDRIV2   Numerical integration, initial value problems, ordinary differential equations, Gear/Adams formulas.

SDRIV3   Numerical integration, initial value problems, ordinary differential equations, implicit equations, sparse Jacobians.

NAG Subprogram Library

D02EAF   Integrates a stiff system of first order differential equations over a range with suitable initial conditions, using a variable order, variable step method implementing the backward differentiation formulas.

D02EBF   Integrates a stiff system of first order ordinary differential equations over a range with suitable initial conditions, using a variable order, variable step method implementing the backward differentiation formulas, and returns the solution at points specified by the user.

D02EGF   Integrates a stiff system of first order ordinary differential equations over a range with suitable initial conditions, using a variable order, variable step method implementing the backward differentiation formulas, until a specified component attains a given value.

D02EHF   Integrates a stiff-system of first order ordinary differential equations over a range with suitable initial conditions, using a variable order, variable step method implementing the backward differentiation formulas, until a user specified function of the solution is zero.

D02EJF   Integrates a stiff system of first order ordinary differential equations over a range with suitable initial conditions, using a variable order, variable step method implementing the backward differentiation formulas, until a user-specified function, if supplied, of the solution is zero, and returns the solution at points specified by the user, if desired.

D02NBF   Forward communication routine for integrating stiff systems of explicit ordinary differential equations when the Jacobian is a full matrix.

D02NCF   Forward communication routine for integrating stiff systems of explicit ordinary differential equations when the Jacobian is a banded matrix.

D02NDF   Forward communication routine for integrating stiff systems of explicit ordinary differential equations when the Jacobian is a sparse matrix.

D02NGF   Forward communication routine for integrating stiff systems of implicit ordinary differential equations coupled with algebraic equations when the Jacobian is a full matrix.

D02NHF   Forward communication routine for integrating stiff systems of implicit ordinary differential equations coupled with algebraic equations when the Jacobian is a banded matrix.

D02NJF   Forward communication routine for integrating stiff systems of implicit ordinary differential equations coupled with algebraic equations when the Jacobian is a sparse matrix.

D02NMF   Reverse communication routine for integrating stiff systems of explicit ordinary differential equations.

D02NNF   Reverse communication routine for integrating stiff systems of implicit ordinary differential equations coupled with algebraic equations.

D02QBF   Integrates a stiff system of first order ordinary differential equations, over a range with suitable initial conditions, using a variable order, variable step Gear method. A variety of facilities for interrupting the calculation are provided. This routine is relatively complicated and is recommended only to experienced users.

D02QDF   Integrates a stiff system of first order ordinary differential equations, over a range with suitable initial conditions, using a variable order, variable step method based on the backward differentiation formulas (BDF). A variety of facilities for interrupting the calculation are provided. This routine is relatively complicated and is recommended to experienced users only.

NMS Subprogram Library

SDRIV2   Numerical integration, initial value problems, ordinary differential equations, Gear/Adams formulas.

Scientific Desk PC Subprogram Library

I1A2   Numerical integration of initial value problems for ordinary differential equations, implicit equations, sparse Jacobians, root finding.

I1A2E   Numerical integration of initial value problems for ordinary differential equations, Gear stiff formulas; easy to use.

I1A2F   Numerical integration of initial value problems for ordinary differential equations, Gear/Adams formulas, root finding.

I1b1:	**Linear multipoint boundary value problems for ordinary differential equations**

CMLIB Library (BVSUP Sublibrary)

BVSUP   Solves boundary value problems for a linear system of ODEs using superposition, orthogonalization, and variable step integration.

NAG Subprogram Library

D02GBF   Solves a general linear two-point boundary value problem for a system of ordinary differential equations using a deferred correction technique.

D02JAF   Solves a regular linear two-point boundary value problem for a single n(th) order ordinary differential equation by a Chebyshev series using collocation and least squares.

D02JBF   Solves a regular linear two-point boundary value problem for a system of ordinary differential equations by a Chebyshev series using collocation and least squares.

D02TGF   Solves a system of linear ordinary differential equations by least-squares fitting of a series of Chebyshev polynomials using collocation.

I1b2:	**Nonlinear multipoint boundary value problems for ordinary differential equations**

Collected Algorithms of the ACM

A569   COLSYS: FORTRAN subroutine for solving nonlinear multipoint boundary value problems for mixed order systems of ordinary differential equations. Based upon spline collocation at Gaussian points using a B-spline basis. Approximate solutions are computed on a sequence of automatically selected meshes until a user-specified set of tolerances is satisfied. (See U. Ascher, J. Christiansen, and R. D. Russell, ACM TOMS 7 (1981) pp. 223–229.)

IMSL MATH/LIBRARY Subprogram Library

BVPFD   Solves a system of differential equations with boundary conditions at two points, using a variable order, variable step-size finite-difference method with deferred corrections.

BVPMS   Solves a system of differential equations with boundary conditions at two points, using a multiple shooting method.

NAG Subprogram Library

D02AGF   Solves the two-point boundary value problem for a system of ordinary differential equations, using initial value techniques and Newton iteration; it generalizes subroutine D02HAF to include the case where parameters other than boundary values are to be determined.

D02GAF   Solves the two-point boundary value problem with assigned boundary values for a system of ordinary differential equations, using a deferred correction technique and a Newton iteration.

D02HAF   Solves the two-point boundary value problem for a system of ordinary differential equations.

D02HBF   Solves the two-point boundary value problem for a system of ordinary differential equations, using initial value techniques (D02PAF) and Newton iteration; it generalizes subroutine D02HAF to include the case where parameters other than boundary values are to be determined.

D02RAF   Solves the two-point boundary value problem with general boundary conditions for a system of ordinary differential equations, using a deferred correction technique and Newton iteration.

D02SAF   Solves a two-point boundary value problem for a system of first order ordinary differential equations with boundary conditions, combined with additional algebraic equations. It uses initial value techniques and a modified Newton iteration in a shooting and matching method.

I1b3:	Eigenvalue (e.g., Sturm-Liouville) multipoint boundary value problems for ordinary differential equations

Collected Algorithms of the ACM

A537   CHARMA: A FORTRAN subprogram for calculating the characteristic values of Mathieu's differential equation for odd or even solutions. (See W. R. Leeb, ACM TOMS 5 (1979) pp. 112–117.)

NAG Subprogram Library

D02AGF   Solves the two-point boundary value problem for a system of ordinary differential equations, using initial value techniques and Newton iteration; it generalizes subroutine D02HAF to include the case where parameters other than boundary values are to be determined.

D02HBF   Solves the two-point boundary value problem for a system of ordinary differential equations, using initial value techniques (D02PAF) and Newton iteration; it generalizes subroutine D02HAF to include the case where parameters other than boundary values are to be determined.

D02KAF   Finds a specified eigenvalue of a regular second order Sturm-Liouville system defined on a finite range, using a Pruefer transformation and a shooting method.

D02KDF   Finds a specified eigenvalue of a regular or singular second order Sturm-Liouville system on a finite or infinite range, using a Pruefer transformation and a shooting method. Provision is made for discontinuities in the coefficient functions or their derivatives.

D02KEF   Finds a specified eigenvalue of a regular or singular second order Sturm-Liouville system on a finite or infinite range, using a Pruefer transformation and a shooting method. It also reports values of the eigenfunction and its derivatives. Provision is made for discontinuities in the coefficient functions or their derivatives.

I1c:	Service routines for ordinary differential equations (e.g., interpolation of solutions, error handling, test programs)

Collected Algorithms of the ACM

A546 SOLVEBLOK: A FORTRAN subprogram for solving almost block diagonal linear systems. Such matrices arise naturally in piecewise polynomial interpolation or approximation and in finite element methods for two-point boundary value problems. (See C. de Boor and R. Weiss, ACM TOMS 6 (1980) pp. 88–91.)

A603 COLROW and ARCECO: FORTRAN subroutines for solving certain almost block diagonal linear systems by modified alternate row and column elimination. Such systems arise when solving boundary value problems for ordinary differential equations. COLROW is designed for systems whose blocks all have the same dimension; ARCECO is designed for systems whose blocks may have different dimensions. (See J. C. Diaz, G. Fairweather, and P. Keast, ACM TOMS 9 (1983) pp. 376–380.)

A648 NSDTST and STDTST: FORTRAN routines for assessing the performance of initial value solvers for stiff or nonstiff systems. (See W. H. Enright and J. D. Pryce, ACM TOMS 13 (1987) pp. 28–34.)

NAG Subprogram Library

D02NRF Enquiry routine for communicating with D02NMF or D02NNF when supplying columns of a sparse Jacobian matrix.

D02NSF Setup routine which must be called by the user, prior to an integrator in the D02N subchapter, if full matrix linear algebra is required.

D02NTF Setup routine which must be called by the user, prior to an integrator in the D02N subchapter, if banded matrix linear algebra is required.

D02NUF Setup routine which must be called by the user, prior to an integrator in the D02N subchapter, if sparse matrix linear algebra is required.

D02NVF Setup routine which must be called by the user, prior to an integrator in the D02N subchapter, if backward differentiation formulas (BDF) are to be used.

D02NWF Setup routine which must be called by the user, prior to an integrator in the D02N subchapter, if the BLEND formulas are to be used.

D02NXF Optional output routine which the user may call, on exit from an integrator in the D02N subchapter, if sparse matrix linear algebra has been selected.

D02NYF	Diagnostic routine which the user may call either after any user-specified exit or after a mid-integration error exit from any of the integrators in the D02N subchapter.
D02NZF	Setup routine which must be called, if optional inputs need resetting, prior to a continuation call to any of the integrators in the D02N subchapter.
D02QQF	Sets up interrupts for use in D02QDF.
D02XAF	Interpolates the system of first order ordinary differential equations from information provided by the Runge–Kutta–Merson routine D02PAF.
D02XBF	Interpolates one component of the solution of a system of first order ordinary differential equations from information provided by the Runge–Kutta–Merson routine D02PAF.
D02XGF	Interpolates the solution of a system of first order ordinary differential equations from information provided by the Adams routine D02QAF or the Gear routine D02QBF.
D02XHF	Interpolates one component of the solution of a system of first order ordinary differential equations from information provided by the Adams routine D02QAF or the Gear routine D02QBF.
D02XJF	Interpolates components of the solution of a system of first order ordinary differential equations from information provided by the integrators in the D02N subchapter (or by the routine D02QDF).
D02XKF	Interpolates components of the solution of a system of first order ordinary differential equations from information provided by the integrators in the D02N subchapter (or by the routine D02QDF). It provides C^1 interpolation suitable for general use.
D02ZAF	Calculates the weighted norm of the local error estimate from inside a MONITR routine called from an integrator in the D02N subchapter.

PORT Subprogram Library

ODESE	Standard error subprogram for the routine ODES1.
ODESH	Default HANDLE routine for ODES. Used to access the results at the end of each integration time step.

I2a1:	**Parabolic partial differential equations**

Collected Algorithms of the ACM

A553	M3RK: A FORTRAN subroutine for solving initial value problems for nonlinear first order systems of ordinary differential equations which originate from semi-discretization of parabolic partial differential equations. M3RK is based on stabilized, explicit three-step Runge–Kutta formulas of order one and two, and degree 2 through 12. (See J. G. Verwer, ACM TOMS 6 (1980) pp. 236–239.)

I2a1a:	Parabolic partial differential equations in one spatial dimension

Collected Algorithms of the ACM

A494 PDEONE: Solution of systems of nonlinear parabolic partial differential equations in one space dimension using the method of lines. (See R. F. Sincovec and N. K. Madsen, ACM TOMS 1 (1975) pp. 261–263.)

A540 PDECOL: A FORTRAN subprogram for solving coupled systems of nonlinear partial differential equations in one space and one time dimension. The solution method uses finite element collocation based upon piecewise polynomials for spatial discretization. The time discretization is performed by general-purpose software for ordinary initial value problems. (See N. K. Madsen and R. F. Sincovec, ACM TOMS 5 (1979) pp. 326–351.)

IMSL MATH/LIBRARY Subprogram Library

MOLCH Solve a system of partial differential equations of the form $U_t = F(x,t,U,U_x,U_{xx})$ using the method of lines with cubic Hermite polynomials.

NAG Subprogram Library

D03PAF Integrates a single linear or nonlinear parabolic partial differential equation in one space variable, using the method of lines and Gear's method.

D03PBF Integrates a system of linear or nonlinear parabolic partial differential equations in one space variable, using the method of lines and Gear's method.

D03PGF Integrates a system of nonlinear parabolic partial differential equations in one space variable, using the method of lines and Gear's method. This routine provides quite general facilities; for simpler versions see D03PAF (for a single equation) or D03PBF (for simple systems).

I2a1b:	Parabolic partial differential equations in two or more spatial dimensions

Collected Algorithms of the ACM

A565 PDETWO/PSETM/GEARB: A FORTRAN package for solving time-dependent coupled systems of nonlinear partial differential equations that are defined over a two-dimensional rectangular region. (See D. K. Melgaard and R. F. Sincovec, ACM TOMS 7 (1981) pp. 126–135.)

A621 BDMG: A FORTRAN subprogram with low storage requirements for two-dimensional nonlinear parabolic differential equations on rectangular spatial domains with mixed linear boundary conditions. (See B. P. Sommeijer and P. J. van der Houven, ACM TOMS 10 (1984) pp. 378–396.)

I2a2:	**Hyperbolic partial differential equations**

Collected Algorithms of the ACM

A540   PDECOL: A FORTRAN subprogram for solving coupled systems
of nonlinear partial differential equations in one space and one
time dimension. The solution method uses finite element collocation based upon piecewise polynomials for spatial discretization.
The time discretization is performed by general-purpose software
for ordinary initial value problems. (See N. K. Madsen and
R. F. Sincovec, ACM TOMS 5 (1979) pp. 326–351.)

A565   PDETWO/PSETM/GEARB: A FORTRAN package for solving
time-dependent coupled systems of nonlinear partial differential
equations that are defined over a two-dimensional rectangular
region. (See D. K. Melgaard and R. F. Sincovec, ACM TOMS 7
(1981) pp. 126–135.)

| **I2b1a:** | **Second order linear elliptic boundary value prob-**
lems |
|---|---|

ELLPACK Program Library

ELLPACK   Solves linear elliptic partial differential equations in general domains in two dimensions and in boxes; a variety of boundary
conditions are handled. Users write programs in the ELLPACK
language (FORTRAN extension), which allows them to declare
elliptic problems and to select from a large library of modules to
solve them numerically. Results can be tabulated or plotted; the
solution is also available as a FORTRAN function for postpro-
cessing.

| **I2b1a1a:** | **Poisson (Laplace) or Helmholtz equation on a rec-**
tangular domain (or topologically rectangular in the
coordinate system) |
|---|---|

Collected Algorithms of the ACM

A527   GMA, GMAS, and KPICK: FORTRAN subroutines for linear
systems arising from five-point discretizations of separable or constant coefficient elliptic boundary value problems on rectangular
domains. A Dirichlet, Neumann, or mixed boundary condition
may be independently specified on each side of the rectangle,
or periodic boundary conditions may be specified on opposing
sides. Implements the generalized marching algorithm. (See
R. E. Bank, ACM TOMS 4 (1978) pp. 165–176.)

A541   FISHPAK: FORTRAN subroutines for solving separable elliptic partial differential equations. Drivers are available for the Helmholtz equation in Cartesian, polar, surface spherical coordinates, cylindrical and interior spherical coordinates. In addition, subprograms for solving systems of linear equations resulting from finite difference approximations to general separable problems are included. (See P. N. Swarztrauber and R. A. Sweet, ACM TOMS 5 (1979) pp. 352–364.)

A543   FFT9: A FORTRAN subroutine for the Dirichlet problem for the Helmholtz equation on a rectangle. The program is based upon 4th and 6th order accurate 9-point finite difference approximations and fast Fourier solution techniques. (See E. N. Houstis and T. S. Papatheodorou, ACM TOMS 5 (1979) pp. 490–493.)

A651   HFFT: FORTRAN routines for solving the Helmholtz equation on bounded two- or three-dimensional rectangular domains. (See R. F. Boisvert, ACM TOMS 13 (1987) pp. 235–249.)

CMLIB Library (FISHPAK Sublibrary)

HSTCRT   Solves the Helmholtz or Poisson equations in two dimensions in Cartesian coordinates on a staggered grid.

HSTCSP   Solves a modified Helmholtz equation in spherical coordinates with axisymmetry using a staggered grid.

HSTCYL   Solves a modified Helmholtz equation in cylindrical coordinates on a staggered grid.

HSTPLR   Solves the Helmholtz or Poisson equation in polar coordinates on a staggered grid.

HSTSSP   Solves the Helmholtz or Poisson equation in spherical coordinates on the surface of a sphere using a staggered grid.

HW3CRT   Solves the Helmholtz or Poisson equation in three dimensions using Cartesian coordinates.

HWSCRT   Solves the Helmholtz or Poisson equation in two dimensions in Cartesian coordinates.

HWSCSP   Solves a modified Helmholtz equation in spherical coordinates with axisymmetry.

HWSCYL   Solves a modified Helmholtz equation in cylindrical coordinates.

HWSPLR   Solves the Helmholtz or Poisson equation in polar coordinates.

HWSSSP   Solves the Helmholtz or Poisson equation in spherical coordinates on the surface of a sphere.

CMLIB Library (VHS3 Sublibrary)

HS3CRT   Sets up and solves the standard seven-point finite difference approximation on a staggered grid to the Helmholtz equation in Cartesian coordinates with a variety of possible boundary conditions.

FPS2H   Solves the Poisson or Helmholtz equation on a two-dimensional rectangle using a fast Poisson solver based on the HODIE finite-difference scheme.

FPS3H   Solves the Poisson or Helmholtz equation on a three-dimensional box using a fast Poisson solver based on the HODIE finite-difference scheme.

I2b1a1b:	**Poisson (Laplace) or Helmholtz equation on a non-rectangular domain**

Collected Algorithms of the ACM

A572   HELM3D: A FORTRAN subroutine for solving the Dirichlet problem for the Helmholtz equation on general bounded three-dimensional regions. Based upon second order accurate finite differences; the resulting linear system of equations is reduced to a capacitance matrix equation that is solved approximately by a conjugate gradient method. (See D. P. O'Leary and O. Widlund, ACM TOMS 7 (1981) pp. 239–246.)

A593   CMMEXP, CMMIMP, and CMMSIX: FORTRAN subroutines for solving the Helmholtz equation on bounded nonrectangular planar regions with Dirichlet or Neumann boundary conditions. Solution is based upon the Fourier method extended to non-rectangular regions using the capacitance matrix method. (See W. Proskurowski, ACM TOMS 9 (1983) pp. 117–124.)

A629   LAPLAC: A FORTRAN subroutine for the interior Dirichlet problem for Laplace's equation on a general three-dimensional domain. Based on integral equation techniques. (See K. E. Atkinson, ACM TOMS 11 (1985) pp. 85–96.)

NAG Subprogram Library

D03EAF   Solves Laplace's equation in two dimensions for an arbitrary domain bounded internally or externally by one or more closed contours, given the value of either the unknown function or its normal derivative (into the domain) at each point of the boundary.

I2b1a2:	**Other separable second order linear elliptic boundary value problems**

Collected Algorithms of the ACM

A527   GMA, GMAS, and KPICK: FORTRAN subroutines for linear systems arising from five-point discretizations of separable or constant coefficient elliptic boundary value problems on rectangular domains. A Dirichlet, Neumann, or mixed boundary condition may be independently specified on each side of the rectangle, or periodic boundary conditions may be specified on opposing sides. Implements the generalized marching algorithm. (See R. E. Bank, ACM TOMS 4 (1978) pp. 165–176.)

CMLIB Library (FISHPAK Sublibrary)

SEPELI   Solves separable elliptic boundary value problems on a rectangle.

SEPX4   Solves separable elliptic boundary value problems on a rectangle with constant coefficients in one direction.

I2b1a3:	**Nonseparable second order linear elliptic boundary value problems**

Collected Algorithms of the ACM

A637   GENCOL: A FORTRAN subprogram for linear second order elliptic problems with general linear boundary conditions on non-rectangular two-dimensional domains. Solves the problem using collocation with bicubic Hermite polynomials. (See E. N. Houstis, W. F. Mitchell, and J. R. Rice, ACM TOMS 11 (1985) pp. 379–412 and 413–415.)

A638   INTCOL and HERMCOL: FORTRAN subprograms for linear second order elliptic problems on rectangular two-dimensional domains. HERMCOL allows general linear boundary conditions while INTCOL requires uncoupled boundary conditions. Problems are solved using collocation with bicubic Hermite polynomials. (See E. N. Houstis, W. F. Mitchell, and J. R. Rice, ACM TOMS 11 (1985) pp. 379–412 and 416–418.)

I2b4:	**Service routines for elliptic boundary value problems**

Collected Algorithms of the ACM

A499   CONOPT: A subprogram which determines the contour scanning path for a two-dimensional region. The path is designed to help accelerate the propagation of edge effects when solving two-dimensional partial differential equations using iterative methods. (See W. Kinsner and E. D. Torre, ACM TOMS 2 (1976) pp. 82–86.)

A625   A FORTRAN subprogram which relates a general two-dimensional domain to a rectangular grid laid over it. (See J. R. Rice, ACM TOMS 10 (1984) pp. 443–452 and 453–462.)

I2b4a:	**Domain triangulation for elliptic boundary value problems** *(search also class P)*

NAG Subprogram Library

D03MAF Places a triangular mesh over a given two-dimensional region. The region may have any shape, including one with holes.

I2b4b:	**Solution of discretized elliptic equations**

Collected Algorithms of the ACM

A512 FACTOR, RHS, and SOLVE: FORTRAN subroutines for solving symmetric positive definite periodic quindiagonal systems of linear equations. (See A. Benson, and D. J. Evans, ACM TOMS 3 (1977) pp. 96–103.)

A527 GMA, GMAS, and KPICK: FORTRAN subroutines for linear systems arising from five-point discretizations of separable or constant coefficient elliptic boundary value problems on rectangular domains. A Dirichlet, Neumann, or mixed boundary condition may be independently specified on each side of the rectangle, or periodic boundary conditions may be specified on opposing sides. Implements the generalized marching algorithm. (See R. E. Bank, ACM TOMS 4 (1978) pp. 165–176.)

A541 FISHPAK: FORTRAN subroutines for solving separable elliptic partial differential equations. Drivers are available for the Helmholtz equation in Cartesian, polar, surface spherical coordinates, cylindrical and interior spherical coordinates. In addition, subprograms for solving systems of linear equations resulting from finite difference approximations to general separable problems are included. (See P. N. Swarztrauber and R. A. Sweet, ACM TOMS 5 (1979) pp. 352–364.)

CMLIB Library (FISHPAK Sublibrary)

BLKTRI Solves block tridiagonal systems of linear algebraic equations arising from the discretization of separable elliptic partial differential equations.

CBLKTR Solves certain complex block tridiagonal systems of linear equations arising from the discretization of separable elliptic partial differential equations.

CMGNBN Solves certain complex block tridiagonal systems of linear equations arising from Helmholtz or Poisson equations in two-dimensional Cartesian coordinates.

GENBUN Solves certain block tridiagonal systems of linear equations arising from Helmholtz or Poisson equations in two Cartesian coordinates.

POIS3D Solves block tridiagonal linear systems of algebraic equations arising from the discretization of separable elliptic partial differential equations in three dimensions.

POISTG   Solves block tridiagonal linear systems of algebraic equations aris-
              ing from the discretization of separable elliptic partial differential
              equations.

                        CMLIB Library (VHS3 Sublibrary)

PSTG3D   Solves certain block tridiagonal systems of linear algebraic equa-
              tions that arise in finite difference approximations to three-dimen-
              sional Helmholtz equations on a staggered grid.

                            NAG Subprogram Library

D03EBF   Uses the Strongly Implicit Procedure to calculate the solution
              to a system of simultaneous algebraic equations of five-point
              molecule form on a two-dimensional, topologically-rectangular
              mesh. (Topological means that a polar grid, for example (r,theta),
              can be used, being equivalent to a rectangular box.)

D03ECF   Uses the Strongly Implicit Procedure to calculate the solution
              to a system of simultaneous algebraic equations of seven-point
              molecule form on a three-dimensional, topologically-rectangular
              mesh. (Topological means that a polar grid, for example, can be
              used if it is equivalent to a rectangular box.)

D03EDF   Solves seven-diagonal systems of linear equations which arise
              from the discretization of an elliptic partial differential equation
              on a rectangular region. This routine uses a multigrid technique.

D03UAF   Performs at each call one iteration of the Strongly Implicit Pro-
              cedure. It is used to calculate on successive calls a sequence of
              approximate corrections to the current estimate of the solution
              when solving a system of simultaneous algebraic equations for
              which the iterative up-date matrix is of five-point molecule form
              on a two-dimensional, topologically-rectangular mesh. (Topolog-
              ical means that a polar grid, for example (r,theta), can be used,
              being equivalent to a rectangular box.)

D03UBF   Performs at each call one iteration of the Strongly Implicit Pro-
              cedure. It is used to calculate on successive calls the sequence of
              approximate corrections to the solution when solving a system of
              simultaneous algebraic equations for which the iterative up-date
              matrix is of seven-point molecule form on a three-dimensional,
              topologically-rectangular mesh. (Topological means that a polar
              grid, for example, can be used if it is equivalent to a rectangular
              box.)

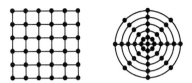

Figure 155.1 Two common computational grids, for rectilinear coordinates and for polar coordinates.

155. Grid Generation

Applicable to   Ordinary differential equations and partial differential equations.

Yields

A grid on which a differential equation may be numerically approximated.

Procedure

When a differential equation is going to be approximated numerically, the points at which the values of the dependent variable will be determined must be specified. This collection of points forms the *grid*, or *mesh*.

The most common computational grids are those in rectilinear coordinates or polar coordinates (see Figure 155.1). These can be used when the domain of a problem naturally fits one of these geometries. For other domains, an appropriate computational grid must be determined. There are many ways in which to construct a grid for a specific equation on a specific domain.

There are many considerations that go into choosing a grid for a specific problem. The grid should be easy to generate, and the algebraic equations used on the grid (usually finite differences or finite elements) must be easy to generate. (On page 581 we have indicated how finite difference approximations may be found on triangular grids.) For finite element methods, it is common to use triangulated grids, or grids composed of simple objects like triangles and rectangles. See example number 3 in the section on finite element methods (on page 662) for an example.

Ideally there should be many grid points where the solution (or its derivatives) are rapidly changing. Some grids naturally lend themselves to grid refinement in certain regions; this can be useful in adaptive techniques.

 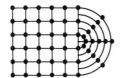 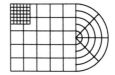

Figure 155.2 A domain, a possible grid on that domain, and a refined grid on that domain.

Example 1

For domains that can be described by combinations of simple geometric regions, a grid may be easy to find. See Figure 155.2 for a simple computational grid for a domain that can be conveniently decomposed into a rectangle and a semicircle. In this figure we have also illustrated how the grid may be modified if it is found that the solution shows great variation in the upper left region of the domain.

Example 2

There are many ways in which a grid may be found for a domain. Figure 155.3, taken from Rice [8], shows six different grids for a single irregularly shaped domain. The first three grids (A,B,C) show different possibilities:

- Grid A is a simple triangulation of the domain.
- Grid B is a uniform rectilinear grid on the domain.
- Grid C is a uniform rectilinear mapping, logically mapped to the domain.

The second three grids (D,E,F) indicate how the the first three grids can adapt to some difficulties near the right boundary.

Notes

[1] One of the greatest obstacles in generating numerical solution to fluid dynamics problems is the difficulty in geometrically describing complex configurations with computational grids.

[2] Conformal mappings are frequently used to construct computational grids (see page 376).

[3] The multigrid method (see page 673) uses a sequence of grids, of varying coarseness, to approximate the solution of a differential equation.

References

[1] L. Abrahamsson, "Orthogonal Grid Generation for Two Dimensional Ducts," *J. Comput. Appl. Math.*, **34**, 1991, pages 305–314.

[2] I. Atlas, R. Manohar, and J. W. Stephenson, "Adaptive Mesh Generation Using Quadratures," *Congr. Numer.*, **62**, 1988, pages 37–45.

[3] I. Atlas and J. W. Stephenson, "A Two-Dimensional Adaptive Mesh Generation Method," *J. Comput. Physics*, **94**, 1991, pages 201–224.

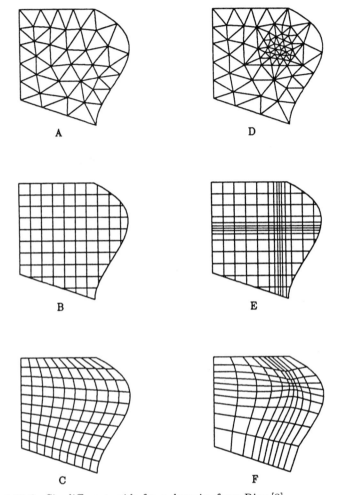

Figure 155.3. Six different grids for a domain, from Rice [8].

[4] J. E. Castillo (ed.), *Mathematical Aspects of Numerical Grid Generation*, SIAM, Philadelphia, 1991.

[5] P. R. Eisman, "Adaptive Grid Generation," *Comp. Meth. Appl. Mech. Eng.*, **64**, 1987, pages 321–376.

[6] W. F. Mitchell, "A Comparison of Adaptive Refinement Techniques for Elliptic Problems," *ACM Trans. Math. Software*, **15**, No. 4, December 1989, pages 326–347.

[7] A. Pardhanani and G. F. Carey, "Optimization of Computational Grids," *Num. Meth. Part. Diff. Eqs.*, **4**, No. 2, 1988, pages 95–117.

[8] J. R. Rice, "Parallel Methods for Partial Differential Equations," in L. H. Jamieson, D. B. Gannon, and R. J. Douglass (eds.), *The Characteristics of Parallel Algorithms*, MIT Press, Cambridge, MA, 1987, pages 209–231.

[9] P. D. Sparis, "A Method for Generating Boundary-Orthogonal Curvilinear Coordinate Systems Using the Biharmonic Equation," *J. Comput. Physics*, **61**, No. 3, 1985, pages 445–462.

[10] J. F. Thompson (ed.), *Special Issue on Numerical Grid Generation*, *Appl. Math. and Comp.*, **10** and **11**, 1982.

[11] J. F. Thompson (ed.), *Numerical Grid Generation*, Proceedings of a Symposium on the Numerical Generation of Curvilinear Coordinate Systems and Their Use in the Numerical Solution of Partial Differential Equations, April 1982, North–Holland Publishing Co., New York, 1982.

[12] J. F. Thompson, Z. U. A. Warsi, and C. Wayne Mastin, *Numerical Grid Generation Foundations and Applications*, North–Holland Publishing Co., New York, 1985.

156. Richardson Extrapolation

Applicable to Approximation techniques for differential equations.

Yields

A procedure for increasing the accuracy.

Procedure

Suppose that a grid with a characteristic spacing h is used to numerically approximate the solution of a differential equation. Then the approximation $u(\mathbf{x}; h)$ at the point $\mathbf{x}$ in the domain will satisfy

$$u(\mathbf{x}; h) = y(\mathbf{x}) + R_m(\mathbf{x})h^m + O(h^{m+1}), \qquad (156.1)$$

where $y(\mathbf{x})$ is the true solution to the differential equation, m is the order of the method, and the other terms represent the error (see page 573).

If the approximation scheme is kept the same, but the characteristic spacing of the grid is changed from h to k, then

$$u(\mathbf{x}; k) = y(\mathbf{x}) + R_m(\mathbf{x})k^m + O(k^{m+1}). \qquad (156.2)$$

Equations (156.1) and (156.2) can be combined to yield the approximation

$$v(\mathbf{x}; h, k) := \frac{k^m u(\mathbf{x}; h) - h^m u(\mathbf{x}; k)}{k^m - h^m} = y(\mathbf{x}) + O(kh^m, hk^m).$$

Note that $v(\mathbf{x}; h, k)$ is one more order accurate than either $u(\mathbf{x}; h)$ or $u(\mathbf{x}; k)$. This process may be iterated to increase the accuracy even more.

In some cases, the order of the method, and hence m in (156.1), will be unknown. The Richardson extrapolation method may still be used, by either estimating m numerically, or by using the Shanks transformation. The Shanks transformation uses three successive terms of the form $A_n = A_\infty + \alpha h^n$ to estimate A_∞ via

$$A_\infty = \frac{A_{n+1} A_{n-1} - A_n^2}{A_{n+1} + A_{n-1} - 2A_n}.$$

This transformation may also be iterated; see Bender and Orszag [1] for details.

Example 1

Given the differential equation

$$\frac{dy}{dx} = y, \quad y(0) = 1,$$

we might choose to approximate the solution by Euler's method

$$u_{n+1;h} = (1 + h)u_{n;h}, \quad u_{0;h} = 1,$$

where $u_{n;h} \simeq y(nh)$, and the step size satisfies $h \ll 1$. Observe that our notation explicitly shows the dependence of the approximation on the grid size. Doing a detailed analysis we can determine that

$$u_{n;h} = y(x) - \left(\frac{x}{2}\right) h + O(h^2), \tag{156.3}$$

where $x = nh$ and hence (here we choose $k = h/2$)

$$u_{2n;h/2} = y(x) - \left(\frac{x}{2}\right) \frac{h}{2} + O(h^2). \tag{156.4}$$

Combining (156.3) and (156.4) results in

$$w_{n;h} := 2u_{n;h} - u_{2n;h/2} = y(x) + O(h^2),$$

which is a numerical approximation that is second order accurate. Since h was reduced by a factor of 2 in going from (156.3) to (156.4), n had to be increased by a factor of 2 to maintain the same physical point, x.

h	u_h	$u_{h;R}$	$u_{h;RR}$	$u_{h;S}$	$u_{h;SS}$
0.200	1.45847				
0.100	1.43792	1.41738		1.41198	
0.050	1.42646	1.41499	1.41420	1.41376	1.41420
0.025	1.42043	1.41441	1.41421	1.41411	
0.012	1.41735	1.41426	1.41421		

Table 156. Numerical approximations to the solution of (156.5). More accurate results are obtained by applying Richardson extrapolation and the Shanks transformation to this data.

Example 2

Suppose we have the differential equation

$$\frac{dy}{dx} = \frac{ty}{t^2 + 1}, \quad y(0) = 1. \tag{156.5}$$

The exact solution to (156.5) is $y(t) = \sqrt{1 + t^2}$. Hence, $y(1) = \sqrt{2} \approx 1.41421$. Approximating (156.5) by use of Euler's method with a step size of h, we can obtain an approximation to the solution at $t = 1$, $u_h \approx y(1)$. As h decreases, this approximation should becomes better.

In Table 156 we show the values of u_h that are obtained when the h's are made successively smaller by a factor of two. Even though the last value is not very close to $\sqrt{2}$, we can improve the accuracy by using transformations. The first application of Richardson extrapolation is defined by (since Euler's method is first order accurate) $u_{h;R} := \dfrac{2u_h - u_{2h}}{2 - 1}$. The second application of Richardson extrapolation is defined by $u_{h;RR} := \dfrac{4u_{h;R} - u_{2h;R}}{4 - 1}$. The first application of the Shanks transformation is defined by

$$u_{h;S} := \frac{u_{2h}u_{h/2} - u_h^2}{u_{2h} + u_{h/2} - 2u_h}$$

The second application then uses the numbers $u_{h;S}$ in the same formula to obtain $u_{h;SS}$. As expected, the transformed values are much closer to the true value of $y(1)$.

Notes

[1] In the example just presented, the quantity $R_1(x)$ could be explicitly determined. However, to utilize this method, this value does not have to be known explicitly.

[2] To numerically approximate the solution to $y' = f(x, y)$, the modified midpoint method determines $y(x + nh)$, given $y(x)$, by

$$z_0 = y(x),$$
$$z_1 = z_0 + hf'(x, z_0),$$
$$z_{m+1} = z_{m-1} + 2hf'(x + mh, z_m), \qquad \text{for } m = 1, 2, \ldots, n-1,$$
$$y(x + nh) \simeq \tfrac{1}{2}[z_n + z_{n-1} + hf'(x + nh, z_n)],$$

where h is a small step size. This method is of second order, but has an error that only involves *even* powers of h. Hence, each Richardson extrapolation of this method increases the order by two. See Press *et al.*. [8] for more details.

[3] Richardson extrapolation is often referred to as *deferred approach to the limit*.

[4] This method also works for non-uniform grids if every interval is subdivided.

[5] Some functions are not well approximated by polynomials, but are well approximated by rational functions (see the section on Padé approximants, page 503). Instead of using a polynomial fit for the error term (as in (156.1)), a rational function approximation could be made—this is the basis of the Bulirsch–Stoer method. See Press *et al.* [8] for more details.

References

[1] C. M. Bender and S. A. Orszag, *Advanced Mathematical Methods for Scientists and Engineers*, McGraw–Hill, New York, 1978, page 369.

[2] J. R. Cash, "On the Numerical Integration of Nonlinear Two-Point Boundary Value Problems Using Iterated Deferred Corrections. II. The Development and Analysis of Highly Stable Deferred Correction Formulae," *SIAM J. Numer. Anal.*, **25**, No. 4, 1988, pages 862–882.

[3] E. Christiansen and H. G. Petersen, "Estimation of Convergence Orders in Repeated Richardson Extrapolation," *BIT*, **29**, 1989, pages 48–59.

[4] P. Deuflhard, "Recent Progress in Extrapolation Methods for Ordinary Differential Equations," *SIAM Review*, **27**, No. 4, December 1985, pages 505–535.

[5] R. Fößmeier, "On Richardson Extrapolation for Finite Difference Methods on Regular Grids," *Numer. Math.*, **55**, 1989, pages 451–462.

[6] E. Isaacson and H. B. Keller, *Analysis of Numerical Methods*, John Wiley & Sons, New York, 1966, pages 372–374.

[7] B. Lindberg, "Compact Deferred Correction Formulas," in J. Hinze (ed.), *Numerical Integration of Differential Equations and Large Linear Systems*, Springer–Verlag, New York, 1982, pages 220–233.

[8] W. H. Press, B. P. Flannery, S. Teukolsky, and W. T. Vetterling, *Numerical Recipes*, Cambridge University Press, New York, 1986, pages 83–86 and 563–568.

[9] L. F. Richardson, "The Approximate Arithmetical Solution by Finite Differences of Physical Problems Involving Differential Equations," *Philos. Trans. Roy. Soc. London*, Ser. A, **210**, 1910, pages 307–357.

157. Stability: ODE Approximations

Applicable to   Ordinary differential equations.

Yields

It is straightforward to determine if a finite difference scheme is stable.

Idea

If a finite difference scheme is stable, then a locally good approximation yields a globally good approximation (provided the differential equation— is well-posed).

Procedure 1

Difference schemes for ordinary differential equations may be stable or unstable. The definition closely parallels the definition for the stability and well-posedness of a differential equation. A stable difference scheme is one in which small changes in the initial and boundary data do not change the solution greatly. An unstable difference scheme is one that shows great sensitivity to the initial and boundary data.

To determine if the difference scheme for an ordinary differential equation is stable (or *zero-stable*), we apply the scheme to the equation $y' = 0$ (which has only a constant solution) and determine if the finite difference approximation stays bounded. Suppose we have the following difference scheme for the first order equation $y' = f(x, y)$:

$$\sum_{j=0}^{p} \alpha_j v_{n+j} - h \sum_{j=0}^{p} \beta_j f(x_{n+j}, v_{n+j}) = 0, \qquad (157.1)$$

where v_n is an approximation to $y(x_n)$ (and $x_n = nh$ for $n = 1, 2, \ldots$). Applying the above scheme to the test equation is equivalent to using

$f(x, y) = 0$ in (157.1). This results in

$$\sum_{j=0}^{p} \alpha_j v_{n-j} = 0. \qquad (157.2)$$

The method is said to be *stable* if all solutions of (157.2) are uniformly bounded for all n and all initial data $\{v_0, v_1, \ldots, v_{p-1}\}$.

The difference equation in (157.2) has solutions of the form $v_n = \lambda^n$. Using $v_n = \lambda^n$ in equation (157.2) results in the characteristic equation for λ

$$\lambda^n \rho(\lambda) = \sum_{j=0}^{p} \alpha_j \lambda^{n-j} = 0. \qquad (157.3)$$

It is easily shown that the method is unstable if any of the roots to (157.3) have magnitudes greater than one, or if there is a multiple root whose magnitude is equal to one.

Procedure 2

Sometimes "stability" is defined in terms of how the approximate solution to the equation $y' = \lambda y$ behaves. Using $f(y, x) = \lambda y$ and then $v_n = \lambda^n$, we are led to the *stability polynomial*. The stability polynomial associated with (157.1) is defined to be $\pi(r; \overline{h}) = \rho(\lambda) - \overline{h}\sigma(\lambda)$, where $\overline{h}$ represents $h\lambda$ and $\rho(x)$ and $\sigma(x)$ represent the first and second characteristic polynomials (see page 574). Using the stability polynomial we have the following definitions (see Lambert [10] for details):

> The method in (157.1) is said to be *absolutely stable* for a given $\overline{h}$ if, for that $\overline{h}$, all the roots of $\pi(r; \overline{h})$ satisfy $|r_s| < 1$ for $s = 1, 2, \ldots, p$, and to be *absolutely unstable* otherwise. An interval (a, b) of the real line is said to be an *interval of absolute stability* if the method is absolutely stable for all $\overline{h} \in (a, b)$.

> The method in (157.1) is said to be *relatively stable* for a given $\overline{h}$ if, for that $\overline{h}$, the roots of $\pi(r; \overline{h})$ satisfy $|r_s| < |r_1|$ for $s = 2, 3, \ldots, p$, and to be *relatively unstable* otherwise. An interval (a, b) of the real line is said to be an *interval of relative stability* if the method is relatively stable for all $\overline{h} \in (a, b)$.

Using these definitions, we define the method in (157.1) is said to be absolutely/relatively stable in a region $\mathcal{R}$ of the complex plane if, for all $\overline{h} \in \mathcal{R}$, the roots of the stability polynomial $\pi(r; \overline{h})$ have the required associated property (defined above).

Using the notion of stability in a region, we define the following types of stability:

- a method has *A-stability* if $\{h\lambda \mid \mathrm{Re}(h\lambda) < 0\} \subset \mathcal{R}$.
- a method has *A(α)-stability* if $\{h\lambda \mid -\alpha < \pi - \arg(h\lambda) < \alpha\} \subset \mathcal{R}$.
- a method has *A_0-stability* if $\{h\lambda \mid \mathrm{Im}(h\lambda) = 0, \quad \mathrm{Re}(h\lambda) < 0\} \subset \mathcal{R}$.

A picture of the region $\mathcal{R}$ is known as a *stability diagram*. When approximating a differential equation on a bounded interval, the limit $n \to \infty$, h fixed, is of interest. The stability diagram will indicate allowable values for h.

Example 1

Euler's method for the ordinary differential equation $y' = f(x, y)$ consists of the approximation: $v_{n+1} - v_n = hf(x_n, v_n)$. To determine if this method is stable, we apply this method to the equation $y' = 0$ to determine the difference scheme

$$v_n - v_{n-1} = 0. \tag{157.4}$$

Using $v_n = \lambda^n$ in (157.4) results in the characteristic equation

$$\rho(\lambda) = \lambda^n - \lambda^{n-1} = 0,$$

which has the roots $\lambda = 1$ and $\lambda = 0$ (with multiplicity $n - 1$). Since the only root with magnitude one, $\lambda = 1$, has multiplicity one, and all the other roots have magnitudes less than one, Euler's method is a stable method.

Example 2

Applying Euler's method to the equation $y' = f(x, y) = \lambda y$ we compute

$$v_{n+1} = v_n + hf_n$$
$$= v_n + h\lambda v_n$$
$$= (1 + \overline{h})v_n.$$

Hence, the region of absolute stability is given by $\mathcal{R} = \{\overline{h} \mid |1 + \overline{h}| \leq 1\}$, see Figure 157.a.

Applying Euler's backwards method to the equation $y' = f(x, y) = \lambda y$ we compute

$$y_{n+1} = y_n + hf_{n+1}$$
$$= y_n + h\lambda y_{n+1}$$
$$= \frac{y_n}{1 - \overline{h}}.$$

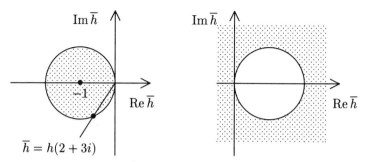

Figure 157. Stability diagrams for Euler's method in (a) and Euler's backwards method in (b). Region of absolute stability is shown shaded.

Hence, the region of absolute stability is given by $\mathcal{R} = \left\{ \overline{h} \; \middle| \; \dfrac{1}{|1 - \overline{h}|} \le 1 \right\}$, see Figure 157.b.

Stability diagrams can be used to determine allowable step sizes. If we were to integrate the ordinary differential equation $y' = (2 + 3i)y$ using Euler's method, then the maximum allowable (real) step size that will produce an absolutely stable method is $h = \frac{4}{13}$, see Figure 157.a. Stability diagrams are also used to qualitatively compare different difference schemes.

Notes

[1] Observe that a difference scheme can be stable and still not be consistent. Stability and accuracy are two entirely different concerns.

[2] For a stability analysis of second order ordinary differential equations, see Gear [6].

[3] Generally, the sequence of methods, {one step methods, iteration methods, implicit methods}, demonstrate progressively better stability. That is, it is generally true that larger step sizes can be taken for implicit methods than for explicit methods.

[4] Karim and Ismail [8] present five different ways in which to determine the stability of a difference scheme. They all lead to the same conclusion, but, on certain classes of equations, some methods are easier to apply than others.

[5] A detailed derivation and example of Euler's method is given on page 653.

[6] To determine if a finite difference scheme for a partial differential equation is stable see either the Courant–Friedrichs–Lewy consistency criterion (page 618) or the Von Neumann stability test (page 621).

[7] There are are many useful theorems in numerical analysis concerning the stability of methods for specific equations. For example, an A-stable method cannot have accuracy $p > 2$. See Dahlquist [3].

[8] A consistent method is called *stiffly stable* if (1) for some constant $D < 0$, all solutions of the difference equation generated by the application of this method to the scalar test equation, $y' = \lambda y$, tend to zero as $n \to \infty$ for all complex λ with Re $\lambda < D$ and for all fixed step sizes h with $h > 0$ and (2) there is an open set S whose closure contains the origin and the method is stable for $h\lambda \in S$. Here, h represents the grid spacing.

[9] There are many other types of stability that have been defined. A partial ordering of some common types of stability is

algebraic stability $\Rightarrow$ Euclidean AN-stability $\Rightarrow$ strong AN-stability
$\Rightarrow$ weak AN-stability $\Rightarrow$ A-stability

See Butcher [2] for details.

References

[1] K. Burrage, "(k,l)-Algebraic Stability of Runge–Kutta Methods," *IMA J. Num. Analysis*, **8**, No. 3, 1988, pages 385–400.

[2] J. C. Butcher, "Linear and Non-Linear Stability for General Linear Methods," *BIT*, **27**, 1987, pages 182–189.

[3] G. Dahlquist, "A Special Stability Problem for Linear Multistep Methods," *BIT*, **3**, 1963, pages 27–43.

[4] K. Dekker and J. G. Verwer, *Stability of Runge–Kutta Methods for Stiff Nonlinear Systems*, North–Holland Publishing Co., New York, 1984.

[5] V. G. Ganzha and R. Liska, "Application of the REDUCE Computer Algebra System to Stability Analysis of Difference Schemes," in E. Kaltofen and S. M. Watt (eds.), *Computers and Mathematics*, Springer–Verlag, New York, 1990, pages 119–129.

[6] C. W. Gear, "The Stability of Numerical Methods for Second Order Ordinary Differential Equations," *SIAM J. Numer. Anal.*, **15**, No. 1, February 1978, pages 188–197.

[7] A. Iserles, "Stability and Dynamics of Numerical Methods for Nonlinear Ordinary Differential Equations," *IMA J. Num. Analysis*, **10**, 1990, pages 1–30.

[8] A. I. A. Karim and G. A. Ismail, "The Stability of Multi-Step Formulae for Solving Differential Equations," *Int. J. Comp. Math.*, **13**, 1983, pages 53–67.

[9] J. D. Lambert, *Computational Methods in Ordinary Differential Equations*, Cambridge University Press, New York, 1973.

[10] J. D. Lambert, "Developments in Stability Theory for Ordinary Differential Equations," in A. Iserles and M. J. D. Powell (eds.), *The State of the Art in Numerical Analysis*, Clarendon Press, Oxford, 1987, pages 409–431.

[11] G. Wanner, "Order Stars and Stability," in A. Iserles and M. J. D. Powell (eds.), *The State of the Art in Numerical Analysis*, Clarendon Press, Oxford, 1987, pages 451–471.

158. Stability:
Courant Criterion

Applicable to   Hyperbolic partial differential equations.

Yields

A statement about whether or not a difference scheme may converge to the exact solution of a hyperbolic equation.

Idea

The "numerical domain of dependence" for a hyperbolic equation must include the actual domain of dependence in order for the numerical approximation of the solution to converge to the true solution.

Procedure

A hyperbolic partial differential equation has characteristics (see page 368). Generally, the dependent variables will satisfy ordinary differential equations along the characteristics. These characteristics will propagate from the curves along which the initial data is given to every point in the domain. Given a specific point at which the solution is desired, the characteristics through that point must be determined.

If a numerical scheme for a hyperbolic equation attempts to compute a numerical approximation to the solution at a point, then all of the relevant characteristics must be present, or the method may not converge to the correct solution.

Example

Suppose we have the wave equation

$$u_{tt} = c^2 u_{xx}, \tag{158.1}$$

for $u(x, t)$ where the constant c represents the wave speed. The initial conditions for (158.1) are assumed to be

$$u(x, 0) = f(x),$$
$$u_t(x, 0) = g(x).$$

We define $v_{n,j} = u(t_n, x_j)$ where $t_n := n\Delta t$ and $x_j := j\Delta x$. If a second order centered difference scheme is used, then (158.1) might be approximated as

$$\frac{u_{n+1,j} - 2u_{n,j} + u_{n-1,j}}{(\Delta t)^2} = c^2 \frac{u_{n,j+1} - 2u_{n,j} + u_{n,j-1}}{(\Delta x)^2},$$

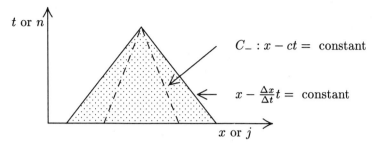

Figure 158.1 Characteristics (indicated by dashed lines) that are included in the numerical domain of dependence (shown shaded).

which can be manipulated into the explicit formula

$$u_{n+1,j} = 2\left[1 - \left(c\frac{\Delta t}{\Delta x}\right)^2\right]u_{n,j} + \left(c\frac{\Delta t}{\Delta x}\right)^2(u_{n,j+1} + u_{n,j-1}) - u_{n-1,j}.$$

$$(158.2)$$

Hence, the value of $u_{n+1,j}$ depends on $\{u_{n,j+k} \mid k = 0, \pm 1\}$ and $u_{n-1,j}$. Applying (158.2) to itself, we see that the value of $u_{n+1,j}$ depends on $\{u_{n-1,j+k} \mid k = 0, \pm 1, \pm 2\}$. Applying (158.2) again, we see that the value of $u_{n+1,j}$ depends on $\{u_{n-2,j+k} \mid k = 0, \pm 1, \pm 2, \pm 3\}$.

In general, the value of $u_{n+1,j}$ will depend on the points $\{u_{0,j+k} \mid k = 0, \pm 1, \ldots, \pm n\}$. These points along the initial curve (where the initial data is given) describe the *numerical domain of dependence*. See Figure 158.1.

The characteristics of (158.1) are the two curves (shown dashed in the figures)

$$C_-: \quad x - ct = x_i,$$
$$C_+: \quad x + ct = x_i,$$

where x_i is any point on the initial curve. Hence, the value of $u(t_n, x_j)$ will depend on the values of $u(0, x_k)$ for $x_k = x_i - ct$ and $x_k = x_i + ct$.

If these values are not included in the numerical domain of dependence, then the numerical approximation will, generally, give the incorrect answer. This is simply because the numerical approximation does not use the data that is important in solving the problem.

The two different possible scenarios are shown in Figure 158.1 and Figure 158.2. In Figure 158.1, the characteristics are included in the numerical domain of dependence (i.e., $\left(\dfrac{\Delta t}{\Delta x}\right)$ is less than one). Because of this, the method *may* converge to the exact solution. In Figure 158.2, the characteristics are not included in the numerical domain of dependence

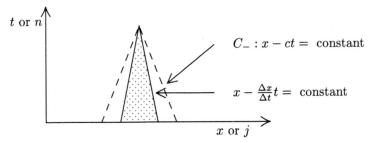

Figure 158.2 Characteristics (indicated by dashed lines) that are not included in the numerical domain of dependence (shown shaded).

(i.e., $\left(\dfrac{\Delta t}{\Delta x}\right)$ is greater than one). Because of this, the method *cannot*, in general, converge to the exact solution of (158.1).

In summary, for this example, if Δx and Δt are chosen so that

(A) $c\dfrac{\Delta t}{\Delta x} > 1$, then the method *cannot* converge to the exact solution.

(B) $c\dfrac{\Delta t}{\Delta x} < 1$, then the method *may* converge to the exact solution.

Notes

[1] This condition is also known as the Courant–Friedrichs–Lewy or CFL condition.

[2] Of course, more complicated hyperbolic problems will require a more detailed analysis.

[3] Another test that can be used to determine the stability of a finite difference scheme for partial differential equations is the Von Neumann stability test (see page 621).

[4] To determine if the difference scheme for an ordinary differential equation is stable, see page 573.

References

[1] J. L. Davis, *Finite Difference Methods in Dynamics of Continuous Media*, The MacMillan Company, New York, 1986, pages 45–47.

[2] D. Gottlieb and E. Tadmor, "The CFL Condition for Spectral Approximations to Hyperbolic Initial-Boundary Value Problems," *Math. of Comp.*, **56**, No. 194, April 1991, pages 565–588.

[3] E. Isaacson and H. B. Keller, *Analysis of Numerical Methods*, John Wiley & Sons, New York, 1966, page 489.

159. Stability: Von Neumann Test

Applicable to   Finite difference schemes for partial differential equations.

Yields

Knowledge of whether the difference scheme is stable.

Procedure

The Von Neumann test determines if the difference scheme for a partial differential equation is stable. For difference schemes with constant coefficients, the test consists of examining all exponential solutions to determine whether they grow exponentially in the time variable even when the initial values are bounded functions of the space variable.

If any of them do increase without limit then the method is *unstable*. Otherwise, it is *stable*.

This test can also be applied to equations with variable coefficients by introducing new, constant coefficients equal to the frozen values of the original ones at some specific point of interest.

Example

If the parabolic equation

$$u_t = u_{xx}$$

is discretized via

$$u_t \simeq \frac{1}{k}\Big[u(x, t + k) - u(x, t)\Big],$$

$$u_{xx} \simeq \frac{1}{h^2}\Big[u(x + h, t) - 2u(x, t) + u(x - h, t)\Big],$$

and $v_{m,n}$ is used to represent $u(mh, nk)$, then the recurrence relation

$$u_{m,n+1} = u_{m,n} + \frac{k}{h^2}\left(u_{m+1,n} - 2u_{m,n} + u_{m-1,n}\right) \qquad (159.1)$$

is obtained. To investigate all possible bounded exponential type solutions, we choose

$$u_{m,n} = e^{im\theta}e^{in\lambda}. \qquad (159.2)$$

Substituting (159.2) into (159.1) results in the relation

$$e^{i\lambda} = 1 - 4\frac{k}{h^2}\sin^2\left(\frac{\theta}{2}\right),$$

which must be satisfied for λ and θ. It can be shown that the imaginary part of λ will be non-negative (and hence the method is stable) if

$$\frac{k}{h^2} \leq \frac{1}{2}.$$

Notes

[1] A stability test for hyperbolic partial differential equations is the Courant–Friedrichs–Lewy consistency criterion (see page 618).

[2] To determine if the difference scheme for an ordinary differential equation is stable, see page 573.

References

[1] J. L. Davis, *Finite Difference Methods in Dynamics of Continuous Media*, The MacMillan Company, New York, 1986, pages 47–50.

[2] P. R. Garabedian, *Partial Differential Equations*, Wiley, New York, 1964, page 469 and page 477.

[3] D. Gottlieb and S. A. Orszag, *Numerical Analysis of Spectral Methods: Theory and Applications*, SIAM, Philadelphia, 1977, pages 48–50.

[4] E. Isaacson and H. B. Keller, *Analysis of Numerical Methods*, John Wiley & Sons, New York, 1966, pages 523–529.

[5] L. Lapidus and G. F. Pinder, *Numerical Solution of Partial Differential Equations in Science and Engineering*, Wiley, New York, 1982, pages 170–179.

IV.B

Numerical Methods for ODEs[*]

160. Analytic Continuation[*]

Applicable to   Initial value ordinary differential equations, a single equation or a system.

Yields

A numerical approximation in the form of a Taylor series.

Idea

If the Taylor series of a function is known at a single point, then the Taylor series of that function may be found at another (nearby) point. This process may be repeated until a particular value is reached.

---

* Some of the methods in this section can be used for partial differential equations as well. These methods are indicated by a star ($*$).

Procedure

Given a system of initial value ordinary differential equations, the method is to replace each dependent variable present by a Taylor series centered at a certain origin. The coefficients in each Taylor series are regarded as unknown quantities. The ordinary differential equations are used to obtain a set of recurrence relations from which the unknown coefficients may be calculated.

Thus, a formal power series solution may be determined to an initial value problem and the series will be convergent in some region about the origin. Then, the truncated power series are evaluated at some point within the region of convergence. At this new point, initial values for the system are obtained from the already obtained Taylor series. Using these initial values, the recurrence relations then yield a second series solution valid in a region about the new origin.

This procedure can be iterated and the solution at a given point may be determined via a sequence of Taylor series. This algorithm is a numerical version of the process of analytic continuation.

Example

Suppose we have the system of ordinary differential equations

$$y' = y^2 + z, \qquad y(0) = 1,$$
$$z' = z^2, \qquad z(0) = 1.$$

This system can be rewritten as the differential/algebraic system

$$a = y^2, \quad b = a + z, \quad c = z^2,$$
$$y' = b, \quad z' = c, \tag{160.1}$$

with $b = 2$, $a = c = y = z = 1$ when $t = 0$. If we define the Taylor series coefficients $\{a_k^{(j)}, b_k^{(j)}, c_k^{(j)}, y_k^{(j)}, z_k^{(j)}\}$ by the expansions

$$a(t) = \sum_{k=0}^{\infty} a_k^{(j)} (t - t_j)^k, \quad b(t) = \sum_{k=0}^{\infty} b_k^{(j)} (t - t_j)^k,$$

$$c(t) = \sum_{k=0}^{\infty} c_k^{(j)} (t - t_j)^k, \quad y(t) = \sum_{k=0}^{\infty} y_k^{(j)} (t - t_j)^k, \tag{160.2}$$

$$z(t) = \sum_{k=0}^{\infty} z_k^{(j)} (t - t_j)^k,$$

then, using (160.2) in (160.1), the following recurrence relations can be
obtained

$$a_k^{(j)} = \sum_{n=0}^{k} y_n^{(j)} y_{k-n}^{(j)}, \quad b_k^{(j)} = a_k^{(j)} + z_k^{(j)},$$

$$c_k^{(j)} = \sum_{n=0}^{k} z_n^{(j)} z_{k-n}^{(j)}, \quad y_k^{(j)} = b_k^{(j)}/(k+1), \tag{160.3}$$

$$z_k^{(j)} = c_k^{(j)}/(k+1).$$

The initial conditions give the starting values: $\{j = 0, t_0 = 0, a_0^{(0)} = c_0^{(0)} = y_0^{(0)} = z_0^{(0)} = 1, b_0^{(0)} = 2\}$. To determine the Taylor series about the point $t_0 = 0$, equation (160.3) is iterated for $k = 1, 2, \ldots, M$. The number of terms in each Taylor series required for a specified numerical accuracy M may be determined dynamically or fixed beforehand (if an appropriate analysis has been done).

Then a new point t_1 is chosen. A Taylor series for each of a, b, c, y, and z is then found about this new point by taking $j = 1$ and determining the initial conditions from.

$$a_0^{(1)} = \sum_{k=0}^{M} a_k^{(0)}(t_1 - t_0)^k, \quad b_0^{(1)} = \sum_{k=0}^{M} b_k^{(0)}(t_1 - t_0)^k, \quad \ldots$$

The recurrence relations in equation (160.3) are then iterated again. This process can be repeated indefinitely.

Notes

[1] In Holubec and Stauffer [5], a Frobenius series is continued instead of a Taylor series. This works particularly well on ordinary differential equations with regular singular points.

[2] A FORTRAN computer program that generates the recurrence relations and then solves the system is described in Corliss and Chang [3].

[3] Sometimes several hundred coefficients are required with this method to obtain an accurate answer. This is especially true when the expansion point for the Taylor series is near a singularity.

[4] Holubec and Stauffer [5] have a discussion on the appropriate step size to take at each stage in the calculation.

[5] Interval bounds (see page 470) for the Taylor series coefficients are discussed in Moore [7].

[6] This technique has been extended to parabolic equations in Chang [1].

References

[1] Y. F. Chang, "Solution of Parabolic Partial Differential Equations," in B. L. Hartnell and H. C. Williams (eds.), *Proceedings of the Sixth Manitoba Conference on Numerical Mathematics*, Utilitas Mathematics Publishing, Winnipeg, Canada, 1977, pages 127–134.

[2] Y. F. Chang, "Solving Stiff Systems by Taylor Series," *Appl. Math. and Comp.*, **31**, 1989, pages 251–269.

[3] G. Corliss and Y. F. Chang, "Solving Ordinary Differential Equations Using Taylor Series," *ACM Trans. Math. Software*, **8**, No. 2, June 1982, pages 114–144.

[4] G. Corliss and D. Lowery, "Choosing a Stepsize for Taylor Series Methods for Solving ODE's," *J. Comput. Appl. Math.*, **3**, No. 4, 1977, pages 251–256.

[5] A. Holubec and A. D. Stauffer, "Efficient Solution of Differential Equations by Analytic Continuation," *J. Phys. A: Math. Gen.*, **18**, 1985, pages 2141–2149.

[6] A. Holubec, A. D. Stauffer, P. Acacia, and J. A. Stauffer, "Asymptotic Shooting Method for the Solution of Differential Equations," *J. Phys. A: Math. Gen.*, **23**, 1990, pages 4081–4095.

[7] R. E. Moore, *Interval Analysis*, Prentice–Hall Inc., Englewood Cliffs, NJ, 1966, Chapter 11.

161. Boundary Value Problems: Box Method

Applicable to   Boundary value problems for ordinary differential equations.

Yields

A numerical approximation of the solution.

Idea

Using finite differences, the solution to a boundary value problem is determined (simultaneously) everywhere on the interval of interest.

Procedure

We will illustrate the procedure on the general second order linear ordinary differential equation. The same technique can be used, with only slight modifications, to systems of higher order ordinary differential equations, with the boundary data given virtually anywhere in the interval of interest.

Given the second order linear ordinary differential equation

$$a(x)y'' + b(x)y' + c(x)y = d(x),$$
$$y(x_L) = y_L, \quad y(x_U) = y_U, \tag{161.1.a--b}$$

we introduce the variable $z(x) = y'(x)$ and write (161.1) as the system

$$\frac{d}{dx}\begin{pmatrix} y \\ z \end{pmatrix} = \begin{pmatrix} z \\ \dfrac{d - cy - bz}{a} \end{pmatrix}. \tag{161.2}$$

Now we choose a grid, not necessarily uniform, on the interval (x_L, x_U), say $x_L = x_1 < x_2 < \cdots < x_N = x_U$. At each one of the grid points, some finite difference scheme is chosen to approximate the equations in (161.2). The scheme used can vary from point to point. For instance, if Euler's method is used for every point, then

$$\begin{pmatrix} y \\ z \end{pmatrix}_{k+1} = \begin{pmatrix} y \\ z \end{pmatrix}_k + (x_{k+1} - x_k)\begin{pmatrix} z \\ \dfrac{d - cy - bz}{a} \end{pmatrix}_k \tag{161.3}$$

to first order, where $y_k = y(x_k)$, $z_k = z(x_k)$, and similarly for $\{a_k, b_k, c_k, d_k\}$. From (161.1.b) the values $y_1 = y_L$ and $y_N = y_U$ are known.

To determine all of the $\{z_k\}$, and the remaining $\{y_k\}$, all of the relations in (161.3) (that is, for $k = 1, 2, \ldots, N$) should be combined into one large matrix equation. First, for ease of notation, define $h_k = x_{k+1} - x_k$, $e_k = d_k/a_k$, $f_k = c_k/a_k$ and $g_k = b_k/a_k$. In these new variables, equation (161.3) may be written as

$$y_{k+1} = y_k + h_k z_k,$$
$$z_{k+1} = z_k + h_k\left(e_k - f_k y_k - g_k z_k\right). \tag{161.4}$$

Combining all of the equations in (161.4) results in

$$\begin{pmatrix} 1 & h_1 & -1 & 0 & 0 & 0 & \cdots \\ h_1 f_1 & -1 + h_1 g_1 & 0 & 1 & 0 & 0 & \cdots \\ 0 & 0 & 1 & h_2 & -1 & 0 & \cdots \\ 0 & 0 & h_2 f_2 & -1 + h_2 g_2 & 0 & 1 & \cdots \\ \vdots & \vdots & & & & & \end{pmatrix}\begin{pmatrix} y_1 \\ z_1 \\ y_2 \\ z_2 \\ y_3 \\ \vdots \\ z_N \end{pmatrix} = \begin{pmatrix} 0 \\ h_1 e_1 \\ 0 \\ h_2 e_2 \\ 0 \\ \vdots \\ h_N e_N \end{pmatrix}.$$

To this matrix equation should be added two more rows, one corresponding to $y_1 = y_L$ and one corresponding to $y_N = y_U$. With these two rows, there results an $2N \times 2N$ matrix equation. This equation can be solved to determine a numerical approximation to the solution at all of the grid points.

Example

The second order linear ordinary differential equation

$$y'' + y = 3,$$
$$y(0) = 3, \quad y\left(\frac{\pi}{2}\right) = 2, \tag{161.5}$$

has the solution $y = 3 - \sin x$. We will use the box method to numerically approximate this solution. Writing (161.5) as a system results in

$$\frac{d}{dx}\begin{pmatrix} y \\ z \end{pmatrix} = \begin{pmatrix} z \\ 3 - y \end{pmatrix}. \tag{161.6}$$

We choose a uniform grid: $x_n = (n-1)h$ for $n = 1, 2, 3, 4$ with $h = \pi/6$. Defining $y_n = y(x_n)$ and $z_n = z(x_n)$, then, using Euler's method, (161.6) may be approximated as

$$y_{n+1} = y_n + h z_n,$$
$$z_{n+1} = z_n + h(3 - y_n). \tag{161.7}$$

Combining all the equations in (161.7) for $n = 1, 2, 3, 4$ results in

$$\begin{pmatrix} 1 & h & -1 & 0 & 0 & 0 & 0 & 0 \\ h & -1 & 0 & 1 & 0 & 0 & 0 & 0 \\ 0 & 0 & 1 & h & -1 & 0 & 0 & 0 \\ 0 & 0 & h & -1 & 0 & 1 & 0 & 0 \\ 0 & 0 & 0 & 0 & 1 & h & -1 & 0 \\ 0 & 0 & 0 & 0 & h & -1 & 0 & 1 \end{pmatrix} \begin{pmatrix} y_1 \\ z_1 \\ y_2 \\ z_2 \\ y_3 \\ z_3 \\ y_4 \\ z_4 \end{pmatrix} = \begin{pmatrix} 0 \\ 3h \\ 0 \\ 3h \\ 0 \\ 3h \end{pmatrix}.$$

Then the following two rows are added, to incorporate the known values of $y(0)$ and $y(\pi/2)$

$$\begin{pmatrix} 1 & 0 & 0 & 0 & 0 & 0 & 0 & 0 \\ 0 & 0 & 0 & 0 & 0 & 0 & 1 & 0 \end{pmatrix} \begin{pmatrix} y_1 \\ z_1 \\ y_2 \\ z_2 \\ y_3 \\ z_3 \\ y_4 \\ z_4 \end{pmatrix} = \begin{pmatrix} 3 \\ 2 \end{pmatrix}.$$

The FORTRAN program in Program 161 numerically approximates the solution to the above equation. Note that this program uses a linear equation solver LSOLVE, whose source code is not listed. The output of the program is

```
HERE IS THE APPROXIMATE SOLUTION:
   3.000    -0.701    2.633    -0.701    2.266    -0.509    2.000    -0.124
HERE IS THE EXACT SOLUTION:
   3.000    -1.000    2.500    -0.866    2.134    -0.500    2.000    0.000
```

The values for y_n are only accurate to one decimal place in this example. Putting more points in the interval would decrease the error, as would using a higher order method in place of Euler's method.

Program 161

```
          DIMENSION ARRAY(8,18),SOLN(8),RHS(8),NROW(100)
          PI=3.1415926
          NPOINT=8
          H=PI/2.* 2./FLOAT(NPOINT-2)
          DO 10 J=1,NPOINT
          DO 10 K=1,NPOINT
10        ARRAY(J,K)=0.0
C CREATE THE MATRIX
          ARRAY(1,1)=1.0
            RHS(1  )=3.0
          ARRAY(NPOINT,NPOINT-1)=1.0
            RHS(NPOINT          )=2.0
          J=1
20        J=J+1
          IF( J .GE. NPOINT ) GOTO 30
C HERE IS THE Y-EQUATION
          ARRAY(J,J-1)=1
          ARRAY(J,J  )=H
          ARRAY(J,J+1)=-1
            RHS(J  )=0
          J=J+1
C HERE IS THE Z-EQUATION
          ARRAY(J,J-2)=H
          ARRAY(J,J-1)=-1
          ARRAY(J,J+1)=1
            RHS(J  )=3.0*H
          GOTO 20
C SOLVE THE MATRIX SYSTEM
30        CALL LSOLVE(NPOINT,ARRAY,SOLN,RHS,NROW,IFSING,NPOINT)
          WRITE(6,5) (SOLN(J),J=1,NPOINT)
5         FORMAT(' HERE IS THE APPROXIMATE SOLUTION:',/,8(1x,F8.3) )
C COMPUTE THE EXACT SOLUTION FOR COMPARISON
          J=1
          DO 40 JJ=1,NPOINT/2
          SOLN(J  )=3.0-SIN( H*FLOAT(JJ-1) )
          SOLN(J+1)=   -COS( H*FLOAT(JJ-1) )
40        J=J+2
          WRITE(6,15) (SOLN(J),J=1,NPOINT)
15        FORMAT(' HERE IS THE EXACT SOLUTION:',/,8(1x,F8.3) )
          END
```

Notes

[1] In our example, if the two rows corresponding to the boundary terms were added to the matrix equation at the correct locations, the resulting matrix will be banded.

[2] This technique is recommended for stiff boundary value problems because many points can be added where the solution undergoes large changes, and different discretization schemes may be used in different regions.

[3] For nonlinear equations or nonlinear boundary conditions, this method can be used iteratively by linearizing the nonlinear terms at each step.

[4] Other techniques for solving boundary value problems include collocation (see page 441), shooting (see page 631), and invariant imbedding (see page 669).

[5] Scott and Watts [7] have a collection of computer programs in FORTRAN for solving two point boundary value problems. Ascher *et al.* [1], Daniel [2], and Mattheij [5] all have discussions of different techniques that can be applied to boundary value problems. Also, see the section beginning on page 586.

References

[1] U. M. Ascher, R. M. M. Mattheij, and R. D. Russel, *Numerical Solution of Boundary Value Problems for Ordinary Differential Equations*, Prentice–Hall Inc., Englewood Cliffs, NJ, 1988.

[2] J. W. Daniel, "A Road Map of Methods for Approximating Solutions of Two-Point Boundary-Value Problems," in B. Childs, M. Scott, J. W. Daniel, E. Denman, and P. Nelson (eds.), *Codes for Boundary-Value Problems in Ordinary Differential Equations*,
Springer–Verlag, New York, 1979, pages 1–18.

[3] J. Gregory and M. Zeman, "Spline Matrices and Their Applications to Some Higher Order Methods for Boundary Value Problems," *SIAM J. Numer. Anal.*, **25**, No. 2, April 1988, pages 399–410.

[4] E. Isaacson and H. B. Keller, *Analysis of Numerical Methods*, John Wiley & Sons, New York, 1966, pages 427–432.

[5] R. M. M. Mattheij, "Decoupling and Stability of Algorithms for Boundary Value Problems," *SIAM Review*, **27**, No. 1, March 1985, pages 1–44.

[6] S. M. Roberts and J. S. Shipman, *Two-Point Boundary Value Problems: Shooting Methods*, American Elsevier Publishing Company, New York, 1972, Chapter 8 (pages 201–231).

[7] M. R. Scott and H. A. Watts, *A Systematized Collection of Codes for Solving Two-Point Boundary Value Problems*, Sandia Report Number 75-0539, Sandia Labs, Albuquerque, NM, 1975.

162. Boundary Value Problems: Shooting Method*

Applicable to   Nonlinear boundary value problems for ordinary differential equations.

Yields

A numerical approximation to the solution.

Idea

Using Newton's method, the correct initial conditions for a boundary value problem can be determined. Knowing the initial conditions, the differential equations can be numerically integrated in a straightforward manner.

Procedure

The general procedure can be illustrated by studying a second order ordinary differential equation. Suppose we wish to numerically approximate the solution $y(x)$ of the equation

$$L(y'', y', y, x) = 0,$$
$$y(0) = 0, \quad y(1) = A,$$
$$(162.1)$$

where A is a given constant. The differential equation $L(\ ) = 0$ may or may not be a linear differential equation. If $z(x; \alpha)$ is defined to be the solution of

$$L(z'', z', z, x) = 0,$$
$$z(0; \alpha) = 0, \quad z'(0; \alpha) = \alpha,$$
$$(162.2)$$

then $y(x)$ will be equal to $z(x; \alpha)$ for one or more values of α. Of course, if $L(\ ) = 0$ were a linear equation, then there would be a single value of α. The parameter α in (162.2) must be determined so that

$$z(1; \alpha) = A.$$

Since (162.2) is an initial value problem, it is straightforward to integrate it numerically from $x = 0$ to $x = 1$. See, for instance, Euler's method (page 653). To use the shooting method, we integrate (162.2) numerically for some arbitrary initial guess for α, say α_0. If $z(1; \alpha_0) = A$, then $y(x) = z(x; \alpha_0)$ and we are done.

If $z(1; \alpha_0) \neq A$, then a new value of α must be chosen, say α_1. Equation (162.2) is then integrated for this new value of α. The process of choosing new values for α is repeated until the value of $z(1;\alpha)$ is sufficiently close to A. If the new α's are chosen well, then $z(1;\alpha)$ will converge to A and a numerical approximation to (162.1) will have been obtained. One way to choose the sequence of α's is by Newton's method

$$\alpha_{n+1} = \alpha_n - \frac{z(1;\alpha_n) - A}{\left.\frac{\partial}{\partial\alpha}z(1;\alpha)\right|_{\alpha=\alpha_n}}. \tag{162.3}$$

A numerical way to implement (162.3) might be

$$\alpha_{n+1} = \alpha_n - \frac{z(1;\alpha_n) - A}{[z(1;\alpha_n + \varepsilon) - z(1;\alpha_n)]/\varepsilon},$$

where ε is a small number.

Example
    Suppose we have the nonlinear second order ordinary differential equation

$$\begin{aligned} y'' + 2(y')^2 &= 0, \\ y(0) = 1, \quad y(1) &= \tfrac{1}{2}. \end{aligned} \tag{162.4}$$

Since (162.4) has no explicit dependence on y, the "dependent variable missing" method (see page 216) can be used to solve this equation exactly. By this technique, the solution of (162.4) is found to be

$$y(x) = 1 + \frac{1}{2}\log\left(1 + \frac{1-e}{e}x\right).$$

Hence, $y'(0) = (1 - e)/2e \simeq -.31606$.
    By use of the shooting method, a computer program should "discover" that $y'(0) \simeq -.31607$. The FORTRAN program in Program 162 utilizes finite differences to determine $y'(0)$ for (162.4). The equation in (162.4) is turned into the two first order ordinary differential equations

$$\frac{dy}{dx} = z,$$

$$\frac{dz}{dx} = -2y^2,$$

and then integrated by the use of Euler's method (see page 653).
    An initial guess of $y'(0) = 0$ is used in the program. The successive approximations of $y'(0)$ appear below:

```
ITERATION NUMBER   0  VALUE OF Y'(0)=   0.
```

```
ITERATION NUMBER   1  VALUE OF Y'(0)=  -0.50000050
ITERATION NUMBER   2  VALUE OF Y'(0)=  -0.49857452
ITERATION NUMBER   3  VALUE OF Y'(0)=  -0.49102421
ITERATION NUMBER   4  VALUE OF Y'(0)=  -0.46366318
ITERATION NUMBER   5  VALUE OF Y'(0)=  -0.40465858
ITERATION NUMBER   6  VALUE OF Y'(0)=  -0.34199798
ITERATION NUMBER   7  VALUE OF Y'(0)=  -0.31799014
ITERATION NUMBER   8  VALUE OF Y'(0)=  -0.31608113
ITERATION NUMBER   9  VALUE OF Y'(0)=  -0.31607109
```

Note that the computer program required a large number of steps in the interval $[0, 1]$ in order to achieve the accuracy shown (this is partly because we used Euler's method, which is of low order).

Program 162

```
            Y0=1.D0
            Y1=.5D0
            YP0=0.D0
C PERFORM A NEWTON ITERATION 9 TIMES
            DO 10 NEWT=1,10
            WRITE(6,5) NEWT-1,YP0
5           FORMAT(' ITERATION NUMBER',I4,' VALUE OF Y''(0)=',F13.8)
10          YP0=FNEWTON(Y0,Y1,YP0)
            END
C THIS FUNCTION PERFORMS ONE NEWTON STEP
            FUNCTION FNEWTON(Y0,Y1,YP0)
            EPS=.000001D0
            YP01=YP0
            YP02=YP0+EPS
            Z1=YAT1(Y0,YP01)
            Z2=YAT1(Y0,YP02)
            FNEWTON=YP0-(Z1-Y1)*EPS/(Z2-Z1)
            RETURN
            END
C THIS FUNCTION DETERMINES Y(1); WHEN Y(0) AND Y'(0) ARE GIVEN
            FUNCTION YAT1(Y0,YP0)
            N=20000
            DX=1.D0/DFLOAT(N)
            Y=Y0
            YP=YP0
C THIS IS THE ACTUAL INTEGRATION LOOP
            DO 10 J=1,N
            Y = Y  + DX * YP
10          YP= YP + DX * ( -2.D0*YP**2 )
            YAT1=Y
            RETURN
            END
```

Notes

[1] If this method is applied to a linear equation, the value of $y'(0)$ will converge to the correct value in a single step.

[2] It is also possible to simultaneous integrate along several rays at once. This is called the *method of multiple shooting*. See Stoer and Bulirsch [8] or Diekhoff *et al.* [1] for details.

[3] A test case that is often used to test computer codes for boundary value problems is *Troesch's problem*

$$\frac{d^2y}{dt^2} - n\sinh ny = 0,$$

$$y(0) = 0, \quad y(1) = 1.$$

See Roberts and Shipman [7] for a solution of this equation.

[4] For a listing of computer software that will implement the method described in this section, see page 586.

References

[1] H.-J. Diekhoff, P. Lory, H. J. Oberle, H.-J. Pesch, P. Rentrop, and R. Seydel, "Comparing Routines for the Numerical Solution of Initial Value Problems in Ordinary Differential Equations in Multiple Shooting," *Numer. Math.*, **27**, 1977, pages 449–469.

[2] H. B. Keller, *Numerical Solutions of Two Point Boundary Value Problems*, SIAM, Philadelphia, 1976, Chapter 1 (pages 1–19).

[3] H. B. Keller and P. Nelson, Jr., "Hypercube Implementations of Parallel Shooting," *Appl. Math. and Comp.*, **31**, 1989, pages 574–603.

[4] R. Lemmert, "The Shooting Method for some Nonlinear Sturm–Liouville Boundary Value Problems," *Z. Angew. Math. Phys.*, **40**, No. 5, 1989, pages 769–773.

[5] P. Marzulli and G. Gheri, "Estimation of the Global Discretization Error in Shooting Methods for Linear Boundary Value problems," *J. Comput. Appl. Math.*, **28**, 1989, pages 309–314.

[6] S. M. Roberts and J. S. Shipman, *Two-Point Boundary Value Problems: Shooting Methods*, American Elsevier Publishing Company, New York, 1972.

[7] S. M. Roberts and J. S. Shipman, "On the Closed Form Solution of Troesch's Problem," *J. Comput. Physics*, **21**, 1976, pages 291–304.

[8] J. Stoer and R. Bulirsch, *Introduction to Numerical Analysis*, translated by R. Bartels, W. Gautschi, and C. Witzgall, Springer–Verlag, New York, 1976.

163. Continuation Method*

Applicable to   Any type of equation at all: algebraic or differential, a single equation or a system.

Yields

A numerical approximation to the solution.

Idea

We embed a given problem into a problem with a continuation parameter σ in it. For one value of σ (say $\sigma = 1$) we obtain the original equations, while for a different value of σ (say $\sigma = 0$) we have an "easier" problem. We solve the simpler problem numerically and then slowly vary the continuation parameter from 0 to 1.

Procedure

After setting up the problem as described above, we define a metric that tells how well a function satisfies the problem when the continuation parameter is between 0 and 1. First, we numerically solve the easier problem (at $\sigma = 0$). Then the continuation parameter σ is increased by a small amount, and a solution is found by using Newton's method (this is accomplished by making the metric as small as possible). We increase σ some more, and repeat this step until we have arrived at $\sigma = 1$.

Example

Suppose we wish to solve the following boundary value problem for $y = y(x)$,

$$y_{xx} + e^y = 0, \qquad y(0) = 1, \quad y(\pi/2) = 0. \tag{163.1}$$

We embed (163.1) into the problem for $v = v(x; \sigma)$,

$$v_{xx} + (1 - \sigma)v + \sigma e^v = 0, \qquad v(0; \sigma) = 1, \quad v(\pi/2; \sigma) = 0. \tag{163.2}$$

Note that when $\sigma = 1$, the problem for $v(x; 1)$ becomes identical to the original problem that we wanted to solve, (163.1). Note also that, when $\sigma = 0$, the problem for $v(x; 0)$ becomes

$$v(x; 0)_{xx} + v(x; 0) = 0, \qquad v(0; 0) = 1, \quad v(\pi/2; 0) = 1,$$

with the solution $v(x; 0) = \cos x$.

The technique is to solve (163.2) numerically on a grid of values from 0 to $\pi/2$. We will start with $\sigma = 0$ and $v(x; 0) = \cos x$ and then increase σ by a small amount and allow $v(x; \sigma)$ to change accordingly.

We choose to solve (163.1) at the $N+1$ grid points: $\{x_n = hn \mid$ for $n = 0, 1, 2, \ldots, N\}$ where $h = \pi/2N$, and we define v_n^σ to be the numerical approximation to $v(x; \sigma)$ at the n-th gridpoint. We take $v_0^\sigma = 1$ and $v_N^\sigma = 0$ so that the boundary conditions to (163.2) are always satisfied.

Now we must define the metric. We choose

$$\varepsilon_n^\sigma = \frac{v_{n+1}^\sigma - 2v_n^\sigma + v_{n-1}^\sigma}{h^2} + (1 - \sigma)v_n^\sigma + \sigma e^{v_n^\sigma}. \qquad (163.3)$$

We choose this metric since, when ε_n^σ is close to zero, (163.2) will be approximately satisfied. This metric was obtained by simply applying a centered second order difference formula to (163.2).

The procedure is now as follows (with $\sigma_0 = 0$, $k = 0$):

(A) Increase σ by a small amount $\delta\sigma$ (i.e., $\sigma_{k+1} = \sigma_k + \delta\sigma$).

(B) Find $\{v_n^\sigma\}$ by making $\varepsilon_n^{\sigma_k} \simeq 0$. This is best accomplished by Newton's method. That is, we keep iterating

$$\begin{pmatrix} v_2^{\sigma_k} \\ v_3^{\sigma_k} \\ \vdots \\ v_{N-1}^{\sigma_k} \end{pmatrix}_{m+1} = \begin{pmatrix} v_2^{\sigma_k} \\ v_3^{\sigma_k} \\ \vdots \\ v_{N-1}^{\sigma_k} \end{pmatrix}_m - J^{-1} \begin{pmatrix} \varepsilon_2^{\sigma_k} \\ \varepsilon_3^{\sigma_k} \\ \vdots \\ \varepsilon_{N-1}^{\sigma_k} \end{pmatrix}_m,$$

where J is the Jacobian matrix defined by $J = \dfrac{\partial(\varepsilon_2^{\sigma_k}, \varepsilon_3^{\sigma_k}, \ldots, \varepsilon_{N-1}^{\sigma_k})}{\partial(v_2^{\sigma_k}, v_3^{\sigma_k}, \ldots, v_{N-1}^{\sigma_k})}$,

until the "difference" between $\begin{pmatrix} v_2^{\sigma_k} \\ v_3^{\sigma_k} \\ \vdots \\ v_{N-1}^{\sigma_k} \end{pmatrix}_{m+1}$ and $\begin{pmatrix} v_2^{\sigma_k} \\ v_3^{\sigma_k} \\ \vdots \\ v_{N-1}^{\sigma_k} \end{pmatrix}_m$ is smaller

than some predefined constant (based on the machine's numerical capabilities).

Note that the Jacobian and the $\{\varepsilon_n^\sigma\}$ all depend on the values of $\{v_n^{\sigma_k}\}_m$. The initial values for $\{v_n^{\sigma_k}\}_0$ will be given by $\{v_n^{\sigma_{k-1}}\}$. If $\delta\sigma$ is small enough, then Newton's method should converge.

(C) If $\sigma_k \neq 1$, go back to step (A).

(D) If $\sigma_k = 1$, then we have found a numerical approximation to the solution of (163.1).

Notes

[1] There are computer codes available that solve (A) through (D). The only input needed for them is the definition of the $\{\varepsilon_n^\sigma\}$.

[2] Continuation methods can be used to track different solution branches of a problem with bifurcations. If the Jacobian ever becomes singular (i.e., $\det J = 0$), a bifurcation point is likely. The null space of the Jacobian will indicate which directions are possible for the different solution branches.

[3] It is not uncommon in practice to find that the iteration in (163.3) *will not* converge unless $\delta\sigma$ is *very* small (at least initially). The better continuation programs available will automatically determine $\delta\sigma$, making it as small as is needed, but also increasing it when possible to speed up the calculation.

[4] Rheinboldt [3] has the FORTRAN listing for a continuation package.

[5] The method of invariant embedding (see page 669) is a specific type of continuation method.

[6] Continuation methods are also known as *homotopy methods*,

References

[1] R. L. Allgower and E. L. Georg, *Numerical Continuation Methods*, Springer-Verlag, New York, 1990.

[2] J. H. Bolstad and H. B. Keller, "A Multigrid Continuation Method for Elliptic Problems with Folds," *SIAM J. Sci. Stat. Comput.*, **7**, No. 4, October 1986, pages 1081–1104.

[3] W. C. Rheinboldt, *Numerical Analysis of Parametrized Nonlinear Equations*, Wiley Interscience, New York, 1986.

[4] W. C. Rheinboldt and J. V. Burkardt, "A Locally Parametrized Continuation Process," *ACM Trans. Math. Software*, **9**, No. 2, June 1983, pages 215–235.

[5] L. T. Watson, "Numerical Linear Algebra Aspects of Globally Convergent Homotopy Methods," *SIAM Review*, **28**, No. 4, December 1986, pages 529–545.

[6] L. T. Watson, S. C. Billups, and A. P. Morgan, "Algorithm 652: HOMPACK: A Suite of Codes for Globally Convergent Homotopy Algorithms," *ACM Trans. Math. Software*, **13**, No. 3, September 1987, pages 281–310.

164. Continued Fractions

Applicable to   Linear second order ordinary differential equations.

Yields

A solution in terms of a continued fraction.

Idea

By finding a simple recurrence pattern, we can express the logarithmic derivative of the solution to an ordinary differential equation in terms of a continued fraction.

Procedure

Suppose we have a linear second order ordinary differential equation in the form

$$y = Q_0(x)y' + P_1(x)y''. \tag{164.1}$$

If (164.1) is differentiated with respect to x, then we obtain

$$y' = Q_1(x)y'' + P_2(x)y''', \tag{164.2}$$

where

$$Q_1 = \frac{Q_0 + P_1'}{1 - Q_0'}, \qquad P_2 = \frac{P_1}{1 - Q_0'}. \tag{164.3}$$

If (164.2) is differentiated with respect to x, then we obtain $y'' = Q_2(x)y''' + P_3(x)y''''$ where $Q_2 = \dfrac{Q_1 + P_2'}{1 - Q_1'}$, $P_3 = \dfrac{P_2}{1 - Q_1'}$. This process can be repeated indefinitely to obtain

$$y^{(n)} = Q_n(x)y^{(n+1)} + P_{n+1}(x)y^{(n+2)} \tag{164.4}$$

with $Q_n = \dfrac{Q_{n-1} + P_n'}{1 - Q_{n-1}'}$, $P_{n+1} = \dfrac{P_n}{1 - Q_{n-1}'}$.

Now, dividing (164.1) by y' produces

$$\frac{y}{y'} = Q_0 + P_1 \frac{y''}{y'}$$

$$= Q_0 + \frac{P_1}{y'/y''}$$

$$= Q_0 + \cfrac{P_1}{Q_1 + P_2\dfrac{y'''}{y''}} \tag{164.5}$$

$$= Q_0 + \cfrac{P_1}{Q_1 + \cfrac{P_2}{Q_2 + P_3\dfrac{y''''}{y'''}}},$$

where we have used (164.3) for the third equality and (164.4) (with $n = 3$) for the fourth equality.

We can extend the continued fraction in (164.5) indefinitely. If it terminates, then it represents the reciprocal of the logarithmic derivative of the solution to (164.1). If it does not terminate, then it will converge if the following three conditions are satisfied:

(A) $P_n \to P$, $Q_n \to Q$ as $n \to \infty$.
(B) The roots $\{\rho_1, \rho_2\}$ of $\rho^2 = Q\rho + P$ are of unequal modulus.
(C) If $|\rho_2| < |\rho_1|$, then $\lim_{n \to \infty} |y^{(n)}|^{1/n} < \begin{cases} |\rho_2|^{-1} & \text{if } |\rho_2| \neq 0, \\ \infty & \text{if } |\rho_2| = 0. \end{cases}$

Example

Suppose we wish to find a continued fraction expansion for the reciprocal of the logarithmic derivative of the equation

$$xy'' - xy' - y = 0. \tag{164.6}$$

Comparing (164.6) with (164.1) we identify $Q_0(x) = -x$, $P_1(x) = x$. Using these values in (164.4), it is easy to show that $Q_n = 1 - x/(n+1)$ and $P_n = x/n$. Using these values, the partial sums for the continued fraction can be evaluated as

$$
\begin{aligned}
&\text{for 1 term:} && -\frac{x^2 + 2}{x}, \\[1em]
&\text{for 2 terms:} && -\frac{x^3 + 5x}{x^2 + 3}, \\[1em]
&\text{for 3 terms:} && -\frac{x^4 + 9x^2 + 8}{x^3 + 7x}, \\[1em]
&\text{for 4 terms:} && -\frac{x^5 + 14x^3 + 33x}{x^4 + 12x^2 + 15}.
\end{aligned}
\tag{164.7}
$$

The information in (164.7) can be used to approximately evaluate y/y'.

Notes

[1] This technique has rarely been extended, with any generality, to any types of differential equations other than linear second order ordinary differential equations. There has been a generalization to "matrix continued fractions" in Risken [7]. In Bellman and Wing [2], continued fractions are used to represent the solution to a Riccati equation.

[2] By taking partial sums of the continued fraction in (164.5), successively better approximations may be found. Rarely, though, can convergence be checked. See Field's paper [3].

[3] Continued fractions have been used recently to obtain high accuracy approximations to eigenvalues and functions of mathematical physics, see Barnett [1] or Gerck and d'Oliveira [4].

References

[1] A. R. Barnett, "High-precision Evaluation of the Regular and Irregular Coloumb Wavefunctions," *J. Comput. Appl. Math.*, **8**, No. 1, 1982, pages 29–33.

[2] R. Bellman and G. M. Wing, *An Introduction to Invariant Imbedding*, John Wiley & Sons, New York, 1975, page 19.

[3] W. L. Ditto and T. J. Pickett, "Exact Solution of Nonlinear Differential Equations Using Continued Fractions," *Nuovo Cimento B*, **105**, No. 4, 1990, pages 429–435.

[4] W. L. Ditto and T. J. Pickett, "Nonperturbative Solutions of Nonlinear Differential Equations Using Continued Fractions," *J. Math. Physics*, **29**, No. 8, '988, pages 1761–1770.

[5] D. A. Field, "Estimates of the Speed of Convergence of Continued Fraction Expansion of Functions," *Math. of Comp.*, **13**, No. 138, April 1977, pages 495–502.

[6] E. Gerck and A. B. d'Oliveira, "Continued Fraction Calculation of the Eigenvalues of Tridiagonal Matrices Arising from the Schroedinger Equation," *J. Comput. Appl. Math.*, **6**, No. 1, 1980, pages 81–82.

[7] F. B. Hilderbrand, *Introduction to Numerical Analysis*, McGraw–Hill Book Company, New York, 1974, pages 494–514.

[8] E. L. Ince, *Ordinary Differential Equations*, Dover Publications, Inc., New York, 1964, pages 178–182.

[9] H. Risken, *The Fokker–Planck Equation*, Springer–Verlag, New York, 1984, Chapter 9.

165.   Cosine Method*

Applicable to   Second order linear autonomous equations of a special form.

Yields

A finite difference scheme from which a numerical approximation to the solution may be obtained.

Idea

An exact representation of the solution is found. This exact representation is discretized to obtain an approximate numerical scheme.

Procedure

Suppose the following second order linear autonomous equation

$$\mathbf{u}'' + A\mathbf{u} = 0,$$
$$\mathbf{u}(0) = \mathbf{u}_0, \qquad \mathbf{u}'(0) = \mathbf{v}_0$$

(165.1)

is given for $\mathbf{u}(t)$, where A is a positive definite symmetric matrix. The solution to (165.1) has the *exact* representation

$$\mathbf{u}(t+k) + \mathbf{u}(t-k) = 2\cos\left(kA^{1/2}\right)\mathbf{u}(t),$$

where k represents a time step. Note that the cosine of a matrix is another matrix. See Moler and Van Loan [2] for how the exponential of a matrix may be computed.

The approximation scheme for (165.1) is based on the use of a rational function to approximate the cosine term:

$$\cos\left(kA^{1/2}\right) \simeq R\left(kA^{1/2}\right) = Q^{-1}\left(kA^{1/2}\right) P\left(kA^{1/2}\right).$$

Once a rational function has been chosen (i.e., P and Q have been picked), we define the approximation to $\mathbf{u}(t_j)$ to be $\mathbf{w}_j$ (where $t_j = jk$). The recurrence relation for $\mathbf{w}_j$ is then given by

$$Q\left(kA^{1/2}\right)(\mathbf{w}_{j+1} + \mathbf{w}_{j-1}) = 2P\left(kA^{1/2}\right)\mathbf{w}_j$$

or

$$\mathbf{w}_{j+1} = 2Q^{-1}\left(kA^{1/2}\right) P\left(kA^{1/2}\right)\mathbf{w}_j - \mathbf{w}_{j-1}$$

The first two values of $\mathbf{w}$ can easily be found (see page 548)

$$\mathbf{w}_0 = \mathbf{u}(0) = \mathbf{u}_0,$$

$$\mathbf{w}_1 = \mathbf{u}(k) = \mathbf{u}(0) + k\mathbf{u}'(0) + \frac{k^2}{2!}\mathbf{u}''(0) + \frac{k^3}{3!}\mathbf{u}'''(0) + \ldots$$

$$= \mathbf{u}(0) + k\mathbf{u}'(0) - \frac{k^2}{2!}A\mathbf{u}(0) - \frac{k^3}{3!}A\mathbf{u}'(0) + \ldots$$

$$= \mathbf{u}_0 + k\mathbf{v}_0 - \frac{k^2}{2!}A\mathbf{u}_0 - \frac{k^3}{3!}A\mathbf{v}_0 + \frac{k^4}{4!}A^2\mathbf{u}_0 + \ldots,$$

(165.2)

where the differential equation itself has been used to compute the higher order derivatives of $\mathbf{u}$. The number of terms kept in this series should correspond to the accuracy of the rational approximation used for the cosine function.

Example

Suppose we have

$$\mathbf{u}'' + \begin{pmatrix} 2 & 1 \\ 1 & 2 \end{pmatrix} \mathbf{u} = \mathbf{0},$$

$$\mathbf{u}(0) = \begin{pmatrix} 1 \\ -1 \end{pmatrix}, \qquad \mathbf{u}'(0) = \begin{pmatrix} 2\sqrt{3} \\ 2\sqrt{3} \end{pmatrix}. \tag{165.3}$$

Here $A = \begin{pmatrix} 2 & 1 \\ 1 & 2 \end{pmatrix}$ is symmetric and positive definite (its eigenvalues are one and three).

The exact solution of the system in (165.3) can be found by converting it into the following first order system

$$\begin{pmatrix} \mathbf{u} \\ \mathbf{v} \end{pmatrix}' = \begin{pmatrix} 0 & I \\ -A & 0 \end{pmatrix} \begin{pmatrix} \mathbf{u} \\ \mathbf{v} \end{pmatrix},$$

$$\begin{pmatrix} \mathbf{u}(0) \\ \mathbf{v}(0) \end{pmatrix} = \begin{pmatrix} \mathbf{u}_0 \\ \mathbf{v}_0 \end{pmatrix} = \begin{pmatrix} 1 \\ -1 \\ 2\sqrt{3} \\ 2\sqrt{3} \end{pmatrix},$$

where I is the two by two identity matrix and $\mathbf{u}' = \mathbf{v}$. The solution of this new system (see page 360) is

$$\begin{pmatrix} \mathbf{u}(t) \\ \mathbf{v}(t) \end{pmatrix} = \begin{pmatrix} \cos t + 2\sin(\sqrt{3}t) \\ -\cos t + 2\sin(\sqrt{3}t) \\ -\sin t + 2\sqrt{3}\cos(\sqrt{3}t) \\ \sin t + 2\sqrt{3}\cos(\sqrt{3}t) \end{pmatrix}.$$

To use the cosine method we need to approximate the cosine function. The (2,2) Padé approximant (see page 503) to the cosine function is

$$\cos(z) \simeq \frac{12 - 5z^2}{12 + z^2},$$

so that

$$Q\left(kA^{1/2}\right) = 12I + k^2 A,$$

$$P\left(kA^{1/2}\right) = 12I - 5k^2 A.$$

From this we obtain our discretization scheme

$$\mathbf{w}_{j+1} = -\mathbf{w}_{j-1} + 2(12I + k^2 A)^{-1}(12I - 5k^2 A)\mathbf{w}_j$$

$$= -\mathbf{w}_{j-1} + \alpha \begin{pmatrix} -15k^4 - 96k^2 + 144 & -72k^2 \\ -72k^2 & -15k^4 - 96k^2 + 144 \end{pmatrix} \mathbf{w}_j \tag{165.4}$$

where $\alpha = \dfrac{2}{3(k^2 + 4)(k^2 + 12)}$.

The FORTRAN program in Program 165 implements the above scheme with $k = .25$. To evaluate $\mathbf{w}_1$, we utilized the first five terms in (165.2). We choose to compare the output from the numerical approximation scheme to the exact solution when t is a multiple of 5. Even for t as large as 30, the results are accurate to two decimal places.

```
AT TIME    5.00   W(J) =    1.6667    1.0993
                  EXACT=    1.6680    1.1007
AT TIME   10.00   W(J) =   -2.8364   -1.1584
                  EXACT=   -2.8373   -1.1592
AT TIME   15.00   W(J) =    0.7422    2.2617
                  EXACT=    0.7403    2.2596
AT TIME   20.00   W(J) =    0.2361   -0.5798
                  EXACT=    0.2413   -0.5749
AT TIME   25.00   W(J) =   -0.2625   -2.2450
                  EXACT=   -0.2680   -2.2504
AT TIME   30.00   W(J) =    2.1371    1.8281
                  EXACT=    2.1386    1.8301
```

Program 165

```fortran
          IMPLICIT DOUBLE PRECISION (A-H,O-Z)
          REAL*8 W(0:1000,2),MAT(2,2),K
          K=.25D0
          TIME=K
          SQRT3=DSQRT(3.D0)
C SET UP THE INITIAL CONDITIONS
          W(0,1)= 1.D0
          W(0,2)=-1.D0
          W(1,1)= 1 + K*2*SQRT3 - K**2/2.D0 - K**3*SQRT3 + K**4/24.D0
          W(1,2)=-1 + K*2*SQRT3 + K**2/2.D0 - K**3*SQRT3 - K**4/24.D0
C SET UP THE MATRIX FOR THE RECURSION
          ALPHA = 2.D0/( 3.D0*(K**2+4)*(K**2+12) )
          MAT(1,1)= ALPHA  * ( - 15*K**4 - 96*K**2 + 144)
          MAT(1,2)= ALPHA  * (           - 72*K**2      )
          MAT(2,1)= MAT(1,2)
          MAT(2,2)= MAT(1,1)
C LOOP IN TIME
          DO 10 J=2,120
          TIME=TIME+K
          W(J,1)= -W(J-2,1) + MAT(1,1)*W(J-1,1) + MAT(1,2)*W(J-1,2)
          W(J,2)= -W(J-2,2) + MAT(2,1)*W(J-1,1) + MAT(2,2)*W(J-1,2)
C COMPUTE THE EXACT SOLUTION ALSO
          IF( MOD(J,20) .NE. 0 ) GOTO 10
          EXACT1=   DCOS(TIME) + 2*DSIN(SQRT3*TIME)
          EXACT2= - DCOS(TIME) + 2*DSIN(SQRT3*TIME)
          WRITE(6,5) TIME,W(J,1),W(J,2),EXACT1,EXACT2
5         FORMAT(' AT TIME',F7.2,'   W(J) =',2F9.4,/,18X,'EXACT=',2F9.4)
10        CONTINUE
          END
```

Notes

[1] This method has been extended to apply to non-homogeneous problems, equations with time dependent coefficients, and second order hyperbolic equations.

[2] Since the iterates in (165.4) do not depend linearly on the step size k, the cosine method is not a multi-step method (as defined on page 573).

References

[1] L. A. Bales, V. A. Douglas and S. M. Serbin, "Cosine Methods for Second-Order Hyperbolic Equations with Time-Dependent Coefficients," *Math. of Comp.*, **45**, July 1985, pages 65–89.

[2] L. A. Bales and V. A. Douglas, "Cosine Methods for Nonlinear Second-Order Hyperbolic Equations," *Math. of Comp.*, **52**, 1989, No. 186, pages 299–319.

[3] J. P. Coleman, "Numerical Methods for $y'' = f(x, y)$ via Rational Approximations for the Cosine," *IMA J. Num. Analysis*, **9**, 1989, pages 145–165.

[4] C. Moler and C. Van Loan, "Nineteen Dubious Ways to Compute the Exponential of a Matrix," *SIAM Review*, **20**, No. 4, October 1978, pages 801–836.

[5] S. M. Serbin, "Some Cosine Schemes for Second-Order Systems of ODE's with Time-Varying Coefficients," *SIAM J. Sci. Stat. Comput.*, **6**, No. 1, 1985, pages 61–68.

[6] S. M. Serbin and A. L. Fisher, "A Post-Processor for the Cosine Method," *SIAM J. Sci. Stat. Comput.*, **9**, No. 1, January 1988, pages 14–23.

166.   Differential Algebraic Equations

Applicable to   Differential algebraic equations, which are differential equations in the form

$$\mathbf{F}(x, \mathbf{y}, \mathbf{y}') = 0. \qquad (166.1)$$

Often, $\mathbf{F}(\ )$ is nonlinear in the $\mathbf{y}'$ term, or $\mathbf{F}(\ )$ contains a collection of differential and algebraic equations. A special subcase of differential algebraic equations is standard ordinary differential equations, in the common form $\mathbf{y}' = \mathbf{f}(x, \mathbf{y})$.

Yields

A numerical approximation to the solution.

Idea

Differential algebraic equations are more difficult to solve than standard ordinary differential equations. These equations are invariably solved exclusively by numerical means. One common numerical technique is to use the backwards Euler method. That is, (166.1) is approximated by

$$\mathbf{F}\left(x_{n+1}, \mathbf{y}_{n+1}, \frac{\mathbf{y}_{n+1} - \mathbf{y}_n}{x_{n+1} - x_n}\right) = 0,$$

and then the resulting system of nonlinear equations is solved for $\mathbf{y}_1$, then $\mathbf{y}_2$, etc.

Many special purpose codes have been written for these systems; see the references. There are, however, a few analytic solution techniques for differential algebraic equations, as the examples show.

Example 1

Algebraic differential equations arise, for instance, in the analysis of mechanical systems. Each component in a mechanical system will have equations of motion, as well as physical constraints (depending on how the given component is attached to other components in the system). It is these physical constraints that become algebraic constraints.

For example, consider a pendulum consisting of a point mass m, under the influence of gravity g, suspended by a massless rod of length l from an attachment point taken to be $x = 0$, $y = 0$. The equations of motion are:

$$\begin{aligned}
x' &= v_x, \\
y' &= v_y, \\
mu'_x &= -x\lambda, \\
mu'_y &= -y\lambda - g, \\
x^2 + y^2 &= l^2.
\end{aligned} \tag{166.2}$$

Here $\lambda(t)$ is the rod tension and $v_x(t)$ and $v_y(t)$ are the x and y velocities.

Example 2

The differential equation

$$y = f(y') = (y')^5 + (y')^3 + y' + 5 \tag{166.3}$$

for $y(x)$ is an example of a differential algebraic equation. It is impossible for (166.3) to be analytically written in the form $y' = g(x, y)$.

However, it is possible to solve differential equations of the form $y = f(y')$ parametrically. The solution may be written as

$$y = f(t), \qquad x = \int t^{-1} f'(t) \, dt + C,$$

where C is an arbitrary constant. Hence, equation (166.3) has the solution

$$x = \tfrac{5}{4}t^4 + \tfrac{3}{2}t^2 + \log t + C,$$
$$y = t^5 + t^3 + t + 5.$$

Example 3

If a differential algebraic equation is of the form $x = f(y')$, then the solution may be written parametrically as

$$x = f(t), \qquad y = \int t f'(t)\, dt + C,$$

where C is an arbitrary constant. Thus, the equation $x = (y')^3 - y' - 1$ has the parametric solution

$$x = t^3 - t - 1,$$
$$y = \tfrac{3}{4}t^4 - \tfrac{1}{2}t^2 + C.$$

Example 4

If a differential algebraic equation is of the form $f(y') = 0$, and there exists at least one real root of $f(k) = 0$, then $y = kx + C$ is a solution (where C is an arbitrary constant). Thus, the equation $(y')^5 - 6(y')^2 - 8 = 0$ has the solution $y = 2x + C$.

Notes

[1] If y is a solution to an algebraic differential equation, then y is called differentially algebraic. If u and v are differentially algebraic functions, then so are $u + v$, uv, u/v, $u \circ v$, u^{-1}, du/dt and $\int_0^t u(s)\, ds$. Hence, all of the elementary functions (such as the rational functions, e^x, $\tan^{-1}$, Bessel functions, etc.) are differentially algebraic. Note that the Gamma function $(\Gamma(x) = \int_0^\infty t^{x-1} e^{-t}\, dt)$ is *not* a differentially algebraic function.

The Shannon–Pour-El–Lipshitz–Rubel theorem roughly states that the outputs of general purpose analog computers are differentially algebraic functions. See Rubel [17].

[2] Differential algebraic equations of the form

$$\mathbf{u}' = \mathbf{f}(\mathbf{u}, \mathbf{v}, t),$$
$$0 = \mathbf{g}(\mathbf{u}, \mathbf{v}, t),$$

are said to be in *semi-explicit form*.

[3] A class of algebraic differential equations that are often studied are systems of the form

$$Ey' = Ay + g(t),$$
$$y(0) = y_0,$$

$$(166.4)$$

where A and E are given matrices. In the cases of interest, A or E (or both) are singular. For example, the system

$$y_2' = y_1 + g(x),$$
$$0 = y_2 + h(x),$$

is an algebraic differential equation in the form of (166.4).

[4] Consider (166.4) when $sE - A$ is a regular matrix pencil (i.e., $\det(sE - A)$ is not identically zero). (If $sE - A$ is not a regular matrix pencil then (166.4) is not well posed.) In this case, non-singular matrices P and Q can be found (see Gantmacher [5]) so that, with $y = Qz = (z_1, z_2)^T$ and $h(t) = Pg(t) = (h_1, h_2)^T$, equation (166.4) then takes the form

$$z_1' + Cz_1 = h_1(t)$$
$$Nz_2' + z_2 = h_2(t)$$

where N is a nilpotent matrix of degree n (i.e., $N^n = 0$ and $N^{n-1} \neq 0$). This is known as Kronecker canonical form. The degree n defines the *index of the problem* in (166.4). The index is equal to the size of the largest Jordan block for the eigenvalue zero ($\lambda = 0$) of $E - \lambda A$. If the index is zero, then E is non-singular and the system is easily solved numerically. Systems with an index greater than 1 are algebraically incomplete which means that the existence and the uniqueness of the solutions are not guaranteed. For example, the equations in (166.2) are of index 3.

As another example, the differential algebraic equations (see Roche [14])

$$y' = f(y, z)$$
$$0 = g(y, z)$$

are of index 1 if $(\partial g/\partial z)^{-1}$ exists and is bounded in the neighborhood of the exact solution.

[5] In Gear and Petzold [7] is the following algorithm in which the index of the problem in (166.4) can be reduced to zero by successive differentiations:

(A) If E is non-singular, go to (F).

(B) Find non-singular matrices P and Q such that $PEQ = \begin{pmatrix} E_{11} \\ 0 \end{pmatrix}$, with E_{11} having full rank.

(C) Make the variable substitution $\mathbf{y} = Q\mathbf{z}$ and multiply the equations from the left by P giving

$$\begin{pmatrix} E_{11} \\ 0 \end{pmatrix} \mathbf{z}' = \begin{pmatrix} F_{11} \\ F_{21} \end{pmatrix} \mathbf{z} + \begin{pmatrix} \mathbf{h}_1(t) \\ \mathbf{h}_2(t) \end{pmatrix}$$

(D) Differentiate the lower part of the system to arrive at the new problem

$$\begin{pmatrix} E_{11} \\ F_{21} \end{pmatrix} \mathbf{z}' = \begin{pmatrix} F_{11} \\ 0 \end{pmatrix} \mathbf{z} + \begin{pmatrix} \mathbf{h}_1(t) \\ -\mathbf{h}_2'(t) \end{pmatrix}$$

(E) If the "E" matrix for the new problem is singular, consider the new problem as the original problem and go to step (B).

(F) Done.

The index of the original problem is equal to the number of times the above loop must be executed.

[6] To indicate how much different the solution to algebraic differential equations can be from standard ordinary differential equations, consider the following theorem in Rubel [15]:

> Given any continuous function ϕ on $(-\infty, \infty)$ and any positive continuous function $\varepsilon(t)$ on $(-\infty, \infty)$, there exists a C^∞ solution of the algebraic differential equation

$$3y'^4 y'' y''''^2 - 4y'^4 y'''^2 y'''' + 6y'^3 y''^2 y''' y''''$$
$$+ 24y'^2 y''^4 y'''' - 12y'^3 y'' y'''^3 - 29y'^2 y''^3 y'''^2 + 12y''^7 = 0$$

> with $|y(t) - \phi(t)| < \varepsilon(t)$ for all $t \in (-\infty, \infty)$.

Hence, *any* continuous function is a "valid" numerical approximation to a solution of the above equation!

[7] A FORTRAN program for approximating the solution to differential algebraic equations of index 1, 2, and 3 is described in Hairer *et al.* [8]. This program is freely available via electronic mail. For a listing of computer software that will implement the method described in this section, see page 586.

References

[1] K. E. Brenan, S. L. Campbell, and L. R. Petzold, *Numerical Solution of Initial-Value Problems in Differential-Algebraic Equations*, North-Holland, Elsevier Pub. Co., Inc., NY, 1989.

[2] R. C. Buck, "The Solutions to a Smooth PDE Can Be Dense in $C[I]$," *J. Differential Equations*, **41**, 1981, pages 239–244.

[3] K. E. Brenan and L. R. Petzold, "The Numerical Solution of Higher Index Differential/Algebraic Equations by Implicit Methods," *SIAM J. Numer. Anal.*, **26**, No. 4, August 1989, pages 976–996.

[4] K. Burrage and L. Petzold, "On Order Reduction for Runge–Kutta Methods Applied to Differential/Algebraic Systems and to Stiff Systems of ODES," *SIAM J. Numer. Anal.*, **27**, No. 2, April 1990, pages 447–456.

[5] F. R. Gantmacher, *The Theory of Matrices*, **I**, **II**, Chelsea Publishing Company, New York, 1959.

[6] C. W. Gear, "Differential Algebraic Equations, Indices, and Integral Algebraic Equations," *SIAM J. Numer. Anal.*, **27**, No. 6, December 1990, pages 1527–1534.

[7] C. W. Gear and L. R. Petzold, "ODE Methods for the Solution of Differential/Algebraic Systems," *SIAM J. Numer. Anal.*, **21**, No. 4, August 1984, pages 716–728.

[8] E. Hairer, C. Lubich, and M. Roche, *The Numerical Solution of Differential-Algebraic Systems by Runge–Kutta Methods*, Springer–Verlag, New York, 1989.

[9] E. Hairer and G. Wanner, *Solving Ordinary Differential Equations, Volume II: Stiff and Differential-Algebraic Problems*, Springer–Verlag, New YorkVerlag, 1991.

[10] M. Hanke, "On a Least-Squares Collocation Method for Linear Differential-Algebraic Equations," *Numer. Math.*, **54**, 1988, pages 79–90.

[11] B. Leimkuhler, L. R. Petzold, and C. W. Gear, "Approximation Methods for the Consistent Initialization of Differential Algebraic Equations," *SIAM J. Numer. Anal.*, **28**, No. 1, February 1991, pages 205–226.

[12] L. Petzold and P. Lötstedt, "Numerical Solution of Nonlinear Differential Equations with Algebraic Constraints. II. Practical Implications," *SIAM J. Sci. Stat. Comput.*, **7**, No. 3, 1986, pages 720–733.

[13] W. C. Rheinboldt, *Numerical Analysis of Parameterized Nonlinear Equations*, Wiley Interscience, New York, 1986, Chapter 10 (pages 183–202).

[14] M. Roche, "Rosenbrock Methods for Differential Algebraic Equations," *Numer. Math.*, **52**, 1988, pages 45–63.

[15] L. A. Rubel, "A Universal Differential Equation," *Bull. Amer. Math. Soc.*, **4**, No. 3, May 1981, pages 345–349.

[16] L. A. Rubel, "Solutions of Algebraic Differential Equations," *J. Differential Equations*, **49**, 1983, pages 441–452.

[17] L. A. Rubel, "Some Mathematical Limitations of the General-Purpose Analog Computer" *Advances in Appl. Math.*, **9**, 1988, pages 22–34.

167. Eigenvalue/Eigenfunction Problems

Applicable to   Sturm–Liouville equations.

Yields

A numerical method for determining the eigenvalues and eigenfunctions of a regular Sturm–Liouville problem.

Idea

The Sturm–Liouville operator can be well approximated numerically by a simple discretization. This leads to a set of simultaneous equations, which can be represented as a matrix eigenvalue problem. The eigenvalues and eigenvectors of this matrix will approximate the eigenvalues and eigenfunctions of the Sturm–Liouville problem.

Procedure

Suppose we wish to numerically approximate the eigenvalues and eigenfunctions of the Sturm–Liouville system (see page 82):

$$(p(x)y')' + q(x)y = \lambda y,$$
$$y(0) = 0, \quad y(1) = 0, \tag{167.1}$$

for $x \in [0, 1]$. We will illustrate how the method of finite differences can be used to approximate the eigenvalues and eigenvectors. Equation (167.1) can be approximated by

$$D_-\left(p_{n+1/2}D_+ u_n\right) + q_n u_n = \lambda_h u_n,$$
$$u_0 = 0, \quad u_N = 0, \tag{167.2}$$

where $n = 1, 2, \ldots, N - 1$; $h = 1/N$; $u_n \simeq y(nh)$; and a function with a subscript of n corresponds to an evaluation at $x = hn$. Also, the forward and backward differencing operators are defined by: $D_- f_n := (f_n - f_{n-1})/h$ and $D_+ f_n := (f_{n+1} - f_n)/h$. It can be shown that (see Keller [8] or Isaacson and Keller [7])

$$|\lambda - \lambda_h| \leq Ch^2,$$

where C is some (unknown) constant. Therefore, for a sufficiently small h, the collection of $\{\lambda_h\}$ will closely approximate the collection of eigenvalues $\{\lambda\}$. The system in (167.2) is equivalent to the linear system of equations

$$A\mathbf{u}_h = h^2 \lambda_h \mathbf{u}_h, \tag{167.3}$$

where $\mathbf{u}_h = (u_1, \ldots, u_{N-1})^{\mathrm{T}}$ and A is the symmetric matrix

$$
\begin{pmatrix}
f_1 & p_{3/2} & 0 & 0 & \cdots & 0 & 0 \\
p_{3/2} & f_2 & p_{5/2} & 0 & \cdots & 0 & 0 \\
0 & p_{5/2} & f_3 & p_{7/2} & & 0 & 0 \\
\vdots & & \ddots & \ddots & \ddots & & \vdots \\
0 & 0 & & p_{N-5/2} & f_{N-2} & p_{N-3/2} & 0 \\
0 & 0 & \cdots & 0 & p_{N-3/2} & f_{N-1} & p_{N-1/2} \\
0 & 0 & \cdots & 0 & 0 & p_{N-1/2} & f_N
\end{pmatrix},
\qquad (167.4)
$$

where $f_m := h^2 q_m - (p_{m-1/2} + p_{m+1/2})$. Hence, the eigenvalues of (167.4), scaled by h^2 (see (167.3)), will approximate the eigenvalues of (167.1). Note that $\mathbf{u}_h$, the eigenvector of (167.4) corresponding to λ_h, is an approximation to the eigenfunction corresponding to λ_h. The eigenvalues and eigenvectors of (167.4) can be computed by standard numerical techniques. As N increases, more eigenvalues and eigenvectors are found and the accuracy of the lower order eigenvalues (and their associated eigenfunctions) increases.

Example

Consider the simple Sturm–Liouville system

$$
\begin{aligned}
y'' + y &= \lambda y, \\
y(0) &= 0, \quad y(1) = 0.
\end{aligned}
\qquad (167.5)
$$

For this system, the eigenfunctions and eigenvalues are given by

$$
\begin{aligned}
y_n(x) &= \sin n\pi x, \\
\lambda_n &= 1 - n^2 \pi^2,
\end{aligned}
\qquad (167.6)
$$

for $n = 1, 2, \ldots$. Hence, the two eigenvalues with the least magnitude are $\lambda_1 = 1 - \pi^2 \simeq -8.86$ and $\lambda_2 = 1 - 4\pi^2 \simeq -38.47$. To utilize the numerical technique presented above, we compare (167.5) with (167.1) to determine that $p(x) = 1$ and $q(x) = 1$.

If $N = 3$ (so that $h = 1/3$), then the matrix in (167.4) is given by

$$
\begin{pmatrix}
-17/9 & 1 & 0 \\
1 & -17/9 & 1 \\
0 & 1 & -17/9
\end{pmatrix}.
\qquad (167.7)
$$

The eigenvalues of the matrix in (167.7) are approximately -1.9 and -.49. When scaled by h^2, the estimates of the smallest eigenvalues of (167.5) become $\lambda_1 \simeq -4.3$ and $\lambda_2 \simeq -17.0$.

For $N = 10$ the estimates are $\lambda_1 \simeq -7.1$ and $\lambda_2 \simeq -30.7$, while for $N = 50$ the estimates are $\lambda_1 \simeq -8.5$ and $\lambda_2 \simeq -36.9$. As N increases, the estimates become better. If a higher order scheme were used to discretize (167.2), then smaller values of N would be required to obtain a given accuracy.

Notes

[1] Of course, Sturm–Liouville systems other than the one in (167.1) can be represented by the simple discretization in (167.2). More complicated boundary conditions may lead to a non-symmetric matrix in (167.3).

[2] Many other techniques have been used to approximate the eigenvalues and eigenfunctions of differential systems. These methods include finite elements, Galerkin methods, invariant embedding, Prüfer substitution, shooting, and variational methods. See page 551 of this book, Keller [8], or Cope [5].

[3] For a listing of computer software that will implement the method described in this section, see page 586.

References

[1] A. L. Andrew, "Correction of Finite Element Eigenvalues for Problems with Natural or Periodic Boundary Conditions," *BIT*, **28**, No. 2, 1988, pages 254–269.

[2] I. Babuška and J. E. Osborn, "Estimates of the Errors in Eigenvalue and Eigenvector Approximation by Galerkin Methods with Particular Attention to the Case of Multiple Eigenvalues," *SIAM J. Numer. Anal.*, **24**, No. 6, December 1987, pages 1249–1276.

[3] P. B. Bailey, M. K. Gordon, and L. F. Shampine, "Automatic Solution of the Sturm–Liouville Problem," *ACM Trans. Math. Software*, **4**, No. 3, September 1978, pages 193–208.

[4] R. Bellman and G. M. Wing, *An Introduction to Invariant Imbedding*, John Wiley & Sons, New York, 1975, Chapter 8 (pages 133–146).

[5] D. Cope, "A Uniformly Convergent Series for Sturm–Liouville Eigenvalues," *Quart. Appl. Math.*, **42**, No. 3, October 1984, pages 373–380.

[6] E. C. Gartland, Jr., "Accurate Approximation of Eigenvalues and Zeros of Selected Eigenfunctions of Regular Sturm–Liouville Problems," *Math. of Comp.*, **42**, No. 166, April 1984, pages 427–439.

[7] E. Isaacson and H. B. Keller, *Analysis of Numerical Methods*, John Wiley & Sons, New York, 1966, pages 434–436.

[8] H. B. Keller, *Numerical Solutions of Two Point Boundary Value Problems*, SIAM, Philadelphia, 1976, Chapter 3 (pages 39–48).

[9] J. P. Leroy and R. Wallace, "Extension of the Renormalized Numerov Method for Second-Order Differential Eigenvalue Equations," *J. Comput. Physics*, **67**, 1986, pages 239–252.

[10] M. D. Mikhailov and N. L. Vulchanov, "Computational Procedures for Sturm–Liouville Problems," *J. Comput. Physics*, **50**, 1983, pages 323–336.

[11] S. Pruess, "On Shooting Algorithms for Calculating Sturm–Liouville Eigenvalues," *J. Comput. Physics*, **75**, 1988, pages 493–497.

168. Euler's Forward Method

Applicable to   Initial value systems of first order ordinary differential equations.

Yields

A numerical marching scheme that is first order accurate.

Idea

A forward difference approximation to a derivative can be easily manipulated into a numerical scheme. The technique in this section is the most elementary finite difference approximation—other techniques are found on page 573.

Procedure

Given the first order system

$$\frac{d}{dt}\mathbf{y}(t) = \mathbf{f}[t, \mathbf{y}(t)],$$

$$\mathbf{y}(t_0) = \mathbf{y}_0,$$

$$(168.1.a\text{--}b)$$

where $\mathbf{y}$ and $\mathbf{f}$ are vectors, we numerically approximate dy/dt by $[\mathbf{y}(t + \Delta t) - \mathbf{y}(t)]/\Delta t$, where Δt is a small *step size*. This numerical approximation is first order accurate. Using this approximation, (168.1.a) can be rewritten as

$$\mathbf{y}(t + \Delta t) \simeq \mathbf{y}(t) + \Delta t\, \mathbf{f}[t, \mathbf{y}(t)]. \qquad (168.2)$$

Hence, to integrate (168.1) we iterate (168.2) and use the initial conditions from (168.1.b) for

$$\mathbf{y}(t_0) = \mathbf{y}_0,$$
$$\mathbf{y}(t_0 + \Delta t) \simeq \mathbf{y}(t_0) + \Delta t\, \mathbf{f}[t_0, \mathbf{y}_0(t_0)],$$
$$\mathbf{y}(t_0 + 2\Delta t) \simeq \mathbf{y}(t_0 + \Delta t) + \Delta t\, \mathbf{f}[t_0 + \Delta t, \mathbf{y}_0(t_0 + \Delta t)],$$
$$\mathbf{y}(t_0 + 3\Delta t) \simeq \mathbf{y}(t_0 + 2\Delta t) + \Delta t\, \mathbf{f}[t_0 + 2\Delta t, \mathbf{y}_0(t_0 + 2\Delta t)],$$
$$\vdots$$

Example

Suppose we want to approximate the value of $y(1)$ when $y(t)$ is defined by

$$\frac{dy}{dt} = \frac{ty}{t^2 + 1}, \qquad y(0) = 1. \tag{168.3}$$

Since this equation is separable, the exact solution is known to be $y(t) = \sqrt{1 + t^2}$. We can use this exact solution to compare the accuracy of the numerical approximation. The FORTRAN code in Program 168 uses Euler's forward method to numerically approximate the solution of (168.3). The code uses a step size of $\Delta t = 0.1$. The output from this program is listed below, with the exact solution alongside for comparison. The error in the calculated value for $y(1)$ is about 1.7%.

```
T= 0.100  Y= 1.00990  EXACT SOLUTION= 1.00499
T= 0.200  Y= 1.02932  EXACT SOLUTION= 1.01980
T= 0.300  Y= 1.05765  EXACT SOLUTION= 1.04403
T= 0.400  Y= 1.09412  EXACT SOLUTION= 1.07703
T= 0.500  Y= 1.13789  EXACT SOLUTION= 1.11803
T= 0.600  Y= 1.18809  EXACT SOLUTION= 1.16619
T= 0.700  Y= 1.24390  EXACT SOLUTION= 1.22066
T= 0.800  Y= 1.30458  EXACT SOLUTION= 1.28062
T= 0.900  Y= 1.36945  EXACT SOLUTION= 1.34536
T= 1.000  Y= 1.43792  EXACT SOLUTION= 1.41421
```

If the number of steps were increased (so the step size decreased) then the accuracy would improve. For example, if (in the above example) NDIV were increased to 100, the calculated value of $y(1)$ would be 1.41672. Hence, the error in the calculated value for $y(1)$ would decrease to about 0.17%.

Program 168

```
          NDIV=10
          TINIT=0.D0
          TEND= 1.D0
          DELTAT=(TEND-TINIT)/DFLOAT(NDIV)
          T=0
          Y=1
C THIS IS THE INTEGRATION LOOP
          DO 10 J=1,NDIV
          T=T + DELTAT
          Y=Y + DELTAT * YPRIME(T,Y)
          EXACT=DSQRT(1+T**2)
          WRITE(6,5) T,Y,EXACT
5         FORMAT(' T=', F6.3,'  Y=', F8.5,'  EXACT SOLUTION=',F8.5)
10        CONTINUE
          END
C THIS FUNCTION SPECIFIES THE DIFFERENTIAL EQUATION
          FUNCTION YPRIME(T,Y)
          YPRIME= T*Y / (T**2+1)
          RETURN
          END
```

Notes

[1] This technique is the easiest to use and program of all the numerical methods presented in this book. A major drawback is that the step size Δt may have to be very small for accurate numerical values.

[2] There is also a method known as *Euler's backward method*. For this implicit method, the difference scheme is given by

$$\mathbf{y}(t + \Delta t) \simeq \mathbf{y}(t) + \Delta t \, \mathbf{f}[t, \mathbf{y}(t + \Delta t)]. \tag{168.4}$$

In general, equation (168.4) will be nonlinear in $\mathbf{y}(t+\Delta t)$. Hence an iterative scheme (such as Newton's method) must be employed to find $\mathbf{y}(t + \Delta t)$.

[3] The stability properties of Euler's forward and backward methods are completely different. Consider applying each method to the scalar differential equation $y' = -cy$, $y(0) = y_0$, where c is a positive constant. For Euler's forward method we have

$$\begin{aligned}
y(t + \Delta t) &\simeq y(t) + \Delta t \, y'(t), \\
&= y(t) - c\Delta t \, y(t), \\
&= (1 - c\Delta t)y(t), \\
&= y_0(1 - c\Delta t)^{t/\Delta t}.
\end{aligned} \tag{168.5}$$

While for Euler's backward method we find

$$\begin{aligned}
y(t + \Delta t) &\simeq y(t) + \Delta t \, y'(t + \Delta t), \\
&= y(t) - c\Delta t \, y(t + \Delta t), \\
&= \frac{y(t)}{1 + c\Delta t}, \\
&= \frac{y_0}{(1 + c\Delta t)^{t/\Delta t}}.
\end{aligned} \tag{168.6}$$

Note that the approximation in (168.5) diverges in an oscillatory fashion when $\Delta t > 2/c$, while the approximation in (168.6) is stable for any value of Δt. In particular, if $c \gg 1$ (so that the problem is stiff, see page 690) then Δt may have to be very small for Euler's forward method to be stable, while a larger value of Δt can be used with Euler's backward method.

[4] As an indication of the different convergence properties of Euler's forward and backward methods, consider the equation: $\dot{y} = -6y + 5e^{-t}$. Figure 168 shows the exact solution ($y = e^{-t}$) and approximations obtained by using Euler's forward method ($\Delta t = .3$ and $\Delta t = .1$) and Euler's backward method ($\Delta t = .3$). On this problem, Euler's backward method is better than Euler's forward method for a fixed step size.

[5] As always, ordinary differential equations of higher order can be written as a system of first order equations (see page 118).

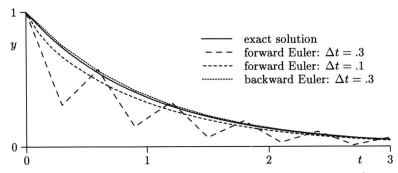

Figure 168. Different numerical techniques applied to $\dot{y} = -6y + 5e^{-t}$.

References

[1] W. E. Boyce and R. C. DiPrima, *Elementary Differential Equations and Boundary Value Problems*, Fourth Edition, John Wiley & Sons, New York, 1986, pages 399–406.

[2] C. W. Gear, *Numerical Initial Value Problems in Ordinary Differential Equations*, Prentice–Hall Inc., Englewood Cliffs, NJ, 1971, pages 10–23.

[3] W. H. Press, B. P. Flannery, S. Teukolsky, and W. T. Vetterling, *Numerical Recipes*, Cambridge University Press, New York, 1986, pages 574–577.

169. Finite Element Method*

Applicable to   Differential equations that arise from variational principles. Principally ordinary differential equations and elliptic partial differential equations.

Yields

A numerical scheme for approximating the solution.

Procedure

The finite element method is one version of the method of weighted residuals (see page 699). The present method is characterized by having "local elements." The finite element method has a specialized vocabulary, several of the terms appearing below will be defined in the example.

Given a differential equation that comes from a variational principle, and a domain in which the equation is to be solved, the steps are as follows:

[1] Discretize the domain into simple shapes (these are the "finite elements"). Define a basis function $\phi_k(\mathbf{x})$ on each of the finite elements. These basis functions should have bounded support.

[2] Assemble the stiffness matrix and the load matrix. These only depend on the finite elements chosen and not on the differential equation to be approximated.

[3] Write the given differential equation as a variational principle. Approximate the unknown in the variational principle by a linear combination of the functions defined on the finite elements; i.e., $u(\mathbf{x}) \simeq u_N(\mathbf{x}) := \sum_{k=1}^{N} c_k \phi_k(\mathbf{x})$. In this last expression, the $\{c_k\}$ are unknown and must be determined.

[4] Construct element stiffness matrices and load vectors, element by element. Then assemble these together into the global stiffness matrix A and the global load vector $\mathbf{f}$.

[5] Relate the minimization in the variational principle to the minimization of the quadratic functional

$$I[u_N] = \mathbf{c}^T A \mathbf{c} - 2\mathbf{c}^T \mathbf{f}. \tag{169.1}$$

When A is symmetric (as it frequently is), the minimization of (169.1) will occur when $\mathbf{c}$ is the solution of the system: $A\mathbf{c} = \mathbf{f}$. In general, A will not be banded or tridiagonal, but it will be sparse. If the original differential equation was nonlinear, then $A = A(\mathbf{c})$ or $\mathbf{f} = \mathbf{f}(\mathbf{c})$.

There is a large literature on the finite element method. We choose to illustrate the basic ideas on simple examples: the first two examples are constant coefficient second order linear ordinary differential equations, the third example is for Laplace's equation. These examples show the major steps involved, without the details that a sophisticated implementation requires.

Example 1

Suppose we have the constant coefficient second order linear ordinary differential equation

$$L[u] := -\frac{d}{dx}\left(p(x)\frac{du}{dx}\right) + q(x)u = f(x) \tag{169.2}$$

on the interval $0 \le x \le 1$. For simplicity, we take $p(x)$ and $q(x)$ to be constants. For this equation, we take the natural boundary conditions

$$u(0) = u(1) = 0. \tag{169.3}$$

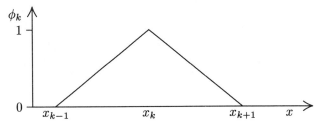

Figure 169.1 The "hat functions" in (169.5).

If we use $I[v]$ to represent the "energy" of the system, then we may form

$$I[v] = \int_0^1 \left[p(v'(x))^2 + qv^2(x) - 2f(x)v(x) \right] \, dx. \tag{169.4}$$

It is straightforward to show that the first variation of $I[v]$ (see page 88) yields (169.2) and (169.3). Hence, $I[v]$ will be minimized when $v = u$.

Now we set up a uniform grid of $N+2$ points on the interval $0 \leq x \leq 1$ (i.e., $x_n = nh$ with $h = 1/(N+1)$ for $n = 0, 1, \ldots, N+1$). We define the interval (x_k, x_{k+1}) to be "finite element number k." We choose as basis functions on the finite elements the functions $\phi_k(x)$ defined by

$$\phi_k(x) = \begin{cases} \dfrac{x - x_{k-1}}{h}, & \text{for } x_{k-1} \leq x \leq x_k, \\ \dfrac{x_{k+1} - x}{h}, & \text{for } x_k \leq x \leq x_{k+1}, \\ 0, & \text{otherwise.} \end{cases} \tag{169.5}$$

These are the "hat functions" shown in Figure 169.1. Note that

$$\phi_k'(x) = \begin{cases} \dfrac{1}{h}, & \text{for } x_{k-1} \leq x \leq x_k, \\ -\dfrac{1}{h}, & \text{for } x_k \leq x \leq x_{k+1}, \\ 0, & \text{otherwise.} \end{cases}$$

Now we approximate the function that minimizes (169.4), $u(x)$, by a linear combination of the $\phi_k(x)$. We take

$$u(x) \simeq u_N(x) := \sum_{k=1}^N c_k \phi_k(x), \tag{169.6}$$

where the unknowns $\{c_k\}$ must be determined. Once the $\{c_k\}$ are known, then the approximation to $u(x)$ at any point can be found from (169.6).

Hence, on finite element k (i.e., for $x_k < x < x_{k+1}$)

$$u_N(x) = c_k\phi_k(x) + c_{k+1}\phi_{k+1}(x),$$
$$u'_N(x) = \frac{-c_k + c_{k+1}}{h}. \tag{169.7}$$

Using $u_N(x)$ for $v(x)$ in (169.4) results in

$$I[u_N] = \sum_{k=0}^{N} \int_{x_k}^{x_{k+1}} \left[p(u'_N)^2 + qu_N^2 - 2fu_N\right] dx,$$
$$:= \sum_{k=0}^{N} \left[I_k^s + I_k^m + I_k^l\right], \tag{169.8}$$

where

$$I_k^s := \int_{x_k}^{x_{k+1}} p(u'_N)^2\, dx = \begin{pmatrix} c_k & c_{k+1} \end{pmatrix} K_k^s \begin{pmatrix} c_k \\ c_{k+1} \end{pmatrix},$$
$$I_k^m := \int_{x_k}^{x_{k+1}} q(u_N)^2\, dx = \begin{pmatrix} c_k & c_{k+1} \end{pmatrix} K_k^m \begin{pmatrix} c_k \\ c_{k+1} \end{pmatrix},$$
$$I_k^l := \int_{x_k}^{x_{k+1}} 2f(x)u_N(x)\, dx$$

by virtue of (169.7). Here K_k^s is the *element stiffness matrix*, and K_k^m is the *element mass matrix*, they are defined by

$$K_k^s = \frac{p}{h}\begin{pmatrix} 1 & -1 \\ -1 & 1 \end{pmatrix}, \qquad K_k^m = \frac{qh}{6}\begin{pmatrix} 2 & 1 \\ 1 & 2 \end{pmatrix}.$$

If p and q were not taken to be constants, then these element matrices would not be so simple. A numerical integration would have been required to find the entries in these matrices.

A numerical integration is required to determine I_k^l. If, on finite element number k, $f(x)$ is approximated by $f(x) \simeq f_k\phi_k(x) + f_{k+1}\phi_{k+1}(x)$, then we find $I_k^l = \left(\mathbf{f}_k^l\right)^{\mathrm{T}}\begin{pmatrix} c_k \\ c_{k+1} \end{pmatrix}$, where the *element load vector* is defined by $\mathbf{f}_k^l = \dfrac{h}{3}\begin{pmatrix} 2f_k + f_{k+1} \\ f_k + 2f_{k+1} \end{pmatrix}$.

The system can now be *assembled*, element by element. That is, we write a single matrix equation representing (169.8). For this example, we find that

$$I[u_N] = \mathbf{c}^{\mathrm{T}}(K + M)\mathbf{c} - 2\mathbf{f}^{\mathrm{T}}\mathbf{c}, \tag{169.9}$$

where $\mathbf{c} = (c_1, c_2, \ldots, c_N)^{\mathrm{T}}$, $\mathbf{f} = \dfrac{h}{6}(f_0 + 4f_1 + f_2,\ f_1 + 4f_2 + f_3,\ \ldots,\ f_{N-2} + 4f_{N-1} + f_N)^{\mathrm{T}}$, and the *global stiffness matrix* K and the *global mass matrix* M are defined by

$$
K = \frac{p}{h}
\begin{pmatrix}
2 & -1 & 0 & 0 & \cdots & 0 & 0 \\
-1 & 2 & -1 & 0 & \cdots & 0 & 0 \\
0 & -1 & 2 & -1 & & 0 & 0 \\
\vdots & & \ddots & \ddots & \ddots & & \vdots \\
0 & 0 & & -1 & 2 & -1 & 0 \\
0 & 0 & \cdots & 0 & -1 & 2 & -1 \\
0 & 0 & \cdots & 0 & 0 & -1 & 2
\end{pmatrix},
$$

$$
M = \frac{qh}{6}
\begin{pmatrix}
4 & 1 & 0 & 0 & \cdots & 0 & 0 \\
1 & 4 & 1 & 0 & \cdots & 0 & 0 \\
0 & 1 & 4 & 1 & & 0 & 0 \\
\vdots & & \ddots & \ddots & \ddots & & \vdots \\
0 & 0 & & 1 & 4 & 1 & 0 \\
0 & 0 & \cdots & 0 & 1 & 4 & 1 \\
0 & 0 & \cdots & 0 & 0 & 1 & 4
\end{pmatrix}.
$$

To minimize the expression in (169.9), $\mathbf{c}$ should be chosen (since $K + M$ is a symmetric matrix in this example) to satisfy the matrix equation $(K + M)\mathbf{c} = \mathbf{f}$. This is a tridiagonal system of equations. It may be solved by standard numerical linear algebra routines.

Example 2

This example shows more of the details for a specific application of the finite element method. Suppose that we wish to approximate the solution of the ordinary differential equation

$$
\begin{aligned}
u'' - u' &= e^x \left(e^{-x} u' \right)' = 0, \\
u(0) &= 2, \quad u(4) = 1 + e^4,
\end{aligned}
\tag{169.10}
$$

whose exact solution is $u(x) = 1 + e^x$. From page 88, we see that the variational principle associated with (169.10) is just $\delta J = 0$ where

$$
J[u] := \int_0^4 e^{-x} \left(u' \right)^2 \, dx.
$$

To use the finite element method on the problem in (169.10), we choose to use three elements: the intervals $[0, 1]$, $[1, 2]$ and $[2, 4]$. We choose the polynomial basis functions on element $[0, 1]$, basis function is $f(x) = \alpha + \beta x + \gamma x^2$,

on element $[1, 2]$, basis function is $g(x) = \delta + \varepsilon x + \zeta x^2$,   (169.11)

on element $[2, 4]$, basis function is $h(x) = \eta + \theta x + \iota x^2$.

After $\{\alpha, \beta, \gamma, \delta, \varepsilon, \zeta, \eta, \theta, \iota\}$ are determined, we will have found an approximate solution, v. The equations needed to satisfy the boundary conditions, and for our approximation, and its first derivative, to be continuous on the interval $[0, 1]$, are

$$
\begin{array}{lll}
\text{boundary conditions:} & f(0) = 2, & h(4) = 1 + e^4, \\
\text{continuity conditions:} & f(1) = g(1), & f'(1) = g'(1), \\
& g(2) = h(2), & g'(2) = h'(2).
\end{array}
\qquad (169.12)
$$

Subject to the constraints in (169.12), we want to minimize $J[v]$. Using our chosen set of finite elements and basis functions, we have

$$
\begin{aligned}
J[v] &= \int_0^1 e^{-x} (f')^2 \, dx + \int_1^2 e^{-x} (g')^2 \, dx + \int_2^4 e^{-x} (h')^2 \, dx \\
&= (4e - 8)\gamma^2 + 4\beta\gamma + (8e^2 - 4e)\zeta^2 + 4e^2\varepsilon\zeta + (e^2 - e)\theta^2 \\
&\quad + 4e^2\iota\theta + (8e^2 - 4e)\iota^2 + (e^2 - e)\varepsilon^2 + (e - 1)\beta^2
\end{aligned}
$$

To minimize this last expression, subject to the constraints in (169.12), we use Lagrange multipliers. The expression obtained after Lagrange multipliers are introduced is differentiated with respect to each of the variables to obtain a linear system of 15 equations (9 equations due to the 9 variables in (169.11) and 6 equations due to the Lagrange multipliers). This system can be solved to determine the basis function on each element:

$$
\begin{aligned}
f(x) &= -3.4508x^2 + 5.3673x + 2, \\
g(x) &= 4.1836x^2 - 9.9014x + 9.6343, \\
h(x) &= 8.8416x^2 - 28.5337x + 28.2666.
\end{aligned}
$$

Figure 169.2 has a comparison of the exact and approximate solutions. At points midway on the elements we find:

$$
\begin{array}{lll}
u(.5) = 2.65, & u(1.5) = 5.48, & u(3) = 21.09. \\
f(.5) = 3.82, & g(1.5) = 4.20, & h(3) = 22.24,
\end{array}
$$

A more accurate approximation could have been obtained by increasing the degree of the basis functions, or by increasing the number of elements.

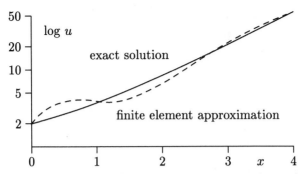

Figure 169.2 Exact solution and finite element approximation to (169.).

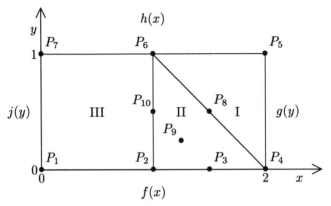

Figure 169.3 Finite elements used in Example 3.

Example 3

Suppose that we want to approximate the solution to

$$\nabla^2 u = 0 \quad \text{in the rectangle} \quad 0 \le x \le 2, \quad 0 \le y \le 1,$$
$$u(x,0) = f(x), \qquad u(1,y) = j(y), \tag{169.13}$$
$$u(x,1) = h(x), \qquad u(2,y) = g(y).$$

For this problem we choose we use three finite elements; two of these elements (I and II) are triangles and one (III) is a square (see Figure 169.3). On the different elements we choose to use the following polynomial functions to represent the solution:

$$u_I = a_{11} + a_{12}x + a_{13}y + a_{14}x^2 + a_{15}y^2 + a_{16}xy,$$
$$u_{II} = a_{21} + a_{22}x + a_{23}y + a_{24}x^2 + a_{25}y^2 + a_{26}xy + a_{27}x^3 + a_{28}y^3,$$
$$u_{III} = a_{31} + a_{32}x + a_{33}y.$$

Now we must specify how the parameters in these approximate solutions are to be determined. Using a subscript to denote evaluation at a node on Figure 169.3, we choose to approximately satisfy the equation and boundary conditions on the individual elements as follows:

On element I:
$$u_I\Big|_{P_4} = g_4, \qquad u_I\Big|_{P_5} = g_5,$$
$$u_I\Big|_{P_6} = h_6, \qquad \nabla^2 u_I\Big|_{P_5} = 0,$$

On element II:
$$u_{II}\Big|_{P_2} = f_2, \qquad u_{II}\Big|_{P_3} = f_3,$$
$$u_{II}\Big|_{P_4} = f_4, \qquad u_{II}\Big|_{P_6} = h_6, \qquad (169.14)$$
$$\nabla^2 u_{II}\Big|_{P_6} = 0,$$

On element III:
$$u_{III}\Big|_{P_1} = f_1, \qquad u_{III}\Big|_{P_2} = f_2,$$
$$u_{III}\Big|_{P_6} = h_6, \qquad u_{III}\Big|_{P_7} = h_7.$$

To connect the elements, we choose the following conditions:

$$u_I\Big|_{P_8} = u_{II}\Big|_{P_8}, \qquad \frac{\partial u_{II}}{\partial n}\Big|_{P_8} = \frac{\partial u_I}{\partial n}\Big|_{P_8},$$
$$u_{II}\Big|_{P_{10}} = u_{III}\Big|_{P_{10}}, \qquad \frac{\partial u_{II}}{\partial n}\Big|_{P_{10}} = \frac{\partial u_{III}}{\partial n}\Big|_{P_{10}}, \qquad (169.15)$$
$$\frac{\partial u_I}{\partial x}\Big|_{P_6} = \frac{\partial u_{III}}{\partial x}\Big|_{P_6}$$

where n stands for the normal.

To actually carry out the solution technique, we choose the boundary conditions: $\{f(x) = x^2,\ g(y) = 4 + y - y^2,\ h(x) = x^2,\ j(y) = y - y^2\}$. For these values, (169.13) has the exact solution: $u(x, y) = x^2 + y - y^2$. Solving the linear equations in (169.14) and (169.15), we obtain the approximate solution:

$$u_I = -2y^2 + (10 - 4x)y + 2x^2 + x - 6,$$
$$u_{II} = -8y^3 + 23y^2 + (8x - 23)y + 24x^3 - 107x^2 + 156x - 72,$$
$$u_{III} = x.$$

Comparing this approximate solution to the exact solution, we determine the maximum errors to be

$$
\begin{array}{lll}
\text{on element I, maximum error} = 1 & \text{at} & x = \tfrac{1}{2},\ y = \tfrac{3}{2} \\[4pt]
\text{on element II, maximum error} = \tfrac{16}{3\sqrt{3}} & \text{at} & x = 1,\ y = 1 - \tfrac{1}{\sqrt{3}} \\[4pt]
\text{on element III, maximum error} = \tfrac{1}{4} & \text{at} & x = 0,\ y = \tfrac{1}{2}
\end{array}
$$

Notes

[1] Nearly every part of the finite element procedure that has been presented in example one can be generalized.

(A) The basis functions do not have to be piecewise linear, but could be piecewise quadratic, cubic, or higher order (they were chosen to be quadratic in example two).

(B) For physically two-dimensional structures, the "finite elements" can be triangles, quadrilaterals, or polygons with more sides (they can be tetrahedrons, cubes, or more complicated structures for three-dimensional structures). However, the smoothness conditions across the boundaries may be difficult to formulate.

(C) Even in one dimension, the "finite elements" do not have to represent intervals of equal length (as in Example 2).

[2] The approximation to the solution in (169.6) will only be C^0, since the basis functions chosen in (169.5) are piecewise linear. The cubic Hermite approximation results in a C^1 approximation by choosing the following two basis functions per finite element:

$$
\eta_k(x) =
\begin{cases}
1 - 3\left(\dfrac{x - x_k}{h}\right)^2 - 2\left(\dfrac{x - x_k}{h}\right)^3, & \text{for } x_{k-1} \le x \le x_k, \\[8pt]
1 - 3\left(\dfrac{x - x_k}{h}\right)^2 + 2\left(\dfrac{x - x_k}{h}\right)^3, & \text{for } x_k \le x \le x_{k+1}, \\[8pt]
0, & \text{otherwise,}
\end{cases}
$$

$$
\zeta_k(x) =
\begin{cases}
(x - x_k)\left(1 + \dfrac{x - x_k}{h}\right)^2, & \text{for } x_{k-1} \le x \le x_k, \\[8pt]
(x - x_k)\left(1 - \dfrac{x - x_k}{h}\right)^2, & \text{for } x_{k-1} \le x \le x_k, \\[8pt]
0, & \text{otherwise,}
\end{cases}
$$

These basis functions are continuous with their first derivatives at the nodes (endpoints of the intervals). See Figure 169.4. Using these functions, an approximation of the form

$$
u(x) \simeq u_N(x) := \sum_{k=1}^{N} d_k \eta_k(x) + e_k \zeta_k(x)
$$

is supposed, where the constants $\{d_k, e_k\}$ must be determined.

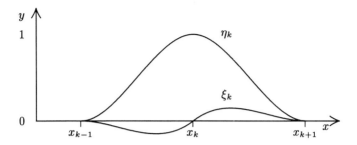

Figure 169.4 The functions for the cubic Hermite approximation.

[3] In higher dimensions, smoother approximations are found analogously. Basis functions are chosen that are continuous (with several of their derivatives) at the nodes of the "finite elements." The nodes could be the vertices of a square (or cube), or some of the vertices and some points along the edges on the square (or cube).

[4] Both Mackerle and Fredriksson [7] and the book edited by Brebbia [3] have comprehensive listings of available software that numerically approximate the solutions of differential equations by finite elements.

[5] Incidentally, by integrating by parts and using the boundary conditions in (169.3), it can be shown that (169.4) is equivalent to $I[v] = (v, L[v]) - 2(f, v)$, where $(g, h) := \int_0^1 g(x)h(x)\,dx$.

[6] In some finite element programs, the discretization errors are controlled by letting the diameter of the largest element h approach zero. This is called the *h-version of the finite element method*. In the *p-version of the finite element method*, the mesh is fixed while the degree of the polynomials on the elements is increased (this is also called the *global element method*). In the *hp-version*, both limits are considered simultaneously. See Babuška [2] for details.

[7] Mackerle [6] contains a very large annotated bibliography.

References

[1] M. B. Allen and M. C. Curran, "Adaptive Local Grid Refinement Algorithms for Finite-Element Collocations," *Num. Methods Part. Diff. Eqns.*, **5**, 1989, pages 121–132.

[2] I. Babuška, "The p and h-p Versions of The Finite Element Method. The State of the Art," in D. L. Dwoyer, M. Y. Hussaini, and R. G. Voigt (eds.), *Finite Elements: Theory and Application*, Springer–Verlag, New York, 1988, pages 199–239.

[3] C. A. Brebbia (ed.), *Finite Element Systems: A Handbook*, Springer–Verlag, New York, 1985.

[4] L. M. Delves and C. Phillips, "A Fast Implementation of the Global Element Method," *J. Inst. Maths. Applics*, **25**, 1980, pages 177–197.

[5] L. Lapidus and G. F. Pinder, *Numerical Solution of Partial Differential Equations in Science and Engineering*, Wiley, New York, 1982.

[6] J. Mackerle, Special Volume—Finite Element Methods: A Guide to Information Sources, *Finite Elements in Analysis and Design*, **8**, No. 1–4, December 1990.

[7] J. Mackerle and B. Fredriksson, *Handbook of Finite Element Software: Supercomputers, Mainframes, Minicomputers, Microcomputers*, Krieger Pub. Co., Melbourne, Fl, 1988.

[8] A. R. Mitchell and R. Wait, *The Finite Element Method in Differential Equations*, Wiley, New York, 1977.

[9] K. C. Rockey, H. R. Evans, D. W. Griffiths, and D. A. Nethercot, *The Finite Element Method — A Basic Introduction for Engineers*, Second Edition, Halstead Press, New York, 1983.

[10] G. Strang and G. J. Fix, *An Analysis of the Finite Element Method*, Prentice–Hall Inc., Englewood Cliffs, NJ, 1973.

[11] G. Strang, *Introduction to Applied Mathematics*, Wellesley–Cambridge Press, Wellesley, MA, 1986, pages 428–445.

[12] P. S. Wang, "FINGER: A Symbolic System for Automatic Generation of Numerical Programs in Finite Element Analysis," *J. Symbolic. Comp.*, **2**, 1986, pages 305–316.

170. Hybrid Computer Methods*

Applicable to   Ordinary differential equations and partial differential equations.

Yields

A numerical approximation to the solution.

Idea

Sometimes the advantages of both digital and analog computers can be used simultaneously on a single differential equation.

Procedure

A hybrid computer is one that combines both digital and analog computing devices. Generally, in such a configuration, the analog computer is used to perform tasks that are very time consuming on a digital computer. The analog computer is constructed, generally by the user, out of capacitors, operational amplifiers, resistors, and other electronic components. The numbers in an analog computer are represented by electrical quantities such as voltage and amperage.

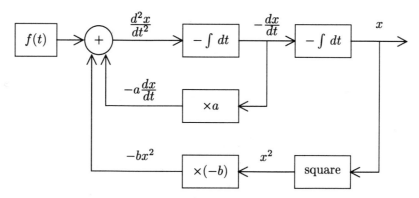

Figure 170. A block diagram for the analog solution of the differential equation
$$\frac{d^2x}{dt^2} + a\frac{dx}{dt} + bx^2 = f(t).$$

As an example of use, a partial differential equation can often be approximated by a large number of ordinary differential equations (see, for example, the method of lines, on page 740, or the Rayleigh–Ritz method, on page 554). Rather than introduce additional approximations in finding solutions of these ordinary differential equations, an analog computer may be used.

In other problems, the analog computer is used to evaluate integrals as they arise. These integrals are often multi-dimensional and would be computationally intensive on a digital computer.

The digital computer is nearly always used to control the solution procedure and to determine the discretization and the overall error.

Example

The block diagram in Figure 170 shows how the differential equation

$$\frac{d^2x}{dt^2} + a\frac{dx}{dt} + bx^2 = f(t)$$

might be solved by an analog computer. Each of the blocks in this figure is easily implemented by electronic components.

The blocks that perform the multiplications will generally have the numerical values of a and $-b$ specified by potentiometers. These values may be changed by adjusting the potentiometers by hand. Or, these values could be changed by a digital computer.

Notes

[1] For an example of a hybrid nonlinear parabolic equation solver, see El-Zorkany and Balasubramanian [3].

[2] Recently, hybrid computers have been introduced that do not require the user to "plug" components together; the specification of the analog part of the machine is performed on the digital part of the machine.

References

[1] J. S. Allison and H. M. Johnson, "Stability of Hybrid Simulation of Dynamic Systems," *Mathematics and Computers in Simulation*, **21**, 1979, pages 289–303.

[2] J. R. Amyot and G. A. Camiré, "Stability of a Class of Hybrid Computer Models of Dynamical Systems," *Mathematics and Computers in Simulation*, **28**, 1986, pages 57–64.

[3] H. I. El-Zorkany and R. Balasubramanian, "Hybrid Solution of Non-Linear P.D.E.'s Based on a Special F.E. Approximation I. One Dimensional Problem," in R. Vichnevetsky (ed.), *Advances in Computer Methods For Partial Differential Equations-II*, IMACS (AICA), North–Holland Publishing Co., New York, 1977, pages 227–234.

[4] H. I. El-Zorkany and R. Balasubramanian, "Hybrid Computer Solution of PDE's Using Laplace–Modified Galerkin Approximation," *Mathematics and Computers in Simulation*, **23**, 1981, pages 304–311.

[5] H. I. El-Zorkany and R. Balasubramanian, "Hybrid Simulation of Multidimensional P.D.E.'s via Solution of Many One Dimensional Problems 1. Theoretical Basis and Hybrid Implementation," *Mathematics and Computers in Simulation*, **25**, 1983, pages 70–76.

[6] P. A. Lawson, "Contraction Mapping Techniques Applied to the Hybrid Computer Solution of Parabolic PDE's," *Mathematics and Computers in Simulation*, **23**, 1981, pages 299–303.

[7] W. Neundorf, "Iterative Block Methods for the Hybrid Computer Solution of the Method of Lines," *Math. and Computers in Simulation*, **23**, No. 2, 1981, pages 142 - 148.

[8] T. Roubíček, "Hybrid Solution of Weakly Formulated Boundary-Value Problems," *Mathematics and Computers in Simulation*, **26**, 1984, pages 11–19.

[9] J. L. Shearer, A. T. Murphy, and H. H. Richardson, *Introduction to System Dynamics*, Addison–Wesley Publishing Co., Reading, MA, Chapter 7 (pages 166–197).

171. Invariant Imbedding*

Applicable to   Most often, two point boundary value problems for ordinary differential equations.

Yields

A new formulation as an initial value problem.

Idea

Invariant imbedding is a type of continuation method (see page 635). For the usual problems that are treated, the length of the interval of interest is considered to be the continuation parameter. Hence the endpoint in a two point boundary value problem is treated as a variable. By differentiating with respect to this variable, an initial value problem can be created.

Procedure

The general technique involves some subtleties, so we choose to illustrate the technique on a class of two point boundary value problems. More details can be found in Casti and Calaba [3]. Suppose we have the system of ordinary differential equations

$$\frac{dx}{dt} = a(t)x(t) + b(t)y(t),$$
$$\frac{dy}{dt} = c(t)x(t) + d(t)y(t) + f(t), \qquad (171.1)$$

with

$$\alpha_1 x(0) + \alpha_2 y(0) = 0,$$
$$\alpha_3 x(T) + \alpha_4 y(T) = 1, \qquad (171.2)$$

on the interval $t \in [0, T]$, where the $\{\alpha_i\}$ are constants and $\{a, b, c, d\}$ are continuous functions. If we think of the endpoint T as being a variable, then the solution to (171.1) and (171.2) can be written, by use of superposition, as

$$x(t) = x(t, T) = u(t, T) + p(t, T),$$
$$y(t) = y(t, T) = v(t, T) + q(t, T), \qquad (171.3)$$

where the functions $\{u, v, p, q\}$ are defined by

$$\frac{du(t, T)}{dt} = a(t)u + b(t)v, \qquad\qquad \alpha_1 u(0, T) + \alpha_2 v(0, T) = 0,$$
$$\frac{dv(t, T)}{dt} = c(t)u + d(t)v + f(t), \qquad \alpha_3 u(T, T) + \alpha_4 v(T, T) = 0,$$
$$\qquad\qquad (171.4)$$

and

$$\frac{dp(t,T)}{dt} = a(t)p + b(t)q, \qquad \alpha_1 p(0,T) + \alpha_2 q(0,T) = 0,$$

$$\frac{dq(t,T)}{dt} = c(t)p + d(t)q, \qquad \alpha_3 p(T,T) + \alpha_4 q(T,T) = 1. \tag{171.5}$$

Using algebraic manipulations, the systems in (171.4) and (171.5) can be written as initial value systems by the introduction of four new variables. Define the functions $\{r,s,m,n\}$ to be the solutions to the following nonlinear ordinary differential equations:

$$r'(t) = b(t)s(t) + [a(t) - \alpha_3 b(t)s(t) - \alpha_4 d(t)s(t)]r(t) - [\alpha_3 a(t) + \alpha_4 c(t)]r^2(t),$$

$$s'(t) = c(t)r(t) + [d(t) - \alpha_3 a(t)r(t) - \alpha_4 c(t)r(t)]s(t) - [\alpha_3 b(t) + \alpha_4 d(t)]s^2(t),$$

$$m'(t) = a(t)m(t) + b(t)n(t) - \Big\{[\alpha_3 a(t) + \alpha_4 c(t)]m(t)$$
$$+ [\alpha_3 b(t) + \alpha_4 d(t)]n(t) + f(t)\Big\}r(t),$$

$$n'(t) = c(t)m(t) + d(t)n(t) + f(t) - \Big\{[\alpha_3 a(t) + \alpha_4 c(t)]m(t)$$
$$+ [\alpha_3 b(t) + \alpha_4 d(t)]n(t) + f(t)\Big\}s(t), \tag{171.6}$$

where $'$ denotes differentiation of a function with respect to its single argument (i.e., the variable t). The initial values for $\{r,s,m,n\}$ are given by

$$\alpha_1 r(0) + \alpha_2 s(0) = 0, \quad m(0) = 0,$$
$$\alpha_3 r(0) + \alpha_4 s(0) = 1, \quad n(0) = 0. \tag{171.7}$$

Note that we must have $\alpha_1 \alpha_4 - \alpha_2 \alpha_3 \neq 0$ if $r(0)$ and $s(0)$ are to be determined from (171.7). Using $\{r,s,m,n\}$, the equations for $\{p,q,u,v\}$ can now be written as

$$\frac{dp(t,T)}{dT} = -\Big\{r(T)[\alpha_3 a(T) + \alpha_4 c(T)] + s(T)[\alpha_3 b(T) + \alpha_4 d(T)]\Big\}p(t,T),$$

$$\frac{dq(t,T)}{dT} = -\Big\{r(T)[\alpha_3 a(T) + \alpha_4 c(T)] + s(T)[\alpha_3 b(T) + \alpha_4 d(T)]\Big\}q(t,T),$$

$$\frac{du(t,T)}{dT}$$
$$= -\Big\{m(T)[\alpha_3 a(T) + \alpha_4 c(T)] + n(T)[\alpha_3 b(T) + \alpha_4 d(T)] + f(T)\Big\}p(t,T),$$

$$\frac{dv(t,T)}{dT}$$
$$= -\Big\{m(T)[\alpha_3 a(T) + \alpha_4 c(T)] + n(T)[\alpha_3 b(T) + \alpha_4 d(T)] + f(T)\Big\}q(t,T). \tag{171.8}$$

The initial conditions for $\{p, q, u, v\}$ may be written as

$$p(t, t) = r(t), \quad q(t, t) = s(t),$$
$$u(t, t) = m(t), \quad v(t, t) = n(t). \tag{171.9}$$

Suppose that the solution of the original system, (171.1) and (171.2), is desired at the set of abscissas $\{t_1, t_2, t_3, \ldots, t_N\}$, where $t_N = T^*$, and T^* is the interval length of interest. The numerical technique is to numerically integrate the equations in (171.6) for $\{r, s, m, n\}$, from $t = 0$ to $t = T^*$. Hence the values of $\{r, s, m, n\}$ will be known at the points $\{t_1, t_2, t_3, \ldots, t_N\}$.

Now fix $t = t_1$ in (171.8) and (171.9). Integrate the resulting equations (with respect to T) from $T = t_1$ to $T = T^*$. This will yield $\{p(t_1, T^*), q(t_1, T^*), u(t_1, T^*), v(t_1, T^*)\}$. If these values are used in (171.3), then $x(t_1, T^*), y(t_1, T^*)$ will be determined. Of course, this is the same as $x(t_1), y(t_1)$. Hence, x and y have been determined at the first point of interest, t_1.

To obtain x and y at $t = t_2$, evaluate (171.8) and (171.9) at $t = t_2$ and integrate the resulting equations with respect to T (from t_2 to T^*). Repeat this for each of $t = t_3, t = t_4, \ldots$.

Example

Suppose we want to turn the boundary value problem

$$\frac{dx}{dt} = 10y, \qquad x(0) = 0,$$

$$\frac{dy}{dt} = 10x, \qquad y(10) = 1,$$

into an initial value problem. Using the above notation, we find that $\{\alpha_1 = \alpha_4 = 1, \alpha_2 = \alpha_3 = 0, a(t) = d(t) = f(t) = 0, b(t) = c(t) = 10, T^* = 10\}$. The system in (171.6) becomes

$$r' = 10(s - r^2),$$
$$s' = 10(s - 1),$$
$$m' = 10(n - mr),$$
$$n' = 10m(1 - s), \tag{171.10}$$

with the initial conditions: $r(0) = 0$, $s(0) = 1$, $m(0) = 0$, $n(0) = 0$. It is easy to see that $n(t) = 0$, $m(t) = 0$, $s(t) = 1$, although these equations could have been integrated if this had not been observed. The system in

(171.8) becomes

$$\frac{dp(t,T)}{dT} = -10r(T)p(t,T),$$

$$\frac{dq(t,T)}{dT} = -10r(T)q(t,T),$$

$$\frac{du(t,T)}{dT} = -10m(T)p(t,T), \qquad (171.11)$$

$$\frac{dv(t,T)}{dT} = -10m(T)q(t,T),$$

with the initial conditions: $p(t,t) = r(t)$, $q(t,t) = s(t)$, $u(t,t) = m(t)$, $v(t,t) = n(t)$. From the above observation, we conclude that $q(t,t) = 1$, $u(t,T) = 0$, $v(t,T) = 0$. Using (171.3) we find: $x(t) = x(t,10) = p(t,10)$ and $y(t) = y(t,10) = q(t,10)$. Let us suppose that we want to know the values of x and y for $t = 2, 4, 6, 8, 10$. The procedure to follow is

(A) Integrate $r(t)$ from $t = 0$ up to $t = 10$ using (171.10). Hence, $r(2)$, $r(4)$, $r(6)$, $r(8)$, $r(10)$ will all be known.

(B) Set $p(2,2) = r(2)$ and $q(2,2) = 1$. Integrate (171.11) for $p(t,T)$ and $q(t,T)$ from $T = 2$ to $T = 10$. Then $\{p(2,10), q(2,10)\}$ will be known and hence $\{x(2), y(2)\}$ will be known.

(C) Set $p(4,4) = r(4)$ and $q(4,4) = 1$. Integrate (171.11) for $p(t,T)$ and $q(t,T)$ from $T = 4$ to $T = 10$. Then $\{p(4,10), q(4,10)\}$ will be known and hence $\{x(4), y(4)\}$ will be known.

(D) Repeat steps (2) and (3) for $t = 6$, $t = 8$, and $t = 10$.

Notes

[1] The paper by Scott [9] lists several different ways in which boundary value problems may be converted into stable initial value problems.

[2] Imbedding methods can be used for more than just boundary value problems. This technique can also be applied to nonlinear variational problems, unconstrained nonlinear control processes, constrained control processes, and Fredholm integral equations. Imbedding methods have also been used in hyperbolic and parabolic partial differential equations.

[3] Wasserstrom [11] discusses how imbedding methods can be analyzed as continuation methods (see page 635).

[4] Other names for the invariant imbedding approach are "field method," "factorization method," "method of sweeps," "compound matrix method," and "Riccati transformation." In this last method, matrix Riccati equations (see page 335) are developed. See Ascher et al. [1] for details.

References

[1] U. M. Ascher, R. M. M. Mattheij, and R. D. Russel, *Numerical Solution of Boundary Value Problems for Ordinary Differential Equations*, Prentice–Hall Inc., Englewood Cliffs, NJ, 1988.

[2] R. Bellman and G. M. Wing, *An Introduction to Invariant Imbedding*, John Wiley & Sons, New York, 1975.

[3] J. Casti and R. Kalaba, *Imbedding Methods in Applied Mathematics*, Addison–Wesley Publishing Co., Reading, MA, 1973.

[4] L. Dieci, M. R. Osborne, and R. D. Russell, "A Riccati Transformation Method for Solving Linear BVPs. I: Theoretical Aspects," *SIAM J. Numer. Anal.*, **25**, No. 5, October 1988, pages 1055–1073.

[5] E. S. Lee, *Quasilinearization and Invariant Imbedding*, Academic Press, New York, 1968.

[6] G. H. Meyer, *Initial Value Methods for Boundary Value Problems: Theory and Application of Invariant Imbedding*, Academic Press, New York, 1973.

[7] G. H. Meyer, "Invariant Imbedding for Fixed and Free Two Point Boundary Value Problems," in A. K. Aziz (ed.), *Numerical Solutions of Boundary Value Problems for Ordinary Differential Equations*, Academic Press, New York, 1975, pages 249–275.

[8] B. S. Ng and W. H. Reid, "A Numerical Method for Linear Two-Point Boundary-Value Problems Using Compound Matrices," *J. Comput. Physics*, **33**, No. 1, October 1979, pages 70–85.

[9] M. R. Scott, "On the Conversion of Boundary–Value Problems into Stable Initial–Value Problems via Several Invariant Imbedding Algorithms," in A. K. Aziz (ed.), *Numerical Solutions of Boundary Value Problems for Ordinary Differential Equations*, Academic Press, New York, 1975, pages 89–146.

[10] M. R. Scott and W. H. Vandevender, "A Comparison of Several Invariant Imbedding Algorithms for the Solution of Two-Point Boundary-Value Problems," *Appl. Math. and Comp.*, **1**, 1975, pages 187–218.

[11] E. Wasserstrom, "Numerical Solutions by the Continuation Method," *SIAM Review*, **15**, No. 1, January 1973, pages 89–119.

172. Multigrid Methods*

Applicable to   Ordinary differential equations and partial differential equations.

Yields

A numerical approximation technique.

Idea

After differential equations are discretized for the purpose of approximating the solution numerically, some linear algebraic operations must be performed. Frequently, a system of linear equations may need to be solved (e.g., see pages 626, 716, and 726). If the system of linear equations is large (e.g., when a fine discretization grid is used), then iterative methods are often used to solve the linear equations.

Multigrid methods are iterative methods for solving systems of linear equations arising from differential equations. Generally, different grids are used, with only a few iterations per grid. The last approximation on one grid becomes the first approximation on the next grid.

Procedure

We sketch the approximation process using the following ordinary differential equation as motivation:

$$u''(x) - \sigma u(x) = -f(x)$$
$$u(0) = 0, \quad u(1) = 0. \tag{172.1}$$

Consider approximating the solution of (172.1) on a uniform grid with a spacing of h (e.g., $x_j = jh$ and $v_j \approx u(x_j)$). Call this grid Ω^h. Using $(v_{j-1} - 2v_j + v_{j+1})/h^2$ as an approximation to $u''(x_j)$, (172.1) can be written as

$$\frac{1}{h^2} \begin{pmatrix} 2+\sigma h^2 & -1 & 0 & & & \\ -1 & 2+\sigma h^2 & -1 & & & \\ 0 & \ddots & \ddots & \ddots & & \\ & & -1 & 2+\sigma h^2 & -1 \\ & & 0 & -1 & 2+\sigma h^2 \end{pmatrix} \begin{pmatrix} v_1 \\ v_2 \\ \vdots \\ v_{N-1} \end{pmatrix} = \begin{pmatrix} f_1 \\ f_2 \\ \vdots \\ f_{N-1} \end{pmatrix}$$
$$\tag{172.2}$$

or simply as $A^h \mathbf{v}^h = \mathbf{f}^h$. (Here a superscript indicates the spacing on the underlying grid.)

The solution of the linear system in (172.2) can be approximated by any of the standard iteration methods, such as Jacobi's method or the Gauss–Seidel method (see Golub and Van Loan [5]). Typically, these iterative methods begin to stall (that is, the convergence rate decreases) when smooth error modes are present. Since a smooth mode on a fine grid looks less smooth on a coarser grid, it is advisable to move to a coarser grid. Iterating on this coarser grid will more effectively reduce the error term. The values on this coarse grid are then fed back to the fine grid.

To illustrate the process, let I_h^{2h} be the linear operator that performs restriction and maps a vector from Ω^h to Ω^{2h}. (For instance, every other value in the vector could be chosen.) Similarly, let I_{2h}^h be the linear operator that performs interpolation and maps a vector from Ω^{2h} to Ω^h. Let us use the term "Relax on" to mean "iterate some number of times using a standard technique such as Gauss–Seidel." Here, then, is how a multigrid method might be implemented:

Relax on $A^h \mathbf{u}^h = \mathbf{f}^h$ with an input initial guess $\mathbf{v}^h$
Find the residual: $\mathbf{r}^h := A^h \mathbf{u}^h - \mathbf{f}^h$
Move to a coarser grid: $\mathbf{f}^{2h} := I_h^{2h} \mathbf{r}^h$
$\quad$ Relax on $A^{2h} \mathbf{u}^{2h} = \mathbf{f}^{2h}$ with the initial guess $\mathbf{v}^{2h} = \mathbf{0}$
$\quad$ Find the residual: $\mathbf{r}^{2h} := A^{2h} \mathbf{u}^{2h} - \mathbf{f}^{2h}$
$\quad$ Move to a coarser grid: $\mathbf{f}^{4h} := I_{2h}^{4h} \mathbf{r}^{2h}$
$\qquad$ Relax on $A^{4h} \mathbf{u}^{4h} = \mathbf{f}^{4h}$ with the initial guess $\mathbf{v}^{4h} = \mathbf{0}$
$\qquad$ Find the residual: $\mathbf{r}^{4h} := A^{4h} \mathbf{u}^{4h} - \mathbf{f}^{4h}$
$\qquad$ Move to a coarser grid: $\mathbf{f}^{8h} := I_{4h}^{8h} \mathbf{r}^{4h}$
$$\vdots$$
$\qquad$ Solve $A^{2^k h} \mathbf{u}^{2^k h} = \mathbf{f}^{2^k h}$ for $\mathbf{u}^{2^k h}$ (which we call $\mathbf{v}^{2^k h}$)
$$\vdots$$
$\qquad$ Revise approximate solution on Ω^{4h}: $\mathbf{v}^{4h} \leftarrow \mathbf{v}^{4h} + I_{8h}^{4h} \mathbf{v}^{8h}$
$\qquad$ Relax on $A^{4h} \mathbf{u}^{4h} = \mathbf{f}^{4h}$ with the initial guess $\mathbf{v}^{4h}$
$\quad$ Revise approximate solution on Ω^{2h}: $\mathbf{v}^{2h} \leftarrow \mathbf{v}^{2h} + I_{4h}^{2h} \mathbf{v}^{4h}$
$\quad$ Relax on $A^{2h} \mathbf{u}^{2h} = \mathbf{f}^{2h}$ with the initial guess $\mathbf{v}^{2h}$
Revise approximate solution on Ω^h: $\mathbf{v}^h \leftarrow \mathbf{v}^h + I_{2h}^h \mathbf{v}^{2h}$
Relax on $A^h \mathbf{u}^h = \mathbf{f}^h$ with the initial guess $\mathbf{v}^h$.

The overall effect is that an approximate solution to the system on the h-grid is input at the top, and a refined approximation to this solution is output at the bottom.

Notes

[1] The multigrid method is applicable to linear algebraic equations. Its importance for differential equations comes about because differential equations can be approximated by solving linear algebraic equations.

References

[1] J. Adams, "Recent Enhancements in MUDPACK, A Multigrid Software Package for Elliptical Partial Differential Equations," *Appl. Math. and Comp.*, **43**, May 1991, pages 79–94.

[2] W. L. Biggs, *A Multigrid Tutorial*, SIAM, Philadelphia, 1988.

[3] A. Brandt, S. McCormick, and J. Ruge, "Multigrid Method for Differential Eigenproblems," *SIAM J. Sci. Stat. Comput.*, **4**, No. 2, 1983, pages 244–260.

[4] C. I. Goldstein, "Multigrid Analysis of Finite Element Methods with Numerical Integration," *Math. of Comp.*, **56**, No. 194, April 1991, pages 409–436.

[5] G. H. Golub and C. F. Van Loan, *Matrix Computations*, Second Edition, The Johns Hopkins University Press, Baltimore, 1989, Chapter 10.

[6] W. Hackbusch and U. Trottenberg, *Multigrid Methods*, Springer–Verlag, New York, 1982.

[7] D. Jespersen, *Multigrid Methods for Partial Differential Equations*, Volume 24 of Studies in Mathematics, Mathematical Association of America, Providence, RI, 1984.

[8] S. McCormick, *Multigrid Methods: Theory, Applications, and Supercomputing*, Marcel Dekker, New York, 1988.

173. Parallel Computer Methods

Applicable to   All types of differential equations.

Yields

Numerical approximations to the solutions.

Idea

Parallel computers may be used to quickly obtain numerical approximations to differential equations.

Procedure

The physical basis for most differential equations is a local and asynchronous model. Hence, it should be possible to numerically approximate a partial differential equation by processors that are loosely coupled.

There are three major ways in which software for differential equations can exploit parallelism: in coding a method so that it can be performed simultaneously on several processors, in splitting variables (in a multivariable system) between processors, and in using parallelism in performing the needed algebraic computations (i.e., solving algebraic systems of equations). We illustrate one parallel technique; it uses the first of these methods.

Example

This example for a two processor MIMD machine is from Iserles and Nørsett [7]. The Butcher-array is a convenient way in which to represent all of the information in a Runge–Kutta method for the equation $\mathbf{y}' = \mathbf{f}(x, \mathbf{y})$, $\mathbf{y}(x_0) = \mathbf{y}_0$ (see page 684). The Butcher array for a four-stage, fourth order Runge–Kutta method is

$$
\begin{array}{c|cccc}
\frac{1}{2} & \frac{1}{2} & 0 & 0 & 0 \\[4pt]
\frac{2}{3} & 0 & \frac{2}{3} & 0 & 0 \\[4pt]
\frac{1}{2} & -\frac{5}{2} & \frac{5}{2} & \frac{1}{2} & 0 \\[4pt]
\frac{1}{3} & -\frac{5}{3} & \frac{4}{3} & 0 & \frac{2}{3} \\[4pt]
\hline
 & -1 & \frac{3}{2} & -1 & \frac{3}{2}
\end{array}
$$

Because of the specific sparsity structure of this Butcher array, we can efficiently implement this technique on two processors. Given the value $\mathbf{y}_n$, to find the approximation at the next time step, $\mathbf{y}_{n+1}$, the steps are as follows:

[1] Use an iteration technique (perhaps Newton–Raphson) to solve the equations

  (A) $\boldsymbol{\xi}_1 = \mathbf{f}\left(t_n + \frac{1}{2}h, \mathbf{y}_n + \frac{1}{2}h\boldsymbol{\xi}_1\right)$ for $\boldsymbol{\xi}_1$ on processor 1,

  (B) $\boldsymbol{\xi}_2 = \mathbf{f}\left(t_n + \frac{2}{3}h, \mathbf{y}_n + \frac{2}{3}h\boldsymbol{\xi}_2\right)$ for $\boldsymbol{\xi}_2$ on processor 2.

[2] Copy the value of $\boldsymbol{\xi}_1$ to processor 2, and copy the value of $\boldsymbol{\xi}_2$ to processor 1.

[3] Use an iteration technique (perhaps Newton–Raphson) to solve the equations

  (A) $\boldsymbol{\xi}_3 = \mathbf{f}\left(t_n + \frac{1}{2}h, \mathbf{y}_n + h\left(-\frac{5}{2}\boldsymbol{\xi}_1 + \frac{5}{2}\boldsymbol{\xi}_2 + \frac{1}{2}\boldsymbol{\xi}_3\right)\right)$ for $\boldsymbol{\xi}_3$ on processor 1,

  (B) $\boldsymbol{\xi}_4 = \mathbf{f}\left(t_n + \frac{1}{3}h, \mathbf{y}_n + h\left(-\frac{5}{3}\boldsymbol{\xi}_1 + \frac{4}{3}\boldsymbol{\xi}_2 + \frac{2}{3}\boldsymbol{\xi}_4\right)\right)$ for $\boldsymbol{\xi}_4$ on processor 2.

[4] Copy $\boldsymbol{\xi}_4$ to processor 1 and then form the estimate at the next time value: $\mathbf{y}_{n+1} = \mathbf{y}_n + h\left(\frac{3}{2}(\boldsymbol{\xi}_2 + \boldsymbol{\xi}_4) - \boldsymbol{\xi}_1 - \boldsymbol{\xi}_3\right)$.

Notes

[1] Many parallel computers can quickly perform matrix operations, such as solving a system of linear equations. Hence, these machines may be used to quickly approximate the solutions to differential equations by using methods (such as finite differences and finite elements) that produce large systems of linear algebraic equations.

  When solving differential equations numerically, it is not uncommon to have large computational needs. For example, a $50 \times 50 \times 50$ grid with 5 degrees of freedom per grid point, such as might be obtained from Euler's equation in fluid dynamics, will lead to matrices of size $N = 625,000$ and a bandwidth $m \approx 25000$. Even though sparse matrix techniques may be used, the complexity of the problem is very high.

However, Rice [14] makes the point that linear algebra approaches are only tangentially relevant to solving partial differential equations and are, in fact, often misleading. Numerical analysis of differential equations begins with the equation itself, not with a discretized version of the equation.

[2] All types of processors have been used to numerically approximate the solutions to differential equations.

(A) By a simple replication of hardware, many Monte Carlo simulations can be performed simultaneously (see pages 721 and 752). This is particularly useful for SIMD machines.

(B) Lattice gas methods (which use cellular automata) are a method of parallel computation; see page 737. Use of cellular automata to numerically approximate the solution of differential equations has also been considered in Boghosian and Levermore [3].

(C) It is also possible to build a specialized VLSI circuit to integrate a specific set of differential equations. A special purpose computer for high-speed, high-precision orbital mechanics computations has been built; see Applegate *et al.* [1]. This was used to demonstrate that the orbit of Pluto was chaotic; see Sussman and Wisdom [16].

It is also possible to construct systolic arrays that solve a class of equations very quickly; see Megson and Evans [10].

[3] The technique described in Garbey and Levine [5] numerically approximates hyperbolic equations by using both characteristics and cellular automata.

References

[1] J. H. Applegate, M. R. Douglas, Y. Gürsel, P. Hunterm, C. L. Seitz, and G. J. Sussman, "A Digital Orrery," *IEEE Trans. Computers*, **C-34**, No. 9, September 1985, pages 822–831.

[2] A. Bellen, R. Vermiglio, and M. Zennaro, "Parallel ODE-solvers with Stepsize Control," *J. Comput. Appl. Math.*, **31**, 1990, pages 277–293.

[3] B. M. Boghosian and C. D. Levermore, "A Cellular Automaton for Burgers' Equation," *Complex Systems*, **1**, 1987, pages 17–29.

[4] T. F. Chan, Y. Saad, and M. H. Schultz, "Solving Elliptic Partial Differential Equations on the Hypercube Multiprocessor," *Appl. Num. Math.*, **3**, 1987, pages 81–88.

[5] M. Garbey and D. Levine, "Massively Parallel Computation of Conservation Laws," *Parallel Computing*, **16**, 1990, pages 293–304.

[6] P. J. van der Houwen and B. P. Sommeijer, "Iterated Runge–Kutta Methods on Parallel Computers," *SIAM J. Sci. Stat. Comput.*, **12**, No. 5, September 1991, pages 1000–1028.

[7] A. Iserles and S. P. Nørsett, "On the Theory of Parallel Runge–Kutta Methods," *IMA J. Num. Analysis*, **10**, 1990, pages 463–488.

[8] A. Lin, "Parallel Algorithms for Boundary Value Problems," *J. Parallel & Distrib. Comput.*, **11**, 1991, pages 284–290.

[9] C. F. Mayo and C. E. Roberts, "Sequential, Parallel and Vector Solution of Ordinary Differential Equations on a Hypercube," *Comp. & Maths. with Appls.*, **18**, No. 9, 1989, pages 797–808.

[10] G. M. Megson and D. J. Evans, "Systolic Arrays for Group Explicit Methods for Solving First Order Hyperbolic Equations," *Parallel Computing*, **16**, 1990, pages 191–205.

[11] C. S. R. Murthy, "Solving Hyperbolic PDE's on Hypercubes," *Comp. & Maths. with Appls.*, **21**, No. 5, 1991, pages 79–82.

[12] C. S. R. Murthy and V. Rajaraman, "A Multiprocessor Architecture for Solving Nonlinear Partial Differential Equations," *Math. and Computers in Simulation*, **30**, 1988, pages 453–464.

[13] J. M. Oretga and R. G. Voigt, "Solution of Partial Differential Equations on Vector and Parallel Computers," *SIAM Review*, **27**, 1985, pages 149–240.

[14] J. R. Rice, "Parallel Methods for Partial Differential Equations," in L. H. Jamieson, D. B. Gannon, and R. J. Douglass (eds.), *The Characteristics of Parallel Algorithms*, MIT Press, Cambridge, MA, 1987, pages 209–231.

[15] W. Schönauer, I. Schnepf, and H. Müller, "PDE Software for Vector Computers," in R. Vichnevetsky and R. S. Stepleman (eds.), *Advances in Computer Methods for Partial Differential Equations*, IMACS, North–Holland Publishing Co., New York, 1984, pages 258–267.

[16] G. J. Sussman and J. Wisdom, "Numerical Evidence that the Motion of Pluto is Chaotic," *Science*, 22 July 1988, pages 433–437.

174. Predictor–Corrector Methods

Applicable to   Ordinary differential equations of the form $y' = f(x, y)$.

Yields

A sequence of numerical approximations.

Idea

To integrate an ordinary differential equation from a point x_n to a new point $x_{n+1} = x_n + h$, a single formula may be used to predict y_{n+1}. Alternately, the value of y_{n+1} could be predicted by one formula, and then that value could be refined by an iterative formula (the "corrector").

Procedure

For the first order ordinary differential equation $y' = f(x, y)$, suppose that the values of x and y are known at the sequence of $m + 1$ points $\{x_{n-m}, \ldots, x_{n-1}, x_n\}$. Then the values of y' are known at those same points (since y' is determined from x and y via $y' = f(x, y)$). An interpolatory polynomial of degree m can be fitted to $m + 1$ values of x and y'. This polynomial can be used to predict the value of y' in the interval (x_n, x_{n+1}). This, in turn, can be used to predict the value of y_{n+1} by a numerical approximation of the relation

$$y_{n+1} = y_n + \int_{x_n}^{x_{n+1}} y'(x) \, dx. \qquad (174.1)$$

Such a formula is called a "predictor."

A modification of this step can be repeated. The values of x and y' are now known at the $m+1$ points $\{x_{n-m+1}, \ldots, x_n, x_{n+1}\}$. A polynomial can be fit through these points, and then the quantity in equation (174.1) can be re-computed. This formula, which furnishes a new estimate of y_{n+1}, is called a "corrector." The corrector may be used repeatedly.

Example

One set of predictor–corrector equations is the Adams–Bashforth predictor formula

$$y_{n+1} = y_n + \frac{h}{24} \left(55y'_n - 59y'_{n-1} + 37y'_{n-2} - 9y'_{n-3}\right), \qquad (174.2)$$

and the Adams–Moulton corrector formula

$$y_{n+1} = y_n + \frac{h}{24} \left(9y'_{n+1} + 19y'_n - 5y'_{n-1} + y'_{n-2}\right), \qquad (174.3)$$

where h is the difference between adjacent x points (The x points are assumed to be equally spaced). These equations are fourth order accurate.

Example

The FORTRAN program in Program 174 uses the method in (174.2) and (174.3) to approximate the solution to the differential equation

$$\frac{dy}{dx} = 1 - x + \frac{y}{x}, \qquad y(1) = 0.$$

Since the solution of (174.3) is given by $y(x) = x(\log x - x + 1)$ (determined by integrating factors), it is easy to see that the values produced:

```
STEP NUMBER=  4  X= 1.60   Y= -0.2080
STEP NUMBER=  5  X= 1.80   Y= -0.3820
STEP NUMBER=  6  X= 2.00   Y= -0.6137
STEP NUMBER=  7  X= 2.20   Y= -0.9054
STEP NUMBER=  8  X= 2.40   Y= -1.2588
STEP NUMBER=  9  X= 2.60   Y= -1.6756
STEP NUMBER= 10  X= 2.80   Y= -2.1570
STEP NUMBER= 11  X= 3.00   Y= -2.7041
STEP NUMBER= 12  X= 3.20   Y= -3.3179
STEP NUMBER= 13  X= 3.40   Y= -3.9991
STEP NUMBER= 14  X= 3.60   Y= -4.7486
```

are all correct to the number of decimal places given.

Note that the program required that the values of y be given for $x = hj$ where $j = 1, 2, 3$. These "starting" values were obtained by using a Runge–Kutta method that was fourth order accurate (these calculations are not shown).

Program 174

```
          REAL*4 X(100),Y(100),YP(100)
C DEFINE THE INITIAL VALUES (FOUND BY RUNGE-KUTTA)
          H=.2
          X(1)= 1.
          Y(1)= 0.
          YP(1)= F(X(1),Y(1))
          X(2)= X(1) + H
          Y(2)=-0.02121
          YP(2)= F(X(2),Y(2))
          X(3)= X(2) + H
          Y(3)=-0.08894
          YP(3)= F(X(3),Y(3))
          X(4)= X(3) + H
          Y(4)=-0.20799
          YP(4)= F(X(4),Y(4))
C HERE IS THE INTEGRATION LOOP
          DO 10 N=4,14
          NP1=N+1
          X(NP1)=   X(N) + H
          Y(NP1)=   PREDIC(X,Y,YP,N,H)
          YP(NP1)= F(X(NP1),Y(NP1))
          Y(NP1)=   CORECT(X,Y,YP,N,H)
          Y(NP1)=   CORECT(X,Y,YP,N,H)
10        WRITE(6,5) N,X(N),Y(N)
5         FORMAT(' STEP NUMBER=',I3,'  X=',F5.2,'   Y=',F8.4)
```

```
          END
C THIS FUNCTION HAS THE PREDICTOR
          FUNCTION PREDIC(X,Y,YP,N,H)
          REAL*4   X(100),Y(100),YP(100)
          PREDIC=Y(N) + H/24.*(55.*YP(N)-59.*YP(N-1)+37*YP(N-2)-9.*YP(N-3))
          RETURN
          END
C THIS FUNCTION HAS THE CORRECTOR
          FUNCTION CORECT(X,Y,YP,N,H)
          REAL*4   X(100),Y(100),YP(100)
          CORECT=Y(N) + H/24.*(9.*YP(N+1)+19.*YP(N)-5.*YP(N-1)+YP(N-2))
          RETURN
          END
C THIS FUNCTION HAS THE RIGHT HAND SIDE OF THE DIFFERENTIAL EQUATION
          FUNCTION F(X,Y)
          F=1.0-X+Y/X
          RETURN
          END
```

Notes

[1] The corrector formula could be iterated as many times as is necessary to insure convergence. This is called *correcting to convergence*. In general, however, if more than two iterations are required, then the step size h is probably too large.

[2] Given the equation $y' = f(x, y)$, let P indicate an application of a predictor, C a single application of a corrector, and E an evaluation of the function f in terms of known values of its arguments. Correcting to convergence can then be represented by $P(EC)^\infty$. See Lambert [8] for an analysis of $P(EC)^m$ and $P(EC)^m E$, where m is a fixed number.

[3] Note that the predictor–corrector method is a finite difference scheme that is not a linear multistep method (as defined on page 573).

[4] To obtain the starting values so that the predictor–corrector pair can be used, Runge–Kutta methods can be used first. This was done in the example above. When this is done, the Runge–Kutta method used should be at least as accurate as the predictor–corrector formula used. See Gear [4] for details.

[5] For the same accuracy, using a predictor–corrector pair to integrate a first order ordinary differential equation generally requires fewer evaluations of the function $f(x, y)$ than a Runge–Kutta method would.

[6] One set of commonly used predictor–corrector equations is "Milne's method"

$$y_{n+1} = y_{n-3} + \frac{4h}{3}\left(2y_n - y'_{n-1} + 2y'_{n-2}\right),$$

$$y_{n+1} = y_{n-1} + \frac{h}{3}\left(y'_{n+1} + 4y'_n + y'_{n-1}\right).$$

These equations are also fourth order accurate. Milne's method is *not* recommended since it is subject to an instability problem, in which the errors do *not* tend to zero as the step size h is made smaller. See Gerald and Wheatley [5] for details.

[7] The Adams–Bashforth formulas are a family of linear multistep methods that are often used as predictors for the equation $y' = f(x, y)$. The k-step fixed-stepsize Adams–Bashforth formula

$$y_n = y_{n-1} + h \sum_{j=1}^{k} \beta_j f(x_{n-j}, y_{n-j}),$$

is equivalent to $y_n = y_{n-1} + \int_{x_{n-1}}^{x_n} p_n(s)\, ds$, where $p_n(x)$ is the unique polynomial of degree $k - 1$ that interpolates $f(x_{n-j}, y_{n-j})$ at x_{n-j} for $j = 1, \ldots, k$.

[8] For a listing of computer software that will implement the method described in this section, see page 586.

References

[1] M. Abramowitz and I. A. Stegun, *Handbook of Mathematical Functions*, National Bureau of Standards, Washington, DC, 1964, formula 25.5.13–25.5.16 (pages 896–897).

[2] W. E. Boyce and R. C. DiPrima, *Elementary Differential Equations and Boundary Value Problems*, Fourth Edition, John Wiley & Sons, New York, 1986, pages 431–438.

[3] R. Bronson, *Modern Introductory Differential Equations*, Schaum Outline Series, McGraw–Hill Book Company, New York, 1973, pages 232–257.

[4] C. W. Gear, "Runge–Kutta Starters for Multistep Methods," *ACM Trans. Math. Software*, **6**, No. 3, September 1980, pages 263–279.

[5] C. F. Gerald and P. O. Wheatley, *Applied Numerical Analysis*, Addison–Wesley Publishing Co., Reading, MA, 1984, pages 314–323.

[6] P. J. Van Der Houwen and B. P. Sommeijer, "Predictor–Corrector Methods for Periodic Second-Order Initial-Value Problems," *IMA J. Num. Analysis*, **7**, 1987, pages 407–422.

[7] A. I. A. Karim and G. A. Ismail, "Nonequidistant Modified Predictor–Corrector Methods for Solving Systems of Differential Equations," *Int. J. Comp. Math.*, **17**, 1985, pages 339–361.

[8] J. D. Lambert, *Computational Methods in Ordinary Differential Equations*, Cambridge University Press, New York, 1973.

[9] P. J. Van Der Houwen and B. P. Sommeijer, "Predictor–Corrector Methods for Periodic Second-Order Initial-Value Problems," *IMA J. Num. Analysis*, **7**, 1987, pages 407–422.

175. Runge–Kutta Methods

Applicable to　　Initial value systems of first order ordinary differential equations.

Yields

A numerical approximation to the solution of an initial value system.

Idea

Given an ordinary differential equation and an initial value, the value of the dependent variable may be found at the next desired value of the independent variable by calculating several intermediate values.

Procedure

Given the first order ordinary differential equation

$$y' = f(x, y), \qquad y(x_0) = y_0, \tag{175.1}$$

the value of $y(x)$ at the point $x_0 + h$ may be approximated by a weighted average of values of $f(x, y)$ taken at different points in the interval $x_0 \le x \le x_0 + h$. The classical Runge–Kutta formula is given by

$$y(x_0 + h) = y(x_0) + \tfrac{1}{6}\left(k_1 + 2k_2 + 2k_3 + k_4\right), \tag{175.2}$$

where

$$\begin{aligned}
k_1 &= hf(x_0, y_0), \\
k_2 &= hf(x_0 + \tfrac{1}{2}h, y_0 + \tfrac{1}{2}k_1), \\
k_3 &= hf(x_0 + \tfrac{1}{2}h, y_0 + \tfrac{1}{2}k_2), \\
k_4 &= hf(x_0 + h, y_0 + k_3).
\end{aligned} \tag{175.3}$$

This approximation to $y(x_0+h)$ is fourth order accurate. After $y(x_0+h)$ has been determined, the same formula may be used to determine $y(x_0 + 2h)$. This process may be repeated.

The Butcher array is a convenient way in which to represent all of the information in a Runge–Kutta method for the equation $\mathbf{y}' = \mathbf{f}(x, \mathbf{y})$, $\mathbf{y}(x_0) = \mathbf{y}_0$. Specifically, the s-stage Runge–Kutta scheme (which uses s intermediate values)

$$\mathbf{y}_{n+1} = \mathbf{y}_n + h \sum_{i=1}^{s} b_i \mathbf{k}_i,$$

$$\mathbf{k}_i := \mathbf{f}\left(x_n + c_i h, \mathbf{y}_n + h \sum_{j=1}^{s} a_{ij} \mathbf{k}_j\right),$$

where $h := x_{n+1} - x_n$, $\sum_i b_i = 1$, and $c_i = \sum_{j=1}^{s} a_{ij}$ for each j, is represented in the tabular form

$$
\begin{array}{c|c}
\mathbf{c} & A \\
\hline
 & \mathbf{b}^{\mathrm{T}}
\end{array}
\quad \text{or} \quad
\begin{array}{c|cccc}
c_1 & a_{11} & a_{12} & \cdots & a_{1s} \\
c_2 & a_{21} & a_{22} & \cdots & a_{2s} \\
\vdots & \vdots & \vdots & \ddots & \vdots \\
c_s & a_{s1} & a_{s2} & \cdots & a_{ss} \\
\hline
 & b_1 & b_2 & \ldots & b_s
\end{array}
$$

Note that an explicit Runge–Kutta scheme has $a_{ij} = 0$ for $j \geq i$ (sometimes these zeros are omitted). See Butcher [5] (page 163) or Dekker and Verwer [9] for details. The explicit method in (175.3) has the Butcher array (with $s = 4$)

$$
\begin{array}{c|cccc}
0 & 0 & 0 & 0 & 0 \\
\frac{1}{2} & \frac{1}{2} & 0 & 0 & 0 \\
\frac{1}{2} & 0 & \frac{1}{2} & 0 & 0 \\
1 & 0 & 0 & 1 & 0 \\
\hline
 & \frac{1}{6} & \frac{1}{3} & \frac{1}{3} & \frac{1}{6}
\end{array}
$$

Example 1

The program in Program 175 calculates a numerical approximation to the solution of the equation

$$
y' = 1 - x + \frac{y}{x}, \quad y(1) = 0, \tag{175.4}
$$

using the method in (175.3). It uses a step size h of .1. The value of $y(2)$ is found to be approximately $y(2) = -.6134$. The exact solution of equation (175.4), determined by integrating factors, is $y(x) = x(\log x - x + 1)$. Hence $y(2) = 2(\log 2 - 1) \simeq -.6137$.

Program 175

```
          H=.1
          X= 1.
          Y= 0.
          DO 10 J=1,10
          Y=RUNGE(X,Y,H)
          X=X+H
10        WRITE(6,88) X,Y
88        FORMAT(' X=',F6.2,' Y=',F7.4)
          END
C THIS FUNCTION PERFORMS ONE INTEGRATION STEP
          FUNCTION RUNGE(X,Y,H)
          FK1=F(X,     Y)
          FK2=F(X+H/2.,Y+H*FK1/2.)
          FK3=F(X+H/2.,Y+H*FK2/2.)
          FK4=F(X+H,    Y+H*FK3)
```

```
RUNGE=H*(FK1+2.*FK2+2.*FK3+FK4)/6.
RETURN
END
C THIS FUNCTION HAS THE RIGHT HAND SIDE OF THE EQUATION
FUNCTION F(X,Y)
F=1.-X+Y/X
RETURN
END
```

Example 2

The derivation of a Runge–Kutta method is instructive since it indicates the arbitrary degrees of freedom that exist in Runge–Kutta methods. Given the equation $y' = f(t, y)$, and using $y_n := y(t_n)$ and $t_n = nh$, to find a 2-stage Runge–Kutta scheme we assume a discrete approximation scheme of the form

$$
\begin{aligned}
y_{n+1} &= y_n + ak_1 + bk_2, \\
k_1 &= hf\left(t_n, y_n\right), \\
k_2 &= hf\left(t_n + \alpha h, y_n + \beta k_1\right),
\end{aligned} \tag{175.5}
$$

We want to find $\{a, b, \alpha, \beta\}$ to make the order of this scheme as high as possible. From (175.5) we can explicitly write y_{n+1} and then find a Taylor series expansion:

$$
\begin{aligned}
y_{n+1} &= y_n + ahf\left(t_n, y_n\right) + bhf\left(t_n + \alpha h, y_n + \beta hf\left(t_n, y_n\right)\right), \\
&= y_n + (a + b)hf_n + h^2\left(\alpha b f_t + \beta b f_y f\right)_n + O\left(h^3\right),
\end{aligned} \tag{175.6}
$$

where a subscript of n denotes evaluation at the point (t_n, y_n). From $y' = f(t, y)$ we can directly construct a Taylor expansion in t to find:

$$
\begin{aligned}
y_{n+1} &= y_n + hf_n + \frac{h^2}{2}\left(\frac{df}{dt}\right)_n + O\left(h^3\right), \\
&= y_n + hf_n + \frac{h^2}{2}\left(f_t + f_y f\right)_n + O\left(h^3\right),
\end{aligned} \tag{175.7}
$$

since $\dfrac{df}{dt} = f_t + f_y \dfrac{dy}{dt} = f_t + f_y f$. Comparing (175.7) to (175.6), we find the 3 equations

$$
a + b = 1, \qquad \alpha b = \tfrac{1}{2}, \qquad \beta b = \tfrac{1}{2},
$$

for the 4 unknowns $\{a, b, \alpha, \beta\}$. Since these equations are undetermined, there are infinitely many second order Runge–Kutta schemes in the form of (175.5).

Fourth order Runge–Kutta methods result in 11 equations for 13 unknowns; 2 of the unknowns may be chosen arbitrarily to achieve some goal. For example, we used a fourth order Runge–Kutta method with a specific sparsity pattern in the Butcher array to allow a parallel implementation in the section on parallel methods (see page 676).

Example 3

To obtain accurate numerical results when using any method, an estimate of the local error must be obtained. This could be done by the standard technique of recomputing the answer with the step size halved; but this requires lots of additional computation. The Runge–Kutta–Fehlberg method is a fifth order method that uses 6 functional evaluations and allows an estimate of the error by re-using the same points:

$$k_1 = hf\left(x_n, y_n\right),$$
$$k_2 = hf\left(x_n + \tfrac{1}{4}h, y_n + \tfrac{1}{4}k_1\right),$$
$$k_3 = hf\left(x_n + \tfrac{3}{8}h, y_n + \tfrac{3}{32}k_1 + \tfrac{9}{32}k_2\right),$$
$$k_4 = hf\left(x_n + \tfrac{12}{13}h, y_n + \tfrac{1932}{2197}k_1 - \tfrac{7200}{2197}k_2 + \tfrac{7296}{2197}k_3\right),$$
$$k_5 = hf\left(x_n + h, y_n + \tfrac{439}{216}k_1 - 8k_2 + \tfrac{3680}{513}k_3 - \tfrac{845}{4104}k_4\right),$$
$$k_6 = hf\left(x_n + \tfrac{1}{2}h, y_n - \tfrac{8}{27}k_1 + 2k_2 - \tfrac{3544}{2565}k_3 + \tfrac{1859}{4104}k_4 - \tfrac{11}{40}k_5\right),$$
$$y_{n+1} = y_n + \left(\tfrac{25}{216}k_1 + \tfrac{1408}{2565}k_3 + \tfrac{2197}{4104}k_k - \tfrac{1}{5}k_5\right)$$
$$\text{error} \approx \tfrac{1}{360}k_1 - \tfrac{128}{4275}k_3 - \tfrac{2197}{75240}k_4 + \tfrac{1}{50}k_5 + \tfrac{2}{55}k_6.$$

Notes

[1] If $f(x, y)$ does not depend on y, then the solution of the initial value problem $y' = f(x)$, $y(x_0) = y_0$, is just the integral $y(x) = y_0 + \int_{x_0}^{x} f(t)\, dt$. The Runge–Kutta method in equation (175.2) then corresponds to the approximation of $y(x)$ by means of Simpson's rule.

[2] There are several Runge–Kutta methods for first order equations. For example, the following scheme for equation (175.1)

$$y(x_0 + h) = y(x_0) + \tfrac{1}{2}\left(k_1 + k_2\right),$$
$$k_1 = hf(x_0, y_0), \tag{175.8}$$
$$k_2 = hf(x_0 + h, y_0 + k_1),$$

is of second order accuracy. A commonly used fourth order accurate method for first order ordinary differential equations (different from the one in (175.3)) is Gill's method; see Abramowitz and Stegun [1] formula 25.5.12.

[3] There are also implicit Runge–Kutta methods, see Burrage and Butcher [4] or Chapter 34 of Butcher [5].

[4] There are also Runge–Kutta methods for ordinary differential equations of orders two through ten. See, for instance, Section 2.4 of Collatz [8] or Abramowitz and Stegun [1]. For example, a Runge–Kutta scheme for the second order equation

$$y'' = g(x, y, y'), \qquad y(x_0) = y_0,\ y'(x_0) = v_0,$$

is given by

$$k_1 = hg\,(x_0, y_0, v_0)\,,$$
$$k_2 = hg\left(x_0 + \tfrac{1}{2}h, y_0 + \tfrac{1}{2}hv_0 + \tfrac{1}{8}hk_1, v_0 + \tfrac{1}{2}k_1\right),$$
$$k_3 = hg\left(x_0 + \tfrac{1}{2}h, y_0 + \tfrac{1}{2}hv_0 + \tfrac{1}{8}hk_1, v_0 + \tfrac{1}{2}k_2\right),$$
$$k_4 = hg\left(x_0 + \phantom{\tfrac{1}{2}}h, y_0 + \phantom{\tfrac{1}{2}}hv_0 + \tfrac{1}{2}hk_3, v_0 + \phantom{\tfrac{1}{2}}k_3\right),$$

$$(175.9)$$

and

$$y(x_0 + h) = y_0 + hv_0 + \tfrac{1}{6}h\,(k_1 + k_2 + k_3)\,,$$
$$y'(x_0 + h) = v_0 + \tfrac{1}{6}\,(k_1 + 2k_2 + 2k_3 + k_4)\,.$$

$$(175.10)$$

This scheme is numerically fourth order accurate.

[5] There are also Runge–Kutta methods for systems of first order ordinary differential equations (see Abramowitz and Stegun [1]). For example, the system

$$y' = m(x, y, z), \qquad z' = n(x, y, z)$$

of ordinary differential equations may be numerically approximated by first calculating

$$k_1 = hm\,(x_0, y_0, z_0)\,,$$
$$l_1 = hn\,(x_0, y_0, z_0)\,,$$
$$k_2 = hm\,(x_0 + h, y_0 + k_1, z_0 + l_1)\,,$$
$$l_2 = hn\,(x_0 + h, y_0 + k_1, z_0 + l_1)\,,$$

$$(175.11)$$

and then the updated values are

$$y(x_0 + h) = y(x_0) + \tfrac{1}{2}(k_1 + k_2),$$
$$z(x_0 + h) = z(x_0) + \tfrac{1}{2}(l_1 + l_2).$$

$$(175.12)$$

This formula is second order accurate. See Dekker and Verwer [9] for details.

[6] The book by Butcher [5] has a very comprehensive account of Runge–Kutta methods (it includes 96 pages of references!).

[7] The Butcher array can represent all multilinear methods for approximating differential equations. For instance:

(A) The backward Euler method $y_{n+1} = y_n + hf(t_n + h, y_{n+1})$ has the Butcher array ($s = 1$)

$$\begin{array}{c|c} 1 & 1 \\ \hline & 1 \end{array}$$

(B) The trapezoidal rule $y_{n+1} = y_n + \dfrac{h}{2}[f(t_n + y_n) + f(t_n + h, y_{n+1})]$ has the Butcher array ($s = 2$)

$$\begin{array}{c|cc} 0 & 0 & 0 \\ 1 & \tfrac{1}{2} & \tfrac{1}{2} \\ \hline & \tfrac{1}{2} & \tfrac{1}{2} \end{array}$$

[8] Pseudo Runge–Kutta methods use not only the stages of the current step, but also the stages of the previous step. For example, for the equation $y' = f(x, y)$ the method has the form:

$$y_{n+1} = y_n + \sum_{i=1}^{s} \alpha_i K_{i,n}$$

$$K_{i,n} = hf \left(x_n + m_i h, y_n + \sum_{j=1}^{s} \bar{\lambda}_{i,j} K_{j,n-1} + \sum_{j=1}^{i-1} \lambda_{i,j} K_{j,n} \right)$$

See Caira *et al.* [7] for details.

[9] To find software that implements this method, see the section beginning on page 586.

[10] To obtain a Runge–Kutta method with a desired order, a minimum number of stages (i.e., function evaluations) are required. From Butcher [5] we have:

desired order:	1	2	3	4	5	6	7	8
minimal number of stages:	1	2	3	4	6	7	9	11

References

[1] M. Abramowitz and I. A. Stegun, *Handbook of Mathematical Functions*, National Bureau of Standards, Washington, DC, 1964, formulae 25.5.6–25.5.12.

[2] P. Bogacki and L. F. Shampine, "Interpolating High-Order Runge–Kutta Formulas," *Comp. & Maths. with Appls.*, **20**, No. 3, 1990, pages 15–24.

[3] W. E. Boyce and R. C. DiPrima, *Elementary Differential Equations and Boundary Value Problems*, Fourth Edition, John Wiley & Sons, New York, 1986, pages 420–423.

[4] K. Burrage and J. C. Butcher, "Stability Criteria for Implicit Runge–Kutta Methods," *SIAM J. Numer. Anal.*, **16**, No. 1, February 1979, pages 30–45.

[5] J. C. Butcher, *The Numerical Analysis of Ordinary Differential Equations*, Wiley, New York, 1987.

[6] J. C. Butcher and J. R. Cash, "Towards Efficient Runge–Kutta Methods for Stiff Systems," *SIAM J. Numer. Anal.*, **27**, No. 3, June 1990, pages 753–761.

[7] R. Caira, C. Costabile and F. Costabile, "A Class of Pseudo Runge–Kutta Methods," *BIT*, **30**, 1990, pages 642–649.

[8] L. Collatz, *The Numerical Treatment of Differential Equations*, Springer–Verlag, New York, 1966, pages 61–77.

[9] K. Dekker and J. G. Verwer, *Stability of Runge–Kutta Methods for Stiff Nonlinear Differential Equations*, North–Holland Publishing Co., New York, 1984, Chapter 3.

[10] D. J. Evans and B. B. Sanugi, "A Nonlinear Runge–Kutta Formula for Initial Value Problems," *SIGNUM Newsletter*, **22**, No. 3, July 1987, pages 27–30.

[11] J. M. Fine, "Low Order Practical Runge–Kutta–Nyström Methods," *Computing*, **38**, 1987, pages 281–297.

[12] L. F. Shampine, "Diagnosing Stiffness for Runge–Kutta Methods," *SIAM J. Sci. Stat. Comput.*, **12**, No. 2, March 1991, pages 260–272.

176. Stiff Equations*

Applicable to   Stiff differential equations; i.e., equations that evolve on more than one scale.

Yields

A numerical approximation technique.

Idea

Since stiff equations evolve on different scales, the techniques used to numerically approximate the solution should change as the different scales become important. This is because the stability aspects of a numerical technique often change as the equation changes (see page 613). Consider, for example, the definition of stiffly stable on page 617—as the eigenvalues of the problem change a method may no longer be stiffly stable.

Procedure

When trying to numerically approximate the solution to a stiff differential equation, the step size used in the discretization process should be variable, becoming very small when needed. The discretization formula should also change in different regions to reflect the different type of local solution (i.e., exponential growth, exponential decay, algebraic growth, etc.)

The step size should be made as small as is needed to obtain a desired accuracy, but it should be increased whenever possible to reduce the total number of computations. The step size should not be allowed to get so large, though, that the discretization technique becomes unstable.

A good choice of step size can be determined by monitoring the change in the solution of the differential equation. For any single step, the change in the function being approximated *and* all of its derivatives should not become too large.

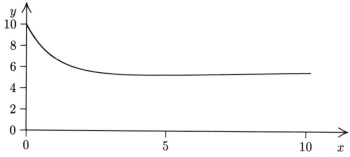

Figure 176. The solution to (176.1) and (176.2) is $y(x) = e^{\varepsilon x} + e^{-x}$.

Example

Suppose we have the ordinary differential equation

$$\frac{d^2 y}{dx^2} + (1 - \varepsilon)\frac{dy}{dx} - \varepsilon y = 0, \tag{176.1}$$

with the initial conditions

$$y(0) = 2, \quad y'(0) = \varepsilon - 1, \tag{176.2}$$

where ε is a small positive number. The solution to (176.1) and (176.2) is

$$y(x) = e^{\varepsilon x} + e^{-x}, \tag{176.3}$$

which has a steep decrease from $x = 0$ to $x \simeq -\log \varepsilon$, and then has a gradual increase, see Figure 176.

When using a simple discretization scheme (such as, say, Euler's method), a small step size is required in the region from $x = 0$ to $x \simeq -\log \varepsilon$ to resolve the exponential decay. After that region, however, the step size should be increased since the solution is no longer rapidly varying.

The FORTRAN program in Program 176 implements this numerical idea. It uses Euler's method and a variable step size, when ε is .01. The parameter TOL determines how much the solution is allowed to change at any step. Note that the change in the solution is defined to also include the change in the value of the derivative. We have chosen TOL = .01.

A few lines of the output of the program are shown below

```
AT TIME=   0.005   DELTAT= 0.0049   Y(T)= 1.9952   EXACT VALUE= 1.9952
AT TIME=   0.317   DELTAT= 0.0049   Y(T)= 1.7307   EXACT VALUE= 1.7312
AT TIME=   0.327   DELTAT= 0.0098   Y(T)= 1.7237   EXACT VALUE= 1.7243
AT TIME=   1.001   DELTAT= 0.0098   Y(T)= 1.3761   EXACT VALUE= 1.3776
AT TIME=   1.021   DELTAT= 0.0195   Y(T)= 1.3691   EXACT VALUE= 1.3707
AT TIME=   1.685   DELTAT= 0.0195   Y(T)= 1.2005   EXACT VALUE= 1.2025
AT TIME=   1.724   DELTAT= 0.0391   Y(T)= 1.1937   EXACT VALUE= 1.1958
```

```
AT TIME=  2.349  DELTAT= 0.0391  Y(T)= 1.1170  EXACT VALUE= 1.1193
AT TIME=  2.427  DELTAT= 0.0781  Y(T)= 1.1105  EXACT VALUE= 1.1129
AT TIME=  2.974  DELTAT= 0.0781  Y(T)= 1.0788  EXACT VALUE= 1.0813
AT TIME=  3.130  DELTAT= 0.1563  Y(T)= 1.0728  EXACT VALUE= 1.0755
AT TIME=  3.599  DELTAT= 0.1563  Y(T)= 1.0613  EXACT VALUE= 1.0640
AT TIME=  9.849  DELTAT= 0.3125  Y(T)= 1.1034  EXACT VALUE= 1.1036
AT TIME= 10.161  DELTAT= 0.3125  Y(T)= 1.1068  EXACT VALUE= 1.1070
```

During the program execution, the step size, DELTAT, has increased from .0049 to .3125. Hence, large steps were taken where the solution is not rapidly changing.

Program 176

```
        IMPLICIT DOUBLE PRECISION (A-H,O-Z)
        TEND=10.D0
        EPSLON=.01D0
        TOL=.01D0
        DELTAT=TEND
        OLDCHG=1.D0
        T=0.D0
        Y=2.D0
        YP=EPSLON-1.D0
C DECREASE THE SIZE OF THE TIME STEP
10      DELTAT=DELTAT/2.D0
20      IF ( DELTAT .GT. .5D0 ) GOTO 10
        CALL STEP(Y,YP,DELTAT,EPSLON,YN,YNP)
        CHANGE= DSQRT((Y-YN)**2 + (YP-YNP)**2)
        IF( CHANGE .GT. TOL          ) GOTO 10
        IF( CHANGE .GT. 2.D0*OLDCHG ) GOTO 10
C STORE AWAY THE NEW VALUES
        T = T + DELTAT
        Y = YN
        YP= YNP
        OLDCHG=CHANGE
        VAL=EXACT(T,EPSLON)
        WRITE(6,5) T, DELTAT, Y, VAL
5       FORMAT(' AT TIME=',F7.3,'  DELTAT=',F7.4,
     1         '  Y(T)=',F7.4,   '  EXACT VALUE=',F7.4)
C INCREASE THE SIZE OF THE TIME STEP
        DELTAT=2.D0*DELTAT
        IF( T .LT. TEND ) GOTO 20
        END
C THIS SUBROUTINE UPDATES Y AND Y' BY EULER'S METHOD
        SUBROUTINE   STEP(Y,YP,DELTAT,EPSLON,YN,YNP)
        IMPLICIT DOUBLE PRECISION (A-H,O-Z)
        YN = Y  + DELTAT*( YP )
        YNP= YP + DELTAT*( EPSLON*Y - YP*(1.D0-EPSLON) )
        RETURN
        END
C THIS FUNCTION COMPUTES THE EXACT SOLUTION TO COMPARE AGAINST
        FUNCTION EXACT(T,EPSLON)
        IMPLICIT DOUBLE PRECISION (A-H,O-Z)
        EXACT=DEXP(EPSLON*T)+DEXP(-T)
        RETURN
        END
```

Notes

[1] In the example shown, we can use the same discretization scheme throughout the region of interest—only the step size needs to be adjusted for efficient computation. In other problems, different discretization schemes will be needed in different regions.

[2] If the new independent variable $\tilde{x} = \varepsilon x$ is introduced, then the solution in (176.3) may be written as $y(x) = e^{\tilde{x}} + e^{-\tilde{x}/\varepsilon}$. In this representation of the solution it is clear that there is a "boundary layer" near $\tilde{x} = 0$; see the section on boundary layers (page 510).

[3] For an example of how the stability of a method may change as the solution of a differential equation evolves, see the stability analysis for Euler's method on page 655. In the example there, as the value of the positive constant c becomes smaller, the step size must also become smaller to ensure stability.

[4] Sometimes non-stiff methods can solve stiff problems, without any special difficulty, except that they can be prohibitively expensive.

[5] Changing the length of the step size leads to accurate solutions to stiff initial value ordinary differential equations and for partial differential equations that may be solved by a marching technique. For boundary value ordinary differential equations, or for elliptic partial differential equations, the analogous technique is to numerically solve the equations on a non-uniform mesh. This mesh should be fine where the solution is rapidly changing, and coarse elsewhere.

[6] It is *not* true that the eigenvalues of the matrix $A(t)$ in the system

$$\frac{d\mathbf{y}}{dt} = A(t)\mathbf{y} \tag{176.4}$$

determine whether the system is stiff or not. For example, the matrix

$$A(t) = \begin{pmatrix} -1 - 9\cos^2 6t + 6\sin 12t & 12\cos^2 6t + \dfrac{9}{2}\sin 12t \\ -12\sin^2 6t + \dfrac{9}{2}\sin 12t & -1 - 9\sin^2 6t - 6\sin 12t \end{pmatrix} \tag{176.5}$$

has the constant eigenvalues -1 and -10, but the solution to (176.4) is

$$\mathbf{y} = C_1 e^{2t}\begin{pmatrix} \cos 6t + 2\sin 6t \\ 2\cos 6t - \sin 6t \end{pmatrix} + C_2 e^{-13t}\begin{pmatrix} \sin 6t - 2\cos 6t \\ 2\sin 6t + \cos 6t \end{pmatrix},$$

where C_1 and C_2 are arbitrary constants. Clearly the exponentials e^{-t} and e^{-10t} are not present in the solution. Also, the solution may blow up as t tends to infinity. Even so, the eigenvalues of the linearized problem are often the most useful piece of information available regarding the conditioning of the system. This example is from Dekker and Verwer [4] (page 11).

[7] If $\eta(t)$ is defined by $\eta = ||\mathbf{y}||^2 = \mathbf{y}^H \mathbf{y}$, then, using (176.4), $\dfrac{d\eta}{dt} = \mathbf{y}^H \left(A + A^H \right) \mathbf{y}$. If $\lambda_{\max}$ represents the largest eigenvalue of $A + A^H$ then $\eta(t) \le \eta_0 e^{\lambda_{\max} t}$. Hence, the eigenvalues of $A + A^H$ allow bounds to be determined for $\mathbf{y}(t)$. For the matrix in (176.5), the eigenvalues of $A + A^H$ are 4 and -26.

[8] An equation is often realized to be stiff only after the differential equation has been numerically integrated. There are tests that can be performed during the integration procedure to determine if the equation is stiff. See, for example, Gear [6] or Shampine [9].

[9] For a recent review of software for stiff equations, see Byrne and Hindmarsh [2] or Chapter 4 of Aiken [1].

[10] For a listing of computer software that will implement the method described in this section, see page 586.

References

[1] R. C. Aiken, *Stiff Computation*, Oxford University Press, New York, 1985, Chapters 3–4 (pages 70–202).

[2] G. D. Byrne and A. C. Hindmarsh, "Stiff ODE Solvers: A Review of Current and Coming Attractions," *J. Comput. Physics*, **70**, 1987, pages 1–62.

[3] K. Dekker and J. G. Verwer, *Stability of Runge–Kutta Methods for Stiff Nonlinear Differential Equations*, North–Holland Publishing Co., New York, 1984.

[4] P. W. Gaffney, "A Performance Evaluation of Some FORTRAN Subroutines for the Solution of Stiff Oscillatory Ordinary Differential Equations," *ACM Trans. Math. Software*, **10**, No. 1, March 1984, pages 58–72.

[5] C. W. Gear, "Automatic Detection and Treatment of Oscillatory and/or Stiff Ordinary Differential Equations," in J. Hinze (ed.), *Numerical Integration of Differential Equations and Large Linear Systems*, Springer–Verlag, New York, 1982, pages 190–206.

[6] W. L. Miranker, *Numerical Methods for Stiff Equations*, D. Reidel Publishing Co., Boston, 1981.

[7] L. F. Shampine, "Stiffness and Nonstiff Differential Equation Solvers, II: Detecting Stiffness With Runge–Kutta Methods," *ACM Trans. Math. Software*, **3**, No. 1, March 1977, pages 44–53.

[8] L. F. Shampine and C. W. Gear, "A User's View of Solving Stiff Differential Equations," *SIAM Review*, **21**, 1979, pages 1–17.

[9] L. Petzold, "Automatic Selection of Methods for Solving Stiff and Nonstiff Systems of Ordinary Differential Equations," *SIAM J. Sci. Stat. Comput.*, **4**, No. 1, March 1983, pages 136–148.

177. Integrating Stochastic Equations

Applicable to   Stochastic differential equations.

Yields

A numerical approximation.

Idea

The "white Gaussian noise" term in a stochastic differential equation can be numerically approximated in many different ways.

Procedure

Suppose we have the stochastic differential equation

$$x' = b(x) + \sigma(x)n(t),$$
$$x(0) = y,$$

(177.1)

where $n(t)$ represents white noise. There exist several numerical approximations for the quantity $x(T)$, where $T = mh$, h is a (small) time step, and T is a fixed time of order one. Three common numerical approximations of (177.1) are

$$\tilde{x}(t_{k+1}) = \tilde{x}(t_k) + b_k h + \sigma_k \sqrt{h}\, \alpha_k,$$
$$\tilde{x}(0) = y,$$

(177.2)

$$\hat{x}(t_{k+1}) = \hat{x}(t_k) + b_k h + \sigma_k \sqrt{h}\, \zeta_k,$$
$$\hat{x}(0) = y,$$

(177.3)

$$\check{x}(t_{k+1}) = \check{x}(t_k) + \left(b - \frac{1}{2}\sigma\frac{\partial\sigma}{\partial x}\right)_k h + \sigma_k \sqrt{h}\, \zeta_k + \frac{1}{2}\left(\sigma\frac{\partial\sigma}{\partial x}\right)_k h\, \zeta_k^2,$$
$$\check{x}(0) = y,$$

(177.4)

where $t_k = kh$ and a subscript of k means evaluation at the k-th point, e.g., $b_k = b(x(t_k))$. The $\{\alpha_k\}$ are independent random variables that take on the values $+1$ and -1 with probability $1/2$, while the $\{\zeta_k\}$ are independent Gaussian random variables with mean 0 and variance 1.

Each of the approximations in (177.2)–(177.4) has a different mean square error for a single step. If $E[\cdot]$ represents the expectation operator, then

$$E[(x(h) - \tilde{x}(h))^2] = O(h),$$
$$E[(x(h) - \hat{x}(h))^2] = O(h^2),$$
$$E[(x(h) - \check{x}(h))^2] = O(h^3).$$

(177.5)

Hence, (177.4) is the most accurate if a sample of $x(T)$ is desired.

However, if the mean of a function of $x(T)$ is required, then each of the three approximations in (177.2)–(177.4) is first order accurate. That is, each of $E[f(\tilde{x}(T))]$, $E[f(\hat{x}(T))]$, and $E[f(\check{x}(T))]$ is equal to $E[f(x(T))] + O(h)$, for general functions f. This next approximation,

$$
\begin{aligned}
z(t_{k+1}) = z(t_k) &+ \left(b - \frac{1}{2}\sigma\frac{\partial\sigma}{\partial x}\right)_k h + \sigma_k\sqrt{h}\,\zeta_k + \frac{1}{2}\left(\sigma\frac{\partial\sigma}{\partial x}\right)_k h\,\zeta_k^2 \\
&+ \left(\frac{1}{2}b\frac{\partial\sigma}{\partial x} + \frac{1}{2}\sigma\frac{\partial b}{\partial x} + \frac{1}{2}\frac{\partial\sigma}{\partial t} + \frac{1}{4}\sigma^2\frac{\partial^2\sigma}{\partial x^2}\right)_k h^{3/2}\zeta_k \\
&+ \left(\frac{1}{2}b\frac{\partial b}{\partial x} + \frac{1}{2}\frac{\partial b}{\partial t} + \frac{1}{4}\sigma^2\frac{\partial^2 b}{\partial x^2}\right)_k h^2,
\end{aligned}
$$
$$
z(0) = y,
$$

(177.6)

has the better error estimate: $E[f(z(T))] = E[f(x(T))] + O(h^2)$. Note that, in (177.6), we have allowed b and σ to be functions of both t and x.

Example

Suppose we have the stochastic differential equation

$$
x' = x + n(t), \qquad x(0) = 1, \tag{177.7}
$$

where $n(t)$ is white noise, and we want to estimate $E[x^2(1)]$. The Fokker–Planck equation corresponding to (177.7) is (see page 254)

$$
\frac{\partial P}{\partial t} = -\frac{\partial}{\partial x}(xP) + \frac{1}{2}\frac{\partial^2}{\partial x^2}(P),
$$

with $P(0, x) = \delta(x - 1)$. By using the method of moments (see page 491), the ordinary differential equation that describes $E[x^2(t)]$ is given by

$$
\frac{d}{dt}E[x^2(t)] = 2E[x^2(t)] + 1, \qquad E[x^2(0)] = 1,
$$

with the solution $E[x^2(t)] = (3e^{2t} - 1)/2$. Therefore, $E[x^2(1)] = (3e^2 - 1)/2 \simeq 10.58$. This is the value that our numerical approximation should produce.

To implement the method in (177.3), the FORTRAN program in Program 177 was constructed. The program takes the results of NTRIAL trials and averages these values together. Note that the program uses a routine called RANDOM, whose source code is not shown, which returns a random value uniformly distributed on the interval from zero to one.

A similar program was written which implemented the methods in (177.2) and (177.4). The results are indicated in Table 177. It should be observed that the numerical results are increasingly accurate when the step size h is decreased.

Table 177:

Numerical comparison of different approximation techniques for equation (177.7).

NTRIAL	h	Eqn. (177.2)	Eqn. (177.3)	Eqn. (177.4)
1000	.25	8.14	8.40	11.19
1000	.2	8.61	8.74	11.11
1000	.1	9.62	9.30	10.59
1000	.05	10.00	10.16	10.87
5000	.25	8.14	8.40	11.19
5000	.2	8.51	8.36	10.60
5000	.1	9.46	9.35	10.59
5000	.05	9.97	10.18	10.90

Program 177

```
C THIS PROGRAM IS A NUMERICAL IMPLEMENTATION OF EQUATION (3)
      NTRIAL=1000
      H=.05
      NTIME=20
      XINIT=1.
      SUMX2=0.
C HERE IS THE INTEGRATION LOOP
      DO 10 NSTEP=1,NTRIAL
      X=XINIT
      DO 20 K=1,NTIME
20    X=X + X*H + SQRT(H)*ZETA()
10    SUMX2=SUMX2 + X**2
      AVERAG=SUMX2/FLOAT(NTRIAL)
      WRITE(6,*) AVERAG
      END
C THIS FUNCTION RETURNS A GAUSSIAN RANDOM VARIABLE
      FUNCTION ZETA()
      DATA  TWOPI/6.2831853/
      Y1=RANDOM( DSEED )
      Y2=RANDOM( DSEED )
      ZETA= SQRT( -2.*ALOG(Y2) ) * COS( TWOPI*Y1 )
      RETURN
      END
```

Notes

[1] Gaussian random variables may be generated from uniformly distributed random variables by the classical technique of Box and Muller [1]. This technique has been used in the function ZETA in Program 177.

[2] Since low numerical accuracy is obtained by this technique, a computer program does not need to work with extended precision arithmetic (such as double precision).

[3] Milśhtein [9] and [8] describes (177.2), (177.3), (177.4), and presents a derivation of (177.6). He also includes a numerically fast implementation of (177.6) using Runge–Kutta methods.

[4] Sun [13] presents a numerical method for approximating the solution to equations of the form $-(pu')' + (q + r\lambda)^2 u = f$, when p, q and r are all functions of the independent variable and both λ and f are random terms.

References

[1] G. E. P. Box and M. E. Muller, "A Note on the Generation of Random Normal Deviates," *Ann. Math. Statistics*, **9**, 1958, pages 610–611.

[2] C.-C. Chang, "Numerical Solution of Stochastic Differential Equations with Constant Diffusion Coefficients," *Math. of Comp.*, **49**, No. 180, October 1987, pages 523–542.

[3] I. T. Drummond, A. Hoch, and R. R. Morgan, "Numerical Integration of Stochastic Differential Equations with Variable Diffusivity," *J. Phys. A: Math. Gen.*, **19**, 1986, pages 3871–3881.

[4] J. Golec and G. Ladde, "Euler-Type Approximation for Systems of Stochastic Differential Equations," *J. Appl. Math. Simulation*, **2**, No. 4, 1989, pages 239–249.

[5] H. S. Greenside and E. Helfand, "Numerical Integration of Stochastic Differential Equations - II," *The Bell System Technical Journal*, **60**, No. 8, October 1981, pages 1927–1940.

[6] R. Janssen, "Discretization of the Wiener-Process in Difference-Methods for Stochastic Differential Equations," *Stochastic Processes and their Applications*, **18**, 1984, pages 361–369.

[7] R. Janssen, "Difference Methods for Stochastic Differential Equations with Discontinuous Coefficients," *Stochastics*, **13**, 1984, pages 199–212.

[8] G. N. Milśhtein, "Approximate Integration of Stochastic Differential Equations," *Theory Prob. Appl.*, **19**, No. 4, 1974, pages 557–562.

[9] G. N. Milśhtein, "A Method of Second-Order Accuracy Integration of Stochastic Differential Equations," *Theory Prob. Appl.*, **23**, No. 2, 1978, pages 396–401.

[10] N. J. Newton, "Asymptotically Efficient Runge–Kutta Methods for a Class of Itó and Stratonovich Equations," *SIAM J. Appl. Math.*, **51**, No. 2, April 1991, pages 542–567.

[11] W. Rümelin, "Numerical Treatment of Stochastic Differential Equations," *SIAM J. Numer. Anal.*, **19**, No. 3, June 1982, pages 604–613.

[12] R. Spigler, "Monte Carlo-Type Simulation for Solving Stochastic Ordinary Differential Equations," *Math. and Computers in Simulation*, **29**, 1987, pages 243–251.

[13] T.-C. Sun, "A Finite Element Method for Random Differential Equations with Random Coefficients," *SIAM J. Numer. Anal.*, **16**, No. 6, December 1979, pages 1019–1035.

178. Weighted Residual Methods[*]

Applicable to   Ordinary differential equations and partial differential equations.

Yields

By introducing approximations, this method changes the numerical calculation of:

(A) an ordinary differential equation to the numerical calculation of a set of algebraic equations,

(B) a partial differential equation to the numerical calculation of a set of ordinary differential equations.

Idea

We approximate the solution by taking a linear combination of an arbitrarily chosen set of functions. The coefficients of the functions, which may be constants or functions themselves, are unknown. We may use any of a number of schemes to find the numerical values for the unknown coefficients.

Procedure

We will illustrate the general technique via a specific example. Suppose we have the following partial differential equation to solve

$$
\begin{aligned}
u_t - N[u] &= 0, &&\text{for } \mathbf{x} \in V, \quad t > 0, \\
u(0, \mathbf{x}) &= v(\mathbf{x}), &&\text{for } \mathbf{x} \in V, &&(178.1.a\text{–}c) \\
u(t, \mathbf{x}) &= f(t, \mathbf{x}), &&\text{for } \mathbf{x} \in S,
\end{aligned}
$$

where $N[\cdot]$ is a differential operator in $\mathbf{x}$ and S is the boundary of V, the region in which we seek the solution.

We choose a $y(t, \mathbf{x})$ and some set of functions $\{u_i(t, \mathbf{x})\}$ with the properties

$$
\begin{aligned}
y(t, \mathbf{x}) &= f(t, \mathbf{x}), &&\text{for } \mathbf{x} \in S, \\
u_j(t, \mathbf{x}) &= 0, &&\text{for } \mathbf{x} \in S,
\end{aligned}
$$

and then form a trial solution by superposition

$$
u_T(t, \mathbf{x}) = y(t, \mathbf{x}) + \sum_{j=1}^{M} c_j(t) u_j(t, \mathbf{x}). \tag{178.2}
$$

Note that the trial solution has been constructed in such a way that it automatically satisfies (178.1.c) but not (178.1.a) or (178.1.b). If we use the trial solution in the original differential equation, (178.1.a), then the right-hand side will not be equal to zero, but will be equal to some residual R_E given by

$$R_E(u_T) = (u_T)_t - N[u_T].\tag{178.3}$$

Instead of this definition of R_E, we might equally well have taken the square of (178.3). Likewise, the boundary condition, (178.10.b), will not be satisfied, but there will be a residue R_B given by

$$R_B(u_T) = v(\mathbf{x}) - \sum_{j=1}^{M} c_j(0)u_j(0, \mathbf{x}).$$

Now we choose M weighting functions, $\{w_j(x)\}$. It is the choice of the weighting functions that defines the method. For example

$$
\begin{aligned}
\text{Galerkin:} \quad & w_j = u_j, \\
\text{Collocation:} \quad & w_j = \delta(\mathbf{x} - \mathbf{x}_j), \\
\text{least squares:} \quad & w_j = \frac{\partial R_E(u_T)}{\partial c_j}, \\
\text{subdomain method:} \quad & w_j = \begin{cases} 1, & x \in V_j, \\ 0, & x \notin V_j, \end{cases}
\end{aligned}
\tag{178.4}
$$

where $\{\mathbf{x}_j \mid j = 1, 2, \ldots, M\}$ is a set of M points in V that must be chosen when collocation is used, and $\{V_j\}$ is a set of disjoint regions whose union is equal to V that must be chosen when the subdomain method is used.

Next, an inner product is defined by

$$(w, z) = \int_V w(\mathbf{x})\, z(\mathbf{x})\, dV,\tag{178.5}$$

or something similar. Then, finally, the unknown coefficients $\{c_j(t)\}$ will be determined from the two conditions

$$
\begin{aligned}
(w_j, R_E(u_T)) = 0, \quad & \text{for } j = 1, 2, \ldots, M, \\
(w_j, R_B(u_T)) = 0, \quad & \text{for } j = 1, 2, \ldots, M.
\end{aligned}
\tag{178.6.a–b}
$$

Condition (178.6.a) generates M simultaneous ordinary differential equations for $\{c_j(t) \mid j = 1, 2, \ldots, M\}$, which will generally be nonlinear. Condition (178.6.b) generates M simultaneous algebraic equations for $\{c_j(0) \mid j = 1, 2, \ldots, M\}$, which will generally be nonlinear.

The procedure is as follows. We solve (178.6.b) for the initial conditions for the $\{c_j(t)\}$. Using (178.6.a) we can then solve the ordinary differential equations to determine the $\{c_j(t)\}$ for all values of t. Using these values in (178.2), we have found an approximation to (178.1).

Example

Suppose we wish to approximate the solution to the equation

$$u_t = N[u] = u^2 + u_{xx}, \quad \text{for } 0 < x < 1, \ t > 0,$$
$$u(0, x) = \sin x = v(x),$$
$$u(t, 0) = 0,$$
$$u(t, 1) = 1.$$

We choose $y(t, x) = x$ and $u_j(t, x) = \sin j\pi x$. Our trial solution then becomes the first M terms in a Fourier sine series

$$u_T(t, x) = x + \sum_{j=1}^{M} c_j(t) \sin j\pi x.$$

Approximating $u(t, x)$ by $u_T(t, x)$ the errors in the equation and the boundary conditions are

$$R_E(u_T) = \sum_{j=1}^{M} c'_j(t) \sin j\pi x - \left[x + \sum_{j=1}^{M} c_j(t) \sin j\pi x \right]^2$$

$$- \sum_{j=1}^{M} j^2 \pi^2 c_j(t) \sin(j\pi x), \quad\quad (178.7.a\text{--}b)$$

$$R_B(u_T) = \sin x - \sum_{j=1}^{M} c_j(0) \sin j\pi x.$$

These two equations are in x and t. Ideally, we would like to have both expressions in (178.7) vanish identically. Since this is not possible (for all x and all t) we choose one of the four methods described in (178.4). Using the chosen method, we will obtain ordinary differential equations for the $\{c_j(t)\}$ and algebraic equations for the $\{c_j(0)\}$. When these equations are satisfied, the expressions in (178.7) will be "close" to zero.

Notes

[1] It is also possible to choose the $\{u_i(t, \mathbf{x})\}$ to satisfy the differential equation in (178.1), but not the boundary conditions. In this case, the integral in (178.5), which defines the inner product, becomes an integral over the boundary.

[2] See the separate sections on least squares method (page 473), finite element method (page 656), Rayleigh–Ritz method (page 554), and collocation (page 441).

[3] Within the Galerkin framework, it is possible to generate finite elements, finite difference, and spectral methods.

[4] This method can also be used to change the numerical calculation of an ordinary differential equation to the calculation of solving algebraic equations. The sequence of steps are the same as for partial differential equations, with the difference that both sets of equations in (178.6) will be algebraic equations. See the finite element method (page 656) for a worked example involving an ordinary differential equation.

References

[1] W. F. Ames, "Ad Hoc Exact Techniques for Nonlinear Partial Differential Equations," in W. F. Ames (ed.), *Nonlinear Partial Differential Equations in Engineering*, Academic Press, New York, 1967, pages 243–261.

[2] L. Collatz, *The Numerical Treatment of Differential Equations*, Springer–Verlag, New York, 1966, pages 408–418.

[3] C. A. J. Fletcher, *Computational Galerkin Methods*, Springer–Verlag, New York, 1984.

[4] D. Gottlieb and S. A. Orszag, *Numerical Analysis of Spectral Methods: Theory and Applications*, SIAM, Philadelphia, 1977.

[5] M. Haque, M. H. Baluch, and M. F. N. Mohsen, "Solution of Multiple Point, Nonlinear Boundary Value Problems by Method of Weighted Residuals," *Int. J. Comp. Math.*, **19**, 1986, pages 69–84.

[6] L. V. Kantorovich and V. I. Krylov, *Approximate Methods of Higher Analysis*, Interscience Publishers, New York, 1958, pages 258–283.

[7] H. Keller, *Numerical Methods for Two-Point Boundary-Value Problems*, Blaisdell Publishing Co., Waltham, MA, 1968, Appendix (pages 173–177).

[8] J. Villadsen and M. L. Michelsen, *Solution of Differential Equation Models by Polynomial Approximation*, Prentice–Hall Inc., Englewood Cliffs, NJ, 1978, Chapter 2 (pages 67–95).

IV.C

Numerical Methods for PDEs

179. Boundary Element Method

Applicable to   Most often linear elliptic partial differential equations, often Laplace's equation. Sometimes parabolic, hyperbolic, or nonlinear elliptic equations.

Yields

An integral equation. The solution of the integral equation is used in an integral representation of the solution.

Idea

The problem of solving a partial differential equation *within* a given domain can be transformed into one solving an equivalent integral equation *on* the boundary of the domain. The unknown in the integral equation will be the "charge density" on the boundary of the domain.

Procedure

Suppose we have Laplace's equation (general linear elliptic equations have results analogous to those listed below)

$$\nabla^2 u(\mathbf{x}) = 0, \tag{179.1}$$

with the Dirichlet or Neumann data

$$u\Big|_S = f(\mathbf{x}) \quad \text{or} \quad \frac{\partial u}{\partial n}\Big|_S = g(\mathbf{x}), \tag{179.2.a–b}$$

where S is the boundary of the domain. Define $\psi(\mathbf{x}; \mathbf{y})$ to be the free space Green's function of (179.1). That is, $\nabla^2 \psi(\mathbf{x}; \mathbf{y}) = \delta(\mathbf{x} - \mathbf{y})$, where $\mathbf{y}$ is an arbitrary point inside the domain. Using Green's theorem, the solution to (179.1) and (179.2) can be represented in any of the following forms:

$$u(\mathbf{x}) = \int_S \sigma(\mathbf{z})\psi(\mathbf{x}; \mathbf{z})\, d\mathbf{z}, \tag{179.3}$$

$$u(\mathbf{x}) = \int_S \mu(\mathbf{z})\frac{\partial \psi(\mathbf{x}; \mathbf{z})}{\partial n}\, d\mathbf{z}, \tag{179.4}$$

$$u(\mathbf{x}) = \int_S \left[\eta(\mathbf{z})\psi(\mathbf{x}; \mathbf{z}) + \zeta(\mathbf{z})\frac{\partial \psi(\mathbf{x}; \mathbf{z})}{\partial n} \right] d\mathbf{z}. \tag{179.5}$$

In these equations, $\sigma(\mathbf{z})$ and $\eta(\mathbf{z})$ represent surface densities of the "single-layer" potential, $\mu(\mathbf{z})$ and $\zeta(\mathbf{z})$ represent the surface densities of the "double-layer" potential, $\mathbf{z}$ represents a point on the boundary, and n represents the outward pointing normal. If $\sigma(\mathbf{z})$, $\mu(\mathbf{z})$, or $\eta(\mathbf{z})$ and $\zeta(\mathbf{z})$ were known, then $u(\mathbf{x})$ could be computed via one of the above three equations. Note there is not a unique way to represent the solution by (179.5); there is a "degree of freedom" in this formulation that may be used for other purposes.

It turns out that the single-layer potential is continuous across the boundary S, while the double-layer potential has a jump of $\mu(\mathbf{y})$. This is because, as $\mathbf{x}$ tends to the boundary point $\mathbf{P}$ from the inside of the domain,

$$u(\mathbf{P}) = -\frac{1}{2}\mu(\mathbf{P}) + \int_S \mu(\mathbf{z})\frac{\partial \psi(\mathbf{P}; \mathbf{z})}{\partial n}\, d\mathbf{z}. \tag{179.6}$$

Using (179.6) a variety of boundary integral equations may be obtained.

For example, using (179.4) to represent the solution to the Dirichlet problem, if we allow the point $\mathbf{x}$ to approach the boundary, we determine from (179.6) that

$$f(\mathbf{y}) = -\frac{1}{2}\mu(\mathbf{y}) + \int_S \mu(\mathbf{z})\frac{\partial \psi(\mathbf{z}; \mathbf{y})}{\partial n}\, d\mathbf{z}.$$

This Fredholm integral equation of the second kind can, in principle, be solved for $\mu(\mathbf{y})$. After $\mu(\mathbf{y})$ is obtained, the value of $u(\mathbf{x})$ may be computed from equation (179.4).

If (179.3) has been used to represent the solution of the Neumann problem, then, after finding the normal derivative of (179.4), the following integral equation for $\sigma(\mathbf{y})$ results

$$g(\mathbf{y}) = -\frac{1}{2}\sigma(\mathbf{y}) + \int_S \sigma(\mathbf{z})\frac{\partial\psi(\mathbf{z};\mathbf{y})}{\partial n}\,d\mathbf{z}.$$

After $\sigma(\mathbf{y})$ is obtained by solving the above integral equation, the value of $u(\mathbf{x})$ may be computed from equation (179.3).

Example

Consider Laplace's equation in the upper half plane, $\nabla^2 u = 0$ for $-\infty < x < \infty$ and $0 < y$, with the boundary conditions

$$u_y(x,0) = 0 \qquad -\infty < x < 0,$$
$$u_y(x,0) - ku(x,0) = 0 \qquad 0 < x < \infty.$$

where k is a constant. The Green's function, $\nabla^2\psi = \delta(x-\xi)\delta(y-\eta)$, in the upper half plane is

$$\psi(x,y;\xi,\eta) = -\frac{1}{2\pi}\log\sqrt{(x-\xi)^2 + (y-\eta)^2} - \frac{1}{2\pi}\log\sqrt{(x-\xi)^2 + (y+\eta)^2},$$

so that, on $y = 0$, we have $\psi(x,0;\xi,\eta) = -\frac{1}{2\pi}\log\left((x-\xi)^2 + \eta^2\right)$. When we actually have Laplace's equation, as we do in this example, (179.3) can be simplified to $u(\mathbf{x}) = -\int_S \frac{\partial u(\mathbf{z})}{\partial n}\psi(\mathbf{x};\mathbf{z})\,d\mathbf{z}$. Using the known values of u_n and ψ in this expression, we find

$$u(\xi,\eta) = \frac{k}{2\pi}\int_0^\infty u(x,0)\log\left((x-\xi)^2 + \eta^2\right)\,dx. \qquad (179.7)$$

If we define $\phi(x) := u(x,0)$, then evaluation of (179.7) at $\eta = 0$ results in

$$\phi(\xi) = \frac{k}{\pi}\int_0^\infty \phi(x)\log|x-\xi|\,dx.$$

After this integral equation is solved for $\phi(x)$, the solution is given by (179.7).

Notes

[1] The representation of the solution in (179.5) would be most appropriate if the boundary conditions were mixed.

[2] This technique has also been applied to the biharmonic equation in several applications. See Ingham and Kelmanson [7] for details.

[3] After the boundary integral equation has been formulated, it is often solved numerically. Some numerical techniques for these equations can be found in Banerjee and Butterfield [1]. In practice one finds that the solution to the original elliptic equation could have been determined by solving a large sparse matrix system, while the boundary element method often requires that a smaller, dense, matrix system be solved to determine the potential. A worked example is shown in Lapidus and Pinder [8].

[4] The principle advantage of the reformulation in this section is that the dimensionality of the problem is reduced. As in the above example, a two-dimensional partial differential equation becomes a one-dimensional integral equation.

[5] For problems in infinite domains, the behavior at infinity is (usually) automatically included in the boundary element formulation. Hence, there is no need for a "remote" boundary simulating an infinite distance. See Margulies [10].

[6] The boundary element method has also been applied to parabolic equations; see Zamani [12] or Duran, Cross, and Lewis [5] for more details. It has also been applied to some hyperbolic equations, see Chapter 12 (pages 191–199) of Brebbia [2]. For an application to nonlinear elliptic equations, see Chapter 4 in Ingham and Kelmanson [7].

[7] The boundary element method and the finite element method have several features in common. See Chapter 9 (pages 141–158) of Brebbia [2] for a general account of the similarities and differences.

[8] The presentation here has been for the *indirect* boundary element method. In this formulation, an integral equation for the potential must be solved, and then the solution to the original equation is given by an integral. It is also possible to directly determine an integral equation whose solution also satisfies the original equation. This is called the *direct* boundary element method. For example, given Laplace's equation, $\nabla^2 \phi = 0$, if we define the Green's function $G(\mathbf{x}; \mathbf{y})$ by $\nabla^2 G = \delta(\mathbf{x} - \mathbf{y})$, then by Green's theorem

$$\frac{1}{2}\phi(\mathbf{y}) = \int (G\nabla^2 \phi - \phi\nabla^2 G)\, dV$$

$$= \int \left(G\frac{\partial \phi}{\partial n} - \phi\frac{\partial G}{\partial n} \right) dS.$$

This integral equation can be solved directly for ϕ.

References

[1] R. Banerjee and P. K. Butterfeld, *Boundary Element Methods in Engineering Science*, McGraw–Hill Book Company, New York, 1981.

[2] C. A. Brebbia, *Boundary Element Techniques in Computer Aided Engineering*, Martinus Nijhoff Publishers, Boston, 1984.

[3] C. A. Brebbia (ed.), *Topics in Boundary Element Research. Volume 1: Basic Principles and Applications*, Springer–Verlag, Berlin, 1984.

[4] C. A. Brebbia (ed.), *Topics in Boundary Element Research. Volume 2: Time-Dependent and Vibration Problems*, Springer–Verlag, Berlin, 1985.

[5] D. Duran, M. Cross, and B. A. Lewis, "A Preliminary Analysis of Boundary Element Methods Applied to Parabolic Partial Differential Equations," in C. A. Brebbia (ed.), *New Developments in Boundary Element Methods*, 1980, pages 179–190.

[6] P. R. Garabedian, *Partial Differential Equations*, Wiley, New York, 1964, Section 9.3 (pages 334–348).

[7] D. B. Ingham and M. A. Kelmanson, *Boundary Integral Equation Analyses of Singular, Potential, and Biharmonic Problems*, Springer–Verlag, New York, 1984.

[8] L. Lapidus and G. F. Pinder, *Numerical Solution of Partial Differential Equations in Science and Engineering*, Wiley, New York, 1982, pages 461–481.

[9] J. Mackerle and T. Andersson, "Boundary Element Software in Engineering," *Adv. Eng. Software*, **6**, 1983, pages 66–102.

[10] M. Margulies, "Exact Treatment of the Exterior Problem in the Combined FEM–BEM," in C. A. Brebbia (ed.), *New Developments in Boundary Element Methods*, Butterworths, London, 1980, pages 43–64.

[11] L. Wardle, "An Introduction to the Boundary Element Method," in J. Noye (ed.), *Computational Techniques for Differential Equations*, North–Holland Publishing Co., New York, 1984, pages 525–549.

[12] N. Zamani, "Some Remarks on the Use of the Boundary Element Method in Transient Heat Conduction Problems," *Mathematics and Computers in Simulation*, **27**, 1985, pages 61–64.

180. Differential Quadrature

Applicable to   Nonlinear partial differential equations, a single equation or a system. Most often, partial differential equations in two independent variables.

Yields

A system of ordinary differential equations whose solution approximates the solution of the original partial differential equation(s).

Idea

All of the derivatives with respect to one or more of the independent variables are replaced by a sum involving the dependent variable.

Procedure

To illustrate the general technique, we show how it works on a class of partial differential equations. Suppose we have the partial differential equation for $u(t, x)$

$$u_t = g(t, x, u, u_x, u_{xx}),$$
$$u(0, x) = h(x),$$
(180.1)

on $t > 0$, $-\infty < x < \infty$. Instead of solving (180.1) for all values of x, we choose a finite set of x values at which the solution will be determined, say $S = \{x_j \mid j = 1, \ldots, N\}$. We now presume that the first derivatives with respect to x, at the points in S, can be written as a linear combination of the values in S. That is

$$u_x(t, x_i) \simeq \sum_{j=1}^{N} a_{ij} u(t, x_j).$$
(180.2)

Viewing (180.2) as the linear transformation $u_x = Au$, it seems natural to approximate $u_{xx} = Au_x = A^2 u$, or

$$u_{xx}(t, x_i) \simeq \sum_{k=1}^{N} \sum_{j=1}^{N} a_{ik} a_{kj} u(t, x_j).$$
(180.3)

Utilizing (180.2) and (180.3) in (180.1) results in the system of ordinary differential equations

$$u_t^i = g\left(t, x_i, u^i, \sum_{j=1}^{N} a_{ij} u^j, \sum_{k=1}^{N} \sum_{j=1}^{N} a_{ik} a_{kj} u^j\right),$$
$$u^i(0) = h(x_i),$$

for $i = 1, \ldots, N$, where $u^i(t) := u(t, x_i)$. These initial value ordinary differential equations may be integrated numerically by any scheme.

Note that this method is similar to the method of lines (see page 740), except that the a_{ij} are *not* chosen in such a way that (180.2) represents a finite difference approximation to the derivative. The a_{ij} are instead chosen so that (180.2) is exact for all polynomials of degree less than or equal to $N - 1$. That is, the a_{ij} satisfy the linear system

$$k (x_i)^{k-1} = \sum_{j=1}^{N} a_{ij} (x_j)^k \qquad (180.4)$$

for $k = 1, 2, \ldots, N$.

Example

We choose to numerically approximate the solution to the nonlinear partial differential equation

$$u_t = u u_x,$$

$$u(0, x) = .2x^2,$$

which has the exact solution $u = .2(x + ut)^2$, or

$$u(t, x) = \frac{[1 - (0.4)tx] - \sqrt{1 - (0.8)tx}}{(0.4)t^2}.$$

The program shown in Program 180 uses twenty x values in the interval from 0 to 1. Note that the source code for the linear equation solver (LSOLVE) is not shown. Some results of the program are shown below:

```
THE TIME IS NOW:      0.5000
HERE IS THE APPROXIMATE SOLUTION AT THIS TIME VALUE:
    0.0005     0.0020     0.0046     0.0083     0.0132     0.0192     0.0264
    0.0348     0.0446     0.0556     0.0681     0.0820     0.0974     0.1143
    0.1328     0.1530     0.1750     0.1985     0.2241     0.2620
HERE IS THE EXACT SOLUTION AT THIS TIME VALUE:
    0.0005     0.0020     0.0046     0.0083     0.0132     0.0192     0.0264
    0.0349     0.0446     0.0557     0.0682     0.0822     0.0977     0.1147
    0.1334     0.1538     0.1760     0.2000     0.2260     0.2540


THE TIME IS NOW:      0.7500
HERE IS THE APPROXIMATE SOLUTION AT THIS TIME VALUE:
    0.0005     0.0021     0.0047     0.0085     0.0135     0.0198     0.0274
    0.0365     0.0470     0.0591     0.0729     0.0885     0.1060     0.1255
    0.1471     0.1712     0.1977     0.2255     0.2687     0.5368
HERE IS THE EXACT SOLUTION AT THIS TIME VALUE:
    0.0005     0.0021     0.0047     0.0085     0.0135     0.0198     0.0275
    0.0365     0.0471     0.0593     0.0732     0.0889     0.1066     0.1263
    0.1484     0.1728     0.2000     0.2301     0.2634     0.3002
```

At $t = .75$, with the last value shown excluded, the relative error in the approximate solution is not more than 4%.

Program 180

```
        DIMENSION X(50),U(50),UNEW(50),A(50,50),CORECT(50)
        DIMENSION SAVE(50,50),COEFF(50,50),RHS(50),NROW(50),SOLN(50)
C SET UP THE PARAMETER VALUES
        N=20
        TIME=0
        DELTAT=.05
        NSTEP=15
C SET UP THE X POINTS
        DO 10 J=1,N
10      X(J)=FLOAT(J)/FLOAT(N)
C SET UP THE COEFFICIENT MATRIX
        DO 20 K=1,N
        DO 20 J=1,N
20      SAVE(K,J)=X(J)**K
C FOR EACH I, DETERMINE A_[IJ]  BY SOLVING A SYSTEM OF EQUATIONS
        DO 40 I=1,N
        DO 30 K=1,N
        RHS(K)=K*X(I)**(K-1)
        DO 30 J=1,N
30      COEFF(J,K)=SAVE(J,K)
        CALL LSOLVE(N,COEFF,SOLN,RHS,NROW,IFSING,50)
        IF( IFSING .NE. 1 ) STOP
        DO 40 J=1,N
40      A(I,J)=SOLN(J)
C SET UP THE INITIAL CONDITIONS
        DO 50 J=1,N
50      U(J)=U0( X(J) )
C THIS IS THE LOOP IN TIME
        DO 100 LOOPT=1,NSTEP
        TIME=TIME + DELTAT
        WRITE(6,5) TIME
C ITERATE EACH ONE OF THE EQUATIONS ONE TIME STEP
        DO 70 J=1,N
        SUM=0
        DO 60 K=1,N
60      SUM=SUM + A(J,K)*U(K)
70      UNEW(J)= U(J) + DELTAT * U(J) * SUM
        DO 80 J=1,N
80      U(J)=UNEW(J)
C WRITE OUT THE APPROXIMATE ANSWER, AND THEN THE EXACT ANSWER
        WRITE(6,*) ' HERE IS THE APPROXIMATE SOLUTION AT THIS TIME VALUE:'
        WRITE(6,15) (U(J), J=1,N)
        DO 90 J=1,N
90      CORECT(J)=EXACT(TIME, X(J) )
        WRITE(6,*) ' HERE IS THE EXACT SOLUTION AT THIS TIME VALUE:'
100     WRITE(6,15) (CORECT(J), J=1,N)
5       FORMAT(' THE TIME IS NOW:',F10.4)
15      FORMAT( 30( 1X, 7(F9.4,1X) / ) )
        END
C THIS FUNCTION HAS THE INITIAL CONDITIONS
        FUNCTION U0(X)
        U0=.2*X**2
        RETURN
        END
C THIS FUNCTION HAS THE EXACT SOLUTION
        FUNCTION EXACT(T,X)
        TEMP=( 1.0 - (0.4)*T*X ) - SQRT( 1.0 - (0.8)*T*X )
```

```
EXACT=TEMP / ( (0.4)*T**2 )
RETURN
END
```

Notes

[1] Note that the coefficient matrix in (180.4) is a Vandermonde matrix.

[2] It is not clear that having the x values uniformly spaced produces the most accurate results. In Bellman, Kashef, and Casti [1], the x values are chosen to be the roots of Legendre polynomials.

[3] In Bellman, Kashef, and Casti [1], a simple error analysis is performed. It is shown, for example, that the error in (180.2) is less than $\dfrac{Kh^{N-1}}{(N-1)!}$ if the mesh has a uniform spacing of h and if $|u^{(N)}(x)| \leq K$ in the domain of interest.

[4] In Civan and Sliepcevich [2] a weighted sum of terms (similar to the approximation in (180.2)) is used to approximate the second derivative terms (such as in (180.3)). This reduces the computational complexity of the coding.

References

[1] R. Bellman, B. G. Kashef, and J. Casti, "Differential Quadrature: A Technique for the Rapid Solution of Nonlinear Partial Differential Equations," *J. Comput. Physics*, **10**, 1972, pages 40–52.

[2] F. Civan and C. M. Sliepcevich, "Solution of the Poisson Equation by Differential Quadratures," *Int. J. Num. Methods Eng.*, **19**, 1983, pages 711–724.

[3] F. Civan and C. M. Sliepcevich, "Differential Quadrature for Multi-Dimensional Problems," *J. Math. Anal. Appl.*, **101**, 1984, pages 423–443.

[4] F. Civan and C. M. Sliepcevich, "On the Solution of the Thomas–Fermi Equation by Differential Quadrature," *J. Comput. Physics*, **56**, 1984, pages 343–348.

[5] G. Naadimuthu, R. Bellman, K. M. Wang, and E. S. Lee, "Differential Quadrature and Partial Differential Equations: Some Numerical Results," *J. Math. Anal. Appl.*, **98**, 1984, pages 220–235.

181. Domain Decomposition

Applicable to   Elliptic second order partial differential equations in non-regularly shaped domains.

Yields

An iterative solution procedure.

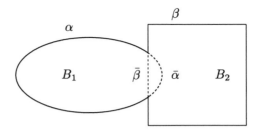

Figure 181.1 The domain for equation (181.1).

Idea

If the geometric domain in which a partial differential equation is to be solved can be written as the union of two (or more) regularly shaped domains, then it may be possible to write a recurrence relation for the solution.

Procedure

Suppose we wish to numerically approximate the solution to the elliptic equation

$$N[u] = F\left(x, y, u, u_x, u_y, u_{xx}, u_{xy}, u_{yy}\right) = 0 \qquad (181.1)$$

in the domain $B = B_1 \cup B_2$ (see Figure 181.1). We presume this is a Dirichlet problem, with the initial data, $f(x, y)$, given on the boundary of B.

Define the part of the boundary of B_1 (∂B_1) that is also a boundary of B to be α; the rest of the B_1 boundary of B_1 will be denoted by $\bar{\alpha}$. Likewise, define the part of the boundary of B_2 (∂B_2) that is also a boundary of B to be β; the rest of the B_2 boundary of B_2 will be denoted by $\bar{\beta}$.

The solution procedure is to first solve (181.1) only in B_1. Then, using this solution, we solve (181.1) only in the domain B_2. This is used to find a new solution of (181.1) in B_1, and then the process is repeated.

Initially, the data on the arc $\bar{\alpha}$ is chosen so that the data on ∂B_1 is piecewise continuous. That is, let $u_1(x, y)$ be the solution of (181.1) with the boundary conditions

$$u_1(x, y) = \begin{cases} f(x, y) & \text{on } \alpha, \\ \phi(x, y) & \text{on } \bar{\alpha}, \end{cases}$$

where $\phi(x, y)$ can be chosen in many different ways. After $u_1(x, y)$ is determined, let $v_1(x, y)$ be the solution of (181.1) with the boundary conditions

$$v_1(x, y) = \begin{cases} f(x, y) & \text{on } \beta, \\ u_1(x, y) & \text{on } \bar{\beta}. \end{cases}$$

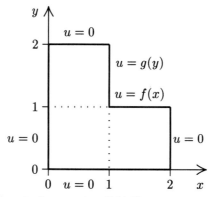

Figure 181.2 The domain for equation (181.5).

Then an iterative sequence of solutions to (181.1) is formed, $\{u_k(x, y), v_k(x, y) \mid k = 1, 2, \ldots\}$ with

$$u_k(x, y) = \begin{cases} f(x, y) & \text{on } \alpha, \\ v_{k-1}(x, y) & \text{on } \overline{\alpha}, \end{cases}$$

$$v_k(x, y) = \begin{cases} f(x, y) & \text{on } \beta, \\ u_k(x, y) & \text{on } \overline{\beta}. \end{cases}$$

Under fairly general conditions, these functions will converge to the solution of (181.1). That is, the limiting $u_k(x, y)$ will be the solution to (181.1) in the region B_1, while the limiting $v_k(x, y)$ will be the solution to (181.1) in the region B_2.

In Kantorovich and Krylov [6], five assumptions are given that are required to assure the convergence of the above sequences. They are:

(A) Equation (181.1), with its boundary conditions, has a unique solution.

(B) If $F[u] = F[u^*] = 0$, and $u^* > u$ on the boundary of the domain, then $u^* > u$ everywhere in the domain.

(C) Within the domain, the solution to (181.1) is bounded by the values of u on the boundary of the domain.

(D) A convergent sequence of uniformly bounded solutions to (181.1) converges to a solution to (181.1).

(E) The boundary data is, at least, piecewise continuous.

Generally, non-pathological examples should satisfy these conditions.

Example

Suppose we want to solve Laplace's equation in the L-shaped region shown in Figure 181.2. For brevity, we define the following portions of the boundary

$$\Gamma_1 = \{x = 2, 0 \le y \le 1\} \cup \{0 \le x \le 2, y = 0\} \cup \{x = 0, 0 \le y \le 1\},$$
$$\Gamma_2 = \{x = 0, 0 \le y \le 2\} \cup \{0 \le x \le 1, y = 0\} \cup \{0 \le x \le 1, y = 2\}.$$

Then the mathematical problem we wish to solve is

$$
\begin{aligned}
\nabla^2 u &= 0, \\
u &= 0, &\text{on } \Gamma_1, \\
u &= 0, &\text{on } \Gamma_2, \\
u &= f(x), &\text{on } \{1 \le x \le 2, y = 1\}, \\
u &= g(y), &\text{on } \{x = 1, 1 \le y \le 2\}.
\end{aligned}
\tag{181.2}
$$

For this example, we break up the original domain into two rectangles, one vertical and one horizontal; the overlap region being the unit square. We start with

$$
\begin{aligned}
\nabla^2 u_1 &= 0, \\
u_1 &= 0, &\text{on } \Gamma_1, \\
u_1 &= f(x), &\text{on } \{1 \le x \le 2, y = 1\}, \\
u_1 &= \phi(x), &\text{on } \{0 \le x \le 1, y = 1\}.
\end{aligned}
\tag{181.3}
$$

Then our iteration sequence becomes

$$
\begin{aligned}
\nabla^2 v_k &= 0, \\
v_k &= 0, &\text{on } \Gamma_2, \\
v_k &= u_{k-1}(1, y), &\text{on } \{x = 1, 0 \le y \le 1\}, \\
v_k &= g(y), &\text{on } \{x = 1, 1 \le y \le 2\},
\end{aligned}
\tag{181.4}
$$

for $k = 1, 2, \ldots$, while

$$
\begin{aligned}
\nabla^2 u_k &= 0, \\
u_k &= 0, &\text{on } \Gamma_1, \\
u_k &= f(x), &\text{on } \{1 \le x \le 2, y = 1\}, \\
u_k &= v_k(x, 1), &\text{on } \{0 \le x \le 1, y = 1\},
\end{aligned}
\tag{181.5}
$$

for $k = 2, 3, \ldots$.

In this case, because of the simple geometry, we can analytically write down the solution to (181.4) and (181.5) by the use of Fourier transforms (see page 299). Note first, if we define $f_n(x) := u_n(x, 1) = \sum_{k=1}^{\infty} f_{nk} \sin k\pi x$,
then $u_n(x, y) = \sum_{k=1}^{\infty} \dfrac{f_{nk}}{\sinh(k\pi/2)} \sinh k\pi y \sin k\pi x$. Similarly, if we define the

expansion $g_n(x) := v_n(1, y) = \sum_{k=1}^{\infty} g_{nk} \sin k\pi y$, then we obtain the result

$$v_n(x, y) = \sum_{k=1}^{\infty} \frac{g_{nk}}{\sinh(k\pi/2)} \sinh k\pi x \sin k\pi y \quad.$$ Using these expansions in

(181.4) and (181.5), we can readily determine that

$$f_{nk} = B_k + \sum_{s=1}^{\infty} A_{ks} g_{n-1,s},$$

$$g_{nk} = C_k + \sum_{s=1}^{\infty} A_{ks} f_{n-1,s},$$ (181.6)

where

$$B_k = \int_1^2 f(x) \sin(k\pi x/2)\, dx,$$

$$C_k = \int_1^2 g(y) \sin(k\pi y/2)\, dy,$$

$$A_{ks} = \frac{2}{\pi} \frac{1}{s^2 + k^2} \left[s \sin\left(\frac{k\pi}{2}\right) \cosh\left(\frac{s\pi}{2}\right) - k \cos\left(\frac{k\pi}{2}\right) \sinh\left(\frac{s\pi}{2}\right) \right].$$

In practice, the two recurrence relations in (181.6) would be iterated until a stationary value was obtained.

Notes

[1] This method is usually implemented numerically, with little analysis done on the equations. For the above example, the equations in (181.3)–(181.5) would be approximated numerically by an elliptic equation package.

[2] This method also works for coupled systems of elliptic equations. For two unknowns, a guess is made for one of the unknowns, and one of the equations is used to solve for the other unknown. Knowing this second unknown, the first unknown is approximated numerically by the other equation, and the process is repeated. See Rice and BoisverttherefG for some examples.

[3] The procedure illustrated in this section is called *Schwarz's method*, it is only one of several different domain decomposition methods (see Glowinski *et al.* [5]).

[4] In Chan, Hou, and Lions [3] it is shown that the convergence rate of the Schwarz alternating procedure, for general second-order elliptic equations, is independent of the aspect ratio for L-shaped, T-shaped, and C-shaped domains.

References

[1] C. Canuto and D. Funaro, "The Schwarz Algorithm for Spectral Methods," *SIAM J. Numer. Anal.*, **25**, No. 1, February 1988, pages 24–40.

[2] T. F. Chan, R. Glowinski, J. Periaux and O. B Widlund (eds.), *Domain Decomposition Methods*, SIAM, Philadelphia, 1989.

[3] T. F. Chan, T. Y. Hou, and P. L. Lions, "Geometry Related Convergence Results for Domain Decomposition Algorithms," *SIAM J. Numer. Anal.*, **28**, No. 2, April 1991, pages 378–391.

[4] L. W. Ehrlich, "The Numerical Schwarz Alternating Procedure and SOR," *SIAM J. Sci. Stat. Comput.*, **7**, No. 3, July 1986, pages 989–993.

[5] R. Glowinski, G. Golub, G. Meurant, and J. Periaux (eds.), *First International Symposium on Domain Decomposition Methods for Partial Differential Equations*, SIAM, Philadelphia, 1988.

[6] L. V. Kantorovich and V. I. Krylov, *Approximate Methods of Higher Analysis*, Interscience Publishers, New York, 1958, Chapter 7 (pages 616–670).

[7] U. Meier, "Two Parallel SOR Variants of the Schwarz Alternating Procedure," *Parallel Comp.*, **3**, No. 3, 1986, pages 205–215.

[8] J. R. Rice and R. F. Boisvert, *Solving Elliptic Problems Using ELLPACK*, Springer–Verlag, New York, 1985, pages 121–135.

182.   Elliptic Equations: Finite Differences

Applicable to   Elliptic partial differential equations.

Yields

A numerical approximation of the solution.

Idea

By use of finite differences, a simultaneous system of equations may be determined. The solution of this algebraic system (which is often a linear system of equations) yields a numerical approximation to the differential equation.

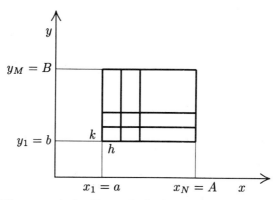

Figure 182.1 The numerical grid on which the problem is to be solved.

Procedure

The method is simply to use finite differences everywhere, and solve the resulting set of simultaneous equations. Since elliptic equations are boundary value problems, the solution at all points in the domain must be determined simultaneously.

We choose to illustrate the method on a second order elliptic equation of the form

$$\alpha u_{xx} + \beta u_{yy} = f(x, y, u, u_x, u_y), \qquad (182.1)$$

where α and β are functions of x and y. We suppose that equation (182.1) applies inside a rectangle with $a \leq x \leq A$, $b \leq y \leq B$, and that the boundary conditions for (182.1) are

$$u(x, y) = \begin{cases} f(y), & \text{on } x = a, \\ g(y), & \text{on } x = A, \\ h(x), & \text{on } y = B, \end{cases} \qquad (182.2)$$

$$\frac{\partial u}{\partial y} + \frac{\partial u}{\partial x} + u^3 = j(x), \quad \text{on} \quad y = b, \qquad (182.3)$$

where $\{f, g, h, j\}$ are all known functions.

We first define a grid that fills the geometric domain (see page 606). For the rectangular geometry given, we choose a rectangular grid with an x spacing of h and a y spacing of k (where $h = (A - a)/(N - 1)$, and $k = (B - b)/(M - 1)$). Here, $N(M)$ is the number of grid points in the x (y) direction. See Figure 182.1.

Let the numerical approximation to $u(x, y)$ be given by v_{ij} (that is, $v_{ij} \simeq u(a + ih, b + jk)$). We can then choose virtually any finite difference approximation to the derivatives appearing in (182.1). For instance, one second order approximation to (182.1) would be

$$\alpha_{ij}\frac{v_{i+1,j} - 2v_{i,j} + v_{i-1,j}}{h^2} + \beta_{ij}\frac{v_{i,j+1} - 2v_{i,j} + v_{i,j-1}}{k^2}$$
$$= f\left(a + ih, b + jk, v_{ij}, \frac{v_{i+1,j} - v_{i-1,j}}{2h}, \frac{v_{i,j+1} - v_{i,j-1}}{2k}\right).$$
$$(182.4)$$

For each i and j, (182.4) represents an algebraic equation among the $\{v_{ij}\}$. Now the boundary conditions must be incorporated. The boundary conditions in (182.2) can be written simply as

$$\begin{aligned} v_{0,j} &= f(b + jk), &\text{for} \quad j = 1, 2, \ldots, M, \\ v_{N,j} &= g(b + jk), &\text{for} \quad j = 1, 2, \ldots, M, \\ v_{i,m} &= h(a + ih), &\text{for} \quad i = 1, 2, \ldots, N. \end{aligned} \qquad (182.5)$$

The boundary condition in (182.3) can be written as

$$\frac{v_{i,1} - v_{i,0}}{k} + \frac{v_{i+1,0} - v_{i,0}}{h} + (v_{i,j})^3 = j(a + ih) \quad \text{for} \quad i = 1, 2, \ldots, N. \qquad (182.6)$$

If equation (182.4) is evaluated for $j = 1, 2, \ldots, M$ and $i = 1, 2, \ldots, N$, and (182.5) and (182.6) are included, there results a simultaneous system of equations for the $\{v_{ij}\}$. There are as many equations as there are unknowns. This system may then be solved numerically.

If the original elliptic equation (182.1) and the boundary conditions are linear in the independent variable, then the resulting system of equations will be linear. For our example, equation (182.6) is not linear (note the $(v_{i,j})^3$ term) since there is a u^3 term in (182.3). The most common type of elliptic systems have linear equations and linear boundary conditions. For this type of elliptic system, a standard linear equation solver may be used. If the system of linear equations is too large to solve directly, an iterative method may be used (see page 726).

Example

Suppose we have the linear elliptic equation

$$(x + 1)u_{xx} + (y + 1)^2 u_{yy} = 1 + u, \qquad (182.7)$$

on $0 \le x \le 1$, $0 \le y \le 1$ with

$$\begin{aligned} u(0, y) &= y, &u(1, y) &= y^2, \\ u(x, 0) &= 0, &u(x, 1) &= 1. \end{aligned} \qquad (182.8)$$

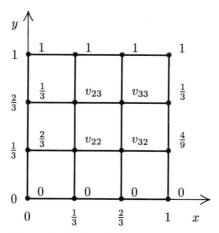

Figure 182.2 The grid on which (182.8) is solved.

If we choose $M = N = 4$ (so that $h = k = 1/3$), then there are 16 points $\{v_{ij} \mid 1 \le i \le 4, 1 \le j \le 4\}$ at which to determine an approximation to $u(x, y)$. The points $\{v_{ij} \mid i = 1 \text{ or } i = 4 \text{ or } j = 1 \text{ or } j = 4\}$ are determined directly by the boundary conditions in (182.8). Hence, the only unknowns that need to be determined are $\{v_{22}, v_{23}, v_{32}, v_{33}\}$. See Figure 182.2. If (182.7) is discretized as

$$(ih + 1)\frac{v_{i+1,j} - 2v_{i,j} + v_{i-1,j}}{h^2}$$
$$+ (jk + 1)^2 \frac{v_{i,j+1} - 2v_{i,j} + v_{i,j-1}}{k^2} = 1 + v_{ij},$$

then the equations for the unknown $\{v_{ij}\}$ may be written as

$$\begin{pmatrix} 57/9 & -16/9 & -4/3 & 0 \\ -25/9 & 25/3 & 0 & -4/3 \\ -5/3 & 0 & 7 & -16/9 \\ 0 & -5/3 & -25/9 & 9 \end{pmatrix} \begin{pmatrix} v_{22} \\ v_{23} \\ v_{32} \\ v_{33} \end{pmatrix} = \begin{pmatrix} -5/9 \\ 24/9 \\ -22/27 \\ 68/27 \end{pmatrix}. \quad (182.9)$$

The equations in (182.9) have the solution (correct to the number of decimal places listed) $v_{22} \simeq 0.0131$, $v_{23} \simeq 0.3791$, $v_{32} \simeq -0.0265$, $v_{33} \simeq 0.3419$.

Notes

[1] The above example is from Ames [1].

[2] The computer language ELLPACK (see Rice and Boisvert [6] is a high level
language that allows linear elliptic problems in two or three dimensions to
be entered in an elementary way. The program generates a discretization
scheme based on user preference. The geometry in two dimensions can be
nearly arbitrary, with holes and other cutouts available. For example, to
solve the problem in the example, the *entire* ELLPACK program would be

```
EQUATION.              (X+1)*UXX+(Y+1)**2*UYY=1.0 + U
BOUNDARY.              U=Y         ON    X=0.0
                       U=Y**2      ON    X=1.0
                       U=0.0       ON    Y=0.0
                       U=1.0       ON    Y=1.0
GRID.                  4 X POINTS
                       4 Y POINTS
DISCRETIZATION.        5 POINT STAR
SOLUTION.              LINPACK BAND
OUTPUT.                TABLE(U)
                       PLOT(U)
END.
```

The use of ELLPACK for two and three dimensional problems is highly
recommended.

[3] Picard iteration (see page 535), Newton's method (see page 500), and Monte
Carlo methods (see page 721) can also be used to numerically approximate
the solution to elliptic problems.

[4] Boisvert and Sweet [4] have a comprehensive listing of currently available
software for solving elliptic problems. For a listing of computer software
that will implement the method described in this section, see page 586.

References

[1] W. F. Ames, *Numerical Methods for Partial Differential Equations*,
Barnes and Noble, Inc., New York, 1969, pages 365–373.

[2] G. Birkhoff, *The Numerical Solution of Elliptic Equations*, SIAM, Philadelphia, 1972.

[3] G. Birkoff and R. Lynch, *Numerical Solution of Elliptic Problems*, SIAM,
Philadelphia, 1984.

[4] R. F. Boisvert and R. A. Sweet, "Mathematical Software for Elliptic Bound-
ary Value Problems," in W. R. Cowell (ed.), *Sources and Development of
Mathematical Software*, Prentice–Hall Inc., Englewood Cliffs, NJ, 1984,
Chapter 9 (pages 200–263).

[5] W. R. Dyksen and C. J. Ribbens, "Interactive ELLPACK: An Interactive
Problem–Solving Environment for Elliptic Partial Differential Equations,"
ACM Trans. Math. Software, **13**, No. 2, June 1987, pages 113–132.

[6] J. R. Rice and R. F. Boisvert, *Solving Elliptic Problems Using ELLPACK*,
Springer–Verlag, New York, 1985.

[7] E. H. Twizell, *Computational Methods of Partial Differential Equations*,
Halstead Press, John Wiley & Sons, New York, 1984, pages 42–80.

183. Elliptic Equations: Monte Carlo Method

Applicable to   Linear elliptic partial differential equations.

Yields

A numerical approximation to the solution of a linear elliptic partial differential equation at a single point.

Idea

Simulation of the motion of a random particle may be used to approximate the solution to linear elliptic equations.

Procedure

The steps for this method are straightforward. First we give an overview, then a more detailed presentation.

First, approximate the given elliptic partial differential equation by a finite difference method. Rewrite the finite difference formula as a recursive function for the value of the unknown at any given point. Then interpret this recursive formula as a set of transition probabilities that determine the motion of a random particle.

Now write a computer program that will allow many (say K) particles to wander randomly around the domain of interest, based on the transition probabilities found from the difference formula. Simulate particles one at a time, with every particle starting off at the same point (say the point $\mathbf{z}$).

(A) If the boundary data are of the Dirichlet type (i.e., the value of the unknown is prescribed on the boundary) then when a particle reaches the boundary, stop that particle and store away the value on the boundary. Begin another particle at the point $\mathbf{z}$.

(B) If the boundary data are not of the Dirichlet type (say Neumann or mixed boundary conditions) then, when the particles reach the boundary, they will be given a finite probability to leave the boundary, and re-enter the domain of the problem. If the particle leaves the boundary, then continue the iteration process. If it does not leave the boundary, then the value at the boundary is stored away, and a new particle is started off at the point $\mathbf{z}$.

The simulation is finished after all K particles have been absorbed into the boundary. If the original elliptic equation has Dirichlet boundary conditions, then an approximation to the solution, at the point $\mathbf{z}$, will be given by the average of all the values obtained (recall, when the particles stop at the boundary they obtain a value).

If the given elliptic equation does not have Dirichlet boundary conditions, then an equation given below will show how to obtain an approximation to the solution. In this latter case, the approximate value of the solution depends on the entire history of the particle.

In more detail, we now describe how the technique may be applied to the linear second order elliptical partial differential equation

$$L[u] = F(x, y), \tag{183.1}$$

with the operator $L[\cdot]$ defined by

$$L[u] := Au_{xx} + 2Bu_{xy} + Cu_{yy} + Du_x + Eu_y,$$

where $\{A, B, C, D, E\}$ are all functions of $\{x, y\}$. The operator $L[\cdot]$ may be discretized to yield the approximation

$$
\begin{aligned}
L[u] \simeq\ & A_{i,j} \left[\frac{v_{i+1,j} - 2v_{i,j} + v_{i-1,j}}{(\Delta x)^2} \right] \\
& + 2B_{i,j} \left[\frac{v_{i+1,j+1} - v_{i,j+1} - v_{i+1,j} + v_{i,j}}{(\Delta x)(\Delta y)} \right] \\
& + C_{i,j} \left[\frac{v_{i,j+1} - 2v_{i,j} + v_{i,j-1}}{(\Delta y)^2} \right] \\
& + D_{i,j} \left[\frac{v_{i+1,j} - v_{i,j}}{\Delta x} \right] + E_{i,j} \left[\frac{v_{i,j+1} - v_{i,j}}{\Delta y} \right],
\end{aligned} \tag{183.2}
$$

where $x_i = x_0 + i(\Delta x)$, $y_j = y_0 + j(\Delta y)$, $v_{i,j} = u(x_i, y_j)$, and a subscript of i, j means an evaluation at the point (x_i, y_j). If the $\{\Gamma_{\cdot,\cdot}\}$ and $Q_{i,j}$ are defined by

$$\Gamma_{i+1,j+1} = \left[\frac{2B_{i,j}}{(\Delta x)(\Delta y)} \right],$$

$$\Gamma_{i+1,j} = \left[\frac{A_{i,j}}{(\Delta x)^2} - \frac{2B_{i,j}}{(\Delta x)(\Delta y)} + \frac{D_{i,j}}{\Delta x} \right],$$

$$\Gamma_{i,j+1} = \left[\frac{C_{i,j}}{(\Delta y)^2} - \frac{2B_{i,j}}{(\Delta x)(\Delta y)} + \frac{E_{i,j}}{\Delta y} \right],$$

$$\Gamma_{i-1,j} = \left[\frac{A_{i,j}}{(\Delta x)^2} \right],$$

$$\Gamma_{i,j-1} = \left[\frac{C_{i,j}}{(\Delta x)^2} \right],$$

$$Q_{i,j} = \left[\frac{2A_{i,j}}{(\Delta x)^2} - \frac{2B_{i,j}}{(\Delta x)(\Delta y)} + \frac{2C_{i,j}}{(\Delta y)^2} + \frac{D_{i,j}}{\Delta x} + \frac{E_{i,j}}{\Delta y} \right],$$

then, using (183.2), equation (183.1) may approximated as

$$Q_{i,j}v_{i,j} = \Gamma_{i+1,j}v_{i+1,j} + \Gamma_{i+1,j+1}v_{i+1,j+1} + \Gamma_{i,j+1}v_{i,j+1}$$
$$+ \Gamma_{i-1,j}v_{i-1,j} + \Gamma_{i,j-1}v_{i,j-1} - F_{i,j},$$

or (dividing through by $Q_{i,j}$ and defining $p_{i,j} := \Gamma_{i,j}/Q_{i,j}$),

$$v_{i,j} = p_{i+1,j}v_{i+1,j} + p_{i+1,j+1}v_{i+1,j+1} + p_{i,j+1}v_{i,j+1}$$
$$+ p_{i-1,j}v_{i-1,j} + p_{i,j-1}v_{i,j-1} - \frac{F_{i,j}}{Q_{i,j}}. \tag{183.3}$$

Since the operator $L[\cdot]$ has been presumed to be elliptic, then Δx and Δy may be chosen small enough so that each of the p's are positive. The p's also add up to one, and we interpret them as probabilities of taking a step in a specified direction. Specifically, equation (183.3) can be interpreted as follows: If a particle is at position (i,j) at step N, then

(A)   with probability $p_{i,j+1}$ the particle goes to $(i, j+1)$ at step $N+1$.
(B)   with probability $p_{i,j-1}$ the particle goes to $(i, j-1)$ at step $N+1$.
(C)   with probability $p_{i+1,j}$ the particle goes to $(i+1, j)$ at step $N+1$.
(D)   with probability $p_{i-1,j}$ the particle goes to $(i-1, j)$ at step $N+1$.
(E)   with probability $p_{i+1,j+1}$ the particle goes to $(i+1, j+1)$ at step $N+1$.

Now, suppose a particle starts at the point $P_0 := \mathbf{z}$, and undergoes a random walk according to the above prescription. After, say, m steps it will hit the boundary. Suppose that the sequence of points that this particle visits is $(P_0, P_1, P_2, \ldots, P_m)$. Then, an unbiased estimator of the value of $u(\mathbf{z})$ for the following elliptic problem

$$L[u] = F(x,y), \quad \text{for all points } x, y \text{ in the domain } R,$$
$$u = \phi(x,y), \quad \text{for all points } x, y \text{ on the boundary } \partial R,$$

is given by

$$u(\mathbf{z}) \simeq \phi(P_m) - \sum_{j=0}^{m} \frac{F(P_j)}{Q(P_j)}.$$

In practice, several random paths will be taken, and the average taken to estimate $u(\mathbf{z})$

$$u(\mathbf{z}) \simeq \frac{1}{K} \sum_{k=1}^{K} \left\{ \phi(P_{m_k}^k) - \sum_{j=0}^{m_k} \frac{F(P_j^k)}{Q(P_j^k)} \right\} \tag{183.4}$$

where $(P_0^k, P_1^k, \ldots, P_{m_k}^k)$, represents the path taken by the k-th random particle.

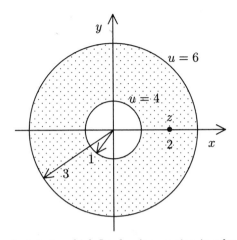

Figure 183. The domain in which Laplace's equation is solved.

Example

Suppose we wish to numerically approximate the solution to Laplace's equation in an annulus. We have $\nabla^2 u = 0$ for $u(r, \theta)$ with the boundary conditions $u(1, \theta) = 4$ and $u(3, \theta) = 6$. (See Figure 183.) We will approximate the value of $u(\mathbf{z})$, when $\mathbf{z} := (r = 2, \theta = 0)$. The exact solution for this problem is $u(r) = 4 + 2 \log r / \log 3$, so that $u(\mathbf{z}) = 4 + \log 2 / \log 3 \simeq 5.261$. To approximate the solution to this problem numerically, we will follow the steps outlined above. We will use the rectangular variables x and y, rather than the polar coordinate variables r and θ.

Using a standard second order approximation to the Laplacian, we find

$$\nabla^2 u \simeq \frac{v_{i+1,j} + v_{i-1,j} + v_{i,j+1} + v_{i,j-1} - 4v_{i,j}}{h^2} = 0, \qquad (183.5)$$

where $v_{i,j} := u(hi, hj)$ and $h \ll 1$. Equation (183.5) can be manipulated into

$$v_{i,j} = \frac{v_{i+1,j}}{4} + \frac{v_{i-1,j}}{4} + \frac{v_{i,j+1}}{4} + \frac{v_{i,j-1}}{4}. \qquad (183.6)$$

We interpret (183.6) probabilistically as follows: If a particle is at position (i, j) at step N, then

(A) with probability $1/4$ the particle goes to $(i, j + 1)$ at step $N + 1$.
(B) with probability $1/4$ the particle goes to $(i, j - 1)$ at step $N + 1$.
(C) with probability $1/4$ the particle goes to $(i + 1, j)$ at step $N + 1$.

(D) with probability $1/4$ the particle goes to $(i-1,j)$ at step $N+1$.

The FORTRAN program in Program 183 was used to simulate the motion of the particles according to the above probability law. The output of that program is given below for $u(r = 2, \theta = 0)$. As more points are taken, the approximation becomes better.

```
NUMBER OF PARTICLES= 1000    APPROXIMATION= 5.3440
NUMBER OF PARTICLES= 2000    APPROXIMATION= 5.3330
NUMBER OF PARTICLES= 3000    APPROXIMATION= 5.3200
NUMBER OF PARTICLES= 4000    APPROXIMATION= 5.3195
NUMBER OF PARTICLES= 6000    APPROXIMATION= 5.3030
NUMBER OF PARTICLES= 7000    APPROXIMATION= 5.3023
NUMBER OF PARTICLES= 8000    APPROXIMATION= 5.2958
NUMBER OF PARTICLES= 9000    APPROXIMATION= 5.2944
NUMBER OF PARTICLES=10000    APPROXIMATION= 5.2914
```

Note that the program uses a routine called RANDOM, whose source code is not given, which returns a random value uniformly distributed on the interval from zero to one.

Program 183

```
           STEP=.10
           SUM=0.
           DO 10 IWALK=1,10000
           X=2.0
           Y=0.0
   20      X=X + SIGN(STEP, RANDOM(DUMMY)-.5 )
           Y=Y + SIGN(STEP, RANDOM(DUMMY)-.5 )
           R=SQRT( X**2+Y**2 )
           IF( R.LT.3 .AND. R.GT.1 ) GOTO 20
   C WHEN A PARTICLE HITS THE BOUNDARY, SUM THE VALUE
           IF( R .LE. 1) SUM=SUM+4
           IF( R .GE. 3) SUM=SUM+6
           IF( MOD(IWALK,1000) .NE. 0 ) GOTO 10
           APPROX=SUM/FLOAT(IWALK)
           WRITE(6,5) IWALK,APPROX
   5       FORMAT(' NUMBER OF PARTICLES=',I5,'   APPROXIMATION=',F7.4)
   10      CONTINUE
           END
```

Notes

[1] If further accuracy is required, the options are:

(A) Increase the number of random particles.
(B) Make the mesh discretization finer (i.e., reduce h).
(C) Do both of the above.

If the number of random particles is not very large then (B) will not help much, and if the mesh is very coarse then (A) will not help much. Generally, the variance of the answer (a measure of the "scatter") decreases as the number of trials to the minus one half power.

[2] Since low numerical accuracy is obtained by this technique, a computer program does not need to work with extended precision arithmetic (such as double precision).

[3] Sadeh and Franklin [8] present several worked examples.

References

[1] V. C. Bhavsar and U. G. Gujar, "VLSI Algorithms for Monte Carlo Solutions of Partial Differential Equations," in R. Vichnevetsky and R. S. Stepleman (eds.), *Advances in Computer Methods for Partial Differential Equations*, IMACS, North–Holland Publishing Co., New York, 1984.

[2] V. C. Bhavsar and J. R. Isaac, "Design and Analysis of Parallel Monte Carlo Algorithms," *SIAM J. Sci. Stat. Comput.*, **8**, No. 1, 1987, pages 573–595.

[3] T. E. Booth, "Exact Monte Carlo Solution of Elliptic Partial Differential Equations," *J. Comput. Physics*, **39**, 1981, pages 396–404.

[4] T. E. Booth, "Regional Monte Carlo Solution of Elliptic Partial Differential Equations," *J. Comput. Physics*, **47**, 1982, pages 281–290.

[5] S. J. Farlow, *Partial Differential Equations for Scientists and Engineers*, John Wiley & Sons, New York, 1982, pages 346–352.

[6] R. Lattès, *Methods of Resolution for Selected Boundary Problems in Mathematical Physics*, Gordon and Breach, New York, 1969, Chapter 8 (pages 158–190).

[7] G. Marshall, "Monte Carlo Methods for the Solution of Nonlinear Partial Differential Equations," *Comput. Physics Comm.*, **56**, 1989, pages 51–61.

[8] E. Sadeh and M. A. Franklin, "Monte Carlo Solution of Partial Differential Equations by Special Purpose Digital Computer," *IEEE Transactions on Computers*, **C-23**, No. 4, April 1974, pages 389–397.

[9] J. Vrbik, "Monte Carlo Simulation of the General Elliptic Operator," *J. Phys. A: Math. Gen.*, **20**, 1987, pages 2693–2697.

184. Elliptic Equations: Relaxation

Applicable to   Elliptic equations, most often Laplace's equations.

Yields

A numerical approximation to the solution.

Idea

The finite difference scheme for an elliptic equation can be interpreted as a local condition on the value of the solution. This local condition leads naturally to an iterative numerical procedure.

Procedure

Given an elliptic equation, choose a finite difference formula to approximate the equation on a grid in the domain of interest. This formula can be manipulated into a relation between the value of the unknown at a point and the values of the unknown at neighboring points. Hence, once values have been assigned to every point in the grid, this formula can be used iteratively to update the value at every point. When the values stops changing (to some specified precision), an approximate solution has been found.

Example

Suppose we want to approximate the solution to Laplace's equation on a square

$$\nabla^2 u = 0,$$
$$u(0, y) = 0, \quad u(1, y) = 0, \quad \text{for } 0 \le y \le 1, \qquad (184.1.a\text{--}c)$$
$$u(x, 0) = 0, \quad u(x, 1) = 1, \quad \text{for } 0 < x < 1.$$

If we choose a grid with a uniform x spacing of Δx and a uniform y spacing of Δy, then (184.1.a) can be discretized as

$$\frac{1}{(\Delta x)^2} \left(v_{i+1,j} - 2v_{i,j} + v_{i-1,j} \right) + \frac{1}{(\Delta y)^2} \left(v_{i,j+1} - 2v_{i,j} + v_{i,j-1} \right) = 0$$

$$(184.2)$$

where $v_{i,j} := u(i\, \Delta x, j\, \Delta y)$, for $i = 1, 2, \ldots, 1/\Delta x$ and $j = 1, 2, \ldots, 1/\Delta y$. Equation (184.2) can be manipulated to yield

$$v_{i,j} = \frac{1}{2(1 + \lambda^2)} \left(\lambda^2 \left(v_{i,j+1} + v_{i,j-1} \right) + v_{i+1,j} + v_{i-1,j} \right) \qquad (184.3)$$

where $\lambda := \Delta y / \Delta x$. From (184.3), we see that $v_{i,j}$ can be replaced by a weighted average of the values at the neighboring points. Note that this is only true for points interior to the boundary.

The numerical technique is this: initialize the values at all points in the grid (one common choice is to use the averaged value of the independent variable on the boundary), then systematically apply (184.3) to all the grid points until the solution converges. In theory, the points to be updated can be chosen in any order. In practice, some choices result in faster convergence.

The FORTRAN program in Program 184 carries out this prescription for the problem in (184.1). In this program, $h = .2$, $k = .2$, and the number of iterative updates required before the approximation did not change more than EPS (set to .0001) was 16. The output from the computer program is given below

```
NUMBER OF ITERATIONS REQUIRED:   16
    0.        1.0000    1.0000    1.0000    1.0000    0.
    0.        0.4545    0.5946    0.5946    0.4545    0.
    0.        0.2234    0.3294    0.3294    0.2234    0.
    0.        0.1097    0.1703    0.1703    0.1098    0.
    0.        0.0454    0.0718    0.0719    0.0454    0.
    0.        0.        0.        0.        0.        0.
```

The symmetry of the solution was to be expected.

The exact solution to (184.1) can be determined by separation of variables (see page 419). The solution is

$$u(x,y) = \frac{4}{\pi} \sum_{n=1}^{\infty} \sin\left[(2n-1)\pi x\right] \frac{\sinh\left[(2n-1)\pi y\right]}{\sinh\left[(2n-1)\pi\right]}.$$

As can be verified, the numerical approximation is accurate to two decimal places.

Program 184

```
        REAL*8 V(6,6)
C INITIALIZE THE GRID
        DO 10 I=2,5
        DO 10 J=2,5
10      V(I,J)=.25D0
C HERE IS THE BOUNDARY DATA
        DO 20 K=1,6
        V(K,1)=0.0D0
        V(K,6)=1.0D0
        V(1,K)=0.0D0
20      V(6,K)=0.0D0
C PERFORM THE ITERATIONS
        EPS=.0001D0
        NUM=0
40      NUM=NUM+1
        IFLAG=0
        DO 30 I=2,5
        DO 30 J=2,5
        VNEW= ( V(I+1,J) + V(I-1,J) + V(I,J+1) + V(I,J-1) ) / 4.D0
        IF( DABS(V(I,J)-VNEW) .GT. EPS )   IFLAG=1
30      V(I,J)=VNEW
C DETERMINE IF ANOTHER ITERATION IS REQUIRED
        IF( IFLAG .EQ. 1 ) GOTO 40
        WRITE(6,5) NUM
5       FORMAT(' NUMBER OF ITERATIONS REQUIRED:', I5)
        DO 50 J=1,6
50      WRITE(6,15) (V(I,7-J),I=1,6)
15      FORMAT( 7(1X,F9.4) )
        END
```

Notes

[1] The equations in (184.2) can be combined into one large system of linear equations, and then iterative methods can be applied to this system. Each different iterative method for a linear system can be interpreted as a relaxation method directly on the grid values.

[2] Depending on the equation to which this method is applied, and on the ordering in which the updated values are obtained, this technique is called

(A) successive over-relaxation (SOR) method,
(B) Jacobi or simultaneous iteration scheme,
(C) Gauss–Seidel or successive iteration scheme,
(D) alternating-direction-implicit (ADI) method,
(E) Liebmann's method.

In the ADI method, the finite difference approximation to Laplace's equation may be written

$$\nabla^2 u \simeq \frac{u_{i,j-1}^{(2n)} - 2u_{i,j}^{(2n)} + u_{i,j+1}^{(2n)}}{(\Delta x)^2} + \frac{u_{i-1,j}^{(2n+1)} - 2u_{i,j}^{(2n+1)} + u_{i+1,j}^{(2n+1)}}{(\Delta y)^2} = 0.$$

The superscripts indicate the iteration number. Hence, the updating is done alternately by rows and columns in the array of values.

[3] This method, when applied to the elliptic equation $L[u] = 0$, can be interpreted as an approximation to the solution of the parabolic equation $u_t = L[u]$. By iterating until the solution stops changing, the steady state solution of the parabolic equation is obtained. This interpretation allows error estimates to be obtained for this method (see Garabedian [4]).

[4] For a listing of computer software that will implement the method described in this section, see page 586.

References

[1] T. F. Chan and H. C. Elman, "Fourier Analysis of Iterative Methods for Elliptic Problems," *SIAM Review*, **31**, No. 1, March 1989, pages 20–49.

[2] S. J. Farlow, *Partial Differential Equations for Scientists and Engineers*, John Wiley & Sons, New York, 1982, pages 304–305.

[3] C. F. Gerald and P. O. Wheatley, *Applied Numerical Analysis*, Addison–Wesley Publishing Co., Reading, MA, 1984, pages 412–417.

[4] P. R. Garabedian, *Partial Differential Equations*, Wiley, New York, 1964, pages 485–492.

[5] E. Isaacson and H. B. Keller, *Analysis of Numerical Methods*, John Wiley & Sons, New York, 1966, pages 463–478.

[6] R. D. Smith, *Numerical Solution of Partial Differential Equations: Finite Difference Methods*, Third Edition, Clarendon Press, Oxford, 1985, Chapter 5 (pages 239–330).

[7] J. M. Vega-Fernández, J. F. Duque-Carrillo, and J. J. Peña-Bernal, "A New Way for Solving Laplace's Problem (The Predictor Jump Method)," *J. Math. Physics*, **26**, No. 3, March 1985, pages 416–419.

185.   Hyperbolic Equations: Method of Characteristics

Applicable to   A single hyperbolic equation, or a system of hyperbolic equations.

Yields

A numerical approximation scheme.

Idea

The method of characteristics (see page 368) can be used directly to create a numerical scheme to integrate hyperbolic equations.

Procedure

To simplify the analysis, we will illustrate the method on the second order hyperbolic partial differential equation

$$au_{xx} + bu_{xy} + cu_{yy} + d = 0. \qquad (185.1)$$

In (185.1), the functions $\{a, b, c, d\}$ are assumed to depend on $\{x, y, u, u_x, u_y\}$. With the usual definition of $p := u_x$ and $q := u_y$, equation (185.1) may be rewritten as the system of equations

$$E_1 := ap_x + bp_y + cq_y + d = 0,$$
$$E_2 := p_y - q_x = 0.$$

If we define $E = E_1 + \lambda E_2$, then E may be written as

$$E = [ap_x + (\lambda + b)p_y] + (cq_y - \lambda q_x) + d = 0.$$

This, in turn, may be written as

$$E = \frac{d}{ds}(p + \mu q) + \left(d - q\frac{d\mu}{ds}\right) = 0, \qquad (185.2)$$

along the curve defined parametrically by

$$\frac{dx}{ds} = a = -\frac{\lambda}{\mu}, \qquad \frac{dy}{ds} = \lambda + b = \frac{c}{\mu}, \qquad (185.3)$$

if such a curve exists. For consistency in (185.3), we must choose μ to satisfy $a\mu^2 - b\mu + c = 0$, that is,

$$\mu_{1,2} = \frac{b \pm \sqrt{b^2 - 4ac}}{2a}. \qquad (185.4)$$

Define $\{\mu_1, \mu_2\}$ to be the distinct real roots given in (185.4) (if the roots are not distinct and real then (185.1) is not hyperbolic), and define $\lambda_i = -a\mu_i$. Then equations (185.2) and (185.3) can be written as

$$\frac{d}{ds}(p + \mu_1 q) = -\left(d - q\frac{d\mu_1}{ds}\right) \quad \text{on the curve } C_1,$$

$$\frac{d}{ds}(p + \mu_2 q) = -\left(d - q\frac{d\mu_2}{ds}\right) \quad \text{on the curve } C_2,$$

$$(185.5)$$

where the characteristic curves C_1 and C_2 are defined by

$$\text{on } C_1: \quad \frac{dx}{ds} = a, \quad \frac{dy}{ds} = \lambda_1 + b,$$

$$\text{on } C_2: \quad \frac{dx}{ds} = a, \quad \frac{dy}{ds} = \lambda_2 + b.$$

$$(185.6.a\text{--}b)$$

These two characteristics curves have slopes that vary from point to point. These curves are generally not orthogonal. Knowing $\{a, b, c, d\}$ allows us to determine $\{\mu_1, \mu_2\}$ and so $\{\lambda_1, \lambda_2\}$ can also be determined. Therefore, the characteristics curves can be calculated numerically.

Now, if $k_1 := p + \mu_1 q$ and $k_2 := p + \mu_2 q$ were known at some common point R (these values arise naturally from (185.5)), then $p(R)$ and $q(R)$ can be found by inverting these relations, that is

$$q(R) = \frac{k_1 - k_2}{\mu_1 - \mu_2},$$

$$p(R) = \frac{\mu_1 k_1 - \mu_2 k_2}{\mu_1 - \mu_2}.$$

$$(185.7)$$

The numerical procedure is now a straightforward application of the method of characteristics. First, the characteristic curves (see (185.6)) are identified, at some point, by evaluating (185.4). Then the equations for k_1 and k_2 are integrated a short distance along the characteristics, by using (185.5). From the values of k_1 and k_2, values for p and q may be determined by use of (185.7). Finally, knowing p and q, the value of $u(x, y)$ can be determined.

In more detail:

(A) Given values at the points P and Q (see Figure 185.a), we will determine the values of all the parameters at a new point R.

(B) Determine R by integrating along characteristic C_1 from P and along characteristic C_2 from Q until the curves intersect.

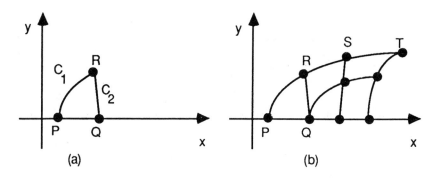

Figure 185. Depiction of the characteristics for a typical calculation.

(C) Integrate $k_1 := p + \mu_1 q$ from P to R and then integrate $k_2 := p + \mu_2 q$ from Q to R. Knowing $\{k_1, k_2\}$ and $\{\mu_1, \mu_2\}$ at R allows $q(R)$ and $p(R)$ to be obtained from (185.7).

(D) Now $du = u_x dx + u_y dy = p\,dx + q\,dy$, and so (approximately) $u(R) = u(P) + p(R)(x_P - x_R) + q(R)(y_P - y_R)$ and also $u(R) = u(Q) + p(R)(x_Q - x_R) + q(R)(y_Q - y_R)$. If these two formulae do not agree on the value of $u(R)$, then an average may be taken.

(E) Now that u and its derivatives are known at the point R, the process can be continued to the points $\{S, T, \ldots\}$ (see Figure 185.b).

Notes

[1] Another way to derive an equation equivalent to (185.2) is to use the relations $dp = u_{xx} dx + u_{xt} dt$ and $dq = u_{xt} dx + u_{tt} dt$ with (185.1). This results in the equation

$$u_{xt}\left(a\left(\frac{dt}{dx}\right)^2 - b\left(\frac{dt}{dx}\right) + c\right) - \left(a\frac{dp}{dx}\frac{dt}{dx} + c\frac{dq}{dx} + d\frac{dt}{dx}\right) = 0.$$

Comparing this equation to (185.2) the characteristic directions μ are immediately seen to correspond to $\dfrac{dt}{dx}$.

[2] This technique also works (in principle) for higher order equations. If the given hyperbolic equation has n independent variables, then the polynomial equation describing the characteristic directions will be of n-th order, and there will be n different characteristic directions. (In our example, there were only two characteristic directions, given by (185.4).)

[3] The procedure presented here can be made more concise by use of matrix notation. With a matrix formulation, equation (185.3) becomes the characteristic polynomial for the eigenvalues of some matrix and the characteristic directions in (185.5) become the eigenvectors of that same matrix.

[4] This technique can be readily modified to work with systems of hyperbolic equations.

[5] This technique is sometimes considered to be superior to using finite differences directly since it utilizes the mathematical structure of the solution. However, this is not always clear: following characteristic surfaces in higher dimensions is difficult, and the method of characteristics has trouble with shocks.

[6] If the independent variables are changed from $\{x,y\}$ to $\{\eta,\zeta\}$ via

$$\zeta = \frac{-b + \sqrt{b^2 - 4ac}}{2x}x + y,$$

$$\eta = \frac{-b - \sqrt{b^2 - 4ac}}{2x}x + y,$$

then (185.1) will have the form $u_{\eta\zeta} = \phi(u, u_\eta, u_\zeta, \eta, \zeta)$. See page 33 for more details.

References

[1] W. F. Ames, *Nonlinear Partial Differential Equations*, Academic Press, New York, 1967, pages 437–461.

[2] J. L. Davis, *Finite Difference Methods in Dynamics of Continuous Media*, The MacMillan Company, New York, 1986, pages 54–67.

[3] P. DuChateau and D. Zachmann, *Applied Partial Differential Equations*, Harper & Row, Publishers, New York, 1989.

[4] R. D. Smith, *Numerical Solution of Partial Differential Equations: Finite Difference Methods*, Third Edition, Clarendon Press, Oxford, 1985, Chapter 4 (pages 175–238).

[5] E. H. Twizell, *Computational Methods of Partial Differential Equations*, Halstead Press, John Wiley & Sons, New York, 1984, pages 116–125.

186. Hyperbolic Equations: Finite Differences

Applicable to   Hyperbolic partial differential equations.

Yields

A numerical approximation scheme.

Idea

Finite differences can be used directly to numerically approximate the solution of a hyperbolic partial differential equation.

Procedure

The technique is to replace all of the derivatives appearing in the given hyperbolic partial differential equation by finite difference approximations. By rearranging the terms in this new equation, an explicit recurrence formula can generally be obtained.

A stability analysis can be performed on this recurrence relation to determine the step sizes that will insure convergence of the numerical approximation to the true solution.

A frequent problem encountered with this method is having enough starting values to begin iterating the recurrence relation. Starting values can generally be obtained by performing manipulations of the original equation.

Example

The hyperbolic equation

$$u_{tt} - \alpha^2 u_{xx} = 0 \qquad (186.1)$$

on the interval $0 < x < L$, for $t > 0$, with the initial and boundary conditions

$$
\begin{aligned}
u(0,t) &= u(L,t) = 0, \\
u(x,0) &= f(x), \\
\frac{\partial u}{\partial t}(x,0) &= g(x),
\end{aligned}
\qquad (186.2)
$$

can be numerically approximated directly by finite differences.

We choose a uniform grid of $M + 1$ points in the x direction (i.e., $x_i = ih$ for $i = 0, 1, 2, \ldots, M$ with $h := L/M$). We choose the step length in the t variable to be k and define $t_j := jk$. We also choose to use the following centered difference formulas for u_{xx} and u_{tt}

$$
\begin{aligned}
u_{tt}(x_i, t_j) &= \frac{u(x_i, t_{j+1}) - 2u(x_i, t_j) + u(x_i, t_{j-1})}{k^2}, \\
u_{xx}(x_i, t_j) &= \frac{u(x_{i+1}, t_j) - 2u(x_i, t_j) + u(x_{i-1}, t_j)}{h^2}.
\end{aligned}
\qquad (186.3)
$$

Each of these formulae is second order accurate. If we define $w_{i,j} := u(x_i, t_j)$, then using (186.3) in (186.1) results in

$$\frac{w_{i,j+1} - 2w_{i,j} + w_{i,j-1}}{k^2} - \alpha^2 \frac{w_{i+1,j} - 2w_{i,j} + w_{i-1,j}}{h^2} = 0.$$

This last equation can be solved for $w_{i,j+1}$ to define the recurrence relation

$$w_{i,j+1} = 2(1 - \lambda^2)w_{i,j} + \lambda^2(w_{i+1,j} + w_{i-1,j}) - w_{i,j-1}, \qquad (186.4)$$

for $i = 1, 2, \ldots, (M-1)$ and $j = 1, 2, \ldots$, where $\lambda := \alpha k/h$. The initial conditions and boundary conditions, from (186.2), can be represented as

$$w_{0,j} = w_{M,j} = 0, \qquad j = 1, 2, \ldots,$$
$$w_{i,0} = f(x_i), \qquad i = 1, 2, \ldots, M. \qquad (186.5)$$

Now comes the problem of starting the recurrence relation off. Suppose we wish to iterate equation (186.4). The values we first compute are the $\{w_{i,2}\}$, but these require knowledge of $\{w_{i,1}\}$, which is not given in (186.5). The procedure for obtaining this data is to perform a Taylor series expansion of $w_{i,1}$. We find that

$$w_{i,1} := u(x_i, t_1)$$
$$= u(x_i, k)$$
$$\simeq u(x_i, 0) + k\frac{\partial u}{\partial t}(x_i, 0) + \frac{k^2}{2}\frac{\partial^2 u}{\partial t^2}(x_i, 0) + \ldots,$$

where this last formula is second order accurate if we retain only the terms shown (higher order approximations can also be obtained). Now u_{tt} is known in terms of u_{xx} from equation (186.1), and $u(x, 0)$ is known in terms of $f(x)$ from equation (186.2). Therefore, equation (186.4) can be simplified to yield

$$w_{i,1} \simeq w_{i,0} + kg(x_1) + \frac{\alpha^2 k^2}{2} f''(x_i). \qquad (186.6)$$

Special Case

The FORTRAN program in Program 186 numerically approximates the solution of the hyperbolic equation

$$u_{xx} - 9u_{xx} = 0, \qquad \text{for} \quad 0 < x < 1, \qquad 0 < t,$$
$$u(0, t) = u(1, t) = 0, \qquad \text{for} \qquad\qquad\qquad 0 < t,$$
$$u(x, 0) = \sin \pi x, \qquad \text{for} \quad 0 \le x \le 1,$$
$$u_t(x, 0) = 0, \qquad \text{for} \quad 0 \le x \le 1.$$

This system has the analytic solution $u(x, t) = \sin \pi x \cos 3\pi t$.

The program utilizes $M = 10$ and the value of k was chosen to be .02. The solution obtained for $t = 1$ at the points $x_i = .1i$ (for $i = 0, 1, \ldots, 10$) is

```
 0.      -0.3082  -0.5862  -0.8069  -0.9485  -0.9973
-0.9485  -0.8069  -0.5862  -0.3082   0.
```

By comparing these values to the exact solution, we observe that the numerical approximation is correct to two decimal places.

Program 186

```
          REAL  W(100,100)
C HERE ARE THE INITIAL VALUES
          ALPHA=3.
          FL=1.
          M=10
          H=FL/FLOAT(M)
          FK=.02
          N=1./FK
          FLAMBD=ALPHA*FK/H
          CONST=2.*(1.-FLAMBD**2)
C SET UP THE INITIAL/BOUNDARY VALUES IN THE MATRIX
          DO 10 J=1,N+1
          W(1,J)=0.
10        W(M+1,J)=0.
          DO 20 I=2,M
          XI=(I-1)*H
          W(I,1)=F(XI)
20        W(I,2)=W(I,1)+FK*G(XI)+FK**2*FPP(XI)/2.
C HERE IS THE RECURRENCE RELATION
          DO 30 J=2,N
          TT=J*FK
          DO 40 I=2,M
40        W(I,J+1)=CONST*W(I,J)+FLAMBD**2*(W(I+1,J)+W(I-1,J))-W(I,J-1)
30        WRITE(6,5) J,TT,(W(K,J+1), K=1,M+1)
5         FORMAT(' AT TIME STEP ',I4,' (T=',F7.3,')'/,4(1X,6(F9.4)/) )
          END
C THESE FUNCTIONS COMPUTE F(X), F''(X) AND G(X)
          FUNCTION F(X)
          F=SIN(3.1415927*X)
          RETURN
          END
          FUNCTION G(X)
          G=0.
          RETURN
          END
          FUNCTION FPP(X)
          FPP=-(3.1415927)**2 * SIN(3.1415927*X)
          RETURN
          END
```

Notes

[1] A stability analysis shows that equation (186.4) is stable if $\lambda < 1$.

[2] If the k^2 term in equation (186.6) had been neglected, then the method would only have been a first order method.

References

[1] R. L. Burden, *Numerical Analysis*, PWS Publishers, Boston, 1985, pages 583–599.

[2] J. L. Davis, *Finite Difference Methods in Dynamics of Continuous Media*, The MacMillan Company, New York, 1986, pages 42–44.

[3] P. DuChateau and D. Zachmann, *Applied Partial Differential Equations*, Harper & Row, Publishers, New York, 1989.

[4] P. R. Garabedian, *Partial Differential Equations*, Wiley, New York, 1964, pages 463–475.

[5] R. A. Renaut-Williamson, "Full Discretisations of $u_{tt} = u_{xx}$ and Rational Approximations to $\cosh\mu$," *SIAM J. Numer. Anal.*, **26**, No. 2, April 1989, pages 338–347.

[6] L. N. Trefethen, "Instability of Difference Models for Hyperbolic Initial Boundary Value Problems," *Comm. Pure Appl. Math*, **37**, No. 3, May 1984, pages 329–367.

187. Lattice Gas Dynamics

Applicable to   Partial differential equations that physically arise from the motion of "particles."

Yields

A numerical approximation methodology.

Idea

Partial differential equations are usually derived from some microscopic dynamical system. It may be possible to simulate the dynamical system directly, without first formulating differential equations.

Procedure

We will illustrate the basic ideas behind this method for the case of a fluid. By considering the interacting particles that make up a fluid, and using continuum theory, the usual Navier–Stokes equation can be derived (see, for example, Hasslacher [10]). This equation describes the evolution of the fluid. To numerically approximate the solution to this equation, the equation is discretized, and the resulting algebraic equations are solved on a computer.

Since a computer will be used to solve a discrete problem, it may be easier (and faster) to directly simulate the motion of the original, discrete particles. The resulting simulation can mimic all of the effects that fluid systems have. By considering only local interaction laws in the simulation,

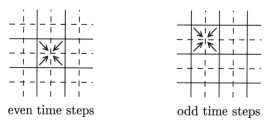

even time steps   odd time steps

Figure 187.1 The 2×2 blocking of the rectilinear array at different time steps.

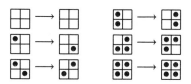

Figure 187.2 All possible motions and interactions on the rectilinear grid in one time step (up to rotations).

we are led to use cellular automata to describe the dynamics of the particles. Methods have been found for constructing cellular automata that are microscopically reversible (and thus support a realistic thermodynamics), obey exact conservation laws, and model continuum phenomena.

Example

We will illustrate one possible set of interaction laws that can be used to simulate gas dynamics; this model goes by the name of HPP. We consider a rectilinear array in which a particle may be present in a cell (indicated by a dot), or it may be absent (indicated by a blank). At each "time step," the grid is considered in 2×2 blocks. The blocking alternates between even and odd time steps (see Figure 187.1). At any time step, a particle in a cell is considered to be moving towards the center of the 2×2 block (see Figure 187.1). Hence, a particle in the upper left corner will move to the lower right corner in one time step. On the next time step, since the blocking has changed, this particle will once again be in the upper left of its new block. Hence, it will continue moving on a diagonal path.

The particles travel straight, with one exception: when exactly two particles coming together from opposite directions collide, they bounce apart in the other two directions. These interactions are particle-conserving, deterministic, and invertible. In Figure 187.2 we have indicated all possible interaction possibilities (up to rotations). With the information presented, it is possible to construct a full-scale simulation of a gas.

Notes

[1] It should be noted that, for some regimes, a lattice gas may fail to well approximate the Navier–Stokes equation and yet be closer to the actual physics than the Navier–Stokes equation itself.

[2] It is possible to amplify the simple example above by having many particles, interaction effects between the different particles, exclusion rules, etc.

[3] The example above is for a rectilinear grid. The articles by Hasslacher [10] describe the use of hexagonal grids.

[4] Papatheodorou and Fokas [12] have shown that "discrete soliton" type behavior is possible in cellular automata.

[5] Using special purpose hardware, simulation in lattice gas dynamics can be performed very quickly. See Margolus, Toffoli, and Vichiniac [11].

References

[1] S. Chen, M. Lee, K. H. Zhao, and G. D. Doolen, "A Lattice Gas Model with Temperature," *Physica D*, **37**, No. 1–3, 1989, pages 42–59.

[2] H. Chen, W. H. Matthaeus, and L. W. Klein, "Theory of Multicolor Lattice Gas: A Cellular Automaton Poisson Solver," *J. Comput. Physics*, **88**, No. 2, 1990, pages 433–466.

[3] G. H. Cottel and S. Mas-Gallic, "A Particle Method to Solve the Navier–Stokes System," *Numer. Math.*, **57**, No. 8, 1990, pages 805–827.

[4] G. D. Doolen, "Lattice Gas Methods for PDE's, Theory, Applications and Hardware," Proc. NATO Advanced Research Workshop, Los Alamos, September 6–8, 1989, North–Holland Publishing Co., New York, 1991.

[5] G. Doolen, B. Hasslacher, U. Frisch, S. Orszag, and S. Wolfram (eds.), *Lattice Gas Methods for Partial Differential Equations*, Addison–Wesley Publishing Co., Reading, MA, 1989.

[6] G. G. Elenin and V. V. Krylov, "Equilibrium Equations for a Multi-Component Nonideal Lattice Gas on Sublattices," *Mat. Model.*, **2**, No. 1, 1990, pages 85–104.

[7] B. Enquist and T. Y. Hou, "Particle Method Approximation of Oscillatory Solutions to Hyperbolic Differential Equations," *SIAM J. Numer. Anal.*, **26**, No. 2, April 1989, pages 289–319.

[8] U. Frisch, B. Hasslacher, and Y. Pomeau, "Lattice Gas Automata for the Navier–Stokes Equation," *Phys. Rev. Let.*, **56**, 1986, pages 1505–1508.

[9] U. Frisch, D. d'Humieres, B. Hasslacher, P. Lallemand, Y. Pomeau, J. P. Rivet, "Lattice Gas Hydrodynamics in Two and Three Dimensions," *Complex Systems*, **1**, No. 4, 1987, pages 649–707.

[10] B. Hasslacher, "Background for Lattice Gas Automata," "The Simple Hexagonal Model," and "The Promise of Lattice Gas Methods," *Los Alamos Science*, 1987, pages 175–186, 187–200, and 211–217.

[11] N. Margolus, T. Toffoli, and G. Vichiniac, "Cellular-Automata Supercomputers for Fluid-Dynamics Modeling," *Phys. Rev. Let.*, **56**, No. 16, 21 April 1986, pages 1694–1696.

[12] T. A. Papatheodorou and A. S. Fokas, "Evolution Theory, Periodic Particles, and Solitons in Cellular Automata," *Stud. Appl. Math.*, **80**, 1989, pages 165–182.

[13] G. Russo, "A Particle Method for Collisional Kinetic Equations. I. Basic Theory and One-Dimensional Results," *J. Comput. Physics*, **87**, No. 2, 1990, pages 270–300.

[14] T. Toffoli, "Cellular Automata as an Alternative to (Rather than an Approximation of) Differential Equations in Modeling Physics," *Physica D*, **10**, 1984, pages 117–127.

[15] T. Toffoli, "Information Transport Obeying the Continuity Equation," *IBM J. Res. & Dev.*, **32**, No. 1, January 1988, pages 29–36.

[16] T. Tonegawa, M. Kaburagi, and J. Kanamori, "Ground State Analysis of the Lattice Gas Model with Two Kinds of Particles on the Triangular Lattice," *J. Phys. Soc. Japan*, **59**, No. 5, 1990, pages 1660–1675.

[17] S. Wolfram (ed.), *Theory and Application of Cellular Automata*, World Scientific, Singapore, 1986.

188. Method of Lines

Applicable to   Elliptic, hyperbolic, and parabolic partial differential equations.

Yields

A system of partial differential equations with one fewer independent variables.

Idea

The basis of the method is substitution of finite differences for the derivatives with respect to one independent variable, and retention of the derivatives with respect to the remaining variables. This approach changes a given partial differential equation into a system of partial differential equations.

Procedure

We will illustrate the general method on a second order elliptic partial differential equation. Say the given equation is

$$A\frac{\partial^2 u}{\partial x^2} + B\frac{\partial^2 u}{\partial x \partial y} + C\frac{\partial^2 u}{\partial y^2} + D\frac{\partial u}{\partial x} + E\frac{\partial u}{\partial y} + Fu = G \qquad (188.1)$$

in a domain Ω, where $\{A, B, C, D, E, F, G\}$ are all functions of x and y. Since (188.1) is assumed to be elliptic, the necessary data for (188.1) are given on the boundary of Ω.

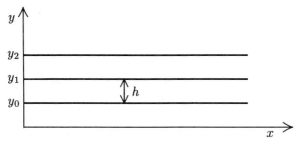

Figure 188.1 Subdivision of the domain to solve (188.1).

If we choose to discretize in the y variable, then we draw lines parallel to the x axis, with a constant distance, h, between adjacent lines. (See Figure 188.1.) Say the lines are specified by

$$y = y_k = y_0 + kh, \qquad k = 0, 1, \dots, N.$$

Then we set $y = y_k$ in (188.1) and use finite differences for the derivatives with respect to y. For example, we can use

$$\left. \frac{\partial u}{\partial y} \right|_{y=y_k} \simeq \frac{1}{h} \left[u_{k+1}(x) - u_k(x) \right],$$

$$\left. \frac{\partial^2 u}{\partial x \partial y} \right|_{y=y_k} \simeq \frac{1}{h} \left[u'_{k+1}(x) - u'_k(x) \right], \qquad (188.2)$$

$$\left. \frac{\partial^2 u}{\partial y^2} \right|_{y=y_k} \simeq \frac{1}{h^2} \left[u_{k+1}(x) - 2u_k(x) + u_{k-1}(x) \right],$$

where $u_k(x)$ is an approximation to $u(x, y_k)$. Using (188.2) in (188.1) (with $y = y_k$) we obtain a first order differential equation involving the unknown functions $\{u_{k-1}, u_k, u_{k+1}\}$. By taking $k = 0, 1, \dots, N$, we obtain a system of first order ordinary differential equations for the $N+1$ unknown functions $\{u_0(x), u_1(x), \dots, u_N(x)\}$.

If (188.1) is elliptic and Ω is convex, then the equations will constitute a two point boundary value system. Any standard (numerical) two point ordinary differential equation system solver can be used to solve this system.

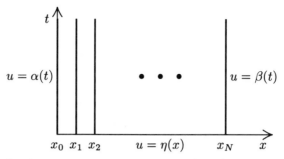

Figure 188.2 Subdivision of the domain to solve (188.3).

Example

Suppose we have the following parabolic equation for $u(x, t)$

$$
\begin{aligned}
u_t &= u_{xx}, \\
u(0, x) &= \eta(x), \\
u(t, 0) &= \alpha(t), \\
u(t, 1) &= \beta(t).
\end{aligned}
\qquad (188.3.a\text{--}d)
$$

If we choose to discretize in the x variable, then we approximate $u(t, x_n)$ by $v_n(t)$, where $x_n = n/N := n\,\Delta x$. Then we can approximate the derivatives with respect to x in (188.3.a) by finite differences to obtain

$$
\frac{d}{dt}v_n(t) \simeq \frac{v_{n+1}(t) - 2v_n(t) + v_{n-1}(t)}{(\Delta x)^2}, \qquad (188.4)
$$

for $n = 1, 2, \ldots, N - 1$. The initial conditions and boundary conditions in (188.3) can be written as

$$
\begin{aligned}
v_n(0) &= \eta(n\,\Delta x), \qquad \text{for } n = 1, 2, \ldots, N - 1, \\
v_0(t) &= \alpha(t), \\
v_N(t) &= \beta(t).
\end{aligned}
\qquad (188.5)
$$

(See Figure 188.2.)

If an explicit scheme (say forward Euler's method) is chosen to numerically approximate (188.4), then the simple formula

$$
v_n(t + \Delta t) = v_n(t) + \frac{\Delta t}{(\Delta x)^2}\left[v_{n+1}(t) - 2v_n(t) + v_{n-1}(t)\right] \qquad (188.6)
$$

results. This formula can be iterated with (188.5) to find a numerical approximation to the solution of (188.3).

If, instead, we choose to discretize (188.3) in the t variable, then we would approximate $u(t_k, x)$ by $w_k(x)$, where $t_k := k \, \Delta t$. Approximating the t derivatives in (188.3) by finite differences, we obtain

$$\frac{w_k(x) - w_{k-1}(x)}{\Delta t} = \frac{d^2}{dx^2} w_k(x), \qquad \cdot \qquad (188.7)$$

with the corresponding initial and boundary conditions

$$
\begin{aligned}
w_0(x) &= \eta(x) \\
w_m(0) &= \alpha(m \, \Delta t), \qquad \text{for } m = 0, 1, \dots \\
w_m(1) &= \beta(m \, \Delta t), \qquad \text{for } m = 0, 1, \dots \, .
\end{aligned}
$$

Note that (188.7) is a constant coefficient ordinary differential equation for the dependent variable $w_k(x)$. Hence, the explicit solution can be obtained and the differential system can be replaced by an algebraic system.

Notes

[1] This method is sometimes called the generalized Kantoravich method.

[2] Observe that the recurrence relation in (188.6) could have been obtained directly by applying finite differences to both the x and t derivatives appearing in (188.3). This is not a clever use of the method of lines.

   A better approach would be to use a computer package to solve the initial value system in (188.4) and (188.5). This package could use an implicit method for the t derivative, and it could adjust the step size as necessary to reduce the error.

[3] For a listing of computer software that will implement the method described in this section, see page 586.

References

[1] M. Berzins, "Global Error Estimation in the Method of Lines for Parabolic Equations," *SIAM J. Sci. Stat. Comput.*, **9**, No. 4, July 1988, pages 687–703.

[2] P. M. Dew and J. E. Walsh, "A Set of Library Routines for Solving Parabolic Equations in One Space Variable," *ACM Trans. Math. Software*, **7**, No. 3, September 1981, pages 295–314.

[3] I. Graney and A. A. Richardson, "The Numerical Solution of Non-Linear Partial Differential Equations by the Method of Lines," *J. Comput. Appl. Math.*, **7**, No. 4, 1981, pages 229–236.

[4] P. Keast and P. H. Muir, "EPDCOL: A More Efficient PDECOL Code," *ACM Trans. Math. Software*, **17**, No. 2, June 1991, pages 153–166.

[5] D. K. Melgaard and R. F. Sincovec, "General Software for Two-Dimensional Nonlinear Partial Differential Equations," *ACM Trans. Math. Software*, **7**, No. 1, March 1981, pages 106–125.

[6] G. H. Meyer, "The Method of Lines for Poisson's Equation with Nonlinear or Free Boundary Conditions," *Numer. Math.*, **29**, 1978, pages 329–344.

[7] M. N. Mikhail, "On the Validity and Stability of the Method of Lines for the Solution of Partial Differential Equations," *Appl. Math. and Comp.*, **22**, 1987, pages 89–98.

[8] S. G. Mikhlin and K. L. Smolitskiy, *Approximate Methods for Solutions of Differential and Integral Equations*, American Elsevier Publishing Company, New York, 1967, pages 263–268.

[9] W. E. Schiesser, *The Numerical Method of Lines*, Academic Press, New York, 1991.

189. Parabolic Equations: Explicit Method

Applicable to   Parabolic partial differential equations.

Yields

An explicit numerical scheme.

Idea

Marching in time is the easiest way to solve a parabolic equation. For this explicit method, the time steps must be small.

Procedure

Suppose we have the parabolic differential equation

$$
\begin{aligned}
u_t &= L(u, \mathbf{x}, t), \\
u(t_0, \mathbf{x}) &= f(\mathbf{x}),
\end{aligned}
\tag{189.1}
$$

for $u(\mathbf{x}, t)$, where $L(u, \mathbf{x}, t)$ is uniformly elliptic. The easiest way to solve (189.1) is by the use of "marching," which is an explicit method.

An explicit numerical approximation is determined by taking a forward difference in the t variable in (189.1), and having no other terms that involve future time values. For example, we can approximate $u(\mathbf{x}, t)$ by $v(\mathbf{x}, t)$ where $v(\mathbf{x}, t)$ satisfies

$$
\begin{aligned}
v(t + \Delta t, \mathbf{x}) &= v(t, \mathbf{x}) + \Delta t \widehat{L}(v(t, \mathbf{x}), \mathbf{x}, t), \\
v(t_0, \mathbf{x}) &= f(\mathbf{x}),
\end{aligned}
\tag{189.2}
$$

and $\widehat{L}(\ )$ is any reasonable finite difference approximation to $L(v(t, \mathbf{x}), \mathbf{x}, t)$ that does not involve $v(t + \Delta t, \mathbf{x})$ (if it did involve this term, then the method would be implicit).

The main drawback of this method is that Δt must often be very small for the method to be stable. If $|\Delta \mathbf{x}|$ is the smallest discretization step in the evaluation of $\widehat{L}(v(t, \mathbf{x}), \mathbf{x}, t)$ then we require

$$\Delta t = O(|\Delta \mathbf{x}|^2)$$

for (189.2) to be a numerically stable technique. More precise restrictions on Δt can be derived from the exact form of $L(v(t, \mathbf{x}), \mathbf{x}, t)$, and the numerical approximation used for the derivatives.

Example
Suppose we want to numerically approximate the solution to the diffusion problem

$$\begin{aligned}
u_t &= u_{xx}, \\
u(t, 0) &= 0, \\
u(t, 1) &= 1, \\
u(0, x) &= 0,
\end{aligned} \qquad (189.3.\text{a--d})$$

for $t \geq 0$ with $0 \leq x \leq 1$. From the method of Fourier series or separation of variables (see page 293 or 419), we find the analytic solution of (189.3) to be

$$u(t, x) = x + \frac{2}{\pi} \sum_{n=1}^{\infty} \frac{(-1)^n}{n} e^{-\pi^2 n^2 t} \sin \pi n x.$$

This exact solution will be used to ascertain the accuracy of the numerical solution.

To numerically approximate the solution to (189.3), we use a grid of N points between 0 and 1: $\{x_n \mid x_n = (n-1)\Delta x, \ n = 1, 2, \ldots, N\}$, where $\Delta x = 1/(N-1)$. We define $v_n(t)$ to be the approximation of $u(t, x)$ at the n-th grid point: $v_n(t) \simeq u(t, x_n)$.

The initial conditions in (189.3.d) can be represented as

$$v_m(0) = 0, \qquad m = 0, 1, 2, \ldots, N,$$

while the boundary conditions in (189.3.b,c) can be represented as

$$v_1(t) = 0, \qquad v_N(t) = 1.$$

The equation (189.3.a) can be discretized as

$$v_m(t + \Delta t) = v_m(t) + \Delta t \left(\frac{v_{m+1}(t) - 2v_m(t) + v_{m-1}(t)}{(\Delta x)^2} \right), \qquad (189.4)$$

which has utilized a centered second order scheme for the u_{xx} term and a first order forward difference scheme for the u_t term.

The FORTRAN program in Program 189 implements the above scheme for $N = 21$ and $\Delta t = .001$. We choose to compare the output from the program to the exact solution, for $t = .1$ and $x = .5$. The exact solution is $u(.1, .5) \simeq .2637$.

Table 189 shows the approximate value of $u(.1, .5)$, for several different choices of N and Δt. From these values we conclude

(A) As N increases, the accuracy of the numerical solution increases.

(B) As Δt deceases, the accuracy of the numerical solution increases.

The difference equation in (189.4) was the example used to demonstrate the Von Neumann stability test (see page 621). It was determined there that the method will be stable only if $\Delta t/(\Delta x)^2$ is less than one.

Table 189:

Approximate value of $u(.1, .5)$ for different N and Δt. The exact value is $u(.1, .5) \simeq .2637$.

N	Δx	Δt	$\Delta t/(\Delta x)^2$	$v(.1, .5)$
5	.25	.05	.80	.6400
5	.25	.01	.16	.2745
11	.10	.005	.50	.2628
11	.10	.001	.10	.2640
21	.05	.001	.40	.2639

Program 189

```
            REAL*8 X(1000),V(1000)
            DELTAT=.001D0
            NTIME=100
            N=21
            DELTAX=1.D0/DFLOAT(N-1)
C INITIALIZE THE GRID
            DO 10 J=1,N
            X(J)=DFLOAT(J-1)*DELTAX
10          V(J)=0.D0
            V(N)=1.D0
            T=0.D0
C THIS IS THE LOOP FOR THE NUMBER OF TIME STEPS
            DO 20 J=1,NTIME
            T=T+DELTAT
C UPDATE THE GRID
            CALL UPDATE(V,DELTAX,N,DELTAT)
C OUTPUT THE ANSWER
20          WRITE(6,5) T, (X(K),V(K),K=1,N)
5           FORMAT(' THE TIME IS=',F8.4,900(/10X,2F12.5) )
            END
C THIS SUBROUTINE INCREMENTS THE SOLUTION BY ONE TIME STEP
            SUBROUTINE  UPDATE(V1,DELTAX,N,DELTAT)
            REAL*8  V1(1000),V2(1000)
            RATIO=DELTAT/DELTAX**2
```

```
          DO 100 J=2,N-1
100       V2(J)=V1(J) + RATIO*( V1(J+1) -2.DO * V1(J) + V1(J-1) )
          DO 200 J=2,N-1
200       V1(J)=V2(J)
          RETURN
          END
```

Notes

[1] For a listing of computer software that will implement the method described in this section, see page 586.

References

[1] P. DuChateau and D. Zachmann, *Applied Partial Differential Equations*, Harper & Row, Publishers, New York, 1989.

[2] J. L. Davis, *Finite Difference Methods in Dynamics of Continuous Media*, The MacMillan Company, New York, 1986, Chapter 4 (pages 167–193).

[3] D. J. Evans and A. R. B. Abdullah, "A New Explicit Method for the Solution of $\frac{\partial u}{\partial t} = \frac{\partial^2 u}{\partial x^2} + \frac{\partial^2 u}{\partial y^2}$," *Int. J. Comp. Math.*, **14**, 1983, pages 325–353.

[4] S. J. Farlow, *Partial Differential Equations for Scientists and Engineers*, John Wiley & Sons, New York, 1982, Lesson 38 (pages 309–315).

[5] C. F. Gerald and P. O. Wheatley, *Applied Numerical Analysis*, Addison-Wesley Publishing Co., Reading, MA, 1984.

[6] R. D. Smith, *Numerical Solution of Partial Differential Equations: Finite Difference Methods*, Third Edition, Clarendon Press, Oxford, 1985, Chapters 2 and 3 (pages 11–174).

[7] E. H. Twizell, *Computational Methods of Partial Differential Equations*, Halstead Press, John Wiley & Sons, New York, 1984, pages 200–265.

[8] W. H. Press, B. P. Flannery, S. Teukolsky, and W. T. Vetterling, *Numerical Recipes*, Cambridge University Press, New York, 1986, pages 635–640.

190. Parabolic Equations: Implicit Method

Applicable to   Parabolic partial differential equations.

Yields

An implicit numerical scheme.

Idea

An implicit scheme will numerically approximate the solution of a parabolic equation and allow large time steps to be taken.

Procedure

Suppose we have the parabolic differential equation

$$
\begin{aligned}
u_t &= L(u, \mathbf{x}, t), \\
u(t_0, \mathbf{x}) &= f(\mathbf{x}),
\end{aligned}
\tag{190.1}
$$

for $u(\mathbf{x}, t)$, where $L(u, \mathbf{x}, t)$ is uniformly elliptic. We desire an implicit difference scheme that will numerically approximate the solution to (190.1). An implicit method is one in which the value of $u(t + \Delta t, \mathbf{x})$ is not determined explicitly by the value of $u(t, \mathbf{x})$, but instead uses both $u(t + \Delta t, \mathbf{x})$ and $u(t, \mathbf{x})$.

For simplicity, we only discuss the case of a single space dimension. The difference scheme will utilize a uniform grid, with a spacing of Δx in the x direction and a spacing of Δt in the t direction. Define $v_{n,j}$ to be an approximation to $u(t_n, x_j)$, where $t_n = n\Delta t$ and $x_j = j\Delta x$.

To discretize (190.1) in t, we choose to use a forward difference in the t variable. That is,

$$
u_t(t_n, x_j) = \frac{v_{n+1,j} - v_{n,j}}{\Delta t}.
$$

Now the x derivatives will be approximated, at any point, by values at time t_n and at time t_{n+1}. That is,

$$
\begin{aligned}
u_x(t_n, x_j) &= (1 - \lambda_1)\frac{v_{n+1,j} - v_{n+1,j-1}}{\Delta x} + \lambda_1 \frac{v_{n,j} - v_{n,j-1}}{\Delta x}, \\
u_{xx}(t_n, x_j) &= (1 - \lambda_2)\frac{v_{n+1,j+1} - 2v_{n+1,j} + v_{n+1,j-1}}{(\Delta x)^2} \\
&\quad + \lambda_2 \frac{v_{n,j+1} - 2v_{n,j} + v_{n,j-1}}{(\Delta x)^2},
\end{aligned}
\tag{190.2}
$$

where λ_1 and λ_2 are any real numbers between zero and one. For any such values, the scheme in (190.2) will be consistent. Note that if $\lambda_1 = \lambda_2 = 1$, there is only dependence on the values at a previous time step and an explicit method is recovered. If neither λ_1 nor λ_2 is equal to one, an implicit difference scheme results.

An implicit scheme often has the advantage that time steps can be taken that are much larger than the time steps that can be taken for an explicit method. More precise restrictions on Δt can be obtained from the form of $L(v, \mathbf{x}, t)$ and the values chosen for λ_1 and λ_2 in (190.2).

Example

Suppose we want to numerically approximate the solution to the diffusion problem

$$u_t = u_{xx},$$
$$u(t,0) = 0,$$
$$u(t,1) = 1,$$
$$u(0,x) = 0,$$

$$(190.3.a\text{--}d)$$

for $t \geq 0$ with $0 \leq x \leq 1$. From the method of Fourier series or separation of variables (see pages 293 or 419), we find the analytic solution to (190.3) is

$$u(t,x) = x + \frac{2}{\pi} \sum_{n=1}^{\infty} \frac{(-1)^n}{n} e^{-\pi^2 n^2 t} \sin \pi n x. \qquad (190.4)$$

This exact solution will be used to determine the accuracy of the numerical solution.

To numerically approximate the solution to (190.3), we use a grid of N points between 0 and 1: $\{x_n \mid x_n = (n-1)\Delta x, \ n = 1, 2, \ldots, N\}$, where $\Delta x = 1/(N-1)$. The initial conditions in (190.3.d) can be represented as

$$v_{0,j} = 0, \qquad j = 0, 1, 2, \ldots, N, \qquad (190.5)$$

while the boundary conditions in (190.3.b,c) can be represented as

$$v_{n,0} = 0, \qquad v_{n,N} = 1, \qquad \text{for } n = 1, 2, \ldots, \qquad (190.6)$$

We choose to discretize the equation with $\lambda_1 = \lambda_2 = 1/2$; this produces the *Crank–Nicolson scheme*. The approximation to (190.3.a) is therefore

$$\frac{v_{n+1,j} - v_{n,j}}{\Delta t} = \frac{1}{2} \frac{v_{n+1,j+1} - 2v_{n+1,j} + v_{n+1,j-1}}{(\Delta x)^2} + \frac{1}{2} \frac{v_{n,j+1} - 2v_{n,j} + v_{n,j-1}}{(\Delta x)^2},$$

which can be manipulated into

$$-\rho v_{n+1,j+1} + (2+2\rho)v_{n+1,j} - \rho v_{n+1,j-1} = \rho v_{n,j+1} + (2-2\rho)v_{n,j} + \rho v_{n,j-1},$$
$$(190.7)$$

where we have defined $\rho = \Delta t/(\Delta x)^2$.

Note that for a given value of n, (190.7) is an algebraic equation for $v_{n+1,j}$ and two of its spatial neighbors. Hence, equation (190.7) cannot be used alone to determine $v_{n+1,j}$. Instead, a system of equations must be solved simultaneously. Utilizing (190.5) and (190.6), this system may be written as

$$
\begin{pmatrix}
1 & 0 & 0 & 0 & \cdots & 0 \\
-\rho & 2+2\rho & -\rho & 0 & \cdots & 0 \\
0 & -\rho & 2+2\rho & -\rho & & 0 \\
\vdots & & \ddots & \ddots & \ddots & \\
0 & 0 & & -\rho & 2+2\rho & -\rho \\
0 & 0 & \cdots & 0 & 0 & 1
\end{pmatrix}
\begin{pmatrix}
v_{n+1,0} \\
v_{n+1,1} \\
v_{n+1,2} \\
\vdots \\
v_{n+1,N-1} \\
v_{n+1,N}
\end{pmatrix}
=
$$

$$
\begin{pmatrix}
0 \\
\rho v_{n,1} + (2-2\rho)v_{n,2} + \rho v_{n,3} \\
\rho v_{n,2} + (2-2\rho)v_{n,3} + \rho v_{n,4} \\
\vdots \\
\rho v_{n,N-2} + (2-2\rho)v_{n,N-1} + \rho v_{n,N} \\
1
\end{pmatrix}.
$$

Because this system of linear equations has a banded matrix of width three, the system can be solved very efficiently.

    The FORTRAN program in Program 190 implements the above scheme with $N = 21$ and $\Delta t = .01$. Note that this program uses a matrix solver, LSOLVE, whose source code is not shown. We choose to compare the output from the program to the exact solution (given in (190.4)), for $t = .1$ and $x = .5$. The exact solution is $u(.1, .5) \simeq .2637$.

    Table 190 shows the approximate value of $u(.1, .5)$, for several different choices of N and Δt. From these values we conclude

(A) As N increases, the accuracy of the numerical solution increases.

(B) As Δt deceases, the accuracy of the numerical solution increases.

Table 190:

        Approximate value of $u(.1, .5)$ for different N and Δt. The exact value is $u(.1, .5) \simeq .2637$.

N	Δx	Δt	$\Delta t/(\Delta x)^2$	$v(.1, .5)$
5	.25	.01	.80	.2526
11	.10	.01	1.00	.2508
11	.10	.005	.50	.2569
21	.05	.01	4.00	.2507

Program 190

```
        DIMENSION  FMAT(100,100),RHS(100),V(100),X(100),NROW(200)
        N=21
        DELTAT=.01
        NTIME=5
        DELTAX=1./DFLOAT(N-1)
        RHO=DELTAT/DELTAX**2
C INITIALIZE THE VECTOR AT T=0
        DO 10 J=1,N
        X(J)=DELTAX*(J-1)
10      V(J)=0
        T=0.
        DO 20 JTIME=1,NTIME
        T=T+DELTAT
C SET UP THE RIGHT HAND SIDE
        RHS(1)=0.
        RHS(N)=1.
        DO 30 J=2,N-1
30      RHS(J)= RHO*V(J-1)+(2.-2.*RHO)*V(J)+RHO*V(J+1)
C SET UP THE MATRIX
        DO 40 J=1,N
        DO 40 K=1,N
40      FMAT(J,K)=0.
        FMAT(1,1)=1.
        FMAT(N,N)=1.
        DO 50 J=2,N-1
        FMAT(J,J-1)=-RHO
        FMAT(J,J )=2.+2.*RHO
50      FMAT(J,J+1)=-RHO
C SOLVE THE MATRIX EQUATION
        CALL LSOLVE(N,FMAT,V,RHS,NROW,IFSING,100)
C PRINT OUT THE ANSWER
20      WRITE(6,5) T, (X(K),V(K),K=1,N)
5       FORMAT(' HERE IS THE SOLUTION AT TIME=',F8.4,/,90(10X, 2F12.5/))
        END
```

Notes

[1] Observe from Table 190 that the numerical method used resulted in reasonable approximations when $\Delta t/(\Delta x)^2$ was as large as 4. Using the Von Neumann test (see page 621), it can be shown that the Crank–Nicolson scheme is unconditionally stable for any value of $\Delta t/(\Delta x)^2$.

[2] Another way to interpret this solution technique is as a sequence of elliptic problems, with one problem being solved at every time step. For example, given the parabolic system

$$
\begin{aligned}
u_t &= L[u] + f(\mathbf{x}, t), & &\text{on } R,\, t > 0, \\
u &= g(\mathbf{x}, t), & &\text{on } \partial R,\, t > 0, \\
u &= u_0(\mathbf{x}), & &\text{on } R \cup \partial R,\, t = 0,
\end{aligned}
\qquad (190.8.a\text{–}c)
$$

we can take a forward difference in t to obtain $u_t(t) \simeq \dfrac{u(t) - u(t - \Delta t)}{\Delta t}$, which allows (190.8) to be rewritten as $\dfrac{u(t) - u(t - \Delta t)}{\Delta t} \simeq L[u(t)] + f(\mathbf{x}, t)$.

This is an elliptic equation for $u(t)$, in which $u(\mathbf{x}, t - \Delta t)$ plays the role of a nonhomogeneous forcing term. Hence, the successive time values of $u(\mathbf{x}, t)$ may be determined by solving a sequence of elliptic problems. The boundary conditions for each elliptic problem come from (190.8.b), while the first value of $u(\mathbf{x}, t)$ is given by $u_0(\mathbf{x})$.

Rice and Boisvert [5] present the template of an ELLPACK program that will numerically approximate the solution of parabolic equations by sequentially solving elliptic equations.

[3] For a listing of computer software that will implement the method described in this section, see page 586.

References

[1] P. DuChateau and D. Zachmann, *Applied Partial Differential Equations*, Harper & Row, Publishers, New York, 1989.

[2] J. L. Davis, *Finite Difference Methods in Dynamics of Continuous Media*, The MacMillan Company, New York, 1986, Chapter 4 (pages 167–193).

[3] S. J. Farlow, *Partial Differential Equations for Scientists and Engineers*, John Wiley & Sons, New York, 1982, Lesson 38 (pages 309–315).

[4] C. F. Gerald and P. O. Wheatley, *Applied Numerical Analysis*, Addison-Wesley Publishing Co., Reading, MA, 1984.

[5] J. R. Rice and R. F. Boisvert, *Solving Elliptic Problems Using ELLPACK*, Springer–Verlag, New York, 1985, pages 111–120.

[6] R. D. Smith, *Numerical Solution of Partial Differential Equations: Finite Difference Methods*, Third Edition, Clarendon Press, Oxford, 1985, Chapters 2 and 3 (pages 11–174).

191. Parabolic Equations: Monte Carlo Method

Applicable to   Linear parabolic partial differential equations.

Yields

A numerical approximation to the solution of a linear parabolic partial differential equation at a single point.

Idea

Simulation of the motion of a random particle may be used to approximate the solution to linear parabolic equations.

Procedure

The steps for this method are straightforward. First we give an overview, then a more detailed presentation.

First, approximate the elliptic part of the given parabolic partial differential equation by a finite difference method. Rewrite the finite difference formula as a recursive function for the value of the unknown at any given point. Then interpret this recursive formula as a set of transition probabilities that determine the motion of a random particle. By creating a finite difference scheme for the time derivative in the differential equation, a natural time scale will be associated with every step of the particle.

Now write a computer program that will allow many (say K) particles to wander randomly around the domain of interest, based on the transition probabilities found from the difference formula. Simulate the particles one at a time, with every particle starting off at the same point (say the point $\mathbf{z}$). If the time step is Δt, and the solution is desired at $t = T$, then the particles will be allowed to wander randomly, but for no more than $M := T/\Delta t$ steps.

(A) If the boundary data are of the Dirichlet type (i.e., the value of the unknown is prescribed on the boundary), then when a particle reaches the boundary, stop that particle and store away the value on the boundary. Begin another particle at the point $\mathbf{z}$.

(B) If the boundary data are not of the Dirichlet type (say Neumann or mixed boundary conditions) then, when the particles reach the boundary, they will be given a finite probability to leave the boundary and re-enter the domain of the problem. If the particle leaves the boundary, continue the iteration process. If it does not leave the boundary, the value at the boundary is stored away, and a new particle is started off at the point $\mathbf{z}$.

(C) For parabolic equations there is also the possibility that the particle will not reach the boundary in M steps. If the particle has not reached the boundary in M steps, then record the position that is finally reached. Using the initial conditions of the problem, there is a value associated with the point reached. Then begin a new particle.

If the parabolic equation was homogeneous, a numerical approximation to the solution at the point $\mathbf{z}$ will be given by an average of the K values stored away. If the given equation was not homogeneous, then an equation given below is used to obtain an estimate of the solution at the point $\mathbf{z}$. In this case, all points on the path that the particle traversed will be utilized.

In more detail, here is how the technique may be applied to the linear parabolic partial differential equation in the domain R

$$
\begin{aligned}
u_t &= L[u] + F(x, y, t), & x, y \in R, \text{ and } t > 0, \\
u &= \phi(x, y, t), & x, y \in \partial R, \text{ and } t > 0, & \quad (191.1.a\text{--}c) \\
u(x, y, 0) &= g(x, y), & x, y \in R,
\end{aligned}
$$

with the operator $L[\cdot]$ defined by

$$
L[u] := A u_{xx} + 2B u_{xy} + C u_{yy} + D u_x + E u_y,
$$

where $\{A, B, C, D, E\}$ are all functions of $\{x, y, t\}$. The operator $L[\cdot]$ may be discretized to yield the approximation

$$
\begin{aligned}
L[u] \simeq A_{i,j} &\left(\frac{v_{i+1,j,n} - 2v_{i,j,n} + v_{i-1,j,n}}{r} (\Delta x)^2 \right) \\
+ 2B_{i,j} &\left(\frac{v_{i+1,j+1,n} - v_{i,j+1,n} - v_{i+1,j,n} + v_{i,j,n}}{(\Delta x)(\Delta y)} \right) \\
+ C_{i,j} &\left(\frac{v_{i,j+1,n} - 2v_{i,j,n} + v_{i,j-1,n}}{(\Delta y)^2} \right) \\
+ D_{i,j} &\left(\frac{v_{i+1,j,n} - v_{i,j,n}}{\Delta x} \right) + E_{i,j} \left(\frac{v_{i,j+1,n} - v_{i,j,n}}{\Delta y} \right),
\end{aligned}
$$

where $x_i = x_0 + i(\Delta x)$, $y_j = y_0 + j(\Delta y)$, $t_n = n(\Delta t)$, $v_{i,j,n} = u(x_i, y_j, t_n)$, and a subscript of i, j, n means an evaluation at the point (x_i, y_j, t_n). If the $\{\Gamma_{\cdot,\cdot,\cdot}\}$ and $Q_{i,j,n}$ are defined by

$$
\Gamma_{i+1,j+1,n} = \left[\frac{2B_{i,j,n}}{(\Delta x)(\Delta y)} \right],
$$

$$
\Gamma_{i+1,j,n} = \left[\frac{A_{i,j,n}}{(\Delta x)^2} - \frac{2B_{i,j,n}}{(\Delta x)(\Delta y)} + \frac{D_{i,j,n}}{\Delta x} \right],
$$

$$
\Gamma_{i,j+1,n} = \left[\frac{C_{i,j,n}}{(\Delta y)^2} - \frac{2B_{i,j,n}}{(\Delta x)(\Delta y)} + \frac{E_{i,j,n}}{\Delta y} \right],
$$

$$
\Gamma_{i-1,j,n} = \left[\frac{A_{i,j,n}}{(\Delta x)^2} \right],
$$

$$
\Gamma_{i,j-1,n} = \left[\frac{C_{i,j,n}}{(\Delta x)^2} \right],
$$

$$
Q_{i,j,n} = \left[\frac{2A_{i,j,n}}{(\Delta x)^2} - \frac{2B_{i,j,n}}{(\Delta x)(\Delta y)} + \frac{2C_{i,j,n}}{(\Delta y)^2} + \frac{D_{i,j,n}}{\Delta x} + \frac{E_{i,j,n}}{\Delta y} \right],
$$

and then u_t is approximated by $\dfrac{u(x,y,t+\Delta t) - u(x,y,t)}{\Delta t}$, the equation in (191.1.a) may be discretized as

$$v_{i,j,n+1} = (\Delta t)\Big[\Gamma_{i+1,j,n}v_{i+1,j,n} + \Gamma_{i+1,j+1,n}v_{i+1,j+1,n} + \Gamma_{i,j+1,n}v_{i,j+1,n}$$
$$+ \Gamma_{i-1,j,n}v_{i-1,j,n} + \Gamma_{i,j-1,n}v_{i,j-1,n}\Big]$$
$$+ \left[1 - Q_{i,j,n}(\Delta t)\right]v_{i,j,n} + (\Delta t)F_{i,j,n}$$

$$(191.2)$$

If we now choose $\Delta t := 1/Q_{i,j,n}$, and define $p_{i,j,n} = \Gamma_{i,j,n}/Q_{i,j,n}$, then (191.2) can be written as

$$v_{i,j,n+1} = p_{i+1,j,n}v_{i+1,j,n} + p_{i+1,j+1,n}v_{i+1,j+1,n} + p_{i,j+1,n}v_{i,j+1,n}$$
$$+ p_{i-1,j,n}v_{i-1,j,n} + p_{i,j-1,n}v_{i,j-1,n} + \frac{F_{i,j,n}}{Q_{i,j,n}}.$$

Note that p's add up to one. We interpret them as probabilities of taking a step in a specified direction. Specifically, if a particle is at position (i, j, n) at step n, then

(A) with probability $p_{i,j+1,n}$ the particle goes to $(i, j + 1)$ at step $n + 1$.
(B) with probability $p_{i,j-1,n}$ the particle goes to $(i, j - 1)$ at step $n + 1$.
(C) with probability $p_{i+1,j,n}$ the particle goes to $(i + 1, j)$ at step $n + 1$.
(D) with probability $p_{i-1,j,n}$ the particle goes to $(i - 1, j)$ at step $n + 1$.
(E) with probability $p_{i+1,j+1,n}$ the particle goes to $(i + 1, j + 1)$ at step $n + 1$.

Now, suppose a particle starts at the point $P_0 := \mathbf{z}$ and undergoes a random walk according to the above prescription. We allow this particle to wander until a time of T has elapsed. If $Q_{i,j,n}$ is constant, then Δt is a constant, and we only need to count the number of steps taken. Either the particle will hit the boundary after, say, N steps, or it will not hit the boundary at all in M steps. Suppose that the sequence of points that this particle visits is $(P_0, P_1, P_2, \ldots, P_N)$, and $N = M$ if the boundary has not been reached. Then, an unbiased estimator of the value of $u(\mathbf{z})$ for the parabolic problem in (191.1) is given by

$$-\sum_{j=0}^{N} \frac{F(P_j)}{Q(P_j)} + \begin{cases} \phi(P_N, t_N), & \text{if the particle reached the boundary,} \\ g(P_N), & \text{if the particle did not reach the boundary.} \end{cases}$$

$$(191.3)$$

In practice, several random paths will be taken, and the average taken to estimate $u(x, y, t)$.

Example

Suppose we wish to numerically approximate the solution to the diffusion equation in the unit square, at a single point. Suppose we have the partial differential equation

$$u_t = \nabla^2 u, \tag{191.4}$$

for $u(t, x, y)$ with the boundary conditions

$$
\begin{aligned}
u(t, x, 0) &= u(t, x, 1) = 0, \\
u(t, 0, y) &= u(t, 1, y) = 0, \\
u(0, x, y) &= 10.
\end{aligned}
\tag{191.5}
$$

The exact solution to (191.4) and (191.5) is

$$u(x, y, t) = \frac{16}{\pi^2} \sum_{n,m=1}^{\infty} \frac{e^{-\left[(2n-1)^2 + (2m-1)^2\right]t}}{(2m-1)(2n-1)} \sin\left[(2m-1)\pi x\right] \sin\left[(2n-1)\pi y\right], \tag{191.6}$$

which was obtained by separation of variables (see page 419). Using (191.6) we can determine that $u(.6, .6, .5) \simeq 5.354$. We choose the point $\mathbf{z} = (.6, .6)$ and try to numerically approximate the solution to (191.4) and (191.5) at the point $\mathbf{z}$ when $t = .5$. We will follow the steps outlined above.

Using the standard second order approximation to the Laplacian, we find

$$\nabla^2 u \simeq \frac{u_{i+1,j,n} + u_{i-1,j,n} + u_{i,j+1,n} + u_{i,j-1,n} - 4u_{i,j,n}}{h^2} = 0,$$

where $u_{i,j,n} := u(hi, hj, n(\Delta t))$ and $h \ll 1$. Using our above approximation to the time derivative, we find that (191.4) may be approximated as

$$\frac{u_{i,j,n+1} - u_{i,j,n}}{\Delta t} = \frac{u_{i+1,j,n} + u_{i-1,j,n} + u_{i,j+1,n} + u_{i,j-1,n} - 4u_{i,j,n}}{h^2},$$

or (defining $\gamma = \Delta t / h^2$)

$$u_{i,j,n+1} = \gamma\left[u_{i+1,j,n} + u_{i-1,j,n} + u_{i,j+1,n} + u_{i,j-1,n}\right] + u_{i,j,n}(1 - 4\gamma). \tag{191.7}$$

If we choose $\gamma = 1/4$, then (191.7) simplifies to

$$u_{i,j,n+1} = \frac{u_{i+1,j}}{4} + \frac{u_{i-1,j}}{4} + \frac{u_{i,j+1}}{4} + \frac{u_{i,j-1}}{4}. \tag{191.8}$$

We interpret (191.8) probabilistically as follows: If a particle is at position (i, j) at step n, then

(A) with probability $1/4$ the particle goes to $(i, j+1)$ at step $n+1$.
(B) with probability $1/4$ the particle goes to $(i, j-1)$ at step $n+1$.
(C) with probability $1/4$ the particle goes to $(i+1, j)$ at step $n+1$.
(D) with probability $1/4$ the particle goes to $(i-1, j)$ at step $n+1$.

The FORTRAN program in Program 191 was used to simulate the motion of the particles according to the above probability law. A total of NSIM random particles were started off. The outcome of that program is given below. As more paths are taken, the approximation becomes better (recall that the exact value is approximately 5.35). Obtaining many decimal places of accuracy requires a very large number of simulations.

```
STEP,DT,M:  0.03000  0.00360  138
AVERAGE AFTER 10000 PARTICLES IS:     4.7320
AVERAGE AFTER 20000 PARTICLES IS:     4.7845
AVERAGE AFTER 30000 PARTICLES IS:     4.7847
```

Note that the program uses a routine called RANDOM, whose source code is not shown, that returns a random value uniformly distributed on the interval from zero to one.

Program 191

```
          NSIM=30000
          TIME=.500
          XHOLD=.60
          YHOLD=.60
C CHOOSE THE STEP LENGTH
          STEP=.02
C THE STEP LENGTH DETERMINES THE TIME STEP
          DT=4.*STEP**2
C DETERMINE THE NUMBER OF TIME STEPS ALLOWED
          M=TIME/DT
          SUM=0.
          DO 30 IWALK=1,NSIM
C START OFF A NEW RANDOM WALK
          X=XHOLD
          Y=YHOLD
          NSTEP=0
10        NSTEP=NSTEP+1
C DETERMINE IF WE HAVE TAKEN M STEPS YET
          IF( NSTEP .GT. M ) GOTO 20
C UPDATE THE POSITION
          X=X + SIGN(STEP,  RANDOM(DUMMY)-.5 )
          Y=Y + SIGN(STEP,  RANDOM(DUMMY)-.5 )
C IF THE PARTICLE ESCAPES THE BOX, START A NEW PARTICLE OFF
          IF( X.GT.1 .OR. X.LT.0 ) GOTO 40
          IF( Y.GT.1 .OR. Y.LT.0 ) GOTO 40
C OTHERWISE TAKE ANOTHER STEP
          GOTO 10
C TIME HAS RUN OUT WITH THE PARTICLE STILL IN THE GRID
20        SUM=SUM+10
40        IF( MOD(IWALK,10000) .NE. 0 ) GOTO 30
          APPROX=SUM/FLOAT(IWALK)
          WRITE(6,5) IWALK,APPROX
30        APPROX=SUM/FLOAT(NSIM)
```

```
      WRITE(6,5) NSIM,APPROX
5     FORMAT(' AVERAGE AFTER',I6,' PARTICLES IS: ',F10.4)
      END
```

Notes

[1] If further accuracy is required, the options are

(A) Increase the number of random particles.
(B) Make the mesh discretization finer (i.e., decrease h).
(C) Do both of the above.

If the number of random particles is not very large, then (B) will not help much, and if the mesh is very coarse, then (A) will not help much. Generally, the variance of the answer (a measure of the "scatter") decreases as the number of trials to the minus one half power.

[2] Since low numerical accuracy is obtained by this technique, a computer program does not need to work with extended precision arithmetic (such as double precision).

[3] Sadeh and Franklin [5] contain several worked examples.

[4] If the time at which the solution is desired is so large that all of the particles end up at the boundaries, then the quantity really being calculated is the steady state solution to the parabolic equation.

[5] If a parabolic equation is interpreted as a Fokker–Planck equation (see page 254), then Itô equations can be associated with the parabolic equation. The Itô equations may be numerically integrated by the technique described on page 695.

[6] Another type of Monte–Carlo approach for parabolic equations, using cellular automata, is described in Boghosian and Levermore [2].

References

[1] V. C. Bhavsar and U. G. Gujar, "VLSI Algorithms for Monte Carlo Solutions of Partial Differential Equations," in R. Vichnevetsky and R. S. Stepleman (eds.), *Advances in Computer Methods for Partial Differential Equations*, IMACS, North–Holland Publishing Co., New York, 1984.

[2] B. M. Boghosian and C. D. Levermore, "A Cellular Automaton for Burgers' Equation," *Complex Systems*, **1**, 1987, pages 17–29.

[3] S. J. Farlow, *Partial Differential Equations for Scientists and Engineers*, John Wiley & Sons, New York, 1982, pages 346–352.

[4] G. Marshall, "Monte Carlo Methods for the Solution of Nonlinear Partial Differential Equations," *Comput. Physics Comm.*, **56**, 1989, pages 51–61.

[5] E. Sadeh and M. A. Franklin, "Monte Carlo Solution of Partial Differential Equations by Special Purpose Digital Computer," *IEEE Transactions on Computers*, **C-23**, No. 4, April 1974, pages 389–397.

[6] E. G. Puckett, "Convergence of a Random Particle Method to Solutions of the Kolmogorov Equation $u_t = \nu u_{xx} + u(1 - u)$," *Math. of Comp.*, **52**, No. 186, April 1989, pages 615–645.

[7] S. Roberts, "Convergence of a Random Walk Method for the Burgers Equation," *Math. of Comp.*, **52**, No. 186, April 1989, pages 647–673.

[8] A. S. Sherman and C. S. Peskin, "A Monte Carlo Method for Scalar Reaction Diffusion Equations," *SIAM J. Sci. Stat. Comput.*, **7**, No. 4, October 1986, pages 1360–1372.

[9] A. S. Sherman and C. S. Peskin, "Solving the Hodgkin–Huxley Equations by a Random Walk Method," *SIAM J. Sci. Stat. Comput.*, **9**, No. 1, January 1988, pages 170–190.

192. Pseudo-Spectral Method

Applicable to   Most commonly, hyperbolic equations with periodic boundary conditions.

Yields

A numerical scheme for calculating the spatial derivatives.

Idea

A numerical finite Fourier transform can be used to obtain difference schemes that are of infinite order.

Procedure

On a uniformly spaced grid $\{x_1, x_2, \ldots, x_N\}$, with $x_{i+1} - x_i = h$, a numerical approximation to $\partial u/\partial x$ at the point x_k that is second order accurate is

$$\left.\frac{\partial u}{\partial x}\right|_{x=x_k} \simeq \frac{1}{2h}\left(u_{k+1} - u_{k-1}\right),$$

where $u_k := u(x_k)$. A numerical approximation that is fourth order accurate is given by

$$\left.\frac{\partial u}{\partial x}\right|_{x=x_k} \simeq \frac{1}{3h}\left(u_{k+1} - u_{k-1}\right) - \frac{1}{6h}\left(u_{k+2} - u_{k-2}\right).$$

A numerical approximation that is sixth order accurate is given by

$$\left.\frac{\partial u}{\partial x}\right|_{x=x_k} \simeq \frac{1}{2h}\left(u_{k+1} - u_{k-1}\right) - \frac{1}{3h}\left(u_{k+2} - u_{k-2}\right) + \frac{1}{30h}\left(u_{k+3} - u_{k-3}\right).$$

Methods of arbitrary high order may be constructed. For higher order methods, more points surrounding the point x_k will be utilized. In the limit, the following centered difference scheme of infinite order accuracy is obtained

$$\left.\frac{\partial u}{\partial x}\right|_{x=x_k} = \sum_{j=1}^{\infty} \frac{2(-1)^{j+1}}{jh}\left(u_{k+j} - u_{k-j}\right). \tag{192.1}$$

Eventually, when implementing methods of progressively higher order, the value of $u(x)$ at a point x_{k+j}, with $k + j > N$, will be required. If we assume that $u(x)$ is periodic, with period Nh, then $u(x_i) = u(x_{i+N})$. By periodicity, then, the value at x_{j+k} is the same as the value at x_{j+k-N}. Hence, methods of arbitrarily high order may be constructed, and only the values $\{u_1, u_2, \ldots, u_N\}$ will be utilized.

Alternately, given $u(x)$ a Fourier transform may be taken to determine

$$\widehat{u}(\omega) = \frac{1}{\sqrt{2\pi}} \int_{-\infty}^{\infty} u(x)e^{i\omega x}\, dx. \tag{192.2}$$

Once determined, $\widehat{u}(\omega)$ may be multiplied by $-i\omega$, and then an inverse transform taken to yield

$$\frac{\partial u}{\partial x} = \frac{-1}{\sqrt{2\pi}} \int_{-\infty}^{\infty} i\omega\widehat{u}(\omega)e^{-i\omega x}\, d\omega. \tag{192.3}$$

An informally derivation of this statement is simple, consider differentiating the formula $u(x) = \dfrac{1}{\sqrt{2\pi}} \int_{-\infty}^{\infty} \widehat{u}(\omega)e^{-i\omega x}\, d\omega$ with respect to x.

Hence, the first derivative at every point in a domain may be computed by taking a Fourier transform, multiplying by $-i\omega$, and then taking an inverse Fourier transform. By discretizing (192.2) and (192.3), the Fourier transforms can be performed by "fast Fourier transforms" (FFTs). The FFT is a fast numerical technique for determining the finite Fourier transform of a function that is defined on a set of equally spaced grid points.

Hence, the derivative at every point in the grid can be computed by taking an FFT, multiplying by the discrete analogue of $i\omega$, and then taking an inverse FFT. This approach yields the same numerical scheme given in (192.1).

Using either technique, a highly accurate finite difference scheme is generated. This scheme may then be used to numerically approximate the u_x term appearing in a differential equation.

Example

Suppose we have the hyperbolic equation for $u(x, t)$

$$\frac{\partial u}{\partial t} = \frac{\partial u}{\partial x}, \tag{192.4}$$

for $t \geq 0$ on $0 \leq x \leq 1$ with the periodic boundary conditions

$$u(0, t) = u(1, t), \tag{192.5}$$

and the initial conditions

$$u(x, 0) = \sin 2\pi x. \tag{192.6}$$

The solution of this system can be determined by the method of characteristics (see page 368) to be $u(x, t) = \sin\left[2\pi(x - t)\right]$. We will use this exact solution to compare against our numerical scheme.

The pseudo-spectral method dictates that we take the derivatives of the periodic component (x in this example) by FFTs. We choose to use a one sided explicit difference scheme for the time derivative term. Of course, a more accurate derivative expression for the $\partial u/\partial t$ term would result in a more accurate numerical approximation (see Gottlieb and Turkel [7]).

A FORTRAN computer program is given in Program 192 that finds a numerical approximation to the solution of (192.4), (192.5), and (192.6). For comparison purposes, the exact solution is also printed out. Note that the program calls a subroutine (called FFT(N,V,SIGNI)), whose source code is not given, to perform the fast Fourier transform. This routine is input a complex-valued vector V and returns the same vector, where the values have been modified by

$$V(k) = \frac{1}{\sqrt{N}} \sum_{j=1}^{N} V(j) \exp\left[2\pi i(j - 1)(k - 1)\frac{\text{SIGNI}}{N}\right]$$

The last few lines of the program output are shown below

```
HERE IS THE SOLUTION AT TIME 0.001000
 X=  0.       Y(APPROX)=  0.0010   Y(EXACT)= -0.0063
 X=  0.1250   Y(APPROX)=  0.7078   Y(EXACT)=  0.7026
 X=  0.2500   Y(APPROX)=  1.0000   Y(EXACT)=  1.0000
 X=  0.3750   Y(APPROX)=  0.7064   Y(EXACT)=  0.7115
 X=  0.5000   Y(APPROX)= -0.0010   Y(EXACT)=  0.0063
 X=  0.6250   Y(APPROX)= -0.7078   Y(EXACT)= -0.7026
 X=  0.7500   Y(APPROX)= -1.0000   Y(EXACT)= -1.0000
 X=  0.8750   Y(APPROX)= -0.7064   Y(EXACT)= -0.7115
```

Program 192

```
      IMPLICIT DOUBLE PRECISION (A-H,O-Z)
      REAL*8    V(100),X(100),EXACT(100)
      COMPLEX*16 VV(100),DERIV(100)
      N=8
      DELTAT=.0001D0
      NTIME=10
      H=1.D0/DFLOAT(N)
      PI=3.141592653589D0
      WO=1.D0/DFLOAT(N/2-1)
C INITIALIZE THE VECTOR WITH THE INITIAL CONDITIONS
      DO 10 J=1,N
      X(J)=DFLOAT(J-1)*H
10    V(J)=DSIN( 2.D0 * PI * X(J) )
C HERE IS THE LOOP IN TIME
      DO 20 LOOP=1,NTIME
      TIME=LOOP*DELTAT
C TAKE THE FOURIER TRANSFORM OF THE V VECTOR
      DO 30 J=1,N
30    VV(J)=V(J)
      CALL FFT(N,VV, 1.D0)
C MULTIPLY BY (I WO)
      NBY2=N/2
      DO 40 J=1,N
40    DERIV(J)= VV(J) * DCMPLX(0.D0,1.D0) * DFLOAT(-NBY2-1+J) * WO
C TAKE THE INVERSE FOURIER TRANSFORM
      CALL FFT(N,DERIV,-1.D0)
C USE THE DERIVATIVE VALUES TO UPDATE THE MESH VALUES
      DO 50 J=1,N
      V(J)=V(J) + DELTAT*DREAL( DERIV(J) )
50    EXACT(J)=DSIN( 2.D0*PI*( X(J)-TIME ) )
20    WRITE(6,5) TIME, (X(K),V(K),EXACT(K),K=1,N)
5     FORMAT(' HERE IS THE SOLUTION AT TIME',F9.6,/,
     1      8(2X,'X=',F8.4,'  Y(APPROX)=',F8.4,'  Y(EXACT)=',F8.4/))
      END
```

Notes

[1] To calculate higher order derivatives, higher powers of $(i\omega)$ should be used to multiply $\hat{u}(\omega)$. See any book on Fourier transforms, for example, Butkov [2].

[2] Note that the method, when applied to partial differential equations, requires that the grid be uniform in every spatial variable in which a FFT is to be taken.

[3] This scheme has also been applied to elliptic and parabolic equations, but the results are not much better than using a relatively low order finite difference scheme.

[4] Comparing this method to finite differences; the pseudo-spectral method (the finite difference method) uses a global (local) interpolation of a function, then an approximation of a derivative is made from this interpolatory function.

[5] Spectral methods are really more general than the limited exposition given here. Theoretically, spectral methods expand the unknown quantities in a series of orthogonal functions; these functions, in turn, result from the solution of a Sturm–Liouville problem. In practice, one considers either a Fourier expansion (as we have done here)–usually for periodic problems–or an expansion in terms of orthogonal polynomials. The Chebyshev polynomials are often used as they are amenable to the fast Fourier transform, but also admit more general boundary values than those allowed in Fourier series.

References

[1] H. F. Ahner, "Walsh Functions and the Solution of Nonlinear Differential Equations," *Am. J. Phys.*, **56**, No. 7, July 1988, pages 628–633.

[2] E. Butkov, *Mathematical Physics*, Addison–Wesley Publishing Co., Reading, MA, 1968.

[3] C. Canuto, M. Y. Hussaini, A. Quarteroni, and T. A. Zang, *Spectral Methods in Fluid Mechanics* Springer–Verlag, New York, 1987.

[4] J. W. Cooley, P. A. W. Lewis, and P. D. Welch, "The Fast Fourier Transform and Its Applications," *IEEE Transactions on Education*, E-12, 1969, pages 27–34.

[5] B. Fornberg, "High-Order Finite Differences and the Pseudospectral Method on Staggered Grids," *SIAM J. Numer. Anal.*, **12**, August 1990, pages 904–918.

[6] D. Gottlieb and S. A. Orszag, *Numerical Analysis of Spectral Methods: Theory and Applications*, SIAM, Philadelphia, 1977.

[7] D. Gottlieb and E. Turkel, "On Time Discretizations for Spectral Methods," *Stud. Appl. Math.*, **63**, 1980, pages 67–86.

[8] B. Mercier, *An Introduction to the Numerical Analysis of Spectral Methods*, Springer–Verlag, New York, 1989.

[9] S. A. Orszag, "Comparison of Pseudospectral and Spectral Approximation," *Stud. Appl. Math.*, **51**, 1972, pages 253–259.

[10] M. Pickering, *An Introduction to Fast Fourier Transform Methods for Partial Differential Equations, with Applications*, John Wiley & Sons, New York, 1986.

[11] E. Tadmor, "Stability Analysis of Finite-Difference, Pseudospectral and Fourier–Galerkin Approximations for Time-Dependent Problems," *SIAM Review*, **29**, No. 4, 1987, pages 525–555.

[12] H. Tal-Ezer, "Spectral Methods in Time for Parabolic Problems," *SIAM J. Numer. Anal.*, **26**, No. 1, February 1989, pages 1–11.

Mathematical Nomenclature

$C^p[a, b]$ The class of functions that are continuous and have p continuous derivatives, on the interval $[a, b]$.

$H(x)$ The Heaviside function or step function, it is defined by

$$H(x) := \int_{-\infty}^{x} \delta(x)\, dx = \begin{cases} 0 & \text{if } x < 0, \\ 1/2 & \text{if } x = 0, \\ 1 & \text{if } x > 0. \end{cases}$$

O We say that $f(x) = O(g(x))$ as $x \to x_0$ if there exists a positive constant C and a neighborhood U of x_0 such that $|f(x)| \leq C|g(x)|$ for all x in U.

o We say that $f(x) = o(g(x))$ as $x \to x_0$ if, given any $\mu > 0$, there exists a neighborhood U of x_0 such that $|f(x)| < \mu|g(x)|$ for all x in U.

p When $z = z(x, y)$, then $p = z_x$; when $y = y(x)$, then $p = y_x$.

q When $z = z(x, y)$, then $q = z_y$.

r When $z = z(x, y)$, then $r = z_{xx}$.

s When $z = z(x, y)$, then $s = z_{xy}$.

t When $z = z(x, y)$, then $t = z_{yy}$.

δ_{ij} The Kronecker delta, it has the value 1 if $i = j$ and the value 0 if $i \neq j$.

ε This is often used to represent a small number, assumed to be much less than one in magnitude.

$\delta(x)$ The delta function, it has the properties that $\delta(x) = 0$ for $x \neq 0$, but $\int_{-\infty}^{\infty} \delta(x)\, dx = 1$.

765

∂S If S is a region or volume, then ∂S denotes its boundary.

$\widetilde{\nabla}$ The space–time gradient operator, it is defined by $\widetilde{\nabla} = [\nabla, \partial/\partial t]$.

∇^2 The Laplacian, it is defined by $\nabla^2 \phi = \mathrm{div}(\mathrm{grad}\,\phi)$.

✡   The vector Laplacian, it is defined by ✡$\mathbf{v} = \mathrm{grad}(\mathrm{div}\,\mathbf{v}) - \mathrm{curl}\,\mathrm{curl}\,\mathbf{v}$.

□   d'Alembert operator, it is defined by $\square = \partial^2/\partial t^2 - \nabla^2$.

$[L, H]$ The commutator of the two differential operators L and H. See page 3.

$\{u, v\}$ The Lagrange bracket of the two independent variables u and v. See page 7.

$[f, g]$ The Poisson bracket of the two functions f and g. See page 10.

$\{y, x\}$ The Schwarzian derivative of y with respect to x. See page 11.

Differential Equation Index

This index only refers to named differential equations that have appeared in the text. A more complete index may be found on page 773.

Index

773